Risk, Systems and Decisions

Health, environment, security, energy, technology are problem areas where man-made and natural systems face increasing demands, giving rise to concerns which touch on a range of firms, industries and government agencies. Although a body of powerful background theory about risk, decision, and systems has been generated over the last several decades, the exploitation of this theory in the service of tackling these systemic problems presents a substantial intellectual challenge. This book series includes works dealing with integrated design and solutions for social, technological, environmental, and economic goals. It features research and innovation in cross-disciplinary and transdisciplinary methods of decision analysis, systems analysis, risk assessment, risk management, risk communication, policy analysis, economic analysis, engineering, and the social sciences. The series explores topics at the intersection of the professional and scholarly communities of risk analysis, systems engineering, and decision analysis. Contributions include methodological developments that are well-suited to application for decision makers and managers.

David M. Berube

Pandemics and Resilience: Lessons we should have learned from Zika

Springer

David M. Berube
North Carolina State University
Raleigh, NC, USA

ISSN 2626-6717 ISSN 2626-6725 (electronic)
Risk, Systems and Decisions
ISBN 978-3-031-25369-0 ISBN 978-3-031-25370-6 (eBook)
https://doi.org/10.1007/978-3-031-25370-6

This Springer imprint is published by the registered company Springer Nature Switzerland AG
The registered company address is: Gewerbestrasse 11, 6330 Cham, Switzerland

Foreword

Too often, the lessons of past public health responses are lost in the scramble to meet the next looming threat. Yet, as the world emerges from the shadow of our most recent pandemic, we know that another potential infectious disease threat is likely to challenge us soon. In this book, Prof. David M. Berube captures the hard-fought insights from the spread of the Zika virus (ZIKV) across large portions of the globe. While much of the world may wish to forget the terrible toll that pandemics can take on all members of the global community, an acceptance of collective amnesia invites us to experience yet again the mistakes we have struggled with during outbreaks of the past decade.

Changes to global ecosystems, social-behavioral practices, and an evolving communication environment make us more vulnerable to a future pandemic. Although highly transmissible respiratory viruses such as COVID-19 are generally the most difficult to stop from achieving pandemic spread, understanding the complexities of outbreaks caused by viruses with other routes of transmission (i.e., vector and sexual transmission in the case of ZIKV) can help us prepare for the pandemic curveballs that may occur in the future.

The ZIKV epidemic occurred in the shadow of the West African Ebola virus outbreak. Outbreak fatigue, caused by attention toward the potential spread of Ebola and the multiyear timeline of the ZIKV epidemic, eroded the ability of public health authorities to maintain long-term preparedness and response efforts. Globally, we face a similar situation with COVID-19.

In this book, readers will find a comprehensive view of the ZIKV epidemic, with meticulous attention to a wide range of issues related to the spread of the virus. In particular, a strong focus on communication is interwoven throughout the book. Misinformation, a potent pandemic accelerator due to its trust-eroding capabilities, was particularly problematic in the ZIKV epidemic. In addition, the way diseases are framed and communicated sets the stage for effective or less effective response efforts. The framing of ZIKV as a problem primarily for pregnant women, who could pass the infection to their fetuses which could then develop congenital disabilities, let critical players—sexual partners, community members, and the larger public—off the hook in protecting the vulnerable from ZIKV disease.

ZIKV also reminds us of long-term recovery needs for populations affected by outbreaks. Often, outbreaks are considered in response, with additional attention paid to preparedness if policymakers are particularly attentive. Yet, recovery is a crucial and often forgotten stage of these disease events. As we have learned with long-COVID, the initial infection can generate health effects only felt in the long run. The social, economic, and political contexts may be with us permanently. Outbreaks and pandemics cannot be considered single events, with only short-term impacts and considerations.

Efforts to meet the threat of ZIKV or leverage its unique capabilities also highlight the potential for new technologies to make meaningful changes across a range of globally impactful diseases. Work to alter vector dynamics shows potential in the fight against mosquito-borne diseases that cause millions of deaths a year. And the potential for new treatments for old scourges such as cancer beckons as we grow our understanding of ZIKV.

If these topics sound familiar after our experience with COVID-19, that's no surprise. Let us learn from the past to chart a better and more resilient future the next time a new disease emerges.

Baltimore, USA

Professor Tara Kirk Sell
Senior Scholar, Johns Hopkins Center for Health Security, Associate Professor, Department of Environmental Health and Engineering, Johns Hopkins Bloomberg School of Public Health

Acknowledgements

Most of my adult life has been spent trying to chase down meanings. For a quarter of a century, my day-to-day life involved inherent geekiness. As a nationally recognized intercollegiate debating coach producing research handbooks on everything from government defense policy to technological breakthroughs, the publisher of research handbooks on over 50 discrete topic areas was one of my occupations. For the next quarter-century, my first academic preoccupation with scientific and technological claims in a set of emerging technologies was embraced with the same level of determination. As a faculty member, my scholarship and instructional duties center on technical and scientific communication and the challenges associated with communication between various stakeholders.

My critical work in science communication started in the field of nanotechnology in the late 80s and 90s, which led to a book (over 500 pages with thousands of references and translated into different languages), a lot of grant money to study the public understanding of nanoscience, and dozens of articles and chapters, many in prestigious technical journals. My approach to writing about nanoscience and nanotechnology as a scholar in science communications demanded reading hundreds of printed and digital articles, books, government reports, etc., and a host of material reported on the Internet. Marjorie Garber calls people like me an "amateur professional [1]." We read everything we can find and ask many questions.

While the newsworthiness of Zika came and went after the outbreak in South America in 2016–2018, the Zika event seemed well suited for a detailed study having both a clear beginning and an end per se (more of that later). While significantly consequential, especially for women of childbearing age, it was not overly apocalyptic and represented the specific infectious disease.

Though my role in nanotechnology continues, other fields have interested me, especially biotechnology, as enabled by nanotechnology and the information sciences developments. Finally, biotechnology and genetic engineering took a turn when cloning and genetically engineered foods became controversial and continued through the development of synthetic biology, most recently CRISPR-Cas9 gene drive technologies. When allowed to become associated with a center on the North

Carolina State University campus interested in synthetic biology, I received a fellowship and a stipend. I began studying synthetic biology and genetic engineering to control mosquitoes as vectors for infectious diseases, emphasizing Zika.

A previous book of mine, Nano-Hype [2], was about exaggeration and hyperbole. This is about how we approach infectious disease as a case history examining various variables impacting comprehension, response, and management. Before laying this out, I edited a major book on Pandemic Communication (2021) [3] that did the same around mostly COVID-19.

This work took six years. I was hit with two bouts of cancer, and onerous treatment regimens only topped off with the COVID pandemic. I escaped infection by COVID but was one of those over 60 with preexisting conditions, wondering what I was in for. After over five years of reading and writing, what was intended as an issue brief became a 500-page+ book with a robust set of references from technical journals and public sources. While I do not claim to be an entomologist or an infectious disease scholar, I read enough to be an "amateur professional." I have learned that the world is unprepared for the threat of contagious disease, which will become more frequent given the broad set of drivers discussed below in this book. Generally, we are woefully underprepared, as our recent experience with COVID-19 has demonstrated.

Neil Postman, in his classic book, Technopoly, warned us that a world in which decision-making was done by the technically literate for the technically illiterate might presage a break in humanity, a technical paternalism, which can be benign. Still, it could also become our worst nightmare [4].

We are in a "post-certain world," if you haven't noticed. The public does not understand most scientific and technological developments. Most do not have the expertise, and many still do not have the interest. Many had self-selected themselves out of the scientific literate world when they decided against another class in mathematics or chemistry. Many others had little choices when they discovered they did not have the means to commit themselves to a world of science, technology, engineering, and mathematics, sometimes called STEM. They were poor. They grew up in incomplete or dysfunctional families and communities. They needed to provide the means for their families to survive monthly. Or they elected to become someone other than a STEM community member, a decision they should not be asked to pay for with their citizenship.

On a different level, we live in a digital world where trust seems more important than truth. The public gets too much scientific information from digital resources, especially news aggregators and social media platforms. News aggregators allow us to select topics in which we are interested. The aggregator will ensure our newsfeeds focus on the descriptors we enter in the aggregator. And if we decide not to use aggregators, then the algorithms designed in the search engines will do that for us.

In 2016, Waddell did a literature review on ZIKV and reported 233 studies and reports up to March 1. However, more than half of the primary literature on ZIKV has been published since 2011 [5], and I have read most of it, nearly 2000 sources.

Anyone claiming to be an expert in science communication must understand as much about the science as possible. This means lots of reading, developing a new vocabulary, reading technical journals and government reports, taking the time and effort to string material together until meaning surfaces, taking a deep breath, sharing what has been written with scientists and engineers, and accepting their input and constructive criticism. It takes time.

Science communication is essential in our globalized world. We are on the precipice of developments that could change our lives, especially in the world of infectious diseases. The public needs help ferret through the reporting and makes sense of the world around them. The alternative, according to Postman, was being led without a chance to participate. Orwell and others remind us that our democracy is vulnerable to dangerous cancers, and a government without transparency and accountability might be the most malignant.

I want to tip my imaginary hat to a colleague, Dr. Roy Schwartzman from UNC-Greensboro. WHO introduced me to the term "propaedeutics." Propaedeutics (*pro* from the Greek for early and *paideutikós* meaning "about teaching") is an introduction to a subject or area of study. It shares a root with the word "encyclopedia." I like the definition found on Wiki and referenced in the *Encyclopedia Americana* (1851).

> … the knowledge necessary or helpful in understanding or practicing art or science, or which unfolds its nature and extent, and the method of learning it. Therefore, it is applied not only to special introductions to branches of study but also to auxiliary sciences, logic, philology, etc., and the comprehensive views of branches of science that facilitate an insight into the relations of the parts. Such a survey can be presented only by one who has studied science in all its ramifications.

Projects such as this book are propaedeutic. I see myself as someone willing to take the time and effort to understand complex science and technology and write about some insights as I compile my offerings authentically and validly.

In addition, my research team in the Public Communication of Science and Technology at NCSU assisted in many ways, which was greatly appreciated. My colleagues have been beneficial, especially Drs. Gina Lane and Elizabeth Sperry from William Jewell College, who never allowed me to forget women's unique roles in this outbreak and how it was manhandled (pun intended). Drs. Jennifer Kuzma, Fred Gould, and Todd Kuiken from the *Genetics Engineering and Society Center* at NCSU heard lots of my complaints and were incredibly patient. I promised them a two-page issue brief, and they got a book. Thanks for waiting.

To my many good friends, Mrs. Sharon Stauffer, Dr. Christopher Cummings, Dr. Jacob Jones, and Dr. Igor Linkov, thank you for encouraging me to strike on. As always, my students are very generous and my best teachers. My doctoral research assistants and MS students, especially Jared Belvin, Ekaterina Bogomoletc, Nicholas Eng, Randall Alex Hammond, Anne Njathi, Ben Whitley, and many others, had much to say about where I was going with this and all my projects.

Special thanks to my undergraduate students Ava Freyaldenjoven, Nicholas Loschin, Charles Smalls, and Ryan Will. They were interested, supportive, and helpful. Ms. Freyaldenjoven helped with Google Flu Trends, and her efforts to locate diagrams, illustrations, and other visualizations were greatly appreciated. Mr. Loschin was an undergraduate student in Natural Resources at NCSU (now as graduate student) and contributed some ideas, especially on the complicated issue of climate change as a driver as an independent study research experience. Mr. Smalls assisted in some difficult copy editing.

One of the perks of being a faculty member at a Research 1 University is access to excellent world-class libraries. This work was assisted substantially by the online search services provided by the North Carolina State University Hunt and Hill Libraries on our Raleigh, North Carolina, campus. Whatever I could not find was found for me by great libraries and my friends at Inter-library Loan. Many resources are critical in a project of this type: PubMed.gov and the National Center for Biotechnology Information (NCBI) in The National Library of Medicine associated with the National Institutes of Health were remarkable.

This book is an academic treatise with over 3500 endnotes from books, government reports, technical articles, popular articles, and web resources, some in Portuguese. This work is comprehensive.

Three books were noteworthy. They were well-written and mainly were not reprinted. Each has its strengths and weaknesses. None were comprehensive.

- Diniz, Deborah wrote Zika: From the Brazilian Backlands to Global Threat in 2016 (London: Zed Books).
- McNeill, Jr., Donald G. wrote Zika: The Emerging Epidemic in 2016 (NY: W. W. Norton).
- Ratna, Kalpish wrote The Secret Life of (sic) Zika Virus in 2015 (New Delhi: Speaking Tiger Publishers).

Many professional journal articles made the work possible. As a seasoned scholar in science communication, it took longer to work through some technical articles on entomology and infectious disease. Too many friends and colleagues worldwide might be mentioned here as I emailed them to verify my understanding of the articles. Special appreciation goes to the following researchers and authors by surname (alpha order). Every one of these works is cited within and worth a read.

- Zach Adelman and Zhijan Tu (2016) took the ideas of genetic manipulation to new levels by introducing the function gene drives may have on controlling mosquito-borne infectious diseases in *Trends in Parasitology*.
- Matthew Aliota and the Global Virus Network (2017) for their review of the literature, "Zika in the Americas, year 2," published in *Antiviral Research*.
- Luke Alphey (2014) was one of the few authors who took the time and effort to provide a highly readable story about the genetic control of mosquitoes in the *Annual Review of Entomology*.
- Patricia Brasil and Karin Nielsen-Saines (2016) had excellent publications in the *New England Journal of Medicine* and *Lancet*. More than any other, these articles seemed to cement the correlation between Zika infection and microcephaly and other infant developmental problems. (Spellcheck nearly killed me: Dr. Brasil is not a country!)
- Christopher Chang et al. (2016), in the *Journal of Autoimmunity*, provided a straightforward summary of the events leading to the crisis designation of Zika.

- João Rafael de Oliveira Dias et al. (2018) wrote an excellent summary of Zika for *Progress in Retinal and Eye Research* and its impact on a handful of vision issues. The piece provided insight and validation throughout.
- Lars Eisen et al. (2009) examined dengue in the Journal of Medical Entomology and made a strong case for proactive vector control.
- Casey A. Klofstad et al. (2019), in *Palgrave Communications*, summarized the work on conspiracy theory as it applies to Zika better than anyone.
- Stephen S. Morse (1995) teased out variables involved in the emergence of infectious diseases in *Emerging Infectious Diseases* eloquently.
- Didier Musso and Duane Gubler (2016) wrote an impressive history for *Clinical Microbiology Reviews*. While brief and outdated, it provided an exciting template for organizing my thoughts on Zika.
- Didier Musso, Albert Ko, and David Baud wrote a remarkable summative review article for the *New England Journal of Medicine* in 2022.
- Will Parks and Linda Lloyd in 2008 wrote a consummate piece on the planning of social mobilization and communication for dengue fever prevention and control published by the WHO.
- Scott Weaver's history of the ZIKV from *Antiviral Research* is one of the most comprehensive written about the virus and its history and biology.

Unsurprisingly, online publications committed to Zika-related information surface. Two are standouts.

- *Zika News* is associated with Precision Vax LLC. PrecisionVax's digital network of vaccine brands offers a broad news network, including the ZikaNews.com (https://www.zikanews.com/). Their first Zika-related post was on August 6, 2017. A small team led by Robert Carlson, MD, linked developments regarding Zika on their news page.
- *Zika Reports.* Governmental and non-governmental organizations have produced a reasonable amount of literature on mosquito vector management. While the vast majority are merely repeating materials created and distributed by the CDC in the USA, a few were outstanding and essential to this project.
- American Mosquito Control Association (AMCA) released a critical best practice for mosquito control report in 2017.
- Association of State and Territorial Health Officials (ASTHO) released Before the Swarm in 2015. Funded through a CDC Cooperative Agreement, it provides a highly readable and brief review of vector management and risk communication.
- Centers for Disease Control (CDC) have an interim response plan published in 2017. It summarizes CDC releases and offers ideal protocols for management to local health officials.
- Council of State and Territorial Epidemiologists (CSTE) provided an insightful understanding of how mosquito boards function, especially in U.S. territories.
- National Association of County and City Health Officials (NACCHO) is the national nonprofit association representing the approximately 2800 local health departments (LHDs) in the United States, including city, county, metro, district, and tribal agencies. NACCHO's letters and flyers were strong sources in

determining how responses were carried out and the federal government's commitment.

- National Institutes of Health (NIH) provided a context for health and infectious disease policy.
- Pan American Health Organization (PAHO) offered a South and Latin American perspective on infectious disease events that affected their region disproportionately. (The Pan American Health Organization [PAHO] works with the countries of the Americas to improve the health and quality of life of their peoples. Founded in 1902, it is the oldest international public health organization. It serves as the regional office for the Americas of WHO and is the specialized health agency of the inter-American system).
- World Health Organization (WHO): There were multiple reports and research papers on Zika as a health crisis, warnings for travelers and pregnant women, and some exciting work on community outreach activities.

Journalists, scientists, and stakeholders made multiple contributions to this work. Thanks to the World Wide Web, it has never been easier to acquire materials from all over the world as digital versions of printed articles, digital articles, blogs, etc. Unfortunately, one of the problems with the World Wide Web involved tracking down papers that were mis-referenced and facts that turned out to be factional at best, fictional sometimes. These are must-reads.

- Lizette Alverez did an incredible job tackling the crisis over releasing genetically engineered mosquitoes in Florida. Her work in the *New York Times* was constructive.
- Karen Atkins has done much of the same in her articles in the *Florida Keys News*, introducing us to the politics of mosquito control in Key Haven and Monroe County, Florida.
- Julie Beck and Adrienne Lafrance did an outstanding job in 2016 covering Zika for *The Atlantic*. Their works were influential, and some commentaries found their way into this book.
- Paula Branswell and her *Neurology* publication helped clarify the discrepancies of reported incidences of Zika and microcephaly in Puerto Rico, a subject unique to her reporting.
- Dr. Peter Hotez is the dean of the National School of Tropical Medicine at Baylor College of Medicine in Houston and an expert on COVID-19. He is cited throughout this book because he is mentioned throughout the popular literature about Zika more than any other entomology and infectious medicine expert.
- Jerome Lessler's 2016 *Science* article on the general history of Zika entitled "Assessing the global threat from Zika virus" may be the most comprehensive in print to date and most cited in the reports on the outbreak. It was fortunate to have been read early in researching this book.
- Maryn McKenna wrote an important article and one of the few on mosquito control infrastructure in the USA for *National Geographic*.
- Alcides Troncoso is best known for his 2010 work on dengue. In 2016, his piece on Zika in the *Asian Pacific Journal of Tropical Biomedicine* and the world pandemic

will provide most readers with a solid introduction to multiple drivers of the ZIKV and its infectious implications.

Thank you so much for my education on this subject and forgive the extensive end noting, but this was a group project in many ways.

A special thank you goes out to Neil Morrison and Donna Liebenberg from Oxitec and Sriram Srinivas and Nel van der Werf, Springer Nature for being patient and considerate.

A final special and gracious thanks to Dr. Tara Kirk Sell for her willingness to write an excellent foreword under a difficult deadline. Thank you again.

This book is dedicated to the children of Zika and their mothers.

References

1. Garber M (2003) Academic instincts. Princeton UP, Princeton, NJ
2. Berube D (2005) Nano-hype: the truth behind the nanotechnology buzz. Prometheus Books, Amherst, NY
3. Berube D (ed) (2021) Pandemic communication and resilience. Springer/NATURE, Cham, Switzerland
4. Postman N (1993) Technopoly: the surrender of culture to technology. Vintage Press, NY
5. Waddell L, Greig J (2016) Scoping review of the Zika virus literature. PLOS One. May 31. http://journals.plos.org/plosone/article?id=10.1371/journal.pone.0156376. Accessed 29 June 2017

Introduction

Pandemics are nearly impossible to predict. However, some features of pandemics are similar, and the approaches are taken to mitigate them share characteristics. How media approaches pandemics has become much opaquer with the arrival of cable broadcast news, social media, and international efforts to spread discord in America and other democratic states as part of a destabilization strategy by China, Russia, Iran, and others. "False news" from other countries has not been isolated to election races and includes misleading and conspiratorial public health postings.

Nevertheless, studying the Zika pandemic in the Americas opened a set of issues that were either willfully ignored or accidentally missed that are important as an international policy is fashioned to deal with the onset of the next pandemic. Indeed, COVID-19, while still active, will be subjected to many postmortems if/when it becomes less virulent and dangerous. It is highly anticipated that there is much more to learn about vaccine hesitancy than ever before once there is the luxury of spending the time and energy on removing ideological and political drivers from conspiracy-laced "fake news" and legitimate concerns over trypanophobia (aversion to needles) and the various side effects correlated to the age of the infected. Since a vaccine strategy for Zika never came to fruition, this debate will remain for another day.

There is a reflective quality to history whereby temporal distance enables scholarly research with more ecological validity. Suppositions can be identified, compared, and separated from others, allowing testable hypotheses to surface. The flurry of journal articles on Zika bottomed out for the time being (2021). However, some research associated with some aspects of the ZIKV pandemic and some associated with the experiments undertaken to reduce vector capacity (Chaps. 8 and 9) should be expected.

After severe acute respiratory syndrome (SARS), Middle East respiratory virus (MERV), the Ebola outbreak, and recurrent re-emergence of chikungunya, cholera, dengue, influenza, and measles across the globe, and before COVID, the world experienced Zika. Its arrival, like the others, serves as a reminder that human beings live in a complex relationship with other organisms in an ecosystem that is often unpredictable [1].

Humans are affected by an impressive diversity of pathogens. For example, in the twenty-first century, there are at least 1407 pathogenic species of viruses, bacteria, fungi, protozoa, and helminths currently recognized. Of this total, 177 (13%) pathogen species are considered emerging or re-emerging. And 816 (58%) are known to be zoonotic. Of the 177 emerging or re-emerging pathogens, 130 (73%) are known to be zoonotic. These emerging and re-emerging zoonoses are associated with a wide range of drivers [2]. A zoonosis is an infectious disease that has jumped from a non-human animal to humans. Zoonotic pathogens may be bacterial, viral, or parasitic, or may involve unconventional agents and can spread to humans through direct contact or food, water, or the environment.

Before moving onward and attempting to answer some of these questions, the subject is situated. The ZIKV is associated predominantly with one species of *Aëdes* mosquito (more on this later). *Aëdes* comes from Greek and means anhedonic or unpleasant. In 1804, Johann W. Meigen discovered Hasselquist's notes and renamed *Culex Aegypti* as *Aëdes Aegypti* (hereafter *Ae. aegypti*). He chose names reflecting the irritation he felt. Over time, we the umlaut disappeared.

The usual common name for *Ae. aegypti* is the "yellow fever mosquito," as it is a principal vector for yellow fever. This species of mosquito is responsible for many other hazardous and debilitating infectious diseases, especially yellow fever, dengue, and chikungunya. Some of the citations used in the text discuss these diseases as a set and do not distinguish much between them though their symptoms may be radically different. It is noted when a source talks about another infectious illness in this text. Sometimes, when it is not, the original discusses how to reduce the interaction with the mosquito regardless of what infectious diseases it may be carrying.

This story is mainly about female *Ae. aegypti* mosquitoes. The females bite for their blood meals which they need to produce viable eggs.

> The weight of a female mosquito, at 1.3 mg, is fugacious. Its lifespan is slightly more than a month. Its habitat is no farther than 50 years from your doorstep. But its flight is buoyant and, at 50–150 cm/sec, has an efficiency more significant than a ramjet's, if we consider the burden of memory on its wing.[3]

It may be a minor and annoying bug, but it has been and remains incredibly dangerous, as shall be seen.

What happened to the Zika virus? It first hit South America just before the Olympics in Rio (2016). It was described as an Earth-changing epidemic in Brazil and a pandemic across South America, the West Indies, and Southern Florida. Zika outbreaks in the past were characterized as mostly annoying but generally harmless sickness, with symptoms like rash, fever, joint pain, and red eyes. These symptoms lasted for seven days [4], and nearly 80% who get infected have no symptoms whatsoever. About one in four infected probably does not even notice they have it [5]. Complaints regarding how problematic ZIKV may be in the second decade of the twenty-first century range from relatively harmless to a worldwide pandemic under the right circumstances.

What happened? Were the original estimations about the outbreak exaggerated? Did the government and public health authorities and the media who made pronouncements about infectious diseases get it all wrong? Did the virus burn itself out, or as Trump once described a virus "just go away"? Did the virus evolve into a less pernicious strand, or did the public sentiment and governmental apprehensions about ZIKV evolve? Or is the modus operandi on the infectious disease industrial complex simply hyperbolic by nature?

On November 18, 2016, the World Health Organization declared that Zika was no longer a global health emergency and should be considered a dangerous mosquito-borne virus, like malaria or yellow fever. One needs to ask why the WHO was taken by surprise and the quick turnaround.

What is Zika?

Ziika means "overgrown" in the Luganda language, a language belonging to the Bantu family [6]. Ziika's spelling was shortened, and an "i" was dropped when the virus was named after the forest in which it was discovered. In 1936, the Uganda Virus Research Institute in Entebbe was established by the Rockefeller Foundation to study yellow fever, a severe problem during the construction of the Panama Canal. Seven miles northeast of the institute is a forested area called "Ziika." It lies along the edge of an extended arm of Lake Victoria and runs parallel with the Entebbe-Kampala Road.

In 1947, wooden platforms were erected in the tree canopy to house rhesus "sentinel" monkeys. By April, there were six platforms. On April 18, Rhesus #766 (an Asian monkey) became ill [7]. A blood sample was taken and injected into two groups of adult mice (standard procedure in viral research in the '40s) and another monkey, Rhesus #771. Rhesus #766 recovered. Rhesus #771 did not appear to get sick, but serum collected after 35 days contained ZIKV antibodies.

The following year, mosquito traps were set up. The catches from January 11 and 12 were ground into a suspension and injected into the brains of mice, and they became seriously ill. Their brains yielded a transmissible agent, and *Aedes Africanus* was established as one of the vectors for ZIKV [8].

It is difficult to determine who was the first to isolate the ZIKV in humans. The earliest published claim involving the isolation of ZIKV from humans comes from MacNamara (1954) [9].

The first alleged deliberate human infection was reported in 1956. W. C. Bearcroft decided to infect himself. He got a mild headache [10].

In 1964, David I. H. Simpson claimed Bearcroft had infected himself with the Spondweni virus, not ZIKV. Simpson claimed he contracted the disease in his work with the virus in the Entebbe laboratory, and he developed the diffuse pink maculopapular rash typical of a contemporary ZIKV infection [11].

ZIKV is a zoonotic disease and a flavivirus (group of RNA viruses, mostly having arthropod vectors, e.g., mosquitoes) and an arbovirus (an arbovirus is a term used to describe a group of viral infections transmitted to humans from a group of insects known as arthropods).

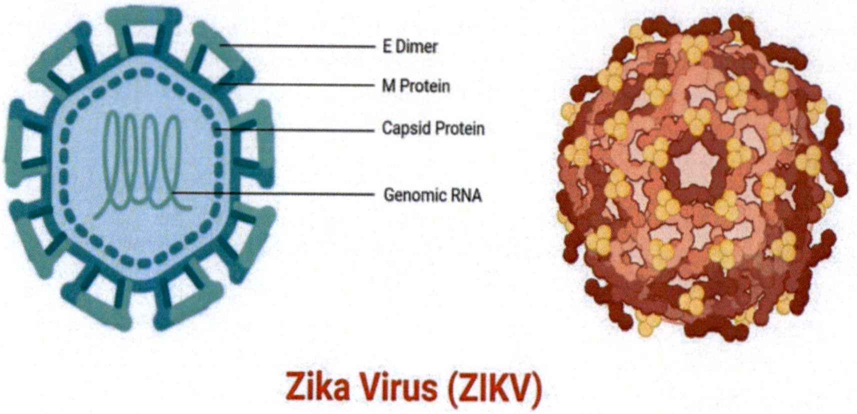

Fig. 1 Zika virus—Structure. Figure designed by Dr. Sagar Aryal, Founder, Microbe Notes. https://microbenotes.com/zika. The mature ZIKV resembles the shape of a golf ball with a smooth surface in which the major surface protein E lies parallel to the viral membrane. The virus is 50 nm in size with 180 copies of E and M protein present in its viral membrane. It consists of a positive-stranded RNA genome with three structural proteins and a lipid envelope

ZIKV is a single-stranded RNA virus distributed throughout much of Africa and Asia [12]. They resemble little balls or spheres. Up close, they are 20-sided polygons called icosahedrons. Twenty-six flaviviruses can cause human disease, but several have produced only laboratory-acquired infections or isolated cases of a disease in man. They share a 10.6 kb, single-stranded, positive-sense RNA genome comprising three structural and seven non-structural genes; the former makes up the vision. Most of the latter participate in genome replication and opposition of host immune defenses. There are sixty-six members in the flavivirus group, of which thirty-one are mosquito borne [13] (Fig. 1).

The literature concludes that ZIKV emerged in Uganda around 1920, most notably between 1892 and 1943 [14].

> …[S]erological and entomological data in the African continent indicated ZIKV infections in Uganda in 1969 and 1970, Nigeria in 1971 and 1975, Sierra Leone in 1972, Gabon in 1975, Central African Republic in 1979, Senegal from 1988 to 1991, and Côte d'Ivoire in 1999.[15]

Two questionable first references appeared in two online Amazon-published books. One by Olmstead claims the first references to human infection from the ZIKV came out of India in 1952 [16]. Otieno reports the first human case was from a man from the United Republic of Tanzania in the 1950s [17]. There is little corroboration between these two authors.

According to Brito and Ordheim et al., in Nigeria in 1979, 31% of 189 serum bank samples were positive by inhibition of hemagglutination for Zika, and 38% were positive by the detection of neutralizing antibodies. In Uganda in 1984, 6% of 132 adult samples had detectable Zika antibodies [18].

Before 2007, ZIKV was of limited public health importance. For nearly half a century, fewer than 20 human infections were documented [19]. In 2007, the first epidemic occurred in Micronesia and the Yap Islands, and another occurred five years later [20]. It arrived in French Polynesia on Tahiti (October 7, 2013), the main island, and some of the islands in the South Pacific (ZIKV outbreak in French Polynesia), attacks occurred in 2014 in New Caledonia (1400 confirmed cases), Cook Islands (50 confirmed and over 900 suspected cases), and Easter Island of Chile (one confirmed and 40 suspected cases). Outbreaks also occurred in Samoa (2015) and American Samoa (2016). Within two months, it had spread to all seventy-six inhabited islands [21].

The phylogenetic tree showed that the ZIKV that emerged in French Polynesia was like that from Yap State 2007 and Cambodia 2010 strains, corroborating the expansion of ZIKV Asian lineage [22], which eventually arrived in Brazil a decade later. Historically, ZIKV remained confined to a narrow equatorial band in Africa and Asia until 2014, when it began to spread eastward, first toward Oceania, and then to South America. Since then, millions of infected individuals have been identified in Brazil, Colombia, and Venezuela, including twenty-five additional countries in the Americas. While the symptoms associated with ZIKV infection are generally mild, consisting of fever, maculopapular rash, arthralgia, and conjunctivitis, there have been reports of more severe reactions related to neurological complications [23].

The third epidemic and the subject of this book began in 2015 in Brazil. Since early 2015, it eventually reached more than forty countries across the Americas, even making it to the Republic of Cabo Verde islands, off the western coast of Africa. More than a million people have become infected [24].

The explosive 2015–2016 pandemic of ZIKV infection occurring throughout South America, Central America, and the Caribbean and potentially threatening the United States is the most recent of four unexpected arrivals of important arthropod-borne viral diseases in the Western Hemisphere over the past 20 years [25]. Autochthonous transmission has been confirmed in 67 countries worldwide and forty-six countries or territories in the Americas [26].

Zika infection cases have been reported from thirty-six countries during this period. Ultimately, ZIKV could have spread globally into other environments where mosquitos can live and breed.

According to the World Health Organization (WHO), as of January 5, 2017, the global risk assessment has not changed. The virus will continue to spread geographically to areas where competent vectors are present. The WHO added that vigilance needs to remain high despite reports of declining cases [27].

After dominating headlines for much of the summer in 2016, news of the mosquito-borne disease that caused life-threatening congenital disabilities fizzled out once temperatures dropped. According to new research by the Centers for Disease Control, it turns out that mosquitoes capable of carrying the virus not only survived the winter, but the threat could be much more widespread than initially anticipated [28]. Public health forecasters at the University of Florida say it's doubtful Zika is finished—although exactly how long the virus may hang around and whether it will spread to other parts of Florida and the U.S. is uncertain. Ira Longini, Biostatistics Professor at the University of Florida's College of Public Health and Medicine, predicted that Zika transmission during the epidemic would continue in South Florida and Texas, the most likely mainland U.S. hot spots [29].

Ae. aegypti, the primary mosquito species known to transmit ZIKV, has now been documented in many counties in the United States. The most concentrated populations are in Southern California, Arizona, Texas, Louisiana, and Florida, but they have been spotted as far north as New Hampshire. This represents a 21% increase in the number of counties that could potentially see a ZIKV outbreak (more on this later).

A close relative, *Ae. Albopictus* is also believed to be a potential carrier, which is even more concerning because it can survive in cooler temperatures than its cousin. The closely related species Ae. Albopictus is often referred to as the "Asian tiger mosquito" [30]. They have been reported in 127 more counties in the USA than previously thought, bringing the total number of counties to 1368. Additionally, 177 counties are home to both species, mainly in Southern California, Arizona, Texas, Florida, and Maryland [31].

What Can We Learn?

ZIKV is unique among the more than one hundred viruses that can cause encephalitis, meningitis, and hemorrhagic disease in humans. The ZIKV can be transmitted through the bite of female mosquito vectors *Ae. aegypti* and *Ae. albopictus*; through sexual contact between humans; congenitally, from a pregnant woman to her fetus; and through blood transfusion [32].

While Zika, or ZIKV as it is abbreviated in the literature as a disease, might seem less impactful than other infectious diseases (though many families in South America and the West Indies may disagree), how it was managed as an issue and a threat is incredibly important as a guide on taking the subsequent outbreak of the following mosquito vector disease. There will be oncoming pandemics involving viruses of all sorts, and it will not end with COVID-19. While keeping humans from spreading infection has proven highly challenging, isolating humans from each other was not a practical option. Human behavior during COVID involved modifications of human behavior and infection mitigation. In general, humanity responded remarkably. However, deterring mosquitoes from seeking a blood meal to reproduce is a different challenge as the *Ae. aegypti* mosquito has been a terror to control. Furthermore, with the increasing regularity given increasing penchants for global travel and development and the seemingly inexhaustible appetites for carbon-based fuels, people find physically closer to this mosquito and its infectious diseases.

Zika is a disease that seems to keep evolving and escalating. The scope and seriousness of this outbreak are unusual in the disease's history. The danger is what might happen later—the fetuses that get microcephaly, the people whose brains and spinal cords swell and possibly sustain damage, the people temporarily paralyzed by

Guillain-Barre, the ones who die because the Guillain-Barre stops their breath (as it can sometimes do), and they are unable to get to a hospital with a respirator.

Zika is an epidemic delay. It will likely take some time for the long-term effects of the epidemic to fully make themselves known. It is safe to say that there will be a generation of kids in some countries with higher-than-usual rates of microcephaly, though it is hard to say just how that will affect their lives. Other neurological complications may linger, or they may be resolved quickly. What about kids who get infected? "Are there going to be long-term cognitive consequences of pediatric exposure? We don't know," Dean Hotez says. But overall, "I think there's going to be a significant amount of long-term [neurological] deficits," he predicts [33].

"ZIKV, for mosquito-borne diseases, is really like no other," said Marcia Castro, Associate Professor of demography at the T. H. Chan School of Public Health who studies vector-borne tropical diseases like Zika. Although it is a flavivirus characterized by a single strand of RNA within a protein envelope, it appears distinct from many of its relatives, such as dengue, yellow fever, West Nile virus, and chikungunya. For one, it has been linked to a higher incidence of Guillain-Barre syndrome, in which the body's immune system attacks nerves, causing paralysis. If it reached the diaphragm and chest walls and the patient was not ventilated, he would die, wide awake, and terrified, staring at the ceiling [34].

ZIKV can also be sexually transmitted and efficiently transferred from a pregnant woman to a fetus, causing congenital disabilities. Furthermore, according to a team of researchers at Harvard Medical School's Center for Virology and Vaccine Research (CVVR) at Beth Israel Deaconess Medical Center, the disease stays in the body for longer than had been previously realized [35].

Chang authored many articles associated with the ZIKV, but little is known about the virus's potential [36]. Pan American Health Organization (PAHO) reports over 1500 papers on Zika [37]. Zika seems to morph and resist common understanding. In the earliest days of this epidemic, Brazilian women gave birth to children with a rare congenital disability, and no one knew why. It is now known why, but each new study seems to turn over a new leaf on a fresh horror, one that wasn't seen coming [38].

The following material on ZIKV is a story, a report, and a historical artifact. While ZIKV remains a potentially crucial viral threat, hopefully it will mutate and move on though there seem to be large naïve populations in Asia, India, and Africa. Nonetheless, more diseases carried by mosquitoes are coming, and some may impact the rich and poor alike. However, Zika decided to take its highest toll among people of color living in places without modern conveniences that make exposure to mosquitoes more manageable. In addition, what ZIKV and especially the mosquitoes who carry it have done was to amplify an essential debate over genetically manipulating the population of mosquitoes is myriads of ways to reduce overall exposure to ZIKV and other viruses these insects spread.

The Chapters

The nineteen chapters examine various issues associated with the Zika pandemic in 2016. Some of the material is technical, but it is all readable. All referenced material is end-noted. Some quoted material was edited for language and idiom consistency and to exclude (sic). While six years have passed since Zika arrived in Brazil, it took a significant amount of time to track down a comprehensive story that fleshes out all the claims and counterclaims. In a world where we expect everything immediately, books like this might seem like historical artifacts when published. Still, the alternative is superficial and unlikely to have much permanence and educational value.

Chapter 1 explains what led me to pursue this work, and in Chap. 2, we understand why someone in the field of science communication was drawn to this subject.

Retelling begins in the following two chapters. In Chap. 3, the re-emergence of Zika and the ZIKV is tracked from Africa to Latin America, and in the next chapter (4), the ebbing process is detailed as well as the most recent outbreaks in the 2020s.

The following two chapters cover what is and is not known about zoonotic infectious diseases. Chapter 5 introduces convergence theory as it tries to explain the "perfect storm" that caused such problems for Latin America in 2015–2016, as well as the main drivers that impact the ZIKV's effect on the people who encounter it lay some blame on the people, this is especially true with the section on climate change and zoonotic diseases. Chapter 6 is about transmission. It examines the related infection and how they move from mosquito to human and back to mosquito again.

Chapters 7 and 8 are entirely dedicated to the effect ZIKV had on children and microcephaly and other developmental disorders they experience. It took two chapters to cover the effects and entertain the debate over the causal connection between the ZIKV and microcephaly. Chapter 9 is relatively short and examines the impact of the ZIKV on adults. Mostly it is about Guillain-Barre syndrome. Then, we return to "zoonotics" and vectors responsible for the ZIKV and potential reservoirs in Chap. 10. Chapter 11 reviews the difficulties in diagnosing ZIKV diseases, the scarcity of treatments, and the failure to produce a vaccine. Chapter 12 reviews traditional twentieth-century vector management tools, primarily larvicides, and insecticides.

The following three chapters are the heart of my fellowship and what drove me to do this research. Chapter 13 examines the utility of Wolbachia in reducing vector strength. Chapter 14 traces the groundbreaking work by the small U.K. company, Oxitec, as it tested its genetically engineered mosquito. The last of these three chapters (15) catalog the development of gene drive technologies, especially CRISPR-Cas9, and how they may be used as a tool to collapse mosquito populations.

Chapter 16 examines the effectiveness travel, and pregnancy warnings may have had on the ZIKV. This chapter tries to make sense of the approach taken by governments and the health industry complex in its approach to the ZIKV.

Chapter 17 takes a much more macroscopic view of pandemics and evaluates how communication theory animates and deepens understanding of health messaging.

Chapter 18 engages the whole idea of engagement as a principle of communication and what roles it can play in pandemic communication.

Chapter 19 tries to answer what should have been learned and what might be done to make the world more resilient when the next epidemic or pandemic hits.

References

1. Chang C et al (2016) The Zika outbreak of the 21st century. J Autoimmunity 68:1–13. February 28. https://www.ncbi.nlm.nih.gov/pubmed/26925496. Accessed 27 Oct 2016
2. Woodhouse MEJ, Gowtage-Sequeria S (2005) Host range and emerging and reemerging pathogens. Emerg Infect Dis 11:12. https://www.ncbi.nlm.nih.gov/pmc/articles/PMC3367654/. Accessed 20 July 2017
3. Ratna K (2017) The secret life of Zika virus. Speaking Tiger Books, New Delhi
4. Alforte A (2016) Zika virus sexually transmitted: More evidence suggested. ITeach-Post. December 24. http://www.itechpost.com/articles/68152/20161224/zika-virus-sexually-transmitted-more-evidence-suggests.htm. Accessed 10 Jan 2017
5. Almendrala A (2016) 4 things to know about Zika's potential spread to the US. The Huffington Post. January 11. http://www.huffingtonpost.com/entry/zika-virus-spread-what-to-know_us_569045afe4b0cad15e64ec4a. Accessed 24 Oct 2016
6. Deniz D (2016) Zika: From the Braziulian Backlands to global threat. Zed books, London
7. McNeil DG (2016) Zika: the emerging epidemic. W. W. Norton & Co., NY
8. Ratna K (2017) The secret life of Zika Virus. Speaking Tiger Books, New Delhi
9. MacNamara FN (1954) Zika virus: a report on three cases of human infection during an epidemic of jaundice in Nigeria. Trans R Soc Trop Med Hyg 48(2). May. 10.1016/0035-9203(54)90006-1
10. Bearcroft WGC (1956) Zika virus infection experimentally induced in a human volunteer. Trans R Soc Trop Med Hyg 50(5):442–448. September
11. McNeil DG (2016) Zika: the emerging epidemic. W. W. Norton & Co., NY
12. Chen H-L, Tang R-B (2016) Why Zika virus infection has become a public health concern? J Chin Med Assoc 79:174–178. http://www.sciencedirect.com/science/article/pii/S1726490116300065. Accessed 10 Apr 2017
13. Hanley KA et al (2013) Fever versus fever: the role of host and vector susceptibility and interspecific competition in shaping the current and future distributions of the sylvatic cycles of dengue virus and yellow fever virus. Infect Genet Evol. March 20. https://www.ncbi.nlm.nih.gov/pubmed/23523817. Accessed 11 June 2019
14. Faye O et al (2016) Molecular evolution of Zika virus during its emergence in the 20th Century. In: Haddow A et al. (eds) Zika virus. China: (SCIRP) Scientific Research Publishing, Wuhan
15. de Oliveira Dias JR et al (2018) Zika and the eye: pieces of a puzzle. Progr Retinal Eye Res. https://doi.org/10.1016/j.preteyeres.2018.04.004. Accessed 6 July 2018
16. Olmstead G (2016) The Zika project. Amazon, Middletown, DE
17. Otieno F (2016) Zika: the emerging epidemic: be on the know. Amazon, Middletown, DE
18. Brito CAA (2017) Zika virus: history and infectology. In: de Fátima Viana Vasco Aragão M (ed) Zika in focus: postnatal clinical, laboratorial and radiological aspects. Springer, Washington, DC; Rodhain F et al (1989) Arbovirus infections and viral haemorrhagic fevers in Uganda: a serological survey in Karamjoa district, 1984. Trans R Soc Trop Med Hyg 83:1989. https://www.ncbi.nlm.nih.gov/pubmed/2559514. Accessed 11 Aug 2018
19. Musso D, Gubler DJ (2016) Zika virus. Clin Microbiol Rev 29. May 9. http://cmr.asm.org/content/29/3/487.abstract. Accessed 18 July 2017
20. Aragão MFVV (2017) Zika virus: an overview. In: de Fátima Viana Vasco Aragão M (ed) Zika in focus: postnatal clinical, laboratorial and radiological aspects. Springer, Washington, DC

21. McNeil DG (2016) Zika: the emerging epidemic. W. W. Norton & Co., NY
22. de Oliveira Dias JR et al (2018) Zika and the eye: pieces of a puzzle. Progr Retinal Eye Res. https://doi.org/10.1016/j.preteyeres.2018.04.004. Accessed 6 July 2018
23. Chang C et al (2016) The Zika outbreak of the 21st century. J Autoimmunity 68:1–13. . February 28. https://www.ncbi.nlm.nih.gov/pubmed/26925496. Accessed 27 Oct 2016
24. Vogel G (2016) Scientific sleuths hunt for Zika-carrying mosquitoes. Science. June 1. http://www.sciencemag.org/news/2016/06/scientific-sleuths-hunt-zika-carrying-mosquitoes. Accessed 9 June 2017
25. Fauci A, Morens D (2016) Zika Virus in the Americas—yet another arbovirus threat. New England J Med 374(7):602–606. February 18. http://www.nejm.org/doi/full/10.1056/NEJMp1600297#t=article. Accessed 4 Sept 2016
26. Kinszer K et al (2017) Reconstruction of Zika virus: introduction in Brazil. Emerg Infect Dis 23:1. January. https://wwwnc.cdc.gov/eid/article/23/1/16-1274_article. Accessed 17 May 2017
27. Lade D (2017) Zika one year later: Is it going away? Sun Sentinel. January 14. http://www.sun-sentinel.com/health/fl-zika-2017-outlook-20170114-story.html. Accessed 19 May 2017
28. Rose J (2016) Is Zika virus still around in the US? A CDC report suggests the threat is still very real. Romper. July 6. https://www.romper.com/p/is-zika-virus-still-around-in-the-us-a-cdc-report-suggests-the-threat-is-still-very-real-68398. Accessed 21 July 2017
29. Lade D (2017) Zika one year later: Is it going away? Sun Sentinel. January 14. http://www.sun-sentinel.com/health/fl-zika-2017-outlook-20170114-story.html. Accessed 19 May 2017
30. OECD (2018) Safety assessment of transgenic organisms in the environment, Volume 8: OECD consensus document of the biology of mosquito Aedes aegypti. Harmonisation of Regulatory Oversight in Biotechnology, OECD Publishing, Paris
31. Rose J (2016) Is Zika virus still around in the US? A CDC report suggests the threat is still very real. Romper. July 6. https://www.romper.com/p/is-zika-virus-still-around-in-the-us-a-cdc-report-suggests-the-threat-is-still-very-real-68398. Accessed 21 July 2017
32. Squiers L et al (2018) Zika virus prevention: U.S. travelers' knowledge, risk perceptions, and behavioral intentions—a national survey. Am J Trop Med Hyg 98(6). June. https://www.ncbi.nlm.nih.gov/pubmed/29737272. Accessed 26 July 2018
33. Beck J (2016) Zika is a delayed epidemic. The Atlantic. April 19. https://www.theatlantic.com/health/archive/2016/04/zika-is-a-delayed-epidemic/478755/. Accessed 5 June 2017
34. McNeil DG (2016) Zika: the emerging epidemic. W. W. Norton & Co., NY
35. Disler M (2017) Studying Zika. Harvard Mag. May 8. http://harvardmagazine.com/2017/05/studying-zika-harvard. Accessed 5 June 2017
36. Chang C et al (2016) The Zika outbreak of the 21st century. J Autoimmunity 68:1–13. February 28. https://www.ncbi.nlm.nih.gov/pubmed/26925496. Accessed 27 Oct 2016
37. Carribean360 (2016) Zika's emergency phase gives way to a long-term public health challenge. October 12. http://www.caribbean360.com/news/zikas-emergency-phase-gives-way-long-term-public-health-challenge. Accessed 15 Mar 2017
38. Beck J Tiny vampires: on living with mosquitoes in the time of Zika. The Atlantic. September 15. https://www.theatlantic.com/health/archive/2016/09/tiny-vampires/500069/. Accessed 8 Mar 2017

Contents

Chapter 1
Why Study Zika?

"After COVID-19, ZIKV became a silent epidemic." Claudia Osorio conducted an analysis of pharmaceutical services regarding preparedness for the epidemic in Campo Grande, the capital city of the State of Mato Grosso do Sul, and the risk perception of two vulnerable women communities located in Sidrolândia, a smaller city in the state, and the city of Rio de Janeiro. She remarked, "lately, it has been impossible to get current data on Zika virus disease. The last information taken was from May 2020, and around 500 people were notified to have had the disease." [1].

Years later, according to experts, despite advances in understanding the transmission, pathogenesis, and clinical progression of the mosquito-borne disease, a lot is still unknown about Zika's emergence as the cause of an unexpected epidemic center in Brazil [2].

According to the ZikAlliance: After years of the epidemic that put Northeast Brazil at its epicenter, the Zika virus and its consequences may be away from the spotlight, but it is still a burden that needs to be addressed.

The world is distracted, not only by shiny objects and unrecognizable sounds but also by the public agenda. Historically, once an epidemic is over, it is soon put out of mind.

Joanna Passos, a mother to a baby affected by the Zika virus (ZIKV) and founder of the Association aBRAÇO a Microcefalia, testified at the 2022 ZikAlliance Consortium Annual Meeting on the implications of putting Zika behind us.

"We women, main and often exclusive caregivers, find ourselves alone, needing to renounce our jobs, lives, dreams, and often ourselves. We live an intense routine of taking care of our children. We cannot have a pap smear or care for our physical health, much less emotional. We were forgotten, invisible," reminding the essential role women have in such a dramatic situation [3].

There is the ZIKV epidemic, an epidemic that mostly had a beginning, a middle, and an end. It is ripe for a post-mortem. Post-mortems demand that most of the variables be settled for the results to be reasonably valid. As such, approaches, activities, and events associated with the ZIKV epidemic of 2016–2018 can be examined

D. M. Berube, *Pandemics and Resilience: Lessons we should have learned from Zika*, Risk, Systems and Decisions, https://doi.org/10.1007/978-3-031-25370-6_1

to discuss and evaluate the feasibility, experience, and general effectiveness of a zoonotic outbreak's treatment and containment strategies.

In terminology, the word "epidemic" is an emotionally charged term. It means different things to different people, and professionals using the term may have an intended meaning quite different from the public's perception of the word.

Interpreting the term "epidemic" could depend on the context in which it is used. The term "outbreak" is considered by epidemiologists as a synonym for "epidemic." Epidemiologists use it in its most general form and define an epidemic as follows: "An epidemic is an occurrence in a community or region of cases of an illness, specified health behavior, or other health-related events clearly over normal expectancy; the community or region, and the period in which cases occur, are specified precisely." [4]. On the other hand, a pandemic is defined more transnationally and internationally, often globally. COVID-19 is a pandemic. Given these definitions, the ZIKV event that is the work's subject is an epidemic unless found in a quoted passage.

While the decline in ZIKV incidence is undoubtedly a positive development, it exposes apparent gaps in understanding of its natural history and epidemiology, which limits the ability to plan for, detect and respond to future epidemics. For example, the short duration of the epidemic and the long lead time needed to investigate comparatively rare congenital impacts have meant maternal cohort studies may be statistically underpowered to assess relative risk and factors associated with ZIKV-related adverse infant outcomes [5].

As aforementioned, events like epidemics are put aside after they are no longer troublesome. It seems as if we want to forget them and the implications and inconveniences. Institutional memories are short and trying to forget an epidemic with all its negativity is understandable but not practical. The ZIKV may appear again among a naïve population or something very much as it will appear instead. Another mosquito transmissible virus is nearly inevitable that being prepared with data-driven strategies should be the foremost goal of the international epidemic management community, and that is not the case.

How the ZIKV epidemic was framed came directly from the top. "It tends to go away, and then you're done," said Anthony Fauci, the National Institute of Allergy and Infectious Diseases director. For most of the population, "it's not scary at all. If you're a woman who's pregnant, who gets infected, particularly in the first trimester, it's overwhelmingly catastrophic." Miscarriages, fetal death, microcephaly, and other congenital disabilities are possible because of Zika infection during pregnancy [6].

ZIKV was framed as a woman's disease, or at least a pregnant woman's condition, and a poor woman and a woman of color's epidemic, mainly because it targeted a woman's reproductive system. Some report that women more than men were victimized by the ZIKV.

Experts tended to organize reports on its effects against the frame of a women's pregnancy. For example, Brasil et al., in their December 16 follow-up on a March study, reported the time of acute ZIKV infection ranged from 6 to 39 weeks of gestation [7]. ZIKV is something new and unique, but "each epidemic is unique," infectious disease expert and COVID talking head and Baylor Dean Peter Hotez also framed the ZIKV as a disease of pregnancy.

"Each epidemic has its little shop of horrors that you must sort out. And Zika has told us what the shop of horrors is. It's vertical transmission to the unborn baby, microcephaly, number one, and number two; you also have some other late neurologic sequelae. Knowing that, you must then plan accordingly." [8]. This perspective will be detailed below.

Like HIV/AIDS, which was initially framed as a gay disease, the ZIKV's framing as a women's disease has had unanticipated consequences. It should be considered how powerful these frames can be when designing a vector management policy (see Chap. 8).

There was a consensus that the ZIKV was especially important. Without any doubt, the ZIKV commanded more basic research because of the nature of the disease–congenital anomalies following an arthropod-borne virus infection. This was being seen for the first time. Also, the transmission mode during pregnancy and how it produced fetal damage was unknown, promoting intensive research on the virus and its biology and disease mechanisms [9]. The Centers for Disease Control and Prevention (CDC) called it "the most difficult" emergency response the agency has ever had to do [6].

It is still not understood how long it can linger in the human body. ZIKV was a virus that had slipped under the radar. While many of the unknowns reported in mid-2016 have been resolved, much to learn about the arbovirus remains.

The context: There are currently 490 known arboviruses, and this vast group contains representatives from several different viral families [10], with Antarctica as the only continent free of endemic arboviruses [11].

In 2015, a search in PubMed turned up just over 200 papers, compared with more than 2500 for chikungunya and more than 14,500 for dengue [12]. In 2017, that increased to over 1800 journal articles [13]. While much of the technical literature was redundant, the ZIKV went from an unknown to a popular subject in infectious disease, entomology, and public health literature.

Factors determining the emergence of flaviviruses and continuing circulation in sylvatic cycles are incompletely understood. Data about the transmission of ZIKV in South American wildlife are critical to inform on the maintenance of the virus in the region, as sylvatic hosts can serve as a source of recurring epidemics and impede efforts for long-term disease prevention. Recognition of wildlife hosts for zoonotic pathogens is essential for evaluating pathogen evolution within wildlife reservoir populations, advancing early detection of outbreaks through identification of sentinel species, mitigating risk at animal–human interfaces, and monitoring threats that zoonotic pathogens might pose to wildlife [14]. In April 2016, researchers in Ecuador and Northeastern Brazil reported the interspecies detection of Zika in monkeys, suggesting a new transmission cycle that could allow the virus to persist. In addition, at least in Brazil, the virus detected in monkeys was identical to the one circulating in humans [15].

Also, there were many "firsts" associated with this outbreak, and for me, these firsts were the main reasons to spend as much time as possible understanding the ZIKV. The virus has killed only 20 people in the Americas since 2015. A host of morbidities (see Chaps. 6 and 7) are associated with this virus, and the full run of effects of infection years after being infected has probably not been seen. Since so many cases were asymptomatic, the true long term and systemic implications may never be known.

In the literature, it is referred to as a neglected tropical disease. Neglected tropical diseases (NTDs) include several parasitic, viral, and bacterial diseases that cause substantial illness for more than one billion people globally. Affecting the world's poorest people, NTDs impair physical and cognitive development, contribute to mother and child illness and death, make it difficult to farm or earn a living, and limit productivity in the workplace. As a result, NTDs trap the poor in a cycle of poverty and disease [16]. And NTDs receive the least attention.

What follows are ten primary considerations motivating me to author this book, as well as one based so immensely on the many references throughout literature. There may be more in your opinion, or even less, but these reasons should be enough for some of us to continue to take this NTD seriously. The ZIKV may return.

1.1 Aegypti and Albopictus Mosquitoes Are Robust

The *Ae. aegypti* and *Ae. albopictus* mosquitoes are nasty bugs. The mosquito *Ae. aegypti* remains the primary vector of viruses responsible for severe diseases such as yellow fever, dengue fever, chikungunya, and ZIKV. This insect is currently subject to biotechnological research and applications (including genetic engineering), aiming to contribute to controlling its population, reduce its capacity to spread diseases, and thus limit its drastic impact on human health. *Ae. albopictus* is a secondary vector for the ZIKV and has demonstrated remarkable adaptability. Over relatively few generations, many populations of this species have evolved to resist cold winters in temperate areas; for example, diapausing eggs have developed cold hardiness. The tiger mosquito has adapted to living in a wide range of human containers, entering private houses, and feeding at night [17].

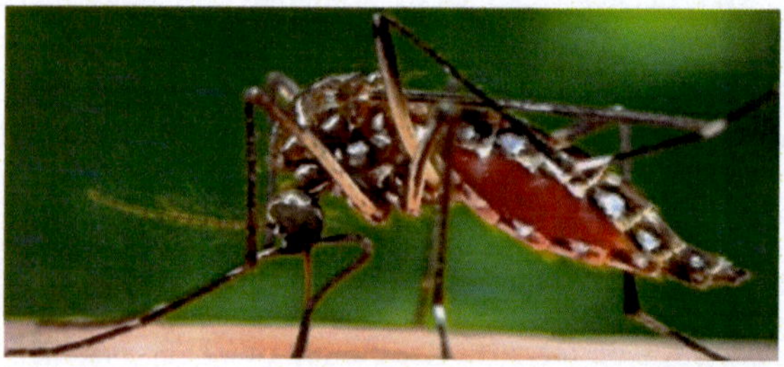

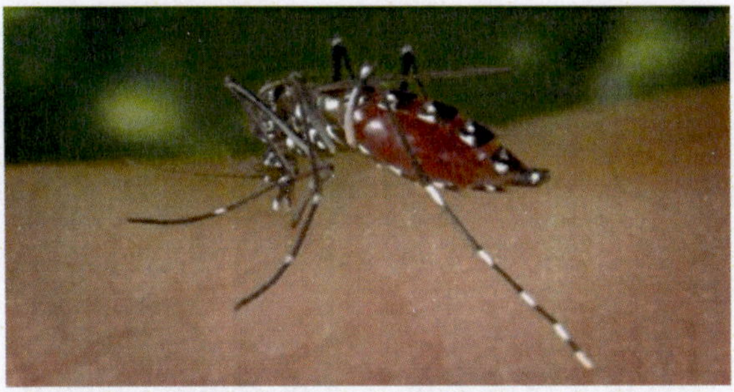

Consequently, this has led to some discussion of novel approaches to mosquito management. With advances in genetic engineering (GE), some techniques involve genetic modification of the carrier species with all the GM baggage of the genetically modified food and genetically modified organism (GMO) debates of the last few decades. In addition, the discovery of efficient engineering techniques and tools involving gene drives, such as CRISPR Cas9, amplifies the GE options (see Chap. 14).

One of the benefits of mosquito management that targets the carriers of the ZIKV would be the coincidentally reduced exposure to the other viruses they may carry, especially dengue fever. The World Health Organization estimates that 390 million dengue infections occur annually, with almost 25% of those severe enough to require hospitalization. The global incidence of the disease has rapidly increased in recent years, and now, roughly half of the people in the world are at risk of catching it [18]. The killer variant is called dengue hemorrhagic fever.

The same mosquitoes transmit chikungunya. Over 2.5 million cases have been reported over the last decade. While chikungunya does not kill as often as dengue, it leaves victims with pain and excessive fatigue. There is no cure for the disease. Treatment is focused on relieving the symptoms. In 2016, 349,936 suspected and 146,914 laboratory-confirmed cases were reported to the PAHO regional office, half the burden compared to the previous year. Countries reporting the most issues were Brazil (265,000 suspected cases), Bolivia, and Colombia (19,000 suspected cases, respectively). [19]

The same mosquitoes also transmit yellow fever. In 2013, yellow fever resulted in about 127,000 severe infections and 45,000 deaths. A modeling study based on African data sources estimated the burden of yellow fever during 2013 was 84,000–170,000 severe cases and 29,000–60,000 deaths [20].

Bot the *Ae. aegypti* and *Ae. albopictus* is tricky to control as well. The adult life expectancy varies from 10 to 35 days for female mosquitoes [21] and 3–6 days for male mosquitoes. However, this is highly dependent on temperature, shorter in tropical regions and more prolonged in temperate climates.

Also, the a*egypti* has developed immunities to many pesticides usually used for vector control. Such resistance has been observed in more than 500 insect species worldwide, including more than 20 Aedes species. Over 400 scientific reports worldwide document insecticide resistance in *Ae. aegypti* [10].

Dried-out *Aedes* eggs, dormant for months, can be revitalized and hatch with a quick moistening rain, Jonathan Day, University of Florida medical entomologist, said.

The United States is hardly immune to these diseases. For example, a dengue outbreak in Key West, also spread by Aedes, survived through the 2009 winter into 2010 [22].

1.2 Underreporting Frequencies and Issues Related to Immunity

ZIKV transmission may occur without an identifiable outbreak since most infections are asymptomatic. Identifying a significant and unreported episode in Cuba in 2017 [23], with cases still being identified in 2018, suggests that ZIKV may still be spreading silently in the Americas [24].

ZIKV has been underreported mainly because of its clinical similarity with dengue and chikungunya [25]. Also, there is significant underreporting when infected asymptomatic and even symptomatic people refuse to visit medical professionals. Estimating the prevalence of ZIKV infections is difficult because many infected individuals do not seek medical attention, and seroprevalence studies are hampered by antigenic cross-reactivity with other flaviviruses [26].

Before moving on, the adaptation of the ZIKV to an urban cycle involving humans and domestic mosquito vectors in tropical areas where dengue is endemic suggests that the incidence of ZIKV infections may be underestimated. There is a high potential for the ZIKV emergence in urban centers in the tropics infested with competent mosquito vectors such as *Ae. aegypti* and *Ae. albopictus* [27]. Also, most of the ZIKV infections were probably missed or incorrectly diagnosed, as suggested by the high prevalence of the ZIKV antibodies found in serosurveys of human populations in Africa and Asia [28].

An exciting study in Martinique provided a peek into the challenges of estimating prevalence. The number of clinically suspect cases in Martinique (French Public Health Institute) was 7600, or 2% of the population, in early March (week 8), and 28,900, or 7.6% of the people, in early June (week 22). This suggests that the proportion of cases that did not seek medical attention was 80–85% and did not change significantly during the study period [26]. See more below.

In addition, there is some concern that the ZIKV mutated. It has been hypothesized that the virus may have evolved to become more neurotropic, exhibit increased replicative capacity, and become more transmissible to humans. Still, causal support for these possibilities is outstanding [29].

Being immune to a virus is a good thing until it's not. That's the lesson from a study that sought to understand the severity of the ZIKV outbreak in Brazil. Experiments on mice suggested previous exposure to dengue or West Nile can worsen a ZIKV infection. "Antibodies you generate from the first infection can facilitate entry of the ZIKV into susceptible cells, exacerbating the disease outcome" [30]. This is called antibody-dependent enhancement. Leftover antibodies from the first episode can sometimes turn traitor, enhancing the growth of the virus inside the body [31]. Citing Dejnirattisai et al. [32], Abbink observed that "dengue-specific antibodies have been reported to increase the ZIKV replication in vitro." [33]. Immunity to DENV might drive more excellent ZIKV replication and have clear implications for disease pathogenesis and future vaccine programs for the ZIKV and DENV [32].

1.3 Sexual Transmission

A significant first involves the fact that this is ***the first mosquito-borne virus that can be sexually transmitted*** [34]. This is a monumental wild card in the ZIKV story.

The ZIKV is unique among arboviruses because it can be transmitted during sexual contact and can cause teratogenic outcomes (development deformations) due to maternal–fetal transmission [24]. "Many viruses are spread through the bite of mosquitoes," Squiers explained. "However, the ZIKV is unique because it also can be transmitted through sex, from a pregnant woman to her baby, and through blood transfusions. This makes it special."

Unlike other tropical viruses, they can be sexually transmitted. "That makes Zika a different breed of dog," Dr. Duane Gubler, a former director of the vector-borne diseases division of the Centers for Disease Control and Prevention, said. If an infected person passes it to a household sexual partner, the virus will be in human blood for up to 20 days at that location for local mosquitoes to pick up and pass on [35].

Although the effect of sexual transmission in areas where the virus is endemic is challenging, estimates are that around 1% of ZIKV infections reported in Europe and the United States were acquired through sexual transmission [24]. The possible sexual transmission from **asymptomatic cases** also increases the risk of the emergence of the ZIKV in Europe in areas where the mosquitoes are present [34].

In humans, it may also be contracted as a sexually transmitted viral infection, like HIV or syphilis appears to be considerably less than HIV or syphilis [36].

Furthermore, this virus strain seems highly associated with a dreadful congenital disability known as microcephaly (see Chap. 6). It carries a thorough list of developmental disorders and another long list of physical abnormalities, and it attacks a vulnerable and dependent population, children. The victims of microcephaly are physically observable and heart-wrenching. Since 2016, it has been learned the damage to fetuses extends beyond heads, measurably more minor than the norm.

1.4 Microcephaly

ZIKV is not only the only member of the genus classified as a teratogen or an agent that intervenes in fetal development. It was also the subject of a contentious debate. While both experimental and epidemiological data now strongly link ZIKV infection to abnormal fetal development and to severe neurological anomalies, like the TORCH pathogens: toxoplasmosis, others (syphilis, parvovirus B19, varicella-zoster), rubella, cytomegalovirus, and herpes simplex virus [37], many claimed the linkage between the ZIKV, and microcephaly was specious.

According to Watson and Mason, the responses at the outbreak of the crisis gave way to a "tyranny of the urgent." [38]. The number of microcephaly cases during the ZIKV outbreak in Brazil was about twenty times higher than normally would be expected, and that became the crux of the debate. Such a congenital malformation has not been seen with any other flavivirus or arthropod-borne virus [39], and what if the causes of microcephaly could be traced to some other reason.

The ZIKV loomed as a **developmental doomsday virus**, attacking the vulnerability of early brain development, and striking at the neurological basis of human potential. There's so much that isn't known. Dr. Peter Jay Hotez, the dean for the National School of Tropical Medicine at Baylor College of Medicine in Houston, said several big questions need to be answered. How does the virus do its damage? What is the full spectrum of damage, from visible microcephaly to less visible neurological changes? And what happens to babies exposed to Zika after birth when the brain is still developing? [40].

Population-level increases in ZIKV-associated congenital disabilities are unlikely to be recognized without ongoing timely and comprehensive surveillance of congenital disabilities that captures all affected fetuses and infants regardless of whether maternal ZIKV exposure or infection was identified.

The approaches and lessons learned from the ZIKV outbreak in the Americas show the necessity of improved and integrated ongoing systems for surveillance of pregnancies, infectious diseases, and congenital disabilities to rapidly address the next emerging health threat affecting pregnant women and infants. Sustained commitment to better monitoring systems for these medically vulnerable populations will more promptly identify serious health threats. It will allow the public health field to affect infant health and prevent congenital disabilities positively.... Population-level increases in ZIKV-associated congenital disabilities are unlikely to be recognized without ongoing timely and comprehensive surveillance of congenital disabilities that captures all affected fetuses and infants regardless of whether maternal ZIKV exposure or infection was identified [41].

"Many viruses, including some similar to Zika, can infect the placenta and the baby's cells," says George Saade, an obstetrician–gynecologist and cell biologist at the University of Texas Medical Branch at Galveston. "This list keeps growing and highlights the risks from viruses that we are not very familiar with."

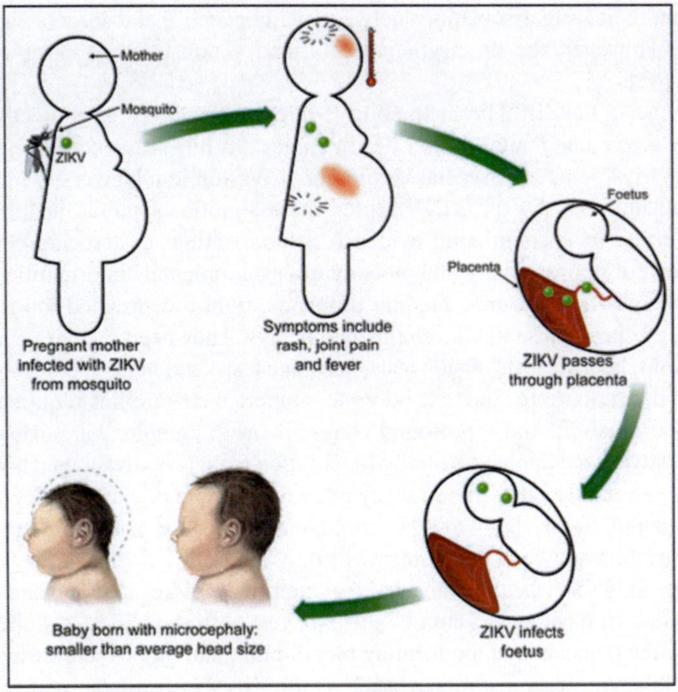

An obstetrician and gynecologist at Emory University School of Medicine in Atlanta, Denise Jamieson, warned, "We are not ready for another emerging infectious disease that may disproportionately affect pregnant women or their fetuses or babies. And we need to be."

1.5 ZIKV Crosses Brain and Placental Barriers

The ZIKV seems to be **the only virus (arbovirus, flavivirus) to cross the placental barrier and stunt the fetus's brain development** [42]. 14.5% of children exposed to the ZIKV during their mother's pregnancy displayed at least one medical issue related to vision, hearing, language, motor skill, or cognitive function [43].

The US has never faced a single mosquito bite that could result in dreadful, permanent, life-altering congenital disabilities. The CDC said about 10 percent of infants of women with confirmed ZIKV infection during pregnancy in US states had ZIKV-associated congenital disabilities. There is limited information about the risk of Zika infection eight weeks before conception or six weeks before the last menstrual period (periconceptional period) [44].

In August 2018, Vital Signs report about 1 in 7 (or 14%) of 1450 babies included had one or more health problems possibly caused by ZIKV. In addition, The CDC says there is information from other viral infections occurring around the time of conception, indicating associations between periconceptional disorders and adverse outcomes. However, the timing of infection and vision in these cases was often unknown [45].

The danger of the ZIKV lies in its ability to cross the body's most protective barrier and cause severe congenital disabilities in babies. "It hits at the core of humanity," said Indira Mysorekar, an associate professor at Washington University [46].

Prenatal infection by the ZIKV represents a significant public health concern worldwide due to accumulating evidence demonstrating an association with the development of microcephaly and other neurodevelopmental abnormalities [47].

These babies risk seizures, feeding problems, tightly contracted joints, thyroid problems, eye problems, and developmental delays. They need regular neurological examinations, hearing tests, vision tests, hormone tests, and many medical back-ups. Providing decent care for these children to support their families requires subspecialists, coordination, and a profound commitment to complex care. That kind of care is not always available, especially for children from less advantaged homes, and there's a concern that Zika, like so many other diseases, may play out like privilege, with poor people more likely to be exposed, either because of less protected living situations or jobs that keep them outside [40].

The human CNS is usually considered an immune-privileged organ and is generally protected from pathogens circulating in the rest of the human body during development by the placenta and the forming blood–brain and blood–cerebrospinal fluid barriers. However, some pathogens, such as the ZIKV, possess the ability to cross the placenta and blood–brain barriers and infect the brain [48].

"We've never seen anything like Zika before, nothing on the scale of what it does to the central nervous system," said Ira Longini, a biostatistics professor from the University of Florida's College of Public Health and Medicine. "It's what makes the virus very serious." [22].

Tang et al. found out that the macrophages were relatively immobilized in the case of a dengue virus infection and remained still occupying one place. On the other hand, those macrophage immune cells infected with the Zika virus migrated. In one mammal infected with Zika, the researchers noticed that the macrophage's cells travel all over the body via the bloodstream. "They are uploading macrophages to other parts of the body," added Tang, asserting that the Zika virus aggressively eliminates the macrophage's ability to perform its typical disease-fighting tasks.

"Now the question is: with the increased ability to spread throughout the body, does the Zika virus also use these infected macrophages to cross the placental barrier, the blood–brain barrier, and the testicular barrier? If we understand how they cross these barriers, we can develop more effective countermeasures to protect people," Hengli Tang concluded [49].

Given that each infant with microcephaly is expected to require $10 million in care over a lifetime, just 100 babies with microcephaly—a tiny fraction of all annual births in the United States—would total $1 billion in costs [50].

1.6 Travel Warnings and Women

ZIKV seems to cause congenital disabilities if a pregnant woman passes the virus to her fetus. This led the WHO, the CDC, and other health organizations to warn that it was essential that pregnant women do not travel to areas with local transmission of the ZIKV. Pregnant women, partners of pregnant women, and couples considering getting pregnant should know how to prevent ZIKV infection both during and after travel." [51].

In January 2016, "out of an abundance of caution," the US Centers for Disease Control and Prevention (CDC) issued its first travel warning. The agency recommended that pregnant women avoid traveling to areas where the virus was circulating. The CDC has issued scads of similar travel warnings, covering more than 50 countries, including some parts of the U.S. [52]. For example, the historic travel warning to stay away from a neighborhood north of downtown Miami marks the first time in 70 years that the U.S. government has issued such a warning [53].

In addition, in the history of public health, countries have advised their populations to postpone planned pregnancies, for example, Brazil, Colombia, El Salvador, and Jamaica [54]. Travel notices for various countries in the Caribbean, Central and South America, Asia, and Africa remained in place two years after the first cases of microcephaly were reported in Brazil.

As Vielot et al. [55] wrote, "there is little hope that they will be lifted until more can be done to prevent the ZIKV infection and its acute and long-term sequelae. Even as the initial epidemic waned, the ZIKV is likely to become endemic in many areas and remains a concern to local populations and travelers [55]. Even in 2018, The CDC continued to issue Travel Alerts warning visitors to Zika endemic areas, such as Mexico, Brazil, India, and the Caribbean Islands, to take precautions [45]. An entire chapter examines the "travel" warning approach.

During the initial phase of the Zika outbreak, the links between the location of microcephaly cases, poverty, public health, the women most affected—the urban poor, the Indigenous in remote areas—and their lack of access to contraceptives and abortion emerged as matters of particular concern. Warnings broadened to include recommendations to avoid pregnancy altogether if exposed to the ZIKV.

Already marginalized women were being asked by governments to avoid pregnancy, apparently without any acknowledgment by these same governments of their role in hindering women's access to contraceptives, sex education, and safe abortion practices in the first place [56].

Advising women to stop having children was unprecedented. Never in history had governments done so. China's one-child policy was different: It was semipermanent and implemented for economic reasons [57]. And the recommended "suggested delays" kept growing. Colombia's health minister asked women to wait 6–8 months. His Jamaican counterpart upped the ante to a year. Then El Salvador proposed two years [57].

1.7 Blue Marble Health

Was the ZIKV epidemic sufficient to demand this much attention? It wasn't much of a U.S. problem, especially if you are inclined not to think of Puerto Rico and the Virgin Islands as part of the U.S. Except for travel-related exposures and mild outbreaks in Texas, California, and Florida, direct exposure via mosquitoes seemed to be minimized.

Some mosquito experts questioned the extent of the emergency. Chris Barker, a mosquito-borne virus researcher at the University of California, Davis School of Veterinary Medicine, told WebMD: "I think the risk for the ZIKV setting up transmission cycles that become established in the continental USA is near zero." Barker expected the ZIKV to go the way of other tropical diseases spread by mosquitoes, such as dengue fever and chikungunya, in the USA, with small clusters of outbreaks in southern states and little activity elsewhere [58]. And, he was generally correct.

Duane J. Gubler, an emerging infectious disease expert at Duke-NUS Medical School in Singapore, put it similarly. "There is no cycle to break," he said "Sustained transmission is impossible, but you get small outbreaks when introductions occur." [59]. This is undoubtedly true in developed countries with mosquito abatement programs and the money to purchase screens and sustained air conditioning, but not so in more impoverished countries and areas within even developed countries. ZIKV was a brown disease affecting primarily people of color and poor people of color. The population of a much whiter North America did not seem to have much to fear.

This nationalistic tone underlies much of the literature on the ZIKV following the 2015–2016 outbreak. There was some odd poetic irony that the first concerns about the virus were associated with the Summer Olympics, another nationalistic symbol.

More than 40 countries in the Americas and several countries in Africa and Asia are Zika-affected areas. A mosquito has transmitted the virus in the US territories of Puerto Rico, the US Virgin Islands, American Samoa, South Florida, FL, and Brownsville, TX. Since 2015, the ZIKV has been reported in more than 5000 travelers returning to the U.S. from affected areas; though underreported, there were several cases of sexual transmission in the U.S. that have also been documented.

Recall what Dean Hotez said. Zika is a "neglected tropical disease" (NTD). With a few exceptions, NTDs are mostly low mortality (measured in years of life lost—YLLs) but high morbidity (measured in years lived with disability—YLDs) [60]. The long-term implications of contracting Zika were also underreported, especially regarding outbreaks of microcephaly and the long-term care of its victims.

Hotez wrote, "this blurring between the health of developed and developing countries has been termed "blue marble health," about the photograph of Earth (and a symbol for peace) taken by the Apollo 17 astronauts, and is meant to foster a global dialogue on the importance of poverty as a key underlying factor for NTDs, regardless of where they occur [60]. The concept of "blue marble health" has been invoked to describe the surprising disease burden of NTDs among the poor in these countries [61].

NTDs are the most common infections globally, thriving in impoverished communities, where they promote and cause poverty. They are the most common afflictions of girls and women and account for an essential yet unmeasured component of the NCDs. The concept of blue marble health recognizes that NTDs are pervasive wherever poverty occurs, including in the world's most devastated nations and emerging market economies and wealthy countries in North America and Europe. It will be essential to incorporate NTDs and blue marble health into the new framework of the sustainable development goals [60].

1.8 Cross-Protection/Cross-Infection

The co-circulation of chikungunya virus, dengue virus, and ZIKV among existing ecological niches in Latin American countries is a significant public health challenge that requires efforts to understand the transmission dynamics and the spectrum of clinical manifestations, health outcomes, and long-term sequelae of those coinfected with any of these emerging arboviruses [62].

Five years after the outbreak, it is still not possible to assess the longer-term effects of ZIKV on subsequent dengue disease risk in humans.

Some population studies showed that previous dengue virus (DENV) immunity correlated with a lower risk of ZIKV infection. The Zika epidemic was followed by several years of low DENV transmission, but in 2019, countries across Latin America reported a significant resurgence of dengue cases.

Dengue virus serotypes 1 to 4 (DENV1 to -4) and Zika virus (ZIKV) are closely related mosquito-borne flaviviruses, and there remains concerned that ZIKV infection or Zika vaccines could enhance subsequent dengue disease [63].

In Katzelnick's Nicaraguan cohort, there was evidence that previous DENV infection may protect against subsequent symptomatic ZIKV disease. Individuals exposed to ZIKV generated both neutralizing and enhancing antibodies for DENV [64].

According to Dr. Aubree Gordon from the University of Michigan's School of Public Health, "If dengue immunity protects from ZIKV, Zika immunity may protect from dengue. And there may be not only cross-protection but also enhancement of disease. Potential cross-protection or enhancement of disease severity will be critical issues to examine in vaccine development and evaluation." [65].

For ZIKV infection after a DENV infection, in vitro testing of sera showed that previous DENV infection led to neutralizing and enhancing responses to ZIKV [64].

Katzelnick et al. found that prior ZIKV infection modulates future dengue disease risk to a similar degree as the previous infection with a DENV serotype. A single ZIKV infection, like one initial DENV infection, increases the probability of symptomatic and severe dengue disease caused by DENV2. Further, one DENV followed by one ZIKV infection also increased the risk of dengue disease, unlike sequential DENV infections, which reduced future risk, suggesting an important essentialness between secondary flavivirus infection and ZIKV versus a DENV serotype [63].

The Katzelnick study serves as a reminder that issues of cocirculation and cross-immunity should remain in view for understanding transmission dynamics and vaccination. In this case, a previous ZIKV infection acts similarly to an earlier DENV infection [64].

Cross-reactivity with dengue is a particular concern. Ferguson et al. [66] indicated that cross-protection and enhancement could shorten the time until epidemics can reoccur and might increase the chances of long-term endemicity [66].

These phenomena implicated vaccine production since if monovalent Zika vaccines induce cross-reactive DENV antibodies such as those observed after natural ZIKV infection, Zika vaccines could increase the risk of subsequent symptomatic and severe dengue disease. The potential cross-protection between dengue and ZIKV is worthy of more research and could substantially change vaccines' research and development landscape [65].

All the research on ZIKV infection outcomes—whether the immunity is elicited by infection or by immunization with flavivirus vaccines—is a matter of debate [24]. Further investigation is needed to determine whether these findings are generalizable across regions in which ZIKV has emerged, whether they apply to severe ZIKV-associated outcomes such as the Guillain–Barré syndrome and congenital disabilities, and whether differing amounts or types of dengue antibodies influence the balance between protection from and enhancement of ZIKV infection [24].

1.9 Long-Term ZIKA

Infectious diseases with an exceptionally high disease burden, sometimes acute, other times chronic, often both. For many, the term "long-term" has been associated with COVID-19 to describe people who seem to hang onto their symptoms much longer than anticipated, months or even longer. For families affected by ZIKV, a difficult road lies ahead, and necessary support services must continue to be provided even though the initial urgency of the outbreak has passed (see Chap. 6).

An editorial commentary by Manuel Martínez-Sellés expertly summarizes some of the long-term complications of Zika. Cardiovascular complications in adults infected with ZIKV have been reported. Human fetal cardiac mesenchymal stromal cells (MSCs can be found in nearly all tissues but mostly located in perivascular niches, playing a significant role in tissue repair and regeneration) can be a target for ZIKV infection, potentially resulting in congenital heart defects during embryological development and cardiovascular events in ZIKV-infected adults [67]. Human fetal cardiac mesenchymal stromal cells can be a target for ZIKV infection [68], potentially resulting in congenital heart defects during embryological development and cardiovascular events in ZIKV-infected adults. Case reports have described adult atrial fibrillation [69] and ZIKV vasculitis as possible causes of stroke in children [70].

Scatularo et al. [71] show that cardiovascular threats of ZIKV may have been overlooked. The heart is usually not involved during ZIKV infections, but cardiovascular manifestations such as myocarditis, pericarditis, heart failure, and arrhythmias may appear [69]. As with other viral infections, an association with hypercoagulable states (tendency to have thrombosis) linked to developing deep venous thrombosis and pulmonary embolism has been reported [72].

Previous studies have reported congenital heart defects in ZIKV-infected infants [73]. Cardiovascular complications in adults infected with ZIKV have also been reported.

In addition, there is the enormous financial and emotional expenses to care for a microcephalic child in countries with minimal child health services (see Chap. 6), as well as a different set of challenges associated with caring for an adult with Guillain-Barré syndrome (see Chap. 7).

1.10 Glioblastoma

Of all the reasons for studying the ZIKV, this was probably the strangest. Some exciting research has been undertaken that suggests that an engineered form of ZIKV or ZIKV pseudovirus may be a possible cancer treatment for one of the most malignant cancers affecting humankind. On the other hand, it must be ensured that ZIKV can be engineered to eliminate its virulence but maintain its oncolytic activity.

The National Cancer Institute (NCI) estimates that nearly 23,000 people will be diagnosed with brain cancer, and about 14,000 will die from the disease this year. (To put that in perspective, more than 230,000 people will get breast cancer, and about 40,000 will die from it.)

One common type is glioma, a broad category of tumors that arise in the brain from glial cells, which are cells wrapped around neurons throughout the central nervous system. The most common and deadly type of glioma is glioblastoma, which can be aggressive in the final stages [74].

Glioblastoma is the most prevalent primary intrinsic brain tumor. Despite surgery, radiation, and chemotherapy, glioblastomas remain lethal [75]. Currently, the survival rate in the U.S. is about eight months.

Doctors can try to treat glioblastoma with surgery to remove the cancerous cells and blast them with radiation and chemotherapy. However, as the disease spreads, it becomes challenging to differentiate healthy brain cells from cancerous glioblastoma stem cells, and the condition is almost impossible to remove altogether [76].

Symptoms include recurring headaches, loss of appetite, blurred vision, vomiting, shifts in personality and the ability to learn, seizures, and the gradual loss of speech [76].

Brain tumor stem cells in the highest-grade tumors, glioblastoma, "borrow" the neural stem cell program essential to brain growth in a developing baby and use this to drive malignant growth and invasion [77].

Delaware Atty Gen. Beau Biden, Sen. Edward Kennedy, and Sen. John McCain had an aggressive brain cancer known as glioblastoma. Bizarrely, it may turn out that the power of the ZIKV may be directed at a type of brain tumor called glioblastoma, cancer which often causes death within a year of diagnosis.

Zika virus (ZIKV) has emerged as a potential anti-tumor therapy in a quest to find better treatments for glioblastoma. As of 2021, no clinical trials have been initiated to validate this therapy. All this research comes from mouse studies.

Nonetheless, oncolytic viruses, including ZIKV, constitute a promising new strategy for glioblastomas treatment. ZIKV is an exciting candidate due to its ability to infect and kill glioblastoma stem cells (GSCs) [78].

Historically, oncolytic virotherapy is a concept with a long history that was first described in 1912, when regression of cervical cancer was seen owing to the administration of the Louis Pasteur rabies vaccine. Clinical trials have been completed using modified versions of herpes simplex, adenovirus, Newcastle disease, reovirus, parvovirus, and poliovirus. Some of these studies demonstrated that a small subset of patients benefited from such virotherapy applications. Still, the statistical significance of these findings must be confirmed in more extensive control trials [79].

A few studies have confirmed that ZIKV selectively infects the fetus's neural stem cells (NSCs). This selective targeting of NSCs has encouraged other scientists and us to investigate whether ZIKV might exert an oncolytic action against GSCs [80].

The recent discovery that ZIKV utilizes the neural cell adhesion molecule (NCAM1 [neural cell adhesion molecule 1]) receptor to enter the cell further suggests that the ZIKV neurotropism could be exploited as a promising strategy. An engineered form of ZIKV, or ZIKV pseudovirus, has recently been developed [81] and holds therapeutic promise for its ability to infect target cells without further replicating, thus avoiding the undesired side effects associated with viral replication and propagation.

In September 2017, a group in California led by Harry Bulstrode from Cancer Research UK, including leading scientists in the brain tumor and virus fields, published compelling and comprehensive evidence that, as expected, Zika selectively attacks tumor stem cells and even prolongs survival in mice with tumors [82].

Moreover, mice with glioblastoma survived substantially longer and at more excellent rates when the tumor was inoculated with a mouse-adapted strain of ZIKV. Results suggest that ZIKV is an oncolytic virus that can preferentially target GSCs; thus, genetically modified strains that further optimize safety could have therapeutic efficacy for adult glioblastoma patients [75].

Kretschmer et al. [79] reported a discovery with a high impact on the future aspects of glioma virotherapy was that Zika virus (ZIKV) showed a specific neurotropism for glial cells. The new emerging ZIKV was shown to cause Guillain–Barré syndrome (GBS), microcephaly, and other congenital brain abnormalities, proving its neurotropic phenotype. An interesting finding was the oncolytic activity of ZIKV against GSCs [79].

When scientists evaluated two strains of the ZIKV on glioblastoma stem cells—which produce new cancer cells and are generally resistant to standard cancer treatments—they discovered that Zika killed those types of cells while leaving other tumor cells intact. This has led scientists to believe that the ZIKV can be a complementary treatment for glioblastoma [83].

"Thus, developing an effective therapy to kill glioblastoma stem cells is urgently needed," Dr. Pei-Yong Shi, a professor of genetics at the University of Texas Medical Branch and author of the study published in the journal mBio told Newsweek. As both the ZIKV and glioblastoma can affect the brain, the team hypothesized that the former could be harnessed to treat the latter [76]. The team injected a combination of glioblastoma stem cells into mice.

Intracerebral injection of ZIKV-LAV (live attenuated ZIKV vaccine candidate) into mice caused no neurological symptoms or behavioral abnormalities. ZIKV-LAV significantly reduced intracerebral tumor growth and prolonged animal survival by selectively killing GSCs within the tumor. Mechanistically, ZIKV infection elicited antiviral immunity, inflammation, and GSC apoptosis. Together, these results further support the clinical development of ZIKV-LAV for GBM therapy [84].

When the team sequenced the RNA (a molecule that helps with gene expressions) of human-derived glioblastoma stem cells exposed to the vaccine and then those which weren't, they found the virus prompted an antiviral reaction in the exposed cells in what is known as an oncolytic response. Before surgery, cancer patients could be given the Zika vaccine to "let the viruses hunt down the GSCs [glioblastoma stem cells] and eliminate them," said virologist Dr. Cheng-Feng Qin of the Chinese Academy of Military Medical Sciences in Beijing, who worked on the study. Shi said: "Scientists may turn the 'bad' side of a devastating pathogen—Zika virus—for potential cancer therapy." [76].

1.11 Nothing to See Here

An issue keeps needling me. Why was the infectious disease community so willing to call an end to Zika per se when so much remains to be addressed. There are considerable uncertainties associated with the ZIKV and approaches to reduce its impact, especially should it rise again or another virus like it.

Some referred to the concern as hysteria, a nineteenth-century term used to marginalize women with psychological problems. Florence Fouque, a World Health Organization (WHO) expert on animals that carry viruses, called the public response to the Zika virus "completely hysterical." She blamed the hysteria on the findings that the virus affects pregnant women and can be sexually transmitted. "It's like AIDS," she told PRI. "People make this link, and that's why they are terrified." At one restaurant in downtown Miami, Florida, where a handful of Zika cases were detected in August 2016, insect repellent was placed on all the tables [58]. Not sure anyone should consider eating at a restaurant that serves repellant, but that's another issue.

In all humility, there was some reason to question the alert status given the Zika epidemic in Latin America. First, there was a questionable relationship between ZIKV and microcephaly (see Chap. 6). According to a report by the New England

Complex Systems Institute (NECSI), there were significant questions about whether the ZIKV is the cause of microcephaly. Preliminary results of a *New England Journal of Medicine* study followed nearly 12,000 pregnant Colombian women infected with the ZIKV.

No cases of microcephaly were reported in their babies as of May 2016. Yet, four cases of microcephaly were reported among women with Zika infection with no symptoms and were therefore not included in the study. The researchers then speculated that there could be four times as many cases of the ZIKV infection unreported for at least 60,000 ZIKV-infected pregnancies in Colombia. Using this data, an analysis revealed the rate of microcephaly to be expected in any area, whether Zika is in the picture or not, which is two cases in 10,000 births.

According to NECSI, "This consistently interprets that there is no direct link between Zika and microcephaly except for random co-occurrence. Note that the base rate of microcephaly in the absence of Zika is 140 per year in Colombia, which is consistent with the approximately 50 microcephaly cases in the first [four] months of 2016, only [four] of which have been connected to Zika. When interpreting Zika as the cause, background cases must be subtracted." [58].

Second, is Zika so bad? There are a lot of viruses out there that are much more problematic. "During the Zika outbreak in Brazil, about 4000 children suffered microcephaly. During a rubella outbreak in the United States between 1963 and 1964—five years before a vaccine was available to prevent it—rubella infected 12.5 million people causing more than 6000 spontaneous abortions, 5000 therapeutic abortions, and 20,000 cases, of permanent congenital disabilities. One out of every 100 births in Philadelphia were complicated by congenital rubella infection during this outbreak. Between 1 and 13% of women infected with the Zika virus will deliver babies with congenital disabilities in the first trimester of pregnancy. First-trimester infections carry a risk of congenital disabilities of 90% for the rubella virus [85].

The main factors preventing these diseases in the United States today are socioeconomic, including lifestyle, housing infrastructure, and good sanitation. If such conditions are maintained, it seems unlikely that local transmission will occur significantly, particularly in the northern states [86].

According to infectious disease expert Dr. Sarah George from Saint Louis University and a Zika vaccine researcher. "We will have another Zika outbreak. We don't know when or where." In the 2016 outbreak, there were more than 5000 Zika cases in the U.S. Most were among people returning from affected countries, but more than 200 cases came from mosquitos in Florida and Texas [87].

1.12 Conclusion

ZIKV was a gift of sorts, the platforms established, and the experience gained must be capitalized on. Our responses and ZIKV research need to evolve into a network capable of rapidly launching a research response to future severe infectious outbreaks caused by emerging pathogens with pandemic potential or the potential to cause severe damage to health and socioeconomics in Latin America [88]. But it didn't.

Accurate estimates of the burden of the Guillain–Barré syndrome and congenital disabilities attributable to ZIKV infection have been hampered by a lack of systematic surveillance of these syndromes before and during the pandemic [24]. Still unknown is whether the ZIKV-associated complications identified during the epidemic were new emerging phenomena—perhaps caused by the virus acquiring enhanced fitness, transmissibility, or disease severity phenotype—or had occurred previously but went undetected because of limited surveillance or infrequent transmission [24].

On February 1, 2016, the WHO gave Zika an international public health emergency designation. Since 2007, this has happened only three times, influenza in 2007 and the polio resurgence, and the Ebola outbreak in 2014. Some have described this reaction as a response to being taken by surprise.

There should be concern. "About 40% of the global population is at risk [from] *Ae. aegypti* circulated arboviruses like the ZIKV," Andrew McKemey, an entomologist, and head of Oxitec Field operations, said [89]. According to Thomas Monath, CEO of the infectious disease division at NewLink Genetics Corp, hundreds of millions are at risk [90].

There remain some potential hotspots. For example, the annual Hajj, the Muslim pilgrimage to Mecca that brings millions of people to Saudi Arabia, has also been postulated as having helped introduce dengue, as it could cause ZIKV infection in 2016 or 2017, and the political destabilization in Venezuela could become a significant contributing factor in the next round of epidemics [61].

In an April eLife study, scientists used similar techniques to map out the worldwide risk of Zika, estimating that some 2.17 billion people may be in the virus's path [91]. The affected areas span many of the Southeastern U.S., including much of Texas and Florida. The researchers estimate that more than 5.4 million births will occur within the next year in parts of the Americas susceptible to ZIKV transmission. This is a cause for concern given the established link between ZIKV and microcephaly. In addition, the researchers found that large parts of India and sub-Saharan Africa—areas populated by 1.42 billion and 453 million people, respectively—are at risk of ZIKV transmission. The map does not account for non-mosquito forms of viral communication (such as sexual transmission). The researchers called for risk estimates to be updated as latest information became available [92].

Considering the widening distribution of the *Ae. aegypti* and *Ae. albopictus* mosquitoes in the Americas and the high mobility of people transiting in and out of the region, the further spread of the ZIKV across the Americas represents a clear and present danger [93]. The CDC's director, Tom Frieden, added his dismay, telling reporters during a press conference in May: "Three months in an epidemic is an eternity." [94].

In December 2021, a Zika outbreak was reported in Uttar Pradesh, India. Indian health authorities say 152 Zika virus cases in Uttar Pradesh as of December 5. The outbreak began in October and represented the first significant outbreak identified in the state [95]. The first-ever case of Zika was reported in Gujarat, India, in 2016, Tamil Nadu in 2017, and Rajasthan and Madhya Pradesh in 2018 (more on India in Chap. 4).

References

1. Fiocruz ZikAlliance (2022) After 6 years, how have social sciences have responded to the Zika epidemic? https://portal.fiocruz.br/en/news/after-6-years-how-social-sciences-have-responded-zika-epidemic. Accessed 10 June 2022
2. Children's National Hospital (2021) Zika 5 years later: still much to learn as 'likely' future outbreak looms. January 21. HealioNews. https://www.healio.com/news/infectious-disease/20210114/zika-5-years-later-still-much-to-learn-as-likely-future-outbreak-looms. Accessed 25 May 2022
3. Fiocruz ZikAlliance (2022) After six years, how have social sciences have responded to the Zika epidemic? https://portal.fiocruz.br/en/news/after-6-years-how-social-sciences-have-responded-zika-epidemic. Accessed 10 June 2022
4. Green MS, Swartz T, Mayshar E, Lev B (2002) When is an epidemic an epidemic? Israel Med Assoc J 4:4–6; Last JM (ed) (1995) A dictionary of epidemiology. Oxford University Press, New York
5. O'Reilly KM et al (2018) Projecting the end of the Zika virus epidemic in Latin America: a modelling analysis. BMC Med 16:180. https://doi.org/10.1186/s12916-018-1158-1168
6. Beck J (2016) Zika is the 'most difficult' emergency health response ever, CDC official says. The Atlantic. June 24. http://www.theatlantic.com/health/archive/2016/06/zika-is-the-most-difficult-emergency-health-response-ever-says-cdc-official/488579/. Accessed 16 Jan 2017
7. Brasil P et al (2016) Zika virus infection in pregnant women in Rio de Janeiro. New Engl J Med 375:2321–2334. December 15. https://doi.org/10.1056/NEJMoa1602412. Accessed 16 Jan 2017
8. Beck J (2016) Zika is a delayed epidemic. The Atlantic. April 19. https://www.theatlantic.com/health/archive/2016/04/zika-is-a-delayed-epidemic/478755/. Accessed 5 June 2017
9. Thomas L (2020) SARS-CoV-2 versus Zika: comparing and contrasting the flow of research. November 25. medRxiv. https://www.news-medical.net/news/20201125/SARS-CoV-2-versus-Zika-Comparing-and-contrasting-the-flow-of-research.aspx. Accessed 11 June 2022
10. OECD (2018) Safety assessment of transgenic organisms in the environment, volume 8: OECD consensus document of the biology of mosquito Aedes aegypti, harmonisation of regulatory oversight in biotechnology. OECD Publishing, Paris
11. Deniz D (2016) Zika: from the Brazilian backlands to global threat. Zed Books, London
12. Enserink M (2015) An obscure mosquito-borne disease goes global. Science 350:6264. November 26. 1-12-1013. http://science.sciencemag.org/content/350/6264/1012. Accessed 10 May 2017
13. Wikan N, Smith DR (2016) Zika virus: history of a newly emerging arbovirus. Lancet Infect Dis 16:7. July. https://www.ncbi.nlm.nih.gov/pubmed/27282424. Accessed 26 July 2018
14. Pandit PS, Doyle MM, Smart KM, Young C et al (2018) Predicting wildlife reservoirs and global vulnerability to zoonotic Flaviviruses. Nat Commun 9(5425). December 21. https://doi.org/10.1038/s41467-018-07896-2
15. WHO (2016) One year into the Zika outbreak: how an obscure disease became a global health emergency. Emergencies. http://www.portal.pmnch.org/emergencies/zika-virus/articles/one-year-outbreak/en/. Accessed 2 Oct 2016
16. CDC (2022) CDC—global health—neglected tropical diseases. https://www.cdc.gov/globalhealth/ntd/index.html. Accessed 23 May 2022

17. Roiz D, Rosa R, Arnold D, Rizzoli A (2010) Effects of temperature and rainfall on the activity and dynamics of host-seeking Aedes albopictus females in Northern Italy. Vector-Borne Zoonotic Dis 10. https://doi.org/10.1089/vbz.2009.0098
18. Corpuz K (2016) Factory-made mosquitoes can reduce disease transmitting populations by 90%. Futurism.com. https://futurism.com/factory-made-mosquitoes-can-reduce-disease-transmitting-populations-by-90/. Accessed 3 Apr 2017
19. WHO (2018) Chikungunya fact sheet. http://www.who.int/en/news-room/fact-sheets/detail/chikungunya. Accessed 19 Sept 2018
20. WHO (2018) Yellow fever factsheet. http://www.who.int/en/news-room/fact-sheets/detail/yellow-fever. Accessed 19 Sept 2018
21. Goindin D et al (2015) Parity and longevity of Aedes aegypti according to temperatures in controlled conditions and consequences on dengue transmission risks. PLoS One 10(8):e0135489. August 10. https://doi.org/10.1371/journal.pone.0135489. Accessed 25 July 2018
22. Lade D (2017) Zika one year later: is it going away? Sun Sentinel. January 14. http://www.sun-sentinel.com/health/fl-zika-2017-outlook-20170114-story.html. Accessed 19 May 2017
23. Grubaugh ND et al (2019) Travel surveillance and genomics uncover a hidden Zika outbreak during the waning epidemic. Cell 178:1057–1071. August 29. https://doi.org/10.1016/j.cell.2019.07.018
24. Musso D, Ko AI, Baud D (2019) Zika virus infection—after the pandemic. N Engl J Med 381:1444–1457. https://doi.org/10.1056/NEJMra1808246
25. Aziz H et al (2017) Zika virus: global health challenge, threat and current situation. J Med Virol 89. https://doi.org/10.1002/jmv.24731/abstract. Accessed 26 June 2017; Lanciotti RS et al (2008) Phylogeny of Zika virus in western hemisphere, 2015. Emerg Infect Dis 22:5. May. https://wwwnc.cdc.gov/eid/article/22/5/pdfs/16-0065.pdf. Accessed 6 July 2018
26. Gallain P et al (2017) Zika virus in asymptomatic blood donors in Martinique. Blood 129:2. January 12. http://www.bloodjournal.org/content/129/2/263?sso-checked=true. Accessed 26 June 2017
27. Musso D, Gubler DJ (2016) Zika virus. Clin Microbiol Rev 29. May 9. http://cmr.asm.org/content/29/3/487.abstract. Accessed 18 July 2017
28. Haddow A et al (2016) Genetic characteristics of Zika virus strains: geographic expansion of the Asian lineage. In: Haddow A et al. (ed) Zika virus. Scientific Research Publishing (SCIRP), Wuhan, China
29. McCall PJ, Kittayapong P (2006) Control of dengue vectors: tools and strategies. Report of the Scientific Working Group on Dengue. TDR/World Health Organization, Geneva. http://www.who.int/tdr/publications/documents/swg_dengue_2.pdf. Accessed 20 July 2017
30. Cunningham A (2017) Getting dengue first may make Zika infection much worse. Science News. March 30. https://www.sciencenews.org/article/getting-dengue-first-may-make-zika-infection-much-worse. Access 11 Apr 2017
31. Beil L (2016) Vaccines may offer defense against dengue, Zika and chikungunya. Science News. June 15. https://www.sciencenews.org/article/vaccines-may-offer-defense-against-dengue-zika-and-chikungunya. Accessed 11 Apr 2017
32. Dejnirattisai W et al (2016) Dengue virus sero-cross-reactivity drives antibody-dependent enhancement of infection with Zika virus. Nat Immunol. https://www.ncbi.nlm.nih.gov/pmc/articles/PMC4994874/. Accessed 26 June 2017
33. Abbink P et al (2016) Protective efficacy of multiple vaccine platforms against Zika virus challenge in rhesus monkeys. Science. https://doi.org/10.1126/science.aah6157. http://science.sciencemag.org/content/early/2016/08/03/science.aah6157. Accessed 16 Jan 2017
34. Fréour T et al (2016) Sexual transmission of Zika virus in an entirely asymptomatic couple returning from a Zika epidemic area, France, April 2016. EuroSurveillance 9(21). June. https://www.ncbi.nlm.nih.gov/pubmed/27311680. Accessed 27 June 2017
35. McNeill D Jr (2016) Predict Zika's spread? It's hard enough to count the cases. New York Times. September 19. https://www.nytimes.com/2016/09/20/health/zika-spread-predictions.html. Accessed 15 May 2017

36. Abramson D, Pitch-Loeb R (2016) U.S. Public's Perception of Zika Risk: awareness, knowledge, and receptivity to public health interventions. NYU Zika Briefing Report #1. https://www.nyu-pir2.org/research. Accessed 19 June 2019
37. Schaub B, Vouga M, Najooullah F, Gueneret M et al (2017) Analysis of blood from Zika virus-infected fetuses: a prospective case series. Lancet Infect Dis 17:520–527
38. Watson A, Mason C (2015) Power of the first hour. Int Feminist J Polit 17:4. https://doi.org/10.1080/14616742.2015.1080908?journalCode=rfjp20. Accessed 11 Apr 2017
39. Department of Molecular Virology and Microbiology, Baylor College of Medicine (2021) Zika virus. Baylor College of Medicine website. https://www.bcm.edu/departments/molecular-virology-and-microbiology/emerging-infections-and-biodefense/specific-agents/zika#:~:text=It. Accessed 30 May 2022
40. Klass P (2016) Doctors brace for Zika babies. The New York Times. September 26. https://www.nytimes.com/2016/09/26/well/family/doctors-brace-for-zika-babies.html?_r=0. Accessed 17 May 2017
41. Yakob L, Walker T (2016) Zika virus outbreak in the Americas: the need for novel mosquito control methods. Lancet 4(3):E148–E149
42. Bernhard B (2018) What happened to Zika? Missouri stops testing most pregnant women as threat drops. St. Louis Post Dispatch. January 6. http://www.stltoday.com/lifestyles/health-med-fit/health/what-happened-to-zika-missouri-stops-testing-mostpregnant-women/article_9d819106-0c3f-5a37-bad1-53ac3f2789f4.html. Accessed 26 Apr 2018
43. Hackett DW (2018) Prenatal Zika exposure lead to 14.5% of infants reporting health issues: Zika cases reported in Puerto Rico, California, and Florida during 2018. PrecisionVaccinations.com. December 13. https://www.precisionvaccinations.com/zika-cases-reported-puerto-rico-california-and-florida-during-2018. Accessed 13 June 2019; Moreira MEL et al (2018) Neurodevelopment in infants exposed to Zika virus in utero. N Engl J Med 379:24. December 13. https://doi.org/10.1056/NEJMc1800098. Accessed 13 June 2019
44. CDC (2019) Data& Statistics on Zika and Pregnancy. March 29. https://www.cdc.gov/pregnancy/zika/data/index.html. Accessed 13 June 2019
45. Hackett DW (2018) Prenatal Zika exposure lead to 14.5% of infants reporting health issues: Zika cases reported in Puerto Rico, California, and Florida during 2018. PrecisionVaccinations.com. December 13. https://www.precisionvaccinations.com/zika-cases-reported-puerto-rico-california-and-florida-during-2018. Accessed 13 June 2019
46. Bernhard B (2017) Researchers in St. Louis making headway against Zika virus. Stltoday.com. July 10. http://www.stltoday.com/lifestyles/health-med-fit/health/researchers-in-st-louis-making-headway-against-zika-virus/article_173e8152-421e-545b-a8d0-3a86a774eb3c.html. Accessed 21 July 2017
47. Morris G et al (2018) Zika virus as an emerging neuropathogen: mechanisms of neurovirulence and neuro-immune interactions. Mole Neurobiol 55:5. May. https://doi.org/10.1007/s12035-017-0635-y. Accessed 7 July 2018
48. Onorati M et al (2016) Zika virus disrupts phospho-TBK1 localization and mitosis in human neuroepithelial stem cells and radial glia. Cell Rep 16. September 6. https://www.ncbi.nlm.nih.gov/pubmed/27568284. Accessed 18 July 2017
49. Caraiman V (2018) Zika virus suppresses the disease-fighting ability of the immune cells, a study revealed. Health Thoroughfare. July 9. https://www.healththoroughfare.com/medicine/zika-virus-suppresses-the-disease-fighting-ability-of-the-immune-cells-a-study-revealed/9908. Accessed 23 July 2018; Tang H et al (2016) Zika virus infects human cortical neural progenitors and attenuates their growth. Cell Stem Cell 18. May 5. https://www.ncbi.nlm.nih.gov/pubmed/26952870. Accessed 14 Apr 2017
50. NACCHO (2016) Joint letter to Zika conferees supporting funding for Zika. June 6. https://www.naccho.org/uploads/downloadable-resources/6-6-zika-coalition-letter-to-house-conferees.pdf. Accessed 28 Aug 2018
51. RTI International (2018) Researchers assess U.S. Travelers' knowledge of Zika virus, willingness to take hypothetical vaccine. Infection Control Today. July 4. https://www.infectioncontroltoday.com/infectious-diseases-conditions/researchers-assess-us-travelers-knowledge-zika-virus-willingness-take. Accessed 26 July 2018

52. Krisch J (2016) The year in Zika. The Scientist. December 16. http://www.the-scientist.com/?articles.view/articleNo/47805/title/The-Year-in-Zika/. Accessed 17 May 2017
53. CNA (2016) The nuclear connection to combatting the Zika virus. CNA website. July. https://cna.ca/news/nuclear-connection-combating-zika-virus-3/. Accessed 10 Apr 2017
54. Jaenisch T et al (2016) Risk of microcephaly after Zika virus infection in Brazil, 2015 to 2016. Bull World Health Organ 95. https://www.ncbi.nlm.nih.gov/pmc/articles/PMC5328112/. Accessed 17 May 2017
55. Vielot NA et al (2018) United States Travelers' concern about Zika infection and willingness to receive a hypothetical Zika vaccine. Am J Trop Med Hygiene 98:6. June. https://www.ncbi.nlm.nih.gov/pubmed/29692314. Accessed 26 July 2018
56. Davies S, Bennett B (2016) A gendered human rights analysis of Ebola and Zika: locating gender in global health emergencies. International Affairs. August 31. https://www.chathamhouse.org/publication/ia/gendered-human-rights-analysis-ebola-and-zika-locating-gender-global-health. Accessed 11 Apr 2017
57. McNeil DG (2016) Zika: the emerging epidemic. W. W. Norton & Co, NY
58. Mercola J (2016) Zika: Brazil admits it's not the virus. Mercola.com. August 16. http://articles.mercola.com/sites/articles/archive/2016/08/16/birth-defects-brazil-not-zika-virus.aspx. Accessed 23 May 2017
59. Sherman A (2017) Is Miami-Dade the no. 1 leader in breaking local Zika transmission? PolitiFact Florida. January 24. http://www.politifact.com/florida/statements/2017/jan/24/carlos-gimenez/miami-dade-no-1-leader-breaking-local-zika-transmi/. Accessed 2 June 2017
60. Hotez PJ (2013) NTDs V.2.0: "blue marble health"—neglected tropical disease control and elimination in a shifting health policy landscape. PLOS Negl Trop Dis 7:11. https://doi.org/10.1371/journal.pntd.0002570. Accessed 17 July 2017
61. Hotez PJ (2016) Neglected tropical diseases in the anthropocene: the cases of Zika, Ebola, and other infections. PLOS Negl Trop Dis. April 8. https://doi.org/10.1371/journal.pntd.0004648. Accessed 17 July 2017
62. Sanchez-Duque JA, Rodriguez-Morales AJ, Trujillo AM, Cardona-Ospina JA, Villamil-Gomez WE (2018) Cocirculation and coinfection associated to Zika virus in the Americas. In: Roderigues-Morales AJ (ed) Current topics in Zika. Intechopen, London. https://doi.org/10.5772/intechopen.77180
63. Katzelnick LC et al (2020) Zika virus infection enhances future risk of severe dengue disease. Science 369:1123–1128. August 28
64. Clapham H (2020) Zika virus increases risk of dengue disease. Science 369(4507):1055–1956. August 28
65. Ward A (2019) A history with dengue fever may provide some protection from symptomatic Zika. ContagionLive. January 28. https://www.contagionlive.com/view/a-history-with-dengue-fever-may-provide-some-protection-from-symptomatic-zika. Accessed 22 May 2022
66. Ferguson NM et al (2016) Countering the Zika epidemic in Latin America. Science.https://doi.org/10.1126/science.aag0219
67. Martinez-Selles M (2022) Editorial commentary: cardiovascular events after Zika virus infection. Trends Cardiovascular Med 32:59–60 citing Rossi F et al (2020) Characterization of Zika virus infection of human fetal cardiac mesenchymal stromal cells. PLOS One 16(1):e0246112. September 17. https://doi.org/10.1371/journal.pone.0246112
68. Rossi F et al (2020) Characterization of Zika virus infection of human fetal cardiac mesenchymal stromal cells. PLoS ONE 15(9):e0239238. https://doi.org/10.1371/journal.pone.0239238pmid:32941515
69. Abdalla LF et al. Atrial fibrillation in a patient with Zika virus infection. Virol J 15(1). https://doi.org/10.1186/s12985-018-0938-2
70. Landais A et al (2017) ZIKA vasculitis: a new cause of stroke in children? J Neurol Sci 383:211–213. December 15
71. Scatularo CE et al (2020) Neglected tropical diseases and other infectious diseases affecting the heart (NET-Heart project). Zika Heart 2020 S1050-1738(20)30147-XEpub ahead of print. https://doi.org/10.1016/j.tcm.2020.11.003

72. Ramacciotti E et al (2019) Zika and Chikungunya virus and risk for venous thromboembolism. Clin Appl Thrombosis/Hemostasis. January 28. https://doi.org/10.1177/1076029618821184.
73. Santana MB et al (2019) Congenital Zika syndrome: is the heart part of its spectrum? Clin Microbiol Infect 25(8):1043–1044; Orifino DHG et al. 24-hour Holter findings in infants within utero exposure to Zika virus: a cross-sectional study. Res Square. https://doi.org/10.21203/rs.2.14894/v1. https://www.researchsquare.com/article/rs-5565/v1. Accessed 26 May 2022
74. Harding A (2015) The facts about Beau Biden's Cancer. Health. June 4. https://www.health.com/news/the-facts-about-brain-cancer. Accessed 23 July 2018
75. Zhu Z et al (2017) Zika virus has oncolytic activity against glioblastoma stem cells. J Exper Med. September 5. http://jem.rupress.org/content/early/2017/09/05/jem.20171093. Accessed 26 July 2018
76. Gander K (2018) Glioblastoma: Zika virus could destroy brain cancer that killed John McCain. Newsweek. September 18. https://www.newsweek.com/glioblastoma-zika-virus-could-destroy-brain-cancer-killed-john-mccain-1125530. Accessed 22 May 2022
77. Bulstrode H (2018) Zika virus as a potential therapy for brain tumours. BrainTumourResearch. July 5. https://www.braintumourresearch.org/media/our-blog/blog-item/our-blog/2018/07/05/zika-virus-as-a-potential-therapy-for-brain-tumours. Accessed 23 July 2018
78. Francipane M et al (2022) Zika virus: a new therapeutic candidate for glioblastoma treatment. Int J Mole Sci 22. https://www.mdpi.com/1422-0067/22/20/10996. Accessed 22 May 2022
79. Kretschmer M, Kadlubowska P, Hoffmann D, Schwalbe B, Auerswald H, Schreiber M (2020) Zikavirus prME envelope pseudotyped human immunodeficiency virus type-1 as a novel tool for glioblastoma-directed virotherapy. Cancers 12.https://doi.org/10.3390/cancers12041000
80. Bulstrode H (2018) Zika virus as a potential therapy for brain tumours. Brain Tumour Research. July 5. https://www.braintumourresearch.org/media/our-blog/blog-item/our-blog/2018/07/05/zika-virus-as-a-potential-therapy-for-brain-tumours. Accessed 23 July 2018; Chen Q, Shi P-Y et al (2018) Treatment of human glioblastoma with a live attenuated Zika virus vaccine candidate. mBio 9:e01683-18. https://doi.org/10.1128/mBio. Accessed 22 May 2022; Kretschmer M, Kadlubowska P, Hoffmann D, Schwalbe B, Auerswald H, Schreiber M (2020) Zikavirus prME envelope pseudotyped human immunodeficiency virus type-1 as a novel tool for glioblastoma-directed virotherapy. Cancers 12. https://doi.org/10.3390/cancers12041000; Rana J, Campos JLS, Leccese G, Francolini M et al (2018) Role of capsid anchor in the morphogenesis of Zika virus. Jo Virol 92(22):eo1174-18; Zhu Z et al (2017) Zika virus has oncolytic activity against glioblastoma stem cells. J Exper Med. September 5. http://jem.rupress.org/content/early/2017/09/05/jem.20171093. Accessed 26 July 2018
81. Kretschmer M, Kadlubowska P, Hoffmann D, Schwalbe B, Auerswald H, Schreiber M (2020) Zikavirus prME envelope pseudotyped human immunodeficiency virus type-1 as a novel tool for glioblastoma-directed virotherapy. Cancers 12. https://doi.org/10.3390/cancers12041000; Rana J, Campos JLS, Leccese G, Francolini M et al (2018) Role of capsid anchor in the morphogenesis of Zika virus. J Virol 92(22):eo1174-18
82. Bulstrode H (2018) Zika virus as a potential therapy for brain tumours. Brain Tumour Research. July 5. https://www.braintumourresearch.org/media/our-blog/blog-item/our-blog/2018/07/05/zika-virus-as-a-potential-therapy-for-brain-tumours. Accessed 23 July 2018
83. Sgobba C (2017) Can the Zika virus actually kill brain cancer? Men's Health. September 6. https://www.menshealth.com/health/a19533571/zika-virus-brain-cancer/. Accessed 10 July 2018
84. Chen Q, Shi P-Y et al (2018) Treatment of human glioblastoma with a live attenuated Zika virus vaccine candidate. mBio 9:e01683-18. https://doi.org/10.1128/mBio. Accessed 22 May 2022
85. Offit P (2017) The anti-vaxxer illness worse than Zika. The Daily Beast. March 26. http://www.thedailybeast.com/articles/2017/03/26/the-anti-vaxxer-illness-plaguing-elementary-schools. Accessed 26 May 2017
86. Moreno-Madriñán M, Turrell M (2017) Factors of concern regarding Zika and other Aedes aegypti-transmitted viruses in the United States. J Med Entomol 54(2). March. https://academic.oup.com/jme/article/54/2/251/2952765/Factors-of-Concern-Regarding-Zika-and-Other-Aedes. Accessed 23 May 2017

87. Boutott C (2018) HealthWatch: vaccine for Zika virus. WeAreGreenBay. June 14. https://www.wearegreenbay.com/health-watch/healthwatch-vaccine-for-zika-virus/1224940177. Accessed 23 July 2018
88. European Commission-CORDIS: Programmes (2016) Addressing the urgent research gaps against the Zika virus and other emerging threats in Latin America. November 29. http://ec.europa.eu/research/participants/portal/desktop/en/opportunities/h2020/topics/sc1-pm-22-2016.html. Accessed 26 June 2017
89. Manaytay A (2017) Houston officials consider use of genetically modified mosquitoes to fight Zika virus. Tech Times. March 21. http://www.techtimes.com/articles/202474/20170321/houston-considers-use-of-genetically-modified-mosquitoes-to-combat-zika.htm. Accessed 21 May 2017
90. McKay B, Loftus P (2016) America's next defense against Zika and other foreign invaders. Wall Street J. December 16. https://www.wsj.com/articles/americas-next-defense-against-zika-and-other-foreign-invaders-1481810402. Accessed 21 May 2017
91. Krisch J (2016) The year in Zika. The Scientist. December 16. http://www.the-scientist.com/?articles.view/articleNo/47805/title/The-Year-in-Zika/. Accessed 17 May 2017; Messina JP et al. Mapping global environmental suitability for Zika virus. eLife 6. April 19. https://elifesciences.org/content/5/e15272. Accessed 19 May 2017
92. Lewis T (2016) Mapping worldwide Zika susceptibility. The Scientist. April 19. http://www.the-scientist.com/?articles.view/articleNo/45898/title/Mapping-Worldwide-Zika-Susceptibility/. Accessed 19 May 2017; Messina JP et al. Mapping global environmental suitability for Zika virus. eLife 6. April 19. https://elifesciences.org/content/5/e15272. Accessed 19 May 2017
93. Hamel R et al (2015) Biology of Zika virus infection in human skin cells. J Virol 89(17). September. https://www.ncbi.nlm.nih.gov/pubmed/26085147. Accessed 2 June 2017
94. Norman A (2016) Romper. What will happen with Zika virus research under president trump? Things don't look good. Romper. November 17. https://www.romper.com/p/what-will-happen-with-zika-virus-research-under-president-trump-things-dont-look-good-22895. Accessed 26 May 2017
95. Crisis24 (2021) India: increased number of Zika virus cases reported in Uttar Pradesh during December 2021. Crisis24 Newsletter. December 13. https://crisis24.garda.com/alerts/2021/12/india-increased-number-of-zika-virus-cases-reported-in-uttar-pradesh-during-december-2021. Accessed 23 Feb 2022

Chapter 2
Epidemic Events Are Communication Events

Pandemics and epidemics are perceived. They are adorned with exaggeration and hyperbole. Responses to them hinge on the public understanding of them. Behavioral adjustments are recommended to the public to reduce the outbreak's spread, such as masking, isolating, and quarantining—however, the recommendations to become integrated into behavior demand well-designed and purposeful communication. Efforts to create and maintain a resilient population demand clear and consistent communication. Recovering from a slapdash approach can be devastating and deadly, as has been observed with COVID-19.

This may seem self-serving from a communication professor, but pandemics and epidemics affect civilizations, societies, communities, and individuals. The implications of an epidemic can be realized from direct experience and indirect experience; indirect experience mainly involves the communication of some sort, whether person to person, printed, broadcasted, or posted on the Internet. While the sources may vary, the media is much the same—an alternative to knowing first-hand.

While scientific research on an epidemic is vital for designing tests, treatments, vaccines, and health policy, much too little effort is spent determining the optimal messaging to reach the highest percentage of the susceptible population. After the horrible communication experiences with the reality of climate change, the democratic nature of the Internet, and the testing, masking, large public settings, and vaccination with COVID-19, it is always the same: Communication is an afterthought. Great science means squat if the public rejects it.

Since COVID-19 appeared, the issue of who can best communicate about pandemics has been questioned. While much of the criticism is mired in politics and ideology, much has been insinuated that the current sources of the messages and the notes they craft and distribute may be inadequate. Everyone who is not reached by this message is a potential victim.

Though many scientists tend to have an elevated level of trust between themselves and the public, that is not universally true. Experts tend to see a public good in their work and technologies and expect the public to embrace them and their wares for the same reasons. Experts lament what is wrong with the public; this vaccine,

D. M. Berube, *Pandemics and Resilience: Lessons we should have learned from Zika*, Risk, Systems and Decisions, https://doi.org/10.1007/978-3-031-25370-6_2

treatment, or behavioral adjustment can save their lives! Then, experts double-down and rephrase and repeat messages in many ways hoping one or more of their efforts will finally influence the public to do what is best for them. What generally follows is a lot of headshaking and sighs from the expert communities as they discover they may not be the optimal senders of these messages. More often, the messages they are sending fail to impress the public.

The data are incomplete on knowing what works and what does not in a world of information overload and new social media platforms found online. Nonetheless, over the last half-century in communication studies, it has been learned that successful communication is a function of many variables such as context and salience. An entire subfield of breakdown in communication deals with health and care communication. Unfortunately, many healthcare professionals have never been exposed to this scholarship, which is reflected in the lack of care to craft effective messaging.

The word "convergence" (see Chap. 5) often describes how some problems demand cross- or even interdisciplinary approaches. Learning to associate a messaging strategy with the appropriate pandemic or epidemic event and the right audience is critical, given nearly half of the world's population is infected with at least one type of vector-borne pathogen [1]. When technical experts sense or guess their way through messaging strategies, they serve neither the technical community nor the public. Errors communicating epidemics can fail to interface life-saving skillsets with those who may need them the most.

Even if a vaccine or vaccines are developed (at warp speed) to protect the uninfected, the public must agree to be vaccinated. But the public avoids vaccines and often postpones or hesitates vaccination for several reasons, some reasonable and others not.

When the public responds to messaging inappropriately and fails to be precautionary, then the epidemic wins. This phenomenon occurs when the messaging is poorly conceived and confronts countermessaging. This happened during the current COVID pandemic and the ZIKV epidemic. The countermessaging in both these cases involved some wild storytelling. While the ZIKV epidemic was not hampered by ineffectual Presidential leadership as has been COVID, it dealt with rationalizations in the form of rumors, innuendos, and falsehoods, often garbed as conspiracy theories.

Pandemics can kill as many people as a devastating war [2]. The annualized expected loss from potential pandemics is more than $60 billion [3]. And one-sixth of the world's infection-associated DALY (disability-adjusted life year) estimate is attributed to vector-borne disease [4].

Thirty-five years ago, there was a belief that the health burden of infectious diseases was close to becoming insignificant, as the means of defense and control, including hygiene, nutrition, drugs, and especially vaccines, had brought about a steady decline in overall mortality. However, in recent decades, it has become clear that the threat persists in a rapidly changing world. Human mortality attributed to infection is projected to remain at 13–15 million deaths annually until at least 2030 [5].

The health burden has direct and indirect consequences. Infectious disease outbreaks that turn into epidemics and potential pandemics can cause massive loss of life and substantial economic disruption [6].

The global implication of infectious disease outbreaks has not gone unnoticed. The intrinsic dynamics of infectious disease outbreaks and the behavioral and policy responses can have an immense economic impact. Pandemics arguably pose more threats to human lives than war, terrorism, or natural disasters. Framed as a risk to economic growth and stability, the danger is equally stark [6].

Now consider how much is invested in pandemic communication. While a pandemic could kill as many people as a devastating war, the resources committed to pandemic prevention and response are a fraction of present security resources.

There are also very few risks with more significant potential for catastrophic economic impact—potentially on a global monetary crisis scale—but the measures to avoid another monetary crisis are entirely different [3].

Perhaps most of all, families wondered why the disease had ravaged Brazil's northeast but much less so in other regions of South America. A host of narratives began circulating in blogs, YouTube videos, and social media that blamed the emergence of the virus on science. These sources argued that various groups—such as pesticide companies, the Eugenics Movement, Monsanto, and Bill Gates—used science to impose an evil agenda upon Latin Americans [7]. This leads to the first confounding variable in health communication related to pandemics: misinformation.

2.1 Conspiracy Theories

Typically used with a derogatory connotation, a "conspiracy" is a contested rhetorical notion applied to specific events or phenomena according to someone's perceptions, point of view, and belief system [8]. They may or may not be informed by any objective information. They are almost impossible to prove and disprove.

Conspiracy theories about a particular event arise specifically from a need for sense-making. This need is vital when the event in question is severe and self-relevant—the exact conditions under which rumors tend to arise [9].

The construction and maintenance of conspiracy theories are often characterized by logical flaws, inconsistent narratives, contradictory rationales, and a lack of evidence. This conspiracist ideation is not limited to any political or cultural entity. Indeed, contemporary trends, especially in the U.S., seem to support the contention that the amount of conspiracy thinking may have grown along with the political association between conspiracies and the current instantiation of the GOP.

Conspiracy theories produced and shared on social and digital media platforms have the power to discursively construct contagious diseases such as Zika, which may fuel misguided public perceptions and impact health policy [10]. This is not unusual. In the past, alarmist predictions, sensationalist rhetoric, rumors, and incorrect information surrounded the severe acute respiratory syndrome (SARS) outbreak, swine flu epidemic, and Ebola crisis.

Uncertainty regarding the origin, transmission, and health consequences of the ZIKV has created a fertile environment for conspiracy theories and pseudoscientific claims. These theories emerged primarily in social media, an active forum for the vaccine refusal community, and have made their way to the mainstream media [11]. This occurs when mainstream media reports on conspiracies found on Internet postings of all sorts as well as on social media. Without assigning motive, mainstream media has become a conduit for conspiracies reported on social media.

Institutional distrust may be widespread among conspiracy theories involving large industries, which are accused of pursuing profits to the detriment of the public's interest [10]. Michel Misse, a sociologist at the Federal University of Rio de Janeiro, said rumors and conspiracy theories played an outsized role in Brazilian society. The crises buffeting the Brazilian government made people question and doubt authority [12].

> V. M. Granmisterio posted a video to YouTube on a channel dedicated to conspiracy theories [13]. The host stated that the virus was spread by a particular mosquito (*Ae. aegypti*), which was accurate. He then, however, argued that in 2012, the company Oxitec had created a genetically modified organism (GMO) mosquito with funding from Bill Gates, whom he then accused of being part of a Eugenics Movement trying to eliminate undesirable races. In the video, the host hypothesized that the GMO mosquito created by Oxitec had evolved extreme pesticide resistance, which enabled it to spread the disease. The host also suggested the Rockefeller Foundation was involved in events because it had the patent on the ZIKV; he indicated that a New World Order conspiracy was using Zika to spread fear so that non-white women would not have children.

Of all the conspiracies reported around the ZIKV, this was the densest. Nonetheless, conspiratorial stories like the above disproportionately impact public understanding.

Shawn Smallman from Portland State offered the disproportionate platforms favoring conspiracies in an interesting paper, a worthy download. "We do know that YouTube videos, WhatsApp postings, and other social media posts discussing conspiracy theories and Zika were viewed millions of times during the epidemic. This number is far higher than the number of views of articles, videos, or links sharing reliable sources of information regarding how people could protect themselves from infection and likely impact some peoples' behavior" [7].

Kloftad et al. (2019) found close to 14% of respondents to the 2016 Cooperative Congressional Election Study blame Zika on genetically modified mosquitos, 7% believe that pharmaceutical companies developed the virus to increase profits, and close to 5% believe that Zika was created by the government to "sicken or kill people." Aggregating these responses into a count of the total number of conspiracy theories indicated, 20% of respondents believe at least one; 7% believe more than one [14].

People believe in conspiracies because they can reduce anxiety and uncertainty and help people come to grips with unfamiliar situations. They make sense of the world. They provide a simple and immediate picture of the problem.

The biggest driver of concern over Zika, belief in Zika conspiracy theories, and faith in multiple Zika conspiracy theories will be underlying conspiracy thinking. Further, since conspiracy theorists are prone to believing in numerous (and sometimes contradictory) conspiracy theories about events [15], it is expected that as individuals' underlying conspiracy thinking increases, people will likely think multiple Zika conspiracy theories are true.

Significantly, conspiracy thinking is also associated with rejecting scientific findings—some crucial results regarding conspiracies on digital media. One of the more extensive research projects on digital conspiracies and Zika involved Wood's [9] analysis of Tweets about Zika from 2015 to 2016 showed that relative to debunking messages, conspiracy theories spread through a more decentralized network are more likely to invoke supposedly knowledgeable authorities in making arguments and ask more rhetorical questions and that Zika conspiracy tweets often invoke supposed experts to suggest a conspiracy, or ask rhetorical questions to cast doubt on official narratives [9]. Another study on Twitter by Dredze et al. [16] documented a "rapid rise in tweets associated with pseudoscientific claims" about Zika vaccines, even before one was available [16].

A study of Facebook postings by Seltzer et al. [17] showed that "misleading" posts about Zika were more popular than accurate ones. An analysis of Instagram found that misleading images were more likely to garner "likes" than images with complete, correct information. Instagram has shown to be a mode of communication for organizations and individuals working to educate the public about the symptoms, prevention methods, and policy developments during this Zika outbreak; however, there is still a large amount of misleading, incomplete, and unclear messaging on Instagram [17].

Claudio Maierovitch, the Brazilian Health Ministry's director of surveillance of infectious diseases, said there was a danger that fear, and scant scientific information would lead people to ignore the government's appeals about protecting themselves from mosquito bites and removing standing water that allows mosquitoes to breed [12].

Finally, in a reversal of conventional scientific reasoning, the evidence against conspiracy theories is often construed as evidence because the evidence is interpreted as arising from the conspiracy in question. This interpretation relies on the notion that the more substantial the proof against a conspiracy, the more the conspirators must want people to believe their version of events. This self-sealing reasoning can engender elaborate theories that are dazzling in their complexity [18].

Conspiracies are a lot like rumors. The more the rumor is argued against, the deeper becomes the rabbit hole from which it sprung. "Rumors are the lifeblood of an epidemic," said Dr. Howard Markel, a medical historian at the University of Michigan [19].

Commenters warn against a premature foreclosing of alternative views, arguing that "as soon as someone has diverging opinions, they get labeled conspiracy theorists and are dismissed as crazy, while their questions and insights are not that crazy at all, and it is incredibly harmful to ridicule conspiracy theorists. When official knowledge, backed by science, politics, and media, is distrusted by various people, they resort to alternative (conspiratorial) explanations" [20].

There are also growing indications that rejection of science is suffused by conspiracist ideation, that is, the general tendency to endorse conspiracy theories, including the specific beliefs that inconvenient scientific findings constitute a "hoax." The latter "convenience" theories illustrate the extent to which rejection of a scientific proposition entails the idea that the relevant evidence is the result of a conspiracy among scientists [18].

Some have argued that giving a stage to dangerous fanatics and the less informed legitimizes their views will lead to undesired tensions between citizens, flawed policymaking, and more significant societal harm. In some situations, this may be true. However, this view seems somewhat at odds with open debate provisions of a free society. Alienation leads to radicalization. Conspiracies may not be objectively valid, but they are expressions of how some people feel about the world in which they live. Epidemics are scary. Engaging conspiracists is not a task for the weak-hearted and those with a quick temperament. However, they can teach us how they feel.

After a quarter-century of working in the public eye, it never fails to astound me that so many people say so many incredibly wrong things for the public interest for many good reasons. Remembering a lecture on "Why good people do bad things?" the motive was interest, whether self-interest, vested interest, or even public interest. Often, allegations that are not true are made by generally good people trying to do good.

Rhetorically dubbed a risk profile shift in an earlier piece of mine, it is when something with a high-risk footprint is associated with another thing with a low-risk impression to make claims of wrongfulness related to the allegation worse [21]. Sometimes, the data used in the shift are incomplete and inaccurate. Sometimes, there is no data at all. Sometimes, it is an error on the part of the allegers; at other times, it is entirely intentional.

Public interest groups are not all equal. To maintain their importance, they need attention, and to get the attention, they make wild allegations. When the first anti-GE seeds and food complaints came from Western Europe, African countries refused GE food under PL 480's Food for Peace program. Shipments were delayed and not forthcoming, and thousands died from malnutrition. Innuendoes and falsehoods in some contexts are more than simply wrong; they are dangerous. It is fortunate the following have not had similar consequences so far.

The possible association of ZIKV infection with microcephaly has spawned several conspiracy theories about how the virus arose and other potential causes of microcephaly. Due to very little knowledge about ZIKV, an aura of mystery has developed about its origin, spread, and consequences. With the lack of scientific explanations, many conspiracy theories about ZIKV have spread explosively [22]. Some of these conspiratorial assumptions are examined below.

2.1.1 Pyriproxyfen

According to a report by the New England Complex Systems Institute (NECSI), a larvicide, not the mosquito-borne ZIKV, was the real cause of the rise in microcephaly cases.

"An alternative cause of microcephaly in Brazil could be the pesticide pyriproxyfen, which is cross-reactive with retinoic acid, which causes microcephaly, and is being used in drinking water" [23]. Dr. Fatima Marinho, the director of information and health analysis at Brazil's Ministry of Health, told the journal Nature that one possibility that has been raised is the pesticide pyriproxyfen, which is applied to drinking water in some parts of Brazil to kill the larvae of the mosquitos that transmit ZIKV and cause microcephaly associated with ZIKV.

They claimed it had been used for 18 months before the spike in reported cases. They contended that spraying had been occurring in Northeast Brazil for 40 years. In 2014, the ministry stopped using an organophosphate to which the *Aedes* larvae had become resistant and switched to pyriproxyfen (also known as Sumilarv).

Pyriproxyfen is an analog for juvenile insect hormone, which is cross-reactive with retinoic acid, known to cause microcephaly [24]. Pyriproxifen is manufactured by Sumitomo Chemical, a Japanese company associated with Monsanto in Latin America for weed control [25].

This group of physicians claims this helps explain the vast preponderance of cases found in Northeast Brazil but not elsewhere, for example, in Southeast Asia and Colombia. The larvicide has been applied directly to water reservoirs in Pernambuco, where the concentration of *Aedes* is high. It acts via endocrine disruption inhibiting the development of adult insect characteristics.

The claim is straightforward: Microcephaly was due to pyriproxyfen, a chemical larvicide. Of course, areas with high uses of pyriproxyfen also had increased cases of Zika because that is where the mosquitoes happen to be [26]. Portal da Saude reports that it is essential to note that some locations that do not use pyriproxyfen also reported cases of microcephaly [27].

This group believes the best defense against Zika is community-based actions and identifies poverty as a key neglected factor in the epidemic. Those areas with poor sanitation and unsafe drinking water suffer the most [28].

Associated with this claim, some blame Brazil for gentrification (moving poor to even more impoverished areas) in its effort to modernize and host an array of international events, such as the 2013 Salesian Youth Movement with a papal visit, the World Cup in 2014, and the Summer Olympics in 2016. Modernization at this speed leads to health and societal harm and has contributed to the increased circulation of the ZIKV [29].

In 2016, there was a similar claim from the PCST (Physicians in Cropped Sprayed Towns) [30], but this time about a larvicide used in mosquito population control. Presumably, this larvicide was used in Northeast Brazil for 40 years, where the highest level of reported cases of microcephaly occurred. Also, Reduas were essential to spreading alarmist misinformation in both the 2016 Zika and 2011 farm cancer belt stories.

The primary complaint of the PCST was that the Brazilian Ministry of Health mostly ignored the role of the chemical model for vector control. This concern was repeated by doctors from the Brazilian Association for Collective Health (ABRASCO) when the ministry reported the Zika-instigated outbreak of microcephaly.

This yields a vociferous debate using some social media tools, with virologists hedging their statements and GM Watch attempting to rescue the premise of the claim that a larvicide may have been responsible for the outbreak of microcephaly.

Agrochemicals, especially pesticides, have a long history of reported health drawbacks. The mass usage of chemical poisons to reduce or eradicate mosquitoes has been conducted in the most vulnerable areas of Latin America. In addition, "[p]yriproxyfen is an analog for juvenile insect hormone which is cross-reactive with retinoic acid, which is known to cause microcephaly." [31].

A 2011 article in La Monde reported health problems around Cordoba, Argentina, related to Monsanto's glycosphate (Roundup). Cancer rates were twice the national average. This was false, but the claim's event spawned a group led by Carlos Nota and Medardo Avila Vasquez, Reduas.

The University Network for Environment and Health (REDUAS) is a coordination between university professionals, academics, scientists, team members in human health at different levels, and other scholars concerned about the harmful effects of human health that generate the environment degraded consequences of productive human activity, mainly when it occurs on a large scale and sustained by an extra-activist vision [32].

The statements made by these groups in 2011 and 2016 could not be sustained. However, it is probably wrongful to conclude these activists are nefarious. Pesticides and larvicides are poisons.

Physicians in the Crop Sprayed Town group admitted they did not complete lab studies or epidemiological research [33]. They argued larvicides **may** have caused the human deformities allegedly linked to Zika. This debate gets stranger as the state's health secretary, João Gabbardo Dos Reis, said that "although we do not indicate that the larvicide pyriproxyfen has a link with microcephaly cases, it is also true that we do not have any strong evidence that it has no links," he said [34].

ABRASCO demanded that urgent epidemiological studies consider this causal link be carried out [35]. In addition, ABRASCO suggested blaming the larvicide may have been driven by the commercial interests of the chemical industry itself, which it warns is deeply integrated into the Latin American ministries of health, among other health organizations. Presumably, ABRASCO went on demand "the suspension of the use of chemicals" (not just pyriproxyfen) and "the adoption of mechanical cleaning methods and environmental sanitation" to drive down mosquito populations. But not before distancing themselves from this claim. Abrasco's coordinator Marcelo Firpo stated: "We did not say that the larvicide [pyriproxifen] is associated with microcephaly." He added that any suggestion that ABRASCO had ever asserted such a link was a "misunderstanding." [36].

Brazilian and U.S. health authorities quickly rejected the Argentine group's assertions and ran contrary to many health authorities in Brazil and internationally. The more probable cause of the rise in suspected microcephaly cases is the mosquito-borne ZIKV rapidly spreading across the Americas [34].

Robinson from GM Watch described the result of this claim as "all hell breaking loose" and did her best to explain the motivation of the Argentinian doctors. They released the statement, which while understandable, seems highly misplaced given the critical health concerns associated with Zika. "But it's undoubtedly no surprise if Argentine physicians, who have had to deal first-hand with the suffering caused by the GMO soy revolution in Argentina with its accompanying pesticide onslaught, should be particularly alert to the role of pesticides in health and development issues in Latin America–and suspicious of the safety claims of chemical corporations. The doctors say their local communities face an exploding health crisis, including children suffering unusual congenital disabilities [37].

Costa and Ko [38] provide the first evidence that exposure to the insecticide pyriproxyfen and vaccines administered during pregnancy was not associated with an increased risk of microcephaly. Despite weak biological plausibility, insecticides and vaccines had been speculated to be potential risk factors at the height of the epidemic [38].

No epidemiological studies show the association between pyriproxyfen and microcephaly's use. The Ministry of Health only uses larvicides recommended by the World Health Organization (WHO). The products undergo a rigorous evaluation process by the World Health Organization Pesticide Evaluation Scheme (WHOPES). Pyriproxyfen is among the products approved by this committee and certified by ANVISA (National Health Surveillance Agency), which assesses the safety of larvicide in Brazil [27].

Associative data for the relationship between Zika and microcephaly come from a case–control study of neonates in eight public maternity hospitals in Recife, Brazil, reporting neither vaccination during pregnancy nor use of larvicide pyriproxyfen was associated with microcephaly [39].

No credible mainstream experts or public health groups lent any credence to that theory and largely leaned toward a link between ZIKV and microcephaly [40].

A study of affected newborns from 14 municipalities in the state of Pernambuco found no statistically significant correlation.

> The overall prevalence of microcephaly was 82 per 10,000 live births in the three municipalities that used the larvicide Bti (Bacillus thuringiensis israelensis) instead of pyriproxyfen and 69 per 10,000 live births in the eleven municipalities that used pyriproxyfen. The difference was not statistically significant. Our results show that the prevalence of microcephaly was not higher in the areas in which pyriproxyfen was used. In this ecological approach, there was no evidence of a correlation between the use of pyriproxyfen in the municipalities and the microcephaly epidemic. [41]
>
> The authors added it should be noted that the mosquito vector control strategies during the past three decades, mainly based on chemical insecticides and larvicides, have proven ineffective. [42]

When people drink water from containers that have been treated with pyriproxyfen, they are exposed to the larvicide–but in tiny amounts that do not harm their health. Moreover, 90–95% of any larvicide ingested is excreted into the urine within 48 h. This product has been used since the late 1990s without being linked to microcephaly [43]. Neurotoxicologist Ian Musgrave commented that "even enormous quantities of pyriproxyfen do not cause the defects seen during the recent Zika outbreak": It is poorly absorbed by humans and quickly broken down. It is used in pet flea collars. He claimed humans need to drink 1000 L daily to feel adverse effects [26].

The WHO has approved pyriproxyfen to combat mosquitoes, and testing has shown it isn't carcinogenic, doesn't cause nervous system damage, or affects reproductive ability [33].

Dr. Frances Collins, the director of the US National Institutes of Health, called the assertion interesting but sketchy. Sadly, the report led to the suspension of one state in Brazil responding to this misinformed allegation. "Let me advise caution in offering credence for this larvicide theory unless and until more data supports it," he said. "The situation in Brazil and elsewhere will not be assisted by attaching unwarranted credibility to this interesting but speculative theory." [33].

An article in the Wall Street Journal associated this report with a decision to discontinue using larvicide to control mosquito outbreaks. In response, one of Brazil's southern states stopped using larvicide. Fortunately, it is one of the least affected by Zika, with only one supported case reported in early 2016. But officials said they would continue to use the larvicide in one of the hardest-hit regions, the northeastern state of Bahia. "If we did not use [Pyriproxifen], we would have more cases yet," said Bahia Health Secretary Fábio Vilas Boas [34].

Brazil's federal government has denied the claim. "Unlike the relationship between the ZIKV and microcephaly, which has had its confirmation shown in tests that indicated the presence of the virus in samples of blood, tissue, and amniotic fluid, the association between the use of pyriproxyfen and microcephaly has no scientific basis." [44].

The Brazilian government has clarified that increased microcephaly cases were also observed in regions where the larvicide, pyriproxyfen, had not been used in its drinking water. Further, the government asserted that WHO approved the use of this larvicide, and there was no scientific proof of its association with microcephaly [45].

2.1.2 *Oxitec Caused the Outbreak*

Another group blames the Oxitec Company stating that the release of the genetically modified mosquitoes in Brazil in 2011 to control the spread of dengue fever is behind the current epidemic of ZIKV (see Chap. 13).

A post from Dorothy (no surname provided) generated some high-octane drama, claiming the Oxitec genetically modified (GM) mosquitoes were likely responsible for Brazil's Zika spread. The post claims the engineering process caused both males and females, which are then separated but are sometimes released. The females bite the post claims truly little is known of the effects this might have on a human population. The post adds protein fragments of herpes virus, *E. coli* bacteria, and cabbage is used in the breeding and may be responsible [46].

This allegation was repeated at one of the Mosquito Control Board meetings in Key Haven, Florida. "It's these chemicals you're spraying that are causing microcephaly," one local environmental activist, Doug Hattendorf, said in September at a meeting in Key West, where town commissioners were deliberating over a proposed resolution supporting the Key Haven project. "There's no proof of Zika causing microcephaly, except what Oxitec says," he added [47].

Christie Wilcox soundly rebutted Dorothy's post. "A new ridiculous rumor is spreading around the Internet. According to conspiracy theorists, the recent outbreak of Zika can be blamed on the British biotech company Oxitec, which some say even intentionally caused the disease as a form of ethnic cleansing or population control." The articles cite a lone Redditor who proposed the connection on January 25th to the conspiracy subreddit [48]. The strongest argument against this claim was made by Wilcox when she compared locations and time frames. The epicenter of the outbreak and the release don't line up—the heart is on the coast rather than inland, where the map points. Epidemiologists have tracked the attack back to where it started. They now say that the Zika outbreak "almost certainly" began in Recife, Brazil, a city almost 400 miles from the nearest Oxitec release location. To claim that anything done in Juazeiro caused an outbreak to occur 400 miles away is the same as claiming that whatever is done in Phoenix, AZ is responsible for disease outbreaks in Los Angeles, CA. Those two cities are nearly 50 miles closer than Juazeiro, BA, and Recife, PE [48].

The timeline is wrong as well. "GM mosquitoes weren't first released in Juazeiro, Bahia (let alone Juazeiro de Norte, Ceará) in 2015. Instead, the announcement by Oxitec was of the published results of a trial that occurred in Juazeiro between May 2011 and Sept 2012—a fact which is clearly stated in the methods and results of the paper that Oxitec was so excited to share. A new control effort employing Oxitec mosquitoes did begin in April 2015, but not in Juazeiro or any of the northeastern states of Brazil where the disease outbreak is occurring" [48].

The WHO also swatted down a theory about genetically modified mosquitoes as the cause. Releasing males that can't reproduce could help Brazil tamp down mosquito populations. But "it's virtually impossible" that GM mosquitoes are transmitting Zika or causing microcephaly, says entomologist William Walton of the University of California, Riverside. "Male mosquitoes don't bite," he says. So, they don't pass what they're carrying on to humans as females do [49].

The WHO concluded: "There is no evidence that genetically modified mosquitoes cause ZIKV disease or microcephaly in Brazil." [43]. On the contrary, Brazilian health officials and scientists have refuted these claims and emphasized that genetically modified mosquitoes have reduced the mosquito population in the affected areas [45].

2.1.3 *Depopulation*

A popular misconception spread across Brazil that vaccines for chickenpox and rubella virus are responsible for the surge in microcephaly cases. A rather bizarre rumor is that this epidemic is a plot by the global elites to depopulate the world [50].

They start with the claim that the ZIKV is a commodity that can be purchased online from the ATCC-LGC (global biological materials resource and standards organizations) for 599 euros, with royalties accruing to the Rockefeller Foundation. Is the ownership of the ZIKV by the Rockefeller Foundation part of that plan of "supernational sovereignty [dominated] by an intellectual elite and world bankers"? [51].

The American Type Culture Collection (ATCC) even has two online forms of the ZIKV that the accidental "scientific researcher," or maybe not "accidental scientific researcher," can purchase. Gwendolyn Olmstead reported that "it appears anyone with a credit card can buy the ZIKV from ATCC [52].

Olmstead is wrong. The ATCC doesn't sell viruses to just anyone. They offer the scientific community credible biological products, advanced model systems, and high-quality services that support complex research in various innovative applications resulting in incredible achievements in basic science, drug discovery, translational medicine, and public health. Their website explains anyone purchasing ATCC materials is ultimately responsible for obtaining the necessary permits and licenses for their receipt. They are heavily regulated. Certain materials require a permit issued by the United States Department of Agriculture (USDA), Animal and Plant Health Inspection Services (APHIS), and Veterinary Services (VS) to transport controlled materials and organisms, and vectors. International customers must get import permits or licenses from their appropriate government agency. Licenses must be provided to the ATCC before shipment.

2.1.4 Weaponizing Viruses

Some marginal scientific communities warn that Zika was a biological warfare experiment gone out of control and not the first of such mishaps [53].

At the same time the U.S. Public Health Service, at last, started trying to eradicate *Ae. aegypti* from the southeast. Historically, another branch of the U.S. government had planned to raise colonies of millions of *Ae. aegypti* mosquitos as biological weapons [54].

> Among these possible insect soldiers, *Ae. aegypti* was "the golden child," writes Jeffrey A. Lockwood in Six-Legged Soldiers, because its disease, yellow fever, was terrible. The Army Chemical Corps, in a 1959 report, notes that yellow fever is "highly dangerous" and that "since 1900, one-third of patients have died." Some parts of the Soviet Union had never been exposed to the disease, making them vulnerable, but they had a suitable climate to support mosquitoes. The Chemical Corps started to experiment with how a brigade of *Ae. aegypti* might be deployed, and what sort of damage they might do.... By 1960, the Chemical Corps was producing 500,000 *Ae. aegypti* every month.... in 1957 and 1958, the Army released *Ae. aegypti* in Avon Park, in the middle of the Florida peninsula. In those same years, in the Panhandle, the Public Health Service had finally started a pilot program to eradicate *Ae. aegypti* in Pensacola, Florida. [54]

Despite the overall failure of the Chemical Corps experiment, this brief bit of insanity seemed to have been imprinted on some military offices. For example, retired US Admiral and NATO commander James Stavridis warns biowar is in sight. He is concerned futuristic warheads laced with diseases such as ZIKV or Ebola could "wipe out swathes of humanity if they fall into terrorists' hands."

He alleges, in theory, a terrorist wouldn't need to create vast amounts of a lethal virus to unleash on the world. Instead, he could create a handful of mosquitoes with a gene for making a toxin and power it with a gene drive (see Chap. 14). Soon all the world's mosquitoes would make the toxin, and every mosquito bite would be lethal [55].

Stavridis adds: "The rise of low-cost synthetic biology technologies, the falling cost of DNA sequencing, and the diffusion of knowledge through the Internet create the conditions for a breakout biological event not dissimilar to the Spanish influenza of roughly a century ago. In that plague, by some estimates, nearly 40% of the world's population was infected, with a 10–20% mortality rate. Extrapolated to our current global population equals more than 400 million dead" [56].

Stavridis' fears regarding zoonoses have not held up to criticism, The debate over the misapplication of synthetic biology is ongoing. However, experts in biosecurity seem to agree warheads would be terrible as delivery systems. Zika would be a ridiculous choice for a pathogen, and Ebola is not that much better. But for the single article in *Foreign Policy*, he seems to stand alone. A close reading of the article suggests the issue is not about zoonoses per se but rather his concerns over synthetic biology [57].

2.1.5 *Vaccines and IPAK (Institute of Pure and Applied Knowledge)*

Some have even blamed child immunizations for the Zika outbreak and the consequent increase in cases of microcephaly [58].

This is a serious virus, and there isn't much known about it yet. But a few things are known: Zika is not caused by vaccines, despite persistent rumors spread by antivaccine zealots. The story that the vaccine, yet to be approved, caused Zika and microcephaly was especially problematic.

As Simas et al. [59] reported, a possible barrier to maternal immunization identified was a rumor that blamed vaccines for microcephaly cases.

> Among vaccine-refusing women surveyed, safety was the main reason not to vaccinate: "I fear my baby will have microcephaly because of the vaccine" (RJ). This reflects the reality of being pregnant after an outbreak that primarily affected expecting mothers. Following probing, women discussed the association between Zika and microcephaly as a cover-up: "I remember pregnant women vaccinating following a campaign, and after that, there were many cases of microcephaly. (. . .) People said it came from mosquitoes, but this is to cover their mistake" (RJ). [59]

Beginning with a post by Jim Stone from the Vaccination Information Network in 2016. The account has been suspended.

> A new TDAP shot Brazil made mandatory at the beginning of 2015 coincides with perfect timing with many newborns born with defects. In late 2014 the Brazilian minister of health announced a new Tdap shot to become mandatory for all expectant mothers as soon as Brazil received it, ending in early 2015. No zika is found in most messed-up babies, but ALL MOTHERS WITH MESSED UP BABIES GOT THAT NEWLY FORMULATED SHOT WHILE PREGNANT (emphasis in original). [60]

A much more rationalized and detailed treatment of the conspiracy came from the Institute for Pure and Applied Knowledge (IPAK). IPAK is a not-for-profit in Pennsylvania that relies on private donations to fund research focused on analyzing peer-reviewed journals related to human health. They also conduct their studies again, typically focused on efforts to reduce human pain and suffering. Sadly, they request to see most of their research because they lack money. The organization's primary goal is to combat biased journals regarding medicine because of the fear that the scientists are gaining profit by miscommunicating information. Overall, they attempt to be a watchdog for scientific research.

James Lyons-Weiler is the founder of IPAK, the organization's leading publisher of scientific journals. He was first known for his research on Ebola, in which he published a book, *Ebola: An Involving Story*, which focused on educating the public on the truth of reports and what went on in Western Africa. His research after that is more centered on the concern of Autism (another book, *Genetic and Environmental Causes of Autism*) and, most recently, a book, *Cures versus Profits: Successes in Translational Research*, which is focused on calling out the medical field and if they are for profit. The rest of what he publishes has a trend of being against vaccines and how they need to be re-evaluated.

In the wake of Zika's spread within Brazil, tons of insecticides were sprayed in and around homes, exposing pregnant women and young children to brain-damaging chemicals. In 2014, the Brazilian Minister of Health mandated that all expectant mothers receive the new Tdap vaccine. At 20 weeks gestation, a vulnerable, developing young life would be exposed to aluminum adjuvant, mercury preservative, formaldehyde, antibiotics, and other chemicals that could damage a fetus' developing brain. Is it only a coincidence that congenital disabilities have spiked in Brazil because of the toxic elements that fetuses have been exposed to because of such vaccines? [53].

After an epidemic in Brazil in 2014, the government started using the tdap vaccine that some believe leads to microcephaly. However, the CDC requires tdap, and the US has not seen the same type of outbreak. Moreover, this study proved the safety of tdap and discussed how the two things were correlated but had no causation. Furthermore, a study by the CDC also demonstrated no link between tdap and microcephaly. If you don't believe me, just google microcephaly and tdap, and the entire first page says no connection. The study the book refers to is this one, where Lyons-Weiler proposes several hypotheses on what is causing the microcephaly and only uses preliminary information.

Be on the lookout for Marco Caceres, an avid believer that ZIKV and microcephaly are unrelated. He can mainly be found at thevaccinereaction.org, but here is one of his many rants against the link between the two.

2.1.6 *Some Others*

One of the more colorful moments involved a false claim made by a group of Argentinian doctors (pyriproxyfen), an incorrect establishment of blame in part toward Monsanto (pyriproxyfen again), and George Takei's (Sat Trek's Mr. Sulu) Facebook page. Takei's mild notoriety on Facebook meant that the study was shared widely, leading to even more widespread concern about the claims made in the item, which was initially published the day before on a website called Second Nexus.

Second, Nexus is not a site devoted to science or medicine. Its "About" page described it as "a New York City-based digital publication bringing newsworthy content in close conjunction with a powerful network of social media platforms." Takei's viral page constituted a "powerful … social media platform." [40].

2.2 Unknowns

Rumsfeld once stated there are known knowns; there are things we know we know. We also know there are known unknowns; we know there are some things we do not know. And they make the public uncomfortable. In a 1927 essay, the legendary horror author H. P. Lovecraft wrote that "the oldest and strongest emotion of mankind is fear, and the oldest and strongest kind of fear is fear of the unknown." [61].

The public does not deal with uncertainty well.

> The human brain, it has been written, is an "anticipation machine, and 'making future' is the most important thing it does." Using past experiences and information about our current state and environment to predict the future allows us to increase the odds of desired outcomes while avoiding or bracing ourselves for future adversity. This ability is directly related to our level of certainty regarding future events – how likely they are, when they will occur, and what they will be like. Uncertainty diminishes how efficiently and effectively we can prepare for the future and thus contributes to anxiety. [62]

So, when confronted with something unknown, they try to make sense of it, and sometimes this goes badly. Hydroxychloroquine and ivermectin, a livestock dewormer, came to mind during the COVID pandemic.

The greatest unknown associated with the ZIKV and Zika was a simple question. Why did Zika hit Brazil, Puerto Rico, and some islands with such ferocity during 2015–2016?

Two hypotheses dominate the literature: (i) evolution for enhanced urban transmission via adaptation to mosquito vectors or enhanced human infection to increase amplification, or (ii) the stochastic introduction of ZIKV into large, naive human populations in regions with abundant *Ae. aegypti* populations, leading to rare, severe infection outcomes for their first recognition [63]. Both may be correct, but it did not stop a slew of suppositions and conspiracy theories like those covered above. Before moving to the actual appearance and spread of the virus and the associated disease, the answers to this question deserve clarity. The following section gets technical; feel free to skip it and return to it later in the book.

2.2.1 Mutations

This is the primary explanation found in the academic and technical literature. According to Lambrechts [64], the virus could have evolved and acquired new features facilitating its emergence in the human population, such as adaptation to the human transmission cycle [64].

Two issues have surfaced about the strain that made its way to Brazil in 2015. First, the strain may have some especially dangerous neurological and fetal pathogenesis (manner of development of a disease). Second, the strain might have had some characteristics that help to explain its rapid dissemination. The consensus is that the strain came from Asia, which seemed to cause minor havoc there, and arrived in Latin America with a medical vengeance.

Viruses come in strains, and different strains behave differently. A Canadian team compared RNA accumulation of two strains, a Thai strain from Canada and the Brazilian strain from Bahi, using (RT-PCR) reverse transcription-polymerase chain reaction analyses. They wanted to determine whether ZIKV increased pathogenicity and rapid ability to spread in tropical areas of the Americas and raise questions regarding whether there is a genetic basis for these changes between the early Asian ZIKV strains and the contemporary Brazilian isolates.

Alpuce-Lazcano et al. [65] reported both cytopathicity (cell disease) and RNA accumulation of the Brazilian ZIKV isolate compared to the Thai isolate could contribute to the increased pathogenicity observed during the Brazilian epidemic, and once compared to other flaviviruses, the virion (a complete infective form of a virus outside a host cell) of the strain that reached Brazil is thermostable and has a more compact surface, which may contribute to its stability in body fluids, such as saliva, urine, or semen [65].

The increased cytopathicity and RNA accumulation of the Brazilian ZIKV isolate compared to the Thai isolate could contribute to the increased pathogenicity observed during the Brazilian epidemic. Whether ZIKV "attenuation" from fetal lethality to sublethality was adaptive or random is currently unknown. At first sight, there is no obvious evolutionary advantage for the virus to cause congenital disabilities rather than abortion. Still, fetal sublethality could be an indirect consequence of the adaptive evolution of another trait influencing viral fitness, such as the increased potential for transmission [64].

Weaver et al. reported that the Asian ZIKV lineage may have adapted to generate higher viremia levels (levels of virus in the bloodstream) in humans, which would lead to more efficient mosquito infection and higher transmission levels and spread. Higher viremia could also enhance transplacental transmission to explain the emergence of microcephaly, or changes in cell tropism could be involved [66].

Another group of researchers says a gene mutation detected after 2012 could have enhanced the virus' power to overcome immunity in the Aedes mosquito, its primary carrier, and increase its power to infect. The evolutionary mutation can increase the ability of Zika to overcome the host mosquito's immune response, reproduce more efficiently and cause more significant infection and epidemics; Shi Peiyong, a virologist at the University of Texas Medical Branch at Galveston and leader of the research team, said [67].

Researchers have tried to explain the process. Here are a few of the most prominent theories. High frequency reassortment can occur when mosquitoes are infected with two viruses, most likely from interrupted feeding and moving from one host to another to complete a blood meal or separate infected bloodmeals one or two days apart. Not only are reassortments able to produce new viruses with new epidemiologies, virulence, and pathologies, but relatively minor mutations are also capable of dramatically affecting vector susceptibility enabling quantum improvements in transmission efficiency in previously inefficient vectors or even new vectors to transmit the virus [68]. According to Braack et al. [68]:

> A case in point is the single amino acid change in the EI envelope glycoprotein (which mediates virus entry by interacting with specific receptors present at the cell surface) of the chikungunya virus (CHIKV). CHIKV had an alanine amino acid (used in the biosynthesis of proteins) at position 226 in EI. This virus was transmitted mainly by *Ae. aegypti*. A single mutation results in a valine amino acid at position 226, greatly enhancing virus replication and transmission by—*Ae. albopictus* led to the widespread dispersal of CHIKV in regions of Asia where *Ae. albopictus* predominated, as well as an outbreak in northern Italy where *Ae. aegypti* did not occur. These events potentially contribute to the survival, spread, and increased incidence of mosquito-borne arboviruses. [68]

A Lawrence Berkeley National Laboratory (Berkeley Lab) researcher, Banumathi Sankaran [69], worked as part of a multiinstitutional team to map a critical viral protein called NS5. NS5 contains two enzyme activities necessary for virus reproduction: One reduces the body's ability to mount an immune response against infection, and the other helps start the genetic replication process [70].

There was serological evidence that ZIKV had circulated in Southeast Asia for many years. Accumulating knowledge [71] suggested that neurovirulent strains of the ZIKV have evolved from more minor pathogenic lineages of the virus. The "journey to pathogenicity" of the Asian strains of ZIKV indicates a combination of factors such as genetic variation in the NS5 gene [72] because of mutations or recombination events, and the reacquisition of an E-154 glycosylation [1] motif could contribute to neurovirulence. Both elements could account for differences in the replication efficiency, neuroinvasiveness, and neurotropism of the Asian ZIKV strains compared to native African ones [71].

This evidence suggests that factors such as genetic variation in the NS5 gene from mutation or recombination events and the reacquisition of an E-154 glycosylation motif could contribute to neurovirulence [73].

Phylogenetic and comparative amino acid analyses by Pettersson et al. [74] revealed four unique amino acid substitutions: One (D683E) was in the viral receptor site, two (T777M and V763M) were found in the viral envelope protein, and the final but possibly the most critical one (S139N) was in the M protein.

One or more of these E mutations could be responsible for either enhanced transmission by the primary epidemic mosquito vector, Aedes aegypti, or altered human tropism that could explain fetal infection and congenital pathogenesis. One of these mutations, encoding an S139N substitution in the transmembrane M protein, has been identified as an essential determinant of viral maturation for the closely related flavivirus DENV [63].

These amino acids play essential roles in virus attachment entry, release, or maturation of the virus during the infectious process. Thus, three unique amino acid substitutions directly associated with virus infection and transmission efficiency occurred just before the emergence and widespread dispersal of the highly epidemic and neurotropic ZIKV [74].

In a new round of genetic sleuthing, Xia et al. [75] claimed they may have pinpointed the single genetic change that has made the ZIKV a fearsome plague to pregnant women and their babies across the Americas, responsible for thousands of cases of microcephaly and other grievous brain abnormalities that sometimes result in death. Once harmless, the ZIKV became lethal after a single genetic mutation took hold around 2013. Some scientists suspect a mutation in the virus made the disease more harmful and triggered the epidemic. It also landed in an ideal climate, where tropical mosquitoes—the Aedes family—helped propel Zika to such large numbers [76]. They claim: The Asian lineage of ZIKV (ZIKV), responsible for the recent epidemics, has fixed a mutation in the NS1 gene after 2012 that enhances mosquito infection [77].

When ZIKV reached Asia, an NS1 V188A substitution occurred, followed by reversion just before the recent outbreaks. If the NS1-188A residue has a selective disadvantage for transmission, this could have happened through a founder effect when a single infected traveler introduced ZIKV to Asia [75].

> According to a study published online in *Science*, a single mutation in the ZIKV made it more virulent, contributing to the increased incidence of microcephaly. Ling Yuan, Ph.D., from the Chinese Academy of Sciences in Beijing, and colleagues conducted a phylogenetic analysis of contemporary epidemic strains of the ZIKV.
>
> The researchers found that a single serine to glutamine substitution (S139N) in the viral polyprotein substantially increased ZIKV infectivity in human and mouse neural progenitor cells. This led to more significant microcephaly in mouse fetuses and higher neonatal mouse mortality.
>
> According to evolutionary analyses, this single S139N substitution arose before the outbreak in French Polynesia in 2013 and has been maintained during the subsequent spread to the Americas. "This functional adaptation makes ZIKV more virulent to human neural progenitor cells, thus contributing to the increased incidence of microcephaly in recent ZIKV epidemics," the authors. [78]

The Chinese researchers also pinpointed when the ZIKV graduated from unwelcome pest status to an international scourge. That change, they surmised, occurred around May 2013, a few months before the start of a two-year outbreak in French Polynesia and three other Pacific islands [79]. When they infected human neural progenitor cells—the forerunners of mature human brain cells—with the Zika strain bearing that single mutation and compared them to the 2010 Zika strain, the mutated version grew and multiplied more prolifically, a ruthless killer of brain cells [79].

"This data and evidence from other viruses like Ebola show us that the smallest of genetic changes can major impact virus behavior," Jonathan Ball, a molecular virologist at the University of Nottingham. He has probed genetic shifts in the Ebola virus added [79].

Changes in the NS1 structure between the African and Asian strains may contribute to increased neuroinvasiveness and different patterns of host gene activation [80].

A set of mouse studies corroborate these findings. Peron and Braga's mouse studies (2017) seemed to draw the same conclusion. Death of precursor cells resulted in reduced cortical thickness in the pups' brains, with reminiscent vacuolized nuclei with apoptotic and autophagic aspects, which was corroborated by gene expression analysis of brain tissue pups born from infected mothers. This may indicate that the virus circulating in Brazil may somehow have undergone some changes rendering the strain more pathogenic [81].

Death of precursor cells resulted in a reduction of the cortical thickness in the brains of the mice pups, with reminiscent vacuolized nuclei with apoptotic and autophagic aspects, which was corroborated by gene expression analysis of brain tissue from pups born from infected mothers [82].

Moore et al.'s [83] mouse study findings support direct neural cell injury by ZIKV and suggest disruption of existing immature neurons, decreased proliferation, and impaired migration due to loss of progenitor cells [83].

Zika, like other RNA viruses, mutates rapidly, so the mechanisms by which it damages brain cells today may not have been present in earlier decades [84]. Recent studies have shown that more genetically diverse circulating and imported ZIKV strains seem to be growing [85]. Pettersson et al. [74] shared the view of Musso and Gubler [86] that genetic changes are the most likely explanation for the dramatic emergence and neuroinvasiveness of ZIKV. Weaver agreed: "Phenotypic changes in Asian lineage ZIKV strains may have led to these disease outcomes associated with the ZIKV." [66].

Very likely, increased incidences of microcephaly (as well as Guillain–Barre) in the current outbreak may be due to specific virulent strains or to a typical pattern of all ZIKV strains that have gone unnoticed low number of cases in previous outbreaks [87]. The ZIKV appears to have mutated in 2000 and becomes able to attack developing brain cells in a fetus [88]. Asian strains of Zika (FSS13025, Cambodia) showed that embryonic stem cell-derived trophoblasts are highly susceptible to ZIKV infection [89]. According to Zhu [90], Zika may spare normal adult brain tissue, even as it seeks and destroys the primitive cells, which, in a fetus, give rise to the brain's diversity of cells (see Chap. 15).

2.2.2 *Naïve Populations*

Back to Lambrechts [64], ZIKV emergence could have simply resulted from the fortuitous introduction of the virus into immunologically naïve human populations in the presence of high densities of *Ae. aegypti*. In this case, novel clinical manifestations would simply result from the increased ability to detect rare complications due to the sheer number of infections. Given its recent rapid spread in immune naïve populations, it is anticipated that ZIKV will continue to spread in the Americas and globally in regions where competent Aedes mosquito vectors are found [64].

In ZIKV endemic areas, most adults have preexisting ZIKV immunity, and new cases primarily occur in children. Introducing ZIKV into immune naïve populations where all ages are susceptible to infection, including women of childbearing age, is the new scenario for ZIKV expansion [91].

ZIKV was transported stochastically to locations with large-enough naive human populations, accompanied by sizeable urban mosquito vector populations, for an explosive outbreak. Under these epidemic conditions, formerly rare diseases were suddenly recognized when prominent enough numbers of infections occurred [63].

If the theory—that Zika blew through Northeastern Brazil in one wave—is correct, it likely means so many people there were infected in 2015 that there were few still vulnerable to the virus in 2016. In some ways, which may be a good sign. It might suggest Zika outbreaks are swift [92].

2.3 Coinfection

ZIKV can circulate simultaneously as other viruses, e.g., dengue and chikungunya, share the same vector. However, there is no evidence that exposure, infection, or immunity to one virus impacts the outcome of a ZIKV infection or provides protection.

A study [93] found an association with more severe symptoms in coinfected patients with malaria [93].

From July 2009 to March 2013, patients from seven healthcare facilities in the Kedougou region presenting with acute febrile illness were enrolled and tested for malaria and arboviral infections, i.e., yellow fever (YFV), West Nile (WNV), dengue (DENV), chikungunya (CHIKV), Crimean Congo hemorrhagic fever (CCHFV), Zika (ZIKV), and Rift Valley fever viruses (RVFV).

They reported that an individual could become coinfected when bitten by a mosquito harboring both the malaria parasite and an arbovirus and a coinfection rate of up to 89% in a sample in Senegal.

While there is hardly a consensus around this issue, it is noteworthy. Co-infection may be problematic on many diverse levels, especially diagnosis and treatment. The issue of coinfection is briefly mentioned here but will be treated in much more detail in Chap. 9 where problems with diagnosing ZIKV are examined.

2.4 Conclusion

The ZIKV virus retains its own unknowns to this day. The primary one reappearing in the literature with an exceedingly high frequency is: Why was microcephaly not recognized in earlier manifestations?

Besides the potential impact of herd immunity and the lack of diagnostics and surveillance in epidemic areas, one plausible hypothesis is that ZIKV has acquired some adaptive mutations to become more virulent to the human fetal brain [94]. The spread was especially pronounced in conjunction with the naïve population in Latin America who had never experienced the ZIKV before.

While the technical research community pondered this question, the public was dealing with their uncertainties complicated by the surfeit of conspiracy theories and rumors abounding in the media, both print and broadcast but primarily in social media and Internet postings. Next, the spread of the ZIKV in 2015–2016 and beyond must be examined.

Endnote

1. Glycosylation is the process by which a carbohydrate is covalently attached to a target macromolecule, typically proteins and lipids.

References

1. National Research Council (2008) Vector-borne diseases: understanding the environmental, human health, and ecological connections, workshop summary (forum on microbial threats). The National Academies Press, Washington, DC
2. GHRF Commission (Commission on a Global Health Risk Framework for the Futures) (2016) The neglected dimension of global security: a framework to counter infectious disease crises. https://www.nap.edu/download/21891. Accessed 21 May 2017
3. Commission on a Global Health Risk Framework for the Future (2016) The neglected dimension of global security: a framework to counter infectious disease crises. The National Academies Press, Washington, DC
4. McGraw E, O'Neill S (2013) Beyond insecticides: new thinking on an ancient problem. Nat Microbiol 11. March. http://www.nature.com/nrmicro/journal/v11/n3/full/nrmicro2968.html. Accessed 22 May 2017
5. Heesterbeek H et al (2015) Modeling infectious disease dynamics in the complex landscape of global health. Science 347. March 13. http://science.sciencemag.org/content/347/6227/aaa4339. Accessed 27 June 2017
6. Sands P, Munduca-Shah C, Dzau VJ (2019) The neglected dimension of global security—a framework for countering infectious-disease crises. N Engl J Med 374:13. https://doi.org/10.1056/NEJMsr1600236
7. Smallman S (2018) Conspiracy theories and the Zika epidemic. J Int Glob Stud 9(2). Works at https://pdxscholar.library.pdx.edu/is_fac. Accessed 1 June 2022
8. Gallie WB (1964) Essentially contested concepts. In: Gallie WB (ed) Philosophy and the historical understanding. Chatto & Windus, London, pp 157–191

9. Wood MJ (2018) Propagating and debunking conspiracy theories on twitter during the 2015–2016 Zika virus outbreak. Cyberpsychol Behavior Soc Netw 2121:485–490. https://doi.org/10.1089/cyber.2017.0669
10. Mitchell S (2019) Population control, deadly vaccines, and mutant mosquitoes: the construction and circulation of Zika virus conspiracy theories online. Can J Commun 44(2):211–237. https://www.proquest.com/docview/2254486919. Accessed 1 June 2022
11. Dredze M, Broniatowski DA, Hilyard KM (2016) Zika vaccine misconceptions: a social media analysis. Vaccine 34:3441–3442
12. Jacobs A (2016) Conspiracy theories about Zika spread through Brazil with the virus. The New York Times. February 16. https://www.nytimes.com/2016/02/17/world/americas/conspiracy-theories-about-zika-spread-along-with-the-virus.html. Accessed 1 June 2022
13. Granmisterio VM (2016, February 5) VIRUS ZIKA Toda la información y la patente de ROCKEFELLER, YouTube video (video no longer available)
14. Klostad CA, Uscinski JE, Connolly JM, West JP (2019) What drives people to believe in Zika conspiracy theories? Palgrave Commun 5(36). https://www.nature.com/articles/s41599-019-0243-8. Accessed 1 June 2022
15. Wood MJ, Douglas KM, Sutton RM (2012) Dead and alive: beliefs in contradictory conspiracy theories. Soc Psychol Personality Sci 3(6):767–773
16. Dredze M, Broniatowski DA, Hilyard KM (2016) Zika vaccine misconceptions: a social media analysis. Vaccine 34(30):3441–3442. https://doi.org/10.1016/j.vaccine.2016.05.008
17. Seltzer EK, Horst-Martz E, Lu M, Merchant RM (2017) Public sentiment and discourse about Zika virus on Instagram. Publ Health 150:170–175
18. Lewandowsky S, Gignac GE, Oberauer K (2013) The role of conspiracist ideation and worldviews in predicting rejection of science. PLoS ONE 8(10):e75637. https://doi.org/10.1371/journal.pone.0075637
19. McNeil D (2016) Zika Virus rumors and theories that you should doubt. The New York Times. February 19. https://www.nytimes.com/interactive/2016/02/18/health/what-causes-zika-virus-theories-rumors.html. Accessed 19 Sept 2018
20. Harambam J, de Grusauskaite KU, Wildt L (2022) Poly-truth, or the limits of pluralism: popular debates on conspiracy theories in a post-truth era. Publ Underst Sci. https://doi.org/10.1177/09636625221092145
21. Berube D (2009) Rhetorical gamesmanship in the nano debates over sunscreens and nanoparticles. J Nanoparticle Res 10:23–37. December 2008. https://doi.org/10.1007/s11051-008-9362-7; Reply from Berube D (2008) NCSU. J Nanoparticle Res 10:265–266. December 2008.https://doi.org/10.1007/s11051-008-9442-8
22. Al-Qahtani AA et al (2016) Zika virus: a new pandemic threat. J Infection Dev Countries 10(3). https://www.ncbi.nlm.nih.gov/pubmed/27031450. Accessed 11 June 2019
23. Mercola J (2016) Zika: Brazil admits it's not the virus. Mercola.com. August 16. http://articles.mercola.com/sites/articles/archive/2016/08/16/birth-defects-brazil-not-zika-virus.aspx. Accessed 23 May 2017
24. New England Complex Systems Institute (2017) New doubts on Zika as cause of microcephaly. Science Daily. June 24. www.sciencedaily.com/releases/2016/06/160624150813.htm. Accessed 18 July 2017
25. Reduas.com (2016) Report from physicians in the crop sprayed town regarding Dengue-Zika microcephaly and massive spraying with chemical poisons. http://reduas.com.ar/report-from-physicians-in-the-crop-sprayed-town-regarding-dengue-zika-microcephaly-and-massive-spraying-with-chemical-poisons/. Accessed 22 Sept 2016
26. Almendrala A (2016) A viral story links the Zika crisis to Monsanto. Don't believe it. The Huffington Post. February 17. http://www.huffingtonpost.com/entry/zika-monsanto-pyriproxyfen-microcephaly_us_56c2712de4b0b40245c79f7c. Accessed 22 Sept 2016
27. Portal da saude (2016) Clarification on the use of larvicide pyriproxifen. February 13. Portal da saude. http://portalsaude.saude.gov.br/index.php/cidadao/principal/agencia-saude/22161-esclarecimento-sobre-o-uso-do-larvicida-pyriproxifen. Accessed 27 Sept 2016

28. GM Watch (2016) Argentine and Brazilian doctors name larvicide as potential cause of microcephaly. GM Watch. February 10. http://www.gmwatch.org/news/latest-news/16706-argentine-and-brazilian-doctors-name-larvicide-as-potential-cause-of-microcephaly. Accessed 23 Sept 2016
29. Herick de Sa T, Reis-Santos B, Rodrigues L (2016) Zika outbreak, mega-events, and urban reform. Lancet 4. September. http://www.thelancet.com/journals/langlo/article/PIIS2214-109X(16)30174-7/fulltext?elsca1=etoc. Accessed 23 Sept 2016
30. Porterfield A (2016) Activists behind Zika virus conspiracy theories, Argentine pesticide birth defect scare. Genetic Literacy Project. March 28. https://www.geneticliteracyproject.org/2016/03/28/activists-behind-zika-virus-conspiracy-theories-argentine-pesticide-birth-defect-scare/. Accessed 29 Sept 2016
31. New England Complex Systems Institute (2016) New doubts on Zika as cause of microcephaly. Science Daily. June 24. www.sciencedaily.com/releases/2016/06/160624150813.htm. Accessed 2 Oct 2016
32. REDUAS: Goals (2016) Reduas.com. http://reduas.com.ar/objetivos/. Accessed 29 Sept 2016
33. Johnson R, Jelmayer R (2016) Brazilian state bans pesticide after Zika claim. Wall Str J. February 15. http://www.wsj.com/articles/brazil-state-bans-pesticide-after-zika-claim-1455584596. Accessed 22 Sept 2016
34. Johnson R, Jelmayer R (2016) Zika claim prompts ban on an insecticide. Wall Str J Europe. February 17. http://www.pressreader.com/belgium/thewallstreetjournaleurope/20160217/281771333249355/TextView. Accessed 25 Sept 2016
35. Nota técnica e carta aberta à população Microcefalia e doenças vetoriais relacionadas ao *Aedes aegypti*: os perigos das abordagens com larvicidas e nebulização química – fumacê. Janeiro de 2016. GT Salud y Ambiente. Asociación Brasileña de Salud Colectiva. ABRASCO. https://www.abrasco.org.br/site/2016/02/notatecnicasobre. Accessed 22 Sept 2016
36. Robinson C (2016) What did Brazilian public health researchers really say about Zika, pesticides, and birth defects? GM Watch. February 26. http://gmwatch.org/news/latest-news/16741-what-did-brazilian-public-health-researchers-really-say-about-zika-pesticides-and-birth-defects. Accessed 25 Sept 2016
37. Robinson C (2016) Zika, microcephaly, and pesticides: Half-truths, hysteria, and vested interests. GM Watch. February 25. http://www.theecologist.org/News/news_analysis/2987284/zika_microcephaly_and_pesticides_halftruths_hysteria_and_vested_interests.html. Accessed 25 Sept 2016
38. Costa F, Co A (2018) Zika virus and microcephaly: where do we go from here? Lancet 18. March. https://doi.org/10.1016/S1473-3099(17)30697-7
39. de Araújo TVB et al (2018) Association between microcephaly, Zika virus infection, and other risk factors in Brazil: final report of a case-control study. Lancet Infectious Dis. 11 Dec 2017. https://doi.org/10.1016/S1473-3099(17)30727-2. Accessed 6 July 2018
40. Second Nexus Staff (2016) Zika virus not to blame? Doctors city man-made cause for birth defect epidemic. February 13. http://secondnexus.com/ecology-and-sustainability/zika-is-not-to-blame/. Accessed 27 Sept 2016
41. Albuquerque MDFP et al (2016) Pyriproxyfen and the microcephaly epidemic in Brazil-an ecological approach to explore the hypothesis of their association. Memórias do Instituto Oswaldo Cruz, Rio de Janeiro. 111:12. December. http://www.scielo.br/scielo.php?script=sci_arttext&pid=S0074-02762016001200774. Accessed 18 July 2018
42. Regis L et al (2014) Characterization of the spatial and temporal dynamics of the dengue vector population established in urban areas of Fernando de Noronha, a Brazilian oceanic island. Acta Tropica 137. September. https://www.ncbi.nlm.nih.gov/pubmed/24832009. Accessed 30 July 2018; de Silva Augusto LG et al (2016) Aedes aegypti control in Brazil. Lancet 387. March 12. https://www.thelancet.com/pdfs/journals/lancet/PIIS0140-6736(16)00626-7.pdf. Accessed 8 Aug 2018
43. WHO (2016) Dispelling rumours around Zika and complications. WHO. http://www.who.int/emergencies/zika-virus/articles/rumours/en/. Accessed 24 Sept 2016

44. Bowater D (2016) Zika virus: Brazil dismisses link between larvicide and microcephaly. Telegraph. February 15. http://www.telegraph.co.uk/news/worldnews/zika/12157747/Zika-virus-Brazil-dismisses-link-between-larvicide-and-microcephaly.html. Accessed 22 Sept 2016
45. Al-Qahtani AA et al (2016) Zika virus: a new pandemic threat. J Infection Dev Countries 10(3). https://www.ncbi.nlm.nih.gov/pubmed/27031450. Accessed 11 June 2019; Mercola J (2016) Zika: Brazil admits it's not the virus. Mercola.com. August 16
46. Genetically modified mosquitoes released in Brazil in 2015 linked to the current Zika epidemic? (2016) https://www.reddit.com/r/conspiracy/comments/42mhii/genetically_modified_mosquitoes_released_in/. Accessed 4 Sept 2016
47. Kolker R (2016) Florida's feud over Zika-fighting GMO mosquitoes. Bloomberg Businessweek. October 6. http://www.bloomberg.com/features/2016-zika-gmo-mosquitos/. Accessed 25 Oct 2016
48. Wilcox C (2016) No, GM mosquitoes didn't start the Zika outbreak. Discovery Magazine. January 31. http://blogs.discovermagazine.com/science-sushi/2016/01/31/genetically-modified-mosquitoes-didnt-start-zika-ourbreak/#.V82RpSgrKUk. Accessed 5 Sept 2016
49. Rosen M (2016) Microcephaly: building a case against Zika. Sci News: 26–29. April 2. https://www.sciencenews.org/article/microcephaly-building-case-against-zika. Accessed 27 Sept 2016
50. Al-Qahtani AA et al (2016). ZIKV: a new pandemic threat. J Infection Dev Countries 10(3). https://www.ncbi.nlm.nih.gov/pubmed/27031450. Accessed 11 June 2019; Mercola J (2016) Zika: Brazil admits it's not the virus. Mercola.com. August 16
51. Kress G (2016) Who owns the ZIKV? Glob Res. September 5. http://www.globalresearch.ca/who-owns-the-zika-virus/5505323. Accessed 24 Sept 2016
52. Olmstead G (2016) The Zika project. Amazon, Middletown, DE
53. The Guardian (2018) GM mosquito trials a health risk in India. The Sunday Guardian. January 7. https://www.sundayguardianlive.com/news/12314-gm-mosquito-trials-health-risk-india. Accessed 27 June 2018
54. Laskow S (2016) While Brazil was eradicating Zika mosquitoes, America made them into weapons. Atlas Obscura. June 22. http://www.atlasobscura.com/articles/while-brazil-was-eradicating-zika-mosquitoes-america-made-them-into-weapons. Accessed 19 May 2017
55. Rathi A (2015) This could be the next weapon of mass destruction. November 20. Quartz Media. https://qz.com/554337/this-could-be-the-next-weapon-of-mass-destruction/. Accessed 29 May 2017
56. Charlton C (2016) December 26. Bio-war in sight. The Sun. https://qz.com/554337/this-could-be-the-next-weapon-of-mass-destruction/. Accessed 3 Apr 2017
57. Stavridis J (2016) Zika is just the first front in the 21st century biowar. Foreign Policy. August 24. https://foreignpolicy.com/2016/08/24/zika-is-just-the-first-front-in-the-21st-century-biowar/. Accessed 30 Oct 2022
58. Worth K (2016) As Brazil confronts Zika, vaccine rumors shape perceptions. FRONTLINE. February 16. https://www.pbs.org/wgbh/frontline/article/as-brazil-confronts-zika-vaccine-rumors-shape-perceptions/, Accessed 12 May 2022
59. Simas C, Paterson P, Lees S, Larson HJ (2021) "From my phone, I could rule the world": critical engagement with maternal vaccine information, vaccine confidence builders, and post-Zika outbreak rumours in Brazil. Vaccine 39:4700–4704
60. Stone J (2016) Small brain disorder in Brazilian babies: caused by Zika virus—or vaccine? Vaccination information network website. https://www.vaccinationinformationnetwork.com/small-brain-disorder-in-brazilian-babies-caused-by-zika-virus-or-vaccine/. Accessed 1 June 2022
61. Lovecraft HP (1927) Supernatural horror in literature. https://www.hplovecraft.com/writings/texts/essays/shil.aspx. Accessed 18 June 2022
62. Grupe DW, Nitschke JB (2013) Uncertainty and anticipation in anxiety. Nat Rev Nanosci 14(7):488–501. July; Opening quote from Gilbert DT (2006) Stumbling on happiness. Random House, NY

63. Weaver SC (2017) Emergence of epidemic ZIKV transmission and congenital Zika syndrome: are recently evolved traits to blame? mBio 8:1. January/February. https://doi.org/10.1128/mBio.02063-16. Accessed 18 Sept 2018
64. Lambrechts L (2021) Did Zika virus attenuation or increased virulence lead to the emergence of congenital Zika syndrome? J Travel Med. March 23. https://doi.org/10.1093/jtm/taab041
65. Alpuce-Lazcano SP et al (2018) Higher cytopathic effects of a Zika virus Brazilian isolate from bahia compared to a Canadian-imported Thai strain. Viruses 10:53. https://www.ncbi.nlm.nih.gov/pubmed/29382068. Accessed 18 July 2018
66. Weaver S et al (2016) Zika virus: history, emergence, biology, and prospects for control. Antiviral Res 130:69–80. https://www.ncbi.nlm.nih.gov/pubmed/26996139. Accessed 30 June 2017
67. XinhuaNet (2018) New research says genetic change possibly behind recent Zika outbreaks. BruDirect.com. January 30. http://brudirect.com/news.php?id=40542. Accessed 27 June 2018
68. Braack L et al (2018) Mosquito-borne arboviruses of African origin: review of key viruses and vectors. Parasites Vectors 11:29. https://www.ncbi.nlm.nih.gov/pubmed/29316963. Accessed 27 June 2018
69. Zhao B et al (2017) Structure and function of the ZIKV full-length NS5 protein. Nat Commun. March 27. https://www.nature.com/articles/ncomms14762.pdf. Accessed 2 July 2017
70. Weiner J (2017) Researchers gain insight into protein critical to ZIKV reproduction. Phys.Org. April 10. https://phys.org/news/2017-04-gain-insight-protein-critical-zika.html. Accessed 2 July 2017
71. Morris G et al (2017) ZIKV as an emerging neuropathogen: mechanisms of neurovirulence and neuro-immune interactions. Mole Neurobiol 55. https://www.ncbi.nlm.nih.gov/pubmed/28601976. Accessed 24 July 2018
72. Flavivirus NS5 protein is the largest viral protein and consists of an N-terminal methyltransferase (MTase) domain (residues 1–262) that is involved in RNA capping and methylations, and a C-terminal RNA-dependent RNA polymerase domain (residues 275–900)
73. de Oliveira Dias JR et al (2018) Zika and the eye: pieces of a puzzle. Progress in retinal and eye research. https://doi.org/10.1016/j.preteyeres.2018.04.004. Accessed 6 July 2018; Morris G et al (2018) ZIKV as an emerging neuropathogen: mechanisms of neurovirulence and neuro-immune interactions. Mole Neurobiol 55:5. May. https://doi.org/10.1007/s12035-017-0635-y. Accessed 7 July 2018
74. Pettersson JH-O, Eldholm V, Seligman SJ et al (2016) How did Zika virus emerge in the Pacific Islands and Latin America? mBio 7(5):e01239-16. https://doi.org/10.1128/mBio.01239-16
75. Xia H et al (2018) An evolutionary NS1 mutation enhances ZIKV evasion of host interferon induction. Nat Commun 414. January 29. https://www.nature.com/articles/s41467-017-02816-2. Accessed 26 July 2018
76. Grinnell A (2018) What happened to Zika? PBS NewsHour. July 6. https://www.pbs.org/newshour/science/what-happened-to-zika. Accessed 24 July 2018
77. Xia H et al (2018) An evolutionary NS1 mutation enhances ZIKV evasion of host interferon induction. Nat Commun 414. January 29. https://www.nature.com/articles/s41467-017-02816-2. Accessed 26 July 2018; Liu Y et al (2017) Evolutionary enhancement of ZIKV infectivity in Aedes aegypti mosquitoes. Nature 545. May 25. https://www.nature.com/articles/nature22365. Accessed 26 July 2018
78. Health Day (2017) The factor that made Zika a mega-outbreak. Health Day. October 9. https://www.neurologyadvisor.com/general-neurology/factor-that-made-zika-mega-outbreak/article/696784/. Accessed 27 June 2018; Yuan L et al (2017) A single mutation in the prM protein of ZIKV contributes to fetal microcephaly. Science: Eaam7120. http://science.sciencemag.org/content/early/2017/09/27/science.aam7120. Accessed 26 July 2018
79. Healy M (2017) Once harmless, the ZIKV became lethal after a single genetic mutation took hold around 2013. LA Times. September 28. http://www.latimes.com/science/sciencenow/la-sci-sn-zika-mutation-microcephaly-20170928-story.html. Accessed 27 June 2018
80. Morris G et al (2018) ZIKV as an emerging neuropathogen: mechanisms of neurovirulence and neuro-immune interactions. Mole Neurobiol 55:5. May. https://doi.org/10.1007/s12035-017-0635-y. Accessed 7 July 2018

81. Peron JPS, Braga PCBB (2017) Zika-related microcephaly in experimental models. Temperature 4:1. https://www.ncbi.nlm.nih.gov/pmc/articles/PMC5388436/. Accessed 8 July 2018
82. Citing Cugola FR et al (2016) The Brazilian ZIKV strain causes birth defects in experimental models. Nat Lett. May 11. https://www.nature.com/articles/nature18296. Accessed 25 July 2018
83. Moore C et al (2016) Characterizing the pattern of anomalies in congenital Zika syndrome for pediatric clinicians. JAMA Pediatr 171:3. https://jamanetwork.com/journals/jamapediatrics/fullarticle/2579543?utm_campaign=articlePDF&utm_medium=articlePDFlink&utm_source=articlePDF&utm_content=jamapediatrics.2016.3982. Accessed 24 July 2018
84. Wapner J (2017) How Zika leads to fetal brain abnormalities. January 17. Newsweek. http://www.newsweek.com/2017/01/27/science-behind-how-zika-leads-fetal-brain-malformations-543541.html. Accessed 30 June 2017
85. Wang Q et al (2016) Molecular determinants of human neutralizing antibodies isolated from a patient infected with Zika virus. Sci Transl Med 8. December 14. https://www.ncbi.nlm.nih.gov/pubmed/27974667. Accessed 30 June 2017
86. Musso D, Gubler DJ (2016) Zika virus. Clin Microbiol Rev 29:487–524. https://doi.org/10.1128/CMR.00072-15
87. Charrel R et al (2016) Background review for diagnostic test development for Zika virus infection. Bull World Health Org 94:574–584D. https://doi.org/10.2471/BLT.16.171207. http://www.who.int/bulletin/volumes/94/8/16-171207.pdf. Accessed 28 Mar 2017
88. Haelle T (2016) Birth defects linked to Zika include more than microcephaly. Everyday Health. http://www.everydayhealth.com/zika/living-with/birth-defects-linked-zika-include-more-than-microcephaly/. Accessed 14 May 2017
89. de Oliveira Dias JR et al (2018) Zika and the eye: pieces of a puzzle. Progr Retinal Eye Res.https://doi.org/10.1016/j.preteyeres.2018.04.004. Accessed 6 July 2018
90. Zhu Z et al (2017) ZIKV has oncolytic activity against glioblastoma stem cells. J Exper Med 214:10. https://doi.org/10.1084/jem.20171093. Accessed 3 May 2018
91. Paul L et al (2016) Dengue virus antibodies enhance Zika virus infection. bioRxiv beta. April 25. http://biorxiv.org/content/early/2016/04/25/050112. Accessed pre-print 15 May 2017
92. Branswell H (2016) Why were there fewer microcephaly cases from Zika last year? Sci Am. March 29. https://www.scientificamerican.com/article/why-were-there-fewer-microcephaly-cases-from-zika-last-year/. Accessed 7 Apr 2017
93. Sow A et al (2016) Concurrent malaria and arbovirus infections in Kedougou, southeastern Senegal. Malaria J 15:47. https://doi.org/10.1186/s12936-016-1100-5?site=malariajournal.biomedcentral.com. Accessed 29 June 2017
94. Yuan L et al (2017) A single mutation in the prM protein of ZIKV contributes to fetal microcephaly. Science: Eaam7120. http://science.sciencemag.org/content/early/2017/09/27/science.aam7120. Accessed 26 July 2018

Chapter 3
Zika Re-emerges

What's new is a black-and-white-striped insect called Aedes, a non-native variety that includes yellow fever mosquitoes and Asian tiger mosquitoes. The yellow fever mosquitoes—technically known as *Ae. aegypti*—are aggressive biters drawn to humans at all hours. They breed in standing water, and their eggs can lie dormant for months or even years on dry surfaces. In addition to yellow fever, they can transmit Zika, dengue fever, and other diseases to humans and pets [1].

Odd title. The truth is, the ZIKV virus is not new at all. It was around the 1950s. Only recently has it re-emerged in Latin America with its current levels of viciousness.

According to the National Institutes of Health in the United States and the United Kingdom, in 2017, ZIKV was classified as a level 2 (moderate risk if inhaled, swallowed, or exposed to the skin) and a level 3 pathogen [2].

While ZIKV has been around for decades, its prevalence in Latin America has garnered much of the world's attention. In 2016, the World Health Organization (WHO) estimated up to four million people could be infected in the Western Hemisphere by the end of the year [3].

ZIKV surfaced as a significant risk factor associated with travel to the American Southern Hemisphere in 2016. However, it was confirmed in May 2015 in the Brazilian states of Bahia and Rio Grande do Norte and was suspected of having been circulating even earlier.

By February 23, 2016, the U.S. CDC had graded 34 countries and territories (American Samoa, Aruba, Barbados, Bolivia, Bonaire, Brazil, Cape Verde, Colombia, Costa Rica, Curacao, Dominican Republic, Ecuador, El Salvador, French Guiana, Guadeloupe, Guatemala, Guyana, Haiti, Honduras, Jamaica, Martinique, Mexico, Nicaragua, Panama, Paraguay, Puerto Rico, Saint Martin, Samoa, Suriname, the Marshall Islands, Tonga, Trinidad and Tobago, Venezuela, and the U.S. Virgin Islands) with a "Level 2" travel health warning.

Under a Level 2 travel warning, travelers are expected to be extra vigilant to protect themselves from mosquito bites. If pregnant, they should NOT travel to areas with an outbreak of Zika [4].

D. M. Berube, *Pandemics and Resilience: Lessons we should have learned from Zika*, Risk, Systems and Decisions, https://doi.org/10.1007/978-3-031-25370-6_3

Two years later, 48 countries and territories in the Western Hemisphere have confirmed autochthonous (indigenous), vector-borne transmission of ZIKV disease, and five countries have reported sexually transmitted ZIKV cases. Overall, ZIKV spread geographically to areas where competent vectors were present (see Chap. 5).

Although only tiny pockets of autochthonous transmission have been documented in the Southern U. S., there was the potential for future vector-borne spread, as both the *Ae. aegypti* and *Ae. albopictus*, competent vectors, are present in multiple regions of the United States [5].

3.1 Zika Surfaces

The recent emergence of Zika virus (ZIKV) across South America, as well as the spread of West Nile virus (WNV) in North America in the early 2000s, reflects the increasing public health risk and pandemic potential posed by flaviviruses (a family of single-stranded enveloped RNA viruses found in arthropods, primarily mosquitoes, and ticks) globally [6].

An article in Science raised another exciting question about Zika: Why has ZIKV emerged as a public health threat in the Americas after being known for decades as a rare and mild tropical disease? [7].

The answer to this question may provide unique insight into the menu of responses to help protect public health when pandemics surface. As the story of this pandemic is written, public health and safety recommendations should become more prevalent as the potential for problematic flaviviruses increases.

First, the story of Zika, the virus, and the diseases must be told. Zika got its name from the Ziika forest (one of the i's seemed to have disappeared once the virus was named). The forest is in Uganda near Lake Victoria and along a road between Entebbe International Airport and Kampala. It encompasses 60 acres of land, remains under the control of the Uganda Virus Research Institute, and, as such, is a protected area.

Enter the Rockefeller Foundation. It funded research in the enzootic (a disease that regularly affects animals in a particular district or a specific season) or sylvatic (certain diseases when contracted by wild animals and the pathogens causing them) cycle of yellow fever virus and some additional arboviruses.

George Dick, Stuart Kitchen, and Alexander Haddow placed six platforms containing caged rhesus monkeys (macaques, an Asian species) in the canopy of the Ziika Forest [8]. They waited to see if any of the monkeys got sick. On April 18, 1947, Rhesus #766 developed a fever [9]. The agent isolated from 766 was called ZIKV (the ZIKV 766 strain) [8].

In January 1948, mosquitoes were collected in the Ziika Forest to isolate the yellow fever virus. Mosquito population analysis of the Zika Forest revealed 58 species of mosquitoes, with the majority belonging to the genera *Aedes* and *Culex* [10]. ZIKV was isolated from a pool of eighty-six collected *Aedes* in the same forest. Africanus mosquitoes [11]: The virus isolated from *Ae. Africanus* was designated ZIKV (E/1 strain) [8]. Eight years later, two other stains were isolated from the same mosquito.

Faye et al. inferred the most likely geographical pathway connecting ZIKV lineages based on these samples. This inference aligns with the first known ZIKV isolation in Uganda in 1947 [12]. These results indicated that ZIKV may have emerged in Uganda between 1892 and 1943, probably around 1920 [11].

3.2 Zika and Humans

There is some disagreement on the first infected human. For example, one source [13] argued the first evidence of human infection by this virus was reported in 1952 when Smithburn demonstrated the presence of neutralizing antibodies in human sera collected from East Africa [13].

However, most seem to agree that Simpson reported the first human ZIKV infection. Simpson described his course of the disease acquired while isolating ZIKV from *Ae. Africanus* mosquitoes in Uganda between 1962 and 1963 [14]. In turn, Simpson reported the first human case in Uganda in 1964 [15], and until 2007, only fourteen sporadic human cases were reported worldwide in some of Africa and Asia [16].

Of course, there is disagreement again. Faye and Haddow disagree. They report the first human ZIKV isolate reportedly came from a 10-year-old Nigerian female in 1954 [17].

Additional serologic studies in the 1950s and 1960s detected ZIKV infections among humans in Egypt, Nigeria, Uganda, India, Malaysia, Indonesia, Pakistan, Thailand, North Vietnam, and the Philippines [18].

According to the academic literature, ZIKV appeared in human populations in the fifties and sixties in Africa and circulated enzootically among unknown vertebrate hosts (presumably non-human primates) [19].

ZIKV lineage from Uganda probably spread to Asia via Malaysia around the mid-fifties. From there, the virus reached Micronesia around 1960, forming what is known as the Asian cluster.

Marchette et al. [20] reported the earliest direct detection of ZIKV in Asia and *Aedes*' first evidence of transmissions. *Aegypti* was in Malaysia in 1969 [20].

Still, there is more disagreement regarding the first human infections in Asia. Early on, Olson et al. [21] claimed they were diagnosed in central Java in Indonesia in 1977 [21]. Weaver et al. [22] agreed and claimed the first human infections in Asia were diagnosed in central Java in Indonesia in 1977 by seroconversion in seven patients who presented themselves with fever, malaise, stomachache, anorexia, and dizziness [22].

After sixty tears since the first reported cases, fewer than 20 human infections have been reported in Asia and Africa, and it wasn't until 2013 that serious complications were reported [23].

At the end of 2016, 48 countries and territories in the Americas have confirmed autochthonous, vector-borne transmission of ZIKV disease since 2015. In addition, five countries in the Americas have reported sexually transmitted Zika cases [24]. Among the places where the ZIKV was widespread were Southern Florida; the

Caribbean, including Puerto Rico; Mexico, Central America, and much of South America; and Southeast Asia, including Vietnam and Thailand [25].

Where did the virus that plagued Brazil come from? Since it is known there are different strains variances of the virus tracing, the source has involved the subclades. The ZIKV from Southeast Asia is distantly related to two African subclades and seems to represent a divergence from a common ancestor that spread throughout Southeast Asia and the Pacific.

Human ZIKV cases were detected in peninsular Malaysia, confirming that ZIKV was active in this region even before 2007 [18]. And it wasn't until 2007 that significant numbers of human patients were reported [26].

Yap, one of the Caroline Islands in the Western Pacific, had a mild outbreak affecting around 900 people in 2007 [27], and this number was reduced to 49 confirmed and 59 probable cases [28]. According to Ratna [29], nearly three-quarters of the entire population on Yap in Micronesia had been infected by ZIKV [29]. Oddly enough, despite the high percentage of Yap's 11,000 or so residents infected with Zika, no neurological complications were reported from the outbreak, according to the WHO [30].

There was another Southeast Asian outbreak of ZIKV in 2013–2015 in New Caledonia, Cook Island, Easter Island in 2014, and Vanuatu, the Solomon Islands, Samoa, and Fiji in 2015 [8]. French Polynesia and New Caledonia outbreak affected 28,000 people (11% of the population) [28]. The responsible mosquito vector was probably *Aedes (Stegomyia) polynesiensis* [28].

This time there were severe symptomologies. In the 2013–2015 outbreak, local officials reported 17 cases of malformations in fetuses and infants hypothesized to be ZIKV associated [31]. This was also the first time Guillain–Barre syndrome was associated with ZIKV [28]. In both these cases, the virus seems to have become locally extinct [32].

In general, the ZIKV remained confined to a narrow equatorial band in Africa and Asia until 2014, when it began to spread eastward, first toward Oceania and then to South America. Since then, millions of infected individuals have been identified in Brazil, Colombia, and Venezuela, including 25 additional countries in the Americas [10].

Initial speculation rested on believing someone could have imported the virus with an undetected infection [33]. The likely principal vector was *Aedes (Stegomyia) hensilli* [28], but virus detection never informed this.

ZIKV is a mosquito-borne flavivirus (a member of the Spondweni serocomplex). What is known today is that it is carried predominantly by the *Ae. aegypti* mosquito (though there is evidence *Ae. albopictus* also take it as was the case in Gabon in 2007 and 2010) [34]. While the two primary mosquito vectors are *Ae. aegypti* and *Ae. albopictus*, several ecologically or geographically distinct mosquito vectors may be responsible for the transmission and maintenance of ZIKV throughout Asia [35].

How did the ZIKV arrive in South America?

The most compelling story comes from Musso, who suggested the virus came to Brazil during the Va'a Canoe World Sprint Championship on August 15, 2014.

The championships opened in Lagoa Rodreigo De Friotas, Rio de Janeiro. Four Pacific countries (French Polynesia, New Caledonia, Cook Islands, and Easter Island) in which Zika circulated in 2014 had teams in this contest [36]. Combined with phylogenetic studies by Al-Qahtani et al. [37] suggest that ZIKV introduction in Brazil may have resulted from this event [37].

Others offer no single event can be identified as a cause. Instead, it should be chalked up to increase general mobility [38].

3.3 Zika in Brazil

There was no evidence of ZIKV circulation in the Western Hemisphere until 2014 [28]. The cases of rash, mild fever, and arthralgia were diagnosed in Northeastern Brazil as early as September 2014 [29].

Generally, the dried rural areas that dominate the interior of Northeast Brazil, called sertáo, were most affected. As of early December 2015, eighteen states in Brazil have confirmed autochthonous virus transmission in the northern, northeastern, southeastern, central-western, and southern regions [8].

In 2016, *Lancet* reported an estimated minimum of 400,000 cases of ZIKV disease reported in 24 states in Brazil and suggested the number of cases could be far higher [39]. Beil in *Science News* said as many as 1.4 million people in Brazil alone may have contracted Zika [40], and the Brazilian Ministry of Health corroborated, estimating that 500,000–1.5 million persons were infected [41].

The story of the ZIKV outbreak in Brazil begins in Recife (capital of Brazil's northeastern state of Pernambuco and the fourth-largest urban area in Brazil), where a team of doctors was trying to explain a spike in cases of microcephaly among newborns. The Oswaldo Cruz Hospital in Recife was probably ground zero. Dr. Vanessa Van Der Linden, who worked there, sounded the alarm after ruling out more specific causation after viewing calcification images in a brain scan [42]. Her mother worked in a different medical center in Recife, telephoned her daughter, and shared notes late summer 2015. At both locations, they shared information on ten cases.

On April 29, 2015, researchers from Salvador (not the country, the most significant state capital in Northeast Brazil) announced ZIKV as the likely etiology of the outbreak. Almost simultaneously, ZIKV was confirmed in 8 of 21 patients from Natal, another city in Northeast Brazil [43].

Over time, Health Secretary for Recife estimated around 50,000 to 100,000 might have been exposed to ZIKV. The Brazilian Ministry of Health estimated between 497,000 and 1,480,000 ZIKV had occurred since the epidemics began. By mid-2016, the corrected case numbers reported out of Brazil dropped to 84,000 cases of Zika. In addition, only about 18 percent of those infected appear symptomatic.

Reports of the disease peaked in the third week of February, with 16,059 cases. In the first week of May, the reports plunged to 2053. The figures demonstrate, once again, the effectiveness of the actions taken against the *Ae. aegypti* mosquito, apart from indicating a different behavior than usual this year. In 2016, the cases started declining earlier than expected, since until then, the diseases transmitted by the *Ae. aegypti* peaked in April [44].

A National Public Health Emergency was declared in Brazil on November 11, and the Pan American Health Organization (PAHO) added its epidemiologic alert on November 17. In December 2015, Brazil estimated 440,000 to 1.3 million suspected cases. In January 2016, a technical note and open letter to the people was released by the ABRASCO (Brazilian Association for Collective Health). The WHO Public Health Emergency was declared on February 1.

In 2016, health officials in Brazil said they expect the number of infections to increase next year as more precautions are introduced dramatically, and more awareness is needed to combat the mosquito [45].

A study by Chopin-Carneiro et al. in 2016 suggested that although susceptible to infection, *Ae. aegypti* and *Ae. albopictus* were unexpectedly low competent vectors for ZIKV presenting the emerging threat. Its potential may be more a function of the high susceptibility of the human populations involved and the proximity to *Aedes* mosquitoes [46].

According to Baylor Dean Peter Hotez, "[t]here are three reasons Zika has slammed this part of Brazil: the presence of the main mosquito species that carries the virus and transmits it to humans, *Ae. Aegypti*; overcrowding and extreme poverty." [47].

ZIKV was not the first and only example of the emergence of a vector-borne disease threatening a new continent. The first was chikungunya from East Africa into the Americas in 2013. This may suggest that other factors, such as the sizeable naïve population for ZIKV and the high densities of human-biting mosquitoes, contribute to the rapid spread of ZIKV during the current outbreak [46].

3.4 Zikv in the USA

There are 2722 total US cases of Zika (8/31/16) [48]. The vast number of infections seems to be travel related. And 23 babies and five pregnancy losses in the mainland USA have been born with Zika-related congenital disabilities [49].

While the hyperbolic claims about the ZIKV will be examined elsewhere, it is essential to recognize that mosquito vector diseases tend to wax and wane along with the climate and weather. For example, the inevitable return of spring and summer raised general expectations among the CDC that ZIKV might be associated with catastrophic health problems even in the U.S. Where one lives on the planet, elevation and average temperatures will impact mosquito populations and their diseases.

The main factors precluding these diseases from occurring in the U.S. at a similar level of ferocity as in Brazil are socioeconomic such as lifestyle, housing infrastructure, and good sanitation. According to Moreno-Madriñán and Turell [50], if such conditions are maintained, local transmission is unlikely to occur significantly, particularly in the northern states [50]. One thing to keep in mind is that the U. S. has everyday creature comforts that severely diminish the ability of the *Ae. aegypti* mosquito, the primary carrier of ZIKV, flourishes and spreads the virus from person to person.

Primarily, this is not a U.S. problem unless Puerto Rico and the American Virgin Islands are not regarded as part of the U.S. (a subject for another treatise). Reporting on ZIKV has oddly treated Puerto Rico, a U.S. territory, as part of the "third world."[1]. Given the state of its economy in 2017, a little more empathy for those Hispanic citizens should be expected.

Second, even if the bulk of the infections occur in the South does not mean the impact of the ZIKV will be felt only there. Considering the multiple transmission modalities and the propensity for privileged populations to intermix through travel, expect Zika implications to be felt from north to south.

While determining all the relevant variables to explain this dispersion has not been established, some suggest both cutbacks in vector control budgets in the U.S. and the lack of adequate and acceptable pesticides may have contributed to the growing dispersion. Not to exaggerate its possible spread, most reports refer to locations where mosquitoes have been observed. That doesn't mean the ZIKV will travel to those cities or bring Zika with it if it does—it just means that the weather conditions would be suitable for it to survive and thrive [51].

Mosquitoes can be a year-round problem in parts of the Southern U.S. and Texas, where hot winters and rainy springs have created conditions for mass breeding [52]. However, to suggest the impacts of ZIKV will only be felt in the Southern U.S. is shortsighted.

At the same time, the Atlanta area tops pest control leader Orkin's 2017 list of top 50 mosquito cities for the fourth year in a row. Atlanta is followed on the list by Washington, DC, and Chicago. Twenty-one metro areas in the are included in the ranking, which is the most of any region in the U.S. [53]. Other areas of concern are pockets in Northern California and Southern California [54], and these species have been found as far north as New York State and Wisconsin [55].

The entire southern half of the continental United States, and much of its east coast, is home to the *Ae. aegypti* mosquitoes. They can carry the ZIKV [56]. In addition, Baylor's Dean Hotez wrote that the same factors responsible for the spread of Zika in Brazil (the main mosquito species, overcrowding, and extreme poverty) are present in the poorest urban areas of coastal Texas, Louisiana, Mississippi, and Alabama, in addition to South Florida, and an area around Tucson [47]. In addition, more than 60% of the counties on the Gulf Coast and Mexican border, the places likeliest to see a ZIKV outbreak, are rated "in need of improvement" for mosquito control, said Dr. Oscar Alleyne, Public Health Adviser to the National Association of County and City Health Officials [57] (Fig. 3.1).

According to Gold and Josephson, as much as 60% of Americans live in regions at risk for the spread of ZIKV [59]. Notwithstanding, while the mosquito species capable of spreading ZIKV live in many parts of the U.S., experts didn't expect an epidemic here; they worried that small clusters of cases were likely, particularly in Florida or Texas, if the insects bite returning travelers and then someone else [60].

Most of the reported occurrences were related to travel and some to sexual transmission between men and women. For example, according to the report, "[t]here

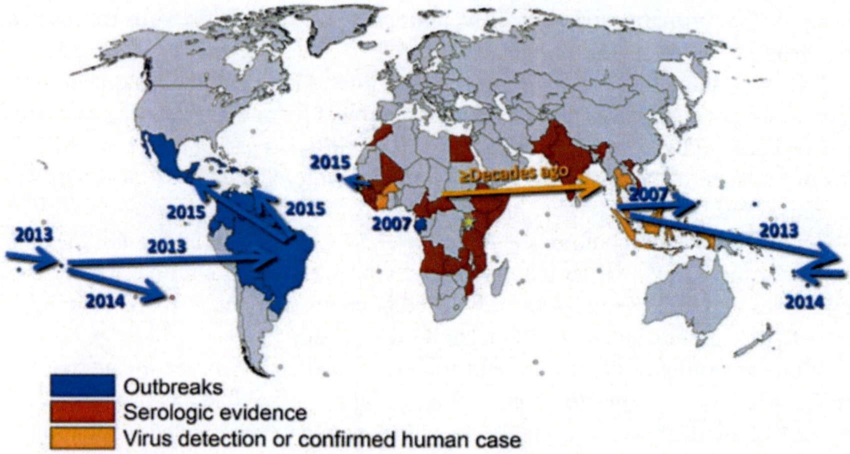

Fig. 3.1 Map showing the known distribution of Zika virus based on serosurveys, virus detection, and laboratory-diagnosed cases. Blue arrows show recent patterns of spread deduced from phylogenetic studies (see this figure). The yellow star shows the location of the Ziika Forest where the virus was discovered in 1947. Weaver et al. [58]

were 24 identified cases of ZIKV in Orange County, California, in 2016, but all contracted the disease outside the county. Specimens of those mosquito types *Aegypti* and *Albopictus* have been identified in 12 cities in Orange County." [61].

However, if the two mosquito species capable of spreading the virus continue their predicted geographical expansion (see Chap. 10), the country could soon be problematic.

Max Moreno-Madrinan, Assistant Professor in public health at Indiana University-Purdue University, and Michael Turell, Independent Research Entomologist, pointed out [62]. For most of the country, ZIKV was travel related from a location with a significant population of mosquitoes. The primary alternative was travel based when the U.S. pulled together a vector control strategy (see Chap. 8).

3.4.1 Zika Everywhere (Travel-Related Reports)

The reported outbreaks did extend across the continental U.S. For many places where mosquitoes are not a significant problem, however, the cause of the explosion was traced to travel in one way or another.

For example, the Connecticut Department of Public Health monitored 30 babies born in or living in Connecticut whose mothers tested positive for ZIKV during pregnancy. Two of those babies had Zika-related congenital disabilities, and another nine "were borderline" for congenital disabilities, the agency reported in a press release on January 30, 2017 [63].

Health officials said that an Alaskan who had traveled abroad contracted the state's third recorded case of ZIKV on April 28, 2017. According to a statement from the Alaska Department of Health and Social Services, the patient tested positive for the mosquito-borne virus after returning from a trip to areas of Central America where the virus is prevalent [64].

And then, reports like those received from the Kapi'olani Medical Center for Women and Children. A Hawai'i Pacific Health hospital identified six mothers who gave birth to babies with microcephaly. Of the six, ZIKV antibodies were detected in three, fifty percent, of the mothers who delivered babies with microcephaly, suggesting the presence of positive ZIKV cases and associated microcephaly in the United States as early as 2009 [65]. It seemed likely the women had either traveled themselves or were impregnated by someone who had traveled to an endemic zone like South America.

In another state, the Illinois Department of Public health says 85 people were infected by the ZIKV statewide, and Dr. Loret De Mola said the numbers would go up. "With winter, people travel and don't tend to travel north. They tend to travel south and more vulnerable areas, which is very common." [66].

Kansas had another travel case. It was reported in Hays County, Kansas. The case was contracted during a trip to Puerto Rico [67].

Maryland reported 170 cases of ZIKV between 2015 and 2016. In 2017, 19 new cases of ZIKV were reported in Maryland, and 13 have been reported in the District of Columbia. There were four Maryland cases in Baltimore, and two were on the Eastern Shore [68].

As of January 20, 2017, Michigan had confirmed 69 cases of ZIKV, including three pregnant women with the ZIKV [25]. All the cases are travel related, health officials say.

The Minnesota Department of Health reported 57 cases of ZIKV in Minnesota in 2016, with five occurring in pregnant women. Thirty-six instances have occurred in females, while 21 males have been infected [69].

The Mississippi State Department of Health reported the first 2017 cases of ZIKV in that state. The two are travel-related cases and involve residents of Warren County who traveled to an area north of Venezuela. There were 23 Mississippi travel-related Zika cases in 2016 [70].

Nebraska reported that an Omaha-area woman was vacationing in Mexico when she began feeling ill, suffering three of the four symptoms characteristic of the virus. At some point, she also realized she was pregnant [71]. Nebraska reported the total number of travel-related cases in Nebraska during 2016–2017 at 21 [72].

In Las Vegas, Nevada [73], the first new non-travel ZIKV case was reported in Clark County. The other 19 reports of ZIKV in Clark County have been from people who traveled to Guatemala, Nicaragua, Dominican Republic, Jamaica, Puerto Rico, Martinique, American Samoa, Brazil, and El Salvador [74].

The travel-related cases were a great concern to a travel hub like New York City. As of October 7, 2016, 617 cases of ZIKV had been identified among New York City (NYC) residents, including 72 points among pregnant women [75]. So far, under 10,000 people have been tested for the virus in New York, with 962 testing positive, including the 325 pregnant women [76]. As early as 2016, at least a dozen babies in New York City had the virus. Five of those babies have stunted brain deformity, also known as microcephaly [77].

New York has reported 155 Zika infections at last count, a figure 56% lower than at the same point the previous year. Only 30 have been in upstate New York, including just one case in Monroe County. Monroe ended last year with 22 Zika cases, far more than any other upstate county [78].

State health officials said that Ohio had 94 confirmed travel-associated Zika cases, one sexual transmission Zika case in 2016, and two points in 2017 [79]. Ohio's Clark County Health Commissioner Charlie Patterson said this is the first suspected case of ZIKV in Clark County and the third in Ohio this year. He said there were 95 last year, 94 of which were contracted overseas, and one was transmitted through sexual contact. None were contracted from infected mosquitoes [80]. And there were many more state reports.

Even the northeastern part of the country was not immune from outbreaks. In addition, some went further to suggest it will eventually become an issue in the northeast. At the end of February, Maine saw its first case of ZIKV, months after the first outbreak in South America. According to the Maine Center for Disease Control, the affected person is older than 65 and had traveled to a ZIKV-affected country on July 18, 2017. And New Hampshire reported its first case of the ZIKV; she was a female who had sexual contact with a man that had traveled to a ZIKV-affected country [81]. "New Englanders are understandably concerned with the new threat that ZIKV brings and the Ae during the primary carrier. *The Aegypti* mosquito is not currently known to be in New England. Over 40 distinct types of mosquitoes in the northeast carry other harmful diseases like eastern equine encephalitis," Mike Peaslee, Technical Manager and Associate Certified Entomologist at Modern Pest Services, said in a press release [81].

3.4.2 Zika Spread (Where Mosquitoes Rule)

In 2016, Peter Jay Hotez, MD, Ph.D., Dean of the National School of Tropical Medicine at Baylor College of Medicine in Houston, offered this warning. "If I were a pregnant woman living on the Gulf Coast or in Florida, in an impoverished neighborhood in a city like Houston, New Orleans, Miami, Biloxi, Miss., or Mobile, Ala., I would be nervous. If mosquitoes carrying the ZIKV reach the United States later this spring or summer, these are the major urban areas where the sickness will spread." [47]. In reality, the spread was less than he intended.

The exact numbers of reported cases were difficult to arrive at when there were asymptomatic cases. The CDC has reported more than 4400 cases of Zika in the U.S., including more than 1000 among pregnant women [82]. Dr. Thomas Frieden, Director of the CDC, said there could be as many as ten undiagnosed cases for every one found, meaning there could be 50,000 infections within the continental U.S.

because four out of five people could be infected with Zika don't show symptoms [83].

The following reviews outbreaks in four central states and territories: California, Florida, Texas, and Puerto Rico. These areas had to worry about mosquito vectors daily. When they returned to their homes, the mosquitoes followed them there.

Concerns have been expressed regarding Southern California and primarily southern states. Active mosquito to human ZIKV transmission was reported in Florida and Texas in February 2017. Mosquitoes did infect about 220 people in pockets of Florida and Texas [83]. As of the week ending January 5, 2018, 48 Zika cases have been reported for 2017, with 323 patients reported for 2015 and 2016. DSHS has also confirmed three Zika cases for 2018, all acquired outside the United States [84].

Cities in South Florida, along the Gulf Coast, and some areas of Arizona and California are already known to host large Aedes aegypti populations. Researchers have identified at least 50 metro areas with meteorological conditions ideal for the species. But even beyond those areas, the CDC has revised one map to extend the distribution of Aedes aegypti, which has been found in pockets of Washington, D.C., New York, and even New England in recent years [85].

The CDC reported figures in January 2018. In the US states, there were 407 symptomatic ZIKV disease cases reported; 398 cases in travelers returning from affected areas; four cases acquired through presumed local mosquito-borne transmission in Florida ($N = 2$) and Texas ($N = 2$), and five cases acquired through sexual transmission.

3.4.3 California

Since 2015, there have been 122 cases of Zika infection in L.A. County, 121 of which were acquired while traveling to countries where the virus is spreading, such as Mexico and Brazil [86]. Since 2015, 86 people in the San Diego region have been diagnosed with the ZIKV, according to statistics kept by county health officials [87]. And a woman in San Diego gave birth to a child with microcephaly [88].

Officials know of six Long Beach residents who have had Zika, none of whom caught the disease locally. Mosquitos of at least one species known to carry the disease have been observed in locales near Long Beach, including Bellflower, Carson, Cerritos, Downey, Hawaiian Gardens, Los Alamitos, and Paramount [89].

A case of sexual transmission associated with Zika infection was found in Fresno County [90]. The California Department of Public Health reported 524 Zika infections statewide on March 17. Of those, 99 involve pregnant women, and five babies have been born with Zika-related congenital disabilities [87].

The California Department of Public Health released an emergency warning on March 31 about two invasive (non-native) mosquito species, *Ae. aegypti* (the yellow fever mosquito) and *Ae. albopictus* (the Asian tiger mosquito), which are known to carry Zika, dengue, chikungunya, and yellow fever, have now been found in ten

California counties, including Fresno, Kern, Imperial, Los Angeles, Madera, Orange, Riverside, San Bernardino, San Mateo, and Tulare as well as in 129 California cities [91].

Overall, in California, there have been 619 cases of the ZIKV since 2015, eight of which were sexually transmitted. "Los Angeles residents already must worry about quakes, mudslides, wildfires, and horrible traffic. But officials worry a new type of mosquito, the *Aedes*, could soon add Zika to that list." [92].

3.4.4 Florida

In 2016, the governor recently declared public health emergencies in four counties due to Zika concerns, despite no evidence of autochthonous transmission in the state [93]. As of October 17, 2016, Florida has experienced 1024 confirmed cases of Zika, according to the Florida Department of Health [94]. At least 836 women have been infected with Zika in Florida, with at least 128 locally contracted points [48]. Five hundred thirty-four travel-related cases of Zika have been reported in Florida (8/25/16), with the highest concentration (49) in Miami-Dade County (9/1/16).

The Florida Department of Health reported trapping Zika-infected mosquitoes [95]. Florida officials said more than 40,000 mosquitoes had been tested since May 2016, and three samples tested positive. "It means that there is a substantial amount of ZIKV in that population because it's not easy to find the virus in the mosquitoes that you trap," said Dr. William Schaffner, Professor of medicine in the division of infectious diseases at Vanderbilt University School of Medicine in Nashville [96].

> Southern states are the most likely spots for Zika to become endemic among local populations of Aedes aegypti mosquitoes. They also tend to be the places with the worst health outcomes and most vulnerable people to disease. Southern states also make up most states that have not expanded Medicaid to all low-income adults under the Affordable Care Act provisions, even with a 100 percent FMAP (federal medical assistance percentage) rate for newly eligible adults. Notably, Florida is among those states that have not expanded Medicaid. Significant populations of highly vulnerable low-income people across the state now lack affordable health coverage. Suppose the virus does gain a foothold in the state. In that case, these people will be the most difficult to monitor and administer preventative services. The bills for complications among uninsured patients will tax the healthcare system's ability to handle Zika. Ironically, many of the most vulnerable in Florida are thousands of Puerto Ricans fleeing the conditions back home and seeking opportunities on the mainland. [97]

By the first week of December 2016, Florida had 27% of all reported cases in the USA (1244). Two hundred and forty-nine were locally acquired. The remainder and most cases, nearly 1000, involved travel with people infected off the mainland. One hundred eighty-five cases involved pregnant women. By year's end, Florida's health department reported 1325 Zika cases, including 1042 travel-related infections and 262 locally acquired ones. Additionally, 21 patients were considered undetermined [98].

According to the health department, as of mid-December 2016, 236 cases resulted from exposure in Miami-Dade. One hundred seven have occurred outside the previously active transmission zones in Wynwood, Little River, and Miami Beach [99]. Virtually, no entomologists believe that the transmission of Zika is limited to a few square miles of downtown Miami and Miami Beach, no matter how vigorously state officials insist it is [100].

Two new Zika travel-related cases in late December were confirmed in the Tampa Bay Area [101]. In January 2017, the Florida Department of Health said two new travel-related cases of the ZIKV were reported in Collier County. There are six new travel-related cases of Zika in Florida, including Miami-Dade, Broward, and Seminole counties [102].

More than 90 Zika infections have been documented outside the four designated neighborhoods near Wynwood in Miami-Dade. Florida's Surgeon General Celeste Philip said, "We will continue to see isolated cases." For nearly five months, Miami-Dade was the only county identified as having an active spread of Zika by mosquitoes [99].

However, only in August, four neighborhoods in Miami and Miami Beach were under federal health warnings, with Wynwood known as the location of the first case of active Zika in the continental USA [103]. In September, the section was expanded to a 4.5-mile area. On October 13, the transmission area was extended to an additional 1 square mile in Miami-Dade [104].

In the USA, the Florida State Department of Health reported no new locally acquired case of ZIKV in the last epidemiological week of 2016 in Miami-Dade County.

"When you create a map and boundaries, you say inside is a danger zone and outside is completely safe," said Joe Furst, Wynwood Business Improvement District Chairperson. "You suffocate a neighborhood doing that." [103]. Unsurprisingly, Gov. Scott lifted the alert on December 9, just in time for the peak travel season. In August, Alvarez reported trips to Miami were down 12.6% compared to a year earlier.

By November, air travel was up again by 3.8%, and on December 9, 2016, Florida Governor Rick Scott made it official when he lifted its Zika alert. Shortly after his announcement, the Centers for Disease Control and Prevention (CDC) lifted its strictest advisory urging pregnant women not to travel to the heart of South Beach. Instead, it suggested caution [103]. State Rep. David Richardson said he met with Ocean Drive business owners about a week ago, and they reported that their revenues were down by about 25% this year, a decline they attribute to Zika [99].

When Scott lifted the ban, he had this to say. "We're going to make sure everybody knows that this state is open for business"—a message with resounding applause. "Tourists want to come here. People want to build their businesses here," he said [99].

There are 48 breeds of mosquitoes in the Florida Keys. In 2009–2011, an outbreak of dengue shocked Key West residents. Its vector was the same mosquito species that has been blamed for the Zika pandemic in the Americas. Nonetheless, mosquitoes are seasonal, and Floridians are aware of their geography Monroe County and Key

West will be the first hit in the next cycle of Zika infections when the rains arrive in early summer. Monroe County includes all the Keys as well as the Everglades [105].

The current approach toward *Ae. aegypti* mosquito control in Key West, Florida, is the integrated pest management and primarily involves source reduction (reducing standing water) and helicopter-delivered larvicides. At the same time, to use of ground-delivered insecticides is used. Backpack sprayers and natural treatments by hand and granules, pellets, and tablets can also treat smaller areas.

Deputy Mayor Alina Hudak said on December 9, 2017, that the county had spent about $22 million fighting ZIKV—primarily to hire additional mosquito control inspectors and workers, to pay for insecticides, equipment, and public education campaigns. Hudak said Florida had sent the county about $12 million to offset the extra expense. The remainder will be additional state and federal grants and an undetermined general revenue from Miami-Dade [99].

In 2017, Florida reported only four travel-related cases and no new local infections. And though the number of locally acquired cases has dwindled with the winter, public health officials have warned that the virus will likely rebound when the temperatures rise and the rainy season kicks in [98].

Florida had 101 cases of the mosquito-borne ZIKV in 2017, with most involving people who brought the virus into the state after being infected elsewhere, according to numbers posted on the state Department of Health website. The total includes 75 "travel-related" infections this year and six cases in which people were exposed in Florida in 2016 and diagnosed in 2017 [106].

A warmer than average winter has mosquito control workers concerned that the ZIKV could still spread as the weather hasn't gotten cold enough to kill off the pests [107]. On December 2, 2018, the State of Florida Health Department reported a year-to-date total of 91 confirmed ZIKV cases as of November 24, 2018. The good news in this Florida Health report is that no cases of locally acquired Zika have been reported during 2018. Moreover, there are no ongoing, active Zika transmission areas in Florida. And the 89 of 91 confirmed Zika cases during 2018 are "travel related." [108].

"Zika continues to pose a threat to pregnant women living in or traveling to Miami-Dade County," said Lyle Petersen, MD, MPH, Director, Division of Vector-Borne Diseases. "Our guidance today strengthens our travel advice and testing recommendations for pregnant women to prevent further the spread of the infection among those most vulnerable." [109].

Zika is here to stay, at least for a while, health and mosquito control officials said, in part because of South Florida's tropical climate and the considerable number of visitors and residents from countries like Brazil, the Dominican Republic, Colombia, and the United States territory of Puerto Rico [103].

3.4.5 Texas

Ten Zika cases have been documented in Texas in 2017 and 320 in 2016 and 2015. About 250 women and children have shown evidence of infection reported to the federal Zika Pregnancy Registry [52]. In addition, Texas has one of the highest unintended pregnancy rates of any state [110].

"There are serious concerns about the possibility of ZIKV infection returning to Texas during the summer months," said Dr. Peter Hotez. "One important reason is that ZIKV transmission occurred in Brownsville and possibly elsewhere in South Texas during the 2016–2017 winter [54].

Presumably, the only homegrown Zika cases in Texas had been in Cameron County (25 miles south of San Antonio) on the border with Mexico [111]. It was reported on November 28. Concerns mounted in Cameron County further since Texas Kids Count reports many women there don't appear to have a doctor. Around a quarter of women in Cameron receive little to no prenatal care. The Texas Medical Associated added that 40.7% of working-age adults in the county were uninsured [112].

On December 9, 2016, health services in Texas confirmed that a woman from Laguna Heights was diagnosed with the first case of ZIKV contracted from a mosquito bite. The woman did not get sick and was tested for Zika during her prenatal care [113]. She reported no recent travel to Mexico or anywhere else with ongoing ZIKV transmission and no other risk factors," the Texas Department of State Health Services said [82].

The Texas Health and Human Services Commission announced on the same day (November 29) that it was re-instating the Medicaid benefit for mosquito repellent due to the new issue, says CBS Dallas-Fort Worth. In Texas, women between the ages of 10 and 45, and those who are pregnant, can pick up free mosquito repellent—up to two cans per month—at participating pharmacies [114].

In Travis County, Texas, 20 people tested positive for ZIKV, including six pregnant women. All cases are travel-associated infections [115]. A baby born to a mother infected with Zika has died in Texas, marking the first known infant death linked to the virus in the U.S. Born in Harris County, Texas, the girl had several Zika-related congenital disabilities, including microcephaly [116].

Although the woman resided in Bexar County, she traveled to Brownsville in November when six locally acquired cases were reported. She reported having had a positive test for genetic material from the ZIKV in her urine (she is not contagious) [117]. The CDC has designated Brownsville as a Zika precautionary zone. They warned pregnant women to avoid traveling to the area [118]. The fact that Zika has finally crossed the Rio Grande is significant because it was the first time that the virus has been recorded in this area after being transmitted by a mosquito bite [117].

"We still don't believe the virus will become widespread in Texas, but there could be more cases, so people need to protect themselves from mosquito bites, especially in parts of the state that stay relatively warm in the fall and winter," John Hellerstedt, Commissioner of the Texas Department of State Health Services said [114]. The agency now recommends testing for all pregnant women in South Texas in Cameron, Hidalgo, Starr, Webb, Willacy, and Zapata counties [119].

The *Ae. aegypti* mosquito happens to be prevalent in Texas, specifically in densely populated areas. The *Ae. aegypti* mosquitoes, which carry the ZIKV, dengue fever, and chikungunya, among other deadly diseases, are common in the Houston region [120]. Dean Hotez said the combination of historically warm weather and the prevalence of *Ae. aegypti* is not good. "I am quite worried that we're still vulnerable to ZIKV transmission," he said [119].

3.4.6 Puerto Rico and U.S. Territories

While some local mosquito control districts and their robust capabilities are described, others are considerably less. Many communities in vulnerable states are not contiguous, leading to potential gaps in vector control. Even where management is most intensive, such as in Puerto Rico, vector resistance to common pesticides is problematic [56].

The World Health Organization declared ZIKV a Public Health Emergency of International Concern on February 1, 2016, after 10,000 Zika infections had been reported in Puerto Rico [121] and another 4000 in other U.S. territories [48]. In addition, 3808 travel cases across the U.S. have been reported to the CDC [122].

In U.S. Territories (American Samoa, Puerto Rico, Virgin Islands), 631 ZIKV disease cases were reported; one case in a traveler returning from affected areas and 630 issues acquired through presumed local mosquito-borne transmission and zero cases acquired through other routes [123].

According to the CDC, 878 pregnant women tested positive for the virus in the 50 states and 1806 territories, including Puerto Rico [122]. There are 25,871 locally acquired cases in U.S. territories.

According to the CDC, over 4800 pregnancies in the U.S. territories had a laboratory result showing confirmed or possible Zika from 2016 to 2018. About one in seven babies had health problems possibly caused by Zika reported, among 1450 babies at least one year old. Only one in three babies had the recommended eye exam, among 1450 babies at least one year old [124].

> Of these cases, 1,450 infants who reached age one or more by Feb. 1, 2018, had some follow-up care after their first 14 days of life. Among these children, 95 percent had at least one physical exam, 76 percent had developmental screening or evaluation, 60 percent had postnatal neuroimaging, and 48 percent had automated auditory brainstem response-based hearing screen or evaluation. Ophthalmic examination was reported in 36 percent of the children. Among all the children with documented follow-up care, 6 percent had at least one Zika-associated congenital disability identified, 9 percent had at least one neurodevelopmental abnormality possibly associated with congenital ZIKV infection, and 1 percent had both. About one in seven children's health problems possibly caused by Zika were reported. [125]

Since Zika appeared in Puerto Rico, doctors believe as many as a million people on the island have been infected. Some 40,000 cases have been confirmed, including some 3200 pregnant women [126].

> It's been less than a year since the first reported case of Zika in Puerto Rico, but doctors in the territory are already at their breaking points.... Since Puerto Rico's decade-long economic crisis began, the commonwealth has hemorrhaged hundreds of thousands of people—usually healthy, younger people seeking work—in a steady mass migration to the mainland. Hundreds of Puerto Rico's doctors were in that flood of people, and perhaps over a thousand physicians have moved away since 2014. The result back on the island is devastating. Not only are the remaining people more likely to be those most vulnerable to Zika—children, older adults, and poorer families and women—many would-be primary-care providers are gone. [97]

Puerto Rico reported 34,963 cases of symptomatic ZIKV disease, with 99.6% occurring through presumed mosquito-borne infection, whereas U.S. states reported 5102 points, 95% of which occurred in travelers returning from ZIKV-affected areas [5]. In May of 2106, Puerto Rico reported its first microcephaly case. Since then, no microcephaly cases have been reported, but federal officials say it is only time.

A report by the CDC showed that neonates may not have microcephaly at birth but may develop microcephaly up to 6–12 months postnatally, illustrating possible false-negative findings for congenital Zika syndrome at the time of delivery [127]. And a Brazilian study in November 2016 found some babies exposed to Zika in the womb who weren't born with microcephaly ended up developing it later. Researchers claim they follow more than 2600 pregnancies (half ongoing in November) [128].

Dr. Alberto de la Vega has seen one-fifth of the pregnant women on the island who tested positive for Zika. "Among those patients, we've had at least 14 or 15 confirmed cases in which severe brain damage, caused by the ZIKV, has occurred," he says. Some of those cases included microcephaly. De la Vega says the risk of brain damage is between 2 and 4% for babies born to mothers infected in the first trimester. But he sees many other problems in his patients infected with Zika, including mothers going into premature labor and a higher number of miscarriages [126].

Medicaid support has been stretched to the limits. "Puerto Rico is entitled to a 55% rate by those calculations, but federal spending is statutorily capped at only 19%. That means that the struggling Puerto Rican government has to foot 81% of the medical bills for many people on the island—including the low-income women and families who face the most risk from Zika complications." [97].

The health crisis in Puerto Rico is a real existential risk to people on the island, even after Congress passed the debt-relief bill PROMESA this summer—and there is a natural and present danger that Zika could cripple the entire health and economic infrastructure [97].

The outbreak peaked during the summer of 2016. By November, the number of new Zika cases in Puerto Rico had dropped dramatically. Yet, health officials worry that the full effect of the outbreak on the island may not be known for months or years, primarily because of continued sexual transmission.

A study published August 19 in JAMA Pediatrics estimates that up to 10,300 pregnant women in Puerto Rico could be infected with Zika. Between 100 and 270 babies could be born with microcephaly through mid-2017 [129].

But, Puerto Rico has reported only 16 cases of congenital disabilities associated with Zika, even though more than 3300 pregnant women have contracted the virus. Several times, that number is believed to have been infected [130].

The CDC estimated in October 2016 that 5900–10,300 pregnant women could have been infected, potentially resulting in 100–270 babies born with abnormally small heads, a condition known as microcephaly, by the middle of this year [83].

US health officials have privately expressed deep concern that Puerto Rico is downplaying the extent of its Zika problem. A multipage document suggests that the dispute has obscured the time of the territory's Zika problem for more than half a year. The record indicated that Puerto Rican official Dr. Miguel Valencia had disregarded the established case definition for identifying affected infants, giving staff different criteria to use—without clearing the change with the CDC. "There is a large discrepancy between the number of cases identified by ZAPSS that meet CDC surveillance case definition and numbers reported by PRDH," according to the document [131].

Why the discrepancy? First, there is the process of counting cases. Jose Cordero, Professor of public health at the University of Georgia, said the CDC has been casting a broad net for possible signs of damage by the virus, including in the offspring of all women infected during pregnancy if there's any sign of the types of congenital disabilities associated with Zika. He said that doesn't appear to be how health authorities in Puerto Rico count cases [130]. "As I understand the conversations I had with Dr. Valencia, he was going in the opposite direction [of the CDC]—that the child must have microcephaly before being affected," Cordero said. Cordero called Valencia "the keeper of all those numbers. And he's certainly keeping them very close to his vest." [130].

> From this April, May, June, July, and August, 1,339 babies will be born from known infected mothers. Of these, 6% are expected to be taken with (Congenital Zika Syndrome) SZC, 80 new babies. So, we're going for 200 babies with SZC.... The problem is that we are not identifying three-quarters of 75% of asymptomatic pregnant women in prenatal care, the so-called "never identified asymptomatic." Puerto Rico has documented 3,339 pregnant women infected, and 55% have symptoms (1,861), and 45% do not (1,478). Then that 45% asymptomatic should be 80% like the rest of Puerto Rico. The total is not 3,339, but 9,305 (the symptomatic 1,861 and the asymptomatic 1,478 now have to be added the "never identified asymptomatic," which is an additional 4,488. This means that 48% of the babies of infected mothers in pregnancy (4,448 out of 9,305) could be born without being identified in our hospitals. Do not forget that between June and November 2016 (6 months), 50 pregnant women were infected per day (20 reported to the system and 30 not written), and in those 180 long days, 9,000 pregnant women were infected in Puerto Rico. Returning to 6% of infection from mother to baby (558/9,305 = 6%), our society will have to give unique services to babies who survive from 558 with SCZ (probably half will not survive pregnancy, or 279 will stay to offer them special services). [132]

Third, the economic situation has also led to the migration of Puerto Ricans elsewhere. Professor Cordero said that some women who know they're going to give birth to a baby with Zika congenital disabilities or who have done so in Puerto Rico might move to the US mainland to access better services for the child. That could also contribute to a lower-the-expected count of affected pregnancies on the island [130].

As a result, after October 2016, the CDC has simply stopped reporting the outcomes of pregnancies in U.S. territories. Their reason: The agency simply said that Puerto Rico wasn't counting cases the same way [130]. Zika pregnancies in Puerto Rico are reported in an entirely separate registry, Puerto Rico Zika Active Pregnancy Surveillance System (ZAPSS).

CDC officials have been working to restore the collaboration with the Department of Health and have refused to specify how the Puerto Rican case counting differs from the CDCs. Earlier this month, Dr. Margaret Honein, Head of the CDC's congenital disabilities branch, told STAT she could not talk about Puerto Rico's data without their permission [131]. Concern about the territory's approach to counting Zika cases strained the relationship between the CDC and Puerto Rico's department of health, which received CDC grants of $9.5 million to establish and operate a pregnancy registry to chart Zika's impact on newborns in the territory [133].

Puerto Rico is facing several challenges. The island's finances are in dire straits, making Zika only one of several significant issues the territorial government must address. "The vulnerability of the Puerto Rican population to Zika is extraordinary," Fauci said. There have already been 1,804 locally transmitted cases on the island, which is susceptible to geography and its ongoing monetary crisis and dense population. Fauci said we should expect Zika to act the same way chikungunya did on the island since the same mosquito spreads them. "When chikungunya hit Puerto Rico, it infected about 20–25% of the population," he said. "So, it is almost certain unless something dramatic changes, 25% of Puerto Rico will get Zika." [134]. In addition, Puerto Rico has a high unintended pregnancy rate of 65% [135].

Puerto Rico may simply be in denial. "They're in denial about the problem," the former official said. "And six months, a year, two years from now, there will be all these babies who aren't learning, and all these problems will come to light." [130].

The CDC urged him. Puerto Rico used an organic larvicide rather than aerial spraying of naled to fight ZIKV [136].

With the rainy season approaching, when disease-carrying mosquitoes become active, Dr. Carmen Zorrilla, who runs the Maternal Infant Studies Center at the University of Puerto Rico hospital, worries many on the island have become complacent. One problem, she says, is that women who have babies with congenital disabilities have been reluctant to go public because, like HIV, Zika carries a stigma. Zorrilla has heard comments like, "Oh, you were not protecting yourself?" "You were not using the mosquito repellent?" or "You were not using condoms?" "You have pregnant women with a viral disease that may cause serious congenital disabilities," Zorrilla says. "And then you're blaming them for getting it." [126].

Puerto Rico declared that the 2016 Zika epidemic is over, saying transmission of the virus that can cause congenital disabilities when pregnant women are exposed has fallen significantly. About ten mosquito-borne disease cases have been reported every four weeks since April 2017, down from more than 8000 cases reported in four weeks at the peak of the epidemic in August 2016, the Puerto Rico Health Department said in a statement [137].

"We are pleased that the peak of the ZIKV outbreak in Puerto Rico has come to a close," CDC Acting Director Dr. Anne Schuchat said. "However, we cannot let our guard down. CDC will continue to focus on protecting pregnant women and work closely with the Department of Health to support comprehensive Zika surveillance and prevention efforts on the island." The CDC is not immediately lifting its advice to pregnant women to avoid traveling to the territory [133].

3.5 International Footprint

The investigators reported that mosquito exposure and lack of access to adequate health care could result in undetected cases of the ZIKV and other illnesses. "Pockets of virus transmission that occur in countries with inadequate reporting can facilitate 'hidden' outbreaks, increasing the risk of infected travelers causing outbreaks in new regions of the world," Grubaugh et al. wrote in 2019 [138].

ZIKV is probably already circulating in other American countries but has not been detected because of misdiagnosis as other arboviruses and a lack of laboratory facilities [8] (Fig. 3.2).

3.5.1 Cuba

The Grubaugh study team hypothesized that Zika outbreaks could be occurring in the Americas without being detected by public health agencies. They discovered an unreported outbreak in Cuba in 2017, a year after peak transmission in neighboring islands.

In 2017, there was a spike in Zika cases in travelers who had been to the Caribbean. Between June 2017 and October 2018, 98% of travel-associated Zika cases reported in Florida and Europe came from Cuba.

The Cuba outbreak likely had 5707 unreported Zika cases, comparable to large outbreaks in other Caribbean countries. More than 99% of those cases occurred in 2017, although the study team called it unusual that the Cuban outbreak had a one-year delay compared with other Caribbean outbreaks.

Although Cuba reported 187 laboratory-confirmed Zika cases in 2016, it said no local Zika cases to the Pan American Health Organization or other international public health agencies in 2017 and 2018 [140].

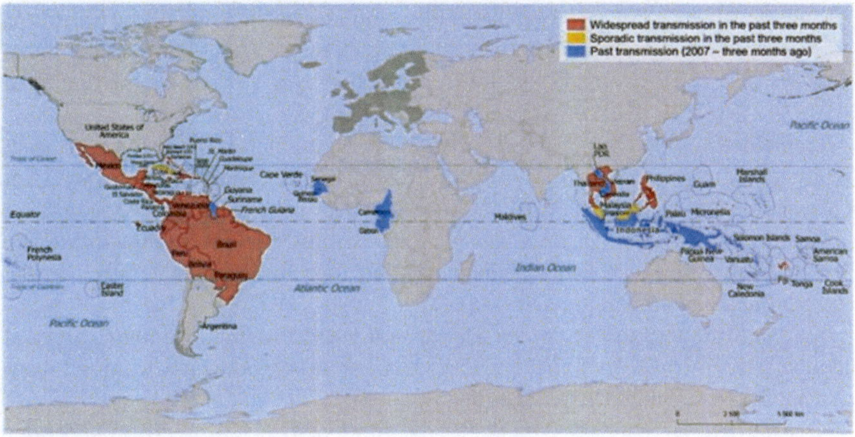

Fig. 3.2 Countries and territories that have reported autochthonous mosquito-borne transmission of ZIKV in the past 3 months. The term "widespread transmission" (red) indicates one of the following three situations: (1) more than ten locally transmitted ZIKV cases reported in a single area, (2) at least two separate areas reporting locally transmitted ZIKV cases, and (3) ZIKV transmission ongoing in an area for more than 3 months. The term "sporadic transmission" (orange) indicates that no more than ten locally transmitted ZIKV cases have been reported in a single area in the past 3 months. The term "past transmission" (blue) indicates that local ZIKV transmission has been reported since 2007, but not in the past 3 months. Source: European Centre for Disease Prevention and Control, current status of ZIKV transmission in the world as of November 23, 2016 (http://ecdc.europa.eu/en/healthtopics/zika_virus_infection/zika-outbreak/pages/zika-countries-with-transmission.aspx). Song et al. [139]

3.5.2 Mexico

On November 20, 2017, a Mexican travel agency out of Puerto Vallarta, there have been only 2742 confirmed cases of Zika across Mexico [141]. A team led by Hernandez-Avila et al. [142] agreed that more infections were expected from Mexico.

They reported that 60,172 symptomatic infections with the Zika virus occurred in Mexico between November 25, 2015, and September 2, 2016. Although this estimate is still conservative, it is about 40 times higher than indicated by the previously reported incidence rate, 1.66 cases per 100,000 population, and almost 30 times higher than the number of confirmed Zika cases reported for Mexico during this time [142].

> The Mexican incidence numbers showed considerable geographical variation, and the numbers for the states of Guerrero (265.3), Chiapas (197.4), Colima (121.6), and Oaxaca (115.6) were closer to those reported in South America and given the problem posed by the dengue virus, also transmitted by *Ae. Aegypti*, in the same areas, we were surprised by the relatively low incidences of symptomatic ZIKV infection we recorded among the pregnant women living in the states of Jalisco, Nayarit, and Tamaulipas. There may well be underreporting of confirmed cases in these states. [142]

Many Mexican states with popular tourist destinations, including Baja California and Sonora, near the Arizona border, continue to see reports of local Zika transmissions. The CDC warned any travel to Mexico is a risk of contracting Zika [143].

It remained very unclear how many congenital Zika syndrome cases had surfaced. The health secretariat has confirmed the first case of microcephaly in a newborn infant in Mexico, linked directly to the ZIKV. The premature baby died during childbirth in November in the state of Oaxaca. The National Institute of Perinatology in Mexico City confirmed the microcephaly diagnosis and found traces of the ZIKV in the amniotic fluid and congenital deformities.

In a press release, the health secretariat said that between November 2015 and January 2017, it confirmed 7634 Zika infections, 4252 of which were pregnant women. Of those, 588 have given birth.

According to Hernandez-Avila et al. [142], the incidence of microcephaly after the Zika virus was introduced in the country was significantly higher than before its introduction. In Mexico, 12.7 million births were registered, and 468 cases of congenital microcephaly were reported before the Zika virus was introduced. The corresponding estimated incidence of microcephaly was 3.7 cases per 100,000 births. In contrast, during the period after the Zika virus was introduced, there were 3.7 million births registered and 428 reported cases of microcephaly, giving an estimated cumulative incidence of 11.5 cases per 100,000 [142].

3.5.3 South America and the Caribbean

Weaver added. “Combine that with large, densely packed cities full of *Ae. Aegypti* mosquitoes, and you have all the fuel you need for a major epidemic like this one.” That, in turn, means a greater likelihood of seeing an increase in rare complications such as microcephaly [144].

More cases and mixed trends in South American and Caribbean countries should be expected. Going into 2017, the cases seemed to be thresholding. At the end of 2016, PAHO reported cases in many South American countries. For example, in Bolivia, autochthonous cases were transmitted in Beni and Pando and the ongoing outbreak in Santa Cruz. In Peru, there was an increase in reported suspected and confirmed cases, particularly in Iquitos [145].

In 2017, PAHO reported an increasing trend of suspected and confirmed cases had been observed in South America, mainly due to increases in the number of reported cases in Argentina, Bolivia (Plurinational State), Brazil, Ecuador, and Peru. An average of 1,246 suspected and confirmed cases were reported weekly in this subregion.

In Argentina, an increase in suspected and confirmed cases was observed related to outbreaks in Formosa, Salta, and Chaco provinces.

There has been an upward trend in Ecuador's suspected and confirmed cases. About 65% (448) of the cases documented in the first 17 weeks of 2017 are from Guayas [146].

They also reported an increasing trend in cases in Anguilla. At the same time, in Saint Martin, a French overseas territory, patients continue to decline with a recent increase in the number of visits to emergency services followed by a further decrease in the following weeks [145]. Local health officials have recorded almost 10,000 notifications of suspected ZIKV cases in Jamaica between January 2016 and March 2017. Figures from the health ministry seen by The Sunday Gleaner show that from the notifications, 7767, or 77%, were determined to be suspected cases of ZIKV. So far, 203 confirmed cases of the virus have been recorded. Congenital syndrome associated with the ZIKV was seen with 50 suspected cases of 170 notifications in Jamaica. Forty-six patients were diagnosed with microcephaly, with 35 non-severe issues, while 11 cases showed severe microcephaly [147].

In 2017, PAHO reported an increase in suspected cases in Turks and Caicos. The decreasing number of reported instances continued in this subregion's other countries/territories, with a weekly average of 330 suspected and confirmed cases [146].

Galveston National Laboratory's Scientific Director Scott Weaver has asked. "So what we don't know, and need to find out, is whether this virus has always had this capacity to produce these outcomes, whether the virus changed, or whether there is something different about the people it infects in the Americas." [144].

Researchers characterize the South American population as immunologically naïve and suggest the unprecedented size and impact of the ZIKV epidemic in the Americas may be the natural result of a random introduction into a large population without preexisting immunity [148].

3.5.4 Asia

There was expected increased infection in Asia as well. Data have been collected on occurrences in nearly every Asian country, including Myanmar, Pakistan, the PDR (Laos), Bangladesh, and the Philippines. Asian countries remain at considerable risk for large Zika outbreaks. There were fears that infections were spreading in Southeast Asia on the scale of Latin America in 2016.

There were some early announcements in Southeast Asia [149]. For example, Thailand also reported ZIKV infections as early as 2013. First reported in Thailand in early 2013, she was a Canadian traveler returning from Southern Thailand. Two additional traveler cases were subsequently identified in a German traveler in November 2013 and a Japanese traveler in July 2014. In September 2016, the Ministry of Public Health reported the first two autochthonous cases of Zika-related microcephaly in Asia [150].

China, with 238,415 travelers and 242 million at-risk residents; Indonesia with 13,865 travelers and 197 million at-risk residents; the Philippines with 35,635 travelers and 70 million at-risk residents; and Thailand with 29,241 travelers and 59 million at-risk residents are all vulnerable [150].

For example, Vietnam's National Institute of Hygiene and Epidemiology discovered a small population of the Aedes aegypti mosquitoes carrying the ZIKV in the tourist town of Nha Trang in the central province of Khanh Hoa. Fifty-six out of 23,682 mosquitoes, or 0.24 percent of the total, tested positive for the ZIKV, according to a study conducted by the institute in March 2015–May 2016 [151]. On April 5, 2016, the National IHR Focal Point of Viet Nam notified WHO of two confirmed cases of locally acquired ZIKV infection [152]. Ho Chi Minh City had four of the seven cases recorded so far, followed by Binh Duong, Khanh Hoa, and Phu Yen with one point each [151]. Ho Chi Minh City had declared a pandemic in An Phu Commune, District 2, and Hiep Thanh Commune, District 12, on October 18, 2016. A four-month-old girl suffering from Zika-related microcephaly was recently reported in the Central Highlands province of Dak Lak; officials say her mother contracted the virus during pregnancy [149].

In 2016, more than 400 people became infected with the virus following the discovery of the first case in August [153]. Singapore reported six cases in the first 12 weeks of 2017, Singapore's National Environment Agency agency's website shows. In 2018, Singapore diagnosed a case and reported it to authorities on January 18. It is the first of its kind in 2018 since the disease was thought to be eradicated from the country in September 2017.

An important note is that a joint sequencing study conducted in 2016 by the health ministry in Singapore and the Agency for Science, Technology, and Research (A* STAR) found that the Zika strains to circulate in Singapore may not have been imported from South America and were like those which have been percolating in Southeast Asia since the 1960s [154].

Hong Kong authorities confirmed one imported case of Zika infection, making it the first of its kind in 2017 and the third since its outbreak in large parts of Latin America and the Caribbean in early 2016 [155].

The Philippine Department of Health (DOH) Secretary Paulyn Jean Ubial announced on February 2, 2017, that there were 57 Zika cases in the country. Seven were pregnant women [156].

All reported Zika cases in Taiwan had been imported from Zika-endemic countries. The first case was detected in January 2016 by airport fever screening in a 24-year-old man returning from Thailand. By May 2017, 14 imported Zika cases had been reported by the Taiwan Ministry of Health and Welfare [150].

By mid-December 2016, Malaysia confirmed its eighth ZIKV case under the current pandemic, and it remains unclear how many were infected locally rather than from travel [157].

The detection of Zika cases in visitors from Zika non-endemic areas to Indonesia, Thailand, Malaysia, Vietnam, and the Philippines suggests the possibility that substantial transmission had been occurring in these Asia countries [158].

In China, an estimated 242 million people live in southern regions where large dengue virus epidemics have occurred previously and hence where the ZIKV has the potential to circulate locally. However, the predicted suitability range for autochthonous transmission by Aedes mosquitoes extends into Northern China, where more than one billion people reside collectively [159]. In 2016. China has recorded 22 had imported Zika cases, seven passing through Hong Kong while returning from South America to the mainland [160].

In 2020, blood samples of 273 healthy individuals were collected from Nanning City, Guangxi Province, China, in March 2019. The team found that 9.5% (26/273) and 1.8% (5/273) of healthy persons were positive for the ZIKV total antibody (IgG and IgM) IgM antibody, respectively, suggesting a more extensive spread.

More than 210 million people live in ZIKV suitable areas in China, and many travelers float between ZIKV-affected areas and China every year [161].

3.5.5 Australia

ZIKV infection was first reported from Indonesia in two Australian travelers, one returning from Jakarta in 2012 and the other from Bali in 2015. Retrospective testing of 103 dengue negative blood samples from a dengue outbreak in Jambi province from December 2014 to April 2015 yielded one ZIKV infection belonging to the Asian lineage. In November 2015, molecular and virological techniques confirmed a Zika case in Sulawesi [150].

3.5.6 Northeast Asia

Only imported Zika cases have been detected in Japan. The first three cases were reported in travelers returning from French Polynesia and Thailand in 2013–2014. Eight Zika cases had been reported in Japanese travelers as of September 2016 [150].

The Republic of Korea reported only travel-related Zika cases. RT-PCR confirmed the first case in March 2016 in a 43-year-old male traveler returning from Brazil. As of May 2017, 19 Zika cases had been reported in Korean travelers; eight returning from the Philippines, four from Vietnam, two from Thailand, one each from Brazil, Dominican Republic, Guatemala, Puerto Rico, and even Russia, and six cases of travel-related Zika have been reported [162].

3.6 Women More Vulnerable

This remains the last important question relevant to this chapter: Are women more vulnerable? Obviously, in terms of the impact, they are vulnerable to the entire array of adult complications (Chap. 7) as well as those associated with pregnancy and the emotional and mental consequences of birthing and rearing a child with congenital development problems (Chap. 6).

The results of a very large-scale study [163] of 28,219 non-pregnant symptomatic cases in Puerto Rico indicated that the incidence of symptomatic infection among females was markedly higher than that among males: 936 versus 576 cases per 100,000. These results, and similar data from Brazil, the Federated States of Micronesia, and Mexico, indicate that, compared with males, females are more likely to be infected, more likely to develop the characteristic symptoms once infected, and more likely to be tested once symptomatic, perhaps because of concern about congenital abnormalities [142].

3.7 Conclusion

A Gallup poll [164] found that 90% of Americans believed they were unlikely to contract ZIKV, and the CDC has confirmed a little more than 5000 cases in the U.S. since January 2015, with a little more than 220 transmitted in the U.S. This was at the height of the pandemic in Brazil. This was the worth profile for the U.S.

On the other hand, the *Ae. aegypti* mosquitoes have been found in over twenty states from coast to coast [165]. Unlike other mosquito species, *Ae. aegypti* mosquitoes bite during the day and night. They carry infectious diseases.

The re-emergence of ZIKV may have been less of a national issue here in the U.S., but it took and continued to take a toll on Latin America and elsewhere. The epidemic of 2015–2016 may be mostly over, but strains mutate, and new viruses surface. Knowing history may just keep it from being repeated. Or, as the American journalist Norman Cousins put it: "History is a vast early warning system." Let's proceed to make certain we are primed to hear the alert should it come.

Endnote

1. "Third World" (used purposefully and sarcastically here) is an outdated and derogatory phrase that has been used historically to describe a class of economically developing nations.

References

1. Werner E (2021) Climate change lets mosquitoes flourish—and feast—in Los Angeles. The Washington Post. September 19. https://www.washingtonpost.com/us-policy/2021/09/19/climate-mosquito-los-angeles/. Accessed 6 June 2022
2. Aziz H et al (2017) Zika virus: global health challenge, threat and current situation. J Med Virol 89. https://doi.org/10.1002/jmv.24731/abstract. Accessed 26 June 2017
3. Hotez P (2016) Will Zika return to the 'old world'? Microbes Infection 18:527–528. http://www.sciencedirect.com/science/article/pii/S1286457916300648. Accessed 27 Sept 2016
4. Almendrala A (2016) An illustrated guide to the Zika outbreak. Huffington Post. February 24. http://www.huffingtonpost.com/entry/guide-to-zika-virus_us_56a272afe4b076aadcc675b6. Accessed 24 Oct 2016
5. Squiers L et al (2018) Zika virus prevention: U.S. travelers' knowledge, risk perceptions, and behavioral intentions—a national survey. Am J Trop Med Hygiene 98(6). June. https://www.ncbi.nlm.nih.gov/pubmed/29737272. Accessed 26 July 2018
6. Pandit PS, Doyle MM, Smart KM, Young C et al (2018) Predicting wildlife reservoirs and global vulnerability to zoonotic Flaviviruses. Nat Commun 9(5425). December 21. https://doi.org/10.1038/s41467-018-07896-2
7. Lessler J et al (2016) Assessing the global threat from Zika virus. Science 353(6300). July 14. http://science.sciencemag.org/content/early/2016/07/13/science.aaf8160. Accessed 19 May 2017
8. Musso D, Gubler DJ (2016) Zika virus. Clin Microbiol Rev 29. May 9. http://cmr.asm.org/content/29/3/487.abstract. Accessed 18 July 2017
9. Edwards SB. The Zika virus. ABDO Publishing, Minneapolis, MN
10. Chang C et al (2016) The Zika outbreak of the 21st century. J Autoimmunity 68:1–13. February 28. https://www.ncbi.nlm.nih.gov/pubmed/26925496. Accessed 27 Oct 2016
11. Faye O et al (2014) Molecular evolution of Zika virus during its emergence in the 20th century. PLoS Neglected Trop Dis 8(1). https://doi.org/10.1371/journal.pntd.0002636. Accessed 4 Apr 2017
12. Dick GWA, Kitchen SF, Haddow AJ (1952) Zika virus I: isolations and serological specificity. Trans R Soc Trop Med Hygiene 46(5):509–520. September. http://www.sciencedirect.com/science/article/pii/0035920352900424. Accessed 14 Oct 2016
13. Fagbami AH (1979) Zika virus infections in Nigeria: virological and seroepidemiological investigations in Oyo state. J Hygiene 83:2. October. https://www.ncbi.nlm.nih.gov/pubmed/489960. Accessed 7 Aug 2018; Smithburn KC (1952) Neutralizing antibodies against certain recently isolated viruses in sera of human beings residing in East Africa. J Immunol 69:2. August. https://www.ncbi.nlm.nih.gov/pubmed/14946416. Accessed 7 Aug 2018
14. Wikan N, Smith DR (2016) Zika virus: history of a newly emerging arbovirus. Lancet Infectious Dis 16:7. July. https://www.ncbi.nlm.nih.gov/pubmed/27282424. Accessed 26 July 2018
15. Simpson DIH (1964) Zika virus infection in man. Trans R Soc Trop Med Hygiene 58:4. July 1. https://www.sciencedirect.com/science/article/pii/0035920364902007. Accessed 7 July 2018
16. Troncoso A (2016) Zika threatens to become a huge worldwide pandemic. Asian Pacific J Trop Biomed 6:6. http://www.sciencedirect.com/science/article/pii/S2221169116302921. Accessed 4 June 2017
17. Faye O et al (2014) Molecular evolution of Zika virus during its emergence in the 20th century. PLOS Neglected Trop Dis 8(1):e2636. https://doi.org/10.1371/journal.pntd.0002636. Accessed 14 Oct 2016; Haddow AD et al (2012) Genetic characterization of Zika virus strains: geographic expansion of the Asian Lineage. PLoS Neglected Trop Dis 6:e1477. https://doi.org/10.1371/journal.pntd.0001477. Accessed 14 Oct 2016

18. Lanciotti R et al (2008) Genetic and Serologic Properties of Zika virus associated with an epidemic, Yap State, Micronesia, 2007. Emerg Infectious Dis 14:8. August. https://wwwnc.cdc.gov/eid/article/14/8/pdfs/08-0287.pdf. Accessed 24 July 2018
19. Aliota M et al (2016) The wMel strain of Wolbachia reduces transmission of Zika virus by Aedes aegypti. Nat Sci Rep 6:28792. https://doi.org/10.1038/srep28792. http://www.nature.com/articles/srep28792. Accessed 16 Jan 2017
20. Marchette NJ, Garcia R, Rudnick A (1969) Isolation of Zika virus from *Aedes aegypti* mosquitoes in Malaysia. Am J Trop Med Hygiene 18:411–415. https://www.ncbi.nlm.nih.gov/pubmed/4976739. Accessed 15 Oct 2016
21. Olson JG, Ksiazek TG, Suhandiman T (1981) Zika virus, a cause of fever in Central Java, Indonesia. Trans R Soc Trop Med Hygiene 75:389–393. https://www.ncbi.nlm.nih.gov/pubmed/6275577. Accessed 15 Oct 2016
22. Weaver S et al (2016) Zika virus: history, emergence, biology, and prospects for control. Antiviral Res 130:69–80. https://www.ncbi.nlm.nih.gov/pubmed/26996139. Accessed 30 June 2017
23. Blümel J et al (2017) Inactivation and removal of Zika virus during manufacture of plasma-derived medicinal products. Transfusion 57. March. https://www.ncbi.nlm.nih.gov/labs/articles/27731495. Accessed 26 June 2017
24. PAHO WHO (2016) Zika—epidemiological update. December 29. http://www.paho.org/hq/index.php?option=com_content&id=11599&Itemid=41691. Accessed 10 Jan 2017
25. Mack J (2017) Michigan has 69 Zika cases so far; officials warn winter travelers. MLive. January 24. http://www.mlive.com/news/index.ssf/2017/01/michigan_has_69_zika_cases_so.html. Accessed 21 May 2017
26. Faye O et al (2014) Molecular evolution of Zika virus during its emergence in the 20th century. PLOS Neglected Trop Dis 8(1):e2636. https://doi.org/10.1371/journal.pntd.0002636. Accessed 14 Oct 2016
27. Hennessey M, Fischer M, Staples JE (2016) Zika virus spreads to new areas—region of the Americas, May 2015–January 2016. MMWR Morbidity Mortality Weekly Rep 65(Early Release):1–4. https://doi.org/10.15585/mmwr.mm6503e1er. http://www.cdc.gov/mmwr/volumes/65/wr/mm6503e1.htm. Accessed 14 Oct 2016
28. Weaver S et al (2016) Zika virus: history, emergence, biology, and prospects for control. Antiviral Res 130:69–80. https://www.ncbi.nlm.nih.gov/pubmed/26996139. Accessed 15 Oct 2016
29. Ratna K (2017) The secret life of Zika virus. Speaking Tiger Books, New Delhi
30. Beck J (2016) What will be the long-term effects of Zika? The Atlantic. April 19. https://www.theatlantic.com/health/archive/2016/04/zika-is-a-delayed-epidemic/478755/. Accessed 4 Feb 2017
31. ECDC (2016) Zika virus disease epidemic: potential association with microcephaly and Guillain-Barré syndrome (first update). Rapid Risk Assessment. January 21. http://ecdc.europa.eu/en/publications/Publications/rapid-risk-assessment-zika-virus-first-update-jan-2016.pdf. Accessed 24 Sept 2016
32. Lessler J et al (2016) Assessing the global threat from Zika virus. Science 353:6300. July 14. http://science.sciencemag.org/content/early/2016/07/13/science.aaf8160. Accessed 19 May 2017
33. Duffy M et al (2016) Zika virus outbreak on Yap Island, Federated States of Micronesia. N Engl J Med 360:24. June 11. https://doi.org/10.1056/NEJMoa0805715#t=article. Accessed 24 Sept 2016
34. Grard G et al (2014) Zika virus in Gabon (Central Africa)—2007: a new threat from *Aedes* albopictus? PLoS Negl Trop Dis 8(2):e2681. https://doi.org/10.1371/journal.pntd.0002681
35. Haddow AD et al (2012) Genetic characterization of Zika virus strains: geographic expansion of the Asian Lineage. PLoS Neglected Trop Dis 6:e1477. https://doi.org/10.1371/journal.pntd.0001477.PDF. Accessed 2 Oct 2016
36. Musso D (2015) Zika virus transmissions from French Polynesia to Brazil. Emerg Infectious Dis 21:10. October. http://www.ncbi.nlm.nih.gov/pmc/articles/PMC4593458/. Accessed 24 Sept 2016

37. Al-Qahtani AA et al (2016) Zika virus: a new pandemic threat. J Infection Dev Countries 10(3). https://www.ncbi.nlm.nih.gov/pubmed/27031450. Accessed 11 June 2019; Mercola J (2016) Zika: Brazil admits it's not the virus. Mercola.com. August 16
38. Faria NF et al (2016) Zika virus in the Americas: early epidemiological and genetic findings. Science 352(6283):345–349. April 15. https://doi.org/10.1126/science.aaf5036. http://science.sciencemag.org/content/early/2016/03/23/science.aaf5036. Accessed 24 Sept 2016
39. Ayres C (2016) Identification of Zika virus vectors and implications for control. Lancet 16. March. http://thelancet.com/journals/laninf/article/PIIS1473-3099(16)00073-6/fulltext. Accessed 12 Jan 2017
40. Beil L (2016) Vaccines may offer defense against dengue, Zika and chikungunya. Sci News. June 15. https://www.sciencenews.org/article/vaccines-may-offer-defense-against-dengue-zika-and-chikungunya. Accessed 11 Apr 2017
41. Paploski IAD et al (2016) Time lags between exanthematous illness attributed to Zika virus, Guillain-Barré syndrome, and microcephaly, Salvador, Brazil. Emerg Infectious Dis 22:8. August. https://wwwnc.cdc.gov/eid/article/22/8/16-0496_article. Accessed 19 July 2017
42. Sifferlin A (2016) How Brazil uncovered the possible connection between Zika and microcephaly. Time. February 1. http://time.com/4202262/zika-brazil-doctors-recife-investigation-outbreak/. Accessed 5 Sept 2016
43. Campos GS, Bandeira AC, Sardi SI (2015) Zika virus outbreak, Bahia, Brazil. Emerg Infectious Dis 21:1885–1886. http://wwwnc.cdc.gov/eid/article/21/10/15-0847_article. Accessed 15 Oct 2016; Zanluca C et al (2015) First report of autochthonous transmission of Zika virus in Brazil. Memórias do Instituto Oswaldo Cruz 110(4):569–572. June 9. http://www.scielo.br/scielo.php?script=sci_arttext&pid=S0074-02762015000400569. Accessed 15 Oct 2016
44. Mendes A (2016) Number of cases of Zika reported in Brazil drops by 87%. June 10. Brazilian Health Ministry-Latest News. http://combateAedes.saude.gov.br/en/latest-news/804-number-of-cases-of-zika-reported-in-brazil-drops-by-87. Accessed 18 Sept 2016
45. BRICS Post (2016) Brazil uses mutant mosquitoes to fight Zika. BRICS Post. November 3. http://thebricspost.com/brazil-uses-mutant-mosquitoes-to-fight-zika/#.WMlVmDvys2w. Accessed 14 Mar 2017
46. Chouin-Carneiro T et al (2016) Differential susceptibilities of Aedes aegypti and Aedes albopictus from the Americas to Zika virus. PLOS Neglected Trop Dis. March 3. https://doi.org/10.1371/journal.pntd.0004543. Accessed 14 Oct 2016
47. Hotez P (2016) Zika is coming. The New York Times. April 8. https://www.nytimes.com/2016/04/09/opinion/zika-is-coming.html?_r=0. Accessed 15 May 2017
48. CDC (2016) Zika virus case counts in the US. http://www.cdc.gov/zika/geo/united-states.html. Accessed 5 Sept 2016
49. Kodjak A (2016) Congress ends spat, agrees to fund $1.1 billion to combat Zika. Health News NPR. September 28. http://www.npr.org/sections/health-shots/2016/09/28/495806979/congress-ends-spat-over-zika-funding-approves-1-1-billion. Accessed 2 Oct 2016
50. Moreno-Madriñán M, Turell M (2017) Factors of concern regarding Zika and other Aedes aegypti-transmitted viruses in the United States. J Med Entomol 54:2. January 3. https://academic.oup.com/jme/article-abstract/54/2/251/2952765/Factors-of-Concern-Regarding-Zika-and-Other-Aedes?redirectedFrom=fulltext. Accessed 17 July 2017
51. Beck J (2016) Which US cities are at risk for a Zika outbreak. The Atlantic. March 17. https://www.theatlantic.com/health/archive/2016/03/which-us-cities-are-at-risk-for-a-zika-outbreak/474081/. Accessed 3 Feb 2017
52. Dart T (2017) 'It's going to hit the poorest people': Zika outbreak feared on the Texas border. The Guardian. April 23. https://www.theguardian.com/world/2017/apr/23/zika-outbreak-rio-grande-valley-texas-border-health. Accessed 9 May 2017
53. Orkin (2017) Orkin releases top 50 mosquito cities list. Orkin website. May 1. https://www.orkin.com/press-room/orkin-releases-top-50-mosquito-cities-list/. Accessed 5 June 2017
54. Medical Xpress (2017) As temperatures heat up, here's what you need to know about Zika virus. Medical Xpress. April 27. https://medicalxpress.com/news/2017-04-temperatures-zika-virus.html. 23 May 2017

55. WISN.com (2017) Mosquito that can spread Zika virus found in Waukesha County. WISN.com. August 7. http://www.wisn.com/article/mosquito-that-can-spread-zika-virus-found-in-waukesha-county/10378747. Accessed 27 June 2018
56. Chante C (2017) CDC urges states to prepare for Zika. The Philippine Star Philstar.com. February 5. http://www.philstar.com/opinion/2017/02/05/1669247/cdc-urges-states-prepare-zika. Accessed 3 Apr 2017
57. McNeil DG Jr (2017) Houston braces for another brush with the peril of Zika. The New York Times. July 17. https://www.nytimes.com/2017/07/17/health/zika-virus-houston-texas.html. Accessed 21 July 2017
58. Weaver SC et al (2017) Zika virus: history, emergence, biology, and prospects for control. Antiviral Res 130:69–80
59. Gold C, Josephson SA (2016) Anticipating the challenges of Zika virus and the incidence of Guillain-Barré syndrome. J Am Med Assoc Neurol 73(8). August. http://jamanetwork.com/journals/jamaneurology/fullarticle/2526494. Accessed 27 June 2017
60. Neergaard L, Swanson E (2016) Poll: some key gaps in Americans' knowledge about Zika virus. AP NORC. April 7. http://www.apnorc.org/news-media/Pages/News+Media/Poll-Some-key-gaps-in-Americans'-knowledge-about-Zika-virus.aspx. Accessed 26 May 2017
61. Graham J (2017) Zika virus could be 'major concern' in Orange County in 5–8 years without proper mosquito abatement, grand jury says. Orange County Register. April 18. http://www.ocregister.com/2017/04/18/zika-virus-could-be-major-concern-in-orange-county-in-5-8-years-without-proper-mosquito-abatement-grand-jury-says/. Accessed 14 May 2017
62. Health Day (2017) Major Zika outbreak in US unlikely: study. Health day. January 6. https://consumer.healthday.com/diseases-and-conditions-information-37/zika-1007/why-major-zika-outbreak-is-unlikely-in-u-s-718278.html. Accessed 14 May 2017
63. New Haven Register (2017) Connecticut monitoring 30 babies for Zika-related birth defects. New Haven Register. January 30. http://www.nhregister.com/general-news/20170130/connecticut-monitoring-30-babies-for-zika-related-birth-defects. Accessed 26 May 2017
64. Klint C (2017) Alaskan gets infected with Zika virus in Central America. Alaska dispatch News. April 28. https://www.adn.com/alaska-news/health/2017/04/28/alaskan-infected-with-zika-virus-in-central-america/. Accessed 5 June 2017
65. Shelton T (2016) Zika virus-infected mothers delivered babies with microcephaly as early as 2009. University of Hawai'i News. December 27. http://www.hawaii.edu/news/2016/12/27/zika-virus-infected-mothers-delivered-babies-with-microcephaly-as-early-as-2009/. Accessed 2 June 2017
66. Fannon E (2016) Digging deeper into the Zika virus medical professionals say the threat of the virus is not slowing down... Illinois homepage.net. November 23. http://www.illinoishomepage.net/news/local-news/digging-deeper-into-the-zika-virus/611790018. Accessed 11 Apr 2017
67. Staff Report (2017) Zika virus confirmed in Hays County. Hays Free Press News Dispatch. January 4. https://issuu.com/haysfreepress/docs/hfp_010417. Accessed 14 May 2017
68. Kelly L (2017) Zika virus target of public awareness campaigns. The Washington Times. May 2. http://www.washingtontimes.com/news/2017/may/2/zika-virus-target-of-public-awareness-campaigns/. Accessed 5 June 2017
69. Minnesota Department of Health (2016) MDH: 57 cases of Zika virus in minnesota this year. CBS Minnesota. November 23. http://minnesota.cbslocal.com/2016/11/23/57-cases-of-zika/. Accessed 22 May 2017
70. Mississippi State Department of Health (2017) Health Dept. confirms 1st Zika cases of 2017. MSDH Web Site. February 7. http://www.msdh.state.ms.us/msdhsite/_static/23,18554,341.html. Accessed 3 Apr 2017
71. Anderson J (2017) Omaha-area resident is state's first pregnant Zika patient. Live Well Nebraska. April 12. http://www.omaha.com/livewellnebraska/omaha-area-resident-is-state-s-first-pregnant-zika-patient/article_0dea9e4e-3fc1-582c-b630-d07c55dd5d81.html. Accessed 3 May 2017

72. Nebraska Department of Health and Human Services (2017) Zika virus. Nebraska Department of Health and Human Services. November 16. http://dhhs.ne.gov/publichealth/CDC/Pages/ZikaVirus.aspx. Accessed 27 June 2018
73. Bowers N (2017) SNHD (Southern Nevada Health District): first non-travel Zika virus case confirmed in Clark County. Las Vegas Now. February 10. http://www.lasvegasnow.com/news/snhd-first-non-travel-zika-virus-case-confirmed-in-clark-county/654766322. Accessed 14 May 2017
74. Jarvis K (2017) First non-travel Zika virus case reported in Clark County. KTNV. February 10. http://www.ktnv.com/news/first-non-travel-zika-virus-case-reported-in-clark-county. Accessed 17 May 2017
75. Whittemore K et al (2017) Zika Virus knowledge among pregnant women who were in areas with active transmission. Emerg Infectious Dis 23:1. January. https://wwwnc.cdc.gov/eid/article/23/1/16-1614_article. Accessed 7 Feb 2017
76. Turner B (2016) Zika virus birth defects growing in numbers in New York. Inquisitor. December 7. http://www.inquisitr.com/3778043/zika-virus-birth-defects-growing-in-numbers-in-new-york/. Accessed 4 June 2017
77. Chi'en A (2016) Zika virus still a concern for New York. December 9. FOX5ny.com. http://www.fox5ny.com/news/222807345-story. Accessed 3 Apr 2017
78. Orr S (2017) Why Zika virus infections are way down in U.S. this summer. USA Today Network. July 12. https://www.usatoday.com/story/news/nation-now/2017/07/13/why-zika-virus-infections-way-down-u-s-summer/474312001/. Accessed 21 July 2017
79. Staff (2017) Among the 62 countries and territories with the Zika virus are Mexico, the Bahamas and Puerto Rico. Vindy.com. http://www.vindy.com/news/2017/apr/03/health-officials-urge-travelers-to-avoid/?print. Accessed 2 June 2017
80. Sweigart J (2017) Zika virus suspected in Clark County. Dayton Daily News. April 29. http://www.daytondailynews.com/news/local/zika-virus-suspected-clark-county/XV3kSkFOsrotZGKBK6fsRM/. Accessed 27 June 2017
81. Searles Z (2017) Zika virus now confirmed in Maine and New Hampshire. The Free Press. July 21. http://usmfreepress.org/2016/03/07/zika-virus-now-confirmed-in-maine-and-new-hampshire/. Accessed 21 July 2017
82. Fox M (2016) Zika virus arrives in South Texas. NBC News. November 28. http://www.nbcnews.com/storyline/zika-virus-outbreak/zika-virus-arrives-south-texas-n689226. Accessed 14 Apr 2017
83. Howell T (2017) Zika largely spares U.S. as virus wreaks havoc worldwide. The Washington Times. January 3. http://www.washingtontimes.com/news/2017/jan/3/zika-largely-spares-us-as-virus-wreaks-havoc-world/. Accessed 15 May 2017
84. Texas Health and Human Services (2018) Zika in Texas. TexasZika.org. http://www.texaszika.org/currentcases.htm. Accessed 26 July 2018
85. Lafrance A (2016) The sneak-attack mosquito. The Atlantic. April 25. https://www.theatlantic.com/science/archive/2016/04/aedes-aegypti/479619/. Accessed 18 May 2017
86. Karlamangla S (2018) L.A. County officials confirm first case of sexually transmitted Zika virus. The LA Times. January 4. http://www.latimes.com/local/california/la-me-ln-zika-sexual-transmission-20180104-story.html, Accessed 27 June 2018
87. McAllister T (2017) 86 people in San Diego area diagnosed with Zika virus since 2015. Times of San Diego. March 23. http://timesofsandiego.com/life/2017/03/23/86-people-in-san-diego-area-diagnosed-with-zika-virus-since-2015/. Accessed 21 May 2017
88. Hayes R (2017) Riverside scientists announce breakthrough in Zika research. ABC 7. March 29. http://abc7.com/health/1st-socal-baby-born-with-zika-defects-renews-travel-warnings/1825136//. Accessed 14 May 2017
89. Edwards A (2017) Zika virus: long beach health officials warn of looming threat. Press Telegram. May 8. http://www.presstelegram.com/health/20170508/zika-virus-long-beach-health-officials-warn-of-looming-threat. Accessed 26 June 2017
90. Staff Reports (2017) Zika virus found in Fresno County. Hanford sentinel. January 20. http://hanfordsentinel.com/selma_enterprise/news/zika-virus-found-in-fresno-county/article_e48b9191-ba3b-5747-87f0-5d7e56b61789.html. Accessed 14 May 2017

91. Street C (2017) Pandemic risk: Zika mosquitos in 129 California cities. Breitbart. April 2. http://www.breitbart.com/california/2017/04/02/pandemic-risk-zika-mosquitos-129-california-cities/. Accessed 2 June 2017
92. CBS (2017) Officials brace for possible cases of Zika with outbreak of new aggressive mosquito in SoCal. http://losangeles.cbslocal.com/2017/03/29/officials-brace-for-possible-cases-of-zika-with-outbreak-of-new-aggresive-mosquito-in-socal/. Accessed 10 Apr 2017
93. Adalja A et al (2016) Genetically modified (GM) mosquito use to reduce mosquito-transmitted disease in the US: a community opinion survey. PLOS Curr Outbreaks. May 25. https://doi.org/10.1371/currents.outbreaks.1c39ec05a743d41ee39391ed0f2ed8d3
94. Atkins K (2016) Wolbachia-infected bugs approved for March trial. Florida Keys News. October 19. http://www.flkeysnews.com/news/local/article109137307.html. Accessed 25 Oct 2016
95. Reuters (2016) Zika: Florida finds local mosquitoes with virus. Newsweek. September 1. http://www.reuters.com/article/us-health-zika-florida-idUSKCN1175S0. Accessed 5 Sept 2016
96. Berkrot B (2016) Florida finds first local mosquitoes with Zika virus. Reuters. September 1. http://www.reuters.com/article/us-health-zika-florida-idUSKCN1175S0. Accessed 25 Sept 2016
97. Newkirk VRII (2016) The American Zika outbreak. The Atlantic. September 13. https://www.theatlantic.com/politics/archive/2016/09/zika-puerto-rico-public-health/499685/. Accessed 26 May 2017
98. Chang D (2017) Zika virus down but not out in Florida as state reports more cases. The Miami Herald. January 30. http://www.miamiherald.com/news/health-care/article129203409.html. Accessed 9 Mar 2017
99. Fletcher J, Chang D (2016) Gov. Rick Scott lifts last Zika zone in Miami Beach, but isolated cases still expected. Miami Herald. December 9. http://www.miamiherald.com/news/health-care/article119893258.html. Accessed 11 Apr 2017
100. McNeill D Jr (2016) Predict Zika's spread? It's hard enough to count the cases. New York Times. September 19. https://www.nytimes.com/2016/09/20/health/zika-spread-predictions.html. Accessed 15 May 2017
101. Lonon S (2016) 2 New Zika virus cases confirmed in Tampa Bay Area. Lakeland Patch. December 28. https://patch.com/florida/lakeland/2-new-zika-virus-cases-confirmed-tampa-bay-area. Accessed 19 May 2017
102. Lolo S (2017) New travel-related cases of Zika virus emerge in Collier County. WINK News. January 28. http://www.winknews.com/2017/01/28/new-travel-related-cases-of-zika-virus-emerge-in-collier-county/. Accessed 19 May 2017
103. Alvarez L (2016) No new local Zika transmissions in Florida, Governor Says. The New York Times. December 9. https://www.nytimes.com/2016/12/09/us/zika-florida-governor-rick-scott.html?_r=0. Accessed 16 Jan 2017
104. CDC (2016) Advice for people living in or traveling to South Florida. October 19. https://www.cdc.gov/zika/intheus/florida-update.html. Accessed 3 Apr 2017
105. Barry-Jester AM (2016) Small Island, big experiment: how a tiny Florida community could influence the way we fight Zika around the world. FiveThirtyEight. https://fivethirtyeight.com/features/zika-mosquito-florida-vote/. Accessed 17 Jan 2017
106. Florida News Service (2017) Florida Zika cases top 100 this year. News4Jax. July 14. https://www.news4jax.com/health/zika-virus/florida-zika-cases-top-100-this-year. Accessed 21 July 2017
107. Russell T (2017) Warm winter fuels Zika fears in Central Florida. WFTV. February 6. http://www.wftv.com/news/local/warm-winter-fuels-zika-fears-in-central-florida/491662850. Accessed 31 May 2017
108. Hackett DW (2018) Florida reports 89 travel related Zika cases. PrecisionVaccinations.com. December 2. https://www.precisionvaccinations.com/zika-infection-during-pregnancy-can-cause-serious-birth-defects-such-microcephaly. Accessed 13 June 2019

109. CDC (2016) CDC updates guidance related to local Zika transmission in Miami-Dade County, Florida. https://www.cdc.gov/media/releases/2016/p1019-zika-florida-update.html. Accessed 3 Apr 2017
110. Guttmacher (2010) State unintended pregnancy rates. https://www.guttmacher.org/sites/default/files/images/unintendedpregnancyratesmap.png. Accessed 1 Feb 2017
111. AP (2017) Houston area may be test zone for genetically modified mosquitoes in Zika fight. Genetic Literacy Project. March 27. https://www.geneticliteracyproject.org/2017/03/27/houston-area-may-test-zone-genetically-modified-mosquitoes-zika-fight/. Accessed 7 Apr 2017
112. Barry-Jester AM (2016) The latest Zika news is more bad news. Five ThiryEight. December 16. https://fivethirtyeight.com/features/the-latest-zika-news-is-more-bad-news/. Accessed 1 Feb 2017
113. Rosa K (2017) First non-travel-related Zika virus in pregnant Texas woman confirmed. ContagionLive. January 29. http://www.contagionlive.com/news/texas-reports-first-non-travel-related-case-of-zika-in-a-pregnant-woman. Accessed 8 Mar 2017
114. Gathany J (2016) Texas reports first Zika case believed to be from local mosquito bite. CBS News. November 29. http://www.cbsnews.com/news/texas-zika-virus-first-case-local-transmission-mosquito-bite/. Accessed 14 Apr 2017
115. CBS Austin (1027) Infant born with microcephaly in Travis Co. diagnosed with Zika virus. CBS Austin. January 6. http://keyetv.com/news/local/infant-born-with-microcephaly-in-travis-co-diagnosed-with-zika-virus. Accessed 16 Jan 2017
116. Phippen JW (2016) The first known Zika-related infant death in the U.S. The Atlantic. August 9. https://www.theatlantic.com/news/archive/2016/08/texas-zika-infant-death/495059/. Accessed 28 May 2017
117. Barlow P (2016) Zika has crossed the Rio Grande … now what? The Independent. December 5. http://www.independent.co.uk/life-style/health-and-families/zika-has-crossed-the-rio-grande-health-a7451436.html. Accessed 16 Jan 2017
118. KRGV (2017) Pregnant woman likely contracted Zika virus in Brownsville. KRGV. January 26. http://www.krgv.com/story/34358408/pregnant-woman-likely-contracted-zika-virus-in-brownsville. Accessed 17 May 2017
119. Lopez A (2017) Texas health officials say we're still 'quite vulnerable' to Zika this mosquito season. Houston Public Media. April 25. https://www.houstonpublicmedia.org/articles/news/2017/04/25/197630/texas-health-officials-say-were-still-quite-vulnerable-to-zika-this-mosquito-season/. Accessed 19 May 2017
120. AP (2017) Houston considers use of genetically modified mosquitoes. The Statesman. News. March 20. http://www.statesman.com/news/state--regional/houston-considers-use-genetically-modified-mosquitoes/zcBceXU5Qzia5WqWVyuu9M/. Accessed 7 Apr 2017
121. Papenfuss M (2016) Activists in Florida block release of GM mosquitoes which could suppress Zika outbreak. International Business Times. August 18. http://www.ibtimes.co.uk/activists-florida-block-release-gm-mosquitoes-which-could-suppress-zika-outbreak-1576655. Accessed 4 Sept 2016
122. CDC (2016) Zika virus. October 6. http://www.cdc.gov/zika/hc-providers/index.html. Accessed 14 Oct 2016
123. CDC (2018) 2017 case counts in the US. Centers for Disease Control and Prevention. January 10. https://www.cdc.gov/zika/reporting/2017-case-counts.html. Accessed 27 June 2018
124. CDC (2018) Zika in babies in US territories. CDC Vitalsigns. August. https://www.cdc.gov/vitalsigns/zika-territories/. Accessed 28 Aug 2018
125. AAFP (2018) Children exposed to Zika in utero need long-term monitoring. August 17. https://www.aafp.org/news/health-of-the-public/20180817mmwr-zika.html. Accessed 14 May 2019
126. Peñaloza M, Allen G (2017) Long-term impact of Zika virus in Puerto Rico unknown. NPR. March 27. http://www.npr.org/sections/health-shots/2017/03/27/519653931/living-with-zika-in-puerto-rico-means-watching-waiting-and-fearing-judgment. Accessed 28 May 2017

127. Van der Linden V et al (2016). Description of 13 infants born during October 2015–January 2016 with congenital Zika virus infection without microcephaly at birth—Brazil. Morbidity Mortality Weekly Rep 65:47. December 2. https://www.cdc.gov/mmwr/volumes/65/wr/mm6547e2.htm. Accessed 19 July 2017
128. Beaubien J (2016) Zika pregnancies and big questions in Puerto Rico. NPR Now. November 29. http://www.npr.org/sections/health-shots/2016/11/29/503592597/zika-pregnancies-and-big-questions-in-puerto-rico. Accessed 3 Feb 2017
129. Ellington S et al (2016) Estimating the number of pregnant women infected with Zika virus and expected infants with microcephaly following the Zika virus outbreak in Puerto Rico, 2016. JAMA Pediatr 170(10):940–945. October. http://jamanetwork.com/journals/jamapediatrics/fullarticle/2545827. Accessed 4 May 2017
130. Branswell H (2017) "They're just hiding": experts say Puerto Rico may be underreporting Zika-affected births. STAT. April 18. https://www.statnews.com/2017/04/18/zika-virus-puerto-rico-pregnancies/. Accessed 4 May 2017
131. Branswell H (2017) Feud erupted between CDC, Puerto Rico over reporting of Zika cases, document shows. STAT. May 1. https://www.statnews.com/2017/05/01/zika-virus-puerto-rico-cdc/. Accessed 5 June 2017
132. Rullán J (2017) El rompecabezas de los bebés con Zika. El Neuvo Día. April 4. http://www.elnuevodia.com/opinion/columnas/elrompecabezasdelosbebesconzika-columna-2307153/. Accessed 31 May 2017
133. Branswell H (2017) Puerto Rico declares its outbreak of Zika virus is over. STAT. June 5. https://www.statnews.com/2017/06/05/puerto-rico-zika-outbreak/. Accessed 26 June 2017
134. Beck J (2016) Zika is the 'most difficult' emergency health response ever, CDC official says. The Atlantic. June 24. http://www.theatlantic.com/health/archive/2016/06/zika-is-the-most-difficult-emergency-health-response-ever-says-cdc-official/488579/. Accessed 16 Jan 2017
135. Beck J (2016) The importance of contraception to the Zika fight. The Atlantic. July 1. https://www.theatlantic.com/health/archive/2016/07/the-importance-of-contraception-to-the-zika-fight/489767/. Accessed 3 Feb 2017
136. AP (2016) Man dies from Zika-related paralysis in Puerto Rico. NBC NEWS. August 19. http://www.nbcnews.com/storyline/zika-virus-outbreak/man-dies-zika-related-paralysis-puerto-rico-n634776. Accessed 1 Feb 2017
137. Steenhuysen J (2017) Puerto Rico declares Zika outbreak over, CDC maintains travel warning. Reuters. June 6. http://www.reuters.com/article/us-health-zika-puertorico-idUSKBN18W2SU. Accessed 19 July 2017
138. Grubaugh ND et al (2019) Travel surveillance and genomics uncover a hidden Zika outbreak during the waning epidemic. Cell 178:1057–1071. August 29. https://doi.org/10.1016/j.cell.2019.07.018
139. Song B-H et al (2017) Zika virus: history, epidemiology, transmission, and clinical presentation. J Neuroimmunol 308:50–64
140. Keet E (2019) An estimated 5700 Zika cases went unreported in 2017 Cuban outbreak. ContagionLive. September 3. https://www.contagionlive.com/view/an-estimated-5700-zika-cases-went-unreported-in-2017-cuban-outbreak. Accessed 25 May 2022
141. Journey Mexico (2017) The Zika virus in Mexico: what you need to know. December 4. Journey Mexico. https://www.journeymexico.com/blog/the-zika-virus-in-mexico-what-you-need-to-know. Accessed 27 June 2018
142. Hernandez-Avila JE, Palacio-Mejia LS, Lopez-Gatell H, Alpuche-Aranda CM (2018) Zika virus infection estimates, Mexico. Bull World Health Organ 96:306–313
143. Buck P (2017) Heading to Mexico? California health officials warn travelers to avoid Zika virus. Sacramento Bee. February 9. http://www.sacbee.com/news/local/health-and-medicine/article131745564.html. Accessed 15 Mar 2017
144. Baker A (2016) Zika hasn't hurt Africa—and that may be the key to beating it. Time. February 12. http://time.com/4219240/zika-africa-origins-microcephaly-vaccine/. Accessed 18 Sept 2016

145. PAHO (2016) Regional Zika epidemiological update (Americas) December 29, 2016. Zika virus—incidence and trends. http://www2.paho.org/hq/index.php?option=com_docman&task=doc_view&Itemid=270&gid=37579&lang=en. Accessed 28 May 2017
146. PAHO (2017) Regional Zika epidemiological update (Americas) May 25, 2017. Zika virus—incidence and trends. http://www.paho.org/hq/index.php?option=com_content&view=article&id=11599%3Aregional-zika-epidemiological-update-americas&catid=8424%3Acontents&Itemid=41691&lang=en. Accessed 28 May 2017
147. Virtue E (2017) Zika attack almost 8,000 suspected cases of the virus in JA 15 months. The Gleaner. April 16. http://jamaica-gleaner.com/article/lead-stories/20170416/zika-attack-almost-8000-suspected-cases-virus-ja-15-months. Accessed 9 June 2017
148. Lessler J et al (2016) Assessing the global threat from Zika virus. Science. July 14. 353:6300.http://science.sciencemag.org/content/early/2016/07/13/science.aaf8160. Accessed 19 May 2017
149. VnExpress (2016) HCMC declares Zika pandemic. VnExpress. October 20. http://e.vnexpress.net/news/news/hcmc-declares-zika-pandemic-3486472.html. Accessed 11 Apr 2017
150. Lim SK, Lim JK, Yoon IK (2017) An update on Zika virus in Asia. Infection Chemother 49:2. June 27. https://www.ncbi.nlm.nih.gov/pmc/articles/PMC5500276/. Accessed 8 Aug 2018
151. Trang P (2016) Zika virus mosquitoes detected in central Vietnam. VnExpress. October 16. http://e.vnexpress.net/news/news/zika-virus-mosquitoes-detected-in-central-vietnam-3484407.html. Accessed 4 June 2017
152. WHO (2016) Zika virus infection—Viet Nam. Disease Outbreak News. April 12. http://www.who.int/csr/don/12-april-2016-zika-viet-nam/en/. Accessed 28 Mar 2017
153. Reuters (2017) Singapore confirms two new cases of Zika virus. Fox News. March 30. http://www.foxnews.com/health/2017/03/30/singapore-confirms-two-new-cases-zika-virus.html. Accessed 30 May 2017
154. Jalil Z (2018) Zika virus re-emerges in Singapore after four-month disappearance. IOL Travel. January 24. https://www.iol.co.za/travel/travel-news/zika-virus-re-emerges-in-singapore-after-four-month-disappearance-12909141. Accessed 27 June 2018
155. Ejinsight (2017) CHP confirms first imported case of Zika virus this year. Ejinsight On the Pulse. April 27. http://www.ejinsight.com/20170427-chp-confirms-first-imported-case-of-zika-virus-this-year/. Accessed 10 May 2017
156. Hinayon R (2017) 57 Zika cases in PH; DOH verifying microcephaly case. ABS-CBN News. February 3. http://news.abs-cbn.com/news/02/03/17/57-zika-cases-in-ph-doh-verifying-microcephaly-case. Accessed 14 May 2017
157. Bhattacharjya S (2016) Malaysia confirms its 8th Zika virus case. International Business Times. December 18. http://www.ibtimes.sg/malaysia-confirms-its-8th-zika-virus-case-5651. Accessed 22 Dec 2016
158. Lim S-K, Lim JK, Yoon I-K (2017) An update on Zika virus in Asia. Infect Chemother 49(2):91–100
159. Bogoch II et al (2016) Potential for Zika virus introduction and transmission in resource-limited countries in Africa and the Asia-Pacific region: a modelling study. Lancet Infectious Dis 16. November. http://www.thelancet.com/journals/laninf/article/PIIS1473-3099(16)30270-5/abstract. Accessed 26 June 2017
160. Mok D, Cheung E (2016) Hong Kong on alert after first case of Zika virus infection confirmed. South China Post. August 26. http://www.scmp.com/news/hong-kong/health-environment/article/2009071/hong-kong-reports-first-zika-virus-infection. Accessed 29 May 2017
161. Zhou CM et al (2020) Emergence of Zika virus infection in China. PLoS Negl Trop Dis 14(5):e0008300. https://doi.org/10.1371/journal.pntd.0008300
162. Icheku V (2017) Is Zika virus is (sic) the definitive culprit in the cases of microcephaly? Zika virus and microcephaly. Lambert Academic Publishing and Middletown, Amazon, Balti, Moldova, DE
163. Lozier M, Adams L, Febo MF, Torres-Aponte J et al (2016) Incidence of Zika virus disease by age and sex—Puerto Rico, November 1, 2015–October 20, 2016. Morb Mortal Wkly Rep 65(44):1219–1223

164. Swift A, Ander S (2017) Zika virus not a worry to Americans. March 27. https://news.gallup.com/poll/207422/zika-virus-not-worry-americans.aspx. Accessed 15 May 2022
165. CDC (2017) Potential range of Aedes aegypti and Aedes albopictus in the United States, 2017. https://www.cdc.gov/mosquitoes/mosquito-control/professionals/range.html. Accessed 15 May 2022

Chapter 4
ZIKV Ebbs

ZIKV ebbed, or did it? To a significant degree, the number of cases reported after the South American pandemic dropped significantly. It declined markedly in the Americas since late 2016; fewer than 30,000 cases were reported in 2018, compared with more than 500,000 cases reported at the peak of the pandemic in 2016 [1].

Here's some good news. Since ZIKA was found in Africa in 1947 and detected in spots in Asia in the 1950s, scientists suspect it has circulated silently on those continents for decades, probably misdiagnosed as mild dengue or other rash-causing fevers. If so, most girls may get it in childhood and become immune before entering their childbearing years [2].

As ZIKV is braced to strike the continental U.S. in one form or another, it's essential to consider where the ZIKV might spread and the implications of seeing clusters of microcephaly cases appear in Europe, the Middle East, Asia Minor, Africa, and Asia [3]."

Globally, however, the actual incidence, prevalence, and geographic distribution of ZIKV are likely underestimated. Some countries report "acute fever and rash," but these infections are not investigated, and the pathogens are not identified.

For example, in Brazil, health authorities initially reported an outbreak of 6000 cases of "exanthematic disease." Still, ZIKV was detected in only a few patients, so the number of actual infections was unknown. In June 2015, the government reported 40,000 cases of the disease, with 24,000 suspected cases of Zika fever, but the exact number of infections is unknown in the absence of routine laboratory testing [4].

ZIKV often presents few symptoms in people who contract it, and many have no symptoms whatsoever. The ZIKV can cause severe complications during pregnancy, leading to congenital disabilities. Additionally, it can cause Guillain–Barre syndrome, a rare sickness that causes the immune system to damage nerve cells, leading to weakened muscles and paralysis.

In 2020, the Guardian reported that 40,000 cases in the U.S. and its territories remain infected with the ZIKV [5].

D. M. Berube, *Pandemics and Resilience: Lessons we should have learned from Zika*, Risk, Systems and Decisions, https://doi.org/10.1007/978-3-031-25370-6_4

4.1 Tracking ZIKV

Tracking infectious diseases is neither facile nor popular. It requires testing. As more was tested, more was learned about the outbreak and the extent of an epidemic. Those responsible for managing the epidemic experience blame, and their incentives to test and track are compromised. It becomes more complicated when many infections are asymptomatic, and the tests are complex, slow, and expensive. Finally, consider the implications when the tested disease looks like other diseases and the testing protocol is challenged to distinguish between them. Welcome to the world of the ZIKV.

4.2 Transmission Challenges

Areas with a sizeable naïve population and abundant *Ae. aegypti* are expected to experience epidemic patterns of ZIKV transmission. However, as areas accumulate with mounting herd immunity, ZIKV tends to spread in smaller outbreaks in the remaining susceptible groups.

Herd immunity, or community immunity, is when a large part of the population of an area is immune to a specific disease. If enough people are resistant to the cause of a disease, such as a virus or bacteria, it has nowhere to go [6]. A caveat: It can go into what is called a reservoir where it can not only infect the reservoir species or not but also mutate for its re-emergence (see Chap. 10).

Some models suggest that the cessation of American ZIKV epidemics could be followed by a relatively low incidence of infection and disease for several decades [7]. Although the susceptible populations in the Americas may be diminishing as future amplifiers of ZIKV, it is anticipated that further transmission may still occur [8]. Nonetheless, the researchers say that Zika virus infection poses "unprecedented challenges to public health" as the first mosquito-borne illness that causes congenital disabilities in infants through perinatal transmission and the first mosquito-borne illness to be sexually transmitted [9].

The framing of ZIKV as a risk by government officials, the media, and other amplification stations [10] has been noteworthy. An amplification station takes a message, in this case, an outbreak message, and crafts around it an amplified, exaggerated set of messages within which a version of the original outbreak message may be found.

The primary amplification station is the media: traditional radio, TV news, newspapers, magazines, documentaries, commercial film, and digital media, with its news reporting more personalized and social. Government officials at all levels, personalities, and interest groups are an alternative amplification station for health-related communication. Together these amplification stations set public agendas and prime and frame outbreaks to a broad class of stakeholders. The media employs exaggeration and hyperbole often to increase readership and viewership, turning nearly every episode into a pandemic. Severe acute respiratory syndrome (SARS) in 2002, Avian flu in 2005, and multidrug resistant tuberculosis (MDR-TB) in 2012.

Some feel it is happening again; the ZIKV implicates genetic engineering this time. Genetic engineering can be defined as modifying genetic material by recombinant DNA techniques, including the deletion, modification, or insertion of genetic material in a genome. For example, a new gene may be added to a genome to add a new function or make a new enzyme. This gene can be from any species, as all organisms use the same DNA coding system [11]. The ZIKV epidemic brought to the fore vector control and a series of individual responses (see Chaps. 12 and 13). These approaches were also amplified, often generating high outrage among public constituencies.

Leslie Lobel, an Israeli physician who has worked with the U.S. Military and the Uganda Virus Research Institute to find a vaccine for Ebola, believes public panic over epidemics can cause more damage than the diseases themselves [12]. Fear is a powerful driver though it is not universally powerful. Under certain circumstances, it is ineffective, if not counterproductive [13].

Lobel suggests the WHO declared an emergency with Zika this time because governments were nervous after failing to deal adequately with Ebola. Similarly, he said the U.S. has suddenly announced almost $2 billion in Zika research funding. It is worried the disease will head up through Mexico once the Northern Hemisphere summer arrives. He believes the money could be spent much more effectively to deal with future, more severe challenges. "There's a lot that might be done that is not being done in a corrective way to control all diseases, not just Zika. This is just the tip of the iceberg. There are going to be a lot more" [12].

This is "pandemic pornography" no less salacious and provocative than photos and videos of barely clothed people end in suggestive and, in some cases, actual sex acts. Pandemic porn is about titillation. Amplification stations often pornographize outbreak-related health risks, using existential descriptors to characterize an outbreak. Time is the worst enemy type of talk and gross exaggerations of the impacts.

Zika is a serious health problem, and many amplification stations have legitimate and sometimes selfless reasons for portraying the outbreak the way they did. For example, Florida House Speaker Richard Corcoran said in a prepared statement on August 29, 2016: "The outbreak of the ZIKV, coupled with the inability of current measures to stop the spread, clearly demonstrate that time is of the essence if we are to beat back the spread of this disease" [14]. Another example came from Alcides Troncoso also in 2016. "The biggest concern is how quickly Zika spreads worldwide and that it could be far more dangerous than previously thought. Zika virus infection, by its explosive potential, has every chance of becoming a global pandemic" [15].

4.3 Underreporting

Given the impact Zika may have on tourism, some underreporting has occurred. This happened in Cuba. According to PAHO, Cuba did not notify anyone of its breakout. Consequently, thousands of Zika virus cases went unreported in Cuba in

2017, according to an analysis of data on travelers to the Caribbean Island. Veiling them may have led to many other cases that year. The Grubaugh study reported that Zika infections peaked in Cuba in the second half of 2017 when the virus waned in mainland North and South America.

Cuba's first case of Zika occurred in March 2016. A PAHO report says the country stopped providing updates on Zika in January 2017. In press reports in May 2017, Cuba said that nearly 1900 infections had been detected up to that point. But Nathan Grubaugh at the Yale School of Public Health and his colleagues estimate that the total cases in 2017 alone would have been more than doubling that at 5700 [16].

The Yale team also sequenced the genomes of Zika viruses retrieved from nine Floridians who traveled to Cuba. This showed that the infection was distinct from Zika infections in Florida. The travel cases revealed that the strains active in Cuba at the time were related to ones previously detected in other Latin American countries.

This is an important discovery, says Duane Gubler at the Duke-NUS Medical School in Singapore. He says Cuba does not report epidemics until they become apparent, and Zika is only mildly symptomatic in adults. "One of the problems we have is that islands that depend on tourism are not forthcoming in immediate reporting," he says [16].

Tourism pressure can be formidable. "Some observers believe Puerto Rico, heavily dependent on tourism, downplayed the scale of its Zika problem. Puerto Rico's not escaping this. They're just hiding," one former US official said of the situation.

The individual, who spoke on condition of anonymity, said months ago, it was clear "dozens and dozens" of babies in Puerto Rico bore the hallmarks of Zika damage. But territorial health officials declined to label most cases of Zika congenital syndrome" [17].

4.4 Anomalies

Many of the predictions about the spread of the ZIKV proved to be incorrect. These events lead to fears of major epidemics and pandemics and significant shifts in geographical, economic, and health impact.

Such breakouts from historic constraints may be temporary, or they may be permanent. They can be adaptive shifts that enable mosquito populations to depart from a previous sylvatic existence to strongly human-adapted coexistence.

Put simply, ZIKV may have been transported stochastically to locations with large enough naive human populations, accompanied by sizeable urban mosquito vector populations, for an explosive outbreak. Under these epidemic conditions, formerly rare diseases can become suddenly recognized when a prominent enough numbers of infections occur [18].

These drivers will be discussed below and may explain the breakout in Brazil and breakouts elsewhere in the future. What happened in Brazil might have been the proverbial "perfect storm." In which case, are there any more "perfect storms" on the horizon?

Oddly enough, studying the appearance of the ZIKV and associated illnesses in Latin America provided some fascinating insight.

4.5 Africa

What surprised many people was that the ZIKV did not have a significant footprint in Africa during the Latin American epidemic. Why not in Africa, especially since Africa was the ancestral origin of what seems like all mosquito and tick-transmitted flaviviruses?

ZIKV has probably been circulating in Africa and nature in a sylvatic cycle for many decades, spilling undetected into human populations across Africa with unknown regularity.

There's now a palpable sense of urgency among scientists to find out more about Zika's potential underappreciated, decades-long trajectory. "If we're trying to understand what would happen with Zika in the Americas ten or 20 years from now, the best place to look for information on the future is Africa," Morris says [19].

Zika might have been causing congenital disabilities in Africa that have gone unnoticed. In Africa, there are a lot of diseases that can lead to conditions such as microcephaly or congenital disabilities, according to Dawn Dudley, who studies Zika in rhesus macaques at the University of Wisconsin–Madison: "ZIKV could have caused those things and would have gone potentially unnoticed" [19].

Here are three examples of African outbreaks: Gabon, Angola, and Tanzania.

GABON: Five human sera tested positive for ZIKV in Gabon in 2007 [20]. Bourgarel, Wauquier, and Gonzalez tell a familiar story. Human incursions in Gabonese forests for exploitation purposes lead to intensified contact between humans and wildlife, thus generating an increased risk of zoonotic diseases. They discuss other drivers, including bushmeat diets by its people and natural resource exploitation as a dominant government promotes the export industry [21]. The Gabon outbreak was traced to the *Ae. albopictus* mosquito which is also present in Southern Europe.

Angola: Angola receives the most significant number of travelers from ZIKV-affected countries in the Americas, probably because of Angola's cultural and historical ties to Brazil. In Angola, there is a historical and ongoing epidemic of yellow fever, an arbovirus also transmitted by *Ae. aegypti* is already straining the public health system [22]. In 2017, Angola announced its first two cases of the virus. "This [ZIKV] strain seems much more virulent than ones we've seen in the past," Donna Patterson told Foreign Policy. "And if it turns from two cases to 20 or more rapidly, we know we have real cause for concern for Angola" [23].

Tanzania: Tanzania's National Institute for Medical Research (NIMR)'s released study findings show that 15.6% of the 533 people whose blood samples were tested have ZIKV. The same study discovered that out of 80 toddlers born with physical disabilities, 43.8% were traced with the virus. NIMR was surveyed in partnership with Bugando Catholic University of Health and Allied Sciences [24].

Many scholars in the field tell researchers to look again at Africa. Wetman (2017) claimed evidence that the ZIKV was widespread across African countries from the 1950s to the 1990s. The proportion of people whose blood contained antibodies to the virus—an indication of prior exposure—ranged from 2% of the population in places like Djibouti to 50% in some parts of Mali. Part of the lack of data, says researchers based in Africa, is that funding has been hard to come by [19].

In East Africa, ZIKV is most probably maintained in a sylvatic cycle with cyclic epizooty involving non-human primates and mosquitoes (Aedes spp.) in the tropical forest. In West Africa and Asia (Southeast Asia), serological surveys indicate a likely silent circulation and antibodies have been detected in various animals, including large mammals such as orangutans, zebra, elephants, water buffalo, and rodents. The virus was isolated in Senegal monkeys (Cercopithecus aethiops and Erythrocebuspatas) [25].

ZIKV has been circulating widely over Africa for decades, with exposure to the virus in at least 25 countries. It has been widely studied, and different things are known about the African strain, including the observation that many more *Aedes* spread it (at least 17 species,) including *Ae. africanus, Ae. aegypti, Ae. albopictus, Ae. apicoargenteus, Ae. luteocephalus, Ae. vitreous, Ae. taylori, Ae. dalzieli, Ae. hirsutus, Ae. metallicus, Ae. unilinaetus, Ae. opok,* and *Ae. furcifer*. This was verified by reverse transcriptase (RT) polymerase chain reaction (PCR) assays [26].

4.6 Strains

There are two main lineages of the virus, African and Asian. Recently, the UTMB team found that only the Asian lineage has been linked with microcephaly. So, what is it about this form of the virus that inflicts such damage? The researchers established a method of investigating how ZIKV alters brain stem cells' production, survival, and maturation using cells donated from three human fetal brains. They focused on the impact of the Asian lineage Zika virus in the first outbreak in North America in late 2015. "We discovered that the Asian lineage Zika virus halted the proliferation of brain stem cells and hindered their ability to develop into brain nerve cells," said Ping Wu, Senior Author of the study and UTMB Professor in the Department of Neuroscience and Cell Biology [27].

On the other hand, others suggest this outbreak is due to one of the African strains, and while involving thousands of cases, they were not linked to neurological disorders [28]. Professor Paul Hunter, an expert in health protection at the University of East Anglia, told *The Independent* that the disease would "almost certainly" now spread to continental Africa in what he predicted could become a major disaster [29].

Others are not so certain. "Given the numerous other viruses related to Zika in Africa—West Nile, Yellow Fever, Wesselsbron, and others—the reinfection in Africa may be as mild as previously reported in the 1940 and 1950s," Academy of Science of Nigeria, Professor Oyewale Tomori said. "Note that Zika is not the only virus originally isolated in Africa, causing mild infection, and had gone on to cause more severe infection in other parts of the world" [30].

Comparing seven ZIKV strains representing the current breadth of viral genetic diversity worldwide, a study by Aubry et al., provided clear experimental evidence that recent African strains are more transmissible and potentially more pathogenic than Asian strains. In their experiments, ZIKV strains of the African lineage were more infectious and were transmitted faster by wild-type *Ae. aegypti* mosquitoes from Colombia, translating into a higher epidemic potential in outbreak simulations. In addition, ZIKV strains of the African lineage were more pathogenic to immunocompromised adult mice. They caused resorption and embryonic death in immunocompetent mouse embryos infected in utero by intraplacental injection [31].

Primarily, the strain implicated in the pandemic in 2016–2017 was Asian and not African. Genetic investigations revealed a distinct strain of ZIKV circulating across much of Asia since 1951. While at the same time, two different genotypes of ZIKV are recognized as the Nigerian cluster and the East African or MR766 prototype cluster.

The African lineage is ancestral to the Asian lineage. Viruses of both lineages vary by less than 4% in their amino acid sequence. A third cluster, the Asian one, seems to have been responsible for the Brazilian outbreak and the pandemic in Latin America [32]. And in contrast to contemporary Asian virus strains, the African ZIKV strains have not been associated with fetal birth defect-like symptoms [33].

The African and Asian lineages are 89% genetically identical, but research at the cellular level has shown that the strains behave differently. For example, compared with the African strain, the Asian lineage viruses significantly influenced the activity of genes involved in DNA replication and repair in neural cells. It also increased the production of tumor suppressor protein p53 by 80%, whereas the African lineage increased production by only 4% [19].

Nutt and Adams and others are less sanguine by this information. According to them, at least 30 laboratory-based studies have suggested that African strains of ZIKV can cause the same, or worse, damage to cells in the central nervous system and reproductive and immune systems as the so-called Asian lineage strains circulating in the Americas. One recent study by virologists at the French National Institute of Health and Medical Research showed that a strain of Zika from the Central African Republic is more than twice as deadly to human neural stem cells as the variant circulating in Latin America [34].

Recent research [31] challenges the profile of the African strains. Surprisingly, a growing body of experimental evidence, both in vitro and in vivo, points toward higher transmissibility and pathogenicity of the African ZIKV strains compared to their Asian counterparts. African ZIKV strains typically cause more productive and lethal infections than Asian strains in cell culture. They are more transmissible by mosquitoes and are associated with more severe pathology in adult mice and mouse embryos [31].

In February 2017, two studies suggested another potential explanation for the lack of observed congenital disabilities associated with Zika in Africa. A research group at the University of Missouri, USA, reported that human trophoblasts making up the early placenta are killed much more rapidly by a Ugandan Zika isolate than by a Cambodian one. "This virus is so destructive of placental cells that pregnancy loss may be occurring very early on," said Coauthor and Obstetrician Danny Schust. "If so, women may not even realize they were pregnant" [35].

There has not been a microcephaly trail seen in Africa. One of the most robust responses is reporting capacity among developed countries versus inferior ones. So, while Zika is believed to have infected 1.5 million in South America, the biggest and only recorded outbreak in Africa was in Gabon in 2007, which infected an estimated 20,000 people. Statistically, that's too low to detect rare complications like microcephaly, says epidemiologist Michael Edelstein [36].

Still, another recent report warns this ZIKV might be problematic in Africa. Seven thousand cases of the strain, probably the Asian theme, the same one that has affected Latin and South America, have been reported in the Republic of Cabo Verde. There have been three cases of microcephaly reported from 170 infected pregnant women. The Johnston article claims this is from an Asian strain that African immunity may not prevent [29].

Also, Guinea-Bissau reported infections during the Latin American outbreak [23]. According to Guinea-Bissau's Ministry of Health, from April to June 2016, six infants in the small West African nation were born with microcephaly after possible exposure to Zika. Clusters of suspected infection in adults were also reported across the country [34].

4.7 Immunity

According to many researchers, Africans may have developed immunity to the ZIKV due to their exposure to the many related viral infections [37]. Rosemary Sang, Head of arbovirology and viral hemorrhagic fevers unit at the Kenya Medical Research Institute, believes that "The risk of disease in Africa is much lower than that in the Americas because of the *Ae. aegypti* mosquitoes have been here for a long time, and Africans may have developed immunity to the ZIKV due to their exposure to the many related viral infections" [37].

According to Albert Ko from the Yale School of Public Health, "Zika came in like a bulldozer, and many people in the Americas who coexist with *Aedes* mosquitoes, which transmit ZIKV, were infected. Now that "there are so many people who've already been exposed to the virus and are presumably immune, it kind of protects indirectly the people who haven't been infected. So that's [what] probably happened" [38].

The WHO is much less optimistic about immunity. As the virus has been detected in parts of Asia and Africa for several decades, some level of endemicity is assumed. However, no one knows whether the presence of the virus over time has resulted in widespread or low-level immune protection or possibly no protection at all [39].

Immunity is still not completely understood. The general understanding of Zika's close relative to dengue is that once someone has had an infection with one type of dengue virus, that person is protected from other diseases. But a 2016 study found that reinfections are possible. So ZIKV immunity might wane over time, perhaps leading to reinfections, said pediatrician and microbiologist Hotez of the Baylor College of Medicine in Houston.

Christian Lindmeier, the World Health Organization's spokesperson on the ZIKV, says that "there is no hard evidence to conclude that East Africans are immune to the virus, per Newsweek… [though] it's possible that parts of Africa or other parts [of the world] might already be quite immunized and therefore would not experience the same sort of outbreak," he said [40].

4.8 Underreporting Again

Some researchers believe that ZIKV outbreaks in Africa may be underreported as the clinical disease symptoms of ZIKV often resemble those of the dengue virus and the chikungunya virus, and precise diagnostic tools are not readily available in rural settings [33]. Moreover, the ZIKV might have been causing congenital disabilities, and congenital disabilities have gone unnoticed. Shaun Morris, a clinician-scientist in the Division of Infectious Diseases at the Hospital for Sick Children in Toronto, found a surprising amount of evidence of the virus's presence on the continent, buried in tables and at the bottom of charts in decades-old papers about other diseases, such as yellow fever or dengue, or in broad studies of mosquito-borne viruses. "Almost all of the papers were studies that incidentally picked up the ZIKV, as opposed to looking specifically at the ZIKV," Morris explained.

It all depends on what is being looked for. Epidemiologists call this the "streetlight effect": the tendency to search for something only where it is easiest to see. When studying infectious diseases, parts of sub-Saharan Africa are poorly illuminated: Doctors are scarce, disease surveillance is weak, and laboratory capacity is severely limited [41]. In addition, Belgian doctors suggested that more than 11% of fevers of unknown origin reported in Cameroon might be traceable to ZIKV.

WHO Regional Director for Europe, Zsuzsanna Jakab, fears Zika might spread to Europe since the mosquitoes in question thrive on Madeira's island and the Black Sea's northeastern coast [29]. For example, the virus has appeared among several patients in Northern Ireland. All of these instances seem to be travel related [42].

The *Ae. albopictus* mosquito species are established in many parts of the EU, primarily around the Mediterranean. On the other hand, the risk of transmission of ZIKV infection is shallow in the EU during the winter as the climatic conditions are not suitable for the activity of the *Ae. albopictus* mosquito [43].

4.9 ZIKA Falls off the Charts

So, what happened to the ZIKV? Intensive mosquito control efforts probably may have had an impact. Travel warnings and personal precautions such as bug spray also might have helped.

Brazil had more than 216,000 probable cases in 2016; as of early September, the new cases for 2017 were around 15,500. Colombia tallied more than 106,000 suspected and confirmed cases from 2015 to 2016. In 2017, recent cases plummeted, with approximately 1700 by mid-October. Mexico went from about 8500 confirmed cases in 2015 and 2016 combined to around 1800 by early October of this year.

The numbers have also dropped in the United States and its territories. In 2016, Puerto Rico, the U.S. Virgin Islands, and American Samoa saw more than 36,000 cases of locally transmitted Zika virus. By 2017, the number had dropped to 665. In Puerto Rico, only 65 cases of local Zika transmission were reported in 2018 through early July, according to the CDC. There have also been two cases in the U.S. Virgin Islands. In the continental U.S. and the other territories, the number of local issues is zero—though there have been 28 cases where Zika exposure occurred during travel [44].

In 2017, the continental U.S. saw only seven cases of local mosquito-borne Zika, down from 224 the previous year [44]. In the 50 states, the CDC counted about 5100 points in 2016. Most were travelers who had been to places where Zika was active, although 224 were locally acquired in Florida and Texas. So far in 2017, only about 300 cases had been reported as of mid-October, mostly from travelers. Local transmission seems to have come to a standstill, with one suspected case in Texas and one case confirmed in Florida [45].

Most experts say the sharp decline in Zika cases was due, at least in part, to herd immunity. When enough people become immune to a virus through vaccination or natural immunity, that disease can't easily travel from person to person, ZIKV may behave like chickenpox. Once the virus is contracted, it won't be contracted again. That means the virus "reservoir"—or humans available to carry the virus—dries up after many people have already been infected [44]. ZIKV infected many people, who are now presumably immune, and those exposed provide indirect protection to people who haven't yet encountered ZIKV. If the mosquito-borne virus can't easily spread, it can't find enough people to infect [46]. A high enough percentage of people were infected across Central and South America that a "herd immunity" developed, making it hard for the virus to continue spreading [47].

Scientists say the virus' swift march through the Americas in 2015 and 2016 provides the best explanation for the drop-off since the blitz produced a robust immune response. Also, 4-in-5 people who contract the virus don't show symptoms, so they probably didn't even know they had it. The CDC said it wasn't surprising to see a downturn in 2017, particularly in Puerto Rico, since a swath of the population is no longer susceptible to the disease [48].

Leading up to the 2016 summer Olympics in Brazil, the ZIKV ravaged Latin America, and photos and videos of patients haunted the news worldwide. A year later, the ZIKV has fallen out of the conversation in the U. S., but not so for the rest of the world [49].

4.10 Is ZIKV Gone?

Several locations may remain susceptible to yearly ZIKV outbreaks, and new regions of the world will likely experience epidemics similar to the one described in South America in 2016 [50]. Christine Kreuder Johnson and researchers from the University of California-Davis who studied around 10,400 bird species and 5400 mammals, ZIKV and yellow fever predict future outbreaks in Southeast Asia, particularly Indonesia and Malaysia, with other hot spots in Europe and Africa [51].

It will probably stick around if the ZIKV behaves like other arboviruses, such as chikungunya and dengue. Arbovirus diseases tend to be cyclical, says Public Health Researcher Ernesto Marques of Pittsburgh. "You have big booms, then they drop. Then a few years later, they come back again," he said.

Experts say despite Zika's slowdown, the disease is relatively new to the Western Hemisphere, so it's difficult to predict what will happen in the coming years. Mosquito-borne diseases such as dengue have been known to go quiet and flare up again, especially in the summer and fall months [48].

"In general, cases of Zika have decreased in most of Central and South America, but the virus is not gone. The mosquitoes carrying ZIKV, and other diseases are still there. The risk for another infection outbreak is still quite prevalent," says Elitza Theel, Director of the Infectious Diseases Serology Laboratory and Codirector of the Vector-Borne Diseases Service Line at Mayo Clinic [49]. Keep in mind that there is no evidence that a Zika outbreak will not return. According to a medical entomologist working at the U.S. Army Medical Research Institute, ZIKV can come back any day [52]. "We're getting a reprieve from ZIKV right now, which is very welcome," Dr. Adams Waldorf, OB-GYN specializing in pregnancy infections and global health at the University of Washington School of Medicine, said. "The problem hasn't gone away in the Americas. It's just, at the moment, a bit underground." ZIKV would need to mutate dramatically to mount a resurgence like the flu virus every year. This scenario is unlikely. ZIKV could also return if it finds new populations that lack herd immunity. Adams Waldorf said that the virus is likely still spreading in communities that have not been exposed to ZIKV yet and lack herd immunity [44].

It's also possible that Zika will find an animal host in the Western Hemisphere, providing the virus a waystation of sorts until the human population is more susceptible again. For example, researchers have detected Zika in capuchin monkeys and common marmosets near humans in Brazil.

All the unknowns make it hard to predict when Zika will re-emerge. There may be epidemics here and there. Then, years later, it pops up "in a place, in a time you can't predict," David Morens, Senior Scientific Adviser to Director of the National Institute of Allergy Infectious Diseases, said. "The ZIKV will be around indefinitely" [53].

Few scientists are naive enough to think they've seen the last of Zika. "The clock is ticking for when we will see another outbreak," says Andrew Haddow, a medical entomologist at the U. S. Army Medical Research Institute of Infectious Diseases in Frederick, Md [54].

Will Zika make a comeback? said, "Dr. Roberta DeBiasi, chief of Pediatric Infectious Diseases and co-director of the Congenital Zika Program at Children's National Health System," said. The question is when. She looks to other flaviviruses for examples. "West Nile sort of disappeared for a few years and then came back with a vengeance six years after that—and then went away for a little bit and returned," DeBiasi said. If Zika behaves more like dengue, the infections will increase seasonally whenever mosquitoes are around. A third option could be a remarkably high number of cases yearly, like Japanese encephalitis. "We will be living with the threat of Zika virus for pregnant women, fetuses, and even small children for a long time," Dr. Adams Waldorf said [44].

The future of Zika could look like the pattern of mosquito-borne West Nile virus, which hit highs of 168 cases in Missouri in 2002 and 9862 issues in the U.S. in 2003 and hasn't reached those numbers since. Or it could be more like the dengue virus, with four subtypes reliably infecting more than one million each year in the Southern Hemisphere [47].

Societal factors, such as poverty and gender, which ultimately shape the emergence and development of infectious diseases, were ignored. The politics of Zika was left out of the discussion [55]. Even though WHO may have downgraded the risks associated with Zika infection when listing viruses in their second annual review for 2018, there was an urgent need for accelerated research and development. Zika was on the shortlist [56].

One of the reasons re-emergence is not wholly unlikely has to do with the possibility of sexual transmission. It significantly increases the ZIKV's capacity to spread, Durigon added, stressing that in his view, medical culture must change, as the profession still focuses on antenatal care for women. "It's no use testing only pregnant women to see if the virus is present and advising only women to use insect repellent and avoid high-risk areas during pregnancy while leaving men to go on with their lives as normal," he said. "Women could be infected by their partners. Doctors aren't paying attention to this possibility" [57].

Another reason may be that the Zika virus mutated so fast in Brazilian patients that different pathogen serotypes could appear shortly, as is already the case with the dengue virus. "Today, there's only one Zika, and people become immune after being infected once," says Edison Luiz Durigon, Professor at the University of São Paulo's Biomedical Science Institute (ICB-USP), "But the virus is constantly mutating, and I wouldn't be surprised if we see Zika 2, 3 and 4 emerging before long…." "What we're seeing is just the tip of the iceberg. We don't know what's underneath," he said [58].

In the summer of 2016 in Miami, it happened where federal health officials issued their first-ever medical travel warning in the U.S. The Zika outbreak in Florida infected nearly 300 people in 2016. Just two people were infected in Florida in 2017.

State health officials shut down their Zika information hotline in January 2018. The WHO counted 12 outbreaks worldwide in 2015, 22 in 2016, and just one in early 2017—three cases in India. The health agency hasn't issued a Zika situation report since March 2017.

The U.S. CDC deactivated its emergency response system for Zika, launched in January 2016 [47].

4.11 ZIKA in the 2000s

Although transmission of ZIKV has declined in the Americas, outbreaks and infection clusters continue to occur in some other regions, such as India and Southeast Asia, where there are large populations of women of childbearing age susceptible to the virus [1].

As of April 2022, there are no outbreaks of Zika worldwide, although a significant outbreak occurred in India in November of 2021. Health experts have warned that a new outbreak could happen at any time, requiring only a single mutation to generate a new variant of the virus.

4.12 India

The Indian subcontinent first reported the presence of Zika in 1952. Musso and Gubler said antibodies to Zika were found in 17% of 196 Indian subjects tested [59]. India has been upgraded to a country where there is "ongoing transmission that is no longer in the new or re-introduction phase but where there is no evidence of interruption," from one which only has the vector with no current transmission to the WHO Zika database, triggering the need for a thorough investigation as most cases remain asymptomatic, the World Health Organization (WHO) warned [60].

Unfortunately, the ZIKV has recently resurged in India with the potential for devastating consequences. Therefore, rapid detection strategies and efficient medicines for the appropriate prevention and efficient control of ZIKV need to be urgently developed [61] and maintained.

An early outbreak was in Tamil Nadu's Krishnagiri district in July 2017 [62], and between September and November 2018, with Bhopal reporting 54, Vidisha 50, Hoshangabad 2, Sehore 21, Sagar 2, and Raisen 1. Two laboratory-confirmed Zika deaths were reported from Madhya Pradesh. In both these cases, death was due to concurrent comorbid conditions (septic shock with dengue encephalitis in one point and multiorgan failure due to septic shock in the other), according to information provided in Rajya Sabha. India's National Institute of Virology (NIV) analyzed the Zika strain from a patient in Gujarat and was close to a Malaysian Zika strain [63].

Mainly, India experienced outbreaks of ZIKV disease in two states: Rajasthan (September–October 2018) and Madhya Pradesh (October–November 2018). The

2018 Rajasthan and Madhya Pradesh outbreaks had 159 and 127 ZIKV positive cases, with 63 and 42 infected pregnant women detected, respectively [64].

The first case surfaced on September 22 when an 85-year-old woman with no travel history tested positive for the disease [65]. These outbreaks led the Rajasthan's Health Department to issue an advisory for pregnant women outside Shastri Nagar not to visit the area. Dr. Neena Valecha, the NIMR director, said the virus appears to be locally transmitted [62].

Complete genome phylogenetic analysis of Jaipur city ZIKV sequences with the known GenBank ZIKV sequences revealed that the outbreak in Jaipur city was caused by the ZIKV belonging to Asian lineage [66]. This was the lineage that reaped havoc in Latin America in 2015–2016.

Recently, reports out of India suggest an outbreak or even an epidemic there. As per the WHO classification scheme on the prevalence of this virus, in 2018, India was in category 2, indicating ongoing transmission of this virus [67].

India detected its first case of ZIKV in November 2016, a postpartum female before she was discharged from the hospital [64]. In February 2017, two additional cases of Zika were confirmed from a densely populated area (Bapunagar) in the city of Ahmedabad in Western Gujarat state. In July 2017, a fourth case was reported from Tamil Nadu [68].

The first four proven cases of the ZIKV from India were reported in 2017. Major outbreaks followed these in Rajasthan and Madhya Pradesh the next year. These outbreaks in India highlighted this disease's spread beyond geographical barriers due to the growing globalization, increased travel, and ubiquitous presence of its vector, the Aedes mosquito [69].

The India Institute of Medical Sciences (AIIMS)-Bhopal reported the prevalence of the Uganda strain of recombinant Zika virus in Madhya Pradesh. The state reported 159 Zika cases since first detected in September. The sequencing took place at AIIMS-Bhopal laboratory, and the report is being submitted to the government, said Director Dr. Sarman Singh. He added that the fully sequenced MP's virus genome is Uganda's strain of Zika, which is not a severe strain [70].

The first case in 2021 of the ZIKV was reported in Kerala on July 9. All state districts have sounded a high alert about the mosquito-borne virus. Kerala has reported 28 cases of Zika virus cases so far, recent data provided by the State Health Department showed. According to top experts, this is a virus of "grave concern" [71]. As of July 27, 2021, the totality of confirmed cases of the current outbreak of Zika infection is 56, all of which are from the state of Kerala in India.

The ZIKV started circulating in Kanpur in September or early October. In October, the number of Zika cases in Kanpur climbed to 123. Infected patients were initially restricted to a 1–2 km radius in colonies like Pokharpur, Adarsh Nagar, Tiwaripur, and Harjinder Nagar and took the infection to their neighborhoods. In Pokharpur, 15 cases have been found. In Adarsh Nagar, nine have been located [72].

At least 89 people, including 17 children, have tested positive for the ZIKV in a surge of cases in the Indian city of Kanpur, its health department said on Monday "There has been a surge in cases of the ZIKV, and the health department has formed several teams to contain the spread," Dr. Nepal Singh, Chief Medical Officer of

Kanpur district in India's most populous state of Uttar Pradesh, told Reuters. The first Zika case in the industrial city of Kanpur was detected on October 23, and the number of cases has increased over the past week. "People are testing positive because we are doing very aggressive contact tracing," said Prasad.

Six pregnant women are among those infected. Kanpur, the epicenter of the outbreak officials, says a massive effort is underway to contain the spread. Four thousand one hundred forty-two people had been randomly tested. Over 2.6% had tested positive. Kanpur has more than 18,000 people living per sq. km, over 700 times the population density in Brazil, where Zika became an epidemic in 2015.

The maximum number of cases recorded in Harjinder Nagar was twenty-seven, and in Tiwaripur, it was fifteen. And they are now being detected in areas beyond 4 km of the city Lucknow, 94 km from Kanpur, and Kannauj, 84 km away, taking the total number of cases in the state to 127.

India has 24 million pregnancies each year. In India, no more than 10% of the population has access to medical technology [73]. The environment in India is ripe for Zika because of factors like the preponderance of the *Ae. aegypti* mosquito is the type that carries the virus. When mosquitoes swarmed in the post-monsoon months, particularly in slum areas with standing water and accumulating trash, it could produce many more infections [59]. Zika virus is known to be circulating in the Southeast Asia Region of India [74]. The Indian government reported three laboratory-confirmed cases of Zika virus disease in the Ahmedabad district of Gujarat in May 2017. Although the first case was detected in November last year, the Ministry of Health and Family Welfare notified the United Nations (UN) health agency on May 15, 2017.

Following the WHO disclosure about the first Zika cases in India, the health ministry has issued a statement—its first since the investigation into suspected cases began "during the post-monsoon period" in 2016. This statement was selective in providing information. It only says that the first case was "further confirmed" on January 3, 2017, without giving the date when it tested positive. The statement says two more cases were confirmed subsequently, without revealing the confirmation dates. Then the ministry said the presence of Zika in Gujarat was reported as part of an answer in parliament on March 17. The WHO was notified of the case on May 15.

The above sequence clarifies that there was no attempt to share information about Zika with the public at any stage [75]. A report in The Hindu states that the "Vibrant Gujarat Summit was a factor in the decision to keep the ZIKV incidence under wraps" [60]. "The rationale was, if you announce ZIKV, there might be a bit of panic," Dr. Soumya Swaminathan, Director-General of the Indian Council of Medical Research, a part of the health ministry, said in a telephone interview. "So, then there was a decision that there was probably no need for a public announcement" [76].

Manmohan Singh said that authorities were increasing their surveillance of the outbreak and eliminating breeding grounds for the mosquitoes that transmit the virus [77].

The 2018 outbreak is considerably more significant, and scientists found mosquitoes infected with the virus for the first time, indicating that it was being transmitted locally. The outbreak comes at an inopportune time for the tourism industry in Rajasthan, which is gearing up for the start of the high season.

Every winter, millions of tourists from other parts of India worldwide travel to Rajasthan, especially its capital, Jaipur, the focus of the outbreak. The city is one of the stops on India's famed "golden triangle" of tourist destinations, including Agra, home of the Taj Mahal, and Delhi, the nation's capital [78].

One hundred fifty-seven laboratory-confirmed cases of ZIKV, including 63 pregnant women, were reported from Rajasthan, India, in 2018 [64].

According to Grubaugh et al., making progress requires a global effort to do large-scale epidemiological investigations and access to historical data to determine the baseline rates of microcephaly and other congenital disabilities currently unavailable in India. Until new evidence suggests otherwise, all ZIKV strains should be considered to have the potential to cause congenital disabilities. Many region-specific factors need to be considered when evaluating the population risks [79].

India has woken up late to Wolbachia as a part of its antidengue/chikungunya arsenal (see Chap. 12), but the country is moving in that direction. The Vector Control Research Center at Puducherry developed a Wolbachia-infected variant for India with Monash University. And the Indian Council of Medical Research (ICMR) is getting ready to conduct field testing of Wolbachia-infected mosquitoes [80]. However, the same reports suggest that transgenic male mosquito solutions like Oxitec were also on the table. While the article addressed dengue and chikungunya, the ZIKV mosquito is the same species.

4.13 Other At-Risk Populations

In an April 2016 eLife study, scientists mapped out the worldwide risk of ZIKV, estimating that some 2.17 billion people may be in the virus's path [81]. The affected areas span a generous portion of the Southeastern U.S., including much of Texas and Florida.

The researchers estimate that more than 5.4 million births will occur within the next year in parts of the Americas susceptible to ZIKV transmission. This is a cause for concern given the established link between ZIKV and microcephaly.

In addition, the researchers found that large parts of India and sub-Saharan Africa—areas populated by 1.42 billion and 453 million people, respectively—are at risk of Zika transmission. The map does not account for non-mosquito forms of viral communication (such as sexual transmission). The researchers called for risk estimates to be updated as the latest information became available [82].

Considering the widening distribution of the Aedes mosquito in the Americas and the high mobility of people transiting in and out of the region, the further spread of ZIKV across the Americas represents a clear and present danger [83]. CDC's

Director, Tom Frieden, added his dismay, telling reporters during a press conference in May: "Three months in an epidemic is an eternity" [84].

4.14 Is ZIKA Still With Us?

While it has primarily faded from mainstream news, Zika is still active in many parts of the world. Even if the mosquito apocalypse that many feared never actually happened, the risk still largely remains for pregnant women and hopeful-to-conceive parents.

Cases declined dramatically in 2017 and virtually disappeared soon after. In the United States, experts believe the epidemic died as people developed greater immunity to the virus. While the virus may remain dormant for years, outbreaks could still occur if the virus mutates or if a large number of young people lack immunity. Ultimately, with most epidemics, "there's never a hard end," said former CDC Official Denise Jamieson [85].

The Zika epidemic has shown signs of a significant slowdown, and the risk to residents and visitors is deemed much lower. However, the region continues to struggle and fight against the *Ae. aegypti* mosquito responsible for its transmission. "Even though the number of cases of Zika has significantly declined from the outbreak of 2016, there is still a need for continued vigilance and action on mosquito-borne diseases, which pose a health security threat, a tourism threat, and an economic threat," stated Dr. C. James Hospedales, Executive Director of the Caribbean Public Health Agency (CARPHA) [86].

The threat of Zika has not diminished: The *Ae. aegypti* mosquito infects around two million Brazilians annually and 300 million globally. Work in life sciences to map global environmental sustainability for the Zika virus suggests over two billion individuals currently inhabit areas with suitable environmental conditions [87]. The Brazilian Ministry of Health has reported continuing transmission of Zika virus across the country, with 6483 cases recorded from January 3, 2021, to January 1, 2022 [88].

But since 2018, no local Zika transmission has been reported in the continental United States; less than 100 travelers have acquired the disease. In 2021, only one traveler contracted Zika, and there have been no confirmed cases of Zika using molecular testing from the U.S. territories.

According to WHO, the ZIKV is present in more than 87 countries. As of 2020, virus activity continues in the Caribbean, most of Latin America, Central Africa, India, Indonesia, Malaysia, Cambodia, and Papua New Guinea, among other places.

The *Ae. aegypti* mosquito is the world's most efficient disease vector, causing more than 50 million infections yearly. "Any virus that can efficiently infect Ae. aegypti also have potential access to billions of humans," wrote Dr. Anthony Fauci [89].

Scientists think the ZIKV is here to stay on the American continent (although, most of the time, fewer people will be infected than during those initial large outbreaks, as more folks develop immunity). Based on the CDC's tracking of infections, the agency expects most Zika-affected pregnancies to have occurred throughout 2017. The CDC plans to work with local health authorities to identify affected families if it has the funding [90].

Nonetheless, there remains an abundance of concern. "About 40% of the global population is at risk [from] this species," Andrew McKemey, Entomologist and Head of Oxitec Field operations, said [91]. According to Thomas Monath, CEO of the infectious-disease division at NewLink Genetics Corp, hundreds of millions are at risk [92].

There remain some potential hotspots. For example, the annual Hajj, the Muslim pilgrimage to Mecca that brings millions of people to Saudi Arabia, has also been postulated as having helped introduce dengue, as it could cause Zika virus infection in 2016 or 2017, and the political destabilization in Venezuela could become a significant contributing factor in the next round of epidemics [93].

4.15 Conclusion

As globalization and climate change increasingly become an aspect of daily life, so are too many new, exotic, and re-emerging vectors and pathogens worldwide.

According to the UN IPCC [94], more than half the global population lives on land that is less than 60 km from a seashore. The population density in coastal areas is expected to increase from 87 persons per km^2 in 2000 to 134 persons per km^2 in 2050. This trend is likely particularly pronounced in tropical developing countries where many mosquito-borne diseases are endemic. Therefore, growing numbers of people will be placed at risk by an increase in mosquito vector populations in coastal zones [95].

Many inhabitants of the outlying areas of large cities in Latin America are marginalized. Some neighborhoods lack electricity, running water, garbage collection, paved streets, sewers, and drains for rain. With this vulnerability, low-income people living in urban areas are more susceptible to suffering a more intense exposure to *Ae. aegypti* [15].

Most efforts to prevent the spread of new diseases tend to focus on vaccine development, early diagnosis, and containment, but that's like treating the symptoms without addressing the underlying cause, says Peter Daszak, Zoologist at the non-governmental organization EcoHealth [96] Alliance in New York, who chaired the IPBES workshop.

Cumulative population-level immunity owing to naturally acquired infection appears to have driven ZIKV to extinction in many regions. However, spatial heterogeneity in ZIKV infection rates during the pandemic may have created pockets of susceptible populations that can sustain transmission in the future [1].

Will Zika make a comeback?

Definitely, DeBiasi said. The question is when. She looks to other flaviviruses for examples. "West Nile sort of disappeared for a few years and then came back with a vengeance six years after that—and then went away for a little bit and came back again," Roberta DeBiasi, Chief of Pediatric Infectious Diseases and Codirector of the Congenital Zika Program at Children's National Health System, said.

If Zika behaves more like dengue, the infections will increase seasonally whenever mosquitoes are around. A third option could be a remarkably high number of cases yearly, like Japanese encephalitis [44].

Still, "we are not finished with Zika," said Dr. Anthony Fauci, Director of the National Institute of Allergy and Infectious Diseases. "Even though it's dramatically down when you look at the number of infections, it doesn't mean they will stay down," Fauci said. "You've got to be careful when dealing with vector-borne diseases. They tend to cycle in and out. It very well could come back" [47].

Although a decline in cases of ZIKV infection has been reported since 2016 in some countries or some parts of countries, vigilance needs to remain high. Seventy-five countries and territories have reported evidence of mosquito-borne ZIKV virus transmission since 2007, 69 with reports from 2015 onward, of which 58 reported an outbreak from 2015 onward [97].

Despite low transmission rates in the past six years, the emergence of new Zika outbreaks remains a looming threat. Of particular concern to health authorities worldwide is the possibility of multiple flavivirus outbreaks, as Zika became endemic in the same regions as dengue viruses [98]. The ZIKV pandemic has waned, but the virus still poses a public health threat, as shown by continued reports of outbreaks in Asia, India, and Africa. Although there are not currently the tools to predict where and when the next large epidemic will happen, the large numbers of susceptible persons residing in Aedes-infested regions make a re-emergence of ZIKV likely [1].

References

1. Musso D, Ko AI, Baud D (2019) Zika virus infection—after the pandemic. N Engl J Med 381:1444–1457. https://doi.org/10.1056/NEJMra1808246
2. McNeil J, Donald G (2017) Houston braces for another brush with the peril of Zika. The New York Times. July 17. https://www.nytimes.com/2017/07/17/health/zika-virus-houston-texas.html. Accessed 21 Jul 2017
3. Hotez P (2016) Will Zika return to the 'Old World'? Microbes and Infection 18:527–528. http://www.sciencedirect.com/science/article/pii/S1286457916300648. Accessed 27 Sept 2016
4. Musso D, Gubler DJ (2016) Zika virus. Clin Microbiol Rev 29. May 9. http://cmr.asm.org/content/29/3/487.abstract. Accessed 18 July 2017
5. The Science Times (2020) Genetically engineered male mosquitos to be released in florida and other parts of US to Curb Zika and dengue Spread. The Science Times. May 6. https://www.theguardian.com/environment/2020/jun/17/genetically-modified-mosquitoes-florida-texas. Accessed 13 May 2022
6. WebMD (2022) What is herd immunity? Herd immunity: what is it and can it end the coronavirus pandemic? (webmd.com). Accessed 25 May 2022
7. Ferguson N et al (2016) Countering the Zika epidemic in Latin America. Science. July 14. http://science.sciencemag.org/content/353/6297/353. Accessed 24 Sept 2016

8. Aliota MT et al (2017) Zika in the Americas, year 2: what have we learned? what gaps remain? a report from the global virus network. Antiviral Res 114. https://www.ncbi.nlm.nih.gov/pubmed/28595824. Accessed 15 May 2019
9. Moore C et al (2016) Characterizing the pattern of anomalies in congenital Zika syndrome for pediatric clinicians. JAMA Pediatrics. 171:3. https://jamanetwork.com/journals/jamapediatrics/fullarticle/2579543?utm_campaign=articlePDF&utm_medium=articlePDFlink&utm_source=articlePDF&utm_content=jamapediatrics.2016.3982. Accessed 24 Jul 2018
10. Kasperson R et al (1988) The social amplification of risk: a conceptual framework. Risk Anal 8(2):177–187
11. Macer D (2013) UNDP/World Bank/WHO. Ethical, legal and social issues of genetically modified disease vectors in public health. Geneva, Switzerland, WHO. http://www.who.int/tdr/publications/tdr-research-publications/seb_topic1/en/. Accessed 8 Aug 2018
12. Watts J (2016) Zika hysteria is way ahead of research into virus, says expert. The Guardian. February 17. https://www.theguardian.com/world/2016/feb/17/zika-hysteria-health-expert-research-leslie-lobel. Accessed 25 Sept 2016
13. Witte K (1992) Putting the fear back into fear appeals: reconciling the literature. Commun Monogr 59:329–349
14. News Service of Florida (2016) Lawmakers call for use of modified mosquitoes in Zika fight. Health News Florida. August 30. http://health.wusf.usf.edu/post/lawmakers-call-use-modified-mosquitoes-zika-fight#stream/0. Accessed 5 Sept 2016
15. Troncoso A (2016) Zika threatens to become a huge worldwide pandemic. Asian Pacific J Tropical Biomedicine. 6:6. http://www.sciencedirect.com/science/article/pii/S2221169116302921. Accessed 4 Jun 2017
16. Baraniuk C (2019) Exclusive: Cuba failed to report thousands of Zika virus cases in 2017. New Scientist. January 8. https://www.newscientist.com/article/2190001-exclusive-cuba-failed-to-report-thousands-of-zika-virus-cases-in-2017/. Accessed 14 May 2019
17. Branswell H (2017) They're just hiding: experts say Puerto Rico may be underreporting Zika-affected births. STAT. April 18. https://www.statnews.com/2017/04/18/zika-virus-puerto-rico-pregnancies/. Accessed 4 May 2017
18. Weaver SC (2017) Emergence of epidemic Zika virus transmission and congenital Zika syndrome: are recently evolved traits to blame? mBio. 8:1. January/February. https://doi.org/10.1128/mBio.02063-16. Accessed 18 Sept 2018
19. Wetman N (2017) The missing pieces: lack of Zika data from Africa complicates search for answer. Nat Med 23(8):904–906
20. Weaver S et al (2016) Zika virus: History, emergence, biology, and prospects for control. Antiviral Res. 130:69–80. https://www.ncbi.nlm.nih.gov/pubmed/26996139. Accessed 15 Oct 2016
21. Bourgarel M, Wauquier N, Gonzalez JP (2010) Emerging viral threats in Gabon: health capacities and response to the risk of emerging zoonotic diseases in Central Africa. Emerg Hum Threats J 3:e7. https://doi.org/10.3134/ehtj.10.163
22. Bogoch II et al (2016) Potential for Zika virus introduction and transmission in resource-limited countries in Africa and the Asia-Pacific region: a modelling study. The Lancet: Infectious Diseases. 16. November. http://www.thelancet.com/journals/laninf/article/PIIS1473-3099(16)30270-5/abstract. Accessed 26 Jun 2017
23. Gramer R (2017) The Zika virus just quietly spread to Southwest Africa. Foreign Policy. January 13. http://foreignpolicy.com/2017/01/13/the-zika-virus-just-quietly-spread-to-west-africa-angola-outbreak/. Accessed 14 May 2017
24. Tanzania Daily News (2016) Africa: Zika virus incidences reach 15.6 percent. Tanzania Daily News. December 16. http://allafrica.com/stories/201612160058.html. Accessed 26 May 2017
25. ECDC (2014) Zika virus infection outbreak, French Polynesia 14 February 2014. Stockholm: European Centre for Disease Prevention and Control. https://www.ecdc.europa.eu/en/publications-data/rapid-risk-assessment-zika-virus-infection-outbreak-french-polynesia. Accessed 10 Jun 2022

26. Charrel R et al (2016) Background review for diagnostic test development for Zika virus infection. Bull World Health Organ 94:574–584D. https://doi.org/10.2471/BLT.16.171207. http://www.who.int/bulletin/volumes/94/8/16-171207.pdf. Accessed 28 Mar 2017
27. UTMB Newsroom (2017) UTMB scientists uncover how Zika virus causes microcephaly. February 16. https://www.utmbhealth.com/support-pages/news. Accessed 9 Jun 2017
28. Broutet N et al (2016) Zika virus as a cause of neurologic disorders. New England J Med 374:16. April 21. http://www.nejm.org/doi/full/https://doi.org/10.1056/NEJMp1602708#t=article. Accessed 24 Sept 2016
29. Johnston I (2016) Zika strain that causes microcephaly found in Africa for the first time, WHO confirms. The Independent. May 20. http://www.independent.co.uk/life-style/health-and-families/health-news/zika-strain-that-causes-microcephaly-found-in-africa-for-the-first-time-who-confirms-a7039641.html. Accessed 15 Sept 2016
30. Wild S (2016) FACTSHEET: Do African countries have to worry about a Zika resurgence? Africa Check. March 10. https://africacheck.org/factsheets/factsheet-do-african-countries-have-to-worry-about-a-zika-resurgence/. Accessed 18 Sept 2016
31. Aubry F et al (2021) Recent African strains of Zika virus display higher transmissibility and fetal pathogenicity than Asian strains. Nature Communications. 12 (916). Recent African strains of Zika virus display higher transmissibility and fetal pathogenicity than Asian strains | Nature Communications. Accessed 23 May 2022
32. Braack L et al (2018) Mosquito-borne arboviruses of African origin: review of key viruses and vectors. Parasites Vectors. 11:29. https://www.ncbi.nlm.nih.gov/pubmed/29316963. Accessed 27 Jun 2018
33. Sheridan MA et al (2018) African and Asian strains of Zika virus differ in their ability to infect and lyse primitive human placental trophoblast. PLoS ONE 13(7):e0200086. https://doi.org/10.1371/journal.pone.0200086
34. Nutt C, Adams P (2017) Zika in Africa—the invisible epidemic? The Lancet. 389:1595–1596
35. Nutt C, Adams P (2017) Zika in Africa—the invisible epidemic? The Lancet. 389. April 22. 1595–1596 and see; Sheridan MA, Balaraman V, Schust DJ, Ezashi T et al (2018) African and Asian strains of Zika virus differ in their ability to infect and lyse primitive human placental trophoblast. PloS ONE, 13(7):e0200086. https://doi.org/10.1371/journal.pone.0200086
36. Baker A (2016) Zika hasn't hurt Africa – and that may be the key to beating it. Time. February 12. http://time.com/4219240/zika-africa-origins-microcephaly-vaccine/. Accessed 18 Sept 2016
37. Ligami C (2016) Scientists say Zika virus unlikely to spread in East Africa. The East African. February 7. http://www.theeastafrican.co.ke/scienceandhealth/Scientists-say-Zika-virus-unlikely-to-spread-in-East-Africa/3073694-3065650-view-printVersion-kvcyxcz/index.html. Accessed 4 Apr 2017
38. Cunningham A (2017) Although the number of Zika cases has fallen, the virus is unlikely to vanish. The Washington Post. November 4. . https://www.washingtonpost.com/national/health-science/although-the-number-of-zika-cases-has-fallen-the-virus-is-unlikely-to-vanish/2017/11/03/dde61900-bfdd-11e7-97d9-bdab5a0ab381_story.html?utm_term=.5914bf5d6746. Accessed 27 Jun 2018
39. WHO (2016) One year into the Zika outbreak: how an obscure disease became a global health emergency. Emergencies. http://www.portal.pmnch.org/emergencies/zika-virus/articles/one-year-outbreak/en/. Accessed 2 Oct 2016
40. Fratti K (2016) Are some people immune to Zika virus? There's no evidence yet. Romper. April 16. https://www.romper.com/p/are-some-people-immune-to-the-zika-virus-theres-no-evidence-yet-9014. Accessed 23 Sept 2016
41. Adams P, Nutt C (2016) A Zika vaccine, but for whom? The New York Times. December 28. https://www.nytimes.com/2016/12/28/opinion/a-zika-vaccine-but-for-whom.html?_r=0. Accessed 10 Jan 2017
42. Dearden L (2016) Zika virus: several patients in Northern Ireland diagnosed with mosquito-borne disease. The Independent. September 14. http://www.independent.co.uk/news/uk/home-news/zika-uk-northern-ireland-latest-diagnosis-virus-people-treated-mosquito-bites-public-health-agency-a7294681.html. Accessed 30 Jun 2017

43. ECDC (2016) Zika virus disease epidemic: Potential association with microcephaly and Guillain-Barré syndrome (first update). Rapid Risk Assessment. January 21. http://ecdc.europa.eu/en/publications/Publications/rapid-risk-assessment-zika-virus-first-update-jan-2016.pdf. Accessed 24 Sept 2016
44. Grinnell A (2018) What happened to Zika? PBS NewsHour. July 6. https://www.pbs.org/newshour/science/what-happened-to-zika. Accessed 24 July 2018
45. Cunningham A (2017) Although the number of Zika cases has fallen, the virus is unlikely to vanish. The Washington Post. November 4. https://www.washingtonpost.com/national/health-science/although-the-number-of-zika-cases-has-fallen-the-virus-is-unlikely-to-vanish/2017/11/03/dde61900-bfdd-11e7-97d9-bdab5a0ab381_story.html?noredirect=on&utm_term=.04ec4c0ba51b. Accessed 28 Jun 2018
46. Cunningham A (2017) As Zika fades from public consciousness, scientists continue to pursue the virus. The Washington Post. December 30. https://www.washingtonpost.com/national/health-science/as-zika-fades-from-public-consciousness-scientists-continue-to-pursue-the-virus/2017/12/29/3fae96dc-e1d3-11e7-bbd0-9dfb2e37492a_story.html?utm_term=.505e6e8a5226. Accessed 27 Jun 2018
47. Bernhard B (2018) What happened to Zika? Missouri stops testing most pregnant women as threat drops. St. Louis Post Dispatch. January 6. http://www.stltoday.com/lifestyles/health-med-fit/health/what-happened-to-zika-missouri-stops-testing-mostpregnant-women/article_9d819106-0c3f-5a37-bad1-53ac3f2789f4.html. Accessed 26 Apr 2018
48. Howell Jr, Tom (2017) Number of Zika virus cases dropped dramatically. Washington Times. December 27. https://www.washingtontimes.com/news/2017/dec/27/number-of-zika-virus-cases-dropped-dramatically/. Accessed 27 Jun 2018
49. University of Notre Dame-Press Release (2018) Fighting mosquito-borne diseases. Press release. https://fightingfor.nd.edu/2017/fighting-mosquito-borne-diseases/. Accessed 10 Jul 2018
50. Magnani DM et al (2017) Neutralizing human monoclonal antibodies prevent Zika virus infection in macaques. Sci Transl Med 9. October 4. https://www.ncbi.nlm.nih.gov/pubmed/28978754. Accessed 27 Jun 2018
51. Blanchard S (2019) Where will the world's next Zika, West Nile or dengue come from? Scientists release a series of 'hot-spot' maps to warn of the regions most likely to harbour killer viruses. The Daily Mail. January 7. https://www.dailymail.co.uk/health/article-6565799/Where-worlds-Zika-West-Nile-dengue-come-from.html. Accessed 13 Jun 2019
52. Kebede E (2017) Zika cases are down, but researchers prepare for the virus's return. Human and Virus. December 13. http://humansandviruses.blogspot.com/2017/12/zika-cases-are-down-but-researchers.html. Accessed 10 Jul 2018
53. Cunningham A (2017) Although the number of Zika cases has fallen, the virus is unlikely to vanish. The Washington Post. November 4. https://www.washingtonpost.com/national/health-science/although-the-number-of-zika-cases-has-fallen-the-virus-is-unlikely-to-vanish/2017/11/03/dde61900-bfdd-11e7-97d9-bdab5a0ab381_story.html?utm_term=.5914bf5d6746. Accessed 27 Jun 2018
54. Cunningham A (2017) As Zika fades from public consciousness, scientists continue to pursue the virus. The Washington Post. December 30. https://www.washingtonpost.com/national/health-science/as-zika-fades-from-public-consciousness-scientists-continue-to-pursue-the-virus/2017/12/29/3fae96dc-e1d3-11e7-bbd0-9dfb2e37492a_story.html?utm_term=.7ff98a23970b. Accessed 27 Jun 2018
55. Ribeiro B, Hartley S (2018) Why Brazil's Zika virus requires a political treatment. The Conversation. February 16. Why Brazil's Zika virus requires a political treatment. Accessed 27 Jun 2018
56. WHO (2018) List of blueprint priority diseases. February. http://www.who.int/blueprint/priority-diseases/en/, Accessed 27 Jun 2018
57. Toledo K, Fundação de Amparo à Pesquisa do Estado de São Paulo (2017) New Zika serotypes may emerge, researcher warns. Science Daily. October 10. https://www.sciencedaily.com/releases/2017/10/171010200120.htm. Accessed 2 Jun 2018

58. Toledo K, Fundação de Amparo à Pesquisa do Estado de São Paulo (2017) New Zika serotypes may emerge, researcher warns. Science Daily. October 10. https://www.sciencedaily.com/releases/2017/10/171010200120.htm. Accessed June '=
59. Najar N (2017) India acknowledges three cases of Zika virus. The New York Times. June 3. https://www.nytimes.com/2017/06/03/world/asia/india-zika-virus.html. Accessed 27 Jun 2017
60. Mitra S (2017) Zika virus cases detected in India: thorough investigation, educating masses is need of the hour. FIRSTPOST. June 6. http://www.firstpost.com/india/zika-virus-cases-detected-in-india-thorough-investigation-educating-masses-is-need-of-the-hour-3522371.html. Accessed 26 Jun 2017
61. Zhang X, Li G, Chen G, Zhu N et al (2020) Recent progresses and remaining challenges for the detection of Zika virus. Med Res Rev Jan 41:2039–2108
62. Press Trust of India. (2018). 18 New Cases of Zika Virus Detected In Rajasthan, Taking Total To 50, NDTV. October 12. https://www.ndtv.com/india-news/10-new-cases-of-zika-virus-detected-in-rajasthan-taking-total-to-42-1931250. Accessed May 21, 2022.
63. Ayub J (2019) African strain of Zika virus in Madhya Pradesh, 159 cases since September. Times of India. January 5. https://timesofindia.indiatimes.com/city/bhopal/african-strain-of-zika-virus-in-madhya-pradesh-159-cases-since-september/articleshow/67396613.cms. Accessed 11 Jun 2019
64. Gupta N, Yadav P, Patil D, Sapkal G (2020) Preparedness of public health-care system for Zika virus outbreak: an Indian perspective. J Infect Public Health 13:949–955
65. Press Trust of India (2018) Zika virus cases in Rajasthan rise to 55. The Economic Times. October 13. https://economictimes.indiatimes.com/news/politics-and-nation/zika-virus-cases-in-rajasthan-rise-to-55/articleshow/66194419.cms?from=mdr. Accessed 21 May 2022
66. Saxena SK, Kumar S, Sharma R, Maurya VK et al (2019) Zika virus disease in India—update October 2018. Travel Med Infect Dis 27:121–122
67. Barman P (2018) Study reveals the spread of Zika virus in Asia. Research Matters. October 4. https://researchmatters.in/news/study-reveals-spread-zika-virus-asia. Accessed 21 May 2022
68. Ishtiaq F (2018) A call to introduce structured Zika Surveillance in India. Trends Parasitol February 34(2):92–95
69. Rackimuthu S et al (2022) Zika virus amid COVID-19 in India: a rising concern. Int J Health Plan Manage 37:556–560
70. Abramson D, Pitch-Loeb R (2016) U.S. public's perception of Zika risk: awareness, knowledge, and receptivity to public health interventions. NYU Zika Briefing Report #1. https://www.nyu-pir2.org/research. Accessed 19 Jun 2019
71. NDTV (2021) Zika virus of grave concern, but not a pandemic: top health expert. NDTV. July 15. https://www.ndtv.com/india-news/zika-virus-not-a-pandemic-but-of-grave-concern-says-top-expert-2487505. Accessed 25 May 2022
72. Barnagarwala T (2021) Inside Uttar Pradesh's Zika outbreak: can India's most populous state contain the virus spread? Scroll.in. November 15. https://scroll.in/article/1010555/inside-uttar-pradeshs-zika-outbreak-can-indias-most-populous-state-contain-the-virus-spread. Accessed 26 Apr 2022
73. Ratna K (2017) The secret life of Zika virus. Speaking Tiger Books, New Delhi
74. Reuters (2017) WHO says India reports cases of Zika virus. US News and World Report. May 27. http://www.reuters.com/article/us-health-zika-india-idUSKBN18N0TI. Accessed 27 Jun 2017
75. Sharma D (2017) By being dishonest about Zika outbreak, government put India at risk. Daily O. June 6. http://www.dailyo.in/technology/zika-virus-india-health-ministry-jp-nadda-world-health-organization/story/1/17656.html. Accessed 19 Jul 2017
76. Najar N (2017) India acknowledges three cases of Zika virus. The New York Times. June 3. https://www.nytimes.com/2017/06/03/world/asia/india-zika-virus.html. Accessed 27 Jun 2017
77. Sharma S (2021) India's latest Zika outbreak sees surge of nearly 100 cases. Reuters. November 8. Accessed 26 Apr 2022

78. Slater J (2018). India wrestles with first significant outbreak of Zika virus. The Washington Post. October 25. https://www.washingtonpost.com/world/asia_pacific/india-wrestles-with-first-significant-outbreak-of-zika-virus/2018/10/25/3eba75f2-d822-11e8-a10f-b51546b10756_story.html. Accessed 21 May 2022
79. Grubaugh ND, Ishtiaq F, Setoh YX, Ko AI (2019) Misperceived risks of Zika-related microcephaly in India. Trends Microbiol May. 27(5). https://doi.org/10.1016/j.tim.2019.02.004
80. The Financial Express (2019) India to include bacteria in its anti-dengue arsenal. The Financial Express. July 19. https://www.financialexpress.com/opinion/india-to-include-bacteria-in-its-anti-dengue-arsenal/1649076/. Accessed 21 May 2022
81. Krisch J (2016) The year in Zika. The Scientist. December 16. http://www.the-scientist.com/?articles.view/articleNo/47805/title/The-Year-in-Zika/. Accessed May 17, 2017; Messina JP et al (2017) Mapping global environmental suitability for Zika virus. eLife. 6. April 19. https://elifesciences.org/content/5/e15272. Accessed 19 May 2017
82. Lewis T (2016) Mapping Worldwide Zika Susceptibility. The Scientist. April 19. http://www.the-scientist.com/?articles.view/articleNo/45898/title/Mapping-Worldwide-Zika-Susceptibility/. Accessed 19 May 2017; Messina JP et al (2017) Mapping global environmental suitability for Zika virus. eLife. 6. April 19. https://elifesciences.org/content/5/e15272. Accessed 19 May 2017
83. Hamel R et al (2015) Biology of Zika virus infection in human skin cells. Journal of Virology. September. 89(17). https://www.ncbi.nlm.nih.gov/pubmed/26085147. Accessed 2 Jun 2017
84. Norman A (2016) Romper. What will happen with Zika virus research under President Trump? things don't look good. Romper. November 17. https://www.romper.com/p/what-will-happen-with-zika-virus-research-under-president-trump-things-dont-look-good-22895. Accessed 26 May 2017
85. Advisory Board (2022) How covid-19 might end: using clues from influenza, HIV, and Zika. March 14. https://www.advisory.com/daily-briefing/2022/03/14/covid-end. Accessed 25 May 2022
86. Garcia G (2017) Zika cases on the decline, but the war on mosquitoes is far from over. CARPHA Release. February 7. https://carpha.org/More/Media/Articles/ArticleID/191/Zika-Cases-on-the-Decline-But-the-War-on-Mosquitoes-is-Far-from-Over. Accessed 8 Jun 2022
87. Lautharte I, Rasul I (2020) From Zika to COVID-19: can we learn lessons across pandemics? International Growth Centre. April 15. https://www.theigc.org/blog/from-zika-to-covid-19-can-we-learn-lessons-across-pandemics/. Accessed 11 Jun 2022
88. TRAVAX (2022) The Brazilian Ministry of health has reported continuing transmission of Zika virus across the country, with 6483 cases recorded from 3 January 2021 to 1 January 2022. January 27. https://www.travax.nhs.uk/outbreaks-index/outbreakitem?newsid=24174. Accessed 10 Aug 2022
89. Goodell J (2020) How climate change is ushering in a new pandemic era. Rolling Stone. December 7. https://www.rollingstone.com/culture/culture-features/climate-change-risks-infectious-diseases-covid-19-ebola-dengue-1098923/. Accessed 6 Jun 2022
90. Diep F (2018) Virus caused a big uptick in serious birth defects in 2016. Pacific Standard. January 25. https://psmag.com/news/zika-virus-caused-a-big-uptick-in-serious-birth-defects-in-2016. Accessed 7 Jul 2018
91. Manaytay A (2017) Houston officials consider use of genetically modified mosquitoes to fight Zika virus. Tech Times. March 21. http://www.techtimes.com/articles/202474/20170321/houston-considers-use-of-genetically-modified-mosquitoes-to-combat-zika.htm. Accessed 21 May 2017
92. McKay B, Loftus P (2016) America's next defense against Zika and other foreign invaders. The Wall Street J December 16. https://www.wsj.com/articles/americas-next-defense-against-zika-and-other-foreign-invaders-1481810402. Accessed 21 May 2017
93. Hotez PJ (2016) Neglected tropical diseases in the anthropocene: the cases of Zika, Ebola, and other infections. PLOS Neglected Tropical Diseases. April 8. http://journals.plos.org/plosntds/article?id=https://doi.org/10.1371/journal.pntd.0004648. Accessed 17 Jul 2017

94. United Nations Intergovernmental Panel on Climate Change (2007) IPCC fourth assessment report: climate change 2007. IPCC, Geneva
95. Ramasamy R, Surendran SN (2012) Global climate change and its potential impact on disease transmission by salinity-tolerant mosquito vectors in coastal zones. Frontiers Physiol. June 19. https://doi.org/10.3389/fphys.2012.00198. Accessed 6 Jun 2022
96. Tollefson J (2020) Why deforestation and extinctions make pandemics more likely. Nature. August 7. 584:175–176
97. WHO (2017) Situation report: Zika virus, microcephaly, guillain-barré syndrome. January 5. http://www.who.int/emergencies/zika-virus/situation-report/05-january-2017/en/. Accessed 1 Jul 2017
98. Castanha PMS, Marques ETA (2021) Zika vaccines: can we solve one problem without creating another one? The Lancet. 21. September. 1199–1199

Chapter 5
Convergence

Convergence has no clear meaning. Some see it as a substitute for scholarship involving different disciplines. Other definitions are grounded in borrowing theory [1], whereby researchers and practitioners borrow methods across disciplines to expose unique features of subjects. Unfortunately, borrowing requires the researcher to deal with all the drawbacks and deficiencies of the borrowed theory and explain them away when used to provide insight into whatever phenomena which led to the decision to borrow the method in the first place.

On an elementary level, when two or more scholars from two or more disciplines work together in a cross-disciplinary effort, there is simple convergence.

Another argument that bothers me is how complicated some research subjects have become associated with researchers calling them "wicked" [2] or even "sticky" [3] or "messy." While the terms are grossly overused, often only in an article title [4], sometimes it is apropos, but those instances are few.

The only time a truly wicked issue demands convergence as a management tool is when the variables that comprise the issue are drawn from disciplines that traditionally have not cooperated. Bringing two or more disciplines to handle an issue that could be mostly done "in-house" merely complicates and confuses the issue even further. Forced convergence is a "Tower of Babel," seldom with shared missions and goals.

In this case, the justification for calling for a convergent solution to this case of infectious diseases is based on the observation that a metaphor, "The Perfect Storm," has risen in many articles and books about ZIKV. The term was used repeatedly by public health providers, government spokespersons, experts in tropical diseases, infectious diseases, land use, development, poverty, and so forth as well as entomologists, virologists, epidemiologists, ecologists, meteorologists, and dozens of others.

A set of circumstances led to the epidemic, including globalization, aviation, shipping, biodiversity, deforestation, climate change, and much more. More importantly, "The Perfect Storm" was also used to describe whether there may be a growing

D. M. Berube, *Pandemics and Resilience: Lessons we should have learned from Zika*, Risk, Systems and Decisions, https://doi.org/10.1007/978-3-031-25370-6_5

tendency to more mosquito vector-associated arboviral infectious diseases over the next few decades if the variables driving epidemics are not resolved.

5.1 Drivers

Drivers make the study of science and technology communication remarkable. Drivers tend to be tangentially associated with what is being driven but vital in designing solutions to what is being driven. Drivers tend to magnify exposure to the ZIKV and the controversy over the ZIKV.

Along with recent epidemics of dengue (DENV) and chikungunya virus (CHIKV), which Ae also transmits. *Ae. aegypti* and *Ae. albopictus* mosquitoes, the emergence of ZIKV suggests an ongoing intensification of environmental and social factors that have given rise to a new regime of arbovirus transmission [5]. These (re)emerging arboviruses are diseases of global movement and change. They share a primary mosquito vector (*Ae. aegypti*) that lives alongside human settlements, lays their eggs in household water containers, and feeds preferentially on humans [5].

For example, there have been two significant amplifying drivers in the case of the threats associated with the ZIKV. First, the Summer Olympics in Rio in 2016 which in terms of functioning as a driver was mostly nonsense (see Chap. 16) [6]. Second, there is a climate change driver, which is not. The primary force behind these drivers was the implications and harmful effects of introducing the *Ae. aegypti* mosquito human–wildlife interaction and increased livestock production, there is a greater probability of zoonotic transmission. In addition, ever-increasing global trade and travel increase the potential for outbreaks of new or resurgent pathogens to become epidemics or pandemics. Globalization drives economic growth but also facilitates the spread of contagion [7]. The risk of zoonotic diseases—infections transmitted from animals to humans—is increasing as we muscle in on the wild. The more we raze habitats for farmland and cities, hunt and trade wildlife, vacation in remote forests, and hike through once-inaccessible caves, the greater the chances of "spillover," as scientists call it, when a virus vaults from a species it doesn't harm into one it does, such as ours [8].

The large-scale connectivity of human populations is a sizeable key factor. Human mobility determines the chance an infection present in one location and will be introduced elsewhere [9]. Increased human population growth in decades ahead and increased international travel and trade are likely to sustain and grow the threat of the further geographical spread of current and new arboviral diseases [10]. Humans occupy the most extensive range of terrestrial habitations. When a species, such as *Ae. aegypti,* evolves abilities to survive together with humans, their territories may expand with human mobility [11].

Other factors associated with the emergence of arboviruses such as ZIKV include genetic changes involving the viruses themselves; climate change; uncontrolled use of insecticides; perturbations of natural systems that are frequently anthropogenic; expansion of the geographic distribution of mosquito vectors; urbanization and local levels of socioeconomic development, deforestation and land reclamation; irrigation projects adaptation to new reservoir/amplification hosts; global growth of human populations with extensive urbanization; lack of effective mosquito control and ineffective or unsustainable vector control; erratic water supplies; commercial transportation and increased travel; and political and military activities that lead to mass human evacuation. Figure 5.1 indicates the interplay between the drivers and mosquito and human contact.

The process by which infectious agents may transfer from animals to humans or disseminate from isolated groups into new populations can be called "microbial traffic" [13].

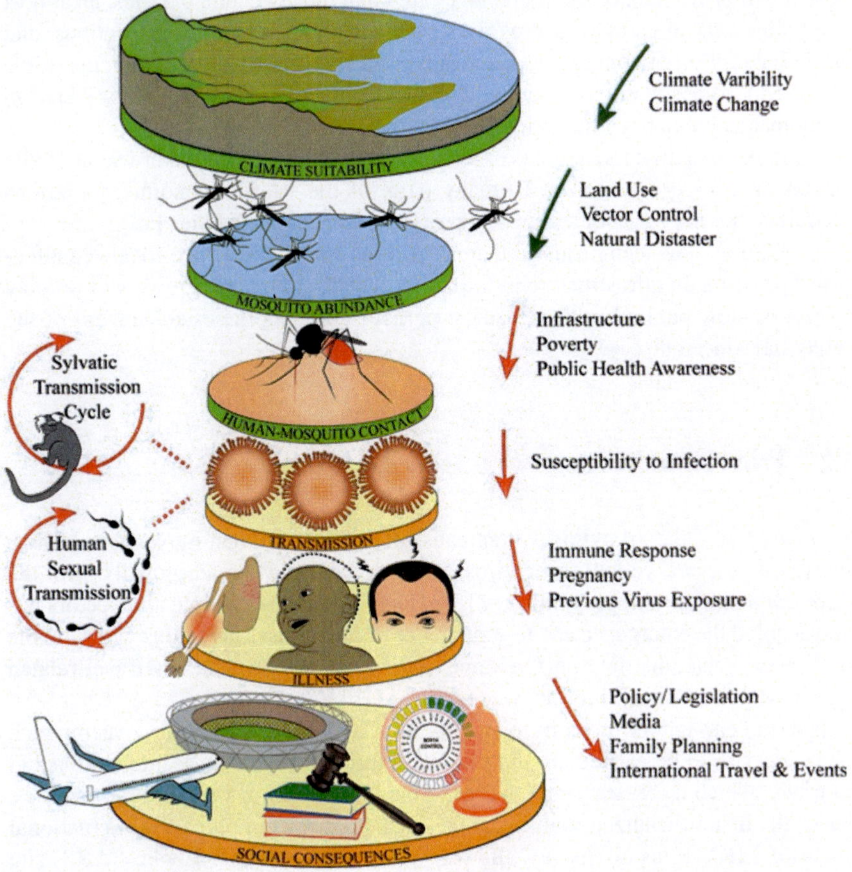

Fig. 5.1 Hierarchy of factors that influence ZIKV transmission, illness, and social consequences. Climate suitability, mosquito abundance, and human–mosquito contact partly determine rates of ZIKV transmission, which causes illness in some cases. Social consequences depend on both actual and perceived risks of illness. Arrows indicate environmental (green), and social(red) changes hypothesized to contribute to the shifting ecology of vector transmission in the Americas. Figure inspired by Plowright et al. [12]. In Ali et al. [5]

Zoonotic introductions may occasionally occur in isolated populations but may go unnoticed so long as the recipients remain isolated. But with increasing movement from rural areas to cities, such isolation is increasingly rare. For example, the migration from rural to urban areas and living under a precarious infrastructure, including the water supply, has contributed to mosquito breeding. In the northeast region of Brazil, over 75% of breeding sites arise due to precarious water storage. In contrast, most breeding sites occur in the southeast region in other domiciliary vessels (jars, plant pots, and roof gutters) [14]. Because *Ae. aegypti* usually uses artificial containers as breeding sites; it does not seem to have specific predators but rather "opportunistic" ones that feed on larvae if they encounter them [15].

The efficient adaptation of *Ae. aegypti* to urban areas makes their control difficult. Only the improvement of sanitation and municipal infrastructure of cities, an almost impossible task in areas with progressive urbanization, budgetary restrictions, and low collaboration of populations, can lead to effective control. *Ae. albopictus* is less adapted to domestic environments than is *Ae. aegypti* but is widely distributed in peridomestic habitats within cities [14].

Human population movements or upheavals caused by migration or war are often crucial in disease emergence. In many parts of the world, economic conditions encourage the mass movement of workers from rural areas to cities [13].

Population movement from rural areas to cities can spread a once-localized infection. The strain on infrastructure in the overcrowded and rapidly growing cities may disrupt or slow public health measures, perhaps allowing the establishment of the newly introduced disease [13].

5.2 Travel and Trade

Ae. aegypti and the arboviruses it spreads often follow human movement: Historically, *Ae. aegypti*, yellow fever virus, and dengue virus spread globally with the slave trade and the shipping industry [5]. Traveling infected people and vectors has also enabled the emergence and re-emergence of flaviviruses in regions with suitable climate and vector distribution. On Monday, February 13, 2016, the first travel-related case in Monroe County, Florida, was reported [16].

Nobody can ignore travel by plane makes it almost inevitable for a vector such as *Ae. aegypti* to be carried around the world through flights [17]. Intercontinental air travel, which increases yearly, has expedited the spread of vector-borne diseases, especially in industrialized countries. Long-distance travel amplifies the international threat of ZIKV transmission by allowing pathogens including Ebola, SARS, and influenza to spread rapidly around the world [5]. Human travel has increased to the point where global movement can occur quickly enough to overcome previous limitations on spreading microbes with a very short incubation period [18].

Tourism and business travel have exposed millions to diseases, especially infectious diseases. In modern times, mosquitoes travel worldwide in cargo airplanes and container ships [19]. Reportedly, the mosquitoes carrying ZIKV can come along with the revelers in their vehicles and luggage [20].

The global number of passengers traveling by air increased from 68.5 million per year in the 1950s to over 2 billion per year in the first decade of the twenty-first century. In 2011, an estimated 2.75 billion people will travel by airplane to and from tropical urban centers [21].

Aircraft passenger numbers have increased nearly 9% per annum in the half-century preceding the mid-2000s. With over 1 billion foreign arrivals in 2016, international tourism trends have continued to grow. The number of tourists traveling to Asia and the Pacific increased by more than 8% in 2016, equaled by more than 8% tourism growth in Africa; these are precisely the regions with the most significant reservoirs of arboviruses in the world, and projections are that the rises in tourism trends will be sustained [10].

According to the most conservative scenario, populations living within the geographical range of ZIKV (during the month where the geographical scope was broadest) were highest in India (1.2 billion people), China (242 million), Indonesia (197 million), Nigeria (179 million), Pakistan (168 million), and Bangladesh (163 million). Geographical variability in the three suitability zones for autochthonous ZIKV transmission on a quarterly and monthly basis is shown for Africa and the Asia–Pacific region [22].

Some countries have unique vulnerabilities. Consider the case of Bangladesh.

A new threat has become introduced with ZIKV infection caused by ZIKV. Millions of Bangladeshi people serve in different countries, including the active areas for ZIKV-associated diseases. If those people get infected and come to Bangladesh during their vacations, they could be responsible for more Zika infections by mosquitoes. As Bangladesh is a densely populated country, there are many overcrowded areas in the big cities where the proper cleaning procedures to eliminate mosquitoes are not maintained adequately. Therefore, if the ZIKV gets introduced in Bangladesh, the condition will worsen within a short period. The latest encounter in America also resulted from the travelers who introduced the virus with them to the countries and further disseminated the infection through a mosquito bite [20].

Since 2004, there has been an increasing number of persons traveling to the Americas from countries where mosquito-borne diseases exist in which humans play a role in the transmission. Thus, the risk for intercountry and intercontinental transmission has increased [23]. 70% of travelers departing from the ecological niche of ZIKV in the Americas for destinations in Africa and the Asia–Pacific region arrived in ten countries: China (238,415 travelers per year), Japan (179,926 travelers per year), Israel (106,365 travelers per year), Australia (96,430 travelers per year), Turkey (90,632 travelers per year), Angola (88,048 travelers per year), South Korea (87,768 travelers per year), United Arab Emirates (70,848 travelers per year), India (67,422 travelers per year), and South Africa (59,318 travelers per year) [22].

Cities in these countries have modern airports through which tens of millions of passengers pass each year, providing the ideal mechanism for transmitting viruses to new cities, regions, and continents where there is little or no effective mosquito control [21].

High-volume rapid movement characterizes travel and other industries in modern society. Modern production methods yield increased efficiency and reduced costs in operations, including food production, where processor uses products of biological origin. Still, they can increase the chances of accidental contamination and amplify the effects of such contamination [13] Shipping contributes to infestation. Despite the change in shipping practices, ships and planes may still bring in large numbers of mosquitoes and infected people [24]. Appearances of both *Ae. aegypti* and *Ae. albopictus* can be traced to shipping. Shipping traffic increased by more than 27% from 1993 to 2006. The middle class is growing in the emerging economies of Asia, Africa, and Latin America, and as disposable income increases, the demand for imported products is directly correlated with shipping [10].

Such opportunity for the movement of disease is not limited to trade and tourism, as there is a steadily increasing number of people visiting friends and relatives, pilgrimages, humanitarian and other volunteer work, and large numbers of politically and economically displaced refugees, all contributing to the potential for infected persons to carry pathogens to other geographical shores and infect local vectors [10].

5.3 Poverty and Subsidence Living

As *Ae. aegypti* and *Ae. albopictus* thrive in stagnant water collections; demand proliferation may be encouraged by human population growth or migratory waves from areas with civil upheaval and uncontrolled slum formation [25].

The lack of adequate water supply in many urban areas makes it necessary to store water in large containers such as clay jars and cisterns, another significant contribution to increased mosquito densities [21]. Public water wells are likely to have a causal link, as they necessitate water storage in individual containers inside houses. Concern that neighbors might consume the water in public wells may also encourage storage in more significant quantities for extended periods [26].

The vast volume of unprocessed solid waste sites in many resource-poor countries and the large volume of untreated sewage that still empties into the waterways surrounding many metropolitan cities—provide excellent breeding sites for mosquitoes [23]. According to a 2016 UN special rapporteur Léo Heller: "There is a strong link between weak sanitation systems and the current outbreak of the mosquito-borne ZIKV, as well as dengue, yellow fever and Chikungunya," and added further that "the most effective way to tackle this problem is to improve the failing services" [27].

Negative, positive feedback comes from the increasing use of technically efficient extractive technologies. Examples include mechanized tube wells chasing increasingly receding groundwater and appropriating water from rural areas by dam-and-transfer schemes favoring the demands of cities and other intensive water users while depleting the natural resource [28].

Periods of economic decline may force individuals into lower socioeconomic status, in which the symptoms of poverty promote ZIKV risk. In 2014, before ZIKV emerged in the Americas, Brazil fell into a recession, and the unemployment rate jumped from 6.8% to 8.5%. The ongoing recession may have increased the risk of ZIKV infection for a new group of vulnerable people [5].

For example, in Brazil, slums are generally situated near metropolitan cities, so the proximity of these socioeconomically distinct regions may also facilitate the spread of ZIKV across social boundaries [5].

Therefore, *Aedes* mosquitoes tend to have generally unhygienic conditions (e.g., irregular garbage collection, promoting *Aedes* breeding sites), and poor house design (e.g., the absence of window screens and water supply) may promote *Aedes* proliferation [26].

Social, political, and economic changes can also impact human exposure to ZIKV, particularly for the urban poor, who often live in areas with inadequate sanitation, infrastructure, and water access [5]. For example, in Brazil, slums are generally situated near metropolitan cities, so the proximity of these socioeconomically distinct regions may also facilitate the spread of ZIKV across social boundaries [5].

Periods of economic decline may force individuals into lower socioeconomic status, in which the symptoms of poverty promote ZIKV risk. In 2014, before ZIKV emerged in the Americas, Brazil fell into a recession, and the unemployment rate jumped from 6.8% to 8.5%. The ongoing recession may have increased the risk of ZIKV infection for a new group of vulnerable people [5].

A mosquito doesn't care who it's biting. Yet the spread of diseases like Zika carries an outsized risk for poor people. That's because "poverty equates to poor quality housing and uncollected garbage and standing water in poor neighborhoods that allow certain insects to breed nearby," wrote Peter Hotez, Dean at Baylor Medical Schools, a pediatrician, and microbiologist at Texas Children's Hospital [29].

> The Zika outbreak has been so widespread in Brazil because many homes don't have screens on windows and doors. That's true, too, in many low-income neighborhoods where the Zika risk is higher in the United States.... Population density, the presence of the mosquito that spreads Zika, and high temperatures all intensify the probability of local transmission—this is why places like Miami, Orlando, and Houston are all considered high risk. As warm weather approaches in the United States, conditions quickly become favorable for mosquito species that can spread Zika, yellow fever, dengue, and other serious illnesses [29].

Socioeconomic drivers are commonly associated with the absence of air-conditioning (thus higher temperatures indoors), the presence of open (i.e., non-screened) windows (more in-and-out access for mosquitoes), and sociocultural practices such as water storage. The latter is an essential sociocultural practice in response to the lack of piped water systems. However, increasing the piped water supply may complicate matters if the collection is intermittent because people tend to store more water when they know that the pool may stop unexpectedly [30].

"Many of these mosquito-borne arboviruses remain quiescent, but some of them must surely have the potential to break out under 'Perfect Storm' conditions to invade other geographical regions with unknown consequences. A range of factors increase this likelihood, including expanding human populations that necessitate clear forests and woodland for cropland and increased sustenance needs. This placed people within reach of wild animal disease cycles" [31].

5.4 Human Population Density and Mosquito Habitat Loss

The more people there are and the more densely they are spaced, the more hosts and easier accessibility for blood-feeding anthropophilic mosquitoes, especially in urban environments. The combination of biodiversity loss, habitat fragmentation, and human contact with forest areas create ideal conditions for introducing known and unknown pathogens into the human population [32].

"Contemporary livelihood and market patterns tend to degrade ecosystems and their services, driving a cycle of degradation in increasingly tightly linked socioecological systems. This contributes to reductions in the natural regulating capacities of ecosystem services to limit disease transfer from animals to humans. It also undermines natural resource availability, compromising measures such as washing and sanitation that may be key to managing subsequent human-to-human disease transmission" [33].

Ecological factors usually precipitate emergence by placing people in contact with a natural reservoir or host for an infection hitherto unfamiliar but usually already present (often a zoonotic or arthropod-borne infection), either by increasing proximity or, often, also by changing conditions to favor an increased population of the microbe or its natural host.

The world has experienced unprecedented population growth in developing tropical countries during the past 50 years. This has resulted in unplanned and uncontrolled urbanization in those countries, deteriorating housing, and inadequate water, sewer, and waste management systems. Finally, increased populations of rodents, mosquitoes, and other animals living in intimate association with crowded human people [34]. The combination of reducing biodiversity, increasing deforestation, unwise land-use policies, and the overabundance of non-degradable plastic containers have increased the likelihood of a new outbreak of infectious diseases.

5.4.1 Biodiversity

Previously, pristine natural areas with high biodiversity were seen as likely sources of new zoonotic pathogens, suggesting that biodiversity could negatively impact human health. Until recently, habitats with naturally high levels of biodiversity were thought to serve as hotspots for the emergence of new zoonotic pathogens, presenting a hazard to humans. This expectation assumed that the diversity of free-living organisms leads to various pathogens. That pathogen diversity per se is a risk factor for zoonotic emergence. But for decades, according to Keesing and Ostfeld, it has also been known that under some conditions, high biological diversity can decrease the transmission of zoonotic diseases that have already become established [35].

The relationship between biodiversity and infectious diseases is complex. It may seem paradoxical initially, but some precepts are clear: Preserved ecosystems act as health promoters, maintaining pathogens in the forest environment. From another perspective, disturbances in highly biodiverse ecosystems facilitate the emergence and spread of new human infections. Biodiversity loss creates new niches that may be occupied by alternative reservoir species, vectors, hosts, and pathogens. In brief, biodiversity loss profoundly alters the infections' dynamics [32].

While some species are going extinct, those that tend to survive and thrive—rats and bats, for instance—are more likely to host potentially dangerous pathogens that can make the jump to humans [36]. A loss in biodiversity usually results in a few species replacing many—and these species tend to be the ones hosting pathogens that can spread to humans. Kate Jones, an ecological modeler at University College in London, and her team [37] compiled more than 3.2 million records from several hundred ecological studies at sites around the world, ranging from native forests to cropland to cities. They found that the populations of species known to host diseases transmissible to humans—including 143 mammals such as bats, rodents, and various primates—increased as the landscape changed from natural to urban, and biodiversity generally decreased.

For example, when biodiversity is lost from ecological communities, large-bodied species with slower life histories are most likely to disappear. In contrast, smaller-bodied species with fast life histories tend to increase abundance. Recent research by Keesing and Ostfeld shows that fast-lived species are more likely to transmit zoonotic pathogens. Together, these processes are likely to increase the number of zoonotic reservoirs when biodiversity is lost from ecological systems [38].

5.4.2 *Deforestation*

Deforestation for timber, new roads, and city sprawl has endangered more than 33 species of New World monkeys. Ratna wrote population density assures a hyperendemic. The urbanization offers the mosquito an evolutionary endorsement. Crowded cities are a semisylvatic habitat. Its grimy tentacles engulf forests, hills, and swamps to replace them with tamped-down garbage landfills [39].

The expansion of cropland and grassland, mainly through deforestation for agricultural and pastoral purposes, increases their interaction potential. These activities involve the felling of trees, which reduces faunal shade, improves air and water temperatures, and increases water pooling. All of these factors promote mosquito breeding and acceleration of larval development from one to two weeks to as quickly as 4.5 days [40]. Areas with less forest cover likely have higher human population densities, urbanization, suitability for *Ae. aegypti*, and ZIKV incidence, so human activities like deforestation and urbanization that decrease forest cover may increase the risk of ZIKV transmission [5].

The establishment of pastures, plantations, or intensive livestock farms close to forest margins may also increase pathogen flow from wildlife to humans. Deforestation and forest fragmentation reshape the dynamics of wildlife communities, possibly leading to the extinction of habitat-specialist species while allowing generalists to thrive. Wildlife species that are hosts of pathogens are relatively more abundant in managed landscapes (e.g., agricultural ecosystems and urban areas) than in adjacent undisturbed sites [41].

The close association of humans with forest habitats may easily promote the emergence of new pathogens. For example, opportunistic and easy human blood-feeding may be provided by the available people living close to the forest's edge than the primates. The emergence of a disease in a particular region is associated with changes that influence people's livelihood strategies, rapid conversion of natural habitats, and urbanization. Factors including host type present may influence host specificity [42].

A study (dos Santos et al.,) done in Brazil looked at the relationship between *Aedes spp.* mosquitoes, specifically *Ae. albopictus*, and the interface between urban areas and the forest edge. The study found that vector density was highest at the forest edge and decreased significantly only 200–300 m into the forest. As the forest edge was close to an urban area, the study identified 75% of engorged mosquitoes at the forest edge as having fed from a human host rather than animal hosts [43]. A similar study (Lourenço-de-Oliveira et al.,) done in Rio de Janeiro, found an absence of *Ae. aegypti* or *Ae. albopictus* 100 m into the forest, again showing that *Aedes spp.* Mosquitoes prefer the edge of forested areas [44].

Finally, even though deforestation may eventually lead to a decrease in microbial diversity, the processes of deforestation are often likely to act to increase the frequency of microbial contact through increased access provided by logging roads [45].

5.4.3 Land Use

Critically, there is increasing evidence that land-use change is a significant driver of emerging infectious diseases. Land-use change has been estimated by the EcoHealth Alliance [46] to be linked to 31% of outbreaks of emerging infectious diseases, including HIV, Ebola, and Zika virus, which are considered connected to anthropogenic changes in tropical rainforests, with 15% of these EIDs linked to agricultural changes [33].

Land development practices and rural and urban land fragmentation increase the likelihood that immunologically naïve humans will encounter infectious vectors at land use interfaces. Land-use change and fragmentation increase the frequency of human-mosquito contact and alters the dynamics of pathogen–vector–host relationships by providing more opportunities for pathogens to expand their geographic range and exploit new habitat niche. Changes in anthropogenic land use due to urbanization, changing agricultural practices, and deforestation has led to the re-emergence of arboviruses, such as ZIKV, across the Americas [40]. Human impacts like land-use change have been linked to emerging infectious diseases of humans in many studies, according to Murray and Daszak [47]. For example, they explored how land-use changes, like deforestation and agricultural conversion, could affect the emergence of zoonotic viruses and presented two hypotheses. In one, the land-use change increases contact between humans and a pool of diverse pathogens without directly affecting the collection of pathogens. On the other hand, land-use change perturbs ecological communities, affecting zoonotic host species, such as rodents or bats, resulting in changes to cross-species transmission rates. These hypotheses are not mutually exclusive. Species that thrive in human-impacted habitats could provide opportunities for spillover based on the diversity of their potential pathogens and their abundance, which might result in more significant contact with humans [38].

Land-use change that modifies the microclimates mosquitoes experience and human density and exposure could have immediate impacts on ZIKV transmission, which might explain the explosive spread of ZIKV in urban centers throughout the Americas [48]. Just as "urban health islands" exacerbate heatwaves, so too can land-use change influence the transmission of infectious diseases. Land cover may affect mosquito habitat by changing local temperature and humidity [49].

Landscape anthropization, especially urbanization, shifts the risk of mosquito-borne pathogen emergence by affecting the mosquito and host communities qualitatively and quantitatively. Landscape disturbances at the urban–forest interface may facilitate anthropophilic mosquito species dispersion into unfavorable habitats, thus modifying vector–host interactions and potentially leading to more contact with sylvatic (i.e., enzootic) reservoirs of zoonotic pathogens [50]. Human modification of the natural environment continues to create habitats in which mosquitoes, vectors of a wide variety of human and animal pathogens, thrive if unabated, with an enormous potential to affect public health negatively. Historic examples of these modifications include impoundments, dams, and irrigation systems that create havens for the mosquitoes that transmit malaria, dengue, and filariasis [51].

Natural mosquito habitats in many regions of the world are abundant even without human environmental modification. Human alteration of the environment, regardless of intent (i.e., clearing land for subsistence agriculture or dam construction for hydroelectric power and recreational use), often exacerbates existing mosquito-associated problems by expanding habitats, creating new habitats, or altering habitats in such a way that limited mosquito populations may explode with the availability of new habitats. Humans affect land-use change has long been recognized as a factor in the exacerbation of the mosquito-borne disease [52].

A larger middle class with increasing disposable income and international traffic to satisfy tourism and import demands will continue to provide an opportunity for the movement of vectors and viruses, including from Africa, where a range of viruses of yet unknown potential for public health impact is known to occur [10].

In addition to direct effects, agricultural development and related activities are associated with sedimentation and runoff. Although well-known as a source of pollutants, rivers, and reservoir sedimentation slows or blocks stream flow and may significantly decrease water depth. Mosquitoes prefer warm, shallow water with little to no flow, suggesting that sedimentation is an ideal process for creating suitable mosquito habitat [51].

Clearing land for subsistence agriculture or dam construction for hydroelectric power and recreational use) often exacerbates existing mosquito-associated problems by expanding habitats, creating new habitats, or altering habitats so that limited mosquito populations may explode with the availability of new habitats [51]. Water is also frequently associated with disease emergence. Infections transmitted by mosquitoes or other arthropods, including some of the most severe and widespread diseases, are often stimulated by the expansion of standing water simply because many mosquito vectors breed in water [13].

A brief aside, one of the most critical agricultural modifications associated with mosquito-borne diseases is the rice paddy. These shallow bodies of water with emergent vegetation and little flow are ideal habitats for the development of immature mosquitoes. In many regions of the world, contiguous rice paddies may cover immense areas resulting in unimaginable mosquito densities [51].

Population growth, urbanization, increasing affluence in middle-income countries, and the associated dietary shifts, including increased demand for animal products, are driving agriculture expansion and changes in animal husbandry—often at the expense of natural ecosystems. Intensive livestock production keeps many animals—often immunosuppressed, with low genetic diversity, and in poor conditions—near one another, making them vulnerable to the emergence and spread of epidemics [41].

Cities like Bangkok, Manila, and Jakarta exploded in population growth, most of it unplanned. The result has been millions of susceptible people moving to the cities and living in shantytowns with inadequate housing and few or no essential services such as water, sewer, and waste management. The resulting crowded human communities and large mosquito populations created ideal conditions for infectious disease transmission [21].

The American region, including Mexico and Central and South America, became highly urbanized in the 1970s. Today, over 75% of the population lives in urban areas, nearly all of which have been reinfested with *Ae. aegypti* [21]. These problems are not isolated to the South American continent. Hotez also points to the poorest urban areas of coastal Texas, Louisiana, Mississippi, Alabama, and South Florida as particularly at-risk. "This could be a catastrophe to rival Hurricane Katrina or other recent miseries that disproportionately affect the poor" [29].

While increased contact between wild animal reservoirs, sylvatic mosquitoes, and human populations due to land-use change may have supported ZIKV emergence, the most crucial role of the sylvatic cycle going forward may be in maintaining a virus reservoir between human epidemics. For example, two Brazilian primate species sampled between July and November 2015, capuchin monkeys and common marmosets, had high rates of ZIKV seroconversion [5].

When the human population migrates from rural areas to cities, the ideal environmental conditions (proliferation of breeding sites) for increasing Aedes mosquitoes are created. On the other hand, poverty complicates the efforts of communities or individuals to carry out effective protection measures. Even when control resources exist, they are inefficiently applied [53].

In general, urban settings provide a variety of potential larval rearing sites for peridomestic *Ae. Aegypti*. In addition, the proximity of people in cities to these sites may facilitate the transmission of these anthroponotic viruses. Furthermore, these viruses are increasingly more likely to arrive in new urban settings because of the increasing globalization and ease of travel and transportation of goods [30].

5.5 Non-degradable Containers

An unexpected driver is a trash, especially trash that allows water to puddle. All over the world, the *Aedes* mosquito population is probably higher today than ever before because nearly all consumer goods are now packed in non-biodegradable plastics or tin containers and refuse collection services and waste disposal practices are grossly deficient [54]. The dispersal and packaging of consumer goods in non-biodegradable plastic have contributed to expanding geographic distribution and increased population densities of *Ae. aegypti* by providing artificial breeding sites for the mosquito [55].

The proliferation of container breeding sites has facilitated the adaptation of *Ae. aegypti* and *Ae. albopictus*, which, in turn, has resulted in an increased incidence of ZIKV, dengue, and other arboviruses in Latin America [40]. Higher human population numbers also mean more waste, creating habitat for container breeders such as *Aedes* [10].

Automobile tires and plastic containers are but two of these containers found in the crowded urban environment and implicated in the high mosquito population densities. Artificial containers are a contributing source of mosquito habitats uniquely associated with man worldwide. These range from bottle caps and flower vases in cemeteries to swimming pools, 55-gallon drums, discarded tires, and junkyards. Virtually any container that can hold water can serve as a mosquito habitat and the source of large mosquito populations. Eliminating many of these sources could significantly reduce mosquito populations and the threat of disease transmission, but this solution is difficult to implement [51].

The proliferation of artificial containers has facilitated the opportunity for an expansion of mosquitoes. There was an explosion in the number of artificial containers that are the ideal larval habitat for this mosquito. These include many non-biodegradable containers, plastics used as household consumer goods, car tires, and many other artificial containers that hold water found in homes [53].

5.6 Climate Change

Climatological variation and ecological perturbation have been pervasive drivers of faunal assembly, structure, and diversification for parasites and pathogens through recurrent events of geographical and host colonization at varying spatial and temporal scales of Earth history [56].

A rise in arthropod-borne emerging infectious disease events due to climate anomalies has been observed during the 1990s. Controlling for reporting efforts, the number of emerging infectious disease events still shows a highly significant relationship with time. The threat of emerging infectious diseases to global health is increasing. This rise corresponds to climate anomalies occurring during the 1990s, supporting hypotheses that climate change may drive the emergence of diseases that have vectors sensitive to changes in environmental conditions such as rainfall, temperature, and severe weather events [57].

Climate change, one of the global environmental changes now underway, is anticipated to have a wide range of impacts on the occurrence of infectious diseases in human populations. Notably, the incubation time of a vector-borne infective agent within its vector organism is typically susceptible to changes in temperature, usually displaying an exponential relationship. [M]osquitoes, perhaps disease-carrying vectors may undergo analogous microevolution which would allow adaptation to altered seasonal patterns associated with global climate change. Climate influences infectious disease transmission by directly affecting the rate of biological processes of pathogens and vectors and influencing their habitats [58]. The IPCC stated, "climate change can affect human health indirectly through changes in the ranges of disease vectors (e.g., mosquitoes) [59]. Both temperature and surface water influence vector-borne infectious disease insect vectors. Of particular importance are vector mosquito species, which spread malaria and viral diseases such as dengue and yellow fever. Mosquitoes need access to stagnant water to breed, and the adults need humid conditions for viability. Warmer temperatures enhance vector breeding and reduce the pathogen's maturation period within the vector organism" [59, 60].

Generally, it seems climate change will significantly change the epidemiology of infectious diseases mainly through changes in microbial and vector geographic ranges. Transmission of infectious diseases is sensitive to climatic conditions, especially diseases like dengue or ZIKV, in which the virus has a life cycle outside the human body. Viruses that insects transfer are exposed to the climatic environment. Climate changes that may affect the transmission of infectious diseases, including temperature, humidity, rainfall, and soil humidity [53]. For example climate change may change the frequency of extreme weather events, such as tropical cyclones, floods, droughts, and hurricanes. It may destabilize and weaken the ecosystem services upon which human society depends. Climate change is also expected to affect the animal, human, and plant health via indirect pathways: It is likely that the geography of infectious diseases and pests will be altered, including the distribution of vector-borne diseases [61].

Most flaviviruses, like ZIKV, are critically dependent on climatic conditions favoring the recruitment and dispersal of their vectors. As such they will likely emerge or even re-emerge in new areas due to projected climate change [62]. Climate change could significantly affect vector-borne diseases in humans affect the reproduction, development, behavior, and population dynamics of the arthropod vectors of these diseases. Climate also can affect the outcome of pathogens in vectors and the population dynamics and ranges of the nonhuman vertebrate reservoirs of many vector-borne diseases [63].

However, climate is incredibly confounding as well as a potentially significant factor in determining: (1) the geographic and temporal distribution of arthropods; (2) characteristics of arthropod life cycles; (3) dispersal patterns of associated arboviruses; (4) the evolution of arboviruses; and (5) the efficiency with which they are transmitted from arthropods to vertebrate hosts. Thus, under the influence of increasing temperatures and rainfall through warming the oceans and altering the natural cycles that stabilize climate, one is inevitably drawn to conclude that arboviruses will continue to emerge in new regions [64].

Since climate change is altering temperature and precipitation patterns across the country, it is critical that public health professionals also prepare for a potential increase in the geographic spread of existing vectors, such as *Ae. aegypti* or *Ae. albopictus*, and potential for new vector-borne diseases [65].

While warmer winter temperatures may expand the geographic range of *Ae, aegypti,* and its arboviruses, warmer spring, summer, and fall temperatures may extend the length of the transmission season in temperate areas, especially in regions where *Ae. aegypti* or *Ae. albopictus* are locally established, such as in Italy and the northeastern U. S. [5]. As a result, climate change and variation may influence the geography of vector transmission and intensify the threat of ZIKV in temperate regions.

Rain events and drought cycles could profoundly influence the spread of vector-borne disease in the United States, not only due to the displacement of populations of humans and disease reservoirs but in terms of vector abundance. Drought conditions could favor closer contact among vectors, reservoirs, and hosts and thus facilitate transmission. Other areas could become too arid or too wet and inhospitable to vectors, thus possibly decreasing the overall disease risk. However, warming temperatures could favor more rapid amplification and dissemination of viruses within reservoirs and vectors. Higher temperatures have been correlated with increased disease incidence at local, regional, and national scales [65]. The consequences of these "greenhouse-effect" rising temperatures will be fundamental and wide-ranging changes in average daily temperatures, precipitation, flood and drought events, and various other climate-related parameters, but with an overall likelihood that these changes will in many geographical regions favor the spread of disease vectors and their abundance [66].

But the grandest impact may be on the emergence of new pathogens from animals. Through intensive agriculture, habitat destruction, and rising temperatures, creatures are being forced to live by the cardinal rule of the climate crisis: adapt or die, which applies to mosquitoes. During the past decade, human and animal pathogenic arboviruses such as West Nile virus (WNV), Chikungunya virus (CHIKV), Rift Valley fever virus (RVFV), and Bluetongue virus (BTV) have merged and caused epidemics in North America, Europe, and the Arabian Peninsula. According to Gould and Higgs [64] their emergence may be attributable to the impact of climate change [67]. Since the West Nile virus debuted in New York City over a decade ago, outbreaks of mosquito-borne diseases, especially the West Nile virus, have become increasingly commonplace. As temperatures reach new highs because of global climate change, mosquitoes that once called the tropics home find the United States just as habitable [68].

It may be an oversimplified assumption that climate change will independently lead to an increased range for this species and a concomitant expansion of the risk of infections worldwide. A range of dynamic factors must be considered when predicting future global distribution trends [15]. The recent resurgence of many of these diseases is a significant cause for concern, but it is facile to attribute this resurgence to climate change. The histories of three such diseases—malaria, yellow fever, and dengue—reveal that climate has rarely been the principal determinant of their prevalence or range; human activities and their impact on local ecology have generally been much more significant. However the main determinants are politics, economics, and human activities [69].

5.6.1 Complex Feedbacks

Climate change is a problematic driver to substantiate, given how complex the relationships are between infectious diseases and temperature/precipitation. Climate influences disease ecology at many levels, and the many nonlinearities and feedbacks in the system create complex dynamics that are not easily modeled or understood. In addition, human factors, including behavior, immunity, and socioeconomic influences, also contribute to the complexity of these relations [70].

The effects of climate change on the epidemiology of mosquito-borne viral diseases are not easily predictable. Although cursory consideration might conclude that increases in temperature and rainfall will produce an increased incidence of arboviral diseases, the ecologic determinants of these diseases interact in complex ways. For examples, the incidence of dengue, yellow fever, and Chikungunya fever, transmitted by Aedes aegypti mosquitoes, sometimes increases during dry seasons because of increased peridomestic water storage [63].

These experts tend to agree extremely high temperatures can increase mosquito mortality, decreasing arboviral disease transmission. Heavy rainfall can wash out mosquito breeding sites. Facile conclusions that higher temperatures and increased precipitation will lead to increased transmission of arboviral diseases must be tempered by a more careful and thoughtful analysis of the interaction of ecologic variables with human behavior [63].

Excessive heat kills insects as effectively as cold does. Nevertheless, within their survivable range of temperatures, mosquitoes proliferate faster and bite more as the air becomes warmer. At the same time, greater heat speeds up the rate at which pathogens inside them reproduce and mature. As whole areas heat up, mosquitoes could expand into formerly forbidden territories, bringing illness with them. Further, warmer nighttime and winter temperatures may enable them to cause more disease for more extended periods in the areas they already inhabit [71].

Nonetheless, many articles claim the following: "The incidence of mosquito-borne diseases, including malaria, dengue, and viral encephalitis, are among those diseases most sensitive to climate. Climate change would directly affect disease transmission by shifting the vector's geographic range, increasing reproductive and biting rates, and shortening the pathogen incubation period. Climate factors strongly determine vector infectivity by affecting pathogen replication, maturation, and the period of infectivity" [72]

How climate change mitigation is approached might impact these relationships as well. For example, Ryan et al., claim that warming temperatures could exceed the upper thermal limits for transmission, estimated at 29.4 °C for *Ae. albopictus* and 34.0 °C for *Ae. aegypti*; we predict, counterintuitively, that partial climate change mitigation could cause more significant increases in exposure risk, particularly for transmission by *Ae. albopictus*, then no mitigation. However, partial mitigation is predicted to decrease the people, geographic areas, and length of seasons of risk for transmission by Ae. aegypti, the primary vector of arboviruses like dengue chikungunya, and ZIKV, in most regions [73].

Some modelers are much more conservative in admitting causality. Transmission largely depends on mosquitoes' ability to survive the extrinsic incubation period, become infectious, and bite new hosts, so differential (unimodal) impacts of temperature on survival, vector competence, and extrinsic incubation period have highly nonlinear effects on transmission. Warmer temperatures do not necessarily translate into more infectious mosquitoes [48]. They added. At warm temperatures, infection and dissemination efficiencies were very robust, but the mortality associated with the warm temperatures resulted in low numbers of mosquitoes that were capable of being infectious [48].

Although past outbreaks have sometimes been associated with extreme climate events and climatic variability, confidence in using these studies for predicting future events is often low, or the results are contentious [63].

The lack of well-designed long-term studies makes it difficult to determine if observed changes in transmission and distribution of vector-borne diseases are related to climate or one or more of the many other global changes concurrently transforming the world, including increased economic globalization, the high speed of international travel and transport of commercial goods, increased population growth, urbanization, civil unrest, displaced [63].

Others express high certainty. Under the influence of increasing temperatures and rainfall through warming of the oceans and alteration of the natural cycles that stabilize climate, one is inevitably drawn to the conclusion that arboviruses will continue to emerge in new regions. Climate is a significant factor in determining: (1) the geographic and temporal distribution of arthropods; (2) characteristics of arthropod life cycles; (3) dispersal patterns of associated arboviruses; (4) the evolution of arboviruses; and (5) the efficiency with which they are transmitted from arthropods to vertebrate hosts [74].

Currently, there is probably not enough information available to construct such threat matrices, but any such exercise would inform the research needed to obtain the required information [75] hence it is worthwhile.

5.6.2 *The Climate Variables*

The impact of climate change on mosquitoes is easy to model, partly because mosquitoes are susceptible to temperature changes and will move to stay in their happy zone. And that comfort zone is expanding [76]. Climate change parameters most often considered for their impact on mosquitoes are temperature, rainfall, and humidity. Primary changes in such parameters, caused principally through the increased emission of greenhouse gases into the atmosphere, can alter the bionomics of mosquito vectors and, therefore, the transmission rates of mosquito-borne diseases [77].

Generally, most if not all, the academic literature mosquito-borne disorders are projected to become increasingly prevalent because their insect carriers, or "vectors," are susceptible to meteorological conditions.

5.6.2.1 Temperature

Global atmospheric temperatures are presently in a warming phase that began 250–300 years ago. Speculations on the potential impact of continued warming on human health often focus on mosquito-borne diseases. Elementary models suggest that higher global temperatures enhance transmission rates and extend their geographic ranges [69]. One of climate change's most prominent effects on the spread of vector-borne illnesses: is rising temperatures that make seasons warmer. That can then expand the niche habitat of disease-carrying mosquitoes)—often farther north—and even speed up the incubation period for that pathogen [78]. "That habitat extension is a combination of climate change and human behavior," Juanita Constible, a climate expert at the NRDC, said. Warmer weather increases the opportunities for exotic vector-borne diseases to enter and establish transmission in the United States by lengthening the transmission season [65].

Mosquito vectors whose optimal survival temperatures are found in lowland areas of the tropics may spread to higher latitudes of the subtropical and temperate zones and the higher altitudes in tropical countries. It has been suggested that a latitudinal range of shift of about 200 km is possible per °C rise in global temperature. A slight temperature rise will favor a limited increase rise in temperature and will also select an increase in the human-biting rate, hasten mosquito development and therefore increase the relative vector density, and reduce the extrinsic incubation period [77].

Instantially, while many try and fail to make it in L.A. one group may prove unstoppable: mosquitoes, which have taken over Southern California and are driving the humans there crazy. According to numerous public health officials, new invasive, disease-bearing species originating from Asia and Africa are thriving in the increasingly long, hot, and humid summers afflicting this region thanks to climate change [79].

How does this happen? "Warming their environment—within their viable range—boosts their reproduction rates and the number of blood meals they take, prolongs their breeding season, and shortens the maturation period for the microbes they disperse," Paul Epstein of Harvard Medical School wrote. In other words, they bite more, breed more, and spread more diseases [80].

Temperature influences vector development rates, mortality, and behavior and controls viral replication within the mosquito. The temperature has been shown to affect the population biology of *Aedes* mosquitoes in the laboratory, while models based on precipitation, temperature, and atmospheric moisture explain much of the intra-annual variation in *Aedes* abundance. Warmer temperatures have been associated with more efficient communication of related flaviviruses and more excellent production of adult mosquitoes [81].

The associations with the different temperature variables confirm that higher temperatures generally favor *Aedes* populations (shown by the positive relationship with minimum temperature), provided they do not exceed harmful upper limits (defined by the optimum maximum temperatures of 33.2–34.2 °C for the various entomological indices) [26].

Different subspecies of the *Aedes* mosquito will react differently to temperature changes brought on by global warming. Climate-driven risk of transmission from both mosquitoes will increase substantially, even in the short term, for most of Europe. Within the next century, nearly a billion people are threatened with new exposure to virus transmission by both *Aedes spp.* in the worst-case scenarios [73].

> By 2050, warming temperatures are expected to dramatically expand *Aedes* transmission risk. For *Ae. aegypti*, significant expansions of one- or two-month transmission risk occur in temperate regions, along with broadening suitability for year-round transmission in the tropics, even into the high-elevation regions previously protected by cooler temperatures. *Ae. albopictus* transmission risk similarly expands substantially into temperate regions, exceptionally high latitude parts of Eurasia and North America. However, warming temperatures are projected to exceed the upper thermal limits to *Ae. albopictus* transmission in many places, significant reductions are projected in regions of seasonal risk (e.g., in North Africa) and year-round suitability (e.g., in northern Australia, the Amazon basin, central Africa, and southern Asia).
>
> By 2080, year-round temperature will be suitable for transmission by *Ae. albopictus* is mainly confined to high elevation regions, southern Africa, and the Atlantic coast of Brazil, during the warmer-adapted *Ae. aegypti* begins to lose some core area of year-round temperature suitability for transmission, especially in the Amazon basin [73].

Aedes, Ae. aegypti is estimated to have a higher thermal optimum for transmission than *Ae. albopictus*. Given the climate limitations on vector distributions, at a minimum, Aedes mosquitoes are projected to shift geographically and seasonally in the face of climate change, with a mix of expansions in some regions and contractions in others and no overwhelming net global pattern of gains or losses. Ecophysiological differences between Aedes vector species are likely to drive differences in thermal niches and, therefore, different distributions of transmission risk [82].

The impacts of global temperature change on disease transmission by mosquito vectors are likely to be broadly similar in coastal and inland areas [77]. Thus, as temperatures move toward the predicted thermal optimum owing to climate change, urbanization, or seasonality, ZIKV could expand north and into longer seasons. As modelers, Tesla et al., noticed strong, unimodal effects of temperature on the number of mosquitoes infected with disseminated infections that became infectious [48].

5.6.2.2 Precipitation

The ebb and flow of rainy seasons correspond with the hatching of millions of mosquitoes—and the spread of diseases they carry. One of the more critical variables involves the relationships between anticipated climate change and precipitation [83]. Climate change alters rainfall which has a direct effect on humidity. An optimal humidity significantly increases mosquito survival.

Furthermore, rainfall, rate of evaporation, and humidity will influence the availability of habitats for oviposition and preimaginal development of the mosquito vectors and therefore influence, the ratio of mosquitoes to humans. An expansion of habitats for preimaginal development because of climate change will tend to increase the vector density of the human population, favoring disease transmission [77].

5.6.2.3 El Niño/Southern Oscillation (ENSO)

Any discussion on climate change associated precipitation would be incomplete without a discussion of El Nino. Suppose ZIKV transmissibility is strongly modulated by longer-term climatic variation (such as El Niño). In that case, the virus may not be able to sustain endemic transmission, resulting in more sporadic but larger-scale epidemics when reseeding of infection coincides with favorable conditions for information [9]. Except for seasonal variability, the Southern Oscillation (ENSO) is the strongest naturally occurring source of climate variability around the globe.

Climate variability associated with El Niño has been associated with includes large epidemics on the Indian subcontinent, in Colombia, Venezuela, and Uganda. However, some epidemics have also occurred in years with no El Niño, and the models have not been validated against new epidemics [49]. Indeed, the literature has suggested a direct relationship between El Niño and increases in vector-borne diseases such as dengue which is transmitted by the same vector as yellow fever, chikungunya, and ZIKV. In the case of ZIKV, the outbreak coincided with an El Niño event [30]. The Zika outbreak from Brazil to Mexico recently coincided with an extreme El Niño event (elevated sea surface temperatures over the tropical eastern Pacific Ocean) and may not be a random coincidence [24].

Lessler et al., believe the warmer temperatures and rainfall resulting from the 2015–2016 El Niño facilitated ZIKV transmission throughout the region and increased the geographic range of *Aedes* mosquitoes. El Nino-associated periods of flooding (which increases mosquito breeding sites) and droughts (which can increase human-mosquito interactions) may facilitate ZIKV transmission [84]. Some have speculated even further that El Niño has played an essential role in spreading Zika in Latin American countries by creating the ideal conditions for the proliferation of mosquitoes [85].

Dr. Anthony Llau from the Global Health Consortium said climate change impacts disease-carrying bugs could be moving north in America, adding "$60 billion a year is the cost of climate change and infectious diseases" [86]. Such has been the case with Dengue fever and ENSO (El Niño Southern Oscillation). ENSO affects dengue occurrence by causing changes in household water storage practices and surface water pooling. Between 1970 and 1995, the annual number of dengue epidemics in the South Pacific was positively correlated with La Niña conditions (i.e., warmer and wetter) [59]. Another case is malaria. For example, the link between malaria and extreme climatic events has long been studied in India. Early last century, the river-irrigated Punjab region experienced periodic malaria epidemics. Excessive monsoon rainfall and high humidity significantly enhanced mosquito breeding and survival. Recent analyses have shown that the malaria epidemic risk increases around fivefold in the year after an El Niño event [59].

A new epidemiological model developed by Caminade and colleagues [52] showed the transmission risk for Zika in South America in 2015 was the highest since 1950. Caminade said their model suggested that temperature conditions related to the 2015–2016 El Niño "played a key role in igniting the outbreak" almost 2 years after the virus' suspected arrival in Brazil from Southeast Asia or the Pacific Islands in 2013 [87]. Caminade's finding raises additional concerns about the impact of large El Niño events on vector-borne diseases' risk in a future warmer, more connected world with increasing levels of drug and insecticide resistance. Flaviviruses, in general, should have a promising future. However, the threat posed by *Ae. albopictus* is not negligible, especially during the warm season in temperate regions, and the overlap of both vector species (*aegypti* and *albopictus)* produces the largest R0 values (how contagious a disease is) [52].

5.6.2.4 Natural Disasters

Climate change has been associated with extreme weather events as well as natural disasters whether associated with climate change or not. Vector-borne disease outbreaks after natural disasters in the U.S. are common. However, since the number and intensity of natural disasters are likely to increase due to climate change, it is essential to understand the effects these disasters will have on vector-borne diseases [65].

Associated in a way with precipitation is flooding. As such, natural disasters such as floods, severe storms, or hurricanes often cause significant public concern about vector-borne disease outbreaks. Members of the public often assume that water-related natural disasters produce more pools of standing water, which leads to more mosquitoes and more cases of vector-borne disease [65].

Spikes in bloodthirsty mosquito populations are common after major storms. And with climate change bringing more storms like Hurricane Florence, and upping the strength, the increased potential for the transmission of vector-borne illnesses like Zika, West Nile, and the chikungunya virus are a top concern for public health experts. "That's one of the hottest topics in climate and health right now," said Juanita

Constible, a climate expert at the Natural Resources Defense Council. "And that's partly because the answer is so dependent on the mosquito and disease in question, and where the mosquito and [pathogen] live" [78].

Also, intensifying floods and droughts resulting from global warming can each help trigger outbreaks by creating breeding grounds for insects whose desiccated eggs remain viable and hatch in still water. As floods recede, they leave puddles. In times of drought, streams can become stagnant pools, and people may put out containers to catch water; these pools and pots, too, can become incubators for new mosquitoes [72].

"The east African countries have had a distinct challenge because of the El Niño phenomenon–they have had flooding every year for several years. There has also been an increase in temperature attributed to climate change. According to Joy Phumaphi, executive secretary of the African Leaders Malaria Alliance, these conditions have increased mosquito populations" [88].

According to Vasquez et al. [89], a spike in ZIKV cases occurred after the earthquake. Patients in the area closest to the epicenter had a delay in seeking care. They expressed concerns over disaster-related problems that might worsen an epidemic: (1) displacement of populations and overcrowding may increase the exposure to vectors and to infected individuals; (2) destruction of property, sewage, and water infrastructure may increase the number of breeding sites; (3) stressful conditions increase the susceptibility of the population to developing the symptomatic disease by changing immune status, leading to more symptomatic viral disease; and (4) health-seeking behavior closest to the epicenter delayed in seeking care [89]. For example in Ecuador, increased local ZIKV transmission occurred following the earthquakes that struck the Manabi Province in April 2016, most likely by destroying infrastructure, contaminating drinking water (leading to makeshift water storage), and forcing people to live outdoors, all of which increase the rate of human contact with ZIKV-infected Ae. aegypti mosquitoes [5].

Also, infectivity may be impacted by ocean surges associated with not only hurricanes and typhoons but also underwater earthquake generated tsunamis and slippages like the Hilina slump in Hawai'i. *Ae. aegypti* and *Ae. albopictus* can oviposit and undergo preimaginal development [90] in collections of brackish water in unused wells, abandoned boats, disposable plastic, and glass food and beverage containers. This brackish water development may be an adaptive response to the almost exclusive application of Aedes larvae control measures (with insecticides such as temephos and Bacillus thuringiensis toxin) to freshwater habitats and the elimination of such habitats in the urban and periurban environment [91].

Rising sea levels can increase the prevalence of many vector-borne diseases in coastal zones. The expansion of brackish water bodies in coastal zones can increase the densities of salinity-tolerant mosquitoes and lead to the adaptation of freshwater mosquito vectors like *Ae. aegypti* and *Ae. albopictus* to salinity. Rising sea levels may therefore act synergistically with global climate change to increase the transmission of mosquito-borne diseases in coastal zones [77].

5.6.3 Variable Interaction

What seems to make the most sense is that these variables will interact. However, [l]inks between climate and diseases with various modes of transmission (vector-, water-, food-, soil-, and airborne) have been identified, with the strongest associations being between climate and mosquito-borne diseases. Variability in precipitation influences habitat availability for *Ae. aegypti* and *Ae. albopictus* larvae and pupae. Temperature further interacts with rainfall as the chief regulator of evaporation, thereby affecting the availability of water habitats. Temperature influences vector development rates, mortality, and behavior and controls viral replication within the mosquito [70]. Together, the conditions for increasingly large populations of mosquito vectors and large outbreaks of infectious diseases seem reasonable.

Warmer temperatures, more frequent droughts, devastating wildfires, and powerful hurricanes have significant implications for public health—and the seasonal birth of millions of mosquitoes, tied to weather patterns, is perhaps the most critical public health risk of them all. Outbreaks can happen in the blink of an eye. The ZIKV carried by *Ae. aegypti* mosquitoes infected over a million people in 2015 alone. Scientists predict that a warming world will invite the spread of more mosquitoes, and more illnesses, threatening a billion more people over the next 60 years [83].

Even wetter conditions may lead to excessive rainfall, wash away larvae and eggs, and reduce the number of small puddles. Of course, as there is *Ae. aegypti*, it can develop indoors in water containers, and its development is, therefore, less dependent on rainfall. *Ae. albopictus*, the alternative dengue vector in mainly periurban and rural settings, tends to undergo larval development in water collections outdoors and is, therefore, more dependent on rain-fed habitats, e.g., water collections in leaf axils, tree holes, and discarded containers [77].

Climate-related changes influence many zoonotic (animal-derived) infections in the density and movement of animal species that act as reservoirs or the expansion of vectors into areas previously inhospitable because of inadequate temperature and humidity [92]. Changes in vector composition due to alterations in the extent and salinity of larval habitats may also modify disease transmission dynamics. Furthermore, genetic changes in pathogens that better adapt to salinity-tolerant mosquitoes can increase disease transmission rates in coastal areas [77]. [L]ike other viruses spread by mosquitos and ticks, ZIKV could soon enjoy a greater reach, thanks to climate change. Last year, a team of researchers mapped the global distribution of Aedes mosquitoes to understand the global human health risk better, noting that mosquitoes are more widely distributed than ever [80]. Species in one part of the world specialize in specific hosts while others feed on various animals. Types of hosts may be chosen due to environmental conditions, such as ecological temperature, which later determines disease outbreaks. The host feeding patterns of mosquitoes are governed by several factors such as innate tendencies, host availability, abundance, defensive behavior of hosts, flight behavior, and feeding periodicity of mosquitoes [93].

5.7 Some Additional Concerns

5.7.1 Climate-Instigated Variables

Given that microbes can adapt to higher temperatures, there is concern that global warming will select microbes with higher heat tolerance that can defeat human endothermy defenses and bring new infectious diseases. There may be a strong possibility that new, previously unknown infectious diseases will emerge from warmer climates as microbes adapt to higher global temperatures that can defeat our endothermic thermal barrier [75].

However, the relationship between climate change and infectious diseases can be complex. Sweltering and dry conditions can reduce mosquito survival as well. "Climate influences dengue ecology by affecting vector dynamics, agent development, and mosquito/human interactions," they wrote, but "although these relationships are known, climate change's impact on transmission is unclear" [94]. On the other hand, precaution has been advised by the vast consensus of climate experts. They argue that climate change is a potentially significant factor that produces short-term effects, such as new epidemics of Zika and even pandemic risk [53]. For instance, a 2°C increase in temperature would simultaneously lengthen the mosquito's lifespan and shorten the extrinsic incubation period of the dengue virus, resulting in more infected mosquitoes for a more extended period [95]. On the other hand, dry seasons favor the cycle. Conversely, excessive rainfall could cause water containers to overflow. A fast water change in rainy seasons may not allow larvae to develop into adults. Still, dry periods may lead to longer water retention times, thus allowing the mosquito life cycle to be completed [24].

5.7.2 The Human Factor

Climate change can also alter how humans interact with the land, altering its use and impacting mosquito population magnitude and species composition [70]. Climate change amplifies the health risks of the poorest people, who suffer a significant increase in diseases by *Ae. aegypti* compared to rich people and less vulnerable population. Changes in human behavior such as massive deforestation, dam construction, the extinction of natural predators, and changes in biodiversity have increased the risk of exposure to mosquitoes [53].

"Urbanization can expand habitats for some species of mosquito that prefer cities, so as people expand into natural areas, those species will go with them." Urban settings have plenty of habitat and food, and mosquitoes lack natural predators in cities [78].

As well, the expansion of vector populations, as a result of climate change, into disease-free areas or areas where disease endemicity is insufficient to elicit good protective immunity will often lead to initial high rates of disease transmission that will decrease in time as the population develops immunity. Similar considerations on population immunity apply to the transmission of mosquito-borne diseases in coastal zones [77].

5.8 Conclusion

"Climate change is going to sicken and kill many people," says Colin Carlson, a biologist at the Center for Global Health and Security at Georgetown. "Mosquito-borne diseases are going to be a big way that happens" [76].

The association between anthropogenic action in the Amazon rainforest, climate change, alterations in vector dynamics, human migration, genetic changes in pathogens, and the poor social and environmental conditions in many Latin American countries can give rise to the "perfect storm" for the emergence and re-emergence of human infectious diseases. In Brazil and other Amazonian countries [32].

A recent study by Sadie Ryan, a medical geographer at the University of Florida, found that heat-loving *Aedes* under a high carbon-emissions scenario (which would cause more severe global warming). *Aegypti* will move into West Africa, putting millions of Africans at risk for dengue, Zika, and other diseases [76]. They will probably find a new home in a lot of places.

One WHO publication predicts that warming of two to three degrees Celsius would put up to 7% more people—several hundred million globally—at risk of malaria, another mosquito-borne disease. "Zika showing up in the Americas is probably more a function of international travel and trade than climate change. But now that the virus has started circulating in the Americas, both climate factors and human behavior will play a role in where it spreads," Kathryn Jacobsen, a professor of global health at George Mason University, said [80].

There is also truly little doubt that ensuing changes in climatic patterns will lead to myriad adverse outcomes, including heat waves, droughts, and increased frequency and violence of significant weather events, and will promote the spread of infectious diseases such as malaria and gastrointestinal infections [96].

Approximately 500 million people in Latin America and the Caribbean are at risk for ZIKV infection because they live in areas less than 2000 m above sea level where competent Aedes vectors are also found [97]. Claudia Codeco, a biologist studying mosquito-borne infections at Rio's Oswaldo Cruz Foundation, said her research suggested similar findings. "ZIKV is not a local problem anymore. This is not a problem in Rio de Janeiro or Brazil. This is already present in 60 countries," she said. "So maybe at this time of the year, we should pay more attention and advise tourists to go where summer is beginning [98].

A creative and organized application of resources is urgently required to control these diseases regardless of future climate change [69]. Whether the nations of the world decide to do anything about climate change, it is clear that climate change and its subdrivers will maintain the state of the ZIKV before the public for some time. Now is the time to move on to the processes of transmission.

References

1. Murray JB, Evers DJ, Janda S (1995) Marketing, theory borrowing, and critical reflection. J Macromark 15:92. https://doi.org/10.1177/027614679501500207
2. Rittel HWJ, Webber MW (1973) Dilemmas in a general theory of planning. Policy Sci 4:155–159; van Woezik AFG, Braakman-Jansen LMA, Kulyk O, Siemons L, van Gemert-Pijnen JEWC (2016) Tackling wicked problems in infection prevention and control: a guideline for co-creation with stakeholder. Antimicrobial Resistance Infect Control 5(20). https://doi.org/10.1186/s13756-016-0119-2
3. von Hippel E (1994) "Sticky information" and the locus of problem solving: implications for innovation. Manage Sci 40(4):429–439, April
4. Imperator PJ (2016) The convergence of a virus, mosquitoes, and human travel in globalizing the Zika epidemic. J Community Health 41:674–679
5. Ali S, Gugliemini O, Harber S, Harrison A, Houle L et al (2017) Environmental and social change drive the explosive emergence of Zika virus in the Americas. PLoS Negl Trop Dis 11(2):e0005135. https://doi.org/10.1371/journal.pntd.0005135
6. Bailey M (2016) Despite Zika hype, bikinis—and no mosquitoes—on Rio's beaches. Stat. August 5. https://www.statnews.com/2016/08/05/zika-olympics-no-mosquiotes/. Accessed 20 Jun 2022
7. Sands P, Munduca-Shah C, Dzau VJ (2019) The neglected dimension of global security—a framework for countering infectious-disease crises. New England J Med 374:13. https://doi.org/10.1056/NEJMsr1600236
8. Smith F (2020) On the hunt for the next deadly virus. National Geographic. June 16. https://www.nationalgeographic.com/science/article/coronavirus-on-the-hunt-for-the-next-deadly-virus. Accessed 25 May 2022
9. Ferguson NM (2016) Countering the Zika epidemic in Latin America. Science. July 14. https://doi.org/10.1126/science.aag0219
10. Braack L et al (2018) Mosquito-borne arboviruses of African origin: review of key viruses and vectors. Parasites and Vectors. 11:29. https://www.ncbi.nlm.nih.gov/pubmed/29316963. Accessed 27 Jun 2018
11. Pergolizzi Jr J et al (2020) The Zika virus: lurking behind the COVID-19 pandemic? J Cli Pharm Therapeutics. October 22. https://doi.org/10.1111/jcpt.13310
12. Plowright RK, Eby P, Hudson PJ, Smith IL, Westcott D, Bryden WL et al (2015) Ecological dynamics of emerging bat virus spillover. Proc R Soc B 282(1798):20,142,124. https://doi.org/10.1098/rspb.2014.2124
13. Morse SS (1995) Factors in the emergence of infectious diseases. Emerg Infect Diseas 1:1. January–March. https://wwwnc.cdc.gov/eid/article/1/1/95-0102_article. Accessed 18 Sept 2018
14. Marcondes C, de Fátima Freire, de Melo Ximenes (2016) Zika virus in Brazil and the danger of infestation by Aedes (Stegomyia) mosquitoes. Revista da Sociedade Brasileira de Medicina Tropical. 49:1. January-February. http://www.scielo.br/scielo.php?script=sci_arttext&pid=S0037-86822016000100004. Accessed 24 Jul 2018
15. OECD (2018) Safety assessment of transgenic organisms in the environment, Volume 8: OECD Consensus Document of the Biology of Mosquito Aedes aegypti, Harmonisation of Regulatory Oversight in Biotechnology, OECD Publishing, Paris
16. Atkins K (2017) First travel-related Zika case of 2017 reported. Florida Keys News. February 15. http://www.flkeysnews.com/news/local/environment/article132829289.html. Accessed 7 Mar 2017
17. Troncoso A (2016) Zika threatens to become a huge worldwide pandemic. Asian Pacific J Tropical Biomed 6:6. http://www.sciencedirect.com/science/article/pii/S2221169116302921. Accessed 4 Jun 2017; Aliota MT et al (2017) Zika in the Americas, year 2: what have we learned? What gaps remain? A report from the Global Virus Network. Antiviral Res 114. https://www.ncbi.nlm.nih.gov/pubmed/28595824. Accessed 15 May 2019

18. Wolfe ND et al (2000) Deforestation, hunting, and the ecology of microbial emergence. Glob Change Hum Health July 1:10. https://doi.org/10.1023/A:1011519513354
19. Edwards SB The ZIKV. Minneapolis, MN, ABDO Publishing
20. Noor R, Ahmed T (2018) ZIKV: epidemiological study and its association with public health risk. J Infect Publ Health. April 28. https://www.sciencedirect.com/science/article/pii/S1876034118300431. Accessed 25 Jul 2018
21. Gubler DJ (2011) Dengue, urbanization, and globalization: the unholy trinity of the 21st century. Tropical Med Health. 39:4. Supplement. https://www.ncbi.nlm.nih.gov/pmc/articles/PMC3317603/. Accessed 17 Jul 2017
22. Bogoch II et al (2016) Potential for ZIKV introduction and transmission in resource-limited countries in Africa and the Asia-Pacific region: a modelling study. The Lancet: Infectious Diseases. 16. November. http://www.thelancet.com/journals/laninf/article/PIIS1473-3099(16)30270-5/abstract. Accessed 26 Jun 2017
23. Calder JAM (2017) Zika virus in the Americas: is it time to revisit mosquito elimination? J Environ Health September 80(2):26–27
24. Moreno-Madriñán M, Turrell M (2017) Factors of concern regarding Zika and other Aedes aegypti-transmitted viruses in the United States. J Med Entomol 54(2). March. https://academic.oup.com/jme/article/54/2/251/2952765/Factors-of-Concern-Regarding-Zika-and-Other-Aedes. Accessed 23 May 2017
25. Vorou R (2016) ZIKV, vectors, reservoirs, amplifying hosts, and their potential to spread worldwide: what we know and what we should investigate urgently. Int J Infect Diseas 48 https://www.sciencedirect.com/science/article/pii/S1201971216310578. Accessed 7 Jul 2018
26. Nagao Y et al (2003) Climatic and social risk factors for Aedes infestation in rural Thailand. Tropical Med Int Health. 8:7. July. http://onlinelibrary.wiley.com/doi/https://doi.org/10.1046/j.1365-3156.2003.01075.x/pdf. Accessed 18 Jul 2017
27. Stone J (2017) How Brazil's Zika epidemic women's everyday plight: watch. Forbes. July 15. https://www.forbes.com/forbes/welcome/?toURL=https://www.forbes.com/sites/judystone/2017/07/15/how-brazils-zika-epidemic-highlights-womens-every-day-plight-human-rights-watch/&refURL=https://www.google.com/&referrer=https://www.google.com/. Accessed 21 Jul 2017
28. Everard M, Johnston P, Santillo D, Staddon C (2020) The role of ecosystems in the mitigation and management of Covid-19 and other Zoonoses. Environ Sci Policy 111:7–17
29. Lafrance A (2016) A Zika catastrophe could rival Hurricane Katrina. The Atlantic. April 11. https://www.theatlantic.com/health/archive/2016/04/zika-katrina/477708/. Accessed 19 May 2017
30. Moreno-Madriñán M, Turell M (2017) Factors of concern regarding Zika and other Aedes aegypti-transmitted viruses in the United States. J Med Entomol 54:2. January 3. https://academic.oup.com/jme/article-abstract/54/2/251/2952765/Factors-of-Concern-Regarding-Zika-and-Other-Aedes?redirectedFrom=fulltext. Accessed 17 Jul 2017
31. Anonymous (2018) Nations must be vigilant against mosquito-borne diseases—review. Horsetalk.co.NZ. January 15. https://www.horsetalk.co.nz/2018/01/15/nations-vigilant-mosquito-borne-diseases/, Accessed 27 Jun 2018
32. Ellwanger JH, Kulmann-Leal B, Kaminski VL, Varverde-Villegas KM et al (2020) Annals of the Brazilian Academy of Sciences 92(1):e20191375. https://doi.org/10.1590/0001-3765202020191375
33. Everard M, Johnston P, Santillo D, Staddon C (2020) The role of ecosystems in mitigation and management of Covid-19 and other Zoonoses. Environ Sci Policy 111:7–17
34. Gubler D, Clark GC (1996) Community involvement in the control of Aedes aegypti. Acta Tropica. 61. https://www.ncbi.nlm.nih.gov/pubmed/8740894. Accessed 26 Jun 2017; Deniz D (2016) Zika: from the Brazilian backlands to global threat. London, Zed books; Aliota MT et al (2017) Zika in the Americas, year 2: what have we learned? What gaps remain? A report from the Global Virus Network. Antiviral Res 114. https://www.ncbi.nlm.nih.gov/pubmed/28595824. Accessed 15 May 2019

35. Keesing F, Ostfeld RX (2020) Impacts of biodiversity and biodiversity loss on zoonotic diseases. PNAS. 118(17). https://doi.org/10.1073/pnas.2023540118; Civitello DJ et al (2015) Biodiversity inhibits parasites: broad evidence for the dilution effect. PNAS. 112:8667–8671; Keesing F et al (2010) Impacts of biodiversity on the emergence and transmission of infectious diseases. Nature 468:647–652
36. Tollefson J (2020) Why deforestation and extinctions make pandemics more likely. Nature. August 7. 584:175–176
37. Gibb R et al (2020) Zoonotic host diversity increases in human-dominated ecosystems. Nature. August 5. 584:398–402
38. Keesing F, Ostfeld RX (2020) Impacts of biodiversity and biodiversity loss on zoonotic diseases. PNAS. 118(17). https://doi.org/10.1073/pnas.2023540118
39. Ratna K (2017) The Secret life of ZIKV. Speaking Tiger Books, New Delhi
40. Weinstein JS, Leslie T, von Fricken M (2020) Spatial associations between land use and infectious disease: Zika virus in Colombia. Int J Environ Res Public Health. February 11. 17:1127. https://doi.org/10.3390/ijerph17041127 www.mdpi.com/journal/ijerph
41. Rulli MC, D'Odorico PD, Galli N, Hayman D (2021) Land-use change and the livestock revolution increase the risk of zoonotic coronavirus transmission from rhinophid bats. Nature Food. 2. June. 409–416
42. Kaddumukasa AM et al (2015) High proportion of mosquito vectors in Zika forest, Uganda, feeding on humans has implications for the spread of new arbovirus pathogens. African J Biotechnol 14(16). April. https://www.ajol.info/index.php/ajb/article/viewFile/117916/107549. Accessed 17 May 2017
43. dos Santos TP, Roiz D, de Abreu FVS, Luz SLB et al (2018) Potential of Aedes albopictus as a bridge vector for enzootic pathogens at the urban-forest interface in Brazil. Emerg Microbes Infect November 28. https://doi.org/10.1038/s41426-018-0194-y
44. Lourenço-de-Oliveira R, Castro MG, Braks MAH, Lounibos LP (2004) The invasion of urban forest by dengue vectors in Rio de Janeiro. Bull Soc Vector Ecol 29:94–100
45. Wolfe ND et al (2000) Deforestation, hunting and the ecology of microbial emergence. Glob Change Hum Health. July. 1. 10. https://doi.org/10.1023/A:1011519513354
46. EcoHealth Alliance (2019) Infectious disease emergence and economics of altered landscapes—IDEEAL. EcoHealth Alliance, New York. https://www.ecohealthalliance.org/wp-content/uploads/2019/09/IDEEAL_report_final.pdf. Accessed 3 Jun 2022
47. Murray KA, Daszak P (2013) Human ecology in pathogenic landscapes: two hypotheses on how land use change drives viral emergence. Curr Opin Virol 3:79–83
48. Tesla B, Demakovsky LR, Mordecai EA, Ryan SJ et al (2018) Temperature drives Zika virus transmission: evidence from empirical and mathematical models. Proc Royal Soc. Proceedings B. 285:20180795. https://doi.org/10.1098/rspb.2018.0795
49. Patz JA, Campbell-Lendrum D, Holloway T, Foley JA (2005) Impact of regional climate change on human health. Nature 438(17):310–317
50. dos Santos TP, Roiz D, Santos de Abreu FV, Luz SLB et al (2018) Potential of Aedes albopictus as a bridge vector for enzootic pathogens at the urban-forest interface in Brazil. Emerg Microbes Infect 7(1):1–8. https://doi.org/10.1038/s41426-018-0194-y
51. Norris D (2004) Mosquito-borne diseases as a consequence of land use change. EcoHealth. 1. https://link.springer.com/article/https://doi.org/10.1007/s10393-004-0008-7. Accessed 27 Jun 2017
52. Caminade C, Turner J, Metelmann S, Hesson JC et al (2016) Global risk model for vector-borne transmission of Zika virus reveals the role of El Niño 2015. PNAS. January 3. 114(1):119–124
53. Troncoso A (2016) Zika threatens to become a huge worldwide pandemic. Asian Pacific J Tropical Biomed. 6:6. http://www.sciencedirect.com/science/article/pii/S2221169116302921. Accessed 4 Jun 2017
54. Raju AK (2003) Community mobilization in Aedes aegypti control programme by source reduction in Peri-Urban District of Lautoka, Viti Levu, Fiji Islands. Dengue Journal. 27. http://apps.who.int/iris/bitstream/handle/10665/163791/dbv27p149.pdf;jsessionid=FC58F0B1A6A1CE84C3FA8EA91A6D7A83?sequence=1. Accessed 9 Aug 2018

55. Gubler D, Clark GC (1996) Community involvement in the control of Aedes aegypti. Acta Tropica. 61. https://www.ncbi.nlm.nih.gov/pubmed/8740894. Accessed 26 Jun 2017
56. Hoberg EP, Brooks DR (2015) Evolution in action: climate change, biodiversity dynamics and emerging infectious disease. Philos Trans R Soc B 370:20130553. https://doi.org/10.1098/rstb.2013.0553
57. Jones KE, Patel NG, Levy MA, Storeygard A et al (2008) Global trends in emerging infectious diseases. Nature 451:990–993
58. Patz JA et al (2003) Climate change and infectious diseases. In: McMichael AJ, Campbell-Lendrum DH, Corvalan CF, Ebi KL, Githeko JD, Woodward A (eds) Climate change and human health. WHO, Geneva, pp 103–132
59. WHO (2003 Climate change and human health risks and responses. Summary. WHO. http://www.who.int/globalchange/climate/summary/en/. Accessed 29 Sept 2016
60. Charrel R et al (2016) Background review for diagnostic test development for Zika virus infection. Bulletin of the World Health Organization. 94:574–584D. https://doi.org/10.2471/BLT.16.171207. http://www.who.int/bulletin/volumes/94/8/16-171207.pdf. Accessed 28 Mar 2017
61. Martin V, Chevalier V, Ceccato P, Anyamba A et al (2008) The impact of climate change on the epidemiology and control of Rift Valley fever. Revue scientifique et technique/ Office international des épizooties. 27(2):413–436
62. Pandit PS, Doyle MM, Smart KM, Young C et al (2018) Predicting wildlife reservoirs and global vulnerability to zoonotic Flaviviruses. Nat Commun 9(5425). December 21. https://doi.org/10.1038/s41467-018-07896-2
63. Gage KL, Burkot TR, Eisen RJ, Hayes JE (2008) Climate and vectorborne diseases. Am J Prev Med 35(5):436–450
64. Gould E, Higgs S (2009) Impact of climate change and other factors on emerging arbovirus diseases. Trans Royal Soc Trop Med Hygiene. 103:2. 109–121; Deniz D (2016) Zika: from the Brazilian Backlands to Global Threat. London: Zed Books
65. ASTHO (2015) Before the Swarm: guidelines for the emergency management of vector-borne disease outbreaks. http://www.astho.org/Programs/Environmental-Health/Natural-Environment/Before-the-Swarm/. Accessed 28 Aug 2018
66. Braack L et al (2018) Mosquito-borne arboviruses of African origin: review of key viruses and vectors. Parasites and Vectors. 11:29. https://www.ncbi.nlm.nih.gov/pubmed/29316963. Accessed 27 Jun 2018
67. Gould E, Higgs S (2009) Impact of climate change and other factors on emerging arbovirus diseases. Trans Royal Soc Tropical Med Hygiene 103(2):109–121
68. Novak S (2018) Scientists race to kill mosquitoes before they kill Us. Sierra Club. January 9. https://www.sierraclub.org/sierra/scientists-race-kill-mosquitoes-they-kill-us and https://www.cdc.gov/zika/vector/range.html. Accessed 27 Jun 2018
69. Reiter P (2001) Climate change and mosquito-borne disease. Environ Health Perspect 109(Supp 1). March. https://www.ncbi.nlm.nih.gov/pmc/articles/PMC1240549/. Accessed 11 Aug 2018
70. Morin C, Comrie AC, Ernst K (2013) Climate and dengue transmission: evidence and implications. Environ Health Perspect 121:11–12. November-December. https://ehp.niehs.nih.gov/1306556/. Accessed 17 Jul 2017
71. Patz JA et al (1996) Global climate change and emerging infectious diseases. J Am Med Assoc 275:3. January 17. https://www.ncbi.nlm.nih.gov/pubmed/8604175. Accessed 19 Jul 2017
72. Epstein PR (2000) Is global warming harmful to health? Sci Am, August 283(2):50–57
73. Ryan SJ, Carlson C, Mordecai E, Johnson L Global expansion and redistribution of Aedes-borne virus transmission risk with climate change. bioRxiv. March 4. https://doi.org/10.1101/172221. Accessed 6 Jun 2022
74. Gould EA, Higgs S (2009) Impact of climate change and other factors on emerging arbovirus diseases. Trans Royal Soc Tropical Med Hygiene. February. 103(2):109–121
75. Casadevall A (2020) Climate change brings the specter of new infectious diseases. J Clin Investigation, February 2020. 130(2):S53–S54

76. Goodell J (2020) How climate change is ushering in a new pandemic era. Rolling Stone. December 7. https://www.rollingstone.com/culture/culture-features/climate-change-risks-infectious-diseases-covid-19-ebola-dengue-1098923/. Accessed 6 Jun 2022
77. Ramasamy R, Surendran SN (2012) Global climate change and its potential impact on disease transmission by salinity-tolerant mosquito vectors in coastal zones. Frontiers Physiol June 19. https://doi.org/10.3389/fphys.2012.00198. Accessed 6 Jun 2022
78. Poon L (2018) Another consequence of climate change: More disease-spreading mosquitoes. Pacific Standard. October 15. https://psmag.com/environment/climate-change-means-more-deadly-mosquitoes. Accessed 6 Jun 2022
79. Werner E (2021) Climate change lets mosquitoes flourish—and feast—in Los Angeles. The Washington Post. September 19. https://www.washingtonpost.com/us-policy/2021/09/19/climate-mosquito-los-angeles/. Accessed 6 Jun 2022
80. Mercer G (2016) The link between Zika and climate change. The Atlantic. February 24. https://www.theatlantic.com/health/archive/2016/02/zika-and-climate-change/470643/. Accessed 23 May 2017
81. Climate Prediction Center (2016) El Nino/Southern Oscillation (ENSO) Diagnostic Discussion. September 8. Nature. 438(17). November. 310-http://www.cpc.ncep.noaa.gov/products/analysis_monitoring/enso_disc_sep2016/ensodisc.pdf. Accessed 18 Sept 2016
82. Ryan SJ, Carlson C, Mordecai E, Johnson L Global expansion and redistribution of Aedes-borne virus transmission risk with climate change. bioRxiv. March 44. https://doi.org/10.1101/172221. Accessed 6 Jun 2022
83. Levy M (2019) How scientists use climate models to predict mosquito-borne disease outbreaks. The Smithsonian. May 10. https://maxglevy.com/2019/05/10/how-scientists-use-climate-models-to-predict-mosquito-borne-disease-outbreaks/. Accessed 6 Jun 2022
84. Lessler J, Chaisson L, Kucirka L, Bi Q et al (2016) Assessing the global threat from Zika virus. Science. August 12. 353(6300). https://doi.org/10.1126/science.aaf8160
85. Charrel R et al (2016) Background review for diagnostic test development for Zika virus infection. Bulletin of the World Health Organization. 94:574–584D. /https://doi.org/10.2471/BLT.16.171207. http://www.who.int/bulletin/volumes/94/8/16-171207.pdf. Accessed 28 Mar 2017; Atif M, Azeem M, Sarwar MR, Bashir A (2016) Zika virus disease: a current review of the literature. Infection. 44(6):695–705
86. Atkins K (2018) How best to kill skeeters? Experts pitch their plans. Florida Keys News. February 23. http://www.flkeysnews.com/news/local/article201838619.html. Accessed 18 Jul 2018
87. Infectious Diseases News (2016) El Niño fueled Zika outbreak in Latin America. Infectious Diseases News. December 22. https://www.healio.com/news/infectious-disease/20161222/el-nino-fueled-zika-outbreak-in-latin-america. Accessed 7 Jun 2022
88. Summers H (2018) Invest in mosquito surveillance to combat malaria, says Bill Gates. The Guardian. April 18. https://www.theguardian.com/global-development/2018/apr/18/commonwealth-nations-malaria-summit-london. Accessed 8 Aug 2018
89. Vasquez D, Palacio A, Nunez J, Briones W et al (2017) Impact of the 2016 ecuador earthquake on Zika virus cases. Am J Publ Health. https://doi.org/10.2105/AJPH.2017.303769
90. A stage in insect development that immediately precedes the imago while imago an insect in its final, adult, sexually mature, and typically winged state
91. Ramasamy R, Surendran SN, Jude PJ, Dharshini S, Vinobaba M (2011) Larval development and of Aedes aegypti and Aedes albopictus in peri-urban brackish water and its implications for transmission of arboviral diseases. PLoS Negl Trop Dis 5:e1369. https://doi.org/10.1371/journal.pntd.0001369
92. Franchini M, Mannucci PM (2015) Impact on human health of climate changes. Euro J Int Med. 26. http://www.sciencedirect.com/science/article/pii/S0953620514003628. Accessed 14 Apr 2017
93. Molaei G et al (2008) Host-Feeding patterns of potential mosquito vectors in connecticut, USA: molecular analysis of bloodmeals from 23 species of Aedes, Anopheles, Culex, Coquillettidia, Psorophora, and Uranotaenia. J Med Entomol. 45:6. https://www.ncbi.nlm.nih.gov/pubmed/19058640. Accessed 17 May 2017

94. Morin C, Comrie A, Ernst T (2014) Climate and Dengue transmission: evidence and implications. Environ Health Perspect 121:11–12. November–December. https://ehp.niehs.nih.gov/1306556/#tab1. Accessed 23 Mar 2017
95. Kyle J, Harris EVA (2008) Global spread and persistence of dengue. Ann Rev Microbiol 62. https://www.ncbi.nlm.nih.gov/pubmed/18429680. Accessed 18 May 2017; Focks D, Barrera R (2007) Dengue transmission dynamics: assessment and implications for control. In: Report of the scientific working group meeting on dengue. Geneva, WHO. http://www.who.int/tdr/publications/documents/swg_dengue_2.pdf. Accessed 18 May 2017
96. Barrett B, Charles J, Temte J (2014) Climate change, human health, and epidemiological transition. Preventive Med. November 28. https://www.ncbi.nlm.nih.gov/pubmed/25434735. Accessed 16 Jan 2017
97. dos Santos T et al (2016) Zika virus and the Guillain–Barré Syndrome—case series from seven countries. New England J Med. August 31. http://www.nejm.org/doi/full/https://doi.org/10.1056/NEJMc1609015#t=article. Accessed 24 Sept 2016
98. Darlington S (2016) How bad is Zika in Rio? CNN Health. June 15. http://www.cnn.com/2016/06/15/health/who-olympics-should-not-be-postponed/. Accessed 18 Sep 2016

Chapter 6
Transmission

Current evidence suggests that the ZIKV can be transmitted through the bite of an infected mosquito and via sexual contact with an infected male partner, blood transfusion, or vertical maternal–fetal transmission prenatally or intrapartum. Thus far, the most common mechanism of information is mosquito-borne. See Fig. 6.1.

The transmission of the virus via amniotic fluid is also discussed briefly in Chap. 2 under mutation and ferocity and in this chapter on children from infected pregnant mothers.

Mosquitoes suck. Nuisance mosquitoes are bothersome in residential or recreational areas. They can have a significant economic impact, as they may reduce property values, slow the economic development of a site, reduce tourism, or affect livestock and poultry production [1]. In this case, they kill and maim.

There are two transmission cycles, one sylvatic involving mosquitoes and non-human primates and the other urban or human. Anti-ZIKV antibodies have been detected in ducks, goats, horses, bats, cows, and carabaos (water buffalo) from Indonesia, showing the widespread circulation of the virus in domestic animals [2]. In Asia, samples collected in 1996 and 1997 from semicaptive orangutans in Borneo, Malaysia, detected anti-ZIKV antibodies [3]. In the Americas, the remarkable diversity of non-human primate species provides the potential to establish a sylvatic ZIKV cycle and more than 200 mosquito species [4].

While the following focuses on the second urban and human cycle, it is essential to note that the sylvatic process can support reservoirs of the virus when the virus is not producing havoc among human populations. Briefly, mosquitoes and primates are born susceptible to ZIKV infection. They are infected at a rate proportional to the number of bites given or received per day and the probability of disease [5].

There seems to be a consensus that the ZIKV can be transmitted when bitten by the mosquito carrying the virus and sexual transmission. It can be passed from a person with Zika before their symptoms start, while they have symptoms, and after signs end. It may also be given by someone infected with the virus but never developed symptoms [6].

D. M. Berube, *Pandemics and Resilience: Lessons we should have learned from Zika*, Risk, Systems and Decisions, https://doi.org/10.1007/978-3-031-25370-6_6

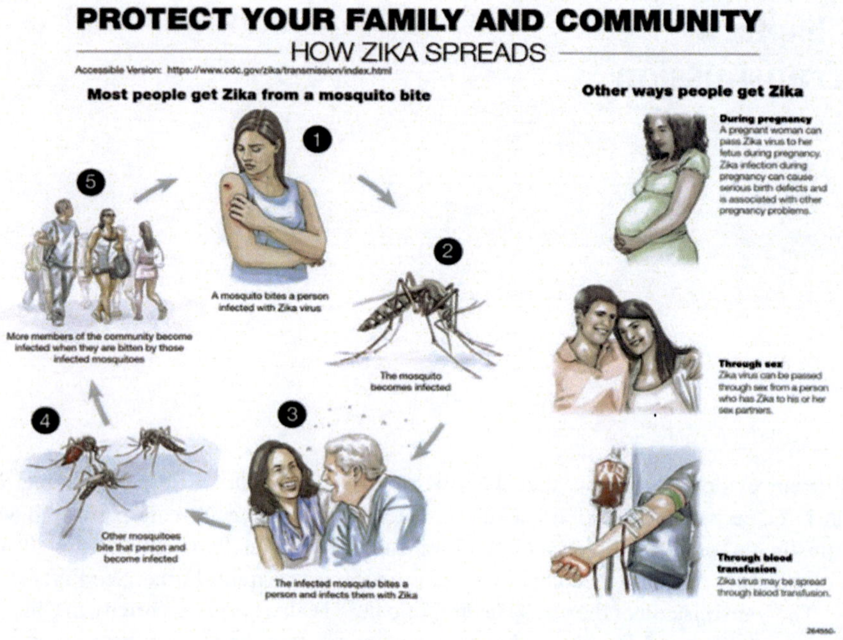

Fig. 6.1. Transmission routes. CDC. Zika Transmission. Zika Virus. https://www.cdc.gov/zika/prevention/transmission-methods.html. Accessed July 3, 2022

According to the American Association of Pharmaceutical Scientists, the ZIKV can live for several hours on hard nonporous surfaces and still be highly contagious under certain conditions. Still, some commonly used disinfectants are highly effective in killing the virus. "Zika can survive on hard, nonporous surfaces for as long as eight hours, longer when the environment contains blood, which is more likely to occur in the real world," said the study's lead researcher S. Steve Zhou, Ph.D. "The good news is that disinfectants such as isopropyl alcohol and quaternary ammonium/alcohol generally kill the virus in this environment and do so in 15 s" [7].

The virus has been isolated from blood, urine, saliva, semen, amniotic fluid, and neurologic tissue [8]. Antibodies to ZIKV become detectable on average nine days after infection [9]. However, the infectiousness of breast milk and urine and data on saliva, animal bites, transplantation, needlestick injury, and laboratory work are inconclusive [10].

6.1 Transmission: Mosquito to Larva

The immature stages of all mosquitoes occur in water; only the adult mosquito lives out of water. Up to 250 are laid singly or in clusters, depending on the species. Those laid on water hatch in 1–5 days, while eggs laid out of water can remain dormant for weeks or months until covered with water. Development stages from egg to adult take 8–14 days, depending on food quality and water temperature. Adult mosquitoes live for one to three months [11].

Vertical transmission involves the virus transfer between mosquitoes, larvae, and eggs. It is known to know *Ae. aegypti* mosquitoes are hardy. Even as mosquito populations dip, these diseases can survive the winter in their eggs. This supports the fear of overwintering, whereby infected eggs may survive the winter. This can mean that once containers fill with water, just a few infected eggs may pass on the infection to others in the clutch upon hatching.

Transovarial transmission involves the transmission of the virus from the infected female mosquito to its offspring. Extraordinarily little has been reported about this phenomenon. It is briefly mentioned in a leaflet size book by Vyas [12].

While there is evidence that dengue, West Nile, and chikungunya can be vertically transmitted, mosquitoes can sometimes pass the viruses to their offspring. It took under 2020 with a new study by Comeau et al., [13] published in the American Journal of Tropical Medicine and Hygiene that demonstrated that ZIKV might also be vertically transmitted.

6.2 Transmission: Mosquito to Mosquito

Until now, there has been no substantial evidence that ZIKV can be transmitted sexually among mosquitoes. However, it is known that female mosquitoes can become infected with arboviruses during hematophagy (blood-feeding, and that females can then vertically transmit viruses to their eggs. Vertical transmission of ZIKV has been observed in experimentally infected *Ae. aegypti* specimens. ZIKV infection of *Ae. aegypti* mosquitoes occurs not only during blood-feeding but also during copulation.

Pereira-Silva et al., showed the presence of ZIKV RNA in previously uninfected mosquitoes of both sexes following copulation with ZIKV-infected mates. These data strongly support the possibility that ZIKV is transmitted in the sexual fluids of mating mosquitoes. This concerns public health because of the venereal transmission of Zika among *Ae. aegypti* mosquitoes could potentially increase mosquito infection rates and thereby increase the spread of the virus [14].

6.3 Transmission: Mosquito to Human (Autochthonous) and Human to Mosquito

When a mosquito bites an infected person during this period of viremia (when the virus is in the blood), it acquires and spreads it to others. The *Aedes* mosquitoes that carry Zika feed primarily on humans, often bite multiple humans in a single meal and live near human habitation. Mosquitoes have sex, and then they die. (The average adult lifespan is 8–15 days for female mosquitoes and 3–6 days for male mosquitoes). Female mosquitoes bite, and research suggests that *Ae. aegypti* can transmit ZIKV for ten weeks with an extrinsic incubation period of 3–12 [15] up to 15 days [16]. One study reported survival of the ZIKV for two weeks in female mosquitoes, suggesting the stability of the virus in the vector host [15].

When a healthy *Aedes* mosquito bites an infected person or other warm-blooded mammals, the virus enters the mosquito with the blood. Then, when this mosquito bites a healthier person, he becomes infected [17].

Kenneth Ratzan, the chief of infectious diseases at Mt. Sinai Hospital in Miami Beach, explained the implications. "We just don't yet have enough people infected with those diseases to serve as a reservoir for the mosquitoes to pick up and transmit the disease by feeding on those already infected, ready to transmit by a mosquito bite the disease to someone susceptible" [18]. Nonetheless, during the first week of ZIKV infection, it is recommended the infected patient should avoid further mosquito bites because the ZIKV can be found in the blood and pass from an infected person to a mosquito. Consequently, an infected mosquito can then spread the virus to another person [19]. However, it seems likely if a mosquito bites a person carrying this virus, the mosquito is expected to pass it on. Some suggest this is possible for up to fifteen other people throughout the mosquito's lifespan [20].

Ae. aegypti mosquitoes carry dengue fever, West Nile virus, yellow fever, Rift valley fever, Murray valley encephalitis, chikungunya, Japanese encephalitis, etc. It is also believed to be responsible for sending the Mayaro virus, which recently appeared in this region for the first time in Haiti [21].

Most sources argue that ZIKV is carried **by** the *Ae. aegypti* mosquito (though there is evidence *Aedes* also has it. For example, *albopictus* as was the case in Gabon in 2007 and 2010) [22]. It was demonstrated that Singapore's *Ae. aegypti* could transmit ZIKV [23]. In contrast, *Ae. aegypti* strains from Kedougou and Dakar (Senegal) were not competent to transmit ZIKV [24]. Similar results recently were observed with *Ae. aegypti* strains from the Americas were susceptible to ZIKV infection but had unexpectedly low transmission potential [25].

As we shall see (Chaps. 11–14), *Ae. aegypti* is exceedingly difficult to control. Although insecticides can drastically reduce the numbers of salt marsh mosquitoes, which are merely a nuisance to humans, the same techniques have been less successful in addressing populations of the far more dangerous *Ae.* aegypti. These mosquitoes live in residential areas (they love closets, for example) and can lay eggs in dry places, where they will sit dormant until the rain. For instance, it is known that large outside drums, which residents use to store rainwater in the tropics, are major breeding sites for *Ae. aegypti*, the dengue vector, ZIKV, chikungunya, and many other viruses. However, when looking at a backyard cluttered with various small objects collecting water, it is difficult to pinpoint which potential breeding sites would be more essential to remove [26].

6.4 Transmission: Sexually Transmitted Zika

Although infected mosquitoes transmit most ZIKV infections, ZIKV transmission has been documented as occurring through sexual contact; ZIKV RNA has been detected in semen, urine, saliva, cerebrospinal fluid, vaginal or cervical secretions, and other body fluids [27]. And this phenomenon seems to set the ZIKV apart from most other infectious diseases. At least twenty-three of the CDC reported Florida cases had been transmitted sexually [28].

The first sexually transmitted case was reported in the female sexual partner of a ZIKV-infected male who was infected in Senegal [29]. According to Wang et al., "it is possible that other transmission routes, such as sexual transmission, may have a greater contribution to the widespread ZIKV in the Americas" [30]. One mathematical modeling study of Barranquilla, Colombia, estimated that as many as 47% of ZIKV cases reported emerged from sexual contact [31].

The *Aedes* mosquitoes carry the virus around, but it's becoming clear that bodily fluids can sometimes transmit it [32]. Indeed, some have commented that mosquitoes and men transmit Zika.

The CDC may have directed men to stop traveling to areas with Zika. Since the ZIKV can be found in semen for longer than six months and is much more easily cleared by a woman, it makes more sense to limit men's travel to Zika endemic areas than women (pregnant or not) [33]. To be somewhat fair, warnings began early in the outbreak, warning males who a ZIKV-bearing mosquito had bitten to refrain from sexual relations fearing the virus would be transmitted to their sexual partner and, in turn, the offspring.

In the August 2017 issue of Emerging Infectious Diseases, Haddow and colleagues reported that four of eight macaques exposed to the virus vaginally developed infections, as did seven of eight that received the virus via the rectum [34].

A mice study concluded that the virus replicated in the vagina for four or five days and seven days in mice with deficient interferon responses. In mice infected with ZIKV, the virus hit these brain regions hard. Nerve cells died, and the parts generated one-fifth to one-half as many new cells as those of uninfected mice. The results might not translate to humans; the mice were genetically engineered to have weak immune systems, making them susceptible to ZIKV [35].

Yockey admitted, "We can't generalize findings in mice to humans, of course, but given the evidence that ZIKV can survive in the human vagina, and the fact that humans are generally more susceptible to the virus than mice, the researchers speculate that ZIKV "introduced into the human vagina is likely to replicate more robustly than in the vaginal cavity of [wild type] mice" [36].

ZIKV has been isolated from semen collected 14 days after symptoms start, while detection of the ZIKV genome was described in semen at 28 and 62 days after the onset of symptoms [15]. A recent report described a patient with symptomatic ZIKV infection in whom the virus was detected in semen for 92 days, supporting recommendations for six months of barrier contraceptive use after symptomatic ZIKV disease

[37]. Viral RNA has been detected for up to 93 days and 11 days after symptom onset in sperm and females' cervical mucus [38].

> Still, another team led by Prisant found comparable results. This time it was a human case study. Doctors at Pointe à Pitre University Hospital in Guadeloupe treated a woman with Zika in May and detected the virus in her genital tract. When she first came in, her blood tested positive for Zika, though her urine did not. Three days after the onset of her symptoms, doctors did a genital swab, a swab around the opening of the cervix, and took a sample of cervical mucus. All three models tested positive for ZIKV RNA. They did another round of tests 11 days after her symptoms started—of blood, urine, and cervical mucus—and while the blood and urine were clear, the virus was still in the mucus [39].

Several cases of sexual transmission of the ZIKV have been documented, related to viral shedding in semen [29]. ZIKV has been isolated from semen collected 14 days after symptoms start, while detection of the ZIKV genome was described in semen at 28 and 62 days after the onset of symptoms [15]. A 2017 report from Gaskell et al. described a patient with symptomatic ZIKV infection in whom the virus was detected in semen for 92 days, supporting recommendations for six months of barrier contraceptive use after symptomatic ZIKV disease [37]. Data from Barzon et al., verify the six months. Barzon et al. suggest ZIKV can survive longer in semen than in the vagina, ***up to six months*** in semen [40]. According to their data, Paz-Bailey's group (2017) noted that 95% of men will have cleared the virus after four months [27]. Atkinson et al. reported in 2016 a study on threshold values for the ZIKV and found positive results (using real-time reverse transcription) at 27 and 62 days after onset of febrile illness. They added: "Although we did not culture infectious virus from semen, our data may indicate the prolonged presence of virus in semen, indicating a prolonged potential for sexual transmission of [Zika]" [41]. In 2016, doctors in Italy reported that two men who had contracted Zika during vacations in January had virus particles in their semen more than 180 days later [42]. Barzon and Nicestri agreed. Zika can use the testicles like a bomb shelter, hunkering down there to escape the onslaught.

While male semen can be a major transmission source, there has also been at least one case of female-to-male transmission [43] compared with dozens of reported cases from men [44]. Murray et al., detected viral shedding in vaginal secretions up to day fourteen. These findings, including the detection of ZIKV RNA up to day fourteen and in erythrocytes up to day 81, are the most extended reported detection duration in this sample type [45]. The duration of ZIKV persistence in the female genital tract and its clearance after the disappearance of the symptoms are unknown though opinion has been coalescing around 11 days. Coyne and Lazear detected ZIKV RNA in the female genital tract in humans and rhesus macaques. This raises the possibility that ZIKV could access the fetal compartment by transvaginal ascending infection and transplacental hematogenous spread [44]. This comment comes from mouse studies. The anatomical and immunological differences between the mouse and human placentas may limit the direct applicability of results from mouse studies to humans [44].

At least one gay man has infected his male partner through anal sex. Also, another man is believed to have infected his female partner through oral sex [46].

Current explanation: Physical barriers such as a blood–testis barrier make immune privilege possible. Cells that form the blood–testis barrier, for instance, secrete factors that suppress immunity to keep immune privilege in check. The immune license allows foreign invaders, including viruses, to find protection from the immune system within the testes. Siemann's team [47] published results outlining that the long gatekeeper cells that protect sperm cells, known as Sertoli cells (somatic cells of the testis that are essential for testis formation and spermatogenesis), succumb to ZIKV. The tightly held space between Sertoli cells—known as the blood-testis barrier—keeps immune cells from reaching sperm. But Zika-infected Sertoli cells attracted cell-eating immune cells known as macrophages that destroyed them, breaking the barrier [48].

The Nicastri et al., study detected ZIKV RNA in urine and saliva 91 days after symptom onset. In semen, up to day 134 might indicate a role played by other non-vector modes of transmission during kissing or vaginal, oral, and anal sex [49]. The CDC website reports that ZIKV is sexually transmitted in vaginal, oral, anal, and shared sex toys.

6.5 Transmission: Vertical (Mother to Child)

Vertical transmission of ZIKV leads to infection of neuroprogenitor cells (progenitor cells of the central nervous system that give rise to many, if not all, of the glial and neuronal cell types that populate the central nervous system) and destruction of brain parenchyma (the functional tissue in the brain). Recent evidence suggests that the timing of disease and host factors may affect vertical transmission [50].

Clearance of virus from amniotic fluid and transient viremia in fetal blood, accompanied by post-mortem isolation of ZIKV from brain tissue, suggests that the virus can infect the fetus, causing severe damage but clearing without leaving an immunological trace [51].

Vertical transmission has been estimated to occur in 26% of fetuses of ZIKV-infected mothers in French Guiana, [52] a percentage like transmission percentages that have been observed for other congenital infections. Among fetuses exposed to ZIKV by vertical transmission, the fetal loss occurred in 14%, and severe complications compatible with congenital Zika syndrome occurred in 21%. In addition, 45% of the fetuses exposed to ZIKV by vertical transmission had no signs or symptoms of congenital Zika syndrome in the first week of life [53]. ZIKV can persist in the placental tissue and umbilical cord blood long after the mother has cleared the active infection in her plasma. According to Stone and shaw (2020), after vertical transmission of ZIKV, the placenta, amniotic fluid, and fetal brain are known to harbor ZIKV [54].

A comprehensive study by Tabata et al., that used first-trimester human placental explants found that several placental cell types might be targeted by ZIKV early in gestation, including cytotrophoblasts and extravillous trophoblast cells [55], which suggests that there may be several routes for ZIKV vertical transmission [56].

Nguyen et al., studied vertical transmission among macaques. They reported that the fetuses of these two first-trimesters' ZIKV pregnancies, as well as two additional late second-/early third-trimester infections, had maternal–fetal ZIKV transmission with vRNA and pathology in fetal tissues at the maternal–fetal interface. While moderate infectious doses of ZIKV are cleared promptly in non-pregnant macaques, initial data from two pregnant macaques infected in the first trimester showed prolonged viremia, like reports of human pregnancies [57]. Nguyen et al. concluded: The results may also suggest that maternal–fetal ZIKV transmission in human pregnancy may be more frequent [58].

Hirsch recently published animal studies on rhesus macaques that have evaluated the role of the placenta in gestational ZIKV infection. Hirsch and colleagues found that even with average fetal growth, a persistent placental infection could lead to uterine and placental vasculitis, leading to a decreased oxygen permeability of the placental villi [59].

In addition, however, consistent vertical transmission in this primate model used by Nguyen et al., may provide a platform to assess risk factors and therapeutic test interventions for interruption of fetal infection. Sapparapu et al., have described the isolation of human monoclonal antibodies that neutralize ZIKV and prevent maternal–fetal transmission in mice [60]. Their next step will be to test mAb in non-human primates, and if vertical transmissions can also be controlled, it could become a prophylactic therapy for at-risk humans [61].

6.6 Transmission: Blood Transfusion

However, as almost more than 130 arboviruses causing human disease can be transmitted through blood transfusion and considering the high rate of asymptomatic subjects during outbreaks, the potential blood-borne transmission of other arboviruses, the widespread of the vectors also in countries with temperate weather, as well as the recently hypothesized new routes of transmission, the potential role of ZIKV in transfusion medicine is plausible [62].

It was reported by Sue Edwards that two patients in Brazil were found to have the ZIKV following blood transfusions [63]. Other researchers suggest that blood transfusions may have contributed to the spread of the disease, as reported earlier in French Polynesia [64].

In 2010, Petersen and Busch warned: "There exists a considerable risk for transfusion transmission of arboviruses due to short periods of asymptomatic viremia in populations with a variable and sometimes extremely high incidence of arboviral infections. Because of the increasing emergence of arboviral disease globally, it is prudent to prepare for both endemic and exotic arboviruses capable of producing large epidemics and subsequent transfusion transmission risk" [65].

> The growing global emergence of arboviral diseases presents unique challenges for transfusion medicine. Arbovirus transfusion risk models suggest that many transfusion transmissions of certain high-incidence arboviruses must occur in epidemic areas. It seems prudent

to prepare for these and other exotic arboviruses capable of producing large epidemics and subsequent transfusion transmission risk. The unexpected should be expected [65].

Massimo and Mannucci, in an editorial in *Blood Transfusion,* commented. The blood-borne transmission of this pathogen raises significant concerns regarding the safety of blood donations and transfusions in those areas characterized by a potential widespread circulation of the virus and its vector [66]. There exists a considerable risk for transfusion transmission of arboviruses due to short periods of asymptomatic viremia in populations with a variable and sometimes extremely high incidence of arboviral infections. Because of the increasing emergence of arboviral disease globally, it is prudent to prepare for both endemic and exotic arboviruses capable of producing large epidemics and subsequent transfusion transmission risk [67].

Although Musso et al. [68] pointed out the potential for ZIKV transmission by blood transfusion during the French Polynesian outbreak, they detected many positive asymptomatic blood donors. From November 2013 to February 2014: 42 (3%) of 1,505 blood donors, although asymptomatic at the time of blood donation, were found positive for ZIKV by PCR [69].

However, no blood transfusion-transmitted cases of ZIKV infection had been confirmed worldwide until a report of a well-defined ZIKV transmission occurred in Brazil [70]. It describes two instances of likely ZIKV transmission by blood transfusion from one person who donated platelets by apheresis (the removal of blood plasma from the body by withdrawing blood) in January 2016 and reported two days later a febrile illness compatible with ZIKV infection. Brazilian authorities announced the first confirmed blood transfusion-mediated transmission cases on 5 February 2016 [71].

In a case study by Barjas-Castro et al., they report a transfusion transmission of ZIKV. Their patient had been transfused with the blood product from an infected donor, in the incubation period after ZIKV infection but before clinical disease onset. They argue it is highly likely that it was caused by the transfusion of a blood product obtained from a donor during the incubation period for ZIKV infection. In this case, the recipient's lack of illness also underscores the characteristics of this disease, in which most infections are asymptomatic [72].

Initially, the F.D.A. barred Puerto Rico from collecting blood donations from the island, fearing viral transmission during blood transfusion, previously seen in Brazil, and asymptomatic French Polynesian blood donors. In response, the usage of blood donations collected from areas of the continental US without active transmission of the ZIKV was recommended.

Six months after the first case of Zika were reported in Puerto Rico, testing using the Hologic assay began in June 2016. The report by Williamson and colleagues provides complementary data to that above for regions of the United States outside of Florida and Puerto Rico [73]. However, with the approval of an investigational blood screen test, Puerto Rico was able to commence local blood donation collection on April 2, 2016.

From January 19 to June 10, 2016, a team led by INSERM (French National Institute of Health and Medical Research) [74] tested 4129 consecutive blood donations and conducted systematic nucleic acid testing for the presence of the ZIKV. The tests allowed the detection of approximately 2% of contaminated blood donations during the outbreak—and their subsequent exclusion to avoid blood-borne transmission. Among these infected samples, the proportion of genuinely asymptomatic cases of Zika disease was approximately 45%, and the balance of patients that did not require medical attention was 80–85% [75]. Molecular sequencing and phylogenetic analysis of ZIKV RNA isolated from the donor and the two patients confirmed the identity of their ZIKV isolates, with nucleotide changes in the envelope gene (codons 11 and 186) shared only by the donor and platelet recipients among available isolates from Brazil [76]. This investigation provides some evidence that ZIKV can be transmitted through platelet transfusion.

Of 466,834 donations screened with individual donor nucleic acid testing (NAT), five were positive by supplemental testing. These donations were collected in Nevada, New York, Arizona, California, and Texas [73]. Several non-commercial entities initiated the development of laboratory-developed tests for diagnostic purposes. Two commercial sponsors experienced in NAT screening tests for blood engaged in developing commercial assays to screen potential donors [73].

On the other hand, plasma derivatives seem safe. Indeed, ZIKV appears to be more sensitive to heat than other closely related viruses. In addition, nanofiltration with a pore size of 40 nm or less removed all ZIKV infectious activity. These findings represent reassuring news regarding ZIKV and the safety of plasma derivatives [77]. Blümel et al. (2017) reported that pasteurization and S/D treatment rapidly inactivated ZIKV, and filters with a pore size of not more than 40 nm removed all infectious ZIKV, demonstrating the effectiveness of these virus reduction strategies used during the manufacture of plasma-derived medicinal products [78].

FDA Commission Robert Califf recommended only using blood from non-Zika-affected areas [79].

> The Pan American Health Organization and the World Health Organization recommend that blood and CSF samples collected at admission be tested using (rRT-PCR). Still, although sensitive, that serologic evaluation for detecting ZIKV, IgM is less specific, as there is antibody cross-reactivity among other flaviviruses (dengue, yellow, fever, and West Nile viruses). If the rRT-PCR is negative or unavailable, the patient should be tested for Dengue, Chikungunya, and ZIKV IgM. If positive for only one of the viruses, the diagnosis of a probable cause can be made [38]

In response to the widespread ZIKV outbreak of 2015–2016 in South and Central America, the U.S. Food and Drug Administration (FDA) published and later updated guidance to screen donors to prevent ZIKV transmission. To keep the transfusion system, safety warnings were posted. People must postpone giving blood for at least 28 days to avoid ZIKV transmission [19].

This guidance inadvertently adversely affected the availability of licensed umbilical cord blood units, as approximately 10% of donors are now ineligible, even as the risk of ZIKV has dramatically decreased [54].

According to a 2016 CDC survey of blood collection centers, there are no known cases of blood transfusion-transmitted ZIKV in the United States or Puerto Rico [80].

6.7 Transmission: Organ Transplantation

Zoonoses represent a problem of rising importance in the transplant population. Although most zoonotic infections develop in the post-transplant period, donor or transfusion-transmitted infections have also been increasingly acknowledged. Case reports describing ZIKV infection in transplant patients are extremely limited.

In 2015 and 2016, ZIKV infection was confirmed among 129 kidney transplants and fifty-eight liver transplants tested in Brazil. Based on this small case series, it was impossible to assess the potential impact of ZIKV in the immunosuppressed SOT recipients, including infectious complications and graft rejection [81].

Many reports on donor-derived or naturally acquired (re) emerging arboviral infections such as dengue, chikungunya, West Nile, tick-borne encephalitis, and ZIKV infection have demonstrated atypical or more complicated clinical courses in immunocompromised hosts [82].

6.8 Transmission Laboratory Exposure

Two infections in laboratories have been reported. Both reports seem problematic for several reasons. Citations involving these works seem to be associated with lists of transmission scenarios without regard to plausibility. Nonetheless, there are two.

Simpson also claims to have published the first proven case of human infection with ZIKV. The single infection reported was a mild febrile episode, accompanied by a generalized pink maculopapular rash, which appeared on the second day before the temperature rose. The virus was isolated and re-isolated in suckling mice and was identified as ZIKV by hemagglutination-inhibition (HI) and neutralization (NT) tests [83].

However, according to Meers, "It could not be decided whether the source of infection had been in the Ziika Forest itself or had been the result of an unrecognized laboratory accident."

Filipe et al., made a second claim. During laboratory work with arboviruses, one of the authors (C. Martens) contracted a benign infection due to an infectious agent that was subsequently found to be the ZIKV. The study of hemagglutination-inhibiting, complement-fixing, and neutralizing antibodies found in the various serum samples collected from the same individual yielded results that seem to be of particular interest concerning the antigenic relationship between yellow fever and ZIKVes [84].

In the case described here, the patient was vaccinated against yellow fever some 11 years before and about two months before developing a ZIKV infection. The antibody response was, therefore, of the anamnestic type already observed by other workers in secondary infections due to group B viruses. Thus, the etiology of current infections could be only ascertained by the isolation of the ZIKV; this contrasts with the first case of a primary laboratory infection with ZIKV [83], which was accurately diagnosed by serological tests.

The conclusion drawn from these observations may serve as a reminder of the care needed in interpreting serological results obtained during epidemiological surveys in areas where several arboviruses are present and in areas where jungle yellow fever exists.

6.9 Transmission: Tears

The most usual form of ZIKV-induced ocular disease is conjunctivitis, which occurs in 10–15% of patients [85].

Finally, researchers have suggested that the eyes could be a "reservoir" of the virus, helping it spread from one person to another [86].

Miner et al., detected abundant viral RNA in tears in mice, suggesting that the virus might be secreted from lacrimal glands or shed from the cornea [87]. In the study, Miner et al., from the Washington University School of Medicine in St Louis infected mice with ZIKV by injecting them under their skin, like how mosquitoes commonly spread infections. They then found a live virus in the mice's eyes and their genetic material in their tears. "There could be a window of time when tears are highly infectious, and people are encountering it and able to spread it" [86].

They also detected abundant viral RNA in tears, suggesting that the virus might be secreted from lacrimal glands or shed from the cornea. This provides a foundation for studying ZIKV-induced ocular disease, defining mechanisms of viral persistence, and developing therapeutic approaches for viral infections of the eye [87].

Given that ZIKV can infect the cornea of mice, human studies might be needed to confirm whether ZIKV analogously infects the human cornea. Suppose corneal infection by ZIKV was established in humans. In that case, widespread ocular disease during epidemics could necessitate testing by eye banks to ensure that ZIKV is not present in the corneas of infected donors [87].

Finally, Swaminathan et al., suggested that in the strange case (below), given the remarkably high level of viremia in Patient 1, infectious levels of the virus may have been present in sweat or tears, of which Patient 2 contacted without gloves. And warns risk for fulminant ZIKV infection may be broader than previously recognized [88].

They added that the transmission of flaviviruses through intact skin or mucous membranes, although uncommon, has been shown in experimental animal models (above) and at least one human case· However, that case was a Boston man with dengue and not ZIKV. There are both flaviviruses [89].

6.10 Transmission: Breast Milk

The ZIKV genome was also detected in breast milk, followed by viral isolation of infective viral particles. There seems to be no evidence that ZIKV can be transmitted this way [90]. Arbovirus transmission via breastfeeding has been previously suggested for dengue, West Nile, and yellow fever, but more information is needed [91].

Dupont-Rouzeyol et al., reported a case from New Caledonia. They report the presence of infective ZIKV particles in breastmilk with substantial viral loads [91].

Colt et al., data are often cited for making this case. It involved a review of only three cases. Two of the three associated newborns had evidence of ZIKV infection. ZIKV was detected in the breast milk of all three mothers. Breast milk detection results were positive in all mothers by RT-PCR, one was positive by culture, and none were tested for ZIKV-specific antibodies [92]. The authors admit, "While ZIKV was detected in the breast milk of all three mothers, the data are insufficient to conclude ZIKV transmission via breastfeeding" [92].

By the Brazilian Protocols for Sexually Transmitted Infections (2020), natural breastfeeding is also ideal for children born to mothers with ZIKV infection. If there is no contraindication to oral feeding, breastfeeding shall be started [93].

6.11 Transmission: Fertility Treatments

Although there is no documented case of ZIKV transmission through fertility treatments (such as donated embryos or gametes), it seems medically possible that this could occur, in that it is unlikely the cryopreservation process would destroy the ZIKV.

The FDA recommended that anonymous donors to fertility clinics be declared ineligible if, in the last six months, they have been diagnosed with the ZIKV, have lived, or traveled to an area with active ZIKV transmission, or have had sex with a male partner who has been diagnosed with the ZIKV or traveled to a place of active ZIKV transmission [94].

6.12 Transmission: An Anomaly

There is the notorious Patient A. On July 12, 2016, the Utah Department of Health (UDOH) was notified by a clinician caring for an adult (patient A) who was evaluated for fever, rash, and conjunctivitis that began on July 1. Patient A had not traveled to an area with ongoing ZIKV transmission; had not had sexual contact with someone who recently traveled; and had not received a blood transfusion, organ transplant, or mosquito bites. Patient A provided care over several days to an elderly male family contact (the index patient) who contracted ZIKV abroad. The index patient developed septic shock with multiple organ failure and died in the hospital on June 25, 2016 [95].

Associated, an outsider case surfaced in Utah where a 38-year-old man contracted the virus at his father's bedside [96]. Testing positive for ZIKV, the father contracted the disease from a mosquito bite while visiting Mexico. He died from complications associated with radiation-induced organ damage from cancer therapy.

The son had not been exposed to mosquitoes carrying ZIKV, hadn't visited Mexico with his father, and hadn't had sex with anyone who may have contracted the disease from exposure to bodily fluids when helping a nurse move his father. "Infectious levels of the virus may have been present in sweat or tears [97], both of which Patient 2 [the son] contacted without gloves," says the paper's authors [88].

The most common interactions with the index patient included kissing, primarily on the cheek ($n = 6$), and assisting in care (e.g., cleaning up vomit, stool, or urine, or wiping tears) ($n = 6$). These activities were performed without PPE. Before the hospitalization of the index patient, patient A had only casual contact (e.g., hugging and kissing) with the index patient [98].

Krow-Lucal et al., studied the case of patient A and did not identify the probable source of his infection and did not identify any additional persons recently infected with ZIKV among family contacts, healthcare workers, or community members. The index patient was unique compared to other persons with ZIKV disease because his illness was fatal, and his relative viral load was estimated to be ˜100,000 times higher than the average level reported [98].

Sankar Swaminathan and colleagues from the University of Utah School of Medicine in Salt Lake City speculated that the man may have had a genetic immune deficiency that just happened to be specific to this virus. In addition, it's not clear why this man suffered so intensely from what is typically a very mild virus. "He'd had dengue in the past; it's possible that the remaining antibodies from that somehow worsened this infection. This can sometimes happen when people get two different strains of dengue—the second dengue infection will be worse" [99].

As to how the son contracted ZIKV, this mystery will live on forever.

6.13 Conclusion

This author understands that ongoing research may discover other relevant transmission routes and debunk some past research on transmissions. Not sure how that could have been prevented. Nonetheless, the material above highlights all the transmission routes documented in one way or another in the literature. The next chapter and saddest covers the transmission from mother to fetus in cases where the mother had been bitten herself or had contracted ZIKV from her male partner.

References

1. American Mosquito Control Association (2017) Best prtactices for integrated mosquito management: a focused update. January. https://www.researchgate.net/publication/315924484_Best_Practices_for_Integrated_Mosquito_Management_A_Focused_Update. Accessed 27 Aug 2018
2. Olson JG et al (1983) A survey for arboviral antibodies in sera of humans and animals in Lombok, Republic of Indonesia. Ann Tropical Med Parasitol 77. https://www.ncbi.nlm.nih.gov/pubmed/6309104. Accessed 7 Jul 2018
3. Wolfe ND et al (2001) Sylvatic transmission of arboviruses among Bornean orangutans. Am J Tropical Med Health 64:5. http://www.ajtmh.org/content/journals/https://doi.org/10.4269/ajtmh.2001.64.310. Accessed 7 Jul 2018
4. Bueno MG et al (2016) Animals in the Zika virus life cycle: what to expect from megadiverse Latin American countries. PLOS Tropical Diseases. December 22. http://journals.plos.org/plosntds/article?id=https://doi.org/10.1371/journal.pntd.0005073. Accessed 7 Jul 2018
5. Althouse BM et al. (2016). Potential for Zika virus to establish a sylvatic transmission cycle in the Americas. PLOS Neglected Tropical Diseases. December 15. https://journals.plos.org/plosntds/article?id=https://doi.org/10.1371/journal.pntd.0005055. Accessed 15 May 2019
6. Lafrance A (2016) Everything you need to know about the Zika virus. The Atlantic. November 10. https://www.theatlantic.com/health/archive/2016/11/zika-cheat-sheet/506822/. Accessed 18 May 2017
7. American Association of Pharmaceutical Scientists (2016) Zika virus can live for hours on hard, nonporous surfaces. Science News. November 15. https://www.sciencedaily.com/releases/2016/11/161115164220.htm. Accessed 11 Apr 2017
8. Lessler J et al (2016) Assessing the global threat from Zika virus. Science. July 14. 353:6300.http://science.sciencemag.org/content/early/2016/07/13/science.aaf8160. Accessed 19 May 2017
9. Rozé BF et al (2016) Zika virus detection in urine from patients with Guillain-Barré syndrome on Martinique, January 2016. EuroSurveillance. 21:9. March 3. http://www.eurosurveillance.org/images/dynamic/EE/V21N09/art21400.pdf. Accessed 19 May 2017
10. Grishott F, Puhan M, Hatz C, Schlagenhauf P (2016) Non-vector-borne transmission of Zika virus: a systematic review. Travel Med Infect Disease. July 7. https://doi.org/10.1016/j.tmaid.2016.07.002
11. Hawai'I Department of Health, Vector Control Branch (2011) Mosquitoes. Bulletin 3. http://health.hawaii.gov/about/files/2013/06/VCB-bulletin_03_11.pdf. Accessed 5 Jun 2017
12. Vyas V (2016) Zika virus: why should we concern (sic). Lambert Academic Publishing, Saarbrüucken, Germany
13. Comeau G, Zinna RA, Scott T, Ernst K, Walker K et al (2020) Vertical transmission of Zika virus in aedes aegypti produces potentially infectious progeny. Am J Trop Med Hyg 103(2):876–883. https://doi.org/10.4269/ajtmh.19-0698
14. Pereira-Silva JW, do Nascimento VA, Mehlcior HCM, Ameida JF et al (2018) First evidence of Zika virus venereal transmission in Aedes aegypti mosquitoes. Memórias do Instituto Oswaldo Cruz. January 113(1):56–61
15. Charrel R et al (2016) Background review for diagnostic test development for Zika virus infection. Bull World Health Organ 94:574–584D. https://doi.org/10.2471/BLT.16.171207. http://www.who.int/bulletin/volumes/94/8/16-171207.pdf. Accessed 28 Mar 2017
16. Aliota M et al (2016) The wMel strain of Wolbachia reduces transmission of Zika virus by Aedes aegypti. Nat: Sci Rep 6:28792. https://doi.org/10.1038/srep28792. http://www.nature.com/articles/srep28792. Accessed 16 Jan 2017
17. Noor R, Ahmed T (2018) Zika virus: epidemiological study and its association with public health risk. J Infect Public Health. April 28. https://www.sciencedirect.com/science/article/pii/S1876034118300431. Accessed 25 Jul 2018; Black WC IV et al (2002) Flavivirus susceptibility in Aedes aegypti. Arch Med Res 33. https://www.ncbi.nlm.nih.gov/pubmed/12234528. Accessed 26 Jun 2017

18. Corsi J (2016) Florida is ground zero for Zika virus in U.S. WND. February 3. http://www.wnd.com/2016/02/florida-is-ground-zero-for-zika-virus-in-u-s/. Accessed 21 July 2017
19. Chen HL, Tang RB (2016) Why Zika virus infection has become a public health concern? J Chin Med Assoc 79:174–178. http://www.sciencedirect.com/science/article/pii/S1726490116300065. Accessed 10 Apr 2017
20. DoctorNDTV (2018) Zika virus may increase the risk of miscarriages, stillbirth: 6 things you need to know about Zika virus. NDTV. July 4. https://www.ndtv.com/health/zika-virus-may-increase-the-risk-of-miscarriages-stillbirth-1877680. Accessed 25 Jul 2018
21. Cayman News Now (2017) Genetic modification project in the Cayman Islands cuts mosquitoes by over 80%. Cayman News Now. January 28. http://www.caribbeannewsnow.com/headline-Genetic-modification-project-in-the-Cayman-Islands-cuts-mosquitoes-by-over-80-percent-33345.html. Accessed 15 Mar 2017
22. Grard G et al (2014) Zika virus in Gabon (Central Africa)—2007: a new threat from *Aedes* albopictus? PLoS Neglected Tropical Diseases. 8:2. https://doaj.org/article/f7cad46b65724a419fe2236184f2c3bc. Accessed 19 Jul 2018; Grard G (2016) In: Andrew H et al (eds), Zika Virus. Wuhan, China, (SCIRP) Scientific Research Publishing
23. Li MI et al (2012) Oral susceptibility of Singapore Aedes (Stegomyia) aegypti (Linnaeus) to Zika virus. PLoS Negl Trop Dis 6:e1792
24. Diagne CT et al (2015) Potential of selected Senegalese Aedes spp. mosquitoes (Diptera: Culicidae) to transmit Zika virus. BMC Infect. Dis 15:492
25. Chouin-Carneiro T et al (2016) Differential susceptibilities of Aedes aegypti and Aedes albopictus from the Americas to Zika Virus. PLOS Negl Trop Dis 10:e0004543
26. Magori K (2018) The incredible value of mosquito surveillance and control programs. BugBitten. June 29. https://blogs.biomedcentral.com/bugbitten/2018/06/29/incredible-value-mosquito-surveillance-control-programs/. Accessed 24 Jul 2018
27. Paz-Bailey G et al (2018) Persistence of Zika Virus in Body Fluids — Final Report. New England J Med 37(13). September 27. https://doi.org/10.1056/NEJMoa1613108
28. Gallagher JJ (2016) Congress halts Zika fund over planned parenthood as cases spread. ABC News. http://abcnews.go.com/Politics/congress-halts-zika-funds-planned-parenthood-cases-spread/story?id=41913670. Accessed 24 Sept 2016
29. Foy B et al (2011) Probable non-vector-borne transmission of Zika virus, Colorado, USA. Emerg Infect Diseas May. 17(5):880–82. https://www.ncbi.nlm.nih.gov/pmc/articles/PMC3321795/. Accessed 14 Oct 2016
30. Wang L et al (2016) From mosquitos to humans: Genetic evolution of Zika virus. Cell Host Microbe 19:561–565. Medline https://doi.org/10.1016/j.chom.2016.04.006. http://www.cell.com/cell-host-microbe/abstract/S1931-3128(16)30142-1. Accessed 15 Oct 2016
31. LaJolla Institute for Allergy and Immunology (2016) LJI researchers strengthen the case for sexual transmission of Zika virus. Life Without Disease Press Release. December 20. https://www.niaid.nih.gov/funded-research/lji-researchers-strengthen-case-sexual-transmission-zika-virus. Accessed May 19, 2017.
32. Fox M (2016) Can I ever get pregnant? and other questions about Zika virus NBC News. August 26. http://www.nbcnews.com/storyline/zika-virus-outbreak/can-i-ever-get-pregnant-other-questions-about-zika-virus-n583231. Accessed 15 Sept 2016
33. RoseWrites (2016) Zika facts you are not being told and CDC'S cover-up. InfoBarrel. October 7. http://www.infobarrel.com/Zika_Facts_You_Are_Not_Being_Told_and_CDCs_Cover-Up_. Accessed 15 May 2017
34. Cunningham, Aimee. (2017). As Zika fades from public consciousness, scientists continue to pursue the virus. The Washington Post. December 30. https://www.washingtonpost.com/national/health-science/as-zika-fades-from-public-consciousness-scientists-continue-to-pursue-the-virus/2017/12/29/3fae96dc-e1d3-11e7-bbd0-9dfb2e37492a_story.html?noredirect=on&utm_term=.42193ac1f6b7. Accessed 7 Jul 2018
35. Rosen M (2016) Zika kills brain cells in adult mice. Science News. August 18. https://www.sciencenews.org/article/zika-kills-brain-cells-adult-mice. Accessed 30 May 2017

36. Yockey LJ et al (2016) Vaginal exposure to Zika virus during pregnancy leads to fetal brain infection. Cell. August 25. 166. http://www.cell.com/fulltext/S0092-8674(16)31053-4. Accessed 14 Apr 2017
37. Gaskell K et al (2017) Persistent Zika virus detection in semen in a traveler returning to the United Kingdom from Brazil, 2016. Emerg Infect Diseas 23:1. January. https://wwwnc.cdc.gov/eid/article/23/1/16-1300_article. Accessed 14 Apr 2017
38. Silva, Gisele S. et al. (2018). Zika virus: Report from the task force on tropical diseases by the world Federation of Societies of intensive and critical care medicine. Journal of Critical Care. 46. August. https://www.ncbi.nlm.nih.gov/pubmed/29779827. Accessed July 26, 2018.
39. Beck J (2016) Can women spread Zika sexually, too? The Atlantic. July 14. https://www.theatlantic.com/health/archive/2016/07/can-women-spread-zika-sexually-too/491165/. Accessed 7 Feb 2017
40. Barzon L et al (2016) Infection dynamics in a traveler with persistent shedding of Zika virus RNA in semen for six months after returning from Haiti to Italy, January 2016. Eurosurveillance. 21(32). https://www.ncbi.nlm.nih.gov/pubmed/27542178. Accessed 14 Apr 2017
41. Atkinson B et al (2016) Detection of Zika virus in Semen. Emerg Infect Diseas 22:5. May. https://wwwnc.cdc.gov/eid/article/22/5/16-0107_article. Accessed 7 Apr 2017
42. Barzon L et al (2016) Infection dynamics in a traveller with persistent shedding of Zika virus RNA in semen for six months after returning from Haiti to Italy. January 2016. Eurosurveillance. August 11. 21:32. http://www.eurosurveillance.org/images/dynamic/EE/V21N32/art22554.pdf. Accessed 4 Apr 2017; Nicastri E et al (2016) Persistent detection of Zika virus RNA in semen for six months after symptom onset in a traveller returning from Haiti to Italy, February 2016. Eurosurveillance. August 11. 21:32. https://www.ncbi.nlm.nih.gov/pubmed/27541989. Accessed 4 Apr 2017
43. Davidson A et al (2016) Suspected female-to-male sexual transmission of Zika virus—New York City. Morbidity and Mortality Weekly Report. July 22. 65:28. 716–717. https://www.cdc.gov/mmwr/volumes/65/wr/mm6528e2.htm. Accessed 5 May 2017
44. Coyne CB, Lazear HM (2016) Zika virus—reigniting the TORCH. Nature Reviews Microbiology. 14. http://www.nature.com/nrmicro/journal/v14/n11/full/nrmicro.2016.125.html. Accessed 14 Apr 2017
45. Murray KO et al (2017) Prolonged detection of Zika Virus in vaginal secretions and whole blood. Emerg Infect Diseas 23:1. January. https://wwwnc.cdc.gov/eid/article/23/1/16-1394_article. Accessed 26 May 2017
46. McNeil D, Saint L, Catherine, St. Fleur N (2016) Short answers to hard questions about Zika virus. The New York Times. July 29. https://www.nytimes.com/interactive/2016/health/what-is-zika-virus.html. Accessed 22 May 2017; Foy BD et al (2011) Probable non-vector-borne transmission of Zika virus, Colorado, USA. Emerging Infectious Diseases. https://wwwnc.cdc.gov/eid/article/17/5/10-1939_article. Accessed 17 Jul 2017; Deckard DT et al (2016) Male-to-male sexual transmission of Zika virus—Texas, January 2016. Morbidity and Mortality Weekly Report. 65:14. April 15. https://www.cdc.gov/mmwr/volumes/65/wr/mm6514a3.htm. Accessed 17 Jul 2017
47. Siemann D et al (2018) Zika virus infects human sertoli cells and modulates the integrity of the in vitro blood-testis barrier model. J Virol 92:10. May. https://www.ncbi.nlm.nih.gov/pubmed/28878076. Accessed 7 Jul 2018
48. Chadradhar S (2018) A private place where HIV, Zika and Ebola hide. Sci Am. January 26. https://www.scientificamerican.com/article/a-private-place-where-hiv-zika-and-ebola-hide/. Accessed 7 Jul 2018
49. Nicastri E et al (2016) Persistent detection of Zika virus RNA in semen for six months after symptom onset in a traveller returning from Haiti to Italy, February 2016. Eurosurveillance. August 11. 21:32. https://www.ncbi.nlm.nih.gov/pubmed/27541989. Accessed 26 May 2017
50. Valentine GC et al (2018) Timing of gestational exposure to Zika virus is associated with postnatal growth restriction in a murine model. Am J Obstetrics Gynecol. June 11. Availabe online. https://www.sciencedirect.com/science/article/pii/S0002937818304927?via%3Dihub. Accessed 26 Jul 2018

51. Ades AE et al (2020) Researching Zika in pregnancy: lessons for global preparedness. Lancet Infect Dis 20:e61-68
52. Pomar L, Vouga M, Lambert V et al (2018) Maternal-fetal transmission and adverse perinatal outcomes in pregnant women infected with Zika virus: prospective cohort study in French Guiana. BMJ 363:k4431
53. Musso D, Ko AI, Baud D (2019) Zika virus infection—after the pandemic. N Engl J Med 381:1444–1457. https://doi.org/10.1056/NEJMra1808246
54. Stone E, Shaz B (2020) Zika virus and its implications on cord blood banking and transplantation. Transfusion 60:889–891
55. "Cytotrophoblast" is the name given to both the inner layer of the trophoblast (also called layer of Langhans) or the cells that live there. The cytotrophoblast is the trophoblastic stem cell. Extravillous trophoblast are one form of differentiated trophoblast cells of the placenta. They are invasive mesenchymal cells which function to establish critical tissue connection in the developing placental-uterine interface
56. Tabata T et al (2016) Zika virus targets different primary human placental cells, suggesting two routes for vertical transmission. Cell Host Microbe. August 10. 20:2. 155–166. http://www.sciencedirect.com/science/article/pii/S1931312816303006. Accessed 5 May 2017
57. Nguyen S et al (2017) Highly efficient maternal-fetal Zika virus transmission in pregnant rhesus macaques. PLOS Pathogens. May 25. http://journals.plos.org/plospathogens/article?id=https://doi.org/10.1371/journal.ppat.1006378. Accessed 25 Jul 2018; Dudley DM et al (2016) A rhesus macaque model of Asian-lineage Zika virus infection. Nature Communications. 7:12204. https://www.nature.com/articles/ncomms12204. Accessed 25 Jul 2018
58. Nguyen S et al (2017) Highly efficient maternal-fetal Zika virus transmission in pregnant rhesus macaques. PLOS Pathogens. May 25. http://journals.plos.org/plospathogens/article?id=https://doi.org/10.1371/journal.ppat.1006378. Accessed 25 Jul 2018
59. Hirsch AJ et al (2018) Zika virus infection in pregnant rhesus macaques causes placental dysfunction and immunopathology. Nat Commun 9:263. https://www.nature.com/articles/s41467-017-02499-9. Accessed 7 Jul 2018
60. Sapparapu G et al (2016) Neutralizing human antibodies prevent Zika virus replication and fetal disease in mice. Nature, 540. December 15. 443–449. http://dx.doi.org/https://doi.org/10.1038/nature20564. Accessed 16 Jan 2017
61. Borden Y (2016) Zika virus: end of transmission? Nat Rev Immunol. November 25. https://doi.org/10.1038/nri.2016.131. http://www.nature.com/nri/journal/v16/n12/full/nri.2016.131.html . Accessed 16 Jan 2017
62. Marono G, Pupella S, Vaglio S, Liumvruno GM, Grazzini G (2016) Zika virus and the never-ending story of emerging pathogens and transfusion medicine. Blood Transfus 14:95–100
63. Edwards SB (2016) The Zika virus. ABDO Publishing, Minneapolis, MN
64. Chattu VK, Kumary S, Jagassar I (2016) Global scenario of Zika virus transmission and prevention: recent updates. Biotecghnol Res 2(3):94–99; Musso D, Nhan T, Robin E, Roche C et al (2014) Potential for Zika virus transmission through blood transfusion demonstrated during an outbreak in French Polynesia, November 2013–February 2014. Euro Surveillance 19(14). https://www.eurosurveillance.org/content/https://doi.org/10.2807/1560-7917.ES2014.19.14.20761. Accessed 9 Jun 2022
65. Petersen LR, Busch MP (2010) Transfusion-transmitted arboviruses. Vox Sang 98:495–503
66. Franchini M, Velati C (2017) Blood safety and zoonotic emerging pathogens: now it's the turn of Zika virus! Blood transfusion. 14. https://www.ncbi.nlm.nih.gov/pubmed/26674809. Accessed 14 Apr 2017
67. Peterson LR, Busch MP (2009) Transfusion-transmitted arboviruses. Vox Sanguinis. 98. https://www.ncbi.nlm.nih.gov/pubmed/19951309. Accessed 28 May 2017
68. Musso D et al (2014) Potential for Zika virus transmission through blood transfusion demonstrated during an outbreak in French Polynesia, November 2013–February 2014. Euro-Surveillance. 19:14. https://www.eurosurveillance.org/content/https://doi.org/10.2807/1560-7917.ES2014.19.14.20761. Accessed 24 Jul 2018

69. Musso D (2015) Zika virus transmissions from French Polynesia to Brazil. Emerg Infect Diseas 21:10. October. http://www.ncbi.nlm.nih.gov/pmc/articles/PMC4593458/. Accessed 24 Sept 2016
70. Magnus MM et al (2018) Risk of Zika virus transmission by blood donationsin Brazil. Hematol Tranfusion Cell Therapy. https://www.sciencedirect.com/science/article/pii/S2531137918300671. Accessed 24 Jul 2018; Ong CW (2016) Zika virus: an emerging infectious threat. Int Med J 46(5):525–530
71. Charrel R et al (2016) Background review for diagnostic test development for Zika virus infection. Bull World Health Organ 94:574–584D. https://doi.org/10.2471/BLT.16.171207. http://www.who.int/bulletin/volumes/94/8/16-171207.pdf. Accessed 28 Mar 2017; Motta I et al (2016) Evidence for transmission of Zika virus by platelet transfusion. New England J Med September 15. http://www.nejm.org/doi/full/https://doi.org/10.1056/NEJMc1607262#t=article. Accessed 26 May 2017
72. Barhas-Castro ML et al (2016) Probable transfusion transmitted Zika virus in Brazil. Transfusion 56(July):1684–1688
73. Marks P, Peterson L (2017) Decision making in the face of uncertainty: the challenge of emerging infectious diseases. Transfusion. 57. March. https://www.ncbi.nlm.nih.gov/pubmed/28345226, Accessed 21 May 2017
74. Gallain P et al (2017) Zika virus in asymptomatic blood donors in Martinique. Blood. January 12. 129:2. http://www.bloodjournal.org/content/bloodjournal/129/2/263.full.pdf. Accessed 23 May 2017
75. Medical Xpress (2017) Martinican blood donors help shed new light on the Zika virus. Medical Xpress. February 15. https://medicalxpress.com/news/2017-02-martinican-blood-donors-zika-virus.html. Accessed 23 May 2017
76. Motta I et al (2016). Evidence for Transmission of Zika Virus by Platelet Transfusion. New England Journal of Medicine. September 15. http://www.nejm.org/doi/full/https://doi.org/10.1056/NEJMc1607262#t=article. Accessed 26 May 2017
77. Marks P, Peterson L (2017) Decision making in the face of uncertainty: the challenge of emerging infectious diseases. Transfusion. 57. March. https://www.ncbi.nlm.nih.gov/pubmed/28345226, Accessed 21 May 2017; Blümel J et al (2016) Inactivation and removal of Zika virus during manufacture of plasma derived medicinal products. Transfusion. 57:3, pt. 2. October. https://www.ncbi.nlm.nih.gov/labs/articles/27731495/. Accessed 21 May 2017
78. Blümel J et al (2017) Inactivation and removal of Zika virus during manufacture of plasma-derived medicinal products. Transfusion. 57, March. https://www.ncbi.nlm.nih.gov/labs/articles/27731495. Accessed 26 Jun 2017
79. Beck J (2016) Zika is the 'most difficult' emergency health response ever, CDC official says. The Atlantic. June 24. http://www.theatlantic.com/health/archive/2016/06/zika-is-the-most-difficult-emergency-health-response-ever-says-cdc-official/488579/. Accessed 16 Jan 2017
80. Anwar S (2016) Zika more threatening than initially believed. ContagionLive: Infectious Diseas Today. April 12. http://www.contagionlive.com/news/zika-more-threatening-than-initially-believed. Accessed 24 Sept 2016
81. Nogueira ML, Estofolete CF, Terzian AC, Mascarin do Vale EP et al (2017) Zika virus infection and solid organ transplantation: a new challenge. Am J Trans 17:791–795
82. Mrzljak A, Novak R, Pendak N, Tabain I et al (2020) Emerging and neglected zoonoses in transplant population. World J Trans March 10(3):47–63
83. Simpson DI (1964) Zika virus infection in man. Trans R Soc Trop Med Hyg 58:335–338
84. Filipe AR, Martins CM, Rocha H (1973) Laboratory infection with Zika virus after vaccination against yellow fever. Archiv fur die gesamte Virusforschung. 43:315–319
85. Miner JJ et al (2016) Zika virus infection in mice causes panuveitis with shedding of virus in tears. Cell Rep. 16. September 20. http://www.cell.com/cell-reports/abstract/S2211-1247(16)31175-5. Accessed 17 Jul 2017
86. Johnston I (2016) Zika virus patients could be spreading the disease through their tears, scientists say. The Independent. September 6. http://www.independent.co.uk/news/science/zika-virus-tears-spread-infection-disease-study-mice-a7228811.html. Accessed 30 Jun 2017

87. Miner JJ et al (2016). Zika virus infection in mice causes panuveitis with shedding of virus in tears. Cell Rep. 16. September 20. http://www.cell.com/cell-reports/abstract/S2211-1247(16)31175-5. Accessed 17 Jul 2017
88. Swaminathan S et al (2016) Fatal Zika virus infection with secondary nonsexual transmission. New England J Med. https://doi.org/10.1056/NEJMc1610613. Accessed 2 Oct 2016
89. Chen LH, Wilson ME (2004) Transmission of dengue virus without a mosquito vector: nosocomial mucocutaneous transmission and other routes of transmission. Clin Infect Dis 39(6):e56-60
90. Besnard M, Lastere S, Teissier A, Cao-Lormeau V, Musso D (2014) Evidence of perinatal transmission of Zika virus, French Polynesia, December 2013 and February 2014. Euro Surveill 19(13):20751
91. Dupont-Rouzeyrol MBA, O'Connor O, Huguon EU, Descloux E (2016) Infectious Zika viral particles in breastmilk. The Lancet. March 1. https://doi.org/10.1016/S0140-6736(16)00624-3
92. Colt S, Garcia-Casal MN, Peña-Rosas JP, Finkelstein JL et al (2017) Transmission of Zika virus through breast milk and other breastfeeding-related bodily-fluids: a systematic review. PLoS Negl Trop Dis 11(4):e0005528. https://doi.org/10.1371/journal.pntd.0005528
93. Duarte G et al (2021) Consensus: Brazilian protocol for sexually transmitted infections 2020: Zika virus infection. J Brazilian Soc Tropical Med 54(Supp. 1):e2020609
94. Armstrong C (2016) CDC updates interim guidance on caring for women with possible exposure to Zika virus. Am Fam Physician 93(10):874–878
95. Brent C, Dunn A, Savage H, Faraji A et al (2016) Preliminary findings from an investigation of Zika virus infection in a patient with no known risk factors—Utah, 2016. Morbidity and Mortality Weekly Report. 65(36):981–982. https://doi.org/10.15585/mmwr.mm6536e4
96. Boseley S (2016) Mystery Zika case in Utah may have been spread via tears or sweat. The Guardian. September 29. https://www.theguardian.com/world/2016/sep/29/mystery-zika-virus-utah-may-spread-through-sweat-tears. Accessed 2 Oct 2016
97. Miner JJ et al (2016) Zika virus infection in mice causes panuveitis with shedding of virus in tears. Cell Rep 12. http://apps.webofknowledge.com/InboundService.do?customersID=ProQuest&mode=FullRecord&IsProductCode=Yes&product=WOS&Init=Yes&Func=Frame&DestFail=http%3A%2F%2Fwww.webofknowledge.com&action=retrieve&SrcApp=Summon&SrcAuth=ProQuest&SID=7EprvsNthrAzgVjiRhu&UT=WOS%3A000383884200013. Accessed 19 Jul 2018
98. Krow-Lucal ER et al (2017) Zika virus infection in patient with no known risk factors, Utah, USA, 2016. Emerg Infect Diseas J 23(8). August. https://doi.org/10.3201/eid2308.170479
99. Beck J (2016) The first documented case of Zika spreading by physical contact. The Atlantic. September 28. https://www.theatlantic.com/health/archive/2016/09/the-first-documented-case-of-zika-spread-by-physical-contact/501923/. Accessed 3 Feb 2017

Chapter 7
Effects on Children: Part 1

The news is disheartening, and researching it was no less so. There are more footnotes to accommodate definitions of some health and medical terms found in this chapter and the next two. The controversy involves whether the link between ZIKV and microcephaly exists or is due to other factors entirely or in part. Some believe it was misreporting, and others claim it is correlative alone. The next chapter will continue the microcephaly issue and offer more medical experimental research involving humans and other mammals.

Enter the ZIKV poster child, the microcephalic child. There probably could not have been a more terrifying symbol and victim designed. Soon after its first appearance, spokespeople, issue handlers, and media outlets employed photographs and film clips of Brazilian microcephalic newborns and warnings about killer mosquitoes. While the list of the infectious diseases mosquitoes carry is frightening, media of all sorts decided to add some visualizations of babies with malformed craniums when reporting on ZIKV. Panic ensued.

In parts of the U.S. where there was the active transmission of ZIKV in 2017, the CDC found an uptick of 21 percent in the number of babies with severe congenital disabilities. The increase amounted to twenty-nine fetuses and infants in Texas, Florida, and Puerto Rico with congenital disabilities affecting the central nervous system and brain, including the brain, eye, joint problems, **microcephaly**, or abnormally small heads. (In a typical six-month span, researchers would expect 140 infants in these regions to have these congenital disabilities, but in June through December of 2016, local health departments reported 169) [1]. The problem was far worse elsewhere.

D. M. Berube, *Pandemics and Resilience: Lessons we should have learned from Zika*, Risk, Systems and Decisions, https://doi.org/10.1007/978-3-031-25370-6_7

7.1 CZS or ZCS Surfaces

Over time, the term "congenital Zika syndrome" (CZS) appeared as a category for children's many associated developmental problems after exposure to ZIKV. Some studies call the congenital disabilities associated with Zika **congenital Zika syndrome** and others **or Zika congenital syndrome** (CZS). Often, they included microcephaly as one symptom of CZS.

ZIKV goes beyond microcephaly, with other symptoms such as visual and hearing impairments and unusual signs and symptoms different from other congenital infections, such as arthrogryposis [1] and no microcephaly, suggesting that the term congenital Zika syndrome is more appropriate [2]. These findings have led some authors to use the term "**congenital Zika syndrome**" (see above), incorporating microcephaly, intracranial calcifications, and other abnormalities, [3] which can include neurologic, ophthalmologic, audiologic, and skeletal findings, are considered congenital Zika syndrome (**CZS**) [4].

The literature does conclude that ZIKV can cause CZS, which the CDC describes as "a unique pattern of congenital disabilities and disabilities found among fetuses and babies infected with ZIKV during pregnancy", including severe microcephaly, decreased brain tissue, eye damage, joints with limited range of motion and too much muscle tone, which restricts body movement after birth [5].

7.2 General Implications

Maternal ZIKV infection during early pregnancy became associated with stillbirth, miscarriage, fetal growth restriction, central nervous system (CNS) abnormalities, and ocular abnormalities [4]. Research indicates a linkage between maternal and fetal infection while the fetus was developing in the placental womb and a lesser body of research on perinatal infection during childbirth and delivery (transplacental transmission).

The ratio of babies affected by congenital disabilities was similar for women with ZIKV infection during pregnancy who experienced symptoms compared to those who did not experience symptoms. Some babies with a possible ZIKV infection during pregnancy might look healthy at birth but can develop long-term health problems as they grow [6].

The CDC reported that the course of ZIKV disease in pregnant women was like that in the general population. No evidence suggests that pregnant women are more susceptible to or experience more severe conditions during pregnancy. There is limited information about the risk of periconceptional ZIKV infection (infection) eight weeks before conception or six weeks before the last menstrual period [6].

[1] Arthrogryposis, also called arthrogryposis multiplex congenita (AMC), involves a variety of nonprogressive conditions that are characterized by multiple joint contractures (stiffness) and involves muscle weakness found throughout the body at birth.

Hoen et al. [7] listed brain abnormalities with or without microcephaly regardless of the presence of additional congenital disabilities and neural tube defects and other early brain malformations, e.g., anencephaly (born without parts of the brain and skull), acrania (absence of a fetal skull), encephalocele (a saclike protrusion or projection of the brain and the membranes that cover it through an opening in the skull), holoprosencephaly (failure of the forebrain, to develop normally), or arhinencephaly (congenital absence of olfactory bulbs and tracts), eye abnormalities, and other consequences of central nervous system dysfunction among fetuses and infants who had neither evident brain abnormalities nor microcephaly [7].

These abnormalities are linked to severe mental retardation and motor disabilities among surviving children [8]. Features included microcephaly, facial disproportionality, cutis gyrata (thickening of the scalp), hypertonia/spasticity (too much/not enough muscle tone), and hyperreflexia (twitching or spastic tendencies), visual impairment, intrauterine growth restriction (IUGR), and irritability. Abnormal neuroimages include fetal intracranial calcifications (hardened brain tissue), ventriculomegaly (areas of the brain filled with fluid), and brainstem hypoplasia (lack of cells in a tissue), and lissencephaly (absence of normal folds in the cerebral cortex). These abnormalities often lead to fetal demise [9].

There seems to be a broad range of ZIKV's brain targets, including the corpus callosum, which facilitates communication between the two hemispheres, the cerebellum, which plays a significant role in the movement, balance, and speech, and the basal ganglia, which are involved in thinking and emotion [10]. Dr. Levine said the images suggest that ZIKV is like a formidable enemy able to damage in three ways: keeping parts of the brain from typically forming, obstructing areas of the brain, and destroying parts of the brain after they start [11].

> Ultrasonographic findings in our cohort showed serious and frequent fetal and central nervous system development problems, affecting 29% of the forty-two women whose fetuses were evaluated by ultrasonography. Abnormalities were noted in the fetuses of women infected at any week of gestation. Fetuses infected in the first trimester had findings suggestive of pathologic change during embryogenesis, but CNS abnormalities were also seen in fetuses infected as late as 27 weeks of gestation. [9]

"And the newest abnormality that's not in the literature is water on the brain or hydrocephalus. That's the newest complication of this that the doctors in Brazil were not seeing in the first 3 to 6 months, but that they are now seeing that the babies are getting to be 6 and 12 months", Dr. William Dobyns, a pediatric neurogeneticist at Seattle Children's Hospital said [12]. Most of the babies in the study had such damage in the cortex, which plays a crucial role in learning, memory, and coordination. It also continues to develop through infancy, suggesting that ZIKV-infected babies who seemed to emerge unscathed might be vulnerable to difficulties as they grow [10].

It may be essential to understand that the disease may not be microcephaly per se, but a new congenital infection presumed to be related to ZIKV, including microcephaly and other malformations. Furthermore, some patients presented an enlarged supratentorial subarachnoid space[2] surrounding the brain and spinal cord, making the head circumference appear bigger than the actual brain volume, masking a tiny brain.

7.3 The Microcephaly Story

According to the WHO, in 2016, at the epidemic's peak, more than 2,000 babies worldwide were born with ZIKV-related microcephaly or other congenital disabilities [13].

The CDC mined the U.S. Zika pregnancy registry to get a clearer picture of how often infections in pregnancy lead to congenital disabilities in infants. According to that study covering entries for 2016, when the researchers only included women with "confirmed" Zika infection during pregnancy, the rate was 10 percent [14]. Some sources report 11 percent [15].

Congenital disabilities were twenty times more likely in pregnant women who had contracted the virus. Be extra precautious if you are pregnant and living in affected areas [16].

The American Academy of Pediatrics cited the typical rate of a ZIKV-free baby born with microcephaly as 0.07 percent, the American baseline. Still, they found roughly four percent of infants or fetuses in the U.S. were diagnosed with microcephaly when their mothers contracted ZIKV while pregnant. Overall, six percent of infants or fetuses born to a ZIKV-infected mother developed abnormalities consistent with the virus [17].

Traces of ZIKV have been found in babies affected by microcephaly's bodily fluids and tissue. Dr. Lyle Petersen, director of the division of vector-borne diseases at the U.S. Centers for Disease Control and Prevention, told a news briefing on March 2, 2016, at the Pan American Health Organization in Washington that numerous lines of evidence are now linking ZIKV with microcephaly. "I don't think there is any question about that anymore", Petersen said [18].

While it has not been proven with 100 percent certainty, ZIKV causes complications such as microcephaly. However, the WHO notes that it is "widely accepted" that the two conditions are linked. In many cases of infants who have had microcephaly, the virus has been confirmed, leading doctors to connect the virus with the complication further [19].

However, there is a significant challenge for researchers attempting to validate microcephaly cases; it was that Brazilian doctors publish little in English journals [20]. Keeping track of developments and research findings became more problematic despite translation software when copies of the articles were attainable. Often, U.S. libraries did not have access to all of them, electronic or otherwise.

---

[2] Space between arachnoid mater [a delicate membrane, one of three, that surrounds the brain and spinal cord] and pia mater [a delicate innermost membrane enveloping the brain and spinal cord]

According to Hoen et al. [21], the risk of congenital disabilities was 12.7% when ZIKV infection occurred in the first trimester, 3.6% when it occurred in the second trimester, and 5.3% when it occurred in the third trimester, and the risk of the congenital Zika syndrome was 6.9%, 1.2%, and 0.9%, respectively [21]. Coyne and Lazear report that ZIKV can induce fetal damage well beyond the first trimester, as infections even late during pregnancy can result in fetal disease and adverse pregnancy outcomes [21].

Brasil et al.'s study attempts to recontextualize the issue of infant congenital Zika syndrome in their December 2016 research published in the New England Journal of Medicine. "Although microcephaly has been widely discussed ZIKV infection, it is important to note that other findings such as cerebral calcifications and fetal growth restriction were present more frequently" [22].

As shall be seen below, these pronouncements did not seem to quell the controversy. There seem to be two reasons for the disagreement: (i) some suggest that ZIKV appears to have mutated in 2000 and become able to attack developing brain cells in a fetus [23] and related to it (ii) before Brazil, there was little evidence of co-occurrence.

A sum of the literature follows, including some from obscure sources. Almost all were peer-reviewed.

7.4 Trimester Distinctions

While most of the technical literature warns pregnant women against infection, especially during the first trimester, this does not mean the dangers are isolated. The highest congenital disability rates appeared when pregnant women had the virus in the first trimester. In eighty-five women with ZIKV symptoms or exposure to the virus in the first trimester, 11 percent of the infants had congenital disabilities. Without the presence of ZIKV, microcephaly occurs between two babies per 10,000 live births, or 0.0002 percent, and twelve babies per 10,000 live births, or 0.0012 percent, in the U.S., according to the CDC [24].

CDC data show that 6 percent of babies carried by ZIKV-infected mothers are born with congenital disabilities. That number would jump to 11 percent if the virus were transmitted to the mother in her first trimester [25]. And when confirmed infection occurred in the first trimester of pregnancy—when the risk Zika poses to a developing fetal brain is highest—15 percent of babies born, or fetuses lost had ZIKV-related defects [26].

[S]ome diseases, like rubella, which can be catastrophic to fetuses whose mothers fall ill in the first trimester, are less severe if a woman is infected later in pregnancy. As a result, conventional wisdom focuses on the first trimester as the most vulnerable time for a fetus. But that does not seem to be the case with the ZIKV. França et al. [25] found that the virus can still cause devastating brain defects among newborn-infected mothers in their last trimester [27].

Branswell [26] reported out when confirmed infection occurred in the first trimester of pregnancy—when the risk Zika poses to a developing fetal brain is highest—15 percent of babies born, or fetuses lost had ZIKV-related defects [26].

Sheridan et al. [5] warned that despite the growing acceptance that ZIKV-associated microcephaly is associated with infection in the first trimester, ambiguity remains as to whether ZIKV can access the fetus via the placenta throughout pregnancy to cause central nervous system abnormalities and whether alternative routes not involving syncytiotrophoblast (the placental barrier between maternal and fetal blood that allows exchanges in nutrients and gases and also represents the endocrine tissue of the human placenta) might exist [27].

Meanwhile, understanding of the biology of ZIKV infection in pregnancy is based on clinically described cases in pregnant women with symptomatic infection. Therefore, there is little knowledge of the effects of mild or asymptomatic ZIKV infections or ZIKV infections in early pregnancy, when women may be unaware of the pregnancy. The risk of adverse events may be higher in symptomatic infections, but mild infections are probably more common and thus may also contribute substantially to the overall burden [28].

Cuevas et al. [29] concluded that the prevalence of microcephaly increased more than fourfold in 2016 compared with 2015, with a ninefold increase in July 2016 (the peak month) compared with July 2015. The temporal association between ZIKV infections and microcephaly, with the peak of reported microcephaly occurring approximately 24 weeks after the peak of the ZIKV outbreak, provides evidence that the most significant risk period is likely the first trimester of pregnancy and early in the second trimester of pregnancy [29].

Nearly half of women infected with the mosquito-borne ZIKV experience some complications with their pregnancy, according to a preliminary study from Brazil. In an article in this week's New England Journal of Medicine, researchers describe the results of a survey of 125 women who displayed symptoms of a ZIKV infection at some stage in their pregnancy. Fifty-five percent of pregnancies had adverse outcomes after maternal infection in the first trimester, 52% after infection in the second trimester, and 29% after infection in the third trimester. The adverse outcomes include miscarriage, calcifications in a baby's brain, babies who are smaller than usual, and brain hemorrhages. There were four cases of microcephaly among the 125 infected women, or 3.4% [28].

Consistent with data from the 50 U.S. states regarding primarily travel-associated ZIKV infections in pregnancy, about one in twenty fetuses or infants had possible ZIKV-associated congenital disabilities. The percentage of infants with possible ZIKV-associated birth defects after infection identified in the first trimester was 8% (95% CI = 5%–12%) in the U.S. territories compared with 15% (95% CI = 8%–26%) in the U.S. states [30].

7.5 Symptoms

Congenital microcephaly is a descriptive diagnosis; it is a condition. It comes from head measurements predominantly. As a clinical finding, in mild cases, it is easily missed [31].

It refers to the size of the cranium, a measurement, and its adverse impacts seem traceable to a genetic condition. One of the problems with tracking the effect has been the lack of standardized criteria for determining its onset. In addition, the literature suggests the condition can develop over time. The absence of a diagnosis at birth does not mean the newborn is safe. The virus' persistence suggests that even healthy babies born to mothers infected with ZIKV may not be out of the woods. Babies born with microcephaly may experience ongoing brain damage from their mothers' ZIKV infection [32].

For example, a follow-up of thirteen infants undertaken by Van der Linden [33] reported thirteen infants with laboratory evidence of congenital ZIKV infection with average head size at birth, including the findings from extensive imaging, neurologic, ophthalmologic, auditory, and orthopedic examinations. Follow-up of these infants has shown that for most, head growth deceleration occurs to the point of microcephaly after birth, and significant neurologic sequelae are evident [M]icrocephaly might not be apparent at birth but can develop after birth in infants with underlying brain abnormalities [33].

The New York Times reported on fifteen of the most severely affected babies born in the northeastern region of Brazil, where ZIKV was most prevalent. "At about 22 months old, these children had the cognitive and physical development of babies younger than six months. They could not sit up or chew and had virtually no language". (Fig. 7.1) [31].

The lifelong consequences of microcephaly could include mental retardation, developmental delays, difficulties with coordination and stature, dwarfism, facial

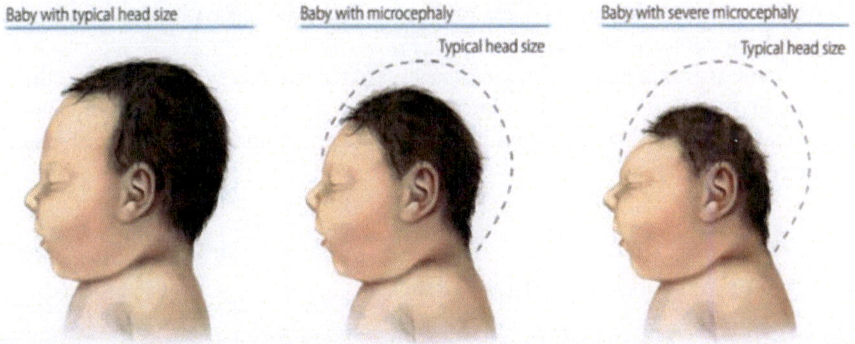

Fig. 7.1 Newborn with normal head size, microcephaly, and severe microcephaly. CDC. Congenital Anomalies of the Nervous System: Microcephaly. https://www.cdc.gov/ncbddd/birthdefects/surveillancemanual/quick-reference-handbook/microcephaly.html. Accessed July 3, 2022

distortions, hyperactivity, and seizures. On the other hand, some children have average intelligence and development [34]. However, for most children with microcephaly, their futures are challenging on many levels for themselves and their caregivers.

Sadly, patients with these serious viral CNS infections during childhood appear to be weakly associated with the later development of schizophrenia and non-affective psychoses [35]. A German study (von der Hagen et al. [36]) of 680 children with microcephaly in Berlin and Dresden found that microcephaly is associated with intellectual impairment in 65% of participants, epilepsy was diagnosed in 43%, and ophthalmological disorders were found in 30%. Brain magnetic resonance imaging revealed abnormalities in 76% of participants. Seventy-two percent did not attend mainstream schools or kindergarten but needed special education. Further clinical findings frequently identified in patients with microcephaly were ophthalmological disorders (30%), facial dysmorphism (19%), anomalies of the oropharynx including cleft palate (13%), and anomalies of the heart (14%), kidneys, and of the urinary tract, as well as of the skeletal system (13% each) and the gastrointestinal tract (9%) [36].

7.6 Long-Term Effects on Children

The first American baby with microcephaly because of the ZIKV was born to a mother in Oahu, Hawaii, who lived in Brazil the previous May, during her first trimester of pregnancy [37].

In a monumental study of Public Health Event Records and Brazil's Live Birth Information System (SINASC) in Brazil, Paixao et al. [38] studied eleven million live-born children from birth to 36 months. Among infants born at term, those with CZS were 14.3 times as likely to die as those without the syndrome, and the risk persisted throughout the first three years of life. This observation and others like it have called for healthcare systems to prepare for an increased burden of adverse pregnancy outcomes in the coming years [28].

The burden of congenital anomalies, diseases of the nervous system, and infectious diseases as recorded causes of death were higher among live-born children with congenital Zika syndrome than among those without the syndrome, and the difference increased with age [38].

7.7 Long-Term Effects on Mothers

The socioeconomic impacts on families with a microcephalic child can be significant. Mothers often are compelled to leave the workforce and devote nearly full-time care to these children.

Microcephaly is a disaster on many levels for mothers of these babies. There are few fears more alarming than a failed pregnancy. Most obstetricians share stories about seamlessly innocent discomforts in a pregnant patient leading her to a doctor's office and the loss of the fetus. Understandably, the anxiety of harboring a new life and a high point in a soon-to-be mother's life hypersensitizes a pregnant woman to her body.

The burden of caring for children affected by ZIKV now falls primarily on their mothers, many of whom are poor, young women of color with little formal education.

In Brazil, they were supposed to get support from the government through subsidies and medical services for their babies; help has been unreliable or difficult to access. These mothers, the invisible victims of ZIKV, report feeling abandoned and helpless. And as their babies grow up, they face a complex reality: They and their children may never escape the poverty into which they were born [39].

The impact of a child with CZS on mothering includes feelings at the time of diagnosis, maternal isolation and mental health, experiences of stigma and prejudice, and exhausting itineraries searching for therapeutic care. The repercussions of CZS have been a significant burden on already vulnerable women, and social inequalities and poverty were important markers in the mothers' reports, says Paula Freitas et al. [40], who interviewed them. Many families affected by CZS already lived in precarious social conditions, which were exacerbated further [40].

Evidence suggests that caring for a child with special healthcare needs can affect many domains of family life, including the caregiver's mental health. A Brazilian study [41] of nine mothers of children aged 5 to 12 months with ZIKV-related microcephaly found that microcephaly was associated with elevated levels of maternal anxiety and low maternal quality of life.

There are significant associated problems with care for microcephalic children, and their socioeconomic condition aggravates these further. Families of children living with CZS face demands for specialized care and a lifelong responsibility for maintenance. For mothers, an accumulated burden of factors influences the daily care of children with significant disabilities [40].

Kotsky et al. found that among primary caregivers of children with evidence of congenital ZIKV infection who do not have childcare support, there was an indirect relationship between child developmental delay and depressive symptoms through economic challenges [42].

In a WhatsApp exchange late last year with L., one of K. Eliza Williamson's research participants in Brazil, she told me about the hardship she was experiencing. Not only was Brazil's rising COVID-19 death toll devastating, especially for structurally vulnerable families like hers, but she was feeling "exhausted" by her daily care duties and "paralyzed" by "everything we go through" as mothers of disabled children in contemporary Brazil [43].

As pointed out by disease expert Dr. Ernesto Marques of the University of Pittsburgh, "Most of these babies are from low socioeconomic status and rely on the public health system to provide care. It is tough to manage those children because they need multiple types of specialists" [44].

Furthermore, the costs associated with treating a microcephalic child can be onerous, especially in a less affluent country where public and private health insurance may be insufficient. "Treatment for microcephalic infants and adults that develop neurological complications can cost millions per patient" [45].

Freitas et al. collected many narratives, most worth reading. There were a few dominant themes in these stories. Coercion to quit a job or dismissal after the child's birth with CZS was a constant in the women's narratives. A painful and difficult choice between the routine of caring for their child and the responsibility to work marked their lives. Unemployment and the unequal burden of caring for a child with CZS added to the narratives revealing their partners' marital infidelity, abandonment, and physical and verbal aggression [40].

Here is a dramatic case in point reported by Agence France-Presse. Taking care of a child with microcephaly requires constant attention, and there are formidable expenses. Young Miguel has been admitted to intensive care eight times so far. His mother says that Miguel could live for 10, 20 years—or two or three. "So, I will bathe him, kiss him, take in his smell. Because at any moment, they could put him in hospital" [46].

Learning a fetus may be stricken by the ZIKV can take quite a toll on the mother. Here is more from the story of young Miguel started above.

When doctors told Miguel's mother the six-month-old fetus she was carrying had severe brain damage caused by the ZIKV, Thamires Ferreira da Silva tried to commit suicide by jumping in front of a bus in Rio de Janeiro. But the bus driver braked in time, and more than two years later, she is raising her son Miguel with the help of her husband Wallace, their families, and medical specialists. Her son, aged two years and four months, suffers from microcephaly and has lissencephaly, where parts of the brain appear smooth. This rare Dandy-Walker syndrome[3] [45] is characterized by deformation of the part of the brain that controls movement, kidney problems, and epilepsy.

In Brazil, families raising kids with CZS are primarily black or brown, poor, and living in neglected areas of the city. Mothers are usually the primary or sole caregivers for their disabled children. Williamson continues her anecdote about L.

> Like most of those directly impacted by ZIKV, she relied primarily on the paltry government assistance paid monthly to disabled people or their caregivers. It was barely enough to cover necessities, let alone any outside support for her son's intensive care needs. Her husband, who was working only sporadically fixing air conditioning units when opportunities arose, had still not bothered to learn how to feed their son properly [43].

In another case, litigation ensured: The mother of a two-year-old boy with congenital disabilities associated with the ZIKV threatened to sue the Ministry of Health in Trinidad and Tobago over its failure to provide comprehensive health care for the toddler. In an associated letter from the toddler's pediatrician, which Dr. Devanand

[3] The symptoms of Dandy Walker syndrome typically include developmental delay, low tone (hypotonia) or later high tone (spasticity), poor coordination and balance (ataxia), and sometimes enlarged head circumference and increased pressure within the skull due to hydrocephalus.

Lakheeram prepared in September 2018, the child cannot sit or stand without full support and has spontaneous and uncoordinated movements of all limbs [47].

In the study by Santos and Farias [48], financial issues were revealed by mothers as a difficulty that generated anguish; moreover, most families suffering from the consequences of ZIKV infection were economically vulnerable [48].

The repercussions of CZS on women's lives in the (Freitas et al. [40]) study were a considerable burden, with important markers of social inequality and poverty revealed in the mothers' reports. Gender inequality also marked the epidemic, an unequal burden of childcare, experiences of stigma, and exhausting therapeutic itineraries in search of medical care for their children.

However, the relationship between parental mental health and child health is likely bidirectional, as parental depression has been associated with harmful parenting practices and risk for child behavior problems [42]. And now to the supposed mystery.

7.8 Epidemiological Mystery

There were many predictions that were made on the brink of an onslaught of cases of ZIKV-associated microcephaly. WHO's Christopher Dye said that based on reports of rash and fever in Northeastern Brazil in early 2016, it was expected that about 1,000 babies would be born with ZIKV-induced microcephaly from late summer onward. Instead, about eighty were recorded in the region [49]. How could this be? Why were the predictions so incorrect? What happened to all the predicted cases of microcephaly? There was an epidemiological problem, if not a mystery.

For example, as of December 2017, 438 were in the state of Pernambuco. Yet, according to the Pan American Health Organization, just 700-odd cases of what is now called congenital Zika syndrome have been confirmed across the rest of the Americas. And nobody can explain the discrepancy.

"We were braced for a large epidemic of microcephaly. We did not see that", said Albert Ko, professor of epidemiology and medicine at the Yale School of Public Health, who has studied the epidemic. "It's still a bit embarrassing that we don't know these answers" [50].

Despite early fears that this mosquito-borne pathogen would cause a widespread, devastating epidemic of congenital disabilities, the occurrence of microcephaly after the second wave of transmission in 2016, when ZIKV spread from northeast Brazil across the country, was substantially lower than that following the first wave of transmission in early 2015. Although the use of sensitive yet non-specific case criteria for microcephaly, among other factors, could have contributed to the initial overestimation of the risk, the progression of the microcephaly epidemic remains puzzling [51].

As of July 2, 2016, almost 90% of the 1,709 confirmed cases of congenital microcephaly or congenital disabilities of the central nervous system reported in Brazil since last November were in a relatively small area: in the coastal hinterland of the country's northeastern tip.

Particularly surprising, says Dr. Fatima Marinho, director of information and health analysis at Brazil's health ministry, is that just three cases have been confirmed in Brazil's second-most populous state, Minas Gerais, which borders the most-affected part of the northeast region. Poor data on the scale and timing of ZIKV outbreaks across Brazil make it hard to tell whether increases in microcephaly elsewhere might have been delayed. Still, ministry scientists now think that the northeast represents a marked outlier [52].

This temporal increase in suspected cases of microcephaly could also be distorted given both raised awareness, with more children than usual being measured and reported, and changing definitions of microcephaly over time. The Latin American Network of Congenital Malformations explained the possibility of over-reporting and misdiagnosis. Their report led to speculation in the international scientific press on the validity of the magnitude of the increase in microcephaly cases [53].

ECLAMC (Latin American Study of Congenital Malformations) found this unsurprising due to the comprehensive media coverage of the possible microcephaly epidemic caused by the ZIKV (ZIKV) should lead to an increased number of microcephaly cases in all Brazilian states resulting from improved notification to SINASC derived from mandatory reporting of cases of microcephaly to the epidemiological surveillance service [54]. Nonetheless, even ECLAMC lamented the increase of 26 times over the estimated risk seen in Pernambuco far exceeds the record of all congenital microcephaly based on previously existing causes [54].

The association between microcephaly and ZIKV has been challenged, and lingering questions remain, including why more than 85 percent of suspected microcephaly cases are confined to Brazil's northeast region [55]. Large numbers of babies with borderline average head sizes were born in Brazil as far back as 2012, two years before the ZIKV is thought to have entered the country, say researchers searching for answers to urgent questions [56].

So, what does this mean? Methods used to estimate prevalence rates of microcephaly have differed over time, as has the working definition of microcephaly until recently, making attribution of microcephaly primarily to a Zika outbreak easier said than proven [57]. Microcephaly has been associated with lack of maternal schooling, living without a partner, smoking during pregnancy, intrauterine growth restriction, vaginal delivery, and the pregnancy being the mother's first. To benchmark the occurrence of microcephaly, a team used the INTERGROWTH-21st standards and the Brazilian Ministry of Health criterion and defined head circumference (HC) > two standard deviations (SDs) below the mean for gestational age (GA) and sex and severe microcephaly defined as HC > 3 SDs below the mean. The prevalence of microcephaly in SL was found to be 3.5%, compared to 3.2% in RP, when using only the last normal menstrual period (LNMP) as the determiner of GA. However, when using either LNMP date or obstetrical ultrasound (OU), the prevalence of microcephaly was reduced in RP (2.5%). The rate of severe microcephaly was higher in SL (0.7%) compared to RP (0.5%). Expected microcephaly and extreme microcephaly rates in an average population are projected to be 0.55% and 0.14%, respectively [58].

Two-thirds of those children are here in the northeast. As of December, 438 were in the state of Pernambuco. Yet just 700-, according to the Pan American Health

Organization, just 700-odd cases of what is now called congenital Zika syndrome have been confirmed across the rest of the Americas can explain the discrepancy [50]. According to Marques et al., as one dose of ZIKV is believed to give immunity, when a second, less intense Zika outbreak followed in 2016, northeastern cities like Recife and Salvador—the capital of Bahia state, which had the most microcephaly cases, at 509—had reached "critical threshold". "That's the proportion of people that need to be infected for the disease to die out", Brady said 63%, in Salvador's case.

In Paraíba, one of the nine States within the epidemic's epicenter, twenty-one medical centers have collaborated via telemedicine since 2012 in a pediatric cardiology network. The network's database currently stores information on more than 100,000 neonates. To support the microcephaly research, the network ran a task force from December 1st to 31st, 2015, and rescued the head circumference of 16,208 neonates. A much higher than expected incidence of microcephaly was observed, varying from 2 to 8% according to the utilized classification criteria. These findings raise questions about the condition's diagnosis and its notification [59].

Hazin [60] led a team that completed head computed tomographies on twenty-three infants from Recife, Brazil. They reported that the global presence of cortical hypogyration and white matter hypomyelination or dysmyelination in all the infants and cerebellar hypoplasia in most of them suggest that ZIKV is associated with a disruption in brain development rather than destruction of the brain. The neuronal and glial proliferation and neuronal migration appear to be affected [60].

But researchers say their information may not be enough to determine whether factors besides ZIKV are involved. Much of the microcephaly raw data comes from routine hospital reports, often incomplete. And lab tests to confirm ZIKV infection are rarely conducted [61]. The limited information on ZIKV infection rates is compounded by the difficulty in the clinical confirmation of microcephaly, evidenced by low confirmation rates in Brazil's independent, temporary microcephaly reporting system in late 2015 [62].

Although ecologic data do not necessarily qualify as an epidemiologic study, data from Brazil regarding the temporal and geographic association between ZIKV infection and the later appearance of infants with congenital microcephaly are compelling [3]. Although evidence suggests that ZIKV can cause microcephaly, the clustering pattern hints that other environmental, socioeconomic, or biological factors could be at play [61].

7.9 Brazil and Microcephaly Cases

Brazilian data on microcephaly remain questionable. Adding to these data complexities, much of the extant microcephaly data from Brazil comes from incomplete hospital reports.

De Araújo et al. [59] investigated this question using population-based surveillance methods from two Brazilian birth cohorts in two cities in Brazil in 2010 before the Zika epidemic manifested itself. The authors used formal working definitions for microcephaly. They found that the prevalence of severe microcephaly in both cities ranged from 0.5 to 0.7% (higher than expected), suggesting microcephaly was endemic in both municipalities before the current Zika epidemic.

In 2015, Brazil had reported a 20-fold increase in microcephaly in regions affected by ZIKV [63]. Over 21 states in Brazil have reported suspected cases of microcephaly across over seven hundred municipalities. After so many such births were recorded in Northeastern Brazil in the last quarter of 2015, the country—and other places where the virus fanned out from Brazil—braced themselves for a similar tsunami in 2016. But it did not materialize. The region's first wave of ZIKV may have been its only wave of ZIKV to date. The discrepancy was plain as day when the data were slotted into a graph [49].

De Magalhães-Barbosa et al. [64] found an overall microcephaly prevalence of 5.6% in the neonatal intensive care unit (NICU) population studied. As the NICU is a unit that admits neonates with a wide range of perinatal conditions, one would expect to find a prevalence of microcephaly higher than in the general population of newborns [64].

The number of cases of congenital microcephaly observed in the northeastern states in the second half of 2015, even considering only those recorded at SINASC (Brazilian Live Birth Information System), exceeds the expected number in most states more than three times. In 2000, SINASC reported that the prevalence of microcephaly in Brazilian newborns was 5.5 cases/100,000 live births; in 2010, it was 5.7 cases/100,000 live births. Over the last three months, it went up to 99.7 per 100.000 live births, corresponding to a 20-fold increase? [59].

As PRI reported in 2016, nobody is entirely sure that the incidence of microcephaly has even risen in Brazil [65]. One author said that before 2015, the incidence was 0.5 cases for every 10,000 live births [31]. On December 26, 2015, Paraíba led Brazil in the rate of reported cases with 82.75 per 10,000 while Pernambuco ranked a close second with 80.38 per 10,000 [20]. Simmons said a minimum norm for Brazil was determined at 2,725 microcephaly cases annually. The minimum incidence for the condition was determined to be 92/100,000 [66].

In February 2016, a retrospective review of microcephaly data from the northeast of Brazil showed undetected seasonal peaks of microcephaly dating back, at least to 2012, and a trend toward an increased number of severe cases starting in 2013 [59].

From 2010 to 2014, Brazil saw an average of 156 babies born yearly with microcephaly. But from October 2015 to January 2016, that number spiked to over 4,000 babies [67]. In Brazil, it was reported that 404/1132 (36%) cases had confirmed microcephaly and central nervous system malformations, and 17/404 (4%) were positive for ZIKV infection [68].

Officially, Brazil has seen more than 84,000 cases of ZIKV and initially reported 3,852 cases of microcephaly [69] among thousands of congenital malformations (nearly 4,000 of which were reported in Pernambuco) [70]. On 27 January, the Brazilian government said that of 4,180 suspected microcephaly cases recorded since October, it has so far confirmed 270 and rejected 462 as false diagnoses [71].

There remains much disagreement on microcephaly rates linked to the ZIKV. Based on confirmed ZIKV-associated cases of microcephaly reported by the Brazilian

Ministry of Health, the northeastern region of Brazil has a ten times larger incidence of confirmed cases compared with the rest of Brazil and other Latin American countries where ZIKV circulates. Though there might be higher rates of microcephaly in Brazil, there remain concerns that the frequencies might both be exaggerated and weakly associated with ZIKV [72].

7.10 Colombia and Microcephaly Cases

Alarming reports of microcephaly caused by the ZIKV came from Colombia's neighbor to the east. Sanz Cortes and Parra Saavedra started monitoring Barrranquilla's pregnancies to see if a similar wave would hit Colombia. It never did. Colombia, with the world's second-largest outbreak, produced only eighty-two [73].

In Colombia, approximately 105,000 suspected cases of ZIKV disease (diagnosed based on clinical symptoms, regardless of laboratory confirmation) were reported from August 9, 2015–to November 12, 2016, including nearly 20,000 pregnant women (1484 (12%)) of these cases confirmed on RT-PCR assay) [29]. But in Colombia, a 2016 study of almost 12,000 pregnant women infected with ZIKV found zero microcephaly cases. If ZIKV is to blame for microcephaly, where are the missing cases? [74].

Bar-Yam et al. reported that the epidemic of Zika in Colombia was expected to result in many cases of microcephaly. However, until epidemiological week 23 (June 11), only six reported cases were reported. The number of cases increased to twenty-one in weeks 24–27, confirming expectations about ZIKV as a cause of microcephaly, though the most recent weeks 28 and 29 are below the expected trend [75]. Therefore, the five additional cases reported this week are the first indication that ZIKV is causing microcephaly in Colombia, while earlier reported cases are consistent with random co-occurrence of microcephaly and ZIKV infections [75].

Nevertheless, using a rate of 1% microcephaly cases for ZIKV infections in the first trimester (inferred from circumstances in French Polynesia) leads to a prediction of 200 total microcephaly cases for Colombia [29]. New England Complex Systems Institute (NECSI) asks if ZIKV is to blame for microcephaly; where are the missing cases? [75]. The NECSI report analyzes the data and shows that the four cases of ZIKV and microcephaly observed until April 28 are just what would be expected due to the background rate of the 60,000 pregnancies. About 20,000 births would already be expected. The expected microcephaly rate for countries with no reported infections of 2 in 10,000 births gives exactly four cases. The study also notes that until April 28, there have been about fifty microcephaly cases in Colombia, of which only four have related to ZIKV. The four cases are expected for the coincidence of ZIKV and microcephaly in identical pregnancies, even if ZIKV is not the cause [74]. NECSI reported five new confirmed ZIKV and microcephaly cases four days later [74].

Pacheco et al [76] reported on 1,850 women who were being tracked whose date of infection with ZIKV was known relative to the start of the pregnancy. Of these, 532, 702, and 616 were infected in the first, second, and third trimesters. Sixteen percent, 29, and 93% (85, 204, and 583) of the pregnancies have concluded. No cases of microcephaly were observed. The total number of pregnancies with ZIKV infections is much more significant, with 11,944 cases with ZIKV symptoms observed in clinical settings. Pacheco et al. [76] reported 1,850 Colombian women, 90% of whom claimed third-trimester infection with the ZIKV but found no apparent anomalies in the fetus [76]. Moore et al. [77] seem to agree. Based on their available data, the most common timing of infection, as determined by maternal symptoms, is the late first and early second trimester. However, they admit some third-semester infections have been reported as well [77].

Cuevas et al. [29] provide preliminary information on cases of congenital microcephaly identified in Colombia during epidemiologic weeks 5–45 (January 31–November 12) in 2016. Colombia's genetic disabilities surveillance system includes reporting microcephaly (International Classification of Disease, 10th Revision code Q02) among live births and pregnancy losses (including spontaneous abortions, pregnancy terminations, and stillbirths) from all reporting areas. Although the microcephaly prevalence in 2016 among infants likely exposed to ZIKV in utero (9.6 per 10,000 live births) in Colombia was not much higher than the median of microcephaly prevalence (6.6 per 10,000 live births) reported by passive surveillance in 17 U.S. states during 2009–2013, the comparison with 2015 Colombia data indicates the magnitude of the increase [29].

As Colombia braced for a wave of congenital disabilities after the ZIKV spread here earlier this year, officials from the remote state of Casanare reported twelve cases of microcephaly in newborn babies—high for the state's small population. Epidemiologists and geneticists descended on the state, only to discover that two of the twelve reported had the defect [78]. News reports asked why microcephaly has not appeared in other Latin American countries with similar climates, such as among the 2,100 pregnant women infected with ZIKV in Colombia? [56].

The popular and peer-reviewed literature provides many explanations. First, Colombia only began official surveillance in August 2015. Second, women in Colombia have access to abortion services, so termination of pregnancy is a viable explanation. Third, the height of the infection was approximately nine months ago (9/16). If, as suspected, ZIKV strikes developing brains in the first trimester, [76] the most incriminating evidence could come when these babies are born [79].

This suggests that there may not be a direct link between ZIKV and microcephaly except for random co-occurrence. Dr. Fatima Marinho, director of information and health analysis at Brazil's health ministry, told the journal Nature, "We suspect that something more than ZIKV is causing the high intensity and severity of cases" [71].

Note that the base rate of microcephaly in the absence of ZIKV is 140 per year in Colombia, which is consistent with the approximately fifty microcephaly cases in the first four months of 2016, only four of which have been connected to ZIKV. When interpreting ZIKV as the cause, background cases must be subtracted [80].

The divergent figures in the two countries, both of which recorded high numbers of infections, underscore the different approaches health authorities have taken to get a handle on a public health crisis after ZIKV hit the region this past year and led to a rash of babies born with abnormally small heads and other defects. Compared to Brazil, Colombia took a far more cautious approach in reporting the incidence of microcephaly—reporting cases only after weeks or months of painstaking research to confirm that a documented congenital disability was indeed an abnormality and then link the defect definitely to ZIKV [78]. International health experts say Colombia's approach is more rigorous than Brazil, which diagnoses a ZIKV-linked defect by clinical assessment if laboratory tests aren't possible. Colombia uses an in-depth process that involves a battery of tests for infants, home visits, genetic screenings, and blood tests for pathogens other than ZIKV—such as toxoplasmosis, a parasitic disease that cats can transmit. It can take months and cost $1,000 to confirm or discard each case [78].

While more complete data from Columbia are waited on, there is some validation for the link between ZIKV and microcephaly in a French Polynesia study using mathematical modeling. The author not only claimed a link between ZIKV infection and microcephaly but also offered solid statistical support for the association between ZIKV infection and microcephaly during the first trimester. It was estimated that the number of microcephaly cases associated with ZIKV was 95 (95% CI 34–191) per 10,000 women infected in the first trimester [81]. This provides the one chance in 1,000 likelihoods that has dominated much of the literature.

Bar-Yam and others reported that until June 11, the cases could be accounted for by background rates. The new cases are consistent with a model in which only first-trimester pregnancy infections cause microcephaly [82]. Preliminary reports suggest that ZIKV-induced fetal abnormalities can occur in all trimesters of pregnancy. However, the most severe manifestations are associated with conditions, especially the first and second trimesters. According to Nielsen-Saines et al., 55% of pregnant women infected in their first trimester had poor pregnancy outcomes, as did 52% of those infected in their second trimester and 29% of those infected in their third trimester. Abnormalities of the central nervous system were seen in fetuses as late as 39 weeks of gestation [32].

Bar-Yam et al. found that the reports from Colombia until June 11, 2016, are consistent with only background cases, while subsequent reports tracked a model in which *only* first-trimester infections result in microcephaly. This team warns that even though they felt the defensible link between infection and microcephaly was most evident in this first trimester, this did not mean there were no consequences of altered pregnancy infections [83].

> There are several possible reasons for the differences between the reported baseline microcephaly prevalences in Brazil and Colombia and the differences in microcephaly increases in the ZIKV outbreaks in the two countries. First, 50–75% of the population of Colombia resides at altitudes above 2,000 meters in areas without active, vector-borne ZIKV transmission. Second, microcephaly is a complex congenital disability to monitor because there are inconsistent definitions, obtaining accurate measurements is challenging, and terminology is inconsistent. Because of these challenges, prevalence estimates vary widely among countries and surveillance systems [even] within the U.S. Third, the reports of microcephaly from Brazil might have served as an early warning. As evidence emerged about the link between ZIKV infection and microcephaly, the Colombian Ministry of Health recommended

in February 2016 advising women to consider delaying pregnancy for six months, which might have affected subsequent birth rates. The number of live births in Colombia during epidemiologic weeks 5–45 decreased by approximately 18,000 from 2015 to 2016 [29].

7.11 What Happened?

The literature supported a host of explanations. However, the consensus seems to have settled on inaccurate baselines and a general frequency of misreporting before and after the epidemic. The remaining literature includes infectious disease cofactors, socioeconomic issues, and less viable explanations.

But scientists in Brazil have been puzzled by the fact that cases of congenital Zika syndrome have clustered in the northeast part of the country and that the expected surge in cases elsewhere in the country has not happened. According to a report by the ministry of health, of 1749 confirmed cases of microcephaly or other central nervous system congenital disabilities, 85% were concentrated northeast, released in late July. A fifth of confirmed cases was in Pernambuco state in northeast Brazil. In east-central Brazil, the country's second most populated east-central Gerais recorded just four malaria cases from ZIKV infection, while Sã Jorge Lopez-Camelo and Ieda Maria Orioli ECLAMC (Latin American Collaborative Study of Congenital Malformations) blame the awareness effect [84].

> …[A] rise in reported cases of microcephaly might be attributable to the intense search for cases of congenital disability and misdiagnoses arising from heightened awareness in the wake of the possible link with ZIKV. The 'awareness' effect is well known and inevitable, they say, and must be revealing cases that would have gone unnoticed under normal circumstances [71].

The discrepancy in numbers between the Brazilian Ministry of Health and this study may reflect underreporting in recent years associated with an even greater incidence of microcephaly than presumed. It is possible that a high prevalence of the non-severe forms of microcephaly had been occurring before the current outbreak. Still, health workers had only notified the live birth information system about the cases of neonates with typical severe phenotypes [85].

7.12 Diagnostic Baselines

Questions remain. What is the baseline prevalence of microcephaly in Brazil before and after the Zika epidemic? Were other risk factors that might play a role in most infants born with microcephaly?

According to the available data, the fraction of first-trimester pregnancies exposed to ZIKV that have confirmed microcephaly is 63%. The number of suspected cases is 289% of the first-trimester ZIKV revealed pregnancies (which in principle is consistent with a first-trimester model of confirmed microcephaly cases), and the fraction of pregnancies exposed to ZIKV that are guaranteed to have both ZIKV and microcephaly is 9.5%. These values are significantly larger than the 1% rate obtained from French Polynesia and used above to model Colombia [65].

One of the most prominent sources on the subject, Juliana Sousa Soares de Araújo of the Heart Circle-Royal Portuguese Hospital in Recife, agrees. "The numbers of microcephaly reported here are much higher than the 6.4 per 10 000 live births reported by the Brazilian live birth information system. The results raise questions about the notification system, the appropriateness of the diagnostic criteria, and future implications for the affected children and their families" [59].

An important paper has surfaced challenging the extreme cases of microcephaly. "The numbers of very extreme cases of microcephaly, for instance, while significantly increasing over the last few months late 2015), are much smaller and until recently fell within the expected ranges for the worldwide reported incidence". The researchers offer these caveats. "It is possible that a high incidence of milder forms of microcephaly occurred well before the current outbreak. Only those extreme cases with classical phenotypes were notified. And as the number of extreme cases increased over these past three or four months, so did the awareness of health professionals who started to notify milder forms" [59]. Infants might be diagnosed with microcephaly when they are globally small—i.e., small for gestational age, without truly isolated microcephaly. This issue deserves attention, especially because in utero growth restriction leading to the birth of small-for-gestational-age infants is also a feature of congenital ZIKV syndrome [86].

The Latin American Network of Congenital Malformations raised the possibility of over-reporting and misdiagnosis. Their report led to the international scientific press speculating the magnitude of the increase in microcephaly cases [87]. In addition, this temporal increase in suspected cases of microcephaly could also be distorted based on awareness, with more children than usual being measured and reported and changing definitions of microcephaly over time [87]. Victora et al. conclude that "the number of suspected cases relied on a secondary specificity screening and therefore overestimated the number of cases by including mostly normal children with small heads" [87].

Some researchers, however, suggest underreporting before ZIKV even arrived, making the increase seem more dramatic than it was. A review of four years' worth of medical records finds far greater numbers of microcephaly cases before the ongoing ZIKV epidemic than had been officially reported. Medical reports of more than 16,000 babies born between 2012 and 2015 at one of the medical centers in the state of Paraiba, Brazil, were examined. Since 2012, Mattos's team [59] found that a vast number of babies—4 percent to 8 percent—appeared to have microcephaly, according to the broadest definitions of the term [88].

> It is possible that a high incidence of milder forms of microcephaly occurred well before the current outbreak. Only those extreme cases, with classical phenotypes, were being notified", the authors wrote: "And as the number of extreme cases increased over these past three or four months, so did the awareness of health professionals who started to notify milder forms. [88]

For example, at the time of the article (08/2016), no cases of microcephaly occurred in all these nearly 12,000 pregnancies in Colombia. However, Pacheco cites four cases of microcephaly with ZIKV in the general population that did not report any ZIKV symptoms, these cases being reported before April 28. This implies they are more unreported, symptomless cases of ZIKV infection. Since there are less than 1 in 12,000 incidences of ZIKV until this point in the epidemic, there should be at least four times as many infected individuals that do not have symptoms for there to be four microcephaly with ZIKV cases, for a total of $5 \times 12,000$ or 60,000 Zika cases [83].

According to the available data, the fraction of first-trimester pregnancies exposed to ZIKV that have confirmed microcephaly is 63%. The number of suspected cases is 289% of the first-trimester ZIKV revealed pregnancies (which in principle is consistent with a first-trimester model of confirmed microcephaly cases), and the fraction of pregnancies exposed to ZIKV that are guaranteed to have both ZIKV and microcephaly is 9.5%. These values are significantly larger than the 1% rate obtained from French Polynesia and used above to model Colombia [83].

Some epidemiologists complain about general underreporting in Brazil over the last five years. In a report by researchers in Paraiba, one of the worst-hit areas, the state saw high numbers of microcephaly cases since 2012, with the condition more common in 2014 than last year, when ZIKV was first recorded in Brazil [70]. The over-reporting has impacted the debate over ZIKV in ways that can only be speculated on at this time. One day soon, when the last chapter is written on this disease, there may be a quantifiable effect of irresponsible statements made in the public interest by initiative-taking but misguided people. While this distraction seemed short-lived and was quickly debunked, the same cannot be said about the misinformation that dominated the debate over the use of genetically engineered mosquitoes (see Chap. 14).

Data from the Brazilian Ministry of Health show many cases of congenital malformations in babies (potentially linked to ZIKV, but also connected to other infections like syphilis and rubella) are concentrated in the northeast of Brazil—particularly in the states of Pernambuco, Bahia, Paraíba, and Maranhão [89].

Data available for the state of Bahia in Brazil have also been analyzed, which includes both ZIKV and microcephaly cases, unlike many other states in Brazil, including Pernambuco, which only reported microcephaly. The results are consistent with a much higher rate of microcephaly than that reported in French Polynesia or Colombia [83].

The Connecticut Department of Public Health announced Monday that it monitors thirty babies born to or living in Connecticut with mothers who had tested positive for ZIKV or flavivirus during their pregnancies. "We must collaborate with pediatricians to monitor these babies for signs of Microcephaly or other ZIKV-related congenital disabilities throughout the first year of life because we have seen that these defects are not necessarily readily apparent at birth", said DPH Commissioner Dr. Raul Pino [90].

There are many reasons for underreporting. First, the pre-ZIKV baseline relied upon was based on a passive surveillance system. Second, the lack of an internationally accepted, standardized approach to measuring and defining microcephaly makes comparing rates across regions and time periods difficult [91]. Other considerations include the diagnostic criteria for microcephaly is relatively un-specific. Simply put, there are multiple criteria used to determine the presence of microcephaly, and reporting during the outbreak may have bundled different symptoms together to produce inflated occurrences. A possible source of discrepancy is the failure to adjust the head size to the infant's weight when defining microcephaly. Therefore, proportionality or lack thereof is essential in ascertaining Brazil's microcephaly [86].

Sarah Boseley, a health editor for the Guardian, offered her assessment in 2017 and defended the data collection process in Brazil.

> The Brazilians have been particularly good at keeping the data apart from anything else, so all the doctors who have seen microcephaly cases, that's cases of babies with small heads, deformed shaped heads, have reported them. That is something that the developing world has not learned yet; it is difficult; it takes people who have been taught to recognize symptoms of any disease, so in Ebola, for instance, that all went under the radar for a long time because you had cases. Still, they were in distant rural places, and nobody recognized them; until you can identify that, you cannot look for them. With ZIKV, it must be said that malformed babies' heads are apparent, but they have been missed. Africa has had ZIKV in the past and must have had microcephaly cases, but we certainly do not know much about that. [92]

Microcephaly is a rare condition. Microcephaly could be challenging to diagnose, as exemplified by a recent German study where authors reported that approximately half of the study population with confirmed microcephaly had not received a prior diagnosis [93].

In the U.S. of America, the prevalence has been estimated to range from 2.0 to 12.0 newborns with microcephaly per 10,000 live births, and the European Surveillance of Congenital Anomalies Center reports 2.9 newborns with microcephaly per 10,000 live births.

The absence of reliable data on microcephaly in Brazil in the years before 2015 makes it difficult to assess the magnitude of the outbreak. Little is known about the previous prevalence of microcephaly in the country's various regions, both in term and preterm neonates. It is possible that a tiny proportion of babies with microcephaly, particularly of the severe type and late in 2015, would have been the result of ZIKV infection [64]. For example, in Brazil, the live birth information system, Public Health Event Records, and Brazil's Live Birth Information System or SINASC reported a prevalence of 0.6 newborns with microcephaly per 10 000 live births in 2010 [86].

ECLAMC (Latin American Collaborative Study of Congenital Malformations) estimated the rate in Brazil (1.98/10,000) may be an underestimation for the Northeastern region, where the prevalence of microcephaly was always higher than that of hospitals in the rest of Brazil [54]. According to Silva et al. [88], the 2010 microcephaly prevalence is estimated at 290 per 10,000 live births (defined as more than two standard deviations below the mean) and 62 per 10,000 live births for severe microcephaly (defined as more than three standard deviations below the mean) [94].

Also, in most cases, tests to confirm ZIKV infection were not conducted [95]. So there are inaccurate baselines and preliminary diagnoses being dealt with as well. In other words, microcephaly might have been diagnosed, but whether the mother had been infected with the ZIKV was, in some cases, not scientifically known.

Meanwhile, understanding of the biology of ZIKV infection in pregnancy is based on clinically described cases in pregnant women with symptomatic infection. Therefore, there is little knowledge of the effects of mild or asymptomatic ZIKV infections or ZIKV infections in early pregnancy, when women may be unaware of the pregnancy. The risk of adverse events may be higher in symptomatic infections, but mild infections are probably more common and thus may also contribute substantially to the overall burden [65].

There is also the "awareness effect". We find what we are looking to find. Nonetheless, there are serious complaints that the cases of microcephaly have been over-reported. Any birth has a minimum probability of 2 in 10,000 microcephaly. Concerns have been expressed that the report from Bahia and elsewhere in Brazil may be excessive. The data from Bahia suggest a much higher rate of microcephaly, implying over 10,000 cases. Other Brazilian states have inconsistent and even higher rates according to reported cases.

Moreover, the occurrence of microcephaly is determined by the head circumference measurement. This is the first problem involved in setting baselines. The definitions of microcephaly used in Brazil have changed and more specific criteria based on WHO InterGrowth standards have been adopted. However, the sensitivity of these more particular criteria may be lower [96].

A customary practice worldwide as well as in Brazil is to use a measuring tape wrapped around the newborn's head right in the delivery room to diagnose microcephaly [81]. In 2016, Brazil reported 6,158 suspected cases of microcephaly (or other brain and spinal cord malformations). But hundreds, if not thousands, of those babies might not have a congenital disability. Doctors have flagged cases with a quick and dirty method: a head circumference of fewer than thirty-two centimeters [79]. Soares de Araújo warns. However, it is not known if a head circumference of 31 m or 32 cm in a term neonate could be within normal limits for this population (referring to Paraíba, Brazil) [59].

The measurement standards change. At the beginning of the epidemic, the inclusion criterion for microcephaly was a head circumference $\leq$ 33 cm. In December 2015, the bar changed to $\leq$ 32 cm, and, in February 2016, it changed again, now to head circumference $\leq$ 31.5 cm for girls and $\leq$ 31.9 cm for boys [59]. In addition, some babies are born with head circumferences at the lower limit of the **normal range** but later progress to microcephaly in about 10% of cases [82].

According to Victoria [53], Brazil's suspected cases relied on a screening test with extremely low specificity. It, therefore, overestimated the actual number of cases by including mostly normal children with small heads [53].

Before 2015, the annual numbers of reported cases of microcephaly in Brazil were consistently below two hundred. Between mid-2015 and Jan 30, 2016, 4,783 suspected cases of microcephaly were reported, including newborn and fetal losses.

Of these, 1103 cases have completed clinical, laboratory, and imaging examinations, and 404 (36.2%) were classified as confirmed cases of microcephaly. Among the confirmed cases, brain abnormalities were detected by imaging in 387 babies, and ZIKV was seen in seventeen babies, including two fetal losses. The remaining 709 cases were discarded, and 3670 suspected cases of microcephaly remain under investigation [53].

By July 3, 130 remained under investigation. "We jumped a little bit too fast into ZIKV. It may be well. Not saying it is not. But just saying there are a few things that do not seem to match very easily in this picture and that we may need to investigate a bit further", pediatric cardiologist Dr. Sandra Mattos said [56]. Mattos has been collecting data on 100,000 newborns in Paraiba in her work on congenital heart disease.

> According to the ECLAMC (Latin American Collaborative Study of Congenital Malformations), the average historical prevalence of microcephaly in Brazil is around two cases per 10,000 births. However, rates in the country's north have typically been higher. The researchers calculate that the maximum number of cases that would have been expected in the northern state of Pernambuco in 2015 is around forty-five. Yet Pernambuco reported twenty-six times that number last year. The report says that even if ZIKV is causing microcephaly, these vast numbers are too high to be credible. [71]

If there has been an over-reporting of microcephaly associated with ZIKV, the next question becomes what should be done. In response to ECLAMPC's suggestions, experts suggest erring toward safety. Other experts believe ECLAMPC's conclusions may not follow from their data. They claim that "prevalence rates of severe microcephaly at birth in Brazil based on the Brazilian Live Birth Information System are underestimated, and the Latin American Collaborative Study of Congenital Malformations estimated that the rate of underreporting is 66%", the researchers continued [82].

> Population-based data from the Brazilian Ribeirão Preto (RP) and São Luís (SL) birth cohort studies, including deliveries by resident mothers, indicate severe microcephaly was much higher than expected in both cities.... The number of cases of microcephaly is grossly underestimated, with an underreporting rate of ~90%... The prevalence of severe microcephaly at birth is expected to be 0.14%. 23 In both cities, most of the severe microcephaly (0.5% in RP and 0.7% in SL) was much higher than expected and higher than previously reported in various national and international studies.... Silva et al.'s (2017) findings suggest that microcephaly was endemic in both municipalities before the circulation of the ZIKV. [94]

Albert Ko at Yale School of Public Health does not quite buy it. "Misdiagnosis is a reasonable hypothesis. But it is not clear that this explanation accounts for the whole story", says Ko, an epidemiologist studying mothers and babies born with ZIKV in the northeast part of Brazil [50].

7.13 Three Theories

These are minority positions. One comes from a loner in the field, another is based on absolute supposition, and the third does not seem to be defended by the expert community. They are included as informational only.

7.14 The Bock Theory

Dr. Randall Bock is a biomedical device consultant and received his M.D. training at the University of Rochester. He has a book published by Drivestraight, Inc. out of Vernal, Utah, touting one employee (April 2022) and a little over $50,000 annual income, so please use careful judgment with what follows. Dr. Bock also has an article in the American Journal of Medicine (IF: 4.965). While his rebuttal has appeared on the fringe, it does not mean his allegations are wrong though they are exceedingly direct.

Of all the literature on the link between ZIKV and microcephaly, the most damning comes from Bock [97].

> The total disregard characterized the 2015 ZIKV microcephaly crash for the scientific method. It is impossible to avoid concluding that there was an intentional effort to present conjecture as scientific fact. The effort to "discover" something new was unrelenting and succeeded fully, sadly rewarding this lack of respect for the scientific method. Rushing to the press cannot be ascribed to a legitimate good-faith reason in the admitted absence of a fatal disease. There was no cluster analysis by experts before the public announcement of microcephaly. There has been little coverage of 2016's data-reconstruction study's contradicting the original "epidemic" (Microcephaly in north east Brazil: a retrospective study on neonates). Clinical laboratory confirmation of ZIKV was an erstwhile scientific impossibility. The neuropediatricians did not use a uniform standard for classifying microcephaly. Conflicts of interest and severe biases abounded and flouting of basic scientific procedures that all physicians know are important all went unchallenged. The paper published by Dr. Luz is not credible on any level. There remains no paper clinically distinguishing ZIKV from dengue.

He starts by reporting the history of the claimed linkage between ZIKV and microcephaly. In 2015 Dr. Kleber Luz, a pediatrician, asserted that his patients with mild clinical dengue were infected with the ZIKV, despite no laboratory confirmation and the impossibility of clinically distinguishing between dengue and ZIKV. There were no published articles in the literature guiding clinicians on how to distinguish dengue from ZIKV. Nothing in the record documents what Dr. Luz found clinically different from mild dengue [97].

Dr. Randall Bock is a biomedical device consultant and received his MD training at the University of Rochester. Further suggestions come from Drs. Gubio Soares Campos and Silvia Sardi's corroborative claims from their RT-PR tests on two dozen samples were suspicious when the Bahia State ministry evaluated five hundred patients from the same cohort. None were positive for ZIKV though some were tested positive for dengue. In addition, Claudia Duarte dos Santos, a virologist from Curitiba, found ZIKV in none of Luz's samples.

Bock adds that the physicians involved leaked their message and became fixed in the public mind, while the careful refutation by expert peers did not receive that same exposure [97].

No medical, scientific, media, or public authority has yet retracted any aspect of the ZIKV–microcephaly theory, six years after the double disappearance that essentially contradicts it.

7.15 Fewer Pregnancies and More Abortions

Another theory is that women in the region who had seen the possible outcome of a ZIKV infection in pregnancy might have avoided or terminated pregnancies in large numbers. It is conceivable that fear of the adverse consequences of ZIKV infection led to fewer conceptions or a more sizable number of pregnancy terminations in 2016. Routinely collected data are not complete enough to determine whether birth rates fell or abortion rates increased in 2016 [98].

But if the maternity wards of hospitals in the region had emptied in 2016, the world would have heard about it by now. "If there were a huge effect, it would have been big news very quickly. It would have been obvious", WHO's Christopher Dye said [49].

7.16 Multicausal Cofactor Theory

ZIKV is but one cause of microcephaly. Silva et al [94] also identified risk factors **associated** with having a microcephalic infant, including maternal schooling level, marital status, smoking during pregnancy, primiparity, vaginal delivery, and intrauterine growth restriction [91].

The spread of an arbovirus such as ZIKV is affected by entomological, environmental, and climatic factors, and, therefore, attack rates might differ between outbreaks. Also, there is a possibility that the risk of microcephaly associated with ZIKV infection will vary in other populations because of genetic factors [63].

7.16.1 Bovine Viral Diarrhea Virus

The first cofactor could be bovine viral diarrhea virus (BVDV). The Brazilian doctor who first reportedly established the link between ZIKV and microcephaly is considering whether another disease, BVDV, may be involved, as BVDV proteins were also detected in the brains of three fetuses with microcephaly. BVDV causes congenital disabilities in cattle but is not known to infect people. Researchers suggested that infection with ZIKV may make it easier for BVDV to infect humans [95].

Adriana Melo and her colleagues at the Federal University of Rio de Janeiro reported on 15 July that they had found BVDV in the brains of three fetuses with microcephaly [87]. Butler admitted they have not ruled out the possibility that their findings might be due to contamination [52].

7.16.2 *Chikungunya*

The second cofactor could be chikungunya, but that claim is less likely. Something that caused a similar illness, likely the chikungunya virus, was probably responsible for the elevated level of fever and rash illnesses Brazil recorded in 2016. A team from the Ministry of Health in Brazil, the PAHO, and Christopher Dye, a WHO epidemiologist, theorized [26]. The problem is the primary candidate is chikungunya, but chikungunya has not been identified as a cause of microcephaly in the past [99].

7.16.3 *Dengue*

Albert Ko at Yale School of Public Health does not quite buy it. "Misdiagnosis is a reasonable hypothesis. But it is not clear that this explanation accounts for the whole story", says Ko, an epidemiologist studying mothers and babies born with ZIKV in the northeast part of Brazil. Ko thinks there is another possible explanation: ZIKV might not be working alone.

Pregnant women who contract ZIKV might not be enough to cause microcephaly in all cases. Some scientists have hypothesized that dengue, another mosquito-spread disease in Brazil, could increase the risk of ZIKV infection. And a new sibling study from researchers at the University of São Paulo, published in February, provided intriguing evidence that genetics could make some babies more susceptible to congenital Zika syndrome. It studied identical twins, where both had microcephaly, and non-identical, where only one did in six of seven cases [50].

Dengue is a complex virus. There are five different versions. Prior exposure to one version of dengue can worsen the illness when exposed to a second version. And what is closely related to dengue? ZIKV. If the dengue theory turns out to be accurate, it could mean the global threat of ZIKV to pregnant women is less dire than scientists initially thought [100].

In the current ZIKV epidemic, some scientists have theorized that dengue and ZIKV may be partners in crime, interacting in some way that allows the ZIKV to cross the placenta. Before the ZIKV arrived in Brazil, it was unknown to cause congenital disabilities [101].

Future studies should also evaluate the influence of potential cofactors and effect modifiers, given the wide geographical variation in risk that Jaenisch et al. observed.

The estimated absolute risk of a predicted confirmed microcephaly case in a baby born to a woman infected during pregnancy, assuming a 50% infection rate, ranged from 0.006% in Paraná State in southern Brazil to 0.99% in Sergipe State in northeast Brazil; the risk was 0.25% in Rio de Janeiro State and 0.54% in Pernambuco State [96].

7.17 Other Causes Altogether

Another possibility is that ZIKV infection during pregnancy is necessary but insufficient for developing microcephaly in newborn infants—in other words, the presence of some other unknown cofactor. Here are two that reappear in the literature.

7.18 Yellow Fever Vaccinations

One alternative cause came from a paper by Butler published in 2106; researchers from Brazilian labs noted a correlation between low vaccination rates for yellow fever and the microcephaly clusters. Because yellow fever and ZIKV are in the same virus family, scientists speculate that the vaccine might provide some protection against ZIKV [52].

In another (de Góes Cavalcanti et al. [102]), pregnant women in regions with high yellow fever vaccination coverage may pose their offspring to lower risk for the development of microcephaly. ZIKV and yellow fever virus, in their urban transmission cycle, both share the same mosquito vector, *Aedes* mosquitoes. Considering the close relation of other arboviruses with ZIKV, such as the yellow fever virus, it might be speculated that the yellow fever vaccine, to some extent, may provide a protective effect against severe disease and sequelae caused by ZIKV infection, such as microcephaly [102].

They concluded that "vaccination against Japanese encephalitis, dengue fever, and yellow fever may produce cross-reactive antibodies. Antibody cross-reactivity occurs between different flavivirus infections, such as yellow fever, Japanese encephalitis, and dengue fever. In a historical study, a volunteer infected with ZIKV virus produced antibodies against yellow fever and ZIKV" [102].

7.19 Socioeconomic Issues

The second alternative phenomenon associated with microcephaly is simply poverty. "[M]alnutrition, previously associated with microcephaly, could intensify when coupled with other etiological factors. Indeed, most of the reported cases have occurred in low-income families" [59].

In addition, the authors noted that other factors were associated with microcephaly back in 2010, including the maternal level of education, marital status, smoking during pregnancy, primiparity (a state in which a woman is bearing a child for the first time and has given birth to an offspring at one time), vaginal delivery, and intrauterine growth restriction—all factors not unique to Brazil or to contracting ZIKV.

The majority of women who have had babies with microcephaly have been young, single, black, poor, and tend to live in small cities or on the outskirts of big ones, Marinho says [52]. Poor Northeasterners, many of them farmworkers, many black and brown, women whose faces and biographies are usually invisible in Brazil's socially stratified world, were the mothers of the babies with microcephaly [20].

Before the ZIKV epidemic, there was a silent endemic of microcephaly associated with these variables linked to poverty and other causes, such as undiagnosed congenital infections and severe microcephaly were grossly underestimated [94].

According to Ricardo Ximenes, a professor of epidemiology and infectious diseases at Pernambuco's state and federal universities, higher levels of microcephaly have been associated with precarious living conditions, poor or open sewage, and irregular water supplies favoring the proliferation of the *Aedes* aegypti mosquito that spreads ZIKV. In Pernambuco, 11% of people do not have a freshwater supply, and only 32% get their sewage treated, according to the state government-controlled sanitation company, Compesa. But as UN and World Health Organization officials argued in 2016, improving Brazil's woeful sanitation is also vital to stopping ZIKV. According to the WHO, thirty-five million people in Brazil lack adequate sanitation, and 3.8 million are without safe drinking water [50].

According to Marinho et al., when the cases first began and were reportedly linked to ZIKV, health officials believed they would see "an explosion of congenital disabilities" across Brazil. But that has not happened. Data compiled by Marinho and colleagues submitted for publication suggest some socioeconomic factors may be involved [95]. Higher levels of microcephaly are associated with precarious living conditions, with poor or open sewage and irregular water supplies favoring the proliferation of the *Aedes* aegypti mosquito that spreads ZIKV. In Pernambuco, 11% of people do not have a freshwater supply, and only 32% get their sewage treated, according to the state government-controlled sanitation company, Compesa [50].

7.20 Conclusion

In the first week of life, neonatal mortality among infants with congenital ZIKV syndrome seems to be as high as 4 to 7% [103].

The spectacular global emergence of ZIKV and the appearance of previously unobserved congenital disabilities caused by infection during pregnancy may thus be due to a combination of extrinsic factors that triggered the human transmission cycle on a large scale, and the propensity of the Asian ZIKV strains to cause fetal abnormalities and not fetal loss [104].

While everyone speculated whether the link between ZIKV and microcephaly was valid, many researchers conducted studies on disability cases and documented other congenital disabilities. Others have developed theories to explain the linkage, which are examined in the following chapter.

References

1. Francis D (2018) Virus caused a big uptick in serious birth defects in 2016. In: Pacific Standard, 25 Jan. Retrieved from https://psmag.com/news/zika-virus-caused-a-big-uptick-in-serious-birth-defects-in-2016. Accessed on 7 Jul 2018
2. van der Linden V et al (2016) Congenital Zika syndrome with arthrogryposis: retrospective case series study. In: The BMJ, 5 Jul. Retrieved from http://www.bmj.com/content/354/bmj.i3899. Accessed on 9 Jun 2017
3. Rasmussen S et al (2016). Zika virus and birth defects — reviewing the evidence for causality. N Engl J Med 374:1981–1987, 19 May. https://doi.org/10.1056/NEJMsr1604338. Accessed on 22 Sept 2016
4. Children's National Hospital (2021) Zika 5 years later: still much to learn as 'likely' future outbreak looms. In: HealioNews, 21 Jan. Retrieved from https://www.healio.com/news/infectious-disease/20210114/zika-5-years-later-still-much-to-learn-as-likely-future-outbreak-looms. Accessed on 25 May 2022
5. de Oliveira Dias JR et al (2018). Zika and the eye: pieces of a puzzle. Prog Retinal Eye Res 66:85–106. https://doi.org/10.1016/j.preteyeres.2018.04.004. Accessed on 6 Jul 2018; Sheridan MA et al (2017) Vulnerability of primitive human placental trophoblast to Zika virus. PNAS 114(9):E1587–E1596, 13 Feb. https://doi.org/10.1073/pnas.1616097114. Accessed on 7 Jul 2018
6. CDC (2022) Zika and pregnancy: data & statistics on zika and pregnancy. Retrieved from https://www.cdc.gov/pregnancy/zika/data/index.html. Accessed on 29 May 2022
7. Hoen B et al (2018) Pregnancy outcomes after zikv infection in French territories in the Americas. N Engl J Med 378(11):985–994, 15 Mar. Retrieved from https://www.ncbi.nlm.nih.gov/pubmed/29539287. Accessed on 27 Jun 2018
8. Miranda-Filho DDB et al. (2016). Initial description of the presumed congenital zika syndrome. Am J Public Health 106(4):598–600.https://doi.org/10.2105/AJPH.2016.303115. Accessed on 23 May 2017
9. Brasil P et al (2016) Zika virus infection in pregnant women in Rio de Janeiro - Preliminary Report. N Engl J Med 375(24):2321–2334. Epub 4 Mar 2016. https://doi.org/10.1056/NEJMoa1602412
10. Belluck P (2016) Brain scans of brazilian babies show array of zika effects. In: The New York Times, 23 Aug. Retrieved from http://www.nytimes.com/2016/08/24/health/zika-a-formidable-enemy-attacks-and-destroys-parts-of-babies-brains.html?_r=0. Accessed on 26 Oct 2016
11. de Oliveira-Szejnfeld PS et al (2016) Congenital brain abnormalities and zika virus: what the radiologist can expect to see prenatally and postnatally. Radiology. 281(1):203–218. https://doi.org/10.1148/radiol.2016161584. Accessed on 26 Oct 2016
12. KING 5. (2017). Zika virus showing new symptoms in infants. In: KING 5, 26 Jan. Retrieved from http://www.king5.com/news/health/childrens-healthlink/zika-virus-showing-new-symptoms-according-to-seattle-childrens-neruogeneticist/393701069. Accessed on 17 May 2017
13. VnExpress (2016) HCMC declares zika pandemic. In: VnExpress, 20 Oct. Retrieved from http://e.vnexpress.net/news/news/hcmc-declares-zika-pandemic-3486472.html. Accessed on 11 Apr 2017
14. Reynolds MR et al (2017) U.S. zika pregnancy registry collaboration. Vital signs: update on zika virus-associated birth defects and evaluation of all US infants with congenital Zika virus exposure — US Zika pregnancy registry. Morb Mortal Wkly Rep 66(13):366–373,

2017. Retrieved from https://www.cdc.gov/mmwr/volumes/66/wr/mm6613e1.htm?s_cid=mm6613e1_w#suggestedcitation. Accessed on 4 May 2017
15. Walker M (2017) Zika: what do we know a year later? In: MedPage Today, 29 Jan. Retrieved from https://www.medpagetoday.com/meetingcoverage/smfm/62809. Accessed on 30 Jun 2017
16. Alvarez M (2017) Protect yourself against Zika and Lyme disease this summer. In: Fox News, 25 Mar. Retrieved from http://www.foxnews.com/health/2017/03/25/protect-yourself-against-zika-and-lyme-disease-this-summer.html. Accessed on 7 Apr 2017
17. Hope M (2017) Texas medicaid to thwart zika pregnancies among low-income homes. In: Breitbart, 26 Apr. Retrieved from http://www.breitbart.com/texas/2017/04/26/texas-medicaid-thwart-zika-pregnancies-among-low-income-homes/. Accessed on 14 May 2017
18. Reuters. (2016) Study shows how zika gets into, damages brain. In: NBC News, 4 Mar. Retrieved from http://www.nbcnews.com/storyline/zika-virus-outbreak/study-shows-how-zika-gets-damages-brain-n531901. Accessed on 30 May 2017
19. Brusie C (2018) How does zika affect toddlers? In: VeryWellFamily, 6 Mar. Retrieved from https://www.verywellfamily.com/how-does-zika-affect-toddlers. Accessed on 1 Apr 2018
20. Deniz D (2016) Zika: from the braziulian backlands to global threat. Zed Books Ltd, London
21. CDC (2022) Infection during pregnancy. Retrieved from https://www.cdc.gov/pregnancy/zika/testing-follow-up/effects-during-pregnancy.html. Accessed on 29 May 2022
22. Brasil P et al (2016) Zika virus infection in pregnant women in Rio de Janeiro. N Engl J Med 375:2321–2334, 15 Dec. https://doi.org/10.1056/NEJMoa1602412. Accessed on 16 Jan 2017
23. Haelle T (2016) Birth defects linked to zika include more than microcephaly. In: Everyday Health.com, 6 Apr. Retrieved from http://www.everydayhealth.com/zika/living-with/birth-defects-linked-zika-include-more-than-microcephaly/. Accessed on 14 Oct 2016
24. Mohney G (2016) Babies exposed to zika virus in first trimester more likely to have birth defects, study says. In: ABC, 14 Dec. Retrieved from http://abcnews.go.com/Health/babies-exposed-zika-virus-trimester-birth-defects-study/story?id=44185819. Accessed on 23 May 2017; Honein MA et al (2016) Birth defects among fetuses and infants of US women with evidence of possible zika virus infection during pregnancy. In: JAMA, 3 Jan. Retrieved from http://jamanetwork.com/journals/jama/fullarticle/2593702. Accessed on 23 May 2017
25. Colli G (2017) Zika virus threat not over yet. In: My Statesman, 27 Jan. Retrieved from http://www.mystatesman.com/news/national/zika-virus-threat-not-over-yet/MVfL2BT4fw2dpTmDsgNSgI/. Accessed on 3 Apr 2017
26. Branswell H (2017) "They're just hiding": experts say Puerto Rico may be underreporting Zika-affected births. In: STAT, 18 Apr. Retrieved from https://www.statnews.com/2017/04/18/zika-virus-puerto-rico-pregnancies/. Accessed on 4 May 2017
27. Lafrance A (2006) The danger of a third-trimester zika infection. In: The Atlantic, 29 Jun. Retrieved from https://www.theatlantic.com/health/archive/2016/06/zika-after-30-weeks/489284/. Accessed on 18 May 2017
28. Johansson MA et al (2016) Zika and the risk of microcephaly. N Engl J Med 375:1–4, 7 Jul. https://doi.org/10.1056/NEJMp1605367#t=article. Accessed on 14 Oct 2016
29. Cuevas E et al (2016) Preliminary report of microcephaly potentially associated with zika virus infection during pregnancy—Colombia, January–November 2016. Morb Mortal Wkly Rep 65(49):1409–1413, 16 Dec. https://doi.org/10.15585/mmwr.mm6549e1. Accessed on 26 Jun 2017
30. Cook M (2016) Zika virus: dangerous but still puzzling. In: BioEdge, 17 Dec. Retrieved from https://www.bioedge.org/bioethics/zika-virus-dangerous-but-still-puzzling/12133. Accessed on 4 Apr 2017; Brasil P et al (2016) Zika virus infection in pregnant women in Rio de Janeiro. N Engl J Med 375:2321–2334, 15 Dec. https://doi.org/10.1056/NEJMoa1602412. Accessed on 16 Jan 2017
31. Ratna K (2017) The secret life of zika virus. Speaking Tiger Books, New Delhi
32. Healy M (2016) New zika findings reveal how virus does its damage. In: Los Angeles Times, 14 Dec. Retrieved from http://www.latimes.com/science/sciencenow/la-sci-sn-zika-antibodies-20161214-story.html. Accessed on 14 May 2017

33. van der Linden V et al (2016) Description of 13 infants born during october 2015–january 2016 with congenital zika virus infection without microcephaly at birth—Brazil. Morb Mortal Wkly Rep 65:1343–1348, 2 Dec. Retrieved from https://www.cdc.gov/mmwr/volumes/65/wr/mm6547e2.htm. Accessed on 19 Jul 2017
34. Mayo Clinic (2016) Complications. Diseases and conditions: microcephaly. Retrieved from http://www.mayoclinic.org/diseases-conditions/microcephaly/basics/complications/con-20034823. Accessed on 25 Oct 2016
35. Dalman C et al (2005) Infections in the CNS during childhood and the risk of subsequent psychotic illness: a cohort study of more than one million Swedish subjects. Am J Psychiatry 165(1):59–65, Jan. Retrieved from https://www.ncbi.nlm.nih.gov/pubmed/18056223. Accessed on 7 Aug 2018
36. von der Hagen M et al (2014) Diagnostic approach to microcephaly in childhood: a two-center study and review of the literature. Dev Med Child Neurol 56(8):732–741, Aug. Retrieved from https://www.ncbi.nlm.nih.gov/pubmed/24617602. Accessed on 7 Aug 2018
37. McNeil DG (2016) Zika: the emerging epidemic. W. W. Norton & Co, NY
38. Paixao ES et al (2022) Mortality from congenital zika syndrome nationwide cohort study in Brazil. N Engl J Med 376:757–767. https://doi.org/10.1056/NEJMoa2101195
39. Fernandes M, Martinelli A (2018) Poor women of color bear the brunt of zika burden in Brazil. In: Huffington Post, 25 Jan 25. Retrieved from https://www.huffingtonpost.com/entry/poor-women-of-color-zika-brazil_us_5a66582ce4b0dc592a0bb36a. Accessed on 27 Jun 2018
40. Freitas PSS et al (2020) How do mothers feel? Life with children with congenital zika syndrome. Int J Gynecol Obstet 148(Suppl 2):20–28
41. dos Santos Oliveira SJG, dos Reis CL, Cipolotti R et al (2017) Anxiety, depression, and quality of life in mothers of newborns with microcephaly and presumed congenital zika virus infection: a follow-up study during the first year after birth. Arch Womens Ment Health 20:473–475
42. Kotzky K et al (2019) Depressive symptoms and care demands among primary caregivers of young children with evidence of congenital zika virus infection in Brazil. J Dev Behav Pediatr 40:344–353
43. Williamson KE (2021) What now? The afterlives of zika and COVID-19. In: Center for the humanities, 11 Nov. Retrieved from https://humanities.wustl.edu/features/eliza-williamson-afterlives-zika-and-covid-19. Accessed on 2 Jun 2022
44. Davies M (2017) Toddlers born with zika virus continue to demonstrate severe developmental issues. In: Jezebel, 14 Dec. Retrieved from https://jezebel.com/toddlers-born-with-zika-virus-continue-to-demonstrate-s-1821302425. Accessed on 27 Jun 2018
45. Newkirk II VR (2016) The American zika outbreak. In: The Atlantic, 13 Sep. Retrieved from https://www.theatlantic.com/politics/archive/2016/09/zika-puerto-rico-public-health/499685/. Accessed on 26 May 2017
46. AFP (2018) 'We've been forgotten': Brazil's zika generation. In: Agence France Presse, 18 Dec. Retrieved from https://mg.co.za/article/2018-12-18-weve-been-forgotten-brazils-zika-generation. Accessed on 14 May 2019
47. Achong D (2019) Mother of zika baby threatens to sue. In: Trinidad & Tobago Guardian, 4 Jan. Retrieved from http://www.guardian.co.tt/news/mother-of-zika-baby-threatens-to-sue-6.2.749862.a92bf371da. Accessed on 14 May 2019
48. Santos JHA, Farias AM (2017). "She is worth five children": the impact of microcephaly on maternity. Proc II Braz Congr Health Sci [in Portuguese] 1:1–12. Campina Grande: Editora Realize
49. Branswell H (2016) Why were there fewer microcephaly cases from Zika last year? In: Scientific American, 29 Mar. Retrieved from https://www.scientificamerican.com/article/why-were-there-fewer-microcephaly-cases-from-zika-last-year/. Accessed on 7 Apr 2017
50. Phillips D (2018) 'There are a lot of unknowns': British scientists set to work on zika vaccine. In: The Guardian, 9 Mar. Retrieved from https://www.theguardian.com/global-development/2018/mar/09/british-scientists-work-on-zika-vaccine-brazil-recife-birth-defects. Accessed on 27 Jun 2018

51. Costa F, Ko, AI (2018) Zika virus and microcephaly: where do we go from here? Lancet Infect Dis 18(3):236–237, Mar. https://doi.org/10.1016/S1473-3099(17)30697-7.
52. Butler D (2016) Brazil's birth-defects puzzle. Nature 535:1–2, 28 Jul. Retrieved from http://nematologia.com.br/wp-content/uploads/2016/07/zykainnature.pdf. Accessed on 26 Jun 2017
53. Victora CG et al (2016) Microcephaly in Brazil: how to interpret reported numbers? Lancet 387(10019):P621–P624, Feb 5. https://doi.org/10.1016/S0140-6736(16)00273-7. Accessed on 26 Jul 2018
54. Lopez-Camelo J, Orioli IM, Castilla E (2015) ECLAMC Final document: summary and conclusions of documents 1–5, 30 Dec. Retrieved from http://rifters.com/real/articles/NS-724-2015_ECLAMC-ZIKA%20VIRUS_V-FINAL_012516.pdf. Accessed on 21 May 2017
55. ECDC (2016) Zika virus disease epidemic: potential association with microcephaly and guillain-barré syndrome (first update). In: ECDC, 21 Jan. Retrieved from http://ecdc.europa.eu/en/publications/Publications/rapid-risk-assessment-zika-virus-first-update-jan-2016.pdf. Accessed on 18 Sep 2016
56. CBC News (2016) Microcephaly cases in Brazil predate zika virus outbreak, study says. Retrieved from http://www.cbc.ca/news/health/microcephaly-brazil-zika-reality-1.3442580. Accessed on 18 Sep 2016
57. First L (2018) Microcephaly in brazil may not just be due to zika virus infection. In: Blog-AAP Gateway, 8 Jan. Retrieved from http://www.aappublications.org/news/2018/01/08/Microcephaly-In-Brazil-May-Not-Just-Be-Due-To-Zika-Virus-Infection-Pediatrics-1-8-18. Accessed on 27 Jun 2018
58. Brooks, Stephanie & Carr, Leah. (2018). Microcephaly rates elevated in Brazil prior to zika virus epidemic. In: 2 Minute Medicine, 5 Jan. Retrieved from https://www.2minutemedicine.com/microcephaly-rates-elevated-in-brazil-prior-to-zika-virus-epidemic/. Accessed on 27 Jun 2018
59. de Araújo JSS et al (2016) Microcephaly in northeast Brazil: a review of 16,208 births between 2012 and 2015. In: Bulletin of the World Health Organization, 19 May. Retrieved from http://www.who.int/bulletin/volumes/94/11/16-170639/en/. Accessed on 20 Jul 2017
60. Hazin AN et al (2016) Computed tomographic findings in microcephaly associated with zika virus. N Engl J Med 374(22):2193–2195, 2 Jun. https://doi.org/10.1056/NEJMc1603617. Accessed on 7 Jul 2018
61. Butler D (2016) Brazil asks whether zika acts alone to cause birth defects. In: Nature News, 25 Jul. Retrieved from http://www.nature.com/news/brazilaskswhetherzikaactsalonetocausebirthdefects1.20309. Accessed on 26 Jun 2017
62. Johansson MA et al (2016) Zika and the risk of microcephaly. N Engl J Med 375:1–4, 7 Jul. https://doi.org/10.1056/NEJMp1605367#t=article. Accessed on 17 May 2017
63. Cauchemez S et al (2016) Association between zika virus and microcephaly in French Polynesia, 2013–15: a retrospective study. Lancet 387(10033):2125–2132. https://doi.org/10.1016/S0140-6736(16)00651-6. Accessed on 14 Oct 2016
64. de Magalhães-Barbosa MC et al (2017) Prevalence of microcephaly in eight south-eastern and Midwestern Brazilian neonatal intensive care units: 2011–2015. Arch Dis Childh 102(8):728–734, Aug. Retrieved from https://www.ncbi.nlm.nih.gov/pubmed/28302630. Accessed on 19 Sep 2018
65. Carless W (2016) Brazil now has doubts that zika alone causes birth defects. In: PRI, 2 Aug. Retrieved from https://www.pri.org/stories/2016-08-02/brazil-now-has-doubts-zika-alone-causes-birth-defects. Accessed on 27 Jun 2017
66. Simmons CH (2016) Establishing base levels of microcephaly in Brazil prior to the arrival of zika viral illnesses. In: Bulletin of the World Health Organization, 8 Feb. Retrieved from http://www.who.int/bulletin/online_first/16-171223.pdf. Accessed on 2 Jun 2017
67. Almendrala A (2016) An illustrated guide to the Zika outbreak. In: Huffington Post, 24 Feb. Retrieved from http://www.huffingtonpost.com/entry/guide-to-zika-virus_us_56a272afe4b076aadcc675b6. Accessed on 24 Oct 2016

68. Chen H-L, Tang R-B (2016) Why zika virus infection has become a public health concern? J Chin Med Assoc 79(4):174–178. Retrieved from http://www.sciencedirect.com/science/article/pii/S1726490116300065. Accessed on 10 Apr 2017
69. Steenhuysen J (2016) Adult mosquitoes can pass zika to their offspring: U.S. study. In: Reuters, 29 Aug. Retrieved from http://www.reuters.com/article/us-health-zika-mosquitoes-idUSKCN1142CD. Accessed on 5 Sep 2016
70. Bowater D (2016) Zika virus: Brazil dismisses link between larvicide and microcephaly. In: The Telegraph, 15 Feb. Retrieved from http://www.telegraph.co.uk/news/worldnews/zika/12157747/Zika-virus-Brazil-dismisses-link-between-larvicide-and-microcephaly.html. Accessed on 22 Sep 2016
71. Butler D (2016) Microcephaly surge in doubt: heightened awareness of Zika virus could help to explain the reported spike in birth defects. Nature 530(7588):13–14, 4 Feb. Retrieved from https://www.researchgate.net/file.PostFileLoader.html?id=56b34ab26307d98bff8b45b0&assetKey=AS%3A325390888390658%401454590642499. Accessed on 27 Oct 2016
72. de Oliveira Melo AS et al (2016) Congenital zika virus infection: beyond neonatal microcephaly. JAMA Neurol 73(12):1407–1416. Retrieved from https://www.ncbi.nlm.nih.gov/pubmed/27695855. Accessed on 23 Jul 2018
73. Molteni M (2017) A clue to the mystery of Colombia's missing zika cases. In: Wired, 31 Jan. Retrieved from https://www.wired.com/2017/01/clue-mystery-colombias-missing-zika-cases/. Accessed on 24 May 2017
74. New England Complex Systems Institute "New doubts on zika as cause of microcephaly". In: Science Daily, 24 Jun. Retrieved from www.sciencedaily.com/releases/2016/06/160624150813.htm. Accessed on 18 Jul 2017
75. Bar-Yam Y et al (2016) Is zika the cause of microcephaly? Status Report June 22, 2016. In: NECSI Report, 22 Jun. Retrieved from http://necsi.edu/research/social/pandemics/statusreport. Accessed on 3 Oct 2016
76. Pacheco O et al (2016) Zika virus disease in Colombia—preliminary report. N Engl J Med 383(6):e44. https://doi.org/10.1056/NEJMoa1604037#t=article. Accessed on 27 Sep 2016
77. Moore CA et al (2016) Characterizing the pattern of anomalies in congenital zika syndrome for pediatric clinicians. JAMA Pediatr 171(3):288–295. Retrieved from https://jamanetwork.com/journals/jamapediatrics/fullarticle/2579543?utm_campaign=articlePDF&utm_medium=articlePDFlink&utm_source=articlePDF&utm_content=jamapediatrics.2016.3982. Accessed on 24 July 2018
78. Muñoz SS (2016) Colombia's careful approach leads to slower reporting of zika-linked defects. In: The Wall Street Journal, 29 Dec. Retrieved from https://www.wsj.com/articles/colombias-careful-approach-leads-to-slower-reporting-of-zika-linked-defects-1483007403. Accessed on 26 May 2017
79. Rosen M (2016) Microcephaly: building a case against zika. In: Science News, 2 Apr. Retrieved from https://www.sciencenews.org/article/microcephaly-building-case-against-zika. Accessed on 27 Sep 2016
80. Bar-Yam Y et al (2016). Is Zika the cause of Microcephaly? Status Report June 27, 2016. In: NECSI Report, 27 Jun. Retrieved from http://necsi.edu/research/social/pandemics/statusreport2.pdf. Accessed on 26 Oct 2016
81. Deniz D (2016) Zika: from the Brazilian backlands to global threat. Zed books, London
82. Aragão MFVV (2017) Zika virus: an overview. Aragão MFVV (ed) Zika in focus: postnatal clinical, laboratorial and radiological aspects. Washington, DC: Springer
83. Bar-Yam Y et al (2016) Determining the rate and week of infection of zika caused microcephaly from Colombian and Brazilian data; Status Report August 1, 2016. In: NECSI Report, 2 Aug. Retrieved from http://necsi.edu/research/social/pandemics/statusreport3. Accessed on 26 Oct 2016
84. Disler M (2017) Studying zika. In: Harvard Magazine, 8 May. Retrieved from http://harvardmagazine.com/2017/05/studying-zika-harvard. Accessed on 5 Jun 2017

85. Wang L et al (2016) From mosquitos to humans: genetic evolution of zika virus. Cell Host Microbe 19(5):561–565. https://doi.org/10.1016/j.chom.2016.04.006. Retrieved from http://www.cell.com/cell-host-microbe/abstract/S1931-3128(16)30142-1. Accessed on 20 Jul 2017
86. Brasil P, Nielsen-Saines K (2016) More pieces to the microcephaly–zika virus puzzle in Brazil. In: The Lancet, 16 Dec. Retrieved from http://www.thelancet.com/journals/laninf/article/PIIS1473-3099(16)30372-3/abstract. Accessed on 7 Apr 2017
87. Victora CG et al (2016) Microcephaly in Brazil: how to interpret reported numbers? Lancet 387(10019):P621–P624, Feb 5. https://doi.org/10.1016/S0140-6736(16)00273-7. Accessed on 25 Sep 2016; Care Press Release (2016) Health investigates 3,670 suspected cases of microcephaly in the country. In: Portal da saude, 2 Feb. Retrieved from http://portalsaude.saude.gov.br/index.php/cidadao/principal/agencia-saude/22032-saude-investiga-3-670-casos-suspeitos-de-microcefalia-no-pais. Accessed on 18 Sep 2016
88. Grens K (2016) Brazil's Pre-zika microcephaly cases. In: The Scientist, 10 Feb. Retrieved from http://www.the-scientist.com/?articles.view/articleNo/45297/title/Brazil-s-Pre-Zika-Microcephaly-Cases/. Accessed on 5 Jun 2017
89. Ribeiro B, Hartley S (2018) Why Brazil's zika virus requires a political treatment. In: The Conversation, 16 Feb. Retrieved from https://theconversation.com/why-brazils-zika-virus-requires-a-political-treatment-91955. Accessed on 27 Jun 2018
90. Cuda A (2017) State monitoring zika patients. In: Connecticut Post, 30 Jan. Retrieved from http://www.ctpost.com/local/article/State-monitoring-Zika-patients-10893895.php. Accessed on 7 Apr 2017
91. Dufort E, White J (2018) Pre-zika microcephaly in Brazil: closer to the elusive baseline and new questions raised. Pediatrics 141(2):e20173811, 5 Jan. Retrieved from http://pediatrics.aappublications.org/content/141/2/e20173811?sso=1&sso_redirect_count=1&nfstatus=401&nftoken=00000000-0000-0000-0000-000000000000&nfstatusdescription=ERROR%3a+No+local+token. Accessed on 7 Jul 2018
92. The Guardian (2017) Zika, drought, conflict: what 2016 meant for the world's poorest – podcast transcript. In: The Guardian, 4 Jan. Retrieved from https://www.theguardian.com/global-development/2017/jan/04/zika-drought-conflict-what-2016-meant-for-the-worlds-poorest-podcast-transcript. Accessed on 19 May 2017
93. Morris G et al (2018) Zika virus as an emerging neuropathogen: mechanisms of neurovirulence and neuro-immune interactions. Mol Neurobiol 55:4160–4184, May. https://doi.org/10.1007/s12035-017-0635-y. Accessed on 7 Jul 2018; von der Hagen M et al (2014) Diagnostic approach to microcephaly in childhood: a two-center study and review of the literature. Dev Med Child Neurol 56(8):732–741, Aug. https://doi.org/10.1111/dmcn.12425. Accessed on 24 Jul 2018
94. Silva AA et al (2018) Prevalence and risk factors for microcephaly at birth in Brazil in 2010. Pediatrics 141(2):e20170589. https://doi.org/10.1542/peds.2017-0589. Accessed on 7 Jul 2018
95. Mercola J (2016) Zika: Brazil admits it's not the virus. In: Mercola.com, 16 Aug. Retrieved from http://articles.mercola.com/sites/articles/archive/2016/08/16/birth-defects-brazil-not-zika-virus.aspx. Accessed on 23 May 2017
96. Jaenisch T et al (2016) Risk of microcephaly after Zika virus infection in Brazil, 2015 to 2016. Bull World Health Organ 95(3):191–198. Retrieved from https://www.ncbi.nlm.nih.gov/pmc/articles/PMC5328112/. Accessed on 17 May 2017
97. Bock R (2022) Investigating zika-microcephaly's crash. Am J Med 135(7):E141–E144. https://doi.org/10.1016/j.amjmed.2022.01.040
98. de Oliveria WK et al (2017) Zika virus infection and associated neurologic disorders in Brazil. N Engl J Med 376(16):1591–1593, 6 Apr. https://doi.org/10.1056/NEJMc1608612#t=article. Accessed on 4 May 2017
99. Fritel X et al (2010) Chikungunya virus infection during pregnancy, reunion, France, 2006. Emerg Infect Dis 16(3):418–425, Mar. Retrieved from https://wwwnc.cdc.gov/eid/article/16/3/09-1403_article. Accessed on 4 May 2017

100. Douclef M (2017) Why didn't zika cause a surge in microcephaly in 2016? In: NPR Now, 30 Mar. Retrieved from http://www.npr.org/sections/goatsandsoda/2017/03/30/521925733/why-didnt-zika-cause-a-surge-in-microcephaly-in-2016. Accessed on 12 Apr 2017
101. Beil L (2016) Vaccines may offer defense against dengue, zika and chikungunya. In: Science News, 15 Jun. Retrieved from https://www.sciencenews.org/article/vaccines-may-offer-defense-against-dengue-zika-and-chikungunya. Accessed on 11 Apr 2017
102. De Góes Cavalcanti LP et al (2016) Zika virus infection, associated microcephaly, and low yellow fever vaccination coverage in Brazil: is there any causal link? J Infect Dev Ctries 10(6):563–566. Retrieved from https://www.ncbi.nlm.nih.gov/pubmed/27367003. Accessed on 21 Jul 2017
103. Musso D, Ko AI, Baud D (2019) Zika virus infection — after the pandemic. N Engl J Med 381:1444–1457. https://doi.org/10.1056/NEJMra1808246
104. Lambrechts L (2021) Did zika virus attenuation or increased virulence lead to the emergence of congenital zika syndrome? J Travel Med 28(5):1–3, 23 Mar. https://doi.org/10.1093/jtm/taab041
105. Coyne CB, Lazear HM (2016) Zika virus—reigniting the TORCH. Nat Rev Microbiol 14(11):707–715. https://doi.org/10.1038/nrmicro.2016.125. Accessed on 14 Apr 2017
106. Shapiro-Mendoza CK et al (2017) Pregnancy outcomes after maternal zika virus infection during pregnancy—U.S. Territories, January 1, 2016–April 25, 2017. Morb Mortal Wkly Rep 66:615–621, 16 Jun. https://www.ncbi.nlm.nih.gov/pubmed/28617773. Accessed on 20 Jul 2017
107. Bar-Yam Y et al (2016) Determining the rate and week of infection of zika caused microcephaly from Colombian and Brazilian data; Status Report August 1, 2016. In: NECSI, 1 Aug. Retrieved from http://necsi.edu/research/social/pandemics/statusreport3.pdf. Accessed on 19 Oct 2016
108. New England Complex Systems Institute (2016) Five new confirmed microcephaly cases in Colombia may be harbingers of epidemic. In: Science News, 28 Jun. Retrieved from https://www.sciencedaily.com/releases/2016/06/160628093035.htm. Accessed on 2 Oct 2016
109. Aragão MFVV et al (2017) Neuroimaging findings of congenital zika syndrome. In: Aragão MFVV (ed) Zika in focus: postnatal clinical, laboratorial and radiological aspects. Washington, DC: Springer
110. Collucci C (2016) Brazil to investigate if other factors act with Zika to cause congenital defects. BMJ 354:i4439, 11 Aug. Retrieved from http://www.bmj.com/content/354/bmj.i4439. Accessed on 5 Jun 2017
111. Bortz K (2018) Microcephaly underreported by about 90% before zika epidemic in Brazil. In: HEALIO: In the Journals. Retrieved from https://www.healio.com/pediatrics/emerging-diseases/news/online/%7B9bbe840f-b355-4c74-a07e-56a6cb639cd2%7D/microcephaly-underreported-by-about-90-before-zika-epidemic-in-brazil. Accessed on 26 Jul 2018
112. de Araújo JSS et al (2016) Microcephaly in northeast Brazil: a review of 16,208 births between 2012 and 2015. In: Bulletin of the World Health Organization, 19 May. Retrieved from http://www.who.int/bulletin/volumes/94/11/16-170639/en/. Accessed on 20 Jul 2017; Miranda-Filho DDB et al (2016). Initial description of the presumed congenital zika syndrome. Am J Public Health 106(4):598–600, Apr. Retrieved from https://www.ncbi.nlm.nih.gov/pubmed/26959258. Accessed on 5 Jun 2017

Chapter 8
Effects on Children, Part 2

While some disagreement was on the linkage between the ZIKV and microcephaly, the research activities and publications grew over time. The consensus seemed to develop around the belief that the linkage was legitimate. The experts tended to coalesce between two overall points of view: (i) epidemiological proof involving a handful of countries and (ii) some hospital observations and postmortems.

Another group remained impressed with the correlations. Below are the most prominent and notable follow-up studies tracking these children after birth. In addition, this chapter visits some of the theoretical work explaining why the linkage might exist, ending with a summary of some of the other congenital disabilities children of mothers with ZIKV visited upon them.

Some of the terminologies in this chapter are challenging, but footnotes should help connote the meanings. Much effort was made to make this section readable, but sometimes technical jargon is arduous to simplify without making it sound simple.

8.1 The Consensus—ZIKV Is Linked to Microcephaly

The overriding consensus based on direct measurements of viral RNA in the amniotic fluid and fetal tissues is that the placenta effectively shuttles the ZIKV to the fetus, where it causes microcephaly [1]. "Virus RNA has been found in placenta, amniotic fluid, and brain tissues of stillborn babies with microcephaly," U. Pittsburg Public Health Researcher Ernesto Marques says [2]. Adibi et al. argue that the placenta directly conveys the ZIKV to the early embryo or fetus. The placenta itself might be mounting a response to the exposure; this response might be contributing to or causing the brain defect [3]. Martines and Bhatnagar found that immune cells called Hofbauer cells[1] located in the placenta may carry the virus into the bloodstreams of developing fetuses [4]. And more of these theories will be discussed below. At

D. M. Berube, *Pandemics and Resilience: Lessons we should have learned from Zika*, Risk, Systems and Decisions, https://doi.org/10.1007/978-3-031-25370-6_8

this point, however, it is worth examining the comments made by government agencies about the association and what, in most conditions, was the basis for these conclusions.

8.1.1 The CDC and the WHO

The CDC does not hedge in its recommendations to women who are pregnant when it comes to ZIKV. For the CDC, the microcephaly debate is over.

> Epidemiologic, clinical, laboratory, and pathologic evidence supports a link between infection during pregnancy and outcomes such as pregnancy loss, fetal microcephaly, intracranial calcifications, and fetal brain and eye abnormalities. The level of risk of these outcomes is not known. Studies suggest that it may be as high as 29%. Still, microcephaly caused by viral destruction of brain tissue is likely part of a spectrum of neurologic damage caused by the ZIKV. This percentage may substantially underestimate the proportion of infants affected. [5]

In January 2016, the CDC reported Brazilian data: "[a]mong the first 35 cases of microcephaly reported to the registry, 74% of mothers reported a rash illness during pregnancy, 71% of infants had severe microcephaly (>3 SD below the mean), approximately half had at least one neurologic abnormality, and among 27 who had neuroimaging studies, all were abnormal." [6].

The CDC claims there is no doubt—ZIKV causes microcephaly. "The Rasmussen study marked a turning point in the ZIKV outbreak. It is now clear that the virus causes microcephaly. We are also launching further studies to determine whether children who have microcephaly born to mothers infected by the ZIKV is the tip of the iceberg of what we could see in damaging effects on the brain and other developmental problems," said Tom Frieden, M.D., M.P.H., Director of the CDC [7].

Based on findings and epidemiologic data, Rasmussen et al. [9] published a special report in the *New England Journal of Medicine* stating that Shephard's criteria [8] (Shephard proposed a set of seven standards for "proof" of human teratogenicity) for establishing that ZIKV infection causes microcephaly had been satisfied [9].

The overriding consensus based on direct measurements of viral RNA in the amniotic fluid and fetal tissues is that the placenta effectively shuttles the ZIKV to the fetus, where it causes microcephaly [9]. The current prediction is that 10–40% of infected pregnancies result in microcephaly [10], with greater risk when exposure occurs early in pregnancy [1].

While flu-like symptoms are the primary symptom for most people infected, there is a reasonably strong relationship between pregnancy and congenital anomalies involving the ZIKV. Despite the limited evidence to date, the Centers for Disease Control and Prevention declared in April that maternal infection with ZIKV is a cause of infant microcephaly [11]. They based their call on the Rasmussen study (see below) in the New England Journal of Medicine [12]. This article is ample proof that uses standard criteria for reviewing evidence for causality in human teratogenicity pioneered by Thomas Shepard in 1994 [13].

However, Director-General of WHO, Margaret Chan, said that although that causal relationship has not been proved, it is "strongly suspected." That is due, in part, to other research that has shown the virus can cross the placental barrier and show up in the amniotic fluid. According to the CDC, retrospective analysis of an earlier outbreak of ZIKV in French Polynesia also suggests an increase in cases of neurological impairment [14].

According to Brasil and Nielsen-Saines (see below), the current prediction is that 10–40% of ZIKV-infected pregnancies result in microcephaly, with greater risk when exposure occurs early in pregnancy [15]. This led to WHO to conclude there is scientific consensus that ZIKV is a cause of microcephaly [16]. According to at least one outlet, this finding amplified the risk: "With the CDC's confirmation of the effects of ZIKV, it is now certain that ZIKV is more dangerous than initially believed." [17].

8.1.2 *Current Point of View*

The severe microcephaly and other brain anomalies observed in many infants are consistent with an infection occurring in the first or early second trimester of pregnancy. Although ecologic data do not necessarily qualify as an epidemiologic study, data from Brazil regarding the temporal and geographic association between ZIKV infection and the later appearance of infants with congenital microcephaly are compelling [18].

Another team went further. "We conclude that the microcephaly epidemic results from congenital ZIKV infection. We recommend that the list of congenital infections normally referred to as TORCH (toxoplasmosis, others [syphilis, varicella-zoster, parvovirus B19], rubella, cytomegalovirus, and herpes) is renamed as TORCHZ and that we prepare for a global epidemic of microcephaly and other manifestations of congenital Zika syndrome." This conclusion was based case–control study in eight public hospitals in Recife, Brazil. Cases were neonates with microcephaly [19].

Rita Driggers, Director of maternal–fetal medicine at Sibley Memorial Hospital in Washington, D.C., said the children in her study must be followed and tested to see if they experience developmental problems. Her concern was that ZIKV children might have tissue scarring in their brains and have underdeveloped brains even though their heads are normal sized. Driggers reported on March 6 a case in which a woman who was infected during her first trimester chose to terminate her pregnancy in week 21. Her ultrasounds hadn't shown that the fetus was microcephalic, but an MRI showed brain abnormalities [20].

8.1.3 Some Studies

The study concludes that regardless of symptoms, women may still pass on the infection and the associated birth defects to their fetuses.

"Base fetuses preliminary data from the U.S. Zika Pregnancy Registry, among 442 completed pregnancies, 6% overall had a fetus or infant with evidence of a ZIKV-related congenital disability, primary microcephaly with brain abnormalities," CDC's Dr. Margaret Honein wrote in the Journal of the American Medical Association [21]. In addition, women who did not remember having symptoms from ZIKV infection were just as likely to have a baby with a congenital disability as women who did have symptoms [22]. Individuals with asymptomatic ZIKV seem susceptible to producing offspring with genetic disabilities. According to Hoen et al. [23], although the rate of complications would be expected to be higher among women with symptomatic infection than among those who were asymptomatic, an observational study involving U.S. women did not show any significant difference in the rate of congenital disabilities between the offspring of women who had symptomatic ZIKV infection and the offspring of women who had asymptomatic ZIKV infection during pregnancy [23].

Secondly, there is some good news. They also found evidence that the fetus can fight off ZIKV infection, perhaps late in pregnancy. Martines, Roosecelis Brasil et al. [24] found evidence of the virus in the placentas of eight women infected during the third trimester who had healthy babies [24].

8.1.3.1 Epidemiological Studies

Regardless of the proceeding reservations, microcephalic babies were born not just in Brazil. Still, in Colombia, Panama, Martinique, and the Cape Verde Islands, not to mention the cluster was discovered retrospectively in French Polynesia. Each cluster had followed a ZIKV outbreak about nine months earlier [25]. While this simplistic association does not constitute scientific proof, it is consistent with the consensus of experts.

Polynesia

Another team investigated the outbreak in French Polynesia. They detected an unusual increase in congenital cerebral malformations and dysfunction in fetuses and newborns in French Polynesia following ZIKV epidemic from October 2013 to March 2014. A retrospective review identified 19 cases, including eight with major brain lesions and severe microcephaly, six with severe cerebral lesions without microcephaly, and five with brainstem dysfunction without visible malformations. Imaging revealed profound neurological lesions (septal and callosal disruption, ventriculomegaly, abnormal neuronal migration, cerebellar hypoplasia, occipital pseudocysts, brain calcifications) [26].

Congenital cerebral malformations are relatively rare in French Polynesia and are not collected in a register of congenital malformations. After 2011, when the prenatal diagnosis unit records started, about 15 congenital cerebral malformations were observed in 2012 and 2013, respectively, and according to PMSI data, between 2001 and 2013, only two cases of brainstem dysfunction were noted in 2009, leading to an average annual incidence rate of 0.34 cases per 10,000 births. Seven cases of congenital microcephaly were reported in the same period, equivalent to an average annual incidence rate of 1.2 cases per 10,000 live births. The present cluster of severe cerebral malformations with a 14-fold increase in congenital microcephaly and 31-fold in brainstem dysfunction was spatially and temporally associated with a large outbreak of ZIKV in French Polynesia [26].

Of the eight microcephaly cases identified in French Polynesia during the 2013–2015 outbreak during the 23-month study period by Cauchemez et al., seven (88%) occurred in the four months from March 1 to July 10, 2014. The timing of these cases was best explained by a period of risk in the first trimester of pregnancy. In this model, the baseline prevalence of microcephaly was two cases (95% CI 0–8) per 10,000 neonates, and the risk of microcephaly associated with ZIKV infection was 95 cases (34–191) per 10,000 women infected in the first trimester [27].

Polynesia estimated that the risk of microcephaly due to ZIKV infection in the first trimester of pregnancy was 0.95% (95% confidence interval, 0.34–1.91) based on eight microcephaly cases identified retrospectively in a population of approximately 270,000 people with an estimated rate of ZIKV infection of 66% [28]. Rasmussen et al. estimated that 1% of pregnant women who became infected in the first trimester of pregnancy had children with microcephaly during the outbreak in French Polynesia in 2013 and 2014 [29].

Brazil

A Brazilian and the UK team led by Barreto de Araujo reported preliminary findings of 91 infected babies, with 30 born with microcephaly in Pernambuco. By testing blood and cerebrospinal fluid, they found that 80% of the mothers who gave birth to microcephalic babies tested positive for ZIKV [30].

In August 2016, França et al. reported their findings. Between November 19, 2015, and February 27, 2015, investigations were completed for 1501 suspected cases reported to the Brazilian Ministry of Health, of whom 899 were discarded. Of the remaining 602 cases, 76 were definite, 54 highly probable, 181 moderately possible, and 291 somewhat probable of congenital ZIKV syndrome [31].

Schuler-Faccini et al. worked off the task force and reported 71% of infants had severe microcephaly (>3 SD below the mean), approximately half had at least one neurologic abnormality, and 27 who had neuroimaging studies, all were abnormal [32].

An observational study of 11 infants in Paraiba, Brazil, with congenital ZIKV infection from gestation to 6 months was done by de Oliveira Melo et al. [33].

> Although variable damage resulted from brain lesions associated with ZIKV congenital infection, a typical pattern of brain atrophy and changes related to disturbances in neuronal migration were observed. Some patients presented with mild brain atrophy and calcifications, and others gave with more severe malformations, such as the absence of the thalamus and lissencephaly. Infection caused by ZIKV during pregnancy causes brain damage, mainly characterized by decreased cortical development and atrophy. [33]

Van der Linden and colleagues also undertake an interesting twin study from Recife. The cases include two sets of identical twins, and both babies in each pair have microcephaly, she said. There are also six sets of fraternal twins, in which one twin has microcephaly, while the other appears unaffected [34]. As Lead Author, van der Linden [35], reported, as has already been reported in different types of congenital infectious disorders in dizygotic twin pregnancies, the virus may affect only one fetus. This article reports two cases of twin pregnancies exposed to the ZIKV, only one of the fetuses was affected by microcephaly and brain damage [35].

Since the identical twins shared one placenta while fraternal twins almost always have separate placentas, Dr. Zatz and other experts suggested that the ZIKV may have penetrated one placenta and not the other. One set of twins has broken the pattern [36].

First and foremost, fraternal twins **usually** do not necessarily have to have different placentas. Secondly, Dr. Vanessa van der Linden, who helped discover that ZIKV causes microcephaly and has treated some of the twins, said one explanation might be that in some fraternal cases, ZIKV crossed both placentas. Still, the twins had genetic differences that influenced why only one became infected or "why the babies reacted differently to the virus." [34]. Thirdly, Dr. Ernesto Marques, Infectious Disease Expert at the University of Pittsburgh and the Oswaldo Cruz Foundation in Recife, Brazil, suggested that an impaired twin was exposed to ZIKV before the mother's body or the placenta developed immune responses against the virus and that the second fetus was infected slightly later. "It should reach both at an equal time," he said. "However, if the virus hit one of the babies before the mother had developed protective immune responses, you have a problem." [34].

They determine why one twin became infected in the womb, while the other did not illuminate how ZIKV crosses the placenta, how it enters the brain, and whether any genetic mutations make a fetus more resistant or susceptible to ZIKV infection.

In 2017, Santos et al. reported in an MCDA twin pregnancy (single placenta), the ZIKV may cause infection in both fetuses, resulting in severe abnormalities in the central nervous system due to neural cell destruction and the disruption. Of the standard development processes of the brain, however, the ZIKV may cause outcomes with different degrees of severity in each fetus [37].

Satterfield-Nash et al. [38] also reported the data from the Zika Outcomes and Development in Infants and Children (ZODIAC) investigation in Brazil that showed that among 19 children aged 19–24 months with ZIKV-associated microcephaly, 100% had a least one adverse health or developmental outcome including feeding challenges, sleeping difficulties, severe motor impairment, vision and hearing abnormalities, and seizures [38].

The ZODIAC study [39] revealed that the findings were consistent for other children who had ZIKV; the other toddlers displayed similar symptoms of being small for their ages, complications such as seizures, more frequent hospital visits, sleeping difficulties, and eating impairments due to issues swallowing. A large majority of the toddlers also had hearing and vision problems, and almost none of the toddlers passed an assessment designed for six month olds [40].

Colombia

Project VEZ (Vigilancia de Embarazadas con Zika) was a surveillance cohort of over 1200 pregnant women with symptomatic ZVD in three cities in Colombia during the peak of the ZIKV outbreak. This cohort constituted nearly 7% of the total pregnant women diagnosed with ZVD in Colombia during the attack [41]. More than 97% of the pregnancies were followed until pregnancy completion, and 4.2% of infants were born with brain or eye defects, including microcephaly. In the subpopulation with confirmed or presumptive ZIKV infection in pregnancy, the proportion with these defects was 8.7%. Overall, the frequencies of preterm delivery and low birth weight were lower in the VEZ cohort than in general population estimates for Colombia [42]. In addition, a recent analysis of 858 cases of microcephaly or central nervous system defects reported to Colombia's congenital disabilities surveillance system between September 2015 and April 2017 classified 503 (59%) as potentially attributable to ZIKV infection [43].

Argentina

There's the Argentine data [44]. They reported that 104 with microcephaly or selected brain abnormalities (cerebral atrophy, cerebellar atrophy, cerebral calcifications, and hydrocephalus ex-vacuo) were detected in 150,057 births in the maternity hospitals of the RENAC (National Network of Congenital Abnormalities Argentina). Comparing this prevalence with the previous period, 2009–2015 (4.1 per 10,000—CI%: 3.1–5.4), a statistically significant increase was detected with a Prevalence Rate Ratio (PRR) of 1.7 (95% CI: 1.2–2.3) [44].

Guatemala

Then there's some Guatemala data as well. Regardless of the case definition used, the estimated background congenital microcephaly of 34–4233 per 10,000 live births in this rural community before and during Guatemala's early ZIKV epidemic was significantly higher than the overall background rate of 3.30 per 10,000 live births reported in Latin America before the ZIKV outbreak [45].

8.1.3.2 Human Studies

Dr. Patricia Brasil and her team did some of the best research on the prevalence of microcephaly among ZIKV-infected mothers and their children [46]. She and her colleagues reported the following:

> By July 2016, 134 ZIKV-affected pregnancies and 73 ZIKV-unaffected pregnancies had been completed, with outcomes known for 125 ZIKV-affected and 61 ZIKV-unaffected pregnancies.... overall adverse effects were 46% among offspring of ZIKV-positive women versus 11.5% among offspring of ZIKV-negative women. Among 117 live infants born to 116 ZIKV-positive women, 42% had grossly abnormal clinical or brain imaging findings, including four infants with microcephaly. [46]

Brasil again: Four infants in the ZIKV-positive group (3.4%) were noted to have microcephaly at birth; two were small-for-gestational-age infants with proportionate microcephaly (i.e., the head size is small but is proportional to the weight and length of the infant), and two had extreme microcephaly (i.e., the head size is small relative to the weight and length of the infant). None of the infants in the control group had microcephaly [46].

Among ZIKV-infected pregnancies, there were five miscarriages (25% of the 20 pregnancies with first-trimester infection), two fetal losses (3% of the 71 pregnancies with second-trimester infection), and two stillbirths (6% of the 34 pregnancies with third-trimester infection). Among 117 live births in the ZIKV-positive cohort, 49 infants (42%) were found to have abnormalities on clinical examination, imaging, or both [46]. According to Dr. Marco Safadi at the Santa Casa Medical School in Sao Paulo, who has been treating and studying cases, 54 miscarriages or newborn deaths have been caused by ZIKV [29].

Brasil and Nielsen-Saines reported preliminary findings from the first case-control study to examine the association between microcephaly and ZIKV infection, done prospectively in the metropolitan region of Recife in Pernambuco state, the hotspot of the microcephaly epidemic in Brazil. Two women with ZIKV infection had miscarriages in the first trimester, and ultrasounds of 12 of the 42 women showed fetuses with some abnormality [47]. Their results highlight the striking magnitude of the association between microcephaly and laboratory-confirmed ZIKV infection: The risk is 50 times higher in all microcephaly cases and more than 100 times higher in patients with brain abnormalities detected by imaging [48].

Dr. Karin Nielsen-Saines, Pediatric Infectious Disease Specialist at UCLA, and a team from the Fundacao Oswaldo Cruz in Brazil followed 88 women in different stages of pregnancy in Rio de Janeiro in 2015–2016. Of the 72 who tested positive for ZIKV, only 42 had ultrasounds performed, but their results were alarming: Twelve of their fetuses showed signs of congenital disabilities, 29%. No such defects were observed in the ultrasounds of ZIKV-negative mothers [49].

Nielsen-Saines was also surprised—and alarmed—to see problems such as fetal death arising in the third trimester of pregnancy, as those risks were thought to be highest in earlier stages of pregnancy. "These babies were just a few weeks from being born," she said. "That's unusual. Syphilis can cause that, but that had not been reported with ZIKV." [49].

CDC Dr. Julu Bhatnagar and colleagues studied tissue samples taken from eight babies with microcephaly who died soon after birth and from the placenta of 44 women, including 22 whose babies were damaged by ZIKV and 22 whose babies seemed healthy even though the mothers had ZIKV infections. They found levels of ZIKV genetic material were 1200 times higher in the brains of the full-term babies killed by ZIKV than in the placenta or fetuses miscarried or aborted before birth. "Our findings show that ZIKV can continue replicating in infants' brains even after birth and that the virus can persist in placentas for longer than we expected." [22]. These US researchers have found evidence of the ZIKV replicating in fetal brains for up to seven months after the mother became infected, and they showed the virus could persist even after birth. "We don't know how long the virus can persist, but it could have implications for babies born with microcephaly and healthy infants whose mothers had ZIKV during their pregnancies," said Dr. Bhatnagar [50].

Interesting associative data for the relationship between ZIKV and microcephaly come from a case–control study of neonates in eight public maternity hospitals in Recife, Brazil:

> The association between microcephaly and ZIKV infection, confirmed by qRT-PCR, capture-IgM ELISA, was strong after controlling for confounders. The association was strong between severe and non-severe microcephaly. None of the other risk factors investigated was associated with microcephaly in multivariable analyses, including larvicide, pyriproxyfen, and vaccine administration during pregnancy. These data support our preliminary findings that the increase in microcephaly prevalence at birth in northeast Brazil was caused by congenital ZIKV infection. [51]

One of the most robust cases for correlation comes from Mlakar et al. in 2016. Following the autopsy of a 29-week-old fetus aborted because of microcephaly, Mlakar et al. [52] recovered from the brain the entire genome of ZIKV. The finding, published this week (February 10) in the New England Journal of Medicine, is considered by some to be the most convincing bit of evidence to date that ZIKV can cause brain malformations [47]. Mlakar et al. [52], in their case report, found severe fetal brain injury associated with ZIKV infection with vertical transmission (mother to child) [52]. Yale's Albert Ko told CNN: "We've seen lesions of a congenital infection in babies in normal size heads, so we're also worried that infection in utero may have downstream important impacts on cognitive developments and other issues." [53].

Still, another 2016 study of brain scans by Soares de Oliveira-Szenfeld et al. [55] examined ultrasound pictures of 45 Brazilian babies whose mothers were infected with ZIKV during pregnancy shows that the virus can inflict severe damage to many different parts of the fetal brain beyond microcephaly, the condition of tiny heads that has become the sinister signature of ZIKV [54].

Brain abnormalities seen in confirmed ($n = 17$) and presumed ($n = 28$) congenital ZIKV infections were similar, with ventriculomegaly in 16 of 17 (94%) and 27 of 28 (96%) infections, respectively; abnormalities of the corpus callosum in 16 of 17 (94%) and 22 of 28 (78%) infections, respectively; and cortical migrational abnormalities in 16 of 17 (94%) and 28 of 28 (100%) infections, respectively [55].

Studies have associated ZIKV with a long list of congenital abnormalities. For example, there was another outbreak of ZIKV in 2013–2015 in French Polynesia and New Caledonia (the largest documented outbreak before the Brazil 2016 outbreak), where local officials reported 17 cases of malformations in fetuses and infants hypothesized to be ZIKV associated [56]. Sixteen babies have been born in the U.S. and have congenital disabilities linked to ZIKV (9/16) [57].

A study of 23 children [58] diagnosed with congenital infection presumably associated with the ZIKV during the Brazilian microcephaly epidemic. Five tested positives for IgM antibodies to ZIKV in cerebrospinal fluid. The other 17 children met the protocol criteria for congenital infection presumably associated with the ZIKV, even without being tested for IgM antibodies (found mainly in blood and lymph fluid, this is the first antibody the body makes when it fights a new infection) to the virus. The authors claim this represents the most extensive and detailed case series of neuroimaging findings in children with microcephaly and presumed ZIKV-related infection.

Van der Linden et al. [58] did a retrospective case study on seven ZIKV-infected children from Brazil. Arthrogryposis was present in the arms and legs of six children (86%) and the legs of one child (14%). Hip radiographs showed bilateral dislocation in seven children, subluxation of the knee associated with genu valgus in three children (43%), and bilateral in two (29%). All presented malformations of cortical development, calcifications predominantly in the cortex and subcortical white matter (especially in the junction between the cortex and white matter), reduction in brain volume, ventriculomegaly, and hypoplasia of the brainstem and cerebellum [58].

For example, Belluck et al. studied 13 babies who did not qualify as microcephalic at birth but were diagnosed as microcephalic months later. Its symptoms include a range of impairments resembling characteristics of cerebral palsy and epileptic seizures, muscle and joint problems, and difficulties swallowing food [59].

Several studies have demonstrated that several ZIKV-induced cytopathic effects may contribute to microcephaly, including the inhibition of neuronal cell proliferation, disturbance of host cell-cycle regulation, and cell death or apoptosis induction [60].

Malformations include agenesis of the vermis, Blake's pouch cyst, and potentially a club foot, including cerebral calcifications, intrauterine growth restriction, and microcephaly [61].

> Eleven of the babies' brain scans taken days or weeks after birth showed significant neurological damage, including improperly formed brain areas, excess fluid in some places, abnormal calcium deposits, or calcification, which probably resulted from brain cell death. But the size of their heads, though small, was not small enough to be considered microcephaly. So, doctors monitored their progress as they grew…. [T] the type of brain damage in the babies who later developed microcephaly "presented the same pattern, but less severe" than those at birth. [59]

Additional evidence for direct transfer and brain damage by ZIKV is the detection of IgM against the viral antigen (but not the viral mRNA) in the cerebral spinal fluid of 30 of the 31 samples analyzed of the babies born with microcephaly [62].

> Severe cerebral damage was found on imaging in most children in this case series with congenital infection presumably associated with the ZIKV. The features most found were brain calcifications in the junction between cortical and subcortical white matter associated with malformations of cortical development, often with a simplified gyral pattern and predominance of pachygyria or polymicrogyria in the frontal lobes. Additional findings were enlarged cisterna magna, corpus callosum abnormalities (hypoplasia or hypogenesis), ventriculomegaly, delayed myelination, and hypoplasia of the cerebellum and the brainstem. [63]

Many infections that target the brain produce clumps of calcium, called calcification. Ventura et al. [64] discovered cerebral calcifications and neuroretinal macular atrophy in a microcephalic ZIKV-infected child [64]. But in ZIKV-infected babies, calcification often occurs in an unusual place: at the intersection of the gray matter of the brain's outer layer, the cortex, and the white matter of the layer just below that.

Oliveira Melo et al. [65] ultrasound examined two fetuses in a microcephaly cluster with results consistent with intrauterine transmission of the virus [65]. Another abnormality seen in most babies' brains involved the ventricles or cavities of the brain becoming so full of cerebrospinal fluid that they "blow up like a balloon." At some point, "like a balloon, these ventricles can pop," Dr. Deborah Levine said. And if they do, "the brain will collapse on itself." [54].

Ultrasounds of pregnant women revealed microcephaly, hardened calcium deposits in the brain, a breakdown of brain tissue, brain swelling, and general poor fetal growth [66]. Samo [67] published a case report evidencing a link between ZIKV infection and hydrops fetalis (a condition in which large amounts of fluid buildup in a baby's tissues and organs, causing extensive swelling (edema) and fetal demise) [67].

8.1.4 Follow-Up Needed

"Although we had a very high number of abnormalities, it's important to emphasize that 71% of the pregnancies seem normal to date," Nielsen-Saines said about their NEJM study. "Women who have ZIKV infection in pregnancy should be followed closely by their obstetricians with serial ultrasound monitoring to follow fetal development," she adds. The women should also be followed through the third trimester because the virus seems to cause problems with the placenta [68].

Another 2016 study in the New England Journal of Medicine [68] found that 42% of women infected with ZIKV during pregnancy gave birth to infants with severe abnormalities. Discrepancy: Anomalies from in utero exposure to ZIKV have shown up well after birth in some cases. The Brazil study had a more extended follow-up period, which may have caught more of the later manifestations [69].

Today, some children born during the outbreak are trying school for the first time—in minimal capacities—while others have died or are struggling to survive, hindered by health and developmental problems. Dr. Epitacio Rolim of the Getulio Vargas Hospital in Recife, where many children with ZIKV-related congenital disabilities are treated, said there are still myriad unknowns. "How much they will learn or live is a huge question mark," said Rolim, who spent hours injecting babies with Botox to ease muscle spasms during a recent afternoon. Beyond developmental delays, around 40% of the children with microcephaly treated at the hospital started showing new physical problems by the time they reached their first birthdays, including dislocated hips, which needed to be repaired surgically [70].

Dr. Liana Ventura, who runs the Altino Ventura Foundation in Recife, said more than 150 of the children with microcephaly getting treatment at the clinic are stable and healthy enough to be candidates for school. "But we can see schools are not prepared; they need caretakers that families can trust, and materials geared toward inclusion," said Ventura, whose family helped found the clinic opened in 2014. Ventura also said 130 children are waiting to get eye exams, glasses, or other optical treatments, which are vital for being able to attend school. But budget shortfalls have made it impossible to hire more doctors or buy supplies. For many children, going to school is out of the question.

According to health figures, in Pernambuco, 159 of the 2513 children diagnosed with microcephaly have died. Paraiba, a neighboring state also hit hard, does not have current figures, but local specialists estimate that 10% of children born with microcephaly have died [70].

While most research published on congenital Zika syndrome tends to warn about infection during the first trimester, concerns extend well into the third trimester. The women should also be followed through the third trimester because the virus seems to cause problems with the placenta. Nielsen-Seines added this commentary.

8.1.5 Toddlers

Babies infected during pregnancy and born with normal head size can develop postnatal-onset microcephaly, suggesting that ZIKV has consequences beyond the gestational period. Some researchers have warned: "Infants who developed ZIKV-induced microcephaly show recurrent seizures, and reports suggest that 60% of normocephalic babies exposed to ZIKV in utero suffer similar symptoms. Cortical dysplasia induced by ZIKV has also been described in babies born with normal head size, a finding that is often underdiagnosed and can be associated with seizures. Thus, it is now clear that focusing only on microcephaly and other physical malformations will lead to underestimating the true magnitude of the consequences of the ZIKV epidemics." [71].

A December 2017 study from the CDC [72] detailed how the first generation of babies born with the ZIKV is now turning two, entering the age of toddlerhood. Doctors have begun to follow their development closely to learn more about how ZIKV might affect children as they grow [40].

Karin Nielsen-Saines, Professor of clinical pediatrics at UCLA's Children's Hospital, reported that 14.5% of infants in the cohort ($n = 131$) had a severe developmental delay of two standard deviations below normal in motor and language or cognitive function or had visual or hearing abnormalities. "This does not include children with moderate impairment, which must be followed over time. This informs the public that ZIKV in pregnancy is not an all-or-nothing phenomenon. There are gradients of disease, and very adverse events occur in at least 15% of children" [73]. In more detail, among 94 children who underwent both neuroimaging and Bayley-III testing (this test assesses infant and toddler development across five domains: Cognitive, Language (Receptive & Expressive), Motor (Gross & Fine), Social-Emotional and Adaptive), neuroradiologists found that ten (11%) had structural abnormalities, five (5%) had nonstructural abnormalities, and 20 (21%) had abnormal results that were limited to a non-specific T_2-weighted hypersignal (T2 hyperintensity is an area of the high intensity on types of magnetic resonance imaging (MRI) scans of the brain of a human or of another mammal that reflects lesions produced primarily by demyelination and axonal loss) on MRI [74].

Nielsen-Saines said six children had microcephaly and were so profoundly impacted that the Bayley test could not be conducted. Roughly 15% had moderate developmental delays, she said.

"It sounds pretty consistent with what we've known. Maybe a little bit higher. But pretty consistent," said Dr. Rita Driggers, Medical Director for maternal–fetal medicine at Sibley Memorial Hospital in Washington, D.C., of the findings [75].

8.1.6 Later in Life

A study [71] suggests that even babies born with normal head perimeter and no detectable malformations at birth are at risk of developing late-stage complications of the infection as they grow. Results suggest that all infants exposed to ZIKV might develop neurological consequences at different stages of life [76].

Their study involved mice exposed to ZIKV and found their subjects had developed frequent seizures during childhood and adolescence. Although these seizures resolved as animals grew into adulthood, animals remained more sensitive to drug-induced attacks in adulthood. They also presented motor deficits throughout their lives and memory impairment. Animals exposed to the virus also showed impaired sociability during adulthood, a hallmark of diseases such as autism and schizophrenia. Surprisingly, viral replication was still active in adult brains long after the acute phase of infection [76]. In addition, mice results from de Oliveira Souza et al. [71] indicate that ZIKV infection during development affected declarative memory and social behavior in adult animals, suggesting that neuropsychiatric disorders might be a further long-term consequence of perinatal ZIKV exposure. This agrees with the association between early-life exposure to several viruses and the development of cognitive impairments and neuropsychiatric disorders later in life [71] (Fig. 8.1).

There have been cases of prolonged viral replication, seizures, and encephalitis in adults exposed to the virus, which means that neurological complications can still occur in adult patients following infection. The late consequences of viral infection in the adult brain remain unknown [76]. Dr. Adams Waldorf has observed the loss of

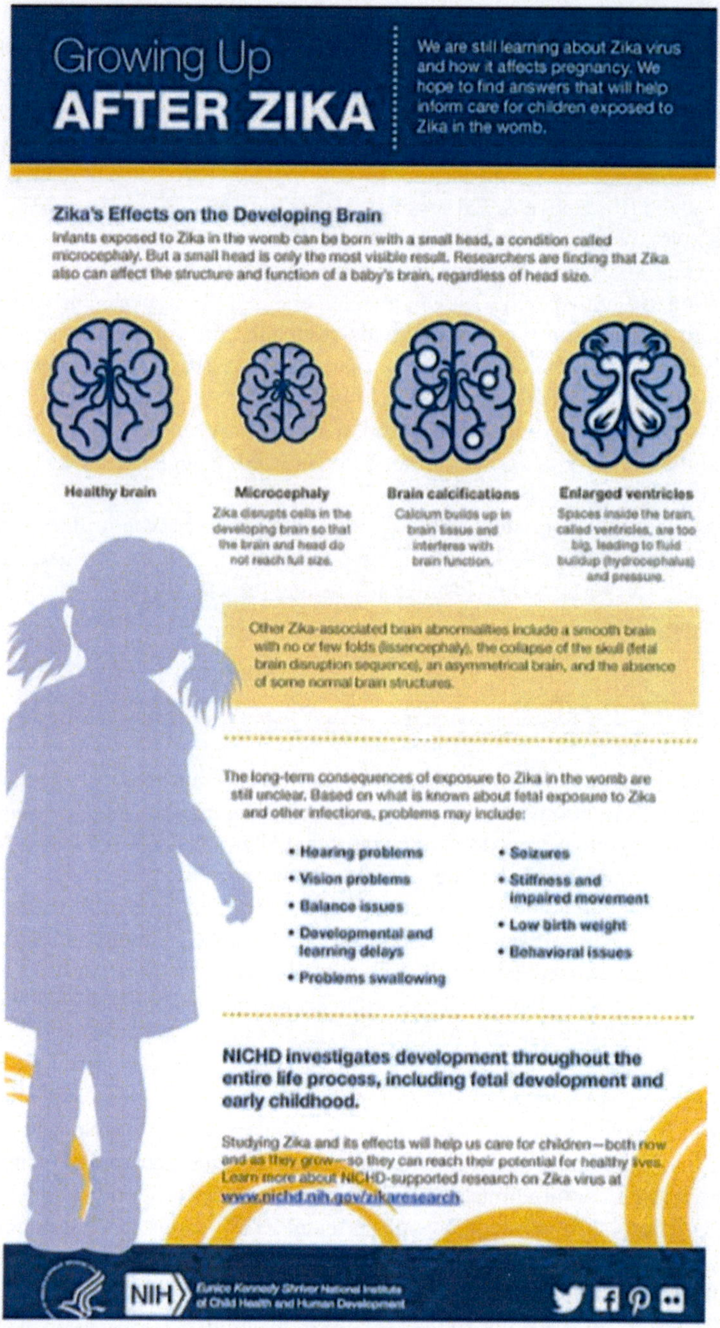

Fig. 8.1 Growing up after Zika. NIH Flicker Gallery. Uploaded February 17, 2017. https://www.flickr.com/photos/nihgov/32111918844/in/photolist-QVBYHb-2iLYt87-2iUt6r3-mGY8ti-G6hTWc-2a3bwxL-2mWVvjC. Accessed July 22, 2022

neural stem cells in the hippocampus of non-human primates, which might be detrimental to long-term memory, social interactions, and abstract reasoning as patients' age [77].

In 2022, Rua et al. did a two-year follow-up of children with CZS and discovered that thirty-eight children (27 males and 11 females; the median age of 4.3 months) with CZS were evaluated. Irritability was present in 50% and 27% of the children at 8 and 24 months. Axial hypertonia was highly prevalent at four months (77%), decreasing to 50% at 24 months. At all ages, spastic tetraparesis (weakness in all four limbs) was the most common motor abnormality. Twenty-seven (71%) participants were diagnosed with epilepsy, and the median age at seizure onset was six months. The most frequent types of seizures were focal seizures and spasms, with spasms being the most frequent in the first year of life (52%) and focal crises being the most frequent in the second year of life (50%) [78].

In the Danielle Oliveira case study [79], a baby born in January to a ZIKV-infected mother was not initially thought to have suffered any ill effects. Still, the virus remained in his blood for at least 67 days after birth [79]. When the infant was examined on day 54, he had no apparent illness or evidence of any immunocompromising condition. However, he showed neuropsychomotor developmental delay by six months, with global hypertonia and spastic hemiplegia, with the dominant right side more severely affected [79].

Suppose a woman is infected with ZIKV while she is pregnant. In that case, scientists don't know how to determine whether the baby will have congenital disabilities or developmental delays related to ZIKV that aren't physically detectable. That's why ZIKV has been described as a "delayed epidemic." [80].

8.2 What Seems to Be Happening?

The overall conclusion between ZIKV infection and CZS: ZIKV infection in pregnancy appears to cause a recognizable pattern of congenital anomalies that is consistent and unique. Although many of the components of this syndrome, such as cognitive, sensory, and motor disabilities, are shared by other congenital infections, five features differentiate CZS from other congenital disorders: (1) severe microcephaly with a partially collapsed skull; (2) thin cerebral cortices with subcortical calcifications; (3) macular scarring and focal pigmentary retinal mottling; (4) congenital contractures; and (5) marked early hypertonia with symptoms of extrapyramidal involvement [81].

Estimates of congenital neurologic defects in infected newborns have ranged from 6 to 42%. The Hoen team [83] included women from French territories in the Americas (French Guiana, Guadeloupe, and Martinique). They were pregnant from March through November of 2016 when it was confirmed that they had been infected with ZIKV (symptomatic). Their study of 555 fetuses and infants found evidence of microcephaly in 5.8%. The rate of severe microcephaly was 1.6% [82]. (Microcephaly (defined as head circumference more than 2 SD below the mean for sex and gestational age) was detected in 32 fetuses and infants (5.8%), of whom 9 (1.6%) had severe microcephaly (more than 3 SD below the mean)) [83].

Primary microcephaly results mainly from the depletion of neural stem/progenitor cells due to centrosomal defects, premature differentiation, and cell death. It is not known to what extent microcephaly results from direct ZIKV infection of developing neural cells versus indirect effects, such as inflammation and altered placental support, which has been shown to affect brain development [84]. There is also some evidence microcephaly may have genetic causes associated with maternal exposures, including radiation, tobacco smoke, alcohol, and viruses.

The link between ZIKV and microcephaly was the subject of some disagreement. The WHO alleged it in February 2016. The basis for the controversy follows. However, as explained above, the consensus has led to the conclusion that there seems to be a relationship between ZIKV outbreaks in 2015–16 and microcephaly.

In April 2016, NIH's Fauci explained at a briefing that in control studies involving injecting monkeys with the virus, preliminary data showed that those pregnant monkeys exhibited viremia (the presence of viruses in the blood) for a significantly more extended period than non-pregnant monkeys. He also reported, "In vitro studies of getting the virus and putting it in neural stem cells [show] that it has a powerful propensity to destroy tissue, which could explain why, besides interfering with the development of the fetus, it might directly attack brain tissue even when the fetus is later on in the period of gestation." [85]. "ZIKV-associated microcephaly differs from cases caused by genetic abnormalities," Dr. Arnold Kriegstein, Director of the Eli and Edythe Broad Center of Regeneration Medicine and Stem Cell Research at UCSF, said. "As with genetic forms, the brain does not grow or develop normally. But in addition, the brain shrinks due to tissue destruction due to the infection. It may be that infections earlier in pregnancy disproportionately affect neural stem cells and later strongly affect more mature cells, including the growing astrocyte cell population." [86].

The timing of acute ZIKV infection ranged from 6 to 35 weeks of gestation (in the March 2016 Brasil study) among the 42 women in whom fetal ultrasonography was performed. In addition, the team found abnormal results on ultrasonography or Doppler studies in 12 cases (29%). Five of the 12 fetuses had intrauterine growth restriction. Cerebral calcifications were noted in four fetuses and other CNS alterations in two fetuses. Abnormal arterial flow in the cerebral or umbilical arteries was seen in four fetuses [61].

Finding traces of ZIKV in newborns may be problematic. Due to the lack of simple laboratory tests and the similar clinical manifestations with dengue fever and yellow fever (cross-reactivity, see VACCINES), miscalculations are anticipated. It is possible that in congenital ZIKV, the infection usually does not last long; alternatively, the virus may be present but at undetectable levels. Even in acquired disorder, PCR (polymerase chain reaction, a technique to produce multiple copies of a piece of DNA) or viral isolation is brutal because the ZIKV viremic (when the virus is present in the blood) period is transient (3–5 days after the onset of symptoms) and viral load is low [87]. There are multiple claims the incubation period for the viremic condition could last longer, from 3 to 12 days [88].

Nonetheless, a retrospective study [26] also found an unusual increase in annual congenital cerebral malformations (twofold), brainstem dysfunction (31-fold), and severe microcephaly (14-fold) among fetuses and newborns, following the French Polynesian ZIKV epidemic. This suggests the linkage was not noticed during the last major outbreak.

Finally, microcephaly can appear later in life. Some babies not born with unusually small heads, the most severe hallmark of brain damage resulting from the virus, have developed microcephaly as they have grown older [59].

The prevalence of microcephaly cases in Southern America (confirmed by laboratory tests) is related to maternal infection with ZIKV in pregnancy, especially in the first and second trimesters. Most studies have linked maternal infection during pregnancy to the development of neonatal microcephaly, and strong evidence for a probable link between ZIKV infection and congenital malformations resulting in severe malformations in children [89].

In 2018, Delaney, Honein, et al. with the CDC analyzed data from 15 U.S. jurisdictions conducting population-based surveillance for congenital disabilities potentially related to ZIKV infection. Among the 2962 infants and fetuses with defects potentially associated with ZIKV infection, there were 2716 (92%) live births. They reported that the prevalence of congenital disabilities strongly linked to congenital ZIKV infection increased significantly, from 2.0 cases per 1000 live births in the first half of 2016 to 2.4 cases in the second half, increasing 29 more cases than expected ($p = 0.009$) [90].

> Data collected from three U.S. population-based congenital disabilities surveillance systems from 2013 and 2014, before introducing ZIKV infection in the World Health Organization's Region of the Americas, showed a baseline prevalence of congenital disabilities potentially related to congenital ZIKV infection of 2.9 per 1000 live births. Based on 2016 data from the U.S. Zika Pregnancy and Infant Registry, the risk for congenital disabilities potentially related to ZIKV infection in pregnancies with laboratory evidence of possible ZIKV infection was approximately 20-fold higher than the baseline prevalence. [90]

After collecting detailed brain images of more than 200 developing fetuses from expecting mothers along the Colombia Caribbean coast, research by a team from Colombia, led by Drs. Sanz Cortes and Parra Saavedra think they have found something close to an explanation: Microcephaly didn't appear in just a few cases; it appeared in the *worst* cases [91].

ZIKV was still causing significant brain damage even in babies without below-average-sized skulls. That means there may be a lot of babies out there who have yet to show signs of congenital ZIKV defects. And if brain scans don't expose them, time will [92].

Delaney et al. [90] concluded that their report provided the first comprehensive data on congenital disabilities (3.0 per 1000 live births) potentially related to ZIKV infection in a birth cohort of nearly 1 million births in 2016 [90].

Congenital disabilities have increased in parts of the U.S. that had local ZIKV transmission in 2016. A new report from the U.S. Centers for Disease Control (CDC) and Prevention said Thursday. Areas with local transmission of ZIKV—Southern Florida, a portion of South Texas, and Puerto Rico—saw a 21% increase in congenital disabilities in the second half of 2016 compared with the first half of that year, according to the study published in CDC's Morbidity and Mortality Weekly Report [93].

A significant increase in congenital disabilities was strongly related to ZIKV during the second half of 2016 compared with the first half in jurisdictions with local ZIKV transmission. Only a tiny percentage of congenital disabilities potentially related to ZIKV had laboratory evidence of ZIKV infection, and most were not tested for ZIKV [90].

While some exciting findings are still left unexplained, this study is one of the few working from a clear benchmark. Given the higher occurrence of congenital disabilities in the American Southern Hemisphere and very little benchmark research, the association between congenital disabilities, especially microcephaly in Brazil and Colombia, has been criticized. However, there is much more known about the association and its mechanics than was known just a few years ago when the outbreak occurred.

It may be essential to note the rarity of ZIKV-associated microcephaly. For example, 11 of 35 infants (31%) with microcephaly reported cases to a Brazil Ministry of Health registry had excessive and redundant scalp skin, which is not typically seen in other forms of microcephaly. These findings suggest an interruption of cerebral growth, but not in that of the scalp skin, after an injury (e.g., viral infection, hyperthermia, or vascular disruption) that occurred after the initial formation of brain structures, followed by partial collapse of the skull [94]. There have been a few interpretations of these results.

Microcephaly is a rare defect estimated to occur in six infants per 10,000 live-born infants in the U.S. ZIKV would not be an occasional exposure among women living in Brazil during the ZIKV outbreak. However, reports of adverse birth outcomes among travelers who spent only a limited time in an area with active ZIKV transmission are consistent with ZIKV being a rare exposure. The fetal brain disruption sequence is rare; only 20 cases were identified in a literature review in 2001 [95].

In parts of the U.S. where there was the active transmission of ZIKV in 2017, there was an uptick of 21% in the number of babies found to have severe birth defects; scientists from the Centers for Disease Control and Prevention have found [96]. The increase amounted to 29 fetuses and infants in Texas, Florida, and Puerto Rico found to have congenital disabilities affecting the central nervous system and brain, including the brain, eye, joint problems, and microcephaly, or abnormally small heads.

8.3 Some Theories and the Placenta

The preponderance of the evidence seems to center on the role of the placenta in this infection. The placenta directly conveys the ZIKV to the early embryo or fetus, or the placenta itself might be mounting a response to the exposure; this response might be contributing to or causing the brain defect [97].

After infection, the virus likely reaches the placenta through viremia. When at the placental barrier, the virus must cross to infect the fetus and be detected in fetal tissue and amniotic fluid. However, the exact mechanism has not yet been established.

> Once across the placental barrier, the virus reaches its target population, neural progenitor cells. The virus disrupts these and related cells, such as radial glia. This may be through inducing autophagy and apoptosis, disrupting cellular development, or disrupting differentiation and proliferation. Neural progenitor cells are highly susceptible to ZIKV infection. These cells are those which differentiate to produce cortical neurons. The fewer neural progenitor cells available, the fewer cortical neurons grown, and therefore, the smaller the cerebral cortex and the smaller the brain, resulting in microcephaly. [98]

There may be other possibilities as well. For example, the virus has neurotropic properties and directly affects and damages the developing brain via the placenta [62]. Another explanation is that ZIKV might be transmitted through semen, potentially giving the virus access to the early unprotected embryo [62]. Another explanation would involve the virus working through the endoplasmic reticulum of the trophoblast to become a sort of cargo of placental exosomes. The virus could be causing localized reactions at the interface of the placenta that can allow the free virus to pass through [62] (see some theories below).

Bayer et al. found data suggesting that ZIKV likely uses alternative strategies to cross the placenta rather than directly infect the placenta [99]. While Bayer et al. rule out trophoblast, Adibi et al. mean placental macrophages may be a route for infection [1]. In 2016, Pylro et al. reported the putative influence of miRNAs on the expression of human genes associated with the symptoms of congenital Zika syndrome [100].

The overriding consensus based on direct measurements of viral RNA in the amniotic fluid and fetal tissues is that the placenta effectively shuttles the ZIKV to the fetus, where it causes microcephaly [1]. "Virus RNA has been found in placenta, amniotic fluid, and brain tissues of stillborn babies with microcephaly," U. Pittsburg Public Health Researcher Ernesto Marques says [2]. Martines and Bhatnagar found that immune cells called Hofbauer cells located in the placenta may carry the virus into the bloodstreams of developing fetuses [4]. Adibi et al. argue that the placenta directly conveys the ZIKV to the early embryo or fetus. The placenta itself might be mounting a response to the exposure; this response might be contributing to or causing the brain defect [3]. They claimed leakage through the trophoblastic plugs that block material blood flow or diffusion of preconceptional viral concentrations into the amniotic and yolk sacs as they form [3].

Next, the death of these neural progenitor cells[2] disrupts brain development and leads to microcephaly, the severe congenital disability seen in babies whose mothers were infected with ZIKV during pregnancy.

American Scientists Bayless and Wysocka [101], who conducted this latest study by infecting human cells with ZIKV in the laboratory, found the virus can infect another type of cell known as well, the cranial neural crest cells—which give rise to skull bones and cartilage—and cause them to secrete signaling molecules that alter their function [102].

The ZIKV might be able to perturb the synthesis or secretion of molecules (i.e., proteins, neuropeptides, non-coding RNAs, or cytokines) within the placental chorionic villi. The chorionic villi (skin of the placenta) have greater exposure to maternal blood than the early embryo, which is embedded within two-fluid sacs and shielded by two membranes from maternal (but not placental) circulation. This finding is exciting since the most frequent symptom in placentas infected with cytomegalovirus was chronic villitis or wide-scale inflammation of the cell layers of the chorionic villi. The degree of placental inflammation has been correlated with the severity of the fetal effects, including microcephaly [103].

Adibi [3] and others have studied the simultaneous over and under-expression of MCPH1-12, CEP63, and CASC5 (mutations in these genes have been causally related to microcephaly) [104]. Should these be secreted within placental exosomes, the expression of the microcephaly proteins might be problematic.

Presumably, Rombi et al. [98] argue that infection with ZIKV results in the down-regulation of various genes involved in cell cycle-related pathways. One of these cell cycle-related genes, Centromere Protein F (CENPF), encodes a kinetochore protein essential for properly aligning fully formed chromosomes during mitosis and recruiting additional proteins that collectively play a part in spindle-pole formation. CENPF has been linked to transcription regulation as well as centrosome maturation [105].

Primary microcephaly phenotypes have previously been affected by the down-regulation of genes involved in various cell cycle stages, including centrosome maturation and spindle-pole formation. This suggests that ZIKV could interact with genes previously implicated in microcephaly [98]. While the world waits for epidemiological data, molecular evidence is coming in.

8.3.1 *The Neural Stem Cell Story*

This theory seems to predominate in the literature. ZIKV readily infects (and kills) one kind of brain cell found in developing embryos, researchers reported online on March 4 in Cell Stem Cell. "It's the first step to show that ZIKV is doing something in the brain," says a study by Tang et al. [106] Coauthor Guo-Li Ming, Neuroscientist at Johns Hopkins School of Medicine. Previous studies have found traces of ZIKV in some damaged fetal brains, but that's just a correlation [2].

Tang et al. offered an entry point to establish a mechanistic link between ZIKV and microcephaly. Their team posits that ZIKV directly infects human cortical neural progenitor cells, a kind of neural stem cell, with high efficiency, resulting in stunted growth of this cell population and transcriptional dysregulation [106]. Neuroepithelial stem cells (NES)[3] and radial glia cells (RGCs)[4] are infected, mitotically impaired, and killed by ZIKV, and early spinal cord stem cells can be infected by ZIKV well [107].

> Neuroepithelial stem cells (NES) lines are derived from primary neuroepithelial cells. These cells constitute the ventricular zone (VZ) of the neural tube and serve as the stem cells of the central nervous system (CNS). Initially, neuroepithelial cells divide symmetrically to expand the stem cell pool. Later, neuroepithelial cells transition into radial glial cells (RGCs), which reside in the ventricular zone (VZ) and inner (iSVZ), and outer subventricular zone (oSVZ).
>
> These cell populations serve as the stem or progenitor cells for neurons and macroglia (i.e., astrocytes and oligodendrocytes) and provide scaffolding for migrating nascent neurons. By comparing NES cells, organotypic fetal brain slices, and the human postmortem tissue in the context of ZIKV infection, we show that ZIKV infects both neocortical (NCX) and spinal cord (SC) NES cells and their fetal homolog, RGCs in the VZ and subventricular zone (SVZ), and to a lesser extent, postmigratory neurons in the cortical plate. ZIKV infection is associated with mitotic impairment, structural disorganization of the proliferative zones, and increased NES cells and RGCs. [108]

While less than 20% of the iPSCs, embryonic stem cells, and neurons became infected, the infection of the neural progenitors was "really striking," said Ming. Up to 90% of the cells contained the virus, and "what is a bit scary to us," she said, "is that we found these progenitor cells can spit out more virus"—with the potential to infect more progenitor cells yet. Neural progenitors "give rise to the larger population of neurons and glial cells of the brain," Ming said. "So, if they are infected and die or have retarded growth, we think that could impact the neurons they will produce." [109]. The developing fetus might be most vulnerable to ZIKV early in the first trimester before a protective zone of mature villous trophoblast[5] has been established [110].

ZIKV, the virus, has ten proteins. Recent research discovered two surface proteins, NS4A and NS4B, were causing microcephaly. Rombi et al. [98] examined three strains of ZIKV in the second trimester of human fetal neural stem cells. They found that these proteins were responsible for creating fetal brain formation and mobilizing these particular ZIKV proteins.

> In addition to affecting genes involved in the cell cycle, ZIKV infection has been shown to significantly downregulate the expression of genes associated with neural stem cells, neuronal cell types, and oligodendrocytes.[6] As well as this downregulation, the infection has been shown to upregulate specific genes, for example, those associated with astrocyte differentiation (astrocytes have a central role in brain development and function) and antiviral genes such as Toll-Like Receptor 3 (TLR3),[7] Viperin (Viperin is an interferon-stimulated gene whose expression inhibits many DNA and RNA viruses including CHIKV, HCMV, HCV, DENV, WNV, SINV, influenza, HIV, and ZIKV) [111], and CCL2.[8] This data suggests that, during infection, ZIKV targets several neural cell types while also inducing differentiation of astrocytes and promoting TLR3 to initiate cellular disorder. [112]

It was confirmed that ZIKV proteins use the energy and nutrients of autophagy, leading to neuronal stem cells being left with metabolic deficits. These proteins controlled the fetal neural stem cells and contracted the size of brain organoids by 65% [113].

"The stem cells are very vulnerable to ZIKV," said Joseph Gleeson from Rockefeller University. The infection in adult mice drastically reduced the creation of new neural stem cells and killed off existing ones—a worrisome finding [114].

"[T]he ZIKV is attacking neural stem cells in the brain—the cells that help the brain develop. In microcephaly, a baby's much smaller than average, Hotez explains.

From what scientists have been able to piece together, it seems that some genes in the virus mutated around the year 2000 and developed the ability to attack the central nervous system of a fetus. "It's every mother's worst nightmare," Hotez says.

8.3.2 Blood Supply Story

ZIKV infection appears to affect oxygen delivery to the fetuses of pregnant monkeys, according to a small study funded by the National Institutes of Health. Researchers also observed a high degree of inflammation in the placenta and lining of the uterus, harming the fetal immune system, and increasing a newborn's susceptibility to additional infections [115]. They claimed leakage through the trophoblastic plugs that block material blood flow or diffusion of preconceptional viral concentrations into the amniotic and yolk sacs as they form [3].

Hirsch et al. [116], studying five rhesus macaques, suggested that the virus induces high levels of inflammation in the blood vessels of the uterus and damages the placental villi. These branch-like growths help transfer oxygen and nutrients from maternal blood to the fetus. Another perspective on one of the reasons for the harmful effects of ZIKV on children comes from Hirsch et al.'s [116] work on macaques. They found that despite seemingly normal fetal growth and persistent fetal-placenta-maternal infection, advanced non-invasive in vivo imaging studies reveal dramatic effects on placental oxygen reserve accompanied by significantly decreased oxygen permeability of the placental villi. The observation of abnormal oxygen transport within the placenta appears to be a consequence of uterine vasculitis and placental villous damage in ZIKV cases. The implication is that in utero ZIKV infection may cause vulnerability in the neonatal innate and adaptive immune responses and thus increase the susceptibility to subsequent sepsis, other infections, and autoimmune diseases [116].

Despite seemingly normal fetal growth and persistent fetal-placenta-maternal infection, advanced non-invasive in vivo imaging studies reveal dramatic effects on placental oxygen reserve accompanied by significantly decreased oxygen permeability of the placental villi. The observation of abnormal oxygen transport within the placenta appears to result from uterine vasculitis and placental villous damage in ZIKV cases. They conclude that this animal model reveals a potential relationship between ZIKV infection and uteroplacental pathology that affects oxygen delivery to the fetus during development [117].

A fascinating study by Streblow et al. (2018) [117] offered a unique hypothesis to help explain childhood disorders associated with ZIKV. ZIKV infection appears to affect oxygen delivery to the fetuses of pregnant monkeys, according to a small study funded by the National Institutes of Health. Researchers also observed a high degree of inflammation in the placenta and lining of the uterus, harming the fetal immune system and increasing a newborn's susceptibility to additional infections [118]. The team found that the virus induces high levels of inflammation in the blood vessels of the uterus and damages the placental villi. These branch-like growths help transfer oxygen and nutrients from maternal blood to the fetus.

8.3.3 The Interferon Story

There is a growing body of collaborative literature. Li [119] found that ZIKV replicated in the embryonic brains of mice and targeted neural progenitor cells. In addition, they had characteristics consistent with the microcephalic phenotype [119].

Cugola et al. [120] used mice gene expression data to confirm a link between the ZIKV outbreak in Brazil and the alarming cases of congenital brain malformations [120]. This team also noted how it was unclear why the virus could not cross the placenta of some of its mice pups but that this result may be due to the robust antiviral immune response of this mouse strain, which secretes significant levels of type I/II interferon, known to confer resistance to ZIKV. Cugola et al. [120] suggest that genetic differences partly explain why some ZIKV-infected pregnant women give birth to newborns without detectable congenital brain malformations [120].

Yockey et al. [122] demonstrate that type I interferon (IFN-α/β) signaling is pathogenic in a mouse model of congenital ZIKV infection, resulting in abnormal placental architecture, intrauterine growth restriction (IUGR), and fetal demise [121]. They implicated type I IFNs as a possible common culprit for virus-associated pregnancy complications. They suggested blockade of type I IFNs as a potential intervention to prevent pregnancy complications in the settings of non-viral interferonopathies [122].

The type I interferons (IFNs), including IFN-β and multiple subtypes of IFN-α, are key antiviral factors that mount a rapid and potent innate defense against viruses [122]. "The type I interferon system is one of the key mechanisms for stopping viral infections," explained Helen Lazear, Virologist at the University of North Carolina at Chapel Hill, who coauthored an editorial accompanying the study. "That same [immune] process is causing fetal damage, and that's unexpected." [123].

Science News reported that pregnant mice were infected vaginally with ZIKV one of two times—one corresponding to the mid-first trimester in humans and the other to the late first trimester. Of the fetuses exposed to infection earlier, those with the interferon receptor died, while those without the receptor continued to develop. For fetuses exposed to infection a bit later in the pregnancy, those with the receptor were much smaller than their receptor-lacking counterparts [124].

Iwasaki and colleagues next added type I interferon to samples of human placental tissue in dishes. After 16–20 h, the placental tissues developed structures that resembled syncytial knots. These knots are widespread in the placentas of pregnancies with such complications as pre-eclampsia and restricted fetal growth. Figuring out which of the hundreds of antiviral proteins made when type I interferon ignites the immune system can trigger placental and fetal damage is the next step, says Iwasaki. That could provide more understanding of miscarriage; other infections that cause congenital diseases, like toxoplasmosis and rubella; and autoimmune disorders that feature excessive type I interferon production, such as lupus [124].

Compared with the pre-epidemic strains from Asian lineage, the epidemic strains isolated after 2012 have undergone an A188V mutation in the NS1 gene that enabled the protein to suppress IFN-β induction [125].

8.3.4 The AXL Story

One of the receptors targeted by ZIKV could be AXL, a protein crowded on the surface of neural stem cells, Nowakowski et al. [126] proposed in a study published online in *Cell Stem Cell.* ZIKV is thought to infect these early-development brain cells preferentially. It could potentially use AXL as an easy entry point, study Coauthor Arnold Kriegstein of the University of California, San Francisco [127].

UCSF researchers determined that the ZIKV preferentially infects brain cells with an abundance of AXL protein, which spans the outer cell membrane of several cell types and serves as a gateway for the invading virus. The fetal brain cells that incorporate this protein and are vulnerable to the virus include neural stem cells and progenitor cells that spin off other types of brain cells and play an especially crucial role in early brain growth and development. Other cells with AXL that the ZIKV infected included microglia, the brain's immune cells, and astrocytes, a fully developed and specialized type of brain cell that supports the signal-conducting neurons. When the researchers blocked or eliminated AXL to laboratory-grown kinds of cells found in the fetal brain, ZIKV was inhibited from infecting the cells [86].

8.3.5 The ANKLE2 Story

An international led by researchers from the University of California at San Francisco and the Baylor College of Medicine conducted systematic comparative analyses of the interactions of proteins from dengue and ZIKV with proteins from the host, both mosquitoes and humans. They discovered new strategies the viruses use to infect their host successfully. For instance, they found that some viral proteins counteract interferon response genes, a human and mosquito defense mechanism, and other viral proteins highjack host proteins and redirect their activities to replicate the virus [128].

The researchers combined their systematic comparative analysis with the fruit fly animal model experiments. They discovered an intriguing mechanism explaining infant microcephaly associated with maternal ZIKV infection.

Using ZIKV-specific PPIs, an NS4A-ANKLE2 interaction was uncovered and shown to impede brain development in Drosophila (fruit flies), suggesting that this connection is, at least in part, responsible for ZIKV-induced microcephaly. They identified a ZIKV-specific interaction between NS4A and ANKLE2, a gene linked to hereditary microcephaly, and showed that ZIKV NS4A causes microcephaly in Drosophila in an ANKLE2-dependent manner [129].

"Mutations in the ANKLE2 gene have been seen to cause microcephaly in humans and fruit flies. Microcephaly caused by lack of ANKLE2 is very similar to the one caused by ZIKV," said Baylor's Dr. Hugo Bellen. The researchers discovered that the ZIKV protein NS4A binds to ANKLE2, the human protein linked to microcephaly. "We found that if we overexpress NS4A in normal flies, the result is the reduced size of the fly's brain. This can be rescued by overexpressing human ANKLE2 in the flies," said Nichole Link, Postdoctoral Associate in the Bellan laboratorty. "These findings also suggest that, if we could develop a drug that could prevent NS4A from binding to ANKLE2, we might be able to prevent ZIKV from causing microcephaly." [128].

8.3.6 *The KINESIN-5 Story*

Edward Wojcik, Professor of biochemistry from LSU, studies microcephaly and blindness and concludes a molecular motor is a target for degradation by an encoded ZIKV protein (Zika protease) [130]. The molecular motor is Kinesin-5, which is required for cell division in humans. Kinesin-5 is a slow homotetrameric motor protein best known for its essential role in the mitotic spindle, limiting the rate at which faster motors can move microtubules. In neurons, experimental suppression of Kinesin-5 causes the axon to grow faster by increasing the mobility of microtubules in the axonal shaft and the invasion of microtubules into the growth cone [131].

Liu et al. [132] proposed that Zika protease targeting of HsEg5[9] may be a key event in the etiology of Zika syndrome and identified Kinesin-5 as a target for the virus and links the infection to microcephaly [130].

8.4 Other ZIKV-Related Implications for Children

CZS seems to include not only cases of microcephaly but also other associated congenital disabilities and secondary implications. A recent study in Rio de Janeiro found that "42% of women infected with ZIKV during pregnancy were found to have given birth to infants with severe abnormalities." [46].

Some researchers have included ZIKV with the disease group known as STORCH, comprising syphilis, toxoplasmosis, other infections, rubella, cytomegalovirus infection, and herpes simplex. It is generally defined as a fatal infection that can cause congenital malformations. ZIKV is now considered one of the STORCH factors, agents that cause intrauterine pathogenesis [133]. The Zika congenital syndrome (see below) is characterized by microcephaly, ventriculomegaly, periventricular calcifications, lissencephaly, hydranencephaly, ocular abnormalities, agenesis of the corpus callosum, placental insufficiency, developmental abnormalities, and fetal loss [134].

ZIKV interferes with the cellular machinery controlling cell division and alters the expression of hundreds of genes responsible for guiding the formation and development of brain cells, according to findings released on January 23, 2017, by Scientific Reports [135]. It functions much like a TORCH pathogen (Toxoplasma Gondii, rubella, cytomegalovirus, and herpes simplex) in terms of having access to the fetal compartment. A placental barrier generally protects the fetus; ZIKV seems to have found a way to overcome it.

> Researchers observed that the pool of human neural stem cells infected by the Brazilian strain of ZIKV was rapidly and wholly depleted compared to non-infected cells. This finding led the group to investigate how ZIKV disrupts the interactome map (or molecular fingerprinting) of infected cells.... The analysis revealed that more than 500 proteins in infected neurospheres had their expression level or status (upregulated vs. downregulated) altered compared to non-infected neurospheres. A number of these altered proteins are generally involved with fixing DNA damage or assuring chromosomal stability. Also, proteins generally required for cell growth were silent in infected neurospheres, which may explain why ZIKV-infected cells die much sooner than their non-infected counterparts. [136]

The absence of microcephaly at birth does not exclude the possibility of delayed development of microcephaly or other ZIKV-related brain and other abnormalities [83]. According to those experts who categorize the neurological implications under the term **CZS**, the actual size of the submerged part of this congenital Zika syndrome "iceberg" is unknown, with its minor changes and without microcephaly, which will probably lead to future problems, such as epilepsy and cognitive and motor impairments [137]. Some warn that normocephalic infants could still present similar but less severe lesions than those found in children with microcephaly [138]. A 2016 prospective study showed fetal ultrasonographic abnormalities in 12 of 42 women (29%) with ZIKV infection during pregnancy; seven of the 42 fetuses (17%) studied had microcephaly, cerebral atrophy, or atrophy, or brain calcifications [139].

According to the American Academy of Family Physicians, about one in seven children age one or older born to women with ZIKV infection during pregnancy have one or more health problems possibly caused by in utero exposure to the virus [140]. According to Rice et al. [141], among children aged ≥ 1 year born in U.S. territories and freely associated states to mothers with laboratory evidence of confirmed or possible ZIKV infection during pregnancy and who had follow-up care reported, 6% had a ZIKV-associated birth defect, 9% had ≥ 1 neurodevelopmental abnormality possibly associated with congenital ZIKV infection, and 1% had both [141] By 18 months of age, these 131 infants exposed to ZIKV in utero and who underwent imaging, neurodevelopmental assessment, sensory organ assessment, or all of these tests, 19 (14%) were found to have a severe neurodevelopmental delay, sensory organ dysfunction, or both.

CDC established pregnancy and infant registries to collect information about pregnant women with laboratory evidence of recent possible ZIKV infection and outcomes in their fetuses and infants. CDC's *Morbidity and Mortality Weekly Report* published a review of cases from the registry by Shapiro-Mendoza et al. [143] from American Samoa, Puerto Rico, Micronesia, the Marshall Islands, and the U.S. Virgin Islands, from January 1, 2016, to April 25, 2017. More than 120 babies have been born with severe congenital disabilities caused by the ZIKV in U.S. territories, the Centers for Disease Control and Prevention reported June 8, 2017. Scientists reviewed 2549 women possibly infected with ZIKV who had completed pregnancies; 2464 resulted in live births, while the others resulted in miscarriage, stillbirth, or abortion. Of the total pregnancies, 1508, or about 60%, had confirmed ZIKV infection. Among the women with confirmed ZIKV infection during the first trimester, 8% had a baby or fetus with ZIKV-associated birth defects. Of pregnancies affected in the second trimester, 5% resulted in congenital disabilities, and 4% were in the third trimester [142].

The U.S. territories reported 3930 pregnancies with laboratory evidence of recent possible ZIKV infection to the registries from January 1, 2016, to May 24, 2017, including 2549 (65%) pregnancies completed on or before April 25, 2017, which resulted in 2464 (97%) live-born infants and 85 (3%) pregnancy losses. Among women with completed pregnancies, 1561 (61%) reported signs or symptoms compatible with ZIKV infection during pregnancy, 966 (38%) were asymptomatic, and symptom information was missing for 22 (1%). Maternal symptoms or positive laboratory test results were identified in the first, second, and third trimesters for 21%, 43%, and 34% of women, respectively; timing of infection was missing or occurred periconceptionally for 41 pregnancies (2%). Of the 2549 completed pregnancies, 122 (5%) resulted in a fetus or infant with possible ZIKV-associated congenital disabilities (5% were symptomatic, and 4% were asymptomatic women). The same percentage of congenital disabilities (5%) was observed among the subset of 1508 (59%) pregnancies with NAT-confirmed ZIKV infections (5% among symptomatic and 7% among asymptomatic women). Among the 122 fetuses or infants that met the surveillance case definition for possible ZIKV-associated birth defects, 108 (89%) were classified as having brain abnormalities and microcephaly [143].

Severe congenital disability caused by ZIKV infection may not be apparent at birth but develop months afterward, confirming that the virus can cause unseen damage to developing babies. Another study "showed that the risk of congenital disabilities did not depend on the presence or the severity of (ZIKV)-related symptoms," Chief Author Dr. Bruno Hoen of the University Medical Center of Guadeloupe told Reuters Health [82]. A pregnant woman who becomes ill from the ZIKV faces a 7% chance that her child will be born with congenital disabilities and that risk jumps to nearly 13% if she becomes ill during the first trimester. These were the conclusions of her (2018) study conducted in French territories in the Americas [82].

The presence of congenital disabilities defined the ZIKV pandemic. This phenomenon was not isolated to South America or a few states in Northern Brazil. For example, areas with local virus transmission—Southern Florida, a portion of South Texas, and Puerto Rico—saw a 21% jump in births with ZIKV-related defects during the second half of 2016 [144]. CDC researchers examined a cohort of nearly 1 million births in 2016 in 14 U.S. states and one territory: Florida (select southern counties), Georgia (select counties in the Atlanta metropolitan area), Hawaii, Illinois, Iowa, Massachusetts, New Jersey, New York (excluding New York City), North Carolina (select regions), Rhode Island, South Carolina, Texas (particular parts), Utah, Vermont, and Puerto Rico. The agency found that three of every 1000 live births in these areas had a birth defect possibly associated with maternal ZIKV infection.

Among the infants with congenital disabilities,

- About half were born with brain abnormalities and microcephaly;
- 20% had neural tube defects and other early brain abnormalities;
- 9% had eye abnormalities without brain abnormalities; and
- 22% had nervous system damage, including joint problems and deafness, without brain or eye abnormalities [144].

The agency warned that because many pregnant women exposed to ZIKV in late 2016 gave birth in 2017, further increases in possible ZIKV-related congenital disabilities could be observed when those 2017 data are analyzed. WHO's Strategy Director Christopher Dye has different concerns. He worries that scientists might be counting only the babies with severe defects. "The concern is that we're seeing the tip of a larger phenomenon," he says. Thousands of babies with less obvious brain damage might be slipping under the radar [145]. For a comprehensive list of potential abnormalities and defects, see Honein et al. [146] and Miranda-Filho et al. [147].

Another problematic issue: Some babies appear fine at birth, only to develop health problems later. What if ZIKV can harm a newborn's still-developing brain like it does a fetal brain? After all, one-way ZIKV does its damage by attacking developing brain cells called neural progenitor cells, and babies retain many of those cells for months after birth [148]. As suggested by the University of Utah, babies exposed to the ZIKV during pregnancy might look normal at the delivery time. Still, afterward, they can develop brain defects or abnormalities. Of the 48 babies chosen for the review, all seemed sound and had moms with affirmed ZIKV disease in pregnancy [149].

An interesting study [150] with preliminary results was discussed at the Annual Meeting of the Pediatric Academic Societies in 2017. Dr. Mulkey's research sought to evaluate the utility of magnetic resonance imaging (MRI) to evaluate fetal brain abnormalities in 48 babies whose mothers had confirmed ZIKV infection during pregnancy. Forty-six women/infant pairs enrolled in the prospective study are Colombian, and two are Washington, D.C. women exposed during travel to a ZIKV hot zone. "Among the 48 study participants, 45 had "normal" fetal MRIs. Three fetuses exposed to ZIKV in the first or second trimester had abnormal fetal MRI: One had heterotopia and an early, abnormal fold on the brain's surface, indications that neurons did not migrate to their anticipated destination during fetal brain development. This pregnancy was terminated at 23.9 gestational weeks. One had parietal encephalocele (https://www.cdc.gov/ncbddd/birthdefects/encephalocele.html), a rare birth defect that results in a sac-like protrusion of the brain through an opening in the skull. According to the CDC, this defect affects one in 12,200 births, or 340 babies, per year. It is not known if this rare finding is related to ZIKV infection. And one had a thin corpus callosum, dysplastic brainstem, heterotopias, significant ventriculomegaly, and generalized cerebral/cerebellar atrophy." [150].

Gulland [151] said that the effects of exposure to ZIKV during pregnancy on children as they get older are unknown. It is too early to tell whether healthy babies will develop problems later, said Professor Fontanet, whose following study would follow children born to ZIKV-infected mothers and compare them with those whose mothers did not become infected during pregnancy. "It is important that we continue to follow up the babies because even if they looked healthy at birth, they might develop difficulties learning how to walk, talk, or read. We need to account for the consequences of ZIKV infection during pregnancy fully," he said [151].

"We don't know the long-term neurological consequences of having ZIKV if your brain looks normal," says Dr. Mulkey, Fetal-Neonatal Neurologist Member of the Children's Congenital ZIKV Program. "The uncertainty about long-term outcomes is what's so scary." [150].

The kinds of congenital disabilities associated with ZIKV, including microcephaly and other brain abnormalities, have increased in parts of the U.S. where mosquitoes were spreading the virus in 2016, including South Florida, according to a 2018 report from the federal Centers for Disease Control and Prevention. "Areas with local spread of ZIKV—including Florida, a portion of southern Texas, and Puerto Rico—saw a 21% spike in these kinds of congenital disabilities during the second half of 2016 compared with births that took place during the first half of that year, CDC officials reported [152]. But researchers said they do not know if the increase is due to the local spread of ZIKV or other factors. Most mothers who delivered babies with birth defects associated with the virus did not have laboratory evidence of infection. CDC officials said that those mothers were never tested, were not tested at the right time, or were not exposed to ZIKV." [152].

They identified 2962 babies and fetuses with congenital disabilities potentially related to ZIKV, including 1456 with brain abnormalities or microcephaly causes abnormally small heads and incomplete brain development. There was no state breakdown. An additional 581 had neural tube defects, 262 had eye abnormalities, and 662 had some other form of central nervous system dysfunction [152].

From the first investigations of microcephaly and subsequent studies in Brazil and elsewhere, it is now clear that ZIKV is a cause of a range of neurologic disorders [153]. To date, 22 countries and territories in the Americas have reported confirmed cases of congenital syndrome associated with ZIKV infection. The most pronounced of which is the reported increase in cases of microcephaly [154]. Screening for microcephaly, a condition in which babies are born with abnormally small heads, isn't enough to detect ZIKV-related congenital disabilities. In other words, an infant whose head circumference appears normal might still have serious health problems from the virus. "Such children would be born with normal-sized heads as cranial growth largely takes up to 30 weeks, yet present important brain damage," França et al. [155] wrote. "Given the huge interest in the epidemic, we believe that underreporting of microcephaly cases is rare, but newborn babies affected late in pregnancy might fail to be reported as their heads will be in the normal range." [155].

8.4.1 *Miscarriages and Stillbirths (Fetal Demise)*

ZIKV infection is associated with congenital disabilities and pregnancy loss. Dudley et al. [156] found that 26% of non-human primates (macaques and marmosets) infected with Asian/American ZIKV in early gestation experienced fetal demise later in pregnancy despite showing few clinical signs of infection. Pregnancy loss due to asymptomatic ZIKV infection may be a common but underrecognized adverse outcome of maternal ZIKV infection. They added: "The timing of infection was an important predictor of fetal demise in these studies. No cases (0/11) of fetal or neonatal death were observed in macaques inoculated after GD (gestational day) 55. In contrast, macaques inoculated on or before GD 55 significantly elevated fetal-demise rate of 37.8%." [156].

A recent study [23] has found a 5.8% miscarriage rate and a 1.6% stillbirth rate in women infected during the first trimester [23]. Sarno et al. [157] reported a case study of a Brazilian woman a few years earlier which they claim provides evidence that, in addition to microcephaly, there may be a link between ZIKV infection and hydrops fetalis and fetal demise [157].

Health officials have speculated that infection in pregnancy can also lead to an increased rate of miscarriages. The U.S. has not previously reported miscarriages in American travelers who have contracted the virus. Still, Dr. Sherif Zaki, CDC's Chief Pathologist, said the agency was aware of at least two such cases (during the first trimester) [158]. "It looks like ZIKV is inhibiting the development of the brain, not just [associated with] small head size, and it's associated with stillbirths," Hotez says [159].

A case report by Samo et al. [160] provides evidence that there may be a link between ZIKV infection and hydrops fetalis and fetal demise in addition to microcephaly. Given the recent spread of the virus, systematic investigation of spontaneous abortions and stillbirths may be warranted to evaluate the risk that ZIKV infection imparts on these outcomes [160].

Neuropathologic examination of infants carried to term that was stillborn or died shortly after birth has shown the frequent presence of arthrogryposis, microphthalmia (birth defect in which one or both eyes did not develop fully, so they are small), and small brains with multiple abnormalities, including thickened leptomeninges (The two innermost layers of tissue that cover the brain and spinal cord.), agyria (complete lissencephaly), ventriculomegaly (the brain ventricles, or fluid-filled cavities, are enlarged due to buildup of cerebrospinal fluid), and parenchymal thinning and calcifications. Histopathology of these cases showed abnormal neuronal migration with polymicrogyria (brain develops too many folds, and they are unusually small), meningeal glioneuronal heterotopia, motor neuron loss, and cerebellar cortical dysplasia [161].

8.4.2 Blindness (Likely Associated with Microcephaly)

Understanding the consequences of ocular infection by viruses has become of more significant concern because of the potential implications of viral persistence and spread [162]. 25.4% of children born to mothers with confirmed or suspected ZIKV infections had eye abnormalities. Researchers suggest that all infants born after ZIKV outbreaks should be universally screened for eye abnormalities, regardless of the presence of microcephaly [163].

Severe eye abnormalities are associated with microcephaly in newborns, ranging from undeveloped retinal vessels to damage to developing retinal vessels, including hemorrhaging. Vision loss can result from malnourished retinal tissue. Microcephaly is also associated with lesions that obstruct vision by decreasing peripheral vision and producing blind spots. Depending upon severity and treatability, this too can cause vision loss [164]. A recent study (2018) by João Rafael de Oliveira Dias et al. suggested that visual impairment most likely results from central nervous system damage [165].

A research team [168] in Salvador, Brazil, studied 29 infants with microcephaly and found that congenital infection due to presumed ZIKV exposure was associated with vision-threatening findings in 35% of the infants examined, which include bilateral macular and periocular lesions (in 7 of 10) as well as optic nerve abnormalities in most cases [166]. Three babies from Brazil with microcephaly and presumed ZIKV intrauterine infection had macular pigment mottling and loss of foveal reflex (affects high-acuity vision), with one manifesting well-defined macular atrophy [167]. Freitas implicates this infection as the cause of chorioretinal scarring (tiny scars, anywhere from a half millimeter to one or two millimeters in size in the back of the eye) and possibly other ocular abnormalities in infants born in Brazil with microcephaly [168].

Another team (da Silva Pone, et al. 2018) examined 112 infants from Rio de Janeiro. Twenty had microcephaly, 31 had other central nervous system abnormalities, and 61 had no significant nervous system findings. Among the 112 mothers, 32 had ZIKV infection in the first trimester, 55 in the second trimester, and 25 in the third trimester.

According to the results: 24 of the 112 infants (21.4%) had sight-threatening eye abnormalities; the most frequent findings were optic nerve and retinal abnormalities. Ten infants (41.7%) with eye abnormalities did not have microcephaly, and eight (33.3%) did not have any central nervous system findings. Fourteen infants with eye abnormalities (58.3%) were born to women infected with ZIKV in the first trimester, eight (33.3%) in the second trimester, and two (8.3%) in the third trimester [169].

Neurologic or eye problems were seen in 12.7% of the babies whose mothers were infected during the first trimester. The rates were 3.6% when mothers were infected during the second trimester and 5.3% during the third. This is "some of the most compelling data that the risk of brain abnormalities, microcephaly, and eye anomalies extends to infections in every trimester of pregnancy," said CDC's Margaret Honein in an editorial [82].

Like de Oliveira Dias et al. [165], others associate ocular issues with the promoted **CZS**. They reported that approximately 29–70% of infants with CZS and microcephaly had severe ocular abnormalities [165]. The ocular findings present in up to 70% of infants with CZS include iris coloboma,[10] lens subluxation, cataract, congenital glaucoma, and especially posterior segment findings. Loss of retinal pigment epithelium, the presence of a thin choroid, a perivascular choroidal inflammatory infiltrate (a blind spot from ischemia-induced scotomas or floaters from vitritis), and atrophic changes within the optic nerve were seen in histologic analyses of eyes from deceased fetuses.

Clinicians in areas with ZIKV should perform ophthalmologic examinations on all microcephalic babies. Because it is still unclear whether the eye lesions occur in the absence of microcephaly, it is premature to suggest ophthalmic screening of all babies born in epidemic areas [168].

8.4.3 Myelin Damage

A team led by di Guardo et al. [170] reported white matter hypomyelination, demyelination, and corpus callosum hypogenesis. Hypoplasia (arrested development) has been recently described in the brains of ZIKV-infected, microcephaly-affected infants from Brazil. Indeed, neuronal and glial cell proliferation and migration pathways within the developing brain appear to be altered during ZIKV infection. Consequently, the role of oligodendrocyte precursor cells (OPCs) in the pathogenesis of myelin damage should receive adequate attention [170].

They and others expressed concern that the pathogenetic characterization of ZIKV-associated/related myelin damage has been largely neglected. ZIKV infection should be considered in patients with acute myelitis (acute inflammation of gray and white matter in one or more adjacent spinal cord segments) living in or traveling from endemic areas. Further study should clarify the spectrum and incidence of neurological associations [171].

Dr. Kristina Adams Waldorf, Professor of global health at the University of Washington, warned that ZIKV could also potentially damage stem cells in the testes. "I am concerned about the fertility of young boys, and men who became infected with ZIKV during this epidemic may be unaware that they have a low sperm count," she said [172].

8.5 Conclusion

Although there have been fewer cases of infection in the past couple of years, ZIKV still poses a risk for pregnant women and their infants, said CDC officials. In addition to the 4800 pregnancies in the U.S. territories and freely associated states described earlier, the CDC noted nearly 2500 pregnancies in the U.S. had laboratory results showing possible or confirmed ZIKV infection [140].

You probably know more about microcephaly than you ever knew before. While the occurrence of microcephaly and other disorders might appear quite low to the reader, please keep in mind the long-term consequences of infection remain unknown. Furthermore, these disorders are incredibly disabling for the infected, and the effect on caregivers should not be discounted. As will be seen next, implications exist for adults who are not necessarily caregivers.

Endnotes

1. These cells are implicated in a wide array of functions important for a successful pregnancy including placental morphogenesis, immune regulation, control of stromal water content, and the transfer of ions and serum proteins across the maternal–fetal barrier.
2. Neural progenitor cells (NPCs) are the progenitor cells of the central nervous system (CNS) that give rise to many, if not all, of the glial and neuronal cell types that populate the CNS.
3. Neuroepithelial cells are stem cells that differentiate into neurons and glia, essential components of the human central nervous system, following the process of neurogenesis.
4. Radial glia are specialized cells in the developing nervous system of all vertebrates.
5. Villous trophoblasts cover the chorionic villi and are involved in the exchange of gas and nutriments between the mother and the fetus.
6. Oligodendrocytes are the myelinating cells of the central nervous system (CNS). They are the product of a cell lineage that undergoes a complex and precisely timed program of proliferation, migration, differentiation, and myelination to finally produce the insulating sheath of axons.
7. Toll-like receptor 3 (TLR3) is an important member of the TLR family, which is an important group of pathogen-associated molecular patterns. TLR3 can recognize double-stranded RNA and induce activation of NF-κB and the production of type I interferons.
8. CL2 is a small cytokine that belongs to the CC chemokine family. CCL2 tightly regulates cellular mechanics and thereby recruits monocytes, memory T cells, and dendritic cells to the sites of inflammation produced by either tissue injury or infection.
9. HsEg5, a kinesin-like microtubule protein, is essential for centrosome separation, mitotic spindle formation, and the postmitotic organization of centrosomes and Golgi complexes.
10. Coloboma of the iris can look like a second pupil or a black notch at the edge of the pupil. This gives the pupil an irregular shape. It can also appear as a split in the iris from the pupil to the edge of the iris.

References

1. Adibi J et al (2016) Placental mechanics in the Zika-Microcephaly relationship. Cell Host Microbe 20. http://www.cell.com/cell-host-microbe/pdf/S1931-3128(16)30268-2.pdf. Accessed 2 Oct 2016
2. Rosen M (2016) Microcephaly: building a case against Zika. Science News, April 2, pp 26–29. https://www.sciencenews.org/article/microcephaly-building-case-against-zika. Accessed 27 Sept 2016
3. Adibi J et al (2016) Teratogenic effects of the Zika virus and the role of the placenta. The Lancet 387. http://thelancet.com/journals/lancet/article/PIIS0140-6736(16)00650-4/fulltext. Accessed 14 Oct 2016
4. Martines RB et al (2016) Notes from the field: evidence of Zika virus infection in brain and placental tissues from two congenitally infected newborns and two fetal losses—Brazil, 2015. Morb Mortal Wkly Rep 65:159–160. http://www.cdc.gov/mmwr/volumes/65/wr/mm6506e1.htm. Accessed 14 April 2017
5. CDC (2017) CDC updates interim guidance on caring for women with possible exposure to Zika virus. CDC Practice Guidelines. http://www.cdc.gov/mmwr/volumes/65/wr/pdfs/mm6512e2.pdf. Accessed 10 June 2022
6. Schuler-Faccini L et al (2016) Possible association between Zika virus infection and microcephaly—Brazil, 2015. Morb Mortal Wkly Rep 65(3). https://www.cdc.gov/mmwr/volumes/65/wr/mm6503e2.htm. Accessed 31 May 2017
7. CDC Media Statement (2016) CDC concludes Zika causes microcephaly and other birth defects. CDC Online Newsroom, April 13. http://www.cdc.gov/media/releases/2016/s0413-zika-microcephaly.html. Accessed 24 Sept 2016
8. Shephard TH (1994) “Proof” of human teratogenicity. Teratology 50(2):97–98
9. Rasmussen S et al (2016) Zika virus and birth defects—reviewing the evidence for causality. N Engl J Med 374: 1981–1987. http://www.nejm.org/doi/10.1056/NEJMsr1604338. Accessed 22 Sept 2016
10. Brasil P et al (2016) Zika virus infection in pregnant women in Rio de Janeiro—preliminary report. N Engl J Med. Published online 4 March 2016. https://doi.org/10.1056/NEJMoa1602412; Jaenisch T et al (2016) Risk of microcephaly after Zika virus infection in Brazil, 2015 to 2016. Bull World Health Organ. E-Published online 30 May 2016. https://doi.org/10.2471/BLT.16.178608; Nishiura H et al (2016) A theoretical estimate of the risk of microcephaly during pregnancy with Zika virus infection. Epidemics 15. http://www.sciencedirect.com/science/article/pii/S1755436516300093. Accessed 15 March 2017
11. Dunleavy B (2016) Zika–microcephaly link confirmed in new study. ContagionLive: Infection Diseases Today, September 20. http://www.contagionlive.com/news/zikamicrocephaly-link-confirmed-in-new-study. Accessed 22 Sept 2016
12. Rasmussen S et al (2016) Zika virus and birth defects—reviewing the evidence for causality. N Engl J Med 374:1981–1987. https://doi.org/10.1056/NEJMsr1604338. http://www.nejm.org/doi/10.1056/NEJMsr1604338. Accessed 22 Sept 2016.
13. Thomas S (1994) “Proof” of human teratogenicity. Teratology 50:97–98. http://onlinelibrary.wiley.com/doi/10.1002/tera.1420500202/pdf. Accessed 2 Oct 2016
14. Maron DF (2016) What would it take to prove the Zika—microcephaly link. Scientific American, January 28. https://www.scientificamerican.com/article/what-would-it-take-to-prove-the-zika-microcephaly-link1/. Accessed 30 May 2022
15. Brasil P et al (2016) Zika virus infection in pregnant women in Rio de Janeiro—preliminary report. N Engl J Med. Published online 4 March 2016. https://doi.org/10.1056/NEJMoa1602412; Jaenisch T et al (2016) Bulletin of the World Health Organization. E-Published online 30 May 2016. https://doi.org/10.2471/BLT.16.178608; Nishiura H et al (2016) A theoretical estimate of the risk of microcephaly during pregnancy with Zika virus infection. Epidemics 2016 (in press). https://doi.org/10.1016/j.epidem.2016.03.0010
16. WHO (2016) Dispelling rumours around Zika and complications. WHO. http://www.who.int/emergencies/zika-virus/articles/rumours/en/. Accessed 24 Sept 2016

17. Anwar S (2016) CDC confirms Zika causes microcephaly and other birth defects. ContagionLive: Infectious Diseases Today, April 14. http://www.contagionlive.com/news/cdc-confirms-zika-causes-microcephaly-and-other-birth-defects. Accessed 24 Sept 2016
18. Kleber de Oliveira W, Cortez-Escalante J, De Oliveira WT et al (2016) Increase in reported prevalence of microcephaly in infants born to women living in areas with confirmed Zika virus transmission during the first trimester of pregnancy—Brazil, 2015. MMWR Morbidity & Mortality Weekly Report 65:242–247. http://www.cdc.gov/mmwr/volumes/65/wr/mm6509e2.htm. Accessed 14 Oct 2016; Reefhuis J, Gilboa SM, Johansson MA et al, Projecting month of birth for at-risk infants after Zika virus disease outbreaks. Emerg Infectious Diseases 22(5). http://wwwnc.cdc.gov/eid/article/22/5/16-0290_article. Accessed 14 Oct 2016; Mitchell AA (2010) Proton-pump inhibitors and birth defects—some reassurance, but more needed. N Engl J Med 363:2161–21663. Accessed http://www.nejm.org/doi/pdf/10.1056/NEJMe1009631. Accessed 14 Oct 2016
19. Barreto de Araujo TV et al (2016) Association between Zika virus infection and microcephaly in Brazil, January to May 2016: preliminary report of a case-control study. Lancet Infect Dis. http://www.thelancet.com/journals/laninf/article/PIIS1473-3099(16)30318-8/abstract. Accessed 24 Sept 2016.
20. Branswell H (2016) Zika infection can damage fetuses even if pregnant women show no symptoms. STAT, June 15. https://www.statnews.com/2016/06/15/zika-infection-asymptomatic-pregnancy/. Accessed 9 March 2017
21. Honein M et al (2016) Birth defects among fetuses and infants of US women with evidence of possible Zika virus infection during pregnancy. JAMA. http://jamanetwork.com/journals/jama/fullarticle/2593702. Accessed 14 April 2017
22. Fox M (2016) Zika virus damages 6 percent of fetuses in U.S. study. NBC News, December 14. http://www.nbcnews.com/storyline/zika-virus-outbreak/zika-virus-damages-6-percent-fetuses-u-s-study-n695666. Accessed 14 April 2017
23. Hoen B et al (2018) Pregnancy outcomes after ZIKV infection in French territories in the Americas. N Engl J Med. https://www.ncbi.nlm.nih.gov/pubmed/29539287. Accessed 27 June 2018
24. Fox M (2016) Zika virus damages 6 percent of fetuses in U.S. study. NBC News, December 14. http://www.nbcnews.com/storyline/zika-virus-outbreak/zika-virus-damages-6-percent-fetuses-u-s-study-n695666. Accessed 14 April 2017; Martines RB et al (2016) Notes from the field: evidence of Zika virus infection in brain and placental tissues from two congenitally infected newborns and two fetal losses—Brazil, 2015. Morb Mortal Wkly Rep 65:159–160. http://www.cdc.gov/mmwr/volumes/65/wr/mm6506e1.htm. Accessed 2 Oct 2016
25. McNeil DG (2016) Zika: the emerging epidemic. W. W. Norton & Co, NY
26. Besnard M et al (2016) Congenital cerebral malformations and dysfunction in fetuses and newborns following the 2013 to 2014 Zika virus epidemic in French Polynesia. Euro-Surveillance 21(13). https://www.ncbi.nlm.nih.gov/pubmed/27063794. Accessed 23 July 2018
27. Cauchemez S et al (2016) Association between Zika virus and microcephaly in French Polynesia, 2013–15: a retrospective study. The Lancet 387. http://thelancet.com/abstract/S0140-6736(16)00651-6. Accessed 3 April 2017
28. Johansson M et al (2016) Zika and the risk of microcephaly. N Engl J Med 375(1):1–4. http://www.nejm.org/doi/full/10.1056/NEJMp1605367#t=article. Accessed 14 Oct 2016; Cauchemez S et al (2016) Association between Zika virus and microcephaly in French Polynesia, 2013–15: a retrospective study. The Lancet 387:2125–2132. http://www.thelancet.com/abstract/S0140-6736(16)00651-6. Accessed 14 Oct 2016
29. Fox M (2016) Zika Virus birth defects may be 'tip of the iceberg', experts say. NBC News, May 1. http://www.nbcnews.com/storyline/zika-virus-outbreak/zika-virus-birth-defects-may-be-tip-iceberg-experts-say-n565631. Accessed 14 April 2017

30. Dunleavy B (2016) Zika–microcephaly link confirmed in new study. ContagionLive: Infection Diseases Today, September 20. http://www.contagionlive.com/news/zikamicrocephaly-link-confirmed-in-new-study. Accessed 22 Sept 2016; Barreto M et al (2016) Zika virus and microcephaly in Brazil: a scientific agenda. The Lancet 387:919–922. http://www.thelancet.com/journals/lancet/article/PIIS0140-6736(16)00545-6/fulltext?rss%3Dyes. Accessed 22 Sept 2016
31. França GV et al (2016) Congenital Zika virus syndrome in Brazil: a case series of the first 1501 livebirths with complete investigation. The Lancet 388. http://www.thelancet.com/journals/lancet/article/PIIS0140-6736(16)30902-3/abstract. Accessed 14 April 2017
32. Schuler-Faccini L et al (2016) Possible association between Zika virus infection and microcephaly—Brazil, 2015. MMWR. Morb Mortal Wkly Rep 65(3):59–62. http://www.cdc.gov/mmwr/volumes/65/wr/mm6503e2.htm. Accessed 14 Oct 2016
33. de Oliveira Melo AS et al (2016) Congenital Zika virus infection: beyond neonatal microcephaly. JAMA Neurol 73(12). https://www.ncbi.nlm.nih.gov/pubmed/27695855. Accessed 23 July 2018
34. Belluck P, Franco T (2017) Clues to Zika damage might lie in cases of twins. The New York Times, May 1, https://www.nytimes.com/2017/05/01/health/zika-twins-transmission-theories.html. Accessed 26 June 2017
35. van der Linden V et al (2017) Discordant clinical outcomes of congenital Zika virus infection in twin pregnancies. Arquivos de Neuro-Psiquiatria 75(6). https://www.ncbi.nlm.nih.gov/pubmed/28658408. Accessed 11 Aug 2018
36. Santos VS et al (2017) Case report: microcephaly in twins due to the Zika virus. Am J Trop Med Hygiene 97(1). https://www.ncbi.nlm.nih.gov/pubmed/28719330. Accessed 26 July 2018
37. Santos VS et al (2017) Case report: microcephaly in twins due to the Zika virus. Am J Trop Med Hygiene 97(1). https://www.ncbi.nlm.nih.gov/pubmed/28719330. Accessed 11 Aug 2018
38. Satterfield-Nash A, Kotzky K, Allen J et al (2017) Health and development at age 19–24 months of 19 children who were born with microcephaly and laboratory evidence of congenital Zika virus infection during the 2015 Zika virus outbreak—Brazil, 2017. Morb Mortal Wkly Rep 66:1347–1351
39. World Health Organization (2016) Zika situation report, February 5. Retrieved from http://apps.who.int/iris/bitstream/10665/204348/1/zikasitrep_5Feb2016_eng.pdf. Accessed 1 April 2018
40. Brusie C (2018) How does Zika affect toddlers? VeryWellFamily, March 6. https://www.verywellfamily.com/how-does-zika-affect-toddlers. Accessed 1 April 2018
41. Ospina ML, Tong VT, Gonzalez M et al (2020) Zika virus disease and pregnancy outcomes in Colombia. N Engl J Med 383:537–545
42. Mercado-Reyes M, Gilboa SM, Valencia D et al (2021) Pregnancy, birth, infant, and early childhood neurodevelopmental outcomes among a cohort of women with symptoms of Zika virus disease during pregnancy in three surveillance sites, Project Vigilancia de Embarazadas con Zika (VEZ), Colombia, 2016–2018. Trop Med Infect Dis 6. https://doi.org/10.3390/tropicalmed6040183
43. Galang RR, Avila GA, Valencia D et al (2020) Etiology of microcephaly and central nervous system defects during the Zika epidemic in Colombia. J Pediatr 222(112–119):e3
44. Tellechea AL, Luppo V, Morales MA et al (2016) Surveillance of microcephaly and selected brain anomalies in Argentina: relationship with Zika virus and other congenital infections. Birth Defects Res. https://doi.org/10.1002/bdr2.1347
45. Rick A-M, Domek G, Cunningham M et al (2017) High background congenital microcephaly in rural Guatemala: implications for neonatal congenital Zika virus infection screening. Glob Health Sci Pract 5(4):686–696
46. Brasil P et al (2016) Zika virus infection in pregnant women in Rio de Janeiro. N Eng J Med 375:2321–2334. https://doi.org/10.1056/NEJMoa1602412. http://www.nejm.org/doi/full/10.1056/NEJMoa1602412. Accessed 16 Jan 2017
47. Haelle T (2016) Birth defects linked to Zika include more than microcephaly. Everyday Health.com, April 6. http://www.everydayhealth.com/zika/living-with/birth-defects-linked-zika-include-more-than-microcephaly/. Accessed 14 May 2017

48. Brasil P, Nielsen-Saines K (2016) More pieces to the microcephaly–Zika virus puzzle in Brazil. The Lancet. http://www.thelancet.com/journals/laninf/article/PIIS1473-3099(16)30372-3/abstract. Accessed 7 April 2017
49. Boodman E (2016) Alarming new data shows high risk of birth defects in Zika-affected pregnancies. STAT, March 4. https://www.statnews.com/2016/03/04/zika-birth-defects-study/. Accessed 8 March 2017
50. Reuters (2016) Zika virus 'can replicate' in foetal brains and after birth. Irish Times, December 13. http://www.irishtimes.com/news/health/zika-virus-can-replicate-in-foetal-brains-and-after-birth-1.2904822. Accessed 30 May 2017
51. de Araújo TVB et al (2018). Association between microcephaly, Zika virus infection, and other risk factors in Brazil: final report of a case-control study. Lancet Infect Dis. https://doi.org/10.1016/S1473-3099(17)30727-2. Accessed 6 July 2018
52. Mlakar J et al (2016) Zika virus associated with microcephaly. N Engl J Med 34(10):951–958. http://www.nejm.org/doi/full/10.1056/NEJMoa1600651. Accessed 14 May 2017
53. LaMotte S (2016) Zika virus might damage vision. CNN, February 11. http://edition.cnn.com/2016/02/11/health/zika-virus-damage-vision/index.html?sr=twcnni021116zika-virus-damage-vision0126PMStoryGalLink&linkId=21181098. Accessed 14 May 2017; Freitas BdP, Sacramento GA (2016) Ocular findings in Infants with microcephaly associated with resumed Zika virus congenital infection in Salvador, Brazil. JAMA Ophthalmol 134(5):529–535. http://jamanetwork.com/journals/jamaophthalmology/fullarticle/2491896. Accessed 14 May 2017
54. Belluck P (2016) Brain scans of Brazilian babies show array of Zika effects. The New York Times, August 23. http://www.nytimes.com/2016/08/24/health/zika-a-formidable-enemy-attacks-and-destroys-parts-of-babies-brains.html?_r=0. Accessed 26 Oct 2016
55. Soares de Oliveira-Szejnfeld P et al (2016) Congenital brain abnormalities and Zika virus: what the radiologist can expect to see prenatally and postnatally. Radiology 281(2). http://pubs.rsna.org/doi/full/10.1148/radiol.2016161584. Accessed 26 May 2017
56. ECDC (2016) Zika virus disease epidemic: potential association with microcephaly and Guillain-Barré syndrome (first update). Rapid Risk Assessment, January 21. http://ecdc.europa.eu/en/publications/Publications/rapid-risk-assessment-zika-virus-first-update-jan-2016.pdf. Accessed 24 Sept 2016
57. Gallagher JJ (2016) Congress halts Zika fund over planned parenthood as cases spread. ABC News. http://abcnews.go.com/Politics/congress-halts-zika-funds-planned-parenthood-cases-spread/story?id=41913670. Accessed 24 Sept 2016
58. Van der Linden V et al (2016) Congenital Zika syndrome with arthrogryposis: retrospective case series study. Br Med J. http://www.bmj.com/content/354/bmj.i3899. Accessed 9 June 2017
59. Belluck P (2016) Microcephaly found in babies of Zika-infected mothers months after birth. The New York Times, November 22. https://www.nytimes.com/2016/11/22/health/zika-microcephaly-babies.html?_r=0. Accessed 9 March 2017
60. Li G et al (2017) Characterization of cytopathic factors through genome-wide analysis of the Zika viral proteins in fission yeast. Proc Natl Acad Sci (2017). www.pnas.org/cgi/doi/10.1073/pnas.1619735114. Accessed 19 May 2017
61. Brasil P et al (2016) Zika virus infection in pregnant women in Rio de Janeiro—preliminary report. N Engl J Med. Published online 4 March 2016. https://doi.org/10.1056/NEJMoa1602412. Accessed 9 March 2017
62. Adibi JL et al (2016) Teratogenic effects of the Zika virus and the role of the Placenta. The Lancet, 387. https://www.ncbi.nlm.nih.gov/pubmed/26952548. Accessed 19 June 2019
63. Aragão MdFV et al (2016) Clinical features and neuroimaging (CT and MRI) findings in presumed Zika virus related congenital infection and microcephaly: retrospective case series study. Br Med J 353:i1901.https://doi.org/10.1136/bmj.i1901. Accessed 6 July 2018
64. Ventura CV et al (2016) Zika virus in Brazil and maculae atrophy win a child with microcephaly. The Lancet. January 7. http://www.thelancet.com/journals/lancet/article/PIIS0140-6736(16)00006-4/fulltext?rss%3Dyes. Accessed September 4, 2016.

65. de Oliveira Melo AS et al (2016) Zika virus intrauterine infection causes fetal brain abnormality and microcephaly: tip of the iceberg? Ultrasound in Obstet Gynecol 47:6–7. https://www.ncbi.nlm.nih.gov/pubmed/26731034. Accessed 19 July 2018
66. Brasil P et al (2016) Zika virus infection in pregnant women in Rio de Janeiro—preliminary report. N Engl J Med. Published online 4 March 2016. https://doi.org/10.1056/NEJMoa1602412
67. Samo M et al (2016) Zika virus infection and stillbirths: a case of hydrops fetalis, hydranencephaly and fetal demise. PLOS; Neglected Trop Dis. http://journals.plos.org/plosntds/article?id=10.1371/journal.pntd.0004517. Accessed 14 Oct 2016
68. Brasil P et al (2016) Zika virus infection in pregnant women in Rio de Janeiro—preliminary report. N Engl J Med. Published online 4 March 2016. https://doi.org/10.1056/NEJMoa1602412. http://www.nejm.org/doi/full/10.1056/NEJMoa1602412#t=article. Accessed 14 Oct 2016.
69. Barry-Jester AM (2016) The latest Zika news is more bad news. Five ThiryEight, December 16. https://fivethirtyeight.com/features/the-latest-zika-news-is-more-bad-news/. Accessed 1 Feb 2017
70. Associated Press (2018) Big questions linger as Brazil's 'Zika babies' grow up. The New York Post, June 16. https://nypost.com/2018/06/06/big-questions-linger-as-brazils-zika-babies-grow-up/. Accessed 18 July 2018
71. de Oliveira Souza IN et al (2018) Acute and chronic neurological consequences of early-life Zika virus infection in mice. Sci Transl Med 10(444):Eaar2749. http://stm.sciencemag.org/content/10/444/eaar2749. Accessed 23 July 2018
72. Centers for Disease Control and Prevention (2017) Overview: Zika virus, August 28. https://www.cdc.gov/zika/about/overview.html. Accessed 1 April 2018
73. Keet E (2019) Severe developmental impairment found in 15% of toddlers exposed to Zika in utero. ContagionLive, December 13. https://www.contagionlive.com/view/severe-developmental-impairment-found-in-15-of-toddlers-exposed-to-zika-in-utero. Accessed 29 May 2022
74. Moreira MEL et al (2018). Neurodevelopment in infants exposed to Zika virus in utero. N Engl J Med
75. Braswell H (2018) Developmental delays persist as Brazil's Zika babies grow up. STAT, December 12. https://www.statnews.com/2018/12/12/zika-babies-brazil-developmental-delays/. Accessed 30 May 2022
76. Clarke J (2018) Zika virus: what does the future hold? Medical.net News, June 18. https://www.news-medical.net/news/20180618/Zika-Virus-What-Does-the-Future-Hold.aspx. Accessed 24 July 2018
77. Grinnell A (2018) What happened to Zika? PBS NewsHour, July 6. https://www.pbs.org/newshour/science/what-happened-to-zika. Accessed 24 July 2018
78. Rua EC et al (2022) Two-year follow-up of children with congenital Zika syndrome: the evolution of clinical patterns. Eur J Pediatr 181:991–999. https://link.springer.com/article/10.1007/s00431-021-04280-z. Accessed 29 May 2022
79. Oliveira D et al (2016) Prolonged shedding of Zika virus associated with congenital infection. N Engl J Med 375(12). http://www.nejm.org/doi/full/10.1056/NEJMc1607583#t=article. Accessed 26 May 2017
80. Lafrance A (2016) Everything you need to know about the Zika virus. The Atlantic, November 10. https://www.theatlantic.com/health/archive/2016/11/zika-cheat-sheet/506822/. Accessed 18 May 2017; Beck J (2016) Zika is a delayed epidemic. The Atlantic, April 19. https://www.theatlantic.com/health/archive/2016/04/zika-is-a-delayed-epidemic/478755/. Accessed 18 May 2017
81. Moore C et al (2016) Characterizing the pattern of anomalies in congenital Zika syndrome for pediatric clinicians. JAMA Pediatr 171(3). https://jamanetwork.com/journals/jamapediatrics/fullarticle/2579543?utm_campaign=articlePDF&utm_medium=articlePDFlink&utm_source=articlePDF&utm_content=jamapediatrics.2016.3982. Accessed 24 July 2018

82. Emery G (2018) Pregnant women with Zika virus have 7% chance of baby with birth defects. Global News, March 14. https://www.reuters.com/article/us-health-zika-pregnancy-risks/birth-defect-rate-pegged-at-7-percent-for-babies-born-to-zika-infected-women-idUSKCN1GQ345. Accessed 27 June 2018
83. Hoen B et al (2018) Pregnancy outcomes after ZIKV infection in French territories in the Americas. N Engl J Med. https://www.nejm.org/doi/full/10.1056/NEJMoa1709481. Accessed 27 June 2018
84. Onorati M et al (2016) Zika virus disrupts phospho-TBK1 localization and mitosis in human neuroepithelial stem cells and radial glia. Cell Rep 16. https://www.ncbi.nlm.nih.gov/pubmed/27568284. Accessed 26 May 2017
85. Anwar S (2016) Zika more threatening than initially believed. ContagionLive: Infectious Diseases Today, April 12. http://www.contagionlive.com/news/zika-more-threatening-than-initially-believed. Accessed 24 Sept 2016.
86. Kurtzman L (2016) Zika in fetal brain tissue responds to a popular antibiotic. UCSF, November 29. https://www.ucsf.edu/news/2016/11/405031/zika-fetal-brain-tissue-responds-popular-antibiotic. Accessed 17 May 2017
87. Miranda-Filho DdB et al (2016) Initial description of the presumed congenital Zika syndrome. Am J Public Health 106(4). https://www.ncbi.nlm.nih.gov/pubmed/26959258. Accessed 23 May 2017; Balm Michaelle N et al (2012). A diagnostic polymerase chain reaction assay for Zika virus. J Med Virol 84(9). https://www.ncbi.nlm.nih.gov/pubmed/22825831. Accessed 23 May 2017
88. Noor, Rashed & Ahmed, Tasnia. (2018). Zika virus: Epidemiological study and its association with public health risk. Journal of Infection and Public Health. April 28. https://www.sciencedirect.com/science/article/pii/S1876034118300431. Accessed 25 July 2018; Matthews L (2017) Understanding the Zika virus: all you need to know. Amazon, Middletown, DE
89. Antoniou E et al (2020). Zika virus and the risk of developing microcephaly in infants: a systematic review. Int J Environ Res Public Health 17:3806. https://doi.org/10.3390/ijerph17113806
90. Delaney A et al (2018) Population-based surveillance of birth defects potentially related to Zika virus infection—15 states and U.S. territories, 2016. Morb Mortal Wkly Rep 67(3). https://www.cdc.gov/mmwr/volumes/67/wr/mm6703a2.htm. Accessed 27 June 2018
91. Sanz Cortes M et al (2018) Clinical assessment and brain findings in a cohort of mothers, fetuses and infants infected with ZIKA virus. Am J Obstet Gynecol 218(4):e1-440. e36. https://doi.org/10.1016/j.ajog.2018.01.012
92. Molteni M (2017) A clue to the mystery of Colombia's missing Zika cases. Wired, January 31. https://www.wired.com/2017/01/clue-mystery-colombias-missing-zika-cases/. Accessed 24 May 2017
93. Xinhua (2018) Zika linked to rise in birth defects: U.S. gov't report. New China, January 26. http://www.xinhuanet.com/english/2018-01/26/c_136925453.htm. Accessed 26 July 2018
94. Rasmussen S et al (2016) Zika virus and birth defects—reviewing the evidence for causality. N Engl J Med 374:1981–1987. http://www.nejm.org/doi/10.1056/NEJMsr1604338. Accessed 29 May 2017
95. Rasmussen S et al (2016) Zika virus and birth defects—reviewing the evidence for causality. N Engl J Med 374:1981–1987. http://www.nejm.org/doi/10.1056/NEJMsr1604338. Accessed 29 May 2017; Corona-Rivera JR et al (2001) Report and review of the fetal brain disruption sequence. Eur J Pediatr 160. https://www.ncbi.nlm.nih.gov/pubmed/11760023. Accessed 29 May 2017
96. Diep F (2018). A new government report find the Zika virus caused a big uptick in serious birth defects in 2016. Pacific Standard, January 25. https://psmag.com/news/zika-virus-caused-a-big-uptick-in-serious-birth-defects-in-2016. Accessed 27 June 2018
97. Adibi JA, Margues Jr TA, Cartus A, Beigi R (2016). Teratogenic effects of the Zika virus and the role of the placenta. The Lancet 387:1587–1590
98. Rombi F, Bayliss R, Tuplin A, Yeoh S (2020) The journey of Zika to the developing brain. Mol Biol Rep 47:3097–3115

99. Bayer A et al (2016) Type III interferons produced by human placental trophoblasts confer protection against Zika virus infection. Cell Host Microbe 19:705–712. https://www.ncbi.nlm.nih.gov/pubmed/27066743. Accessed 2 Oct 2016
100. Pylro VS et al (2016) ZIKV—CDB: a collaborative database to guide research linking SncRNAs and ZIKA virus disease symptoms. PLOS: Negl Trop Dis 10(6):e0004817. https://doi.org/10.1371/journal.pntd.0004817
101. Bayless N et al (2016) Zika virus infection induces cranial neural crest cells to produce cytokines at levels detrimental for neurogenesis. Cell Host Microbe 20(4). https://www.ncbi.nlm.nih.gov/pubmed/27693308. Accessed 30 May 2017
102. Reuters (2016) The science of Zika revealed: study finds killer virus infects neural cells related to skull formation. Daily Mail, September 29. http://www.dailymail.co.uk/health/article-3814158/Study-finds-Zika-infects-neural-cells-related-skull-formation.html. Accessed 30 May 2017; Bayless NL et al (2016) Zika virus infection induces cranial neural crest cells to produce cytokines at levels detrimental for neurogenesis. Cell Host Microbe 20. https://www.ncbi.nlm.nih.gov/pubmed/27693308. Accessed 2 July 2017
103. Mostoufi-Zadeh M, Driscoll SG, Biano SA, and Kundsin RB (1984) Placental evidence of cytomegalovirus infection of the fetus and neonate. Arch Pathol Lab Med 108:403–500. https://www.ncbi.nlm.nih.gov/pubmed/6324716. Accessed 14 Oct 2016; Gabrielli L et al (2009) Histological findings in foetuses congenitally infected by cytomegalovirus. J Clin Virol 46(suppl 4):S16–21. https://www.ncbi.nlm.nih.gov/pubmed/19879801. Accessed 14 Oct 2016; Adibi J et al (2016) Teratogenic effects of the Zika virus and the role of the placenta. The Lancet 387. http://thelancet.com/journals/lancet/article/PIIS0140-6736(16)00650-4/fulltext. Accessed 14 Oct 2016
104. Gilmore E, Walsh C (2013) Genetic causes of microcephaly and lessons for neuronal development. Wiley Interdiscip Rev Dev Biol 2(4):461–78. http://onlinelibrary.wiley.com/doi/10.1002/wdev.89/abstract. Accessed 14 Oct 2016; Homem CC, Repic M, Knoblich JA (2015) Proliferation control in neural stem and progenitor cells. Nat Rev Neurosci 16(11):647–659. https://www.ncbi.nlm.nih.gov/pubmed/26420377. Accessed 14 Oct 2016; Pulvers JN, Journiac N, Arai Y, Nardelli J (2015) MCPH1: a window into brain development and evolution. Front Cell Neurosci 9:92. https://www.ncbi.nlm.nih.gov/pubmed/25870538. Accessed 14 Oct 2016
105. Barkovich AJ et al (2012) A developmental and genetic classification for malformations of cortical development: update 2012. Brain J Neurol 135:1348–1369
106. Tang H et al (2016) Zika virus infects human cortical neural progenitors and attenuates their growth. Cell Stem Cell 18. https://www.ncbi.nlm.nih.gov/pubmed/26952870. Accessed 14 April 2017.
107. Onorati M et al (2016) Zika virus disrupts phospho-TBK1 localization and mitosis in human neuroepithelial stem cells and radial glia. Cell Rep 16. https://www.ncbi.nlm.nih.gov/pubmed/27568284. Accessed 18 July 2017
108. Onorati M et al (2016) Zika virus disrupts phospho-TBK1 localization and mitosis in human neuroepithelial stem cells and radial glia. Cell Rep 16. https://www.ncbi.nlm.nih.gov/pubmed/27568284. Accessed 27 May 2017
109. Williams R (2016) Zika infects neural progenitors. The Scientist, March 4. http://www.the-scientist.com/?articles.view/articleNo/46841/title/Zika-Infects-Adult-Neural-Progenitors-Too/. Accessed 1 July 2017.
110. Sheridan MA et al (2017) Vulnerability of primitive human placental trophoblast to Zika virus. Proc Natl Acad Sci 114(9). https://www.ncbi.nlm.nih.gov/pubmed/28193876. Accessed 25 July 2018
111. Panayiotou C, Lindqvist R, Kurhade C, Vonderstein K, Pasto J et al (2018) Viperin restricts Zika virus and tick-borne encephalitis virus replication by targeting NS3 for proteasomal degradation. J Virol 92(12):e00501-18. https://doi.org/10.1128/JVI.00501-18. Print 2018 June 15
112. Romi F, Bayliss R, Tuplin A, Yeoh S (2020) The journey of Zika to the developing brain. Mol Biol Rep 47:3097–3115

113. Cervantes M (2017) Zika virus updates: researchers discover damaging proteins. Counsel & Heal.com, January 7. http://www.counselheal.com/articles/29328/20170107/zika-virus-updates-researchers-discover-damaging-proteins.htm. Accessed 15 March 2017; Li G et al (2017) Characterization of cytopathic factors through genome-wide analysis of the Zika viral proteins in fission yeast. Proc Natl Acad Sci. www.pnas.org/cgi/doi/10.1073/pnas.1619735114. Accessed 15 March 2017
114. Li H et al (2016) Zika virus infects neural progenitors in the adult mouse brain and alters proliferation. Cell Stem Cell 19:593–598. http://www.cell.com/cell-stem-cell/pdf/S1934-5909(16)30252-1.pdf. Accessed 4 April 2017
115. NIH News Release (2018) Zika infection during pregnancy may disrupt fetal oxygen supply. NIH News Release, January18. https://www.nih.gov/news-events/news-releases/zika-infection-during-pregnancy-may-disrupt-fetal-oxygen-supply. Accessed 30 May 2022
116. Hirsch AJ, Smith JL, Haese NN, Broeckel RM, Parkins CJ, et al. (2017) Correction: Zika Virus infection of rhesus macaques leads to viral persistence in multiple tissues. PLOS Pathogens 13(4): e1006317. https://doi.org/10.1371/journal.ppat.1006317
117. Hirsch AJ et al (2018) Zika virus infection in pregnant rhesus macaques causes placental dysfunction and immunopathology. Nat Commun 9:263. https://doi.org/10.1038/s41467-017-02499-9
118. NIH (2018) Zika infection during pregnancy may disrupt fetal oxygen supply. NIH. https://www.nih.gov/news-events/news-releases/zika-infection-during-pregnancy-may-disrupt-fetal-oxygen-supply. Accessed 7 July 2018
119. Li C et al (2016) Zika virus disrupts neural progenitor development and leads to microcephaly in mice. Cell Stem Cell 19(1):120–126. https://doi.org/10.1016/j.stem.2016.04.017
120. Cugola FR et al (2016) The Brazilian Zika virus strain causes birth defects in experimental models. Nat Lett. https://www.nature.com/articles/nature18296. Accessed 25 July 2018
121. Casazza R, Lazear H (2018) Antiviral immunity backfires: pathogenic effects of type I interferon signaling in fetal development. Sci Immunol 3(19):eaar3446. https://www.ncbi.nlm.nih.gov/pubmed/29305463. Accessed 23 July 2018
122. Yockey L et al (2018) Type I interferons instigate fetal demise after Zika virus infection. Sci Immunol 3:eaao1680. https://www.sciencenews.org/article/key-virus-fighter-implicated-pregnancy-woes. Accessed 6 July 2018
123. Cunningham A (2018) A key virus fighter is implicated in pregnancy woes. Science News 193(2). https://www.sciencenews.org/article/key-virus-fighter-implicated-pregnancy-woes. Accessed 6 July 2018; Yockey L et al (2018) Type I interferons instigate fetal demise after Zika virus infection. Sci Immunol 3:eaao1680. https://www.sciencenews.org/article/key-virus-fighter-implicated-pregnancy-woes. Accessed 6 July 2018
124. Cunningham A (2018) A key virus fighter is implicated in pregnancy woes. Science News 193(2). https://www.sciencenews.org/article/key-virus-fighter-implicated-pregnancy-woes. Accessed 6 July 2018
125. Xia H et al (2018) An evolutionary NS1 mutation enhances Zika virus evasion of host interferon induction. Nat Commun 414. https://www.nature.com/articles/s41467-017-02816-2. Accessed 26 July 2018
126. Nowakowski T et al (2016). Expression analysis highlights AXL as a candidate Zika virus entry receptor in neural stem cells. Cell Stem Cell 18(5). http://www.cell.com/cell-stem-cell/abstract/S1934-5909(16)00118-1. Accessed 30 May 2017
127. Rosen M (2016) Zika structure mapped for first time. Science News, March 31. https://www.sciencenews.org/article/zika-structure-mapped-first-time. Accessed 30 May 2017
128. Mickey A (2018) New insights into Zika's microcephaly link, dengue infection. BCM News, December 13. https://www.bcm.edu/news/insight-zika-microcephaly-dengue-traits. Accessed 30 May 2022
129. Shah PS et al (2018) Comparative flavivirus-host protein interaction mapping reveals mechanisms of dengue and Zika virus pathogenesis. Cell 175:1931–1945

130. Henderson E (2021) Study discovered how microcephaly and blindness may develop in Zika-infected fetuses. News Med Life Sci. https://www.news-medical.net/news/20210406/Study-discovers-how-microcephaly-and-blindness-may-develop-in-Zika-infected-fetuses.aspx. Accessed 29 May 2022; Liu L et al (2021) Interorganelle interactions between the ER and mitotic apparatus facilitates Zika protease cleavage of human Kinesin-5 and contributes to distinct mitotic defects. iScience. https://doi.org/10.1016/j.isci.2021.102385
131. Kahn OI, Sharma V, Gonzalez-Billault C, Baas PW (2015) Effects of kinesin-5 inhibition on dendritic architecture and microtubule organization. Mol Biol Cell 26:66–77
132. Liu L et al (2021) Interorganelle interactions between the ER and mitotic apparatus facilitates Zika protease cleavage of human Kinesin-5 and contributes to distinct mitotic defects. iScience. https://doi.org/10.1016/j.isci.2021.102385
133. Coyne CB, Lazear HM (2016). Zika virus—reigniting the TORCH. Nat Rev Microbiol 14. http://www.nature.com/nrmicro/journal/v14/n11/full/nrmicro.2016.125.html. Accessed 14 April 2017
134. Soares de Oliveira-Szejnfeld P et al (2016) Congenital brain abnormalities and Zika virus: what the radiologist can expect to see prenatally and postnatally. Radiology 281(2). http://pubs.rsna.org/doi/full/10.1148/radiol.2016161584. Accessed 26 Oct 2016; Coyne CB, Lazear HM (2016) Zika virus—reigniting the TORCH. Nat Rev Microbiol 14. http://www.nature.com/nrmicro/journal/v14/n11/full/nrmicro.2016.125.html. Accessed 14 April 2017
135. D'Or Institute for Research and Education. (2017). Network of molecular interactions in brain cells infected by Zika virus reveals new therapeutic targets. Science News, January 23. https://www.sciencedaily.com/releases/2017/01/170123094726.htm. Accessed 11 April 2017
136. D'Or Institute for Research and Education (2017) Network of molecular interactions in brain cells infected by Zika virus reveals new therapeutic targets. Science News, January 23. https://www.sciencedaily.com/releases/2017/01/170123094726.htm. Accessed 11 April 2017; Garcez Patricia et al (2017) Zika virus disrupts molecular fingerprinting of human neurospheres. Nature. https://www.nature.com/articles/srep40780. Accessed 14 April 2017
137. Aragão MFVV et al (2017) Neuroimaging findings of congenital Zika syndrome. In: Zika in focus: postnatal clinical, laboratorial and radiological aspects (Aragão MFVV, ed). Springer, Washington, DC
138. Aragão MFVV et al (2017) Neuroimaging findings of congenital Zika syndrome. In: Zika in focus: postnatal clinical, laboratorial and radiological aspects (Aragão MFVV, ed). Springer, Washington, DC; Aragão MdFV et al (2016) Clinical features and neuroimaging (CT and MRI) findings in presumed Zika virus related congenital infection and microcephaly: retrospective case series study. Br Med Jo 353:i1901. https://doi.org/10.1136/bmj.i1901. Accessed 6 July 2018; de Oliveira Melo AS et al (2016) Congenital Zika virus infection: beyond neonatal microcephaly. JAMA Neurol 73(12). https://www.ncbi.nlm.nih.gov/pubmed/27695855. Accessed 23 July 2018
139. Brasil P et al (2016) Zika virus infection in pregnant women in Rio de Janeiro—preliminary report. N Engl J Med 374(22). http://www.nejm.org/doi/full/10.1056/NEJMoa1602412#t=article. Accessed 12 April 2017
140. AAFP (2018) Children exposed to Zika in utero need long-term monitoring, August 17. https://www.aafp.org/news/health-of-the-public/20180817mmwr-zika.html. Accessed 14 May 2019
141. Rice ME et al (2018) Vital signs: Zika-associated birth defects and neurodevelopmental abnormalities possibly associated with congenital Zika virus infection—U.S. territories and freely associated states, 2018. Morb Mortal Wkly Rep 67(31). https://www.cdc.gov/mmwr/volumes/67/wr/mm6731e1.htm?s_cid=mm6731e1_w. Accessed June 13, 2019.
142. Leonard K (2017) Zika caused serious birth defects in 120 babies in US territories, CDC says. Washington Examiner, June 8. http://www.washingtonexaminer.com/zika-caused-serious-birth-defects-in-120-babies-in-us-territories-cdc-says/article/2625370. Accessed 27 June 2017
143. Shapiro-Mendoza CK et al (2017). Pregnancy outcomes after maternal Zika virus infection during pregnancy—U.S. territories, January 1, 2016–April 25, 2017. Morb Mortal Wkly Rep. https://www.ncbi.nlm.nih.gov/pubmed/28617773. Accessed 20 July 2017

144. AAFP (2018) Local Zika transmission in U.S. Tied to rise in birth defects. AAFP.org, January 30. Accessed 26 April 2018. https://www.aafp.org/news/health-of-the-public/20180130mmwrzika.html. Accessed 27 June 2018
145. Rosen M (2016) Microcephaly: building a case against Zika. Science News, March 4 (reprinted April 2). https://www.sciencenews.org/article/microcephaly-building-case-against-zika. Accessed 2 June 2017
146. Honein M et al (2016) Birth defects among fetuses and infants of US women with evidence of possible Zika virus infection during pregnancy. JAMA. http://jamanetwork.com/journals/jama/fullarticle/2593702. Accessed 14 May 2017
147. Miranda-Filho DdB et al (2016) Initial description of the presumed congenital Zika syndrome. Am J Public Health 106(4). https://www.ncbi.nlm.nih.gov/pubmed/26959258. Accessed 23 May 2017
148. Neergard L (2017) Zika virus: the threat continues. News Tribune, July 6. http://www.newstrib.com/free/zika-virus-the-threat-continues/article_72fd02a2-6290-11e7-ab20-a39d99f70d7b.html. Accessed 21 July 2017
149. Roy P (2017) Zika virus may produce abnormalities in fetus: 1 out of 10 babies with prenatal exposure in danger. Science Times, May 7. http://www.sciencetimes.com/articles/14495/20170507/zika-virus-may-produce-abnormalities-in-fetus-1-out-of-10-babies-with-prenatal-exposure-in-danger.htm. Accessed 27 June 2017
150. Mulkey S (2017) Three of 48 fetuses exposed to Zika in utero had abnormal fetal magnetic resonance imaging. Children's National, May 4. https://childrensnational.org/news-and-events/childrens-newsroom/2017/three-of-48-fetuses-exposed-to-zika-in-utero-had-abnormal-mri. Accessed 10 March 2023
151. Gulland A, Majid A (2018) Infection with Zika in early weeks of pregnancy poses high risk to baby. The Telegraph, March 14. https://www.telegraph.co.uk/news/2018/03/14/infection-zika-early-weeks-pregnancy-poses-high-risk-baby/. Accessed 27 June 2018
152. Chang D (2018) More birth defects reported in states with local spread of Zika, including Florida. Miami Herald, January 25. https://www.miamiherald.com/news/health-care/article196616159.html. Accessed 27 June 2018
153. de Oliveria WK et al (2017) Zika virus infection and associated neurologic disorders in Brazil. N Engl J Med. http://www.nejm.org/doi/full/10.1056/NEJMc1608612#t=article. Accessed 4 May 2017
154. PAHO WHO (2016) Zika—epidemiological update, December 29. http://www.paho.org/hq/index.php?option=com_content&id=11599&Itemid=41691. Accessed 10 Jan 2017
155. Lafrance A (2006) The danger of a third trimester Zika infection. The Atlantic, June 29. https://www.theatlantic.com/health/archive/2016/06/zika-after-30-weeks/489284/. Accessed 18 May 2017
156. Dudley DM et al (2018) Miscarriage and stillbirth following maternal Zika virus infection in nonhuman primates. Nat Med. https://www.nature.com/articles/s41591-018-0088-5. Accessed 24 July 2018
157. Sarno M et al (2016) Zika virus infection and stillbirths: a case of hydrops fetalis, hydranencephaly and fetal demise. PLoS Negl Trop Dis 10(2):e0004517. http://journals.plos.org/plosntds/article?id=10.1371/journal.pntd.0004517. Accessed 25 July 2018
158. Branswell H (2016) Exclusive: miscarriages reported in pair of American women with Zika virus. STAT, February 10. https://www.statnews.com/2016/02/10/zika-american-women-miscarriages/. Accessed 9 March 2017
159. Prasad R (2017) GM mosquito trials to control dengue, chikungunya launched. The Hindu, January 25. http://www.thehindu.com/sci-tech/health/GM-mosquito-trials-to-control-dengue-chikungunya-launched/article17093840.ece. Accessed 14 May 2017
160. Samo M et al (2016) Zika virus infection and stillbirths: a case of hydrops fetalis, hydranencephaly and fetal demise. PLOS Negl Trop Dis. http://journals.plos.org/plosntds/article?id=10.1371/journal.pntd.0004517. Accessed 31 May 2017
161. Aliota MT et al (2017) Zika in the Americas, year 2: what have we learned? What gaps remain? A report from the Global Virus Network. Antiviral Res 114. https://www.ncbi.nlm.nih.gov/pubmed/28595824. Accessed 15 May 2019

162. Miner JJ et al (2016) Zika virus infection in mice causes panuveitis with shedding of virus in tears. Cell Rep 16. http://www.cell.com/cell-reports/abstract/S2211-1247(16)31175-5. Accessed 17 July 2017
163. Hackett DW (2018) Children born During Zika outbreaks should have eye exams. Precisionvaccinations.com, September 25. https://www.precisionvaccinations.com/zika-virus-caused-optic-nerve-and-retina-abnormalities-regardless-microcephaly-diagnosis. Accessed 13 June 2019
164. Ho F (2016) Focus on eyes: how the Zika virus damages eyes. Florida Today, December 27. http://www.floridatoday.com/story/life/wellness/2016/12/27/focus-eyes-zika-virus-damages-eyes/95822986/. Accessed 14 May 2017.
165. de Oliveira Dias JR et al (2018). Zika and the eye: pieces of a puzzle. Progress in Retinal and Eye Research. https://doi.org/10.1016/j.preteyeres.2018.04.004. Accessed 6 July 2018
166. Freitas BdP, Sacramento GA (2016) Ocular findings in Infants with microcephaly associated with resumed Zika virus congenital infection in Salvador, Brazil. JAMA Ophthalmol 134(5):529–535. http://jamanetwork.com/journals/jamaophthalmology/fullarticle/2491896. Accessed 14 May 2017
167. Ventura CV et al (2016) Zika virus in Brazil and maculae atrophy in a child with microcephaly. The Lancet. http://www.thelancet.com/journals/lancet/article/PIIS0140-6736(16)00006-4/fulltext?rss%3Dyes. Accessed 4 Sept 2016
168. Jambol L, Debra G (2016) Zika virus infection and the eye. JAMA Ophthalmol 134(5):535–536. http://jamanetwork.com/journals/jamaophthalmology/article-abstract/2491895. Accessed 14 Oct 2016
169. MedicalXpress (2017) Which infants exposed to Zika virus infection in pregnancy should have eyes examined? MedicalXpress, July 17. https://medicalxpress.com/news/2017-07-infants-exposed-zika-virus-infection.html. Accessed 21 July 2017
170. Di Guardo G, Baleeiro Beltrão Braga P, Schatzmann Peron JP (2016) Zika virus-associated brain damage: animal models and open issues. Emerg Microb Infect 5:e106. https://doi.org/10.1038/emi.2016.103
171. Mécharles S et al (2016) Acute myelitis due to Zika virus infection. The Lancet 387. https://www.thelancet.com/journals/lancet/article/PIIS0140-6736(16)00644-9/fulltext. Accessed 9 Aug 2018
172. Children's National Hospital (2021) Zika 5 years later: still much to learn as 'likely' future outbreak looms. HealioNews, January 21. https://www.healio.com/news/infectious-disease/20210114/zika-5-years-later-still-much-to-learn-as-likely-future-outbreak-looms. Accessed 25 May 2022

Chapter 9
Effects on Adults

This is a brief chapter and reflects what is known about ZIKV today. There have been few adult mortalities from ZIKV and we know only one major problem for adults is associated with Guillain-Barré syndrome (GBS). Nineteen countries and territories have reported confirmed cases of Guillain-Barré syndrome associated with ZIKV infection [1].

However, it would be short-sighted to suggest that all the effects on adults are known. There are good reasons to believe that some long-term implications may surface over time as children and adults mature.

9.1 Research Findings

ZIKV did not seem to have significantly impacted adult populations compared to its effects on fetuses and young children (see Fig. 9.1). The symptoms seem minimal. However, acute inflammatory demyelinating polyneuropathy, acute motor axonal neuropathy, and the Miller-Fisher syndrome (a subset of the Guillain-Barré syndrome characterized by ophthalmoplegia [paralysis of the extraocular muscles that control the movements of the eye], ataxia [a group of disorders that affect coordination, balance, and speech], and areflexia [the absence of deep tendon reflexes]) have been observed with ZIKV-associated Guillain-Barré syndrome. Still, the relative proportions of these subtypes vary across studies and regions [2].

9.1.1 Guillain-Barré Syndrome

Guillain-Barré Syndrome (GBS) is an immune-mediated progressive flaccid paralysis. GBS damages parts of nerves and damage causes tingling, muscle weakness, and paralysis. GBS most often affects the nerve covering (myelin sheath). This damage is

D. M. Berube, *Pandemics and Resilience: Lessons we should have learned from Zika*, Risk, Systems and Decisions, https://doi.org/10.1007/978-3-031-25370-6_9

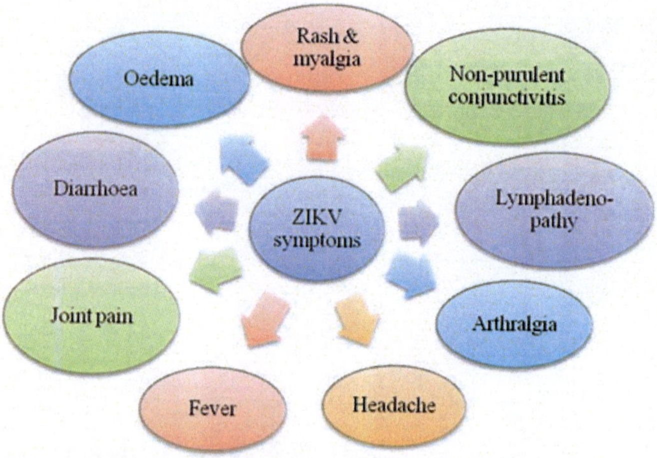

Fig. 9.1 Symptoms of Zika infection. Kumari et al. [3]

demyelination (any condition that causes damage to the protective covering (myelin sheath) that surrounds nerve fibers in the brain, the nerves leading to the eyes (optic nerves), and the spinal cord). It causes nerve signals to move more slowly. Damage to other parts of the nerve can cause the nerve to stop working [4]. GBS is an acute, immune-mediated polyradiculoneuropathy (a slowly developing autoimmune disorder in which the body's immune system attacks the myelin that insulates and protects the body's nerves) typically occurring after minor viral and bacterial infections and characterized by rapidly progressive symmetrical weakness, areflexia, sensory disturbances, and involvement of cranial nerves that can cause death [5].

GBS described a broad spectrum of acute autoimmune neuropathies [6]. Death is rare and is usually caused by respiratory failure, autonomic dysfunction, or deep vein thrombosis [7]. The syndrome begins as the body's immune system attacks peripheral nerves, often causing weakness or tingling in the lower extremities. In severe cases, paralysis can result. Medical complications are common in patients with severe GBS. Venous stasis in paralyzed limbs places patients at risk for deep vein thrombosis [8].

According to Barbi [9], GBS has a mortality rate of approximately 5%, and 20% of the affected patients suffer a significant disability. Moreover, whereas GBS mortality occurs in around 5% of the cases, this number rises to 10–35% if the GBS-affected person is a pregnant woman [9].

Recovery can take weeks, months, or years. Most people survive and recover completely. In some people, mild weakness may persist [4]. According to the National Organization for Rare Disorders, the prognosis is favorable. However, ZIKV-associated Guillain-Barré syndrome results in higher morbidity and more frequent cranial neuropathy [10].

Death from GBS is rare in countries with intensive care facilities, occurring in fewer than 5% of patients. The overall prognosis of GBS is favorable. Not all patients with GBS show complete recovery; however, most patients have a good outcome on a one-year follow-up [11].

Most people recover completely. However, many have mild, residual effects such as foot drops or abnormal sensations for more than two years.

Scientists estimate that while most recover, 3–5% of people die from complications [12]. The clinical outcomes of patient samples with ZIKV and the Guillain-Barré syndrome were favorable. Cor-Lamoneau et al. recommend that their findings be considered in intensive care bed decision-making [13]. GBS is the leading cause of acute ascending muscle paralysis, with an annual incidence of 1–2 per 100,000 persons worldwide [14].

Persistent fatigue and pain may be problematic. Fewer than 15% have a substantial long-term disability requiring a walker or wheelchair.

However, that is not universally true. NBC reported in July 2016, a man in his forties from the San Juan metro area died from Guillain-Barré [15]. While he was obese, he had no other health conditions [16]. Even in countries with advanced health systems, despite immunotherapy, around 5% of patients with the syndrome die [17].

The potential cost of missing the signs and symptoms of early GBS can be severe, as some patients experience rapid worsening within hours to a few days of onset [8]. An epidemic of GBS associated with ZIKV would stress hospital and intensive care resources nationally. For example, up to 20% of patients with severe GBS cannot walk independently for six months after symptoms. A surge in cases may saturate acute inpatient rehabilitation units [8].

9.1.1.1 Linkage

It was first associated with ZIKV in the 2014 outbreak in French Polynesia. A case-control study of the French Polynesia ZIKV outbreak in 2013–2014 confirmed a link between ZIKV infection and Guillain-Barré syndrome (1 per 4000 infections) [13]. During the outbreak in French Polynesia, a 20-fold increase in the incidence of GBS was observed and concerns about an association between ZIKV infection and GBS were first raised [18]. Benoît Rozé et al. [19] from the University Hospital of Martinique reported two cases of Guillain-Barré syndrome who had concomitant ZIKV viruria (living viruses in the urine). This viruria persisted for longer than 15 days after symptom onset. The cases occurred in Martinique in January 2016, beginning the ZIKV outbreak [19].

However, the best evidence for an association seems to come from a metanalysis by Krauer et al. [20].

> For GBS, we assessed thirty-six studies that addressed questions about one or more causality dimensions. In several countries in the Americas and the Pacific region, a temporal association has been found with symptoms of ZIKV infection preceding the onset of GBS. In these countries, surveillance reports of cases of GBS followed the pattern of reports of ZIKV-like illness. During a ZIKV outbreak in French Polynesia in 2013–14, scientists estimated that around one in four thousand people with ZIKV infection developed GBS. The odds of having

had a recent ZIKV infection were more than thirty times higher in patients with GBS than those without in a hospital-based study in French Polynesia. [20]

The information generated from studies of the outbreak in Colombia corroborated these findings. From November 2015 through March 2016, clusters of Guillain-Barré syndrome cases were observed during the outbreak of ZIKV infection in Colombia. A total of sixty-six patients (97%) had symptoms compatible with ZIKV infection before the Guillain-Barré syndrome. The median period between the onset of symptoms of ZIKV infection and symptoms of the Guillain-Barré syndrome was seven days (interquartile range from 3 to 10). Among the 68 patients with the Guillain-Barré syndrome, 50% were found to have bilateral facial paralysis on examination [21]. The increase in Guillain-Barré syndrome cases during the ZIKV outbreak in Colombia and the absence of such an increase, while the epidemiologic dengue virus and chikungunya virus were circulating within the region in previous years, evidence of the link between ZIKV infection and the Guillain-Barré syndrome [21].

The frequencies reported in the literature are not consistent. For example, according to Monel [22], the frequency of Guillain-Barré syndrome was estimated to be one per 5000 cases of ZIKV infection (or 0.5%) [22], a factor less than Barbi and others. According to Noor and Ahmed [23] and Besnard and associated (2018) [23], GBS incidents increase five times when active ZIKV infection prevails [23].

Epidemiological evidence from an earlier outbreak of GBS set the stage for claims associated with ZIKV and GBS. If conditions were to mimic the French Polynesian episode [8], then as many as 30,000 cases of ZIKV-associated GBS might be expected (population 318 million; 60% of the people at risk for infection; 66% infection rate in those at risk; 0.24 cases/1000 ZIKV infections), representing a roughly tenfold increase from the baseline incidence of GBS nationally [8]. In Venezuela, the occurrence of GBS demonstrated a gender bias for men. Since women more than men reported ZIKV infections, which might be because women are more aware of the risks because of the pregnancy link, reporting might have been skewed. However, men are 28% more likely to develop Guillain-Barré [24].

Morris, Lanciotti, Parra, and many others have tried to explain the linkage. They report that while it is possible that the rapid onset of GBS following ZIKV infection could be driven by molecular mimicry before the start of symptoms, profound immune dysregulation resulting from direct neuroinvasiveness and prolonged illness as evidenced by the presence of replicating ZIKV in the cerebral spinal fluid of GBS sufferers extending well beyond the symptomatic and viremic phases could also play a pathophysiological role. In addition, there seems to be some evidence indicating that immune responses evoked by acute ZIKV infection, particularly in terms of the type 1 interferon response, differ from that of other mosquito-borne viruses, which may contribute to the emergence of GBS [25].

The accumulated evidence suggests a link between ZIKV infection and illness and GBS. The World Health Organization stated, "sufficient evidence to conclude that ZIKV is a cause of congenital abnormalities and is a trigger of GBS." This conclusion was based on a systematic review of the evidence published on 30.05.2016. Since then, the body of evidence has grown, leading to this systematic review update with new evidence published from 30.05.2016 to 18.01.2017, update one. This review confirms previous conclusions that ZIKV is a cause of congenital abnormalities, including microcephaly, and is a trigger of GBS. The transition to living systematic review techniques and methodology provides a proof of concept for using these methods to synthesize evidence about an emerging pathogen such as ZIKV [26].

During the ZIKV epidemic, Gorgora-Rivera et al. [27] indicated that increases in GBS would occur among those with antecedent symptomatic ZIKV. ZIKV may be associated with phenotypic presentations of GBS. ZIKV-associated GBS has been associated with more dysautonomia (a group of medical conditions caused by problems with the autonomic nervous system), facial nerve palsy, and more rapid onset of clinical GBS signs [27].

The explanation of available evidence from ZIKV infection and GBS outbreaks is that ZIKV virus infection triggers GBS [20].

9.1.1.2 Incidences

GBS has usually seen in about every 5000–10,000 infections like ZIKV [24]. This estimate represents an annual incidence of 5.5–8.7 cases/100,000 population, 3.2–5.1 times the baseline incidence. Associated healthcare resource needs will increase accordingly. Estimated long-term care needs in 2016 were predicted to be 3–5 times greater than in years with no ZIKV transmission [7].

A 2016 study in the New England Journal of Medicine evaluated over 160,000 confirmed and suspected cases of ZIKV and over 1400 cases of Guillain-Barré in six countries in South America and found a close association between increases in both cases [28]. Following the ZIKV outbreak in northeastern Brazil, a cluster of GBS cases was identified and reported by Do Rosio et al. [29] Brazil, Venezuela, El Salvador, and Martinique reported cases of Guillain-Barré seemingly associated with ZIKV. Guillain-Barré has been linked to the ZIKV [30]. An association between ZIKV circulation and an increased incidence of GBS has been reported in many different countries, in Southern America and the West Indies. ZIKV is the only circulating flaviviruses that add weight to this presumed association in some countries [31].

At the population level, 11 countries in Latin America (Brazil, Colombia, El Salvador, French Guiana, Honduras, Venezuela, Suriname), the Caribbean (Dominican Republic, Jamaica, Martinique), and French Polynesia have reported an increase in GBS cases during outbreaks of Zika infection [20]. From April 1, 2015, to March 31, 2016, 164,237 confirmed and suspected cases of ZIKV disease and 1474 cases of the Guillain-Barré syndrome were reported in Bahia, Brazil; Colombia; the Dominican Republic; El Salvador; Honduras; Suriname; and Venezuela [32]. In July 2015, Brazil reported 76 cases of ZIKV infection with neurologic syndromes in Bahia State, and 42 of those cases were confirmed to have GBS. In late January 2016, 104 cases of GBS were reported from El Salvador, two from Martinique, one from New Zealand, and 255 from Venezuela [33].

During the weeks of ZIKV transmission, "there were significant increases in the incidence of the Guillain-Barré syndrome, as compared with the pre-ZIKV baseline incidence, in (Brazil's) Bahia State (an increase of 172%), Colombia (211%), the Dominican Republic (150%), El Salvador (100%), Honduras (144%), Suriname (400%), and Venezuela (877%)," Brazilian doctors wrote. "When the incidence of ZIKV disease increased, so did the incidence of the Guillain-Barré syndrome." [34].

Weekly reports of hospital cases reveal that GBS incidence was markedly higher in the northeast region in 2015 and 2016 and in other areas of Brazil in 2016 than in the years before the ZIKV epidemic (2010–2014) [35]. From January 1 to July 31, 2016, 56 cases of GBS were reported in Puerto Rico; evidence of ZIKV or flavivirus infection was found in 34 (61%) of these. As in other locations, GBS cases in Puerto Rico are anticipated to increase with ongoing ZIKV transmission [7]. In the end, Puerto Rico reported thirty-four cases of Guillain-Barré [36].

However, Ernesto Marques of the University of Pittsburgh stresses that "it's important that people don't think that if you get ZIKV, you will get Guillain-Barré." The chance is much less than 1% [37]. Nonetheless, instances of Guillain-Barré seem to have skyrocketed in countries hit by ZIKV epidemics, health officials, and researchers from Latin America reported [24].

Systematic reviews and meta-analyses of studies of ZIKV and GBS by Barbi et al. [38] have been published. In a recent meta-analysis, from a total pooled number of 164,651 ZIKV-infected individuals, 1513 developed ZIKV-associated GBS, 1.23% (95% CI = 1.17–1.29%) [38].

9.1.1.3 Chikungunya Is a Cofactor

Like the underreporting of microcephaly incidence, they were reporting GBS has been problematic. This problem was addressed in an important letter to the New England Journal of Medicine in 2017. PAHO and some research leaders identified possibilities for underreporting overall, with one explicitly explaining the drop in GBS cases.

They argued, "in 2016, infections attributed to ZIKV and linked to an increase in the incidence of GBS were caused by another arbovirus that *Aedes* also transmits. *Aegypti* mosquitoes since there was herd immunity against ZIKV infection after widespread infection in 2015." They also suggested chikungunya was to blame. "Chikungunya virus was introduced into Brazil in 2014 and has caused successively larger epidemics in the northeast region in 2015 and 2016. Chikungunya is a cause of GBS, and some chikungunya infections were misclassified as ZIKV infection in Pernambuco in 2016."

The team believed this reason was most plausible. "The cases in the first year, back in 2015, were ZIKV cases. And that is why we saw the microcephaly in 2015. But in 2016, it was chikungunya, not ZIKV, which is why we saw Guillain-Barré, but not microcephaly." [39].

9.1.2 Male Infertility

Claims associated with male infertility rose as well. There have been very few studies linking ZIKV to infertility in men. In 2–17, Moley says a CDC study is underway in men in Puerto Rico examining a link between ZIKV, sperm motility, and a decrease in testosterone levels [40].

According to Meinhardt [41], ZIKV virus infections affect developing fetuses in pregnant women and can be a threat to fertility in men. Whether this threat could be managed or mitigated remains uncertain [41]. His research was drawn from a mice study. He showed the virus' surprise effect on male fertility, including lower testosterone levels, sperm counts, and, in some animals, dramatically shrunken testicles [42]. In mice, infection of spermatozoa is fast and extensive, and spermatogenesis eventually ceases.

> Prompted by previous studies in mice, in which the highest ZIKV load was found in the testes, the detection of ZIKV in semen, and data that show male-to-female and male-to-male transmission in humans in a longitudinal study, initial damage to spermatogenesis was observed just 14 days after infection, with germ cell loss and impairment of Sertoli cell (cells that nurture the immature sperm) function indicated by decreased levels of serum inhibin B, a Sertoli cell functional marker. [43]

If the virus persisted, progressive damage was observed until, on day twenty-one, spermatogenesis was wholly eradicated. Most germ cells were lost, with only the robust Sertoli cells persisting in the tubules. ZIKV infection has been identified as a previously unknown threat to male fertility in mice.

> In mice, infection of spermatozoa is fast and extensive, and spermatogenesis eventually ceases. Data indicate a high infectious viral load in semen and established sexual transmission from men to women. However, sperm in the ejaculate suggests a different, less radical pattern of spermatogenic damage in humans compared with mice. Thus, fertility in men could be at risk owing to the transmission of the disease by ZIKV-infected spermatozoa during natural conception or ART. In contrast, infertility by non-obstructive azoospermia—as seen in mice—could be less of a human problem. [43]

Another study led by Fikrig, and colleagues (2017) [44] "demonstrated how the ZIKV replicates in and damages testes," said first author Ryuta Uraki. The persistence of the virus in a storage compartment known as the epididymis, which conveys sperm from the testicle to the urethra, is consistent with the reported cases of male-to-female sexual transmission, he explained. Reduced testicular size—testicular atrophy—indicates a potentially long-term effect on male fertility. "These results suggest that infection can cause a reproductive deficiency in males," Uraki noted [44].

At the same time, physician-scientist Kelle Moley is researching the impact of ZIKV on men. Doctor Moley examined the reproductive systems of ZIKV-infected mice as well. "By day 21, we saw no germ cells, so basically, this would imply that it would lead to infertility if it had the same effect," said Dr. Moley.

9.1.3 Epilepsy

On epilepsy, the CDC for Disease Control and Prevention warned that cases of epilepsy caused by the virus might be misdiagnosed or underreported.

ZIKV's effects on a developing brain are like those of other central nervous system infections associated with epilepsy, according to the CDC, whose article cited two research reports conducted in Brazil in 2016 that suggest a link [45]. However, because epilepsy symptoms vary, the condition is often diagnosed based on caregivers' symptom descriptions rather than from a medical examination. If parents and healthcare professionals do not recognize the symptoms of epilepsy in children, ZIKV-linked cases may not be diagnosed and reported appropriately [46].

Seizures have been reported after ZIKV infection in adult patients, and studies have shown that 50–60% of microcephalic babies due to ZIKV congenital infection develop a spontaneous epileptic activity. Seizures associated with changes in sleep electroencephalographic patterns were found in these babies, although in some cases, electrographic seizures were identified even in the absence of other clinical manifestations. In addition, follow-up studies performed with normocephalic babies exposed in utero to ZIKV suggest seizures are clinical manifestations in ~ 50% of babies. A high incidence (~ 95%) of seizures was found in neonatal mice exposed to ZIKV [47].

In two 2016 case series reports [48], seizures and epilepsy were reported in some infants with probable congenital ZIKV infection. In a case series of forty-eight infants from Brazil with congenital ZIKV syndrome, 50% had clinical seizures. Among forty-eight babies from Brazil with probable congenital ZIKV infection, "50% reportedly had clinical seizures," according to physicians and public health experts Pastula, Yeargin-Allsopp, and Kobau [49]. They studied ZIKV at the CDC and are primarily responsible for the ZIKV-epilepsy connection.

According to the same CDC team, besides, forty-eight infants already cited, seven of another group of 13 ZIKV-exposed babies in Brazil were also diagnosed with epilepsy. The team noted that the finding is not overly surprising since the types of brain abnormalities seen in ZIKV-affected newborns have been linked to seizures and epilepsy in the past [50].

In another case series, 4 of 13 infants in Brazil with laboratory evidence of congenital ZIKV infection had cortical malformations, all had subcortical and or/basal ganglia intracerebral calcifications, 92% had decreased brain volume, and 83% had ventriculomegaly on neuroimaging. Seven (54%) were diagnosed as having epilepsy [51].

In a case series 6 of 19 children with previous congenital cytomegalovirus infection, 17 (89%) had abnormal brain neuroimaging (e.g., intracerebral calcifications, ventriculomegaly, white matter abnormalities, and cortical malformations, such as schizencephaly (rare congenital (present from birth) brain malformation in which abnormal slits or clefts form in the cerebral hemispheres of the brain) or polymicrogyria [a condition characterized by abnormal development of the brain before birth]), and epilepsy eventually developed in 7 (41%) of these 17 (at a mean [range] age of 20.7 [2–37] months). [52]

Also, the brain damage done to children should they reach adulthood may manifest in a series of problems, such as "epilepsy, personality changes, depression, and dementia." [53]. Epilepsy is associated with considerable morbidity and costs, and early recognition and treatment of epilepsy may mitigate some adverse outcomes associated with developmental delay [54].

9.1.4 Acute Disseminated Encephalomyelitis (ADEM)

ZIKV may be associated with ADEM (acute disseminated encephalomyelitis), according to a small study presented at the American Academy of Neurology's 68th Annual Meeting in Vancouver, Canada, from April 15 to 21, 2016 [55]. Presumably, two people from Brazil developed a neurological disorder called acute disseminated encephalomyelitis (ADEM) [56]. However, little is reported on this occurrence and whether additional cases have surfaced.

ADEM is a brief but intense attack of inflammation in the brain and spinal cord that damages the myelin and can be difficult to distinguish from multiple sclerosis because of common symptoms (loss of vision, weakness, numbness, and loss of balance) [57].

In a cohort follow-up of six patients in Pernambuco, Brazil, Ferreira (2016) reported that the symptom presentation was fever followed by a cutaneous rash. Some patients also had pruritus, myalgia, arthralgia, and conjunctival hyperemia. Neurologic manifestations appeared from zero to 15 days after the first clinical symptoms. MRI white matter lesions were in two acute disseminated encephalomyelitis cases and elevated protein concentration with average cell count in four Guillain-Barré syndrome cases. Blood and chronic fatigue syndrome molecular tests for arboviruses were positive only for ZIKV. After hospital discharge, five patients had sustained motor dysfunction, one patient had low visual acuity, and another had a cognitive decline [58].

"At present, it does not seem that ADEM cases are occurring at a similarly high incidence as the GBS cases, but these findings from Brazil suggest that clinicians should be vigilant for the possible occurrence of ADEM and other immune-mediated illnesses of the central nervous system," said James Sejvar, MD, with the Centers for Disease Control and Prevention in Atlanta and member of the American Academy of Neurology. "Of course, the remaining question is 'why'–why does ZIKV appear to have this strong association with GBS and potentially other immune/inflammatory diseases of the nervous system? Hopefully, ongoing investigations of ZIKV and immune-mediated neurologic disease will shed additional light on this important question." [55].

9.1.5 Encephalitis

There has been a smattering of case reports linking ZIKV to other adult neurological disorders. For example, a woman infected with ZIKV died in Brazil from brain inflammation or encephalitis early this year [59]. Other adults have been hospitalized with the spinal cord, brain, and surrounding tissue inflammation [60].

Carteaux et al.'s case study of central nervous system infection with ZIKV associated with meningoencephalitis in an adult concluded clinicians should be aware that ZIKV may be associated with meningoencephalitis [60]. Li et al.'s [61] result from a mice study demonstrated blood-borne ZIKV administration could lead to pronounced evidence of ZIKV infection in adult neural stem cells, leading to cell death and reduced proliferation. The authors' research suggested that ZIKV infection can enter the adult brain, leading to neuropathology in mammals [61].

9.1.6 Other Inflammatory Diseases

The list of inflammatory diseases affecting the brain and nervous systems attributable to ZIKV is extensive though the frequencies of these disorders have not been reported.

Researchers found that ZIKV can cause other severe neurological problems, such as encephalitis (inflammation of the brain), meningoencephalitis (inflammation of the brain and the membrane that covers the brain and spinal cord), myelitis (inflammation of the spinal cord), neuromyelitis optica (inflammation of the spinal cord and the optic nerve), and acute disseminated encephalomyelitis (a rare autoimmune disease with inflammation in the brain and spinal cord) [62]. Other autoimmune disorders, such as thrombocytopenic purpura (a blood disorder characterized by a decrease in the number of platelets in the blood0, have also been associated with ZIKV infection [2].

A team from Oregon State and Harvard [63] suggested a reason. Weeks after the virus disappears from the bloodstream, it lingers in the lymph nodes and the central nervous system of rhesus monkeys. At the same time, the virus disappeared from the monkeys' bloodstreams after ten days. But it remained for as long as 42 days in cerebrospinal fluid, which circulates throughout the brain, and up to 72 days in the lymph nodes. That could help explain why ZIKV infection can cause neurological problems in infants (see the previous chapter) and adults [64].

Early in gestation, before the brain has developed into a complex organ with specialized zones, it is comprised entirely of neural progenitor cells. With the capability to replenish the brain's neurons throughout its lifetime, these are the brain's stem cells. In healthy individuals, neural progenitor cells eventually become fully formed neurons. It is thought that at some point along with this progression, they become resistant to ZIKV, explaining why adults appear less susceptible to the disease. But current evidence suggests that ZIKV targets neural progenitor cells, leading to loss of these cells and reduced brain volume. This closely mirrors what is seen in microcephaly, a developmental condition linked to ZIKV infection in developing fetuses that results in a smaller than normal head and a wide variety of developmental disabilities [65].

"ZIKV can cause damage in the adult brain," said Joseph Gleeson, adjunct professor, head of the laboratory of pediatric brain disease at New York's Rockefeller University, and an author of the Li study [66]. Gleeson reported that ZIKV affects the specific cells (stem cells known as neural progenitors) in adults as it does in fetuses [67].

9.2 Comorbidity Considerations

Individuals with compromised immunity were more susceptible to severe symptoms and maladies than healthier people. Sujan Shresta, an associate professor at La Jolla Institute for Allergy and Immunology, pointed out that people with immune deficiencies could prove similarly vulnerable. "We have HIV patients, cancer patients. If for some reason these people are affected by ZIKV, there is a chance that it will get into the brain," she said [68]. "Based on our findings, getting infected with ZIKV as an adult may not be as innocuous as people think," said Joseph Gleeson, head of the Laboratory of Pediatric Brain Disease at Rockefeller University. He warns specific adult brain cells may be vulnerable to infection as well. Among these are populations of cells that serve to replace lost or damaged neurons throughout adulthood critical to learning and memory. Coauthor Yale's Marco Muotri believes that if the virus gets into adult brains, it could have "consequences for memory and learning. Similarly, the virus could trigger an inflammatory response accumulating over time, damaging the brain." [69]. Noor et al. warned patients with underlying sickle cell disease can show life-threatening conditions if diagnosed with ZIKV, chikungunya, and dengue [70].

Few ZIKV fever-related fatalities among adult populations have been recorded. There are two reasons: First, earlier outbreaks were less egregious, and second, diseases associated with ZIKV were misdiagnosed.

However, two such occurrences appeared in the literature: one in a 15-year-old Colombian female with sickle cell disease and a concurrent ZIKV infection and another in a 70-year-old Puerto Rican male who developed severe thrombocytopenic purpura [71] as a rare complication of ZIKV infection [72].

9.3 Social Poverty

"Too often, these diseases affect only the poorest among the poor, in remote, isolated areas that local governments pay very little attention to," computational biologist Guido Núñez-Mujica and a researcher at the Engineering School of the University of Development, UDD in Santiago, Chile says. "And while many of these diseases kill in dramatic and terrible ways, many others are slow and invisible, so the real magnitude of the problem is hidden." [73].

The physiological effects are only one facet of the epidemic's long-term impact on adult communities. The expenses associated with combatting an infectious disease, caring for those who grow ill, and long-term assistance for those suffering from developmental difficulties. Children with microcephaly or other developmental disordered are going to need specialized care—"not only a doctor or a pediatrician," Marcos Espinal, director of infectious diseases for the Pan American Health Organization, says, "it's going to require a neurologist, but it's also going to require special neurosurgeons and so on for special situations." [74]. These consequences are expected to substantially add to the affected countries' short-term and long-term economic burdens [75]. For example, even the cautious warning from the CDC to restrict travel to affected countries came with a cost for the tourism industry in the West Indies and South America.

ZIKV could be a real economic drain, especially since it is expected to hit impoverished areas hardest and in locations with some of the poorest public health and vector control funding. Mosquito control requires outreach and educational programs to get local populations to help reduce the places where mosquitoes lay their eggs, expensive larvicides and insecticides used to manage people, and the costs of distributing these to where they will maximize impacts, and efforts to maintain mosquito management programs annually.

On a different level, reducing birth rates from women delaying pregnancies will have social effects on these countries. The result will be short-term, with women compensating for delayed pregnancies after the threat settles. However, if birth rates decline, there will be an associated impact on growth and development goals in these developing economies as fewer workers enter the labor market after a decade and a half.

9.4 Conclusions

Extraordinarily little is still known about the long-term impact of ZIKV on the adult communities in the West Indies and South America. As children with developmental disorders age, their disabilities tend to worsen. There is also no idea what impact someone who had shown no, or mild symptoms might express as they age.

How frequently do asymptomatic or second- and third-trimester infections lead to central nervous system diseases? Is ZIKV-induced GBS the specific immune-mediated disease, or is there a direct virus invasion? What are the long-term sequelae of intrauterine ZIKV infection? What is the reason for the substantial size, severity, and unexpected complications of the recent ZIKV outbreak in the Americas compared with what has been seen with this virus in the past? Is it a result of immunological enhancement from prior dengue virus exposure in this population, or has there been a critical sequence change in the virus genome, making it more neurovirulent? Or a proliferation or increase in the range of mosquitos in the area?

And a broader question: How many congenital disabilities presently of the "unclear cause" will be found to be virus-induced? [76].

It is a dangerous gamble to think ZIKV is behind us with so many unanswered questions. Many of the 2015–2016 pandemic results might be yet to surface. And over the next few decades, ZIKV might appear elsewhere. Next, to examine is the effectiveness of the warnings posted by health providers and government agencies, how effective they may have been, and what should have been learned.

References

1. PAHO WHO (2016) Zika—epidemiological update, December 29. http://www.paho.org/hq/index.php?option=com_content&id=11599&Itemid=41691. Accessed 10 Jan 2017
2. Musso D, Ko AI, Baud D (2019) Zika virus infection—after the pandemic. N Engl J Med 381:1444–1457. https://doi.org/10.1056/NEJMra1808246
3. Kumari S, Kumar S, Dhamija I (2016) Zika virus: an informative note. Crit Rev Pharmaceut Sci 5(1):9. http://earthjournals.in/crps_167.pdf. eISSN 2319-1082
4. MedlinePlus (2016) Guillain-Barré syndrome. Medline Plus, October 5. https://medlineplus.gov/ency/article/000684.htm. Accessed 23 May 2017
5. de Oliveira Dias, João Rafael et al (2018) Zika and the eye: pieces of a puzzle. Progr Retinal Eye Res. https://doi.org/10.1016/j.preteyeres.2018.04.004. Accessed 6 July 2018
6. Wakerley BR, Uncini A, Yuki N (2014) Guillain-Barré and Miller Fisher syndromes: new diagnostic classification. Nat Rev Neurol 10:537–544. https://www.ncbi.nlm.nih.gov/pubmed/25072194. Accessed 11 April 2017
7. Dirlikov E et al (2016) Guillain-Barré syndrome and healthcare needs during Zika virus transmission, Puerto Rico, 2016. Emerg Infect Dis 23(1). https://www.ncbi.nlm.nih.gov/pubmed/27779466. Accessed 11 April 2017
8. Gold C, Josephson SA (2016) Anticipating the challenges of Zika virus and the incidence of Guillain-Barré syndrome. J Am Med Assoc Neurol 73(8). http://jamanetwork.com/journals/jamaneurology/fullarticle/2526494. Accessed 27 June 2017
9. Barbi L et al (2018) Prevalence of Guillain-Barré syndrome among Zika virus infected cases: a systematic review and meta-analysis. Braz J Infect Dis 22(2). https://www.ncbi.nlm.nih.gov/pubmed/29545017. Accessed 15 May 2019
10. Muñoz LS, Parra B, Pardo CA (2017) Neurological implications of Zika virus infection in adults. J Infect Dis 216(Suppl 10):S897–S905; Dirlikov E, Major CG, Medina NA et al (2018) Clinical features of Guillain-Barré syndrome with vs without Zika virus infection, Puerto Rico, 2016. JAMA Neurol 75:1089–1097
11. National Organization for Rare Disorders (2017) Guillain-Barré syndrome. Rare Diseases.org. https://rarediseases.org/rare-diseases/guillain-barre-syndrome. Accessed 15 May 2017
12. Sanders L, Rosen M (2016) Scientists track Zika's link to neurological disorder. Science News, April 2, pp 26–29. https://www.sciencenews.org/article/microcephaly-building-case-against-Zika. Accessed 27 Sept 2016
13. Cao-Lormeau V-M et al (2016) Guillain-Barré syndrome outbreak associated with Zika virus infection in French Polynesia: a case-control study. The Lancet 387:1531–1539
14. Jadah RHS (2020) Guillain-Barre syndrome and Miller Fisher variant in Zika virus disease. In: Current concepts in Zika research (Rodriguez-Morales AJ, ed). https://www.intechopen.com/chapters/72694. Accessed 1 June 2022
15. AP (2016) Man dies from Zika-related paralysis in Puerto Rico. NBC News, August 19. http://www.nbcnews.com/storyline/Zika-virus-outbreak/man-dies-Zika-related-paralysis-puerto-rico-n634776. Accessed 14 April 2017

16. AP (2016) Man dies from Zika-related paralysis in Puerto Rico. NBC News, August 19. http://www.nbcnews.com/storyline/Zika-virus-outbreak/man-dies-Zika-related-paralysis-puerto-rico-n634776. Accessed 1 Feb 2017
17. Campisi J (2016) Zika dangers could threaten more than just infants, scientists say. Miami Herald, October 12. http://www.miamiherald.com/news/nation-world/national/article107702257.html. Accessed 15 March 2017; Soares de Oliveira-Szejnfeld P et al (2016) Congenital brain abnormalities and Zika virus: what the radiologist can expect to see prenatally and postnatally. Radiology 281(2). http://pubs.rsna.org/doi/full/10.1148/radiol.2016161584. Accessed 26 Oct 2016
18. Paploski IAD et al (2016) Time lags between exanthematous illness attributed to Zika virus, Guillain-Barré syndrome, and microcephaly, Salvador, Brazil. Emerg Infect Dis 22(8). https://wwwnc.cdc.gov/eid/article/22/8/16-0496_article. Accessed 19 July 2017
19. Rozé B et al (2016) Zika virus detection in urine from patients with Guillain-Barré syndrome on Martinique, January 2016. Euro Surveillance 21(9). https://www.ncbi.nlm.nih.gov/pubmed/26967758. Accessed 11 Aug 2018
20. Krauer F et al (2017) Zika virus infection as a cause of congenital brain abnormalities and Guillain-Barre syndrome: systematic review. PLoS Med 14(1). http://journals.plos.org/plosmedicine/article?id=10.1371/journal.pmed.1002203. Accessed 27 June 2018
21. Parra B et al (2016) Guillain-Barré syndrome associated with Zika virus infection in Colombia. N Engl J Med 375(16). https://www.nejm.org/doi/full/10.1056/NEJMoa1605564. Accessed 9 Aug 2018
22. Monel B et al (2017) Zika virus induces massive cytoplasmic vacuolization and paraptosis-like death in infected cells. EMBO J 36(12). https://www.ncbi.nlm.nih.gov/pubmed/28473450. Accessed 17 July 2017
23. Noor R, Ahmed T (2018) Zika virus: epidemiological study and its association with public health risk. J Infect Public Health. https://www.sciencedirect.com/science/article/pii/S1876034118300431. Accessed 25 July 2018; Besnard M et al (2014) Evidence of perinatal transmission of Zika virus, French Polynesia, December 2013, and February 2014. Euro-Surveillance 19(13). http://www.eurosurveillance.org/images/dynamic/EE/V19N13/art20751.pdf. Accessed 19 May 2017
24. Fox M (2016) As Zika spread, paralyzing Guillain-Barré syndrome skyrocketed. NBC News, September 1. http://www.nbcnews.com/storyline/Zika-virus-outbreak/study-finds-strong-link-between-Zika-paralyzing-guillain-barr-syndrome-n641276. Accessed 10 Jan 2017
25. Morris G et al (2018) Zika virus as an emerging neuropathogen: mechanisms of neurovirulence and neuro-immune interactions. Mol Neurobiol 55(5). https://link.springer.com/article/10.1007/s12035-017-0635-y. Accessed 7 July 2018; Lanciotti R et al (2008) Genetic and serologic properties of Zika virus associated with an epidemic, Yap State, Micronesia, 2007. Emerg Infect Dis 14(8). https://wwwnc.cdc.gov/eid/article/14/8/pdfs/08-0287.pdf. Accessed 24 July 2018; Rozé B et al (2016) Zika virus detection in urine from patients with Guillain-Barré syndrome on Martinique, January 2016. EuroSurveillance. https://www.eurosurveillance.org/content/10.2807/1560-7917.ES.2016.21.9.30154. Accessed 24 July 2018; Parra B et al (2016) Guillain-Barré syndrome associated with Zika virus infection in Colombia. N Engl J Med 375(16). https://www.nejm.org/doi/10.1056/NEJMoa1605564. Accessed 24 July 2018
26. Counotte M, Egli-Gany D, Wang J, Low N (2018). Zika virus infection as a cause of congenital brain abnormalities and Guillain-Barré syndrome: from systematic review to living systematic review [version 1; referees: 1 approved with reservations]. f1000 Research, February 15. https://f1000research.com/articles/7-196/v1. Accessed 23 July 2018
27. Gongora-Rivera F, Grijalva I, Infante-Valenzuela A, Camara-Lemarroy C et al (2020) Zika virus infection and Guillain-Barre syndrome in Northeastern Mexico: a case-control study. PLoS ONE 15(3):e0230132. https://doi.org/10.1371/journal.pone.0230132
28. Steenhuysen J (2016) Study finds strong link between Zika and Guillain-Barre. Reuters-Health News, August 31. http://uk.reuters.com/article/us-health-Zika-guillainbarre-idUKKCN1162X8. Accessed 18 Sept 2016

29. do Rasário MS et al (2016) Guillain-Barrė syndrome after Zika infection in Brazil. Am J Trop Med Hygiene. https://www.ncbi.nlm.nih.gov/pubmed/27645785. Accessed 11 April 2017
30. WHO (2016) Dispelling rumours around Zika and complications. WHO. http://www.who.int/emergencies/Zika-virus/articles/rumours/en/. Accessed 24 Sept 2016
31. WHO (2016) Mosquito control: can it stop Zika at source? Emergencies, February 17. http://www.who.int/emergencies/Zika-virus/articles/mosquito-control/en/. Accessed 2 Oct 2016
32. dos Santos T et al (2016) Zika virus and the Guillain-Barré syndrome—case series from seven countries. N Engl J Med. http://www.nejm.org/doi/full/10.1056/NEJMc1609015. Accessed 12 April 2017
33. Chen H-L, Tang R-B (2016) Why Zika virus infection has become a public health concern? J Chin Med Assoc 79:174–178. http://www.sciencedirect.com/science/article/pii/S1726490116300065. Accessed 10 April 2017
34. Fox M (2016) As Zika spread, paralyzing Guillain-Barré syndrome skyrocketed. NBC News, September 1. http://www.nbcnews.com/storyline/Zika-virus-outbreak/study-finds-strong-link-between-Zika-paralyzing-guillain-barr-syndrome-n641276. Accessed 14 April 2017
35. de Oliveria WK et al (2017) Zika virus infection and associated neurologic disorders in Brazil. N Engl J Med. http://www.nejm.org/doi/full/10.1056/NEJMc1608612#t=article. Accessed 4 May 2017
36. Dirlikov E, Major CG, Mayshack M et al (2016) Guillain-Barré syndrome during ongoing Zika virus transmission—Puerto Rico, January 1–July 31, 2016. Morb Mortal Wkly Rep 65:910–914. https://doi.org/10.15585/mmwr.mm6534e1
37. Laruren S, Rosen M (2016) Scientists track Zika's link to neurological disorder. Science News, April 2, pp 26–29. https://www.sciencenews.org/article/microcephaly-building-case-against-Zika. Accessed 27 Sept 2016
38. Barbi L, Coelho AVC, Alencar LCA, Crovella S (2018) Prevalence of Guillain-Barre syndrome among Zika virus infected cases: a systematic review and meta-analysis. Brazil J Infect Dis (An official publication of the Brazilian Society of Infectious Diseases). https://doi.org/10.1016/j.bjid.2018.02.005. Accessed 1 June 2022
39. Branswell H (2016) Why were there fewer microcephaly cases from Zika last year? Scientific American, March 29. https://www.scientificamerican.com/article/why-were-there-fewer-microcephaly-cases-from-Zika-last-year/. Accessed 7 April 2017
40. McFadden M (2017) Doctors warn of a Zika virus resurgence. WNDU, April 28. http://www.wndu.com/content/news/Doctors-warn-of-a-Zika-virus-resurgence-420761043.html. Accessed 5 June 2017
41. Meinhardt A (2017) Infection: a new threat on the horizon—Zika virus and male fertility. Nat Rev Urol 14. http://www.nature.com/nrurol/journal/v14/n3/full/nrurol.2016.265.html#access. Accessed 31 May 2017
42. Sin LH (2017) Meet three women on the front lines of Zika vaccine testing. Tampa Bay Times, January 13. http://www.tampabay.com/news/health/meet-three-women-on-the-front-lines-of-Zika-vaccine-testing/2309421. Accessed 2 June 2017
43. Meinhardt A (2017) A new threat on the horizon—Zika virus and male fertility. Nat Rev Urol. https://www.nature.com/nrurol/journal/v14/n3/full/nrurol.2016.265.html. Accessed 26 June 2017
44. Zashef Z (2017) Zika virus harms testes, says study. Yale News, February 22. http://news.yale.edu/2017/02/22/Zika-virus-harms-testes-says-study. Accessed 31 May 2017; Uraki R et al (2017) Zika virus causes testicular atrophy. Sci Adv 3(2). http://advances.sciencemag.org/content/3/2/e1602899.full. Accessed 31 May 2017
45. Chang D (2017) CDC urges doctors to screen for Zika-related epilepsy in infants born to infected moms. Miami Herald, April 17. http://www.miamiherald.com/news/health-care/article145029239.html. Accessed 5 May 2017
46. Kegel M (2017) Zika virus may trigger epilepsy, CDC researchers caution in report. Epilepsy News Today, April 21. https://epilepsynewstoday.com/2017/04/21/Zika-epilepsy-cdc-warn-report/. Accessed 17 May 2017

47. de Oliveira Souza IN et al (2018) Acute and chronic neurological consequences of early-life Zika virus infection in mice. Sci Transl Med 10(444):Eaar2749. http://stm.sciencemag.org/content/10/444/eaar2749. Accessed 23 July 2018
48. Moura da Silva AA et al (2016) Early growth and neurologic outcomes of infants with probable congenital Zika virus syndrome. Emerg Infect Dis 22(11). https://wwwnc.cdc.gov/eid/article/22/11/16-0956_article. Accessed 10 May 2017; van der Linden V et al (2016) Description of 13 infants born during October 2015–January 2016 with congenital Zika virus infection without microcephaly at birth: Brazil. Morb Mortal Wkly Rep 65(47). https://www.cdc.gov/mmwr/volumes/65/wr/mm6547e2.htm. Accessed 28 May 2017
49. Pastula DM, Yeargin-Allsopp M, Kobau R (2017) Enhanced epilepsy surveillance and awareness in the age of Zika. JAMA Neurol. https://doi.org/10.1001/jamaneurol.2017.0215
50. Health Day News (2017) Zika virus may trigger epilepsy in infants. Neurology Adviser. Health Day News, May 22. https://www.neurologyadvisor.com/epilepsy/epilepsy-linked-to-Zika-virus-infection-in-infants/article/651567/. Accessed 27 June 2018
51. Pastula DM, Yeargin-Allsopp M, Kobau R (2017) Enhanced epilepsy surveillance and awareness in the age of Zika. JAMA Neurol. http://jamanetwork.com/journals/jamaneurology/article-abstract/2618389. Accessed 21 July 2017; van der Linden V et al (2016) Description of 13 infants born during October 2015–January 2016 with congenital Zika virus infection without microcephaly at birth: Brazil. Morb Mortal Wkly Rep 65(47). https://www.cdc.gov/mmwr/volumes/65/wr/mm6547e2.htm. Accessed 21 July 2017
52. Pastula DM, Yeargin-Allsopp M, Kobau R (2017) Enhanced epilepsy surveillance and awareness in the age of Zika. JAMA Neurol. http://jamanetwork.com/journals/jamaneurology/article-abstract/2618389. Accessed 21 July 2017; Suzuki T et al (2008) Epilepsy in patients with congenital cytomegalovirus infection. Brain Dev 30. https://www.ncbi.nlm.nih.gov/pubmed/18215482. Accessed 21 July 2017
53. The Independent (2016) Zika may cause brain damage in adults too. The Independent, September 2. https://www.google.com/search?q=Zika+may+cause+brain+damage+in+adults+too&oq=Zika+may+cause+brain+damage+in+adults+too&aqs=chrome..69i57j69i60.935j0j8&sourceid=chrome&ie=UTF-8. Accessed 15 May 2017
54. Pastula DM, Yeargin-Allsopp M, Kobau R (2017) Enhanced epilepsy surveillance and awareness in the age of Zika. JAMA Neurol. http://jamanetwork.com/journals/jamaneurology/article-abstract/2618389. Accessed 21 July 2017
55. American Academy of Neurology Press Release (2016) Zika virus may now be tied to another brain disease, April 11. https://www.aan.com/PressRoom/Home/PressRelease/1451. Accessed 27 June 2017
56. Edwards SB, The Zika virus. ABDO Publishing, Minneapolis, MN
57. National Multiple Sclerosis Society (2016) Zika virus linked to Acute Disseminated Encephalomyelitis, a disorder than can be confused with MS. News, April 11. http://www.nationalmssociety.org/About-the-Society/News/Zika-Virus-Linked-to-Acute-Disseminated-Encephalom. Accessed 5 June 2017
58. Ferreira MLB (2016) Zika Virus may now be tied to another brain disease. Press Release. In: AAN 68th annual meeting abstract, April 10. https://www.aan.com/PressRoom/Home/PressRelease/1451. Accessed 5 June 2017
59. Soares CN et al (2016) Fatal encephalitis associated with Zika virus infection in an adult. J Clin Virol 83:63–65. https://www.ncbi.nlm.nih.gov/pubmed/27598870. Accessed 4 April 2017
60. Carteaux G et al (2016) Zika virus associated with meningoencephalitis. N Engl J Med 374:1595–1596. http://www.nejm.org/doi/full/10.1056/NEJMc1602964. Accessed 4 April 2017
61. Li H et al (2016) Zika virus infects neural progenitors in the adult mouse brain and alters proliferation. Cell Stem Cell 19. https://www.ncbi.nlm.nih.gov/pubmed/27545505. Accessed 19 May 2017
62. Madialdea-Carrrera (2017) Four reasons why we shouldn't forget about Zika. The Conversation, April 13. http://theconversation.com/four-reasons-why-we-shouldnt-forget-about-Zika-75257. Accessed 21 May 2017

63. Aid M et al (2017) Zika virus persistence in the central nervous system and lymph nodes of rhesus monkeys. Cell 169(4). http://www.sciencedirect.com/science/article/pii/S009286741730421X. Accessed 14 May 2017
64. Hamers L (2017) Zika hides out in body's hard-to-reach spots. Science News, April 27. https://www.sciencenews.org/article/Zika-hides-out-bodys-hard-reach-spots. Accessed 14 May 2017
65. Fenz K (2016) Zika infection may affect adult brain cells, suggesting risk may not be limited to pregnant women. Newswire, August 18. http://newswire.rockefeller.edu/2016/08/18/Zika-infection-may-affect-adult-brain-cells-suggesting-risk-may-not-be-limited-to-pregnant-women/. Accessed 18 April 2017
66. McKay B (2016) Study suggests Zika may damage adult brains. The Wall Street Journal, August 18. https://www.wsj.com/articles/study-suggests-Zika-may-damage-adult-brains-1471536001. Accessed 21 May 2017
67. Ramkumar A (2016) Zika may cause brain damage in adults, too. Bloomberg Business, August 19. https://www.bloomberg.com/news/articles/2016-08-19/Zika-may-cause-brain-damage-in-adults-too. Accessed 29 May 2017
68. D'Amora D (2016) Zika may be a threat to the adult brain, too. Mother Jones, September 21. http://www.motherjones.com/environment/2016/09/science-Zika-threat-adult-brain. Accessed 4 April 2017
69. Wagner D (2016) San Diego scientists find Zika may infect adult brains too. KPBS, August 18. http://www.kpbs.org/news/2016/aug/18/san-diego-scientists-find-Zika-may-infect-adult-br/. Accessed 29 June 2017
70. Noor R, Ahmed T (2018) Zika virus: epidemiological study and its association with public health risk. J Infect Public Health. https://www.sciencedirect.com/science/article/pii/S1876034118300431. Accessed 25 July 2018; Sikka V et al (2016) The emergence of Zika virus as a global health security threat: a review and a consensus statement of the INDUSEM Joint Working Group (JWG). J Glob Infect Dis 8. https://www.ncbi.nlm.nih.gov/pubmed/27013839. Accessed 25 July 2018
71. Dirlikov E et al (2016) Update: ongoing Zika virus transmission—Puerto Rico, November 1, 2015–April 14, 2016. Morb Mortal Wkly Rep. https://www.cdc.gov/mmwr/volumes/65/wr/mm6517e2.htm. Accessed 23 May 2017
72. May M (2016) A comprehensive systems biology approach to studying Zika virus. PLOS One. http://journals.plos.org/plosone/article?id=10.1371/journal.pone.0161355. Accessed 23 May 2017
73. Dvorksky G (2014) Why "neglected diseases" are becoming a global danger. GIZMODO, November 20. http://io9.gizmodo.com/why-neglected-diseases-are-becoming-a-global-danger-1660705505. Accessed 4 April 2017
74. Beck J (2016) What will be the long-term effects of Zika? The Atlantic, April 19. https://www.theatlantic.com/health/archive/2016/04/Zika-is-a-delayed-epidemic/478755/. Accessed 4 Feb 2017
75. Al-Qahtani AA et al (2016) Zika virus: a new pandemic threat. J Infect Dev Countries 10(3). https://www.ncbi.nlm.nih.gov/pubmed/27031450. Accessed 11 June 2019; Mercola J (2016) Zika: Brazil admits it's not the virus. Mercola.com, August 16
76. Roos R (2016) Zika virus—a public health emergency of international concern. JAMA Neurol 73(12). http://jamanetwork.com/journals/jamaneurology/fullarticle/2557229?widget=personalizedcontent&previousarticle=2618389. Accessed 30 May 2017

Chapter 10
Vectors and Reservoirs

The primary vector for COVID-19 may turn out to be bats or pangolins, and there remains some concern that it was an accidental release from a Chinese lab in Wuhan [1]. The vector is the mosquito for infectious diseases such as yellow fever, dengue, chikungunya, and ZIKV.

With the emergence or re-emergence of numerous mosquito-borne diseases in recent years, effective emergency vector control response methods have become necessary to reduce human infections. Current vector control practices often vary significantly between jurisdictions and are executed independently and on different spatial scales. Several types of surveillance information (e.g., number of human infections or adult mosquitoes) trigger the implementation of control measures. While this patchy implementation of control measures likely alters control efficacy, it remains the primary approach to controlling mosquito populations [2].

Vector control is any method to limit or eradicate vectors, particularly mosquitoes, which transmit disease pathogens. Disease control reduces an infectious disease's incidence, prevalence, morbidity, or mortality to a locally acceptable level or, if possible, its elimination or eradication. To be sustainable, a vector control strategy must limit the spread of resistance to the method within target mosquito populations [3]. In addition, the plan must be tolerable to humans and other species within the ecosystem.

Vector-borne diseases transmitted by insect vectors such as mosquitoes occur in over one hundred countries. More than 80% of the world's population is at risk of vector-borne disease, with half at risk from two or more conditions. Many of these diseases are concentrated in the poorest communities in tropical and subtropical regions. They cause unacceptable mortality and morbidity and impede economic growth [4]. In the United States, there are over 170 species of mosquito, and there are 3500 species found in the world.

No mosquitoes mean no disease. This was the common sensibility engaged by U.S. Surgeon General William Gorgas and Fred Soper from the Rockefeller Foundation when they eliminated breeding sites to mitigate yellow fever among workers building the Panama Canal in the early twentieth century. According to the Focks model [5],

D. M. Berube, *Pandemics and Resilience: Lessons we should have learned from Zika*, Risk, Systems and Decisions, https://doi.org/10.1007/978-3-031-25370-6_10

the mosquito population will need to be taken well below the disease transmission threshold to claim that population reduction/crash would be sufficient to mitigate disease.

Morrison wrote the goal for ZIKV should be to reduce adult populations or their interactions with humans below that which can sustain an epidemic; it is unrealistic to expect to eradicate either the vector or the viruses [6]. However, with a few exceptions, vector control methods have been largely unsuccessful because of the absence of a sustained commitment of resources and the inability to scale up and successfully apply interventions over large geographical areas and modern megacities [7].

There are different approaches involving genetics and species population controls. Some are potentially irreversible, and others are not geographically self-limiting. The appeal of genetically modifying insects to curb their numbers is that it avoids using indiscriminate pesticides, which can harm other wildlife and, in some cases, threaten human health [8].

However, there is a clear need for novel efficacious approaches, given that existing strategies such as insecticides and larval biological control have proven unsustainable and ineffective at halting disease spread [9].

Nonetheless, distinguishing between approaches with wholly different risk profiles is challenging given the historical debates over genetically modified seeds, cloning, etc. In addition, even in the case of the ZIKV carrying *Aedes aegypti* mosquito, there is wholly non-specific fearmongering.

If environmentalists fight genetically modified mosquitoes as hard as they have fought naked, they may be stuck with ZIKV permanently. Accordingly, former CDC Director Thomas Frieden said during a conference in Miami in 2016 that the ZIKV will become endemic to this hemisphere if the *Ae. aegypti* mosquito is not eradicated [10].

10.1 Primary Vector

Some of the closest ecological relationships between humans and other species are with arthropod vectors of disease, which often depend on human blood and artificial breeding sites for survival [11].

ZIKV has been isolated from a wide range of mosquitoes, including at least seventeen species of *Aedes*, the malaria vector *Anopheles gambiae*, two species of *Eretmapodites*, and *Mansonia uniformis*. Despite some earlier conflicting reports, it has been shown that *Culex quinquefasciatus*, and possibly other *Culex* species, are not effective vectors, and *An. gambiae* and *Anopheles stephensi* [12].

There are around 3600 mosquitoes, and about one hundred spread human diseases. Anne Schuchat from the CDC said *Ae. aegypti*, the mosquito species that primarily transmits the virus, is present in about thirty states (April 2016).

The primary vector of ZIKV is the *Ae. aegypti* mosquito, which lays eggs in temporary pools of water. Both natural (e.g., tree holes) and human-made (e.g., plant pot saucers) containers can serve as mosquito breeding sites, potentially bringing mosquitoes closer to areas of human habitation [13]. *Ae. aegypti* does not just spread ZIKV and "about 40% of the global population is at risk [from] this species," said Andrew McKemey, an entomologist, and the head of field operations for Oxitec. "It's kind of the rat of the mosquito world." [14].

The word "*Aedes*" comes from Greek and means unpleasant or cruel. The *Aegypti* species are found in Africa, Africa, and Asia. Half of the planet's population lives in areas where it is present. According to UC-Riverside entomologist Omar Akbari, "*Ae. aegypti* is probably the most dangerous animal in the world." Mr. Bill Gates has blogged that they are even more dangerous to humans than humans. Mosquitoes kill nearly fifteen times more people than snakes and 72,000 more than sharks. According to WHO, mosquitoes kill 700,000 people annually [15].

Ae. aegypti mosquitoes carry dengue fever, West Nile virus, yellow fever, Rift valley fever, Murray valley encephalitis, chikungunya, Japanese encephalitis, etc. It is also believed to be responsible for carrying the Mayaro virus, which recently appeared in this region for the first time in Haiti [16].

It is very opportunistic and shows a remarkable ability to adapt to changing environments, especially those created by changes in how humanity inhabits the planet. Over the years, it has exploited these opportunities, increases in international travel and trade, and rapid unplanned urbanization, with exceptional efficiency. They can breed nearly anywhere water collects, even the screw tops from plastic water bottles. The mosquitoes can also breed in the microbial stew in septic tanks, toilet tanks, and shower stalls. Construction sites that use tires and clogged rain gutters offer additional opportunities to breed in large numbers [17].

Females will also oviposit on the inner wall of various artificial containers such as tanks, vases, jars, tires, drums, buckets, pots, cans, scrap metal, and gutters, distributed inside houses or in their yards. The variability in the preference of the diverse types of containers as sites for oviposition by female *Ae. aegypti* depends on the availability of artificial containers, urbanization degree, and season [18].

These mosquitoes live in residential areas (they love closets, for example) and can lay eggs in dry places, where they will sit dormant until the rain comes. For instance, it is known that large outside drums, which residents use to store rainwater in the tropics, are major breeding sites for *Ae. aegypti*, the dengue vector, ZIKV, chikungunya, and many other viruses. However, when looking at a backyard cluttered with various small objects collecting water, it is difficult to pinpoint which potential breeding sites would be more essential to remove [19].

Ae. aegypti are known as sip-feeders because they like to take little glugs of blood from various sources, which increases the likelihood that any single mosquito will transmit disease to multiple people. They spend time together in closets, under beds, and behind washing machines, often emerging around dawn and dusk for blood meals from people who usually will not even notice they have been bitten until it is too late. Tests conducted by Jong and Knols [20] demonstrated that *Ae. aegypti* prefers to bite the head and upper part of the trunk of persons lying prone or supine but will often chew on the lower legs beneath tables and when the host is seated.

According to the OECD, females can take a blood meal required for egg development. The females bite. A single human blood meal can produce 100–200 eggs. And a single female can make five batches of eggs [21].

Finally, a team of Chinese scientists led by Jin Lin may have uncovered significant research findings that have reported that LTRIN, a protein obtained from the salivary glands of *Ae. aegypti* facilitates the transmission of the ZIKV. LTRIN modulates the host immune responses to a mosquito bite, providing chances for the ZIKV to spread [22].

10.2 How Do We Know?

To evaluate whether a given species can transmit a virus, researchers feed insects on infected blood in the lab and collect saliva from them a week or so later. The species is considered a "competent" vector if the saliva contains an infective virus. Not all lab-competent vectors spread disease, however. That depends on several factors, such as how often the species bites, whether it feeds primarily on humans or other animals, and how long it lives. To confirm that a species is transmitting disease, researchers must find virus-infected mosquitoes in the wild.

Lab tests can be misleading, however. "There is a classic discordance between what you see in the lab and what happens in the wild," Oliver Brady, an entomologist at the University of Oxford in the United Kingdom, says. "*Albopictus* and *aegypti* are both highly competent in the lab" as vectors for dengue. "But in Europe, where we have widespread *albopictus* and almost no *aegypti*, you don't have huge dengue outbreaks." [23].

After an extensive mapping exercise, Kraemer et al. [24] demonstrated that *Aedes* distributions to be the widest ever recorded, now extensive in all continents, including North America and Europe. Today both *Ae. aegypti* and *Ae. albopictus* are present in most Asian cities and large parts of the Americas [24].

According to Waddell [26], eighteen species of mosquitos were positive for ZIKV during epidemiological sampling in Africa and Asia from 1956 to 2015. Eight were evaluated experimentally for vector competence, *Ae. aegypti* and *Ae. Africanus* were investigated most often.

Recently studies have reported significant differences in susceptibility for ZIKV infection between wild mosquito populations of *Ae. aegypti*, *Ae. albopictus* and other *Aedes* species [27]. Moreover, at least ten *Aedes* species are known to harbor ZIKV [28]. *Ae. albopictus*, a species of particular interest as a possible vector for ZIKV in North America, was evaluated for vector competence in one study and as a naturally infected vector of ZIKV in two other studies [26].

There are hundreds of mosquito species and evaluating them is difficult and costly. So far, only a few species have been assessed to see if they transmit ZIKV. However, computational tools called decision trees could help predict which mosquitoes can transmit a virus based on shared traits, such as a mosquito's geographic range or the symptoms of a virus.

University of Georgia ecologist Evans et al. [29] used decision trees to create a model that predicts which mosquitoes are potential carriers of ZIKV and should therefore be prioritized for testing. The model considered all known viruses that belong to the same family as ZIKV and the mosquitoes that carry them.

Evans et al. predicted that thirty-five species may be able to take the ZIKV seven of which are found in the United States. On average, closely related viruses of the Flaviviridae family are vectored by over nine mosquito species. Thus, because ZIKV may be associated with multiple mosquito species, it may be considered necessary to develop a more comprehensive list of potential ZIKV vectors [29].

Evans et al. expressed concern over conclusions regarding vector competence of mosquitoes for ZIKV. "Our model predicts that fewer than one-third of the potential mosquito vectors of ZIKV have been identified, with over twenty-five additional mosquito species worldwide that may have the capacity to contribute to transmission. The continuing focus in the published literature on two species known to transmit the ZIKV virus (*Ae. aegypti* and *Ae. albopictus*) ignores the potential role of other vectors, potentially misrepresenting the spatial extent of risk. Models predicted four species to be competent vectors—*Ae. vexans, Culex quinquefasciatus, Cx. pipiens*, and *Cx. Marsalis*—are found throughout the continental United States" [29].

Vector-based solutions assume how the virus spreads can be controlled in the case of the *Ae. aegypti* mosquito has been identified as the primary carrier or vector of ZIKV. However, the presence of the virus does not automatically make the species an efficient vector for the disease. While the vectorial competence of *Ae. aegypti* is well established, that of *Ae. albopictus* is not [28]. The isolation of a virus from a mosquito is not evidenced that it is a vector of the virus. To demonstrate that a mosquito is a vector, it must be shown capable of transmission [30].

10.2.1 Debates Over Aegypti

Native to Africa, globalization has aided the *Aedes*' travels to nearly every continent and the proliferation of disease. Except for yellow fever, these diseases do not have vaccines or cures. The *Aedes* assists in killing tens of thousands of people each year and infects many, many more. As its footprint has widened, methods of fighting it have weakened [31].

Most importantly to this project, ZIKV involves a mosquito species, *Ae. aegypti,* and it traveled to the Americas via slave and trading ships during the seventeenth to nineteenth centuries. These ships carried freshwater reservoirs on board and could maintain breeding colonies of *Ae. aegypti,* so it is probable that the species was introduced to the rest of the world via this means [32].

They are the most important vector of human arboviruses. They were introduced from Egypt, hence the name, and are not endemic to the U.S. or the Americas. They live for 2–4 weeks, but their eggs can survive for extended periods in a dry state (several months) and are known to reintroduce large numbers after a cold, dry winter [33]. They were responsible for the yellow fever outbreaks in the sixties and today carry dengue, yellow fever, chikungunya, and dengue alone infects up to one hundred million people worldwide each year. *Ae. aegypti*, which bites during daylight, is uniquely domestic among mosquito vectors: Tt mates, feeds, rests, and lays eggs in and around human habitation [6]. In addition, they display a strong preference for feeding on humans.

These mosquitoes were responsible for Napoleon's troops being routed in Haiti at the beginning of the nineteenth century when General Victor-Emmanuel LeClerc lost a third of his soldiers to Yellow Fever.

In another mosquito-driven moment in history, Aegypti mosquitoes killed 22,000 men who built the Panama Canal at the end of the same century [34].

> *Ae. aegypti* has a cosmopolitan distribution between the 40° N and 40° S latitudes and is phenotypically polymorphic, varies in gene frequencies as detected by biochemical and molecular genetic markers, and exhibits variation in vector competence for arboviruses. In sub-Saharan Africa, *Ae. aegypti* appears as a sylvan race or subspecies, *Ae. aegypti formosus* oviposits primarily in tree holes.
>
> A light-colored domestic race, *Ae. aegypti* is distributed in tropical and subtropical regions outside Africa. This race displays an oviposition preference for artificial containers (e.g., tires and discarded jars) associated with human habitats—electrophoretic analysis of allozyme variation among populations … of *Ae. aegypti* identified eight genetic groups, including sylvan races (*Ae. aegypti formosus*) in West and East Africa and domestic races (*Ae. aegypti aegypti*) in East Africa, southeast U.S., southwest U.S.-Mexico, Central-South America, the Caribbean, and Southeast Asia-Pacific. [35]

Ae. aegypti, the "container-breeding" mosquito, is strongly associated with humans and highly anthropophilic, predominating in densely populated urban areas. They are commonly found indoors, breeding in artificial containers, with females needing to feed on blood to produce eggs. *Ae. aegypti* females bite many people—about half of their diet is human blood, says University of Southern Mississippi aquatic ecologist Don Yee to Science News. In the tropics, they seek out human environments; he adds, "The adults are resting on the walls or the ceiling. They are hanging around the bathroom." [36].

There is some minor disagreement on whether *Ae. aegypti* was responsible for ZIKV, but it is an outsider's testimony. Fiona Hunter, a Canadian medical entomologist, remarked: "No mosquito infection data supports ZIKV transmission by *Ae. aegypti* in Brazil." This statement also appeared on a PowerPoint slide during her presentation at the Zika Symposium: 2016 International Congress of Entomology held on September 26, 2016, in Orlando, Florida [37]. Remarks like hers have aggravated an ongoing debate about whether an aggressive effort to crash the *Ae. aegypti* population is sufficient to reduce sufficiently the occurrence of Zika.

This debate remains important as Gardner, Chen, and Sarkar noted: "If *Ae. aegypti* is the only competent ZIKV vector, then the risk is geographically restricted, in North America to Florida, Louisiana, and Texas. However, if *Ae. albopictus* is a capable vector, then there is the risk of autochthonous transmission cycles in Canada, Chile, much of Western Europe, and South and East Asia; and for all these areas, the risk compounds that from flights originating in other regions historically endemic for ZIKV virus" (more below) [28].

The average lifespan of an *Ae. aegypti* mosquito is two weeks [38]. In their lifespan, they travel less than 150 m (164 yards), according to the CDC, though the WHO reports an average flight range of four hundred meters (437 yards). Another minority source argues the lifespan ranges from fifteen to thirty days, and the flight range is four hundred meters [39].

Aegypti like people's blood and live and breed close to humans. *Ae. aegypti* mosquito, admirably adapted to urbanized areas, especially peridomestic, often breeds in artificial containers and almost exclusively bites people [40].

They can breed in a space the size of a bottle cap. "Even a tiny amount of water, in a candy wrapper, tossed somewhere, can be a breeding ground for *Aedes* eggs," says Margareth Capurro, a professor of biosciences at the University of São Paulo.

Ae. aegypti mosquitoes lay their eggs every three or four days. A single female can lay around four hundred eggs in her lifetime.

The eggs are also resistant to drought and can survive for more than a year, with larvae emerging when the eggs encounter water [41]. Stella Fogleman, the director of emergency preparedness and response for the Los Angeles County Department of Public Health, coyly commented: "Their eggs can lie dormant for a year, like a sea monkey." [42]. David Severson at the University of Notre Dame discovered a population of *Ae. aegypti* mosquitoes have spent the past four winters underground in Washington, D.C.'s Capitol Hill neighborhood [43].

"The *Ae. aegypti* is a domesticated mosquito that likes people a lot, and it is spread ZIKV in Miami to more than 280 people." [44]. A recent report from the CDC found that *Ae. aegypti*—which can transmit the ZIKV, dengue, and chikungunya viruses—could find suitable breeding habitats in 75% of the contiguous United States [45]. "We are near-certain that ZIKV will reach all areas inhabited by the (supposed) principal vector, *Ae. aegypti*. Probably all countries in the Americas, except Canada. … Maybe to the Mediterranean," said Christopher Dye, the strategy director in the Office of the WHO [46].

Finally, "sometime during the past decade, ZIKV underwent adaptive evolution (or genetic drift resulting from a population bottleneck and not the result of selection) that resulted in more efficient transmission by *Ae. aegypti* and perhaps other closely related mosquito vectors (in the *Aedes (Stegomyia)* subgenus) such as *Ae. hensilii* incriminated in Yap [47] or *Ae. polynesiensis* suspected as a vector in French Polynesia [48].

Recently, a lab at Colorado State, postdoc fellow Claudia Rückert discovered *Ae. aegypti* was capable of coinfection. *Ae. aegypti* can spread multiple viruses with one bite: ZIKV, dengue, and chikungunya. According to a corroborating study conducted at the University of California, Berkeley, 20% of individuals infected with one arbovirus were coinfected with another [49].

Of all mosquitoes, *Ae. aegypti* is the most efficient at transmitting ZIKV in a given location and time for various reasons: It feeds off human blood; it can bite several people during a single meal, flitting inconspicuously from one victim to another; its bite goes almost unnoticed; human dwellings are its habitat, and it feeds only in the daytime [50]. Paul Reiter, a medical entomologist at the Institut Pasteur in Paris, said when it comes to *Ae. aegypti* mosquitoes, there is just extraordinarily little we can do about it. The mosquito wins." [51].

Vector status is not static. Therefore, in addition to the broader geographic region supporting potential vectors, Evans et al. suggest that rural and urban areas could serve as habitats for potential vectors of ZIKV [29].

This phenomenon of some mosquitoes being able to lay eggs that can survive variable periods of complete drying out and hatch at the subsequent inundation by rain or floods or humans watering a garden, and even have staggered hatching events spread over multiple inundations, is probably the single most crucial biological trait that (in combination with solid anthropophilic) has allowed such effective global spread of *Ae. aegypti* and *Ae. albopictus* and their associated diseases of DENV, CHIKV, YFV, and others [12].

10.2.2 Debates Over Albopictus

As mentioned above, second mosquito species has surfaced as a likely vector for ZIKV, *Ae. albopictus*. The spread of the Asian tiger mosquito (*Ae. albopictus*) across densely populated urban areas in the U.S. has established a new risk landscape. Manore et al. [52] developed a model that demonstrates under specific but realistic conditions, fifty percent of introductions by infectious travelers to a high human, high mosquito density city could initiate a local transmission, and 10% of which could result in one hundred or more people infected. Despite the propensity for *Ae. albopictus* to bite non-human vertebrates, Manore et al. demonstrate that local virus transmission and human outbreaks may occur when vectors feed on humans even 40% of the time [52].

While *Ae. albopictus* is an inferior vector capacity, it remains a concern. The opportunistic feeding on nonhuman vertebrates, high human-biting propensity, and high probability of human-vector interactions makes *Ae. albopictus* as a potential vector of zoonotic pathogens and arboviruses in endemic habitats [53].

The establishment of *Ae. albopictus* in locations with cooler climates has been aided by its ecological plasticity, with eggs able to undergo diapause (dormancy) like its *Ae. aegypti* sisters, one possible explanation for populations persisting through winters that are too cold for adult survival [54].

The land areas with environmental conditions suitable for *Ae. albopictus* populations are expected to increase from 5 to 16% in the next two decades and 43–49% by the end of the century. Presently, about one-third of the total human population of fifty-five million in the Northeastern U.S. reside in urban areas where *Ae. albopictus* is present. This number is predicted to double to about 60% by the end of the century, encompassing all major urban centers and placing over thirty million people under the threat of dense *Ae. albopictus* infestations [55].

Unlike mosquito species traditionally encountered in the Northeast, *Ae. albopictus* larvae prefer small, artificial container habitats which are ubiquitous and diffusely distributed in urban areas and nearby parkland. Additionally, many of these container habitats are located within private residential backyards that might be inaccessible to mosquito control personnel. *Ae. albopictus* range expansion in the northeast threatens to present challenges far exceeding the resources likely to be available to combat them unless new and effective control strategies are developed [55].

Ae. albopictus is seen in wider geographical distribution, including subtropical and temperate climates; it is resilient and aggressive, is long-lived (4–9 weeks), can survive in rural and urban environments, and can survive through cold winters. *Ae. albopictus* has also colonized almost every Mediterranean country and has adapted to temperate regions, spreading further north in Europe and United States, becoming a concern for populations in temperate climates [56]. *Ae. albopictus* is expected to spread broadly throughout Europe, reaching broad areas of France and Germany. Locations in the Northern U.S. and highland regions of South America and East Africa are also projected to establish *Ae. albopictus* over the next 30 years [24].

The invasive mosquito *Ae. albopictus* is often considered a poor vector of human pathogens, owing to its catholic feeding behavior. However, it was recently incriminated as a significant vector in several Chikungunya epidemics outside its native range. Kamgang et al.'s [57] analysis of ingested blood in outdoor-resting females showed that *Ae. albopictus* preferentially fed on humans rather than on available domestic animals (95% of the blood meals contained human blood) [57].

Data from Kek et al. [58] corroborate Kamgang and others. *Ae. albopictus* primarily fed on humans (83.2%) even in semiurban-rural settings, such as nature reserves where the human density is low, and the availability of other vertebrate hosts is relatively high. Unsurprisingly, all positive bloodmeals in urban areas were of humans, the most abundant host in residential estates [58].

A North Carolina study by Richards et al. [59] reported *Ae. albopictus* fed predominantly on mammalian hosts (83%). Shared mammalian hosts included humans (24%), cats (21%), and dogs (14%). However, a notable proportion (7%) of bloodmeals also was taken from avian hosts. However, when feeding indices were time weighted, *Ae. albopictus* fed preferentially upon humans. Of forty human bloodmeals, 32 (80%) were from a single human, whereas eight (20%) were multiple bloodmeals taken from more than one human host [59].

Although *Ae. albopictus* preferentially bites mammals, females can also feed on most groups of vertebrates, both cold- and warm-blooded, including reptiles, birds, and amphibians. Such feeding plasticity, which has been found to vary according to mosquito populations' geographic origin, maximizes *Aedes*'s fitness (fecundity and survival). *Albopictus* enhances the risk that it may propagate zoonotic pathogens from wildlife or domestic animals to humans [57].

Richards et al. [59] derived time-weighted feeding indices, using estimates of the outdoor exposure of residents and their pets, and found that *Ae. albopictus* took a more considerable proportion of bloodmeals than would be expected from humans relative to dogs and cats. The results support the assertions of Franco-Estrada and Craig [60] that *Ae. albopictus* is highly anthropophilic. Still, that host abundance and availability significantly impact the host-feeding patterns of this peridomestic mosquito species [59].

Ae. albopictus, easily identified by its stripey white legs and daytime biting habits, arrived in Texas in 1985. It is much more tolerant of cold temperatures, thrives more in the suburbs than in the city, and now lives in 40 U.S. states [61]. They hitched a ride with a shipment of used tires from Japan [62].

Eighty-three percent of female *Ae. albopictus* mosquitoes in suburban areas of Singapore had fed on humans [63], but some were reactive to shrews, swine, dogs, cats, turtles, and other hosts in rural settings. In urban areas, all positive blood meals were from humans [64]. In its native tropical range, *Ae. albopictus* feeds exclusively on humans in Indonesia. However, some studies on the feeding preferences of *Ae. albopictus* have indicated a preference for mammals in suburban areas of the U.S. [65]. Additionally, studies conducted in Thailand have reported that, *Ae. albopictus* feed on humans, swine, buffalo, dogs, and chickens. In contrast, recent investigations say that *Ae. albopictus* feeds only on humans, with a few (0.6%) double-host blood meals between humans and swine/cat/dog. In Cameroon, 95% of reported blood meals contained human blood [66]. In temperate Japan, *Ae. albopictus* primarily feed on mammals, with a high propensity for humans, birds, and amphibians/reptiles.

Studies conducted at a tire dump in Missouri, USA, reported that *Ae. albopictus* will feed on birds (17%) but prefer mammals (64%), with 8.2% of those mammalian feedings obtained from humans. A follow-up study conducted in other tire yards and surrounding vegetation of rural and urban habitats in Missouri, Florida, Indiana, Illinois, and Louisiana, USA, concluded that *Ae. albopictus* showed a strong preference for mammals (0.94%), with up to 8% human-derived blood meals, while also detecting avian (1%) and reptilian (5%) blood meals. An additional study in suburban landscapes of North Carolina, USA, reported that *Ae. albopictus* feeds predominately on mammalian hosts (83%) but also on birds (7%), amphibians (2%), and reptiles (2%) [67].

According to Oxitec (Chap. 15), *Ae. albopictus* can spread disease but is relatively inefficient at disease transmission. *Ae. aegypti* feeds almost exclusively on humans. It lives in and around the home and can take multiple bites to get a blood meal. These features make it highly effective at spreading disease once a person has entered an area carrying the virus. *Ae. albopictus*, by contrast, is an aggressive biter and will tend to secure its blood meal more quickly from various sources. It is less selective about what it bites—so it will bite dogs, birds, etc., which are unaffected and do not function as a source for onward transmission to people. The two species can occupy the same habitat, but *Ae. aegypti* is predominately restricted to areas of human habitation, whereas *Ae. albopictus* is not and tends to live more at the fringes of town, in forested areas, underbrush, etc. [68].

Ae. albopictus can carry ZIKV, but it also tends to bite humans much less than the *Ae. albopictus* is less fickle and nibbles on animals as well. According to Stan Cope, president of the American Mosquito Control Association, "If [an *Ae. albopictus* mosquito] happens to pick up ZIKV from a human, that mosquito is much more likely to bite than something other than a human, and the virus will not survive in those other animals." [69].

Grard et al. reported in 2014 that they had found ZIKV in *Ae. albopictus* was collected in the wild in Gabon from 2007 to 2010 [70]. In a 2013 study, Singaporean researchers successfully infected the Asian tiger mosquito with ZIKV [71]. "The detection of ZIKV RNA from five adults *Ae. albopictus* reared from eggs collected during the 2015 outbreak in Camaçari, Bahia, Brazil, is consistent with the potential for vertical or sexual transmission of ZIKV by *Ae. albopictus*; however, evidence supporting this was not conclusive," they wrote [72]. Another first reported in April 2016 involved researchers at a government laboratory in Mexico who reported detecting the ZIKV in female *Ae. albopictus* mosquitoes collected in the wild, as opposed to experimentally infected—another first for the Western Hemisphere. *Ae. albopictus*, also known as the Asian tiger mosquito, is an invasive species that continues to expand geographically well beyond Asia and has adapted to flourish in various habitats strongly associated with humans. As the mosquito can survive the winter in temperate climates, its ability to carry the ZIKV could significantly expand the map of areas at risk of ZIKV transmission [73]. Furthermore, A 2017 study by Smartt et al. [72] reported that field-collected *Ae. albopictus* eggs from Camçari, Bahi, Brazil, may contain ZIKV RNA that requires further tests for infectious ZIKV.

The Smartt study involved grinding up the mosquitoes that grew from eggs—male and female—they found genetic pieces of ZIKV. For the study, Smartt and her colleagues collected mosquitoes in Brazil, hatched the insects' eggs, and found Asian tiger males tested positive for ZIKV genetic makeup but not the live virus. It is unclear yet whether Asian tigers can transmit the devastating disease to humans, but the finding proves more research is needed into other possible carriers of ZIKV, Smartt added [74]. Smartt and colleagues said [v]ertical and sexual transmission of ZIKV had been seen in several *Aedes* species, but not *Ae. albopictus*. "Detecting ZIKV RNA fragments without finding live ZIKV suggests that either the female parent was not itself infected with live ZIKV or it was not able to transfer live ZIKV to her eggs," Smartt said "ZIKV RNA in field-collected eggs from mosquitoes where there is current [ZIKV] transmission is concerning," Smartt and colleagues concluded. "Samples of mosquitoes, including those resulting from field-collected eggs that are returned to the laboratory from regions with [ZIKV], must be treated with the potential the resulting adult mosquitoes or their offspring might be positive for [ZIKV virus] RNA [75]. Finally, Smartt et al. [76] concluded: "The detection of ZIKV RNA from five adult *Ae. albopictus* reared from eggs collected during the 2015 outbreak in Camaçari, Bahia, Brazil, is consistent with the potential for vertical or sexual transmission of ZIKV by *Ae. albopictus*; however, evidence supporting this was not conclusive." [76].

Albopictus is found throughout Southern Europe. *Albopictus*, the Asian tiger mosquito for its white stripes, is considered the most invasive mosquito species [77]. *Albopictus* can adapt to new environments and has been associated with dengue, chikungunya, and dirofilariasis. It is an opportunistic feeder and, unlike *Aegypti,* does not feed exclusively on humans. If the *Aegypti* population is significantly reduced, there is some expressed concern that *Albopictus* may move into the vacated ecosystem [78].

Now the question is how well the virus lives in the bodies of the Asian tiger mosquito. Simply finding a virus in a mosquito does not necessarily mean the mosquito spreads the virus. The virus must replicate in the insect's salivary glands to be transmitted in a bite [61].

"But most people believe *Aedes albopictus* will not be as efficient [a vector] as *Ae. aegypti* because it also feeds on other mammals and birds," said Peter Hotez, the dean of the National School of Tropical Medicine at Baylor College of Medicine. In other words, if the tiger mosquito is getting a good portion of its blood meals from horses and dogs, that means fewer opportunities for infected mosquitoes to spread disease among humans [79].

Since its first detection in Brazil in the 1980s, *Ae. albopictus* has geographically expanded to approximately 60% of Brazilian municipalities across urban, periurban, and rural environments, and sometimes at the border of the Atlantic and Amazon forests [80]. Over ten months, a team led by Ricardo Lourenço-de-Oliveira, an entomologist at the Oswaldo Cruz Foundation (Fiocruz) in Rio de Janeiro, collected more than 1500 mosquitoes, identified them, and tested pooled samples of the same sex and species for the presence of ZIKV and other viruses. Nearly half were *Ae. aegypti*, and most of the rest were *Culex quinquefasciatus*, another common mosquito in urban Brazil. Roughly, 5% were distinct species. A species called *Ae. albopictus*, widely known as the Asian tiger mosquito, which can also transmit ZIKV in the lab and has been found infected with the virus in Mexico and Gabon, made up only 2% of the catch, Lourenço-de-Oliveira says. They found the ZIKV virus in three sets of female *Ae. aegypti* mosquitoes, but none of the other species [81].

MosquitoMate (Chap. 12) is awaiting permits from the U.S. EPA to sell a related mosquito species, known as the "Asian tiger mosquito," infected with Wolbachia as a pest control service. Though these mosquitoes also can carry viruses, experts consider them less of a threat for triggering outbreaks than *Ae. Aegypti* [82].

10.2.3 Debates Over **Aegypti** *and* **Albopictus**

Entomological studies have demonstrated that Brazilian and other American populations of *Ae. aegypti* and *Ae. albopictus* mosquitoes are competent to ZIKV, but they present various levels of susceptibility [3]. According to Wong et al. [83], data from their previous and current studies suggest that both *Ae. aegypti* and *Ae. albopictus* will have a significant role in the transmission of ZIKV [83].

"[At least i]n the U.S., the range of *Ae. albopictus* occupies more states than *Ae. aegypti*," Fauci said. But he doubts it will spread Zika much. He noted that *Ae. aegypti* has usually been the main spreader of Zika [84].

These species have solid ecological plasticity. *Ae. albopictus* has been called "the most invasive mosquito globally" and has "relatively cold-hardy and long-lived eggs," allowing it to survive in cooler temperatures than *Ae. aegypti*. Concerns are that this species could fill the ecological niche left when *Ae. aegypti* were eliminated or substantially reduced. In addition, its spread would expose even more populations to infection.

Both *Ae. aegypti* and *Ae. albopictus* are resilient. They can survive extended periods of trans-oceanic transport as dry eggs in car tires or other containers [12]. At present, aside from its tropical Asian home, *Ae. albopictus* can be found in temperate Asian countries, tropical and temperate Americas, Europe, the Middle East, the Pacific islands, Australia, and Africa. With the increased global urban expansion and human-made developments, Aegypti and Albopictus species have expanded their geographic range and become associated with urban landscapes. *Aedes Albopictus* is less adapted to domestic environments than is *Ae. aegypti* but is widely distributed in peridomestic habitats within cities [64].

A study by Gubler [85] has shown that although they are susceptible to ZIKV infection, American populations of both *Ae. aegypti* and *Ae. albopictus* collected from Florida have unexpectedly low levels of vector competence (the mosquito's ability to transmit ZIKV biologically). However, *Ae. aegypti* (and to a lesser extent *Ae. albopictus*) is thought to have an elevated level of vectorial capacity (a measurement of the efficiency of ZIKV transmission) because it primarily bites humans, infects several persons during a single blood meal, and lives in close association with humans [86].

10.2.4 *Debates Over Interspecific Mating Between Aegypti and* Albopictus

Suppose there is an attempt to reduce the *Ae. aegypti* population only to have male *Ae. albopictus* mosquitoes cross breed with female *aegypti* mosquitoes, producing infertile male *aegypti* might be problematic. In 2015, Bargielowski, Blosser, and Lorunibus [87] reported that *aegypti* females were more susceptible to interspecific insemination than *Albopictus* females. However, Leahy and Craig [88], Harper and Paulson [89], and Lee et al. [90] reported interspecific mating does not produce viable offspring.

There is have solid data that reducing viable male *Ae. aegypti* populations, in general, should have a corresponding effect on the entire viable offspring populations of *Ae. aegypti.* It is not likely female *Ae. aegypti* mosquitoes could be satirized by another mosquito species, such as *Ae. albopictus.* Honório et al. [91] found that *Ae. albopictus* males from Brazil inseminated *Ae. aegypti* females at meager rates, regardless of their population origin [91].

Research by Marceloa et al. [92] further suggests that *Ae. aegypti* males were not species-specific in mating. If released into the field as practiced in genetically modified mosquito techniques, they may mate with both *Ae. aegypti* and *Ae. albopictus* females, hence reducing populations of both species by producing infertile eggs [92]. The debate over whether the offspring of interspecific mating is viable has not been settled. Other studies on interspecific mating between *Ae. aegypti* and *Ae. albopictus* reported both viable [93] and nonviable offspring [94].

Nasci et al. [95] researched interspecific mating between *Ae. albopictus* males and *Ae. aegypti* females in the field using mark-release-recapture techniques. By three days after the release of virgin *Ae. aegypti* females into a field site containing only *Ae. albopictus*, 100% of the captured females were inseminated. Laboratory investigations indicated that male *Ae. albopictus* were very proficient at inseminating *Ae. aegypti* females and that *Ae. aegypti* males rarely inseminated *Ae. albopictus* females, especially if *Ae. aegypti* females were available [95].

10.2.5 Debates Over Culex

Some continue to insist that *Culex* are ZIKV vectors as well [96]. Recently, a controversial report originating from China [97], and a second report from Brazil [98], suggested that *Culex quinquefasciatus* could also potentially transmit the virus.

Also, in 2016, a team from Beijing reported laboratory results demonstrating the potential role of *Cx. p. quinquefasciatus* as a vector of ZIKV in China. Because there are pretty different vector management strategies required to control Aedes (Stegomyia) species and *Cx. p. quinquefasciatus*, an integrated approach may be required should a Zika epidemic occur [99]. They add: Compared with *Ae. aegypti* and *Ae. albopictus*, *Cx. p. quinquefasciatus* has much greater densities in human habitats. Although *Cx. p. quinquefasciatus* is an opportunistic blood feeder, humans are its common host in urban China. Although *Cx. p. quinquefasciatus* may not be the primary vector of ZIKV (*Ae. aegypti* occupies that role) owing to its high density, human blood-feeding behavior, and vector competence shown herein, its potential importance in the transmission of ZIKV cannot be ignored, especially in urban areas and towns [99].

Brazilian researchers found comparable results. Researchers led by Ayres analyzed 456 female *Culex* mosquitoes, divided into 80 "pools" or sample groups of between one and ten mosquitoes each. They found ZIKV-infected insects in three of these pools. Ayres said the research proved that *Culex* can transmit ZIKV and could have played a role in Brazil's rapid spread of the disease [100].

Guedes et al. [101] searched for evidence of other mosquito species, including *Culex quinquefasciatus*, playing a role in the Brazilian outbreak. To evaluate this hypothesis, they compared the vectorial competence of laboratory-reared *Ae. aegypti* and *Culex quinquefasciatus*. ZIKV has been found in the midgut, salivary glands, and saliva of artificially fed *Culex quinquefasciatus*. The present study indicates that *Culex quinquefasciatus* mosquitoes may be involved in ZIKV transmission in Recife [101].

Amraoul et al. concluded that *Culex* mosquitoes are unlikely to be involved in the rapid spread of ZIKV based on their study of two laboratory colonies of *Culex quinquefasciatus* and *Culex pipiens*. In both cases, they could not transmit ZIKV either up to 21 days post an infectious blood meal or 14 days post intrathoracic inoculation [102].

Sérgio Bessa Luz, the director of the federal Fiocruz Amazônia research center in Manaus, cautioned that restraint was in order: "This is preliminary data," he said. He said that even if the virus is in the saliva, "it's not definitive" that it can transfer ZIKV effectively to humans. "We can't yet conclude [*Culex* is] a vector." [103].

In addition, *Culex quinquefasciatus* is found on all islands and is the most common night-biting mosquito in Hawaii [104]. In California, *Culex* has 53 species and is the second-largest genus of mosquitoes [105]. In Florida, *Culex quinquefasciatus* is found in all 67 counties [106]. *Culex quinquefasciatus* is found as far north as Iowa and Indiana in the United States, although window screens and other factors protect people [23].

While the consensus is that the *Ae. aegypti* mosquito is the primary vector to humans worldwide, the *Culex* mosquitoes transmit several viruses related to ZIKV, and it would not be shocking if both *Culex* and could spread ZIKV. Gubler agrees that *Culex* is a plausible carrier. He notes that several ZIKV relatives spread by *Culex* mosquitoes, including the West Nile virus, target the nervous system, which ZIKV also seems to do [23].

The two mosquitoes are radically different: The *Aedes* feeds during the day; the *Culex* at dawn and dusk. *Aedes* likes fresh water; *Culex* likes dirty water. *Aedes* specializes in feeding on people; *Culex* prefers birds [103]. In addition, *Aedes* and *Culex* do not share the same breeding sites. Amraoul et al. vector control should focus on larval and adult habitats specific to *Aedes* mosquitoes to control ZIKV vectors efficiently [102].

Just because a mosquito can carry a disease does not decide whether it is "an important vector," he said. "The main remaining question regarding *Culex* is how often they feed consecutively on humans, which is likely far less than *(Aedes) aegypti*," Scott Weaver, the scientific director of the Galveston National Laboratory, said [103]. In addition, though *Culex* mosquitoes have an even broader territory than *Ae. aegypti* and are common across Europe and the U.S. they carry disease less efficiently than *Aedes* because they tend to bite fewer humans in a row, limiting how many people they infect.

Before moving on, concerns have been expressed about vector competence research for neglecting other mosquito species, such as *Culex perfuscus*, which is abundant and has been implicated in spreading other arboviruses closely related to ZIKV, including West Nile Virus.

In addition, it has been reported researchers have isolated ZIKV virus strains from *Anopheles constant*, *Mansonia iniformis*, and *Aedes albopictus*. Ayres argues that "vector control strategies must be directed at all potential vectors [107].

These vectors are not equally concerning unless one or more refills the niche left decimated from a successful vector strategy against *Ae. aegypti*. Although *Culex. quinquefasciatus* may not be the primary vector of ZIKV (*Ae. aegypti* occupies that role) owing to its high density, human blood-feeding behavior, and vector competence shown herein. The potential importance in transmitting ZIKV cannot be ignored, especially in urban areas and towns" [99].

"Some populations [of *Culex*] have already adapted to living side-by-side with humans and are as efficient at biting humans as *Ae. aegypti*. Larvae will develop in sewers and pit latrines, and adults live in people's houses." Were a *Culex* mosquito to become infectious, there is also no telling whether it would prefer humans over the birds and other animals it has usually been known to feed on, Tulane's Dawn Wesson added. Several studies have concluded that *Culex quinquefasciatus*, the southern house mosquito, feeds minimally on humans [108]. "ZIKV is a loose cannon," said Dina Fonseca, an entomologist at Rutgers University. Until it is proven that the virus will not infect *Culex* mosquitoes or any other reservoir animals—like the birds that exacerbated the West Nile virus—scientists just do not know how much of a threat ZIKV poses [108].

Peter Hotez, the dean of the National School of Tropical Medicine at Baylor College of Medicine, told The Atlantic. That would be particularly worrisome for public health officials in North America because *Culex* mosquitoes are heartier than their tropical cousins and already live as far north as Canada. "If that turns out to be the case," Hotez added, "then we're all totally screwed." [108]. However, Hotez was skeptical that *Culex* mosquitoes were going to start spreading ZIKV in a significant way. "So far, every place we have seen ZIKV has been a place where you have *Ae. aegypti* mosquitoes," he said. "There is no reason one would have to speculate that another mosquito vector is involved." [109].

If further research confirms the early results detected ZIKV in *Culex* mosquitoes [107], many countries could be forced to rethink their mosquito control campaigns radically. In Brazil, those efforts have focused almost exclusively on eradicating a different kind of mosquito—the *Aedes aegypti.* Ayres said that if *Culex* mosquitoes prove to be a significant vector, it would require an overhaul of mosquito control strategies: "The strategies that we are using to control *Ae. aegypti* are useless for *Culex* because they are completely distinct species." [103].

10.3 Debates Over Reducing Vector Densities

We are witnessing a teaching and a learning moment if we can set aside concerns regarding how practical a speciescide population crash approach may be. ZIKV has opened the consideration of whole novel approaches to mosquito management [110].

Four issues overlap: (i) How effective are current pesticides? Since the 1980s, vector control efforts have been cut back or hampered by a lack of effective pesticides, allowing the *Aedes* species to return to North and South America and the Caribbean. (ii) How pervasive is the carrier? Revised vector surveillance estimates by the CDC show that *Ae. aegypti* is present over a wider geographic area in the US than previously thought, as far North as New York and Northern California. This is a function of climate shifts coupled with urbanization and travel. (iii) How has policy toward mosquito eradication changed? This is mainly dealt with below when examining the role of contemporary mosquito control agencies and state and federal pest control strategies. (iv) How has the ZIKV impacted the debate? Interestingly, the ZIKV carrying *Ae. aegypti* mosquito is enabling an essential conversation about the role genetic engineering may play in the future. Here seems to be an unimportant species that does not occupy an important niche in an ecosystem [111].

Out of the more than 3500 mosquito species, only around four hundred can transmit diseases like malaria and West Nile virus to people, and most do not feed on humans. Many of the complaints found below are associated with mosquitoes that carry deadly diseases such as *Ae. aegypti* and *Ae. albopictus* is heavily inflated.

10.3.1 Benefits from Mosquitoes

Nonetheless, it is essential to understand the general concerns about the roles played by mosquitoes in the ecosystem. Genetically engineering mosquitoes to die off could risk species that rely on them, including threatened amphibians, bats, and birds.

As Yvonne-Marie Linton, the research director at the Walter Reed Biosystematics Unit, which curates Smithsonian's U.S. National Mosquito Collection put it: "Mosquitoes do not deserve such a bad rap and form an essential source of biomass in the food chain—serving as food for fish as larvae and for birds, bats, and frogs as adult flies—and some species are important pollinators [112].

They live in almost every habitable continent except the permanently frozen areas and serve various functions in different ecological conditions. In diverse ecological habitats, the mosquito is someone's prey, or it preys upon other organisms. Complete eradication of mosquitoes may halt the evolutionary process. Ecological processes disrupted by extinction or species decline may also lead to cascading and catastrophic coextinctions [113]. But no one advocates the complete end of the mosquito population, just those carrying diseases and participating in human bloodmeals.

The oldest known mosquito with anatomy like modern species was found in seventy-nine million years old Canadian amber in the Cretaceous period. An even older sister species with more primitive features was found in Burmese amber, which is 90–100 million years old.

10.3.1.1 Animals

Aquatic turtles each have a diversity of animals, including mosquito larvae. The Red-Eared Slider Turtle has an enormous capacity to consume mosquito larvae. Juvenile turtles can eat more than 500 3rd and 4th instar (larval stage of development) mosquitoes daily, which is why they are often used in stormwater catch basins, holding ponds, and water storage tanks though hardly practical for some locations such as roadside ditches [114].

According to Kirchner [115], the Eastern Red-Spotted Newt eats mosquito eggs and larvae but lives nearly exclusively in forests.

The Center for Food Safety was concerned about bats that depend on mosquitoes for their diet, predominantly brown bats. Brown bats eat them as well though they seem to prefer small flies [116]. Recent research at the University of Wisconsin [117] showed that bats eat many types of mosquitoes. While bats may play an essential role in suppressing agricultural pests, studies that found bats consume ten mosquitoes per minute or 1000 per hour came from an enclosure experiment not under natural conditions [118]. Even Wray et al. admit that molecular studies have generally found that mosquitoes represent a small portion of the diet of wild bats and that the taxonomic richness of mosquitoes consumed by bats is low. According to entomologist Michael Doyle, a bat would have to eat thousands of them to equal a couple of moths [119]. Less than 2% of their gut content is mosquitoes [120]. This was due to their small size, poor detectability by low-frequency echolocation, and variable field metabolic rates and because *Ae. aegypti* mosquitoes are active during the day, whereas bats are crepuscular [121]. If you are expending energy," says medical entomologist Janet McAllister of the Centers for Disease Control and Prevention in Fort Collins, Colorado, "are you going to eat the 22-oz filet-mignon moth or the 6-oz hamburger mosquito?" [120].

Caribou select paths during migration to face the wind to escape swarming mosquitoes. Conceivably, a change in traditional paths could have significant consequences in an Arctic valley where caribou each lichen, transport nutrients, feed wolves, and generally alter the ecology [120]. The mosquito species inhabiting Arctic valleys are not *Ae. aegypti*, at least not yet.

10.3.1.2 Birds

"The number of migratory birds that nest in the tundra may drop by more than 50% without mosquitoes to eat," says Bruce Harrison from North Carolina's Department of Environment and Natural Resources.

An intensive seven-year diet study conducted at PMCA headquarters in Edinboro, PA, failed to find a single mosquito among the five hundred diet samples collected from parent martins bringing beakfuls of insects to their young. Therefore, due to the temporal separation in activity periods and that the mosquito is likely to form only a tiny part of the bird's diet, it is unlikely that a single species such as *Ae. aegypti* would significantly impact insectivorous birds [121].

Nighthawks, Purple Martins, Eastern Bluebirds, Red-Eyed Vireos, Yellow Warblers, Downy Woodpeckers, House Wrens, Baltimore Orioles, and Hummingbirds consume adult mosquitoes.

10.3.1.3 Insects

Insects, including dragonflies, eat mosquitoes on the wing, and dragonfly larvae eat mosquito larvae.

The Center for Food Safety expressed concerns over amphibians, including mosquitoes in their diet [122]. The CVM concluded that it is unlikely that extinguishing a single species would have a significant impact on predatory amphibians [121].

10.3.1.4 Plants

On the east coast of the USA, pitcher plants use mosquito larvae symbiotically. Midges chew up carcasses trapped in the plant, and mosquito larvae devour waste products, releasing nitrogen the plant uses to grow [120].

Also, mosquitoes are pollinators, and nectar from flowering plants, not blood, is used for energy. Like bees or butterflies, mosquitoes transfer pollen from flowers as they feed on nectar, fertilizing plants and allowing them to form seeds and reproduce [123].

According to Peach and Gries [124], mosquitoes are generally considered nectar thieves that do not contribute to the pollination of the flowers they visit. They consume nectar as legitimate pollinators do but without effectively transferring pollen between inflorescence, essentially "cheating" by reaping the rewards without providing any pollination service [124]. A case can be made that nectar thieves reduce the number of visits by legitimate pollinators and may adversely affect plant fitness [125].

Despite feeding on plant nectar, mosquitoes likely transfer pollen to some extent, although there is little scientific information on this [121]. There is limited information on the pollination of plant species by mosquitoes and no reports that *Ae. aegypti* is a pollinator for any essential plant species.

Janet McAllister of the Center for Disease Control and Prevention in Colorado added that pollination from mosquitoes is not crucial for crops on which humans depend. However, at least one orchid species seems to depend on mosquito pollination [126].

For example, *Aedes* spp. mosquitoes, including *Ae. aegypti*, are effective visitors to the Blunt Leaf (*Platanthera obtusata*) orchid. Small moths also pollinate it [127]. Another study examined this species of orchid, also known as a bog orchid) and found the eight subspecies of Aedes, including neither *Ae. aegypti* nor *Ae. albopictus* as pollinators [128]. A study by Lahnodere et al. observed this orchid was visited by various mosquito species, both sexes, which belonged to the Aedes group [129]. A second plant species, *Silene otitis*, called the Spanish catchfly, is native to Europe and the Transcaucasus area and was introduced to Xinjiang in China. According to observations by Brantjes and Leemans [130], mosquitoes might be an important pollinator due to high activity among the flowers on most evenings [130]. Here again, mosquitoes and moths were copollinators.

According to John Dexter from Columbia over one hundred years ago, this is the only case reported in which mosquitoes seem to be of primary importance—as agents of pollination [131].

Mosquitoes are not considered significant pollinators; an exception exists in subarctic regions of Northern Canada and Russia, where they play a significant role in pollination. In other areas, bees and butterflies outdo the mosquitoes in pollination. Hence, the results of mosquito extinction might not be felt in all areas, but it would take a toll in subarctic regions where plants rely on them for pollination [113].

10.3.1.5 Wetland Ecosystems

Mosquito larvae feed on detritus that floats and clogs the water surface preventing the detritus from choking off nitrogen and oxygen necessary for the plants below. In this sense, mosquitoes are essential to a functional wetland ecosystem, processing detritus, and aquatic microbes and eventually providing a link between aquatic and terrestrial systems when they emerge [113].

Mosquitoes are generally crucial for pond ecosystems. Mosquitoes and tadpoles consume a wide range of organic materials, including algae, detritus, and fecal pellets.

Their presence could significantly influence ecosystem structure and processes through direct consumption of resources and the export of stored biomass from the aquatic to terrestrial biome at metamorphosis [132].

Schafer also addressed mosquito value [133] and considered mosquitoes' contribution to wetland biodiversity in various wetland types, from forests to meadows to shallow water.

Allie Mokany, the strategic director of the Australian Department of Agriculture, Water, and the Environment (2007), claims mosquito larvae increase metaphyton dominance through the preferential consumption of edible phytoplankton and bacteria over the less palatable metaphyton algal mats [134].

Global wetland loss means constructed urban wetlands are an increasingly valuable resource for conservation and a priority for ecologically minded developers. However, preferences for managing urban wetlands for protection often conflict with management to reduce potential mosquito risks, such as nuisance biting and pathogen transmission. As constructed urban wetlands become more common, so does the need for routine maintenance and management of threatened and invasive species. Future design and management of urban wetlands must consider mosquitoes' impact on humans [135].

10.3.1.6 Parasites, Pathogens and Symbionts

Mosquitoes act not only as vectors/hosts of these organisms but also as intermediate or definitive hosts in the life cycle of hose parasites/pathogens. Indeed, it seems a variety of parasites, pathogens, and symbionts are dependent on mosquitoes for their survival.

Eradication of mosquitoes from nature may have an excellent selection pressure on these organisms. Most of them may become extinct if their vectors/hosts are eradicated [113].

10.3.2 Consequences of Eradicating Ae. aegypti

Generally, scientists acknowledge that the ecological scar left by a missing mosquito would heal quickly as other organisms filled the niche [120]. Eradicating *Ae. aegypti* would have a minimal impact on the environment. It is an invasive species with no apparent role in the food chain.

In addition, since *Ae. aegypti* did not coevolve in Florida or even in America. It is unlikely that the other species rely heavily on or moderately on *Ae. aegypti* as a food item or provider of ecosystem services. Furthermore, *Ae. aegypti* does not constitute a keystone species. [T]here appears to be no specific predator that preys on *Ae. aegypti* but rather predators that are generally opportunistic and feed on larvae when they encounter them. From a review of the scientific literature conducted in PubMed, no papers were identified in which a predator was found to depend on *Ae. aegypti* alone as a food source [121].

10.3.3 Minimal Impacts

There would be some impacts, but those seem minimal. Unlike other types of mosquitoes, *Ae. aegypti* is not a food source for birds and fish. [136] For example, mosquitofish are stocked in rice fields and swimming pools as pest control could experience significant population losses. Some species of insect, spider, salamander, lizard, and frog might also lose a primary food source [120]. Mosquito-eating birds would probably switch to other insects that post-mosquitoes might repopulate their niches [120]. Others like Cathy Curby from the Alaska Fish & Wildlife Services say that midges are a more important food source [120].

Mosquitoes do not play critical roles in many environments except the Arctic tundra as a food resource for birds. Biologists believe in most other systems; the ecosystem could survive the loss [137].

10.3.4 Ethics

Mother nature is unbiased toward all her creations. Every creation of natural selection is no less important than others and has a role to play in nature, and the mosquito is no exception. Mosquito eradication doctrine may also raise a bioethical [113] question—do humans have the right to decide which insect, creation of organic evolution and an integral part of different ecosystems, deserve to live on planet Earth?

10.4 Debates Over Reservoir Hosts

Most arboviruses cause zoonoses that usually depend on nonhuman animal species for natural maintenance. Many animal species are host reservoirs (hosts of an infection in which the infectious agent multiplies and develops and on which the agent is dependent for survival in nature) of arboviruses; humans, with few exceptions, are dead-end or accidental hosts (hosts from which infectious agents are not transmitted to other susceptible hosts) [138].

Two distinct species of flaviviruses, dengue virus (DENV) and yellow fever virus (YFV), originated in sylvatic cycles maintained in non-human primates, and forest-dwelling have emerged repeatedly into sustained human-to-human transmission by *Ae. aegypti* mosquitoes [139]. Arboviruses do regularly occur in new transmission cycles. Among these are YFV, well-known for its ability to move from a jungle cycle into a devastating, albeit transient, urban cycle in humans, and DENV, which has emerged from an enzootic cycle on four separate occasions to establish ecologically distinct human transmission cycles [139].

> A significant determinant of ZIKV stability in the Americas will be its ability to establish an enzootic, non-human primate-hosted transmission cycle. A recent modeling study (Atlhouse et al., 2016) demonstrated a high probability of enzootic establishment across a wide range of biologically plausible parameters, such as host and vector population sizes, host birthrates, and the ZIKV force of infection. A sylvatic cycle of ZIKV would make future elimination efforts in the Americas practically impossible and paints a dire picture for the epidemiology of ZIKV and our ability to end the ongoing outbreak of congenital Zika syndrome.... Suppose the virus establishes a sylvatic cycle in the Americas. In that case, mosquito control and even herd immunity from vaccination or limitation of transmission from demographic turnover will not suffice to eradicate it from the region. [140]

The available data indicate two cycles of ZIKV: a sylvan cycle involving nonhuman primates and forest-dwelling mosquitoes and an urban/suburban cycle involving humans and *Ae. aegypti* and, to a lesser extent, *Ae. albopictus* [141].

Ferguson et al. warn: If a sylvatic reservoir for Zika is established in the Americas, background levels of human exposure may increase [142].

The wide distribution of the virus in animal hosts and vectors favors the emergence of recombinants. There is insufficient information regarding the animal reservoirs and amplification hosts, including domestic animals, and the vectors of ZIKV, as well as the vector capacity of the genus *Aedes* and genus *Anopheles*. This represents a public health emergency. Understanding of these factors must improve because they will define the transmission dynamics and geographic distribution of ZIKV and indicate the timing and scale of environmental public health interventions [143].

The ever-present hazard is that novel human pathogens emerge from livestock, wild mammals, and bird reservoirs [144]. Reservoir hosts are carriers of a pathogen and often do not get the disease. It carries a subclinical infection and is asymptomatic and non-lethal to the host. Reservoir hosts are often used as another reason to attack the panoply of vectors of infection. Many species of mosquitoes associated with ZIKV may not favor humans for a blood meal but the effort to crash the *Ae. aegypti* population may not be sufficient to eradicate ZIKV if other species can find transferable ZIKV in animals serving as reservoir hosts.

The virus has been isolated in monkeys, and antibodies have been detected in domestic sheep, goats, horses, cows, ducks, rodents, bats, orangutans, and carabaos [143]. ZIKV seropositivity in rodents, bats, ducks, horses, and sizeable domestic livestock [26], zebras, orangutans, elephants, and rodents [145]. The cell lines from rabbits, pigs, and chickens also show rapid replication of ZIKV [146].

Dr. Marco Safadi thinks ZIKV may stay around for years, citing a published study that showed monkeys had been infected with Zika in Brazil. That means they can function as a reservoir, as they do with yellow fever. Even if a vaccine is developed and people get vaccinated, or if enough of the population gets infected to confer widespread immunity, the monkeys will be there for the mosquitoes to bite and carry the infection to people in the future. That will make Zika a problem for years to come [147].

Some evidence suggests that men shed infectious ZIKV in semen for weeks or months after the acute infection is resolved, potentially acting as long-term transmission reservoirs [148]. Froeschl et al. found ejaculate samples in a ZIKV-positive male after 77 days and in whole blood samples until 101 days. They believe they may have identified possible localizations of ZIKV replication in the human male reproductive tract [149].

10.4.1 Birds: Bulbuls

For ZIKV, there is some speculative evidence on bulbuls. It comes predominantly from a Canadian blogger called RoseWrites, and she publishes her blogs on InfoBarrel. She claims she has a background in nursing and was a frontline worker during the SARS outbreak. She sells Zika-related products on Zazzle, the profits from which she claims goes to Zika research. While she is an outsider, her comments are often footnoted in the peer-reviewed literature—more from her below concerning the side effects of Wolbachia.

Some species of birds are called amplifying hosts, meaning that when they get infected by a WNV-carrying mosquito, the virus can replicate to such high numbers in the blood that a new mosquito coming along for a snack will be able to pick up the virus as well. [B]irds can have up to a billion times more viral particles in their blood than similarly infected humans. [B]irds maintain these high blood virus levels for 15 days. During this time, a single "super amplifier" bird can infect 100 s of mosquitoes [150].

The bulbuls are a popular pet in Southeast Asia. The bulbul appears to be the most likely amplifying reservoir host [151]. In 1971, 15% of birds evaluated in Uganda, primarily bulbuls, had Zika.

Red-whiskered bulbuls have been introduced across most of the planet, especially in Southeast Asia, where one bird could infect hundreds of mosquitoes. The Red-whiskered Bulbul has established itself in Australia, Los Angeles, Hawaii, and Florida in the U.S., Mauritius, Assumption Island, and the Mascarene Islands.

10.4.2 Bats

Concerning the neotropical chiropteran model, viral RNA was found in different tissues of fruit bats at 28 dpi, suggesting ZIKV can infect bats which may serve as virus reservoirs. The bats did not show any signs of disease [152].

10.4.3 Primates

ZIKV can spread between people via sexual contact. Still, the virus likely dwells in some wild animal population—probably non-human primates—before being picked up by mosquitos and transmitted to humans [153].

Scientists usually worry that animal diseases could spill over to humans. But "spill-back" of ZIKV into monkeys in South America could be just as dangerous. Phylogenetic evidence indicates ZIKV frequently passes between non-human primates and humans in Africa, and numerous studies in Africa and Asia show serologic evidence for ZIKV infection in nonhuman primates [154].

In the wild, animals can function as reservoirs for ZIKV between human outbreaks. A small number of black-striped capuchin monkeys and common marmosets in a region of Brazil with high numbers of human cases were already found to carry the virus, the first such report among Western Hemisphere monkeys [155].

According to disease ecologist Barbara Han of the Cary Institute of Ecosystem Studies in Millbrook, N.Y., "If the disease gets established in monkeys or other wild primates, the animals may serve as reservoirs for future human outbreaks. That could make it nearly impossible to get rid of the virus. Han and colleagues calculated the risk of ZIKV entering South American primate populations using criteria that include species range, body size, and diet. Two contenders on her list of at-risk species—black-striped capuchin monkeys and common marmosets—had been found by other researchers to be infected with ZIKV matching the human strain in Brazil. The finding indicates the spillback has already started. Capuchins are of particular concern because they are often kept as pets and used to attract tourists. The possibility for close contact with humans is already there," Han said [156].

10.4.4 Humans

Even asymptomatic individuals can serve as virus reservoirs that can disseminate the virus if mosquitos take blood meals from those individuals, allowing those mosquitos to spread the virus to others who may develop a more severe infection [157].

Transmission persists at low but steady levels with no symptoms [158]. However, the virus may stay in acute or chronic stages in humans or as a zoonotic infection in reservoir hosts, which can be a source for future outbreaks [159].

10.4.4.1 Testes, Cerebrospinal Tissue and Lymph Nodes

Campos et al. [160] investigated the persistence of ZIKV in serum samples of seventy-seven subjects infected by the virus between eighteen months and three years before the start of this study. The subjects included children with microcephaly and their parents who have detected viremia in healthy carriers up to three years after symptom onset. Humans acting as potential viral reservoirs have significant implications for the current understanding of ZIKV infection [160]. The results reported in this study show that ZIKV can persist in asymptomatic infected individuals and potentially transmit or perpetuate the virus as viral reservoirs. It is well documented that ZIKV can continue in human body fluid with RNA detection in serum, urine, saliva, and cerebrospinal fluid, and for more extended periods in semen (188 days) [160].

Scientists believe the prostate or testes serve as a reservoir, according to Lazear et al. [161]. "We looked for evidence of ZIKV in the mouse testes mainly as an afterthought due to mounting evidence of sexual transmission and were surprised that viral levels were the highest we saw in any tissue" [161].

Mansuy [162] describes a French case study from a traveler to Brazil. The viral load in the semen was roughly 100,000 times that of his blood or urine more than two weeks after symptom onset. The reason for this difference is unknown and needs investigation. Perhaps the virus can replicate in the male genital tract, fill a specific genital reservoir, or both [162].

Nicastri et al. [163] collected data suggesting the virus could replicate specifically in the male genital tract and may persist in semen, with implications for potential male-to-female sexual transmission, even in the absence of haematospermia. They concluded "[b]ecause of prolonged detection of ZIKV RNA and isolation of the replication-competent virus in semen, the testes are considered an immune-privileged replication site for ZIKV" [163]. The virus can persist even if it cannot replicate in the male reproductive tract because the testes are immunologically privileged, and the immune response is restricted to enable sperm survival [164].

The presence of ZIKV in semen is a significant challenge, as is the possible teratogenicity of the virus. More than 80% of infected people are probably asymptomatic, making them an enormous potential reservoir. Hence, pregnant women in areas where the ZIKV virus is widespread should protect themselves not only from mosquitoes but also from the infectious virus in the semen of their partners—at least during pregnancy [162].

Aid et al. [165] studied the persistence of ZIKV in infected rhesus monkeys. Prolonged viral shedding has been reported in semen, suggesting the presence of anatomic viral reservoirs. They showed that ZIKV could persist in cerebrospinal fluid (CSF) and lymph nodes (LN) of infected rhesus monkeys for weeks after the virus has been cleared from peripheral blood, urine, and mucosal secretions. In contrast with the rapid virologic control in peripheral blood, viral persistence was observed in the CSF for up to 42 days and in LN and colorectal tissues for up to 72 days, suggesting the presence of multiple anatomic sanctuaries for the virus. These data indicate that viral persistence may result from the stimulation of pathways that promote the survival of ZIKV-infected cells and block the recruitment of immune cells to the site of infection. ZIKV-specific antibodies were not detected in CSF, despite robust and prolonged responses in peripheral blood, suggesting an additional mechanism for viral persistence in immune-privileged sites. These data indicate that persistent or occult neurologic and lymphoid disease may occur following clearance of peripheral virus in ZIKV-infected individuals [165].

"As exciting as it is to get added information, it also raises questions," Fauci continued. Among them, "if the virus can linger in the central nervous system even after people recover, will we start seeing some subtle effects later, or will it just be an inconsequential phenomenon clinically? Right now, we do not know the answer to that because we do not have enough experience with the long-term effects of Zika." [166].

10.5 Debates Over Jurisdiction

Vector control is a critical element of the effort to combat the spread of mosquito-borne disease. Novel mosquito control technologies have gained greater attention as an element of this effort; however, there has been some confusion concerning FDA's and EPA's respective authority over such mosquito-related products. Jurisdictional issues associated with government policy in the U.S. is known.

Given the public health implications of mosquito control, FDA is providing this guidance to clarify the regulatory oversight of mosquito-related products, including but not limited to those produced through biotechnology [167].

The US Food and Drug Administration, evaluating Oxitec's skeeters since 2011, transferred approval power to the EPA. In short, Oxitec's bugs have been deemed more like a pesticide—used to suppress wild mosquito populations—than a drug used to prevent disease. And that promises to speed up the company's ability to get its product on the ground, especially in hurricane-prone areas, where big storms can exacerbate mosquito-borne diseases [168].

Since the Federal Insecticide, Fungicide, and Rodenticide Act's definition of pesticide was amended in 1975, EPA has registered as pesticides articles that control the population of mosquitoes by killing them or interfering with their reproduction, which is consistent with FDA's and EPA's general agreement that articles or categories of articles that control the population of mosquitoes are most appropriately regulated as pesticides [167].

Given this history, FDA clarified that the phrase "articles (other than food) intended to affect the structure, or any function of the body of man or other animals" in the Food, Drug, and Cosmetics Act's drug definition [21 U.S.C. 321 (g)(1)(C)] does not include articles intended to function as pesticides by preventing, destroying, repelling, or mitigating mosquitoes for population control purposes [167].

The FDA oversees technologies for sterilizing and controlling animal populations but giving it responsibility for gauging the environmental impact of a mosquito raised eyebrows on both sides of the debate. "Without relevant expertise, not surprisingly, the FDA has been ill-equipped to review the application expeditiously," wrote attorney John Cohrssen and physician and former FDA official Henry Miller in a Forbes op-ed last January. In draft guidance released in January 2017, FDA explains that products intended to reduce mosquito populations should be considered pesticides. That means that, if finalized, the direction routes future mosquito strains to EPA for review. (Designer mosquitoes intended to reduce disease transmission, such as those that spread the insect parasite Wolbachia, would still be considered "new animal drugs" and would fall to FDA) [169].

The travails for Oxitec (Chap. 13) do not seem to be over any time soon though they are carrying out some experiments. Six organizations—some based locally, some internationally—argue the FDA did not consider the impact the experiment could have on endangered species living in the Florida Keys [170]. Friends of the Earth, Foundation Earth, International Center for Technology Assessment, Florida Keys Environmental Coalition, and Food and Water Watch.

"Threats to endangered and protected species are detailed in numerous comments to FDA, including comments from the U.S. Fish and Wildlife Service requesting more data. Under the Endangered Species, FDA is required to consult FWS on these potential impacts to protected species," George Kimbrell, from the Center for Food Safety, said in the statement. "If FDA does not cure these violations within 60 days, the listed organizations intend to file suit against the responsible agencies and officials to enforce the Endangered Species Act," he said [170].

The switch from FDA to EPA oversight moves vector collapse strategies forward. That is because the EPA is required by federal law to review new pesticides "as expeditiously as possible," which the statute defines as within 12 months after applying [171].

10.6 Conclusion

Vector control programs can be effectively crippled by the proliferation of backyard non-degradable trash containers in today's throw-away society, inadequate community involvement in eliminating or controlling mosquito developmental sites, and lack of access to households for insecticide application [172].

Vector control can be proactive and reactive, or both. Proactive vector control refers to measures implemented before knowledge of elevated vector or disease risk based on surveillance results in the present time. Reactive vector control for actions implemented only in direct response to increased risk based on vector, virological/serological, or disease surveillance results in the present time [172].

In the case of dengue, Eisen et al. [172] argue an effective control strategy should include a strong focus on locally appropriate proactive control approaches while still retaining the logistical capacity for implementing reactive emergency vector control measures in an event outbreak [172].

Eisen et al.'s primary complaint was that vector control activities are initiated only in response to laboratory-confirmed dengue cases in some settings. They are reactive. Unless laboratory confirmation is achieved rapidly, this system will result in the response activity, using a fire-fighting analogy, arriving on the scene when the building has already been reduced to ashes. The fire has moved on to another structure [172].

Therefore, an improved understanding of the potential wildlife host involvement in transmitting emerging flaviviruses is essential. For instance, data about the transmission of ZIKV in South American wildlife are critical to inform on the maintenance of the virus in the region, as sylvatic hosts can serve as a source of recurring epidemics and impede efforts for long-term disease prevention [173].

For emerging flaviviruses such as ZIKV, there is growing concern for establishing sylvatic cycles due to spillback transmission from humans to wildlife in geographical regions predicted with high ZIKV host diversity [173].

Another concern, supported by the detection of ZIKV RNA in Brazilian monkeys in proximity to humans [174], is the possibility that ZIKV will establish a zoonotic cycle in the Americas. This would be akin to introducing yellow fever to the Americas in the seventeenth century and would serve as a focus for future spillover infection to humans [175].

The next chapter will review vector management in the twentieth century, how it is practiced, what it involves, and where it may be going.

References

1. Korman C (2021) The mysterious case of the COVID-19 lab-leak theory. The New Yorker, October 12. https://www.newyorker.com/science/elements/the-mysterious-case-of-the-covid-19-lab-leak-theory. Accessed 14 May 2022
2. Schwab SR et al (2018) The importance of being urgent: the impact of surveillance target and scale on mosquito-borne disease control. Epidemics 23. https://www.ncbi.nlm.nih.gov/pubmed/29279187. Accessed 11 Aug 2018
3. OECD (2018) Safety assessment of transgenic organisms in the environment, Volume 8: OECD consensus document of the biology of mosquito *Aedes aegypti*, harmonisation of regulatory oversight in biotechnology. OECD Publishing, Paris
4. Alonzo P et al (2017) Renewed push to strengthen vector control globally. WHO Commentary, June 6. http://www.who.int/mediacentre/commentaries/strengthenvectorcontrol/En/. Accessed 26 June 2017
5. Focks D et al (2000) Transmission thresholds for dengue in terms of *Aedes aegypti* pupae per person with discussion of their utility in source reduction efforts. Am J Trop Med Hygiene 62(1). https://www.ncbi.nlm.nih.gov/pubmed/10761719. Accessed 27 May 2017
6. Morrison A et al (2008) Defining challenges and proposing solutions for control of the virus vector *Aedes aegypti*. PLOS Med 5(3). http://journals.plos.org/plosmedicine/article?id=10.1371/journal.pmed.0050068. Accessed 26 May 2017

7. Lambrechts L et al (2015) Assessing the epidemiological effect of Wolbachia for dengue control. Lancet Infect Dis 15. http://thelancet.com/journals/laninf/article/PIIS1473-3099(15)00091-2/fulltext. Accessed 19 May 2017
8. Chakradhar S (2015) Buzzkill: regulatory uncertainty plagues rollout of genetically modified mosquitoes. Nat Med 21(5):416–418. http://www.nature.com/nm/journal/v21/n5/full/nm0515-416.html. Accessed 3 April 2017
9. Dutra H et al (2016) Wolbachia blocks currently circulating Zika virus isolates in Brazilian *Aedes aegypti* mosquitoes. Cell Host Microbe 19. https://www.ncbi.nlm.nih.gov/pubmed/27156023. Accessed 14 April 2017
10. Iannelli J (2016) Miami beach asks FDA for emergency permission to release anti-Zika GMO mosquitoes. Miami New Times, November 1. http://www.miaminewtimes.com/news/miami-beach-asks-fda-for-emergency-permission-to-release-anti-zika-gmo-mosquitoes-8893201. Accessed 15 May 2017
11. Brown JE et al (2013) Human impacts have shaped historical and recent evolution in *Aedes aegypti*, the dengue and yellow fever mosquito. Evolution 68(2). https://www.ncbi.nlm.nih.gov/pubmed/24111703. Accessed 7 Aug 2018
12. Braack L et al (2018) Mosquito-borne arboviruses of African origin: review of key viruses and vectors. Parasites and Vectors 11(29). https://www.ncbi.nlm.nih.gov/pubmed/29316963. Accessed 10 July 2018
13. CSTE (Council of State and Territorial Epidemiologists) (2018) 2018 Zika preparedness resources toolkit. CSTE, Atlanta, GA. https://cdn.ymaws.com/www.cste.org/resource/resmgr/zika/Zika_Virus_Preparedness_Reso.pdf. Accessed 8 Aug 2018
14. Lafrance A (2016) Genetically modified mosquitoes: what could possibly go wrong? The Atlantic, April 26. https://www.theatlantic.com/technology/archive/2016/04/genetically-modified-mosquitoes-zika/479793/. Accessed 18 May 2017
15. Editors (2017) Can bacteria-infected mosquitoes stop spread of dengue, Zika? The Guardian, January 4. https://guardian.ng/features/health/can-bacteria-infected-mosquitoes-stop-spread-of-dengue-zika/. Accessed 14 May 2017
16. Cayman News Now (2017) Genetic modification project in the Cayman Islands cuts mosquitoes by over 80 percent. Cayman News Now, January 28. http://www.caribbeannewsnow.com/headline-Genetic-modification-project-in-the-Cayman-Islands-cuts-mosquitoes-by-over-80-percent-33345.html. Accessed 15 March 2017
17. WHO (2016) Mosquito control: can it stop Zika at source? Emergencies, February 17. http://www.who.int/emergencies/zika-virus/articles/mosquito-control/en/. Accessed 2 Oct 2016
18. Rubio A, Cardo MV, Vezzani D (2011) Tire-breeding mosquitoes of public health importance along an urbanisation gradient in Buenos Aires, Argentina. Mem Inst Oswaldo Cruz 106(6):678–684
19. Magori K (2018) The incredible value of mosquito surveillance and control programs. BugBitten, June 29. https://blogs.biomedcentral.com/bugbitten/2018/06/29/incredible-value-mosquito-surveillance-control-programs/. Accessed 24 July 2018
20. Jong R, Knols BG (1996) Selection of biting sites by mosquitoes. In Bock GR, Cardew G. (eds) Olfaction in Mosquito-Host Interactions, UK: Ciba Foundation. 1–331
21. OECD (2018) Safety assessment of transgenic organisms in the environment, Volume 8: OECD consensus document of the biology of mosquito *Aedes aegypti*, harmonisation of regulatory oversight in biotechnology. OECD Publishing, Paris; Lehane MJ (1991) Biology of blood-sucking insects, 1st edn. Chapman and Hall, London
22. XinhuaNet (2018) Chinese scientists find facilitator in Zika transmission. XinhuaNet, March 7. http://www.xinhuanet.com/english/2018-03/07/c_137020572.htm. Accessed 7 July 2018
23. Vogel G (2016) Scientific sleuths hunt for Zika-carrying mosquitoes. Science, June 1. http://www.sciencemag.org/news/2016/06/scientific-sleuths-hunt-zika-carrying-mosquitoes. Accessed 9 June 2017
24. Kraemer MUG, Siski M, Duda K, Mylne A et al (2015) The global distribution of the arbovirus vectors *Aedes aegypti* and *Ae. albopictus*. eLife 4:e08347. https://doi.org/10.7554/eLife.08347

25. Lambrechts L, Paaijmans KP, Fansiri T, Carrington LB, Kramer LD (2011) Impact of daily temperature fluctuations on dengue virus transmission by *Aedes aegypti*. PNAS 108:7460–7465
26. Waddell L, Greig J (2016) Scoping review of the Zika virus literature. PLOS One, May 31. http://journals.plos.org/plosone/article?id=10.1371/journal.pone.0156376. Accessed 29 June 2017
27. Costa-da-Silva AL et al (2017) Laboratory strains of *Aedes aegypti* are competent to Brazilian Zika virus. PLOS One. http://journals.plos.org/plosone/article?id=10.1371/journal.pone.0171951. Accessed 26 June 2017
28. Gardner L, Chen N, Sakar S (2016) Global risk of Zika virus depends critically on vector status of *Aedes albopictus*. Lancet Infect Dis 26. http://thelancet.com/journals/laninf/article/PIIS1473-3099(16)00176-6/fulltext. Access 27 June 2017
29. Evans M et al (2017) Data-driven identification of potential Zika virus vectors. eLife. https://elifesciences.org/content/6/e22053. Accessed 10 May 2017
30. Musso D, Gubler DJ (2016) Zika virus. Clin Microbiol Rev 29. http://cmr.asm.org/content/29/3/487.abstract. Accessed 18 July 2017
31. Brown K (2016) Genetically modified mosquitoes could wipe out the world's most deadly viruses. If we let them. Fusion.net, September 19. http://fusion.net/story/347298/oxitec-genetically-modified-mosquitoes/. Accessed 5 May 2017
32. OECD (2018) Safety assessment of transgenic organisms in the environment, Volume 8: OECD consensus document of the biology of mosquito *Aedes aegypti*, harmonisation of regulatory oversight in biotechnology. OECD Publishing, Paris; Christophers SR (1960) *Aedes aegypti* (L.) The yellow fever mosquito. Its life history, bionomics and structure. Cambridge University Press, Cambridge; Tabachnick WJ (1991) Evolutionary genetics and arthropod-borne disease: The yellow fever mosquito. Am Entomol 37. https://academic.oup.com/ae/article-abstract/37/1/14/2389211?redirectedFrom=fulltext. Accessed 25 July 2018
33. Chang C et al (2016) The Zika outbreak of the 21st century. J Autoimmunity 68:1–13. https://www.ncbi.nlm.nih.gov/pubmed/26925496. Accessed 27 Oct 2016
34. Levine R (2016) A short history of the mosquito that transmits Zika virus. OUPblog, February 22. https://blog.oup.com/2016/02/history-zika-virus-mosquito/#sthash.BGxvZ2QH.dpuf. Accessed 27 June 2017
35. Black WC IV et al (2002) Flavivirus susceptibility in *Aedes aegypti*. Arch Med Res 33. https://www.ncbi.nlm.nih.gov/pubmed/12234528. Accessed 26 June 2017
36. Machemer T (2021) Genetically modified mosquitoes take flight to fight invasive species in Florida. Smithsonian Magazine, May 17. https://www.smithsonianmag.com/smart-news/genetically-modified-mosquitoes-take-flight-fight-invasive-species-florida-180977748/. Accessed 31 May 2022
37. RoseWrites (2016) A crime against humanity: how the CDC and WHO are promoting the global spread of Zika. InfoBarrel, October 17. http://www.infobarrel.com/A_Crime_Against_Humanity_How_the_CDC_and_WHO_Are_Promoting_the_Global_Spread_of_Zika. Accessed 20 July 2017
38. Kwan N (2016) Zika arrives in US: debunking top myths about the virus. FOX News, August 5. http://www.foxnews.com/health/2016/08/05/zika-arrives-in-us-debunking-top-myths-about-virus.html. Accessed 21 July 2017
39. Matthews L (2017) Understanding the Zika virus: all you need to know. Amazon, Middletown, DE
40. Maciel-de-Freitas R et al (2014) Undesirable consequences of insecticide resistance following *Aedes aegypti* control activities due to a dengue outbreak. PLOS One. http://journals.plos.org/plosone/article?id=10.1371/journal.pone.0092424. Accessed 10 May 2017
41. PAHO (2016) Zika virus infection step-by-step guide to risk communication and community engagement. http://iris.paho.org/xmlui/handle/123456789/18599. Accessed 27 May 2017
42. Edwards A (2017) Zika virus: long beach health officials warn of looming threat. Press Telegram, May 8. http://www.presstelegram.com/health/20170508/zika-virus-long-beach-health-officials-warn-of-looming-threat. Accessed 26 June 2017

43. Steenhuysen J (2016) Zika mosquitoes' habits may foil U.S. elimination efforts. Reuters, February 3. http://www.reuters.com/article/us-health-zika-mosquitoes-insight-idUSKCN0VC04B. Accessed 19 May 2017
44. Allen G (2017) Bacteria-infected mosquitoes tested as a way to control population. NPR Now, April 20. http://www.npr.org/2017/04/20/524833658/bacteria-infected-mosquitoes-tested-as-a-way-to-control-its-population. Accessed 3 May 2017
45. Novak S (2018) Scientists race to kill mosquitoes before they kill Us. Sierra Club, January 9. https://www.sierraclub.org/sierra/scientists-race-kill-mosquitoes-they-kill-us. Accessed 10 July 2018, and https://www.cdc.gov/zika/vector/range.html. Accessed 10 July 2018
46. Lafrance A (2016) Tracing Zika back to patient zero. The Atlantic, May 16. https://www.theatlantic.com/health/archive/2016/05/zika-mutations-patient-zero/482739/. Accessed 18 May 2017
47. Ledermann JP et al (2014) *Aedes hensilli* as a potential vector of Chikungunya and Zika Viruses. PLoS Negl Trop Dis 8(10):e3188. https://doi.org/10.1371/journal.pntd.0003188
48. Weaver S et al (2016) Zika virus: history, emergence, biology, and prospects for control. Antiviral Res 130:69–80. https://www.ncbi.nlm.nih.gov/pubmed/26996139. Accessed 30 June 2017
49. Small N (2016) CSU researchers identify mosquito that transmits Zika, Dengue with single bite. Rocky Mountain Collegian, December 7. https://collegian.com/2016/12/csu-researchers-identify-mosquito-that-transmits-zika-yellow-fever-with-single-bite/. Accessed 2 June 2017
50. Deniz D (2016) Zika: from the Brazilian backlands to global threat. Zed books, London
51. Enserink M (2014) Crippling virus Set to conquer western hemisphere. Science 344. http://science.sciencemag.org/content/344/6185/678. Accessed 7 Aug 2018
52. Manore CA et al (2017). Defining the risk of Zika and Chikungunya virus transmission in human population centers of the Eastern United States. PLOS Negl Trop Dis 11(1):e0005255. http://journals.plos.org/plosntds/article?id=10.1371/journal.pntd.0005255. Accessed 8 Aug 2018
53. Little EE, Harriott O, Akaratovic K, Kiser J et al (2021) Host interactions of *Aedes albopictus*, an invasive vector of arboviruses, in Virginia, USA. PLoS Negl Trop Dis 15(2):E0009173. https://doi.org/10.1371/journal.pntd.0009173. eCollection 2021 Feb
54. Kraemer MU, Reiner RC, Brady OJ, Messina JP et al (2019) Past and future spread of the arbovirus vectors *Aedes aegypti* and *Aedes albopictus*. Nat Microbiol 4:854–863
55. Rochlin I, Ninivaggi D, Hutchinson ML, Farajollahi A (2013) Climate change and range expansion of the Asian Tiger Mosquito (*Aedes albopictus*) in Northeastern USA: implications for public health practitioners. PLoS ONE 8(4):e60874. https://doi.org/10.1371/journal.pone.0060874
56. Medlock J, Leach S (2015) Effect of climate change on vector-borne disease risk in the UK. Lancet Infect Dis 15(6). https://www.sciencedirect.com/science/article/pii/S1473309915700915. Accessed 7 July 2018
57. Kamgang B, Nchoutpouen E, Simard F, Paupy C (2012) Notes on the blood-feeding behavior of *Aedes albopictus* (Diptera: Culicidae) in Cameroon. Parasites & Vectors 5(57). https://www.ncbi.nlm.nih.gov/pubmed/22433236. Accessed 7 Aug 2018
58. Kek R et al (2014) Feeding host range of *Aedes albopictus* (Diptera: Culicidae) demonstrates its opportunistic host-seeking behavior in rural Singapore. J Med Entomol 51(4). https://www.ncbi.nlm.nih.gov/pubmed/25118424. Accessed 7 Aug 2018
59. Richards SL et al (2006). Host-feeding patterns of *Aedes albopictus* (Diptera: Culicidae) in relation to availability of human and domestic animals in suburban landscapes of Central North Carolina. J Med Entomol 43(3). https://www.ncbi.nlm.nih.gov/pubmed/16739414. Accessed 11 Aug 2018
60. Fran Counter Estrada JG, Craig GB Jr (1995) Biology, disease relationships, and control of *Aedes albopictus*. Pan Am Health Organization. Tech Paper No. 42. 1995. World Health Organization, Geneva
61. Fox M (2017) Zika found in common backyard Asian tiger mosquito. NBC News, April 14. http://www.nbcnews.com/storyline/zika-virus-outbreak/zika-found-common-backyard-asian-tiger-mosquito-n746646. Accessed 14 May 2017

62. Kambhampati S, Black WC, Rai Karamjit S (1991) Geographic origin of the US and Brazilian *Aedes albopictus* inferred from allozyme analysis. Heredity 67(Pt. 1). https://www.nature.com/hdy/journal/v67/n1/pdf/hdy199167a.pdf. Accessed 17 May 2017; Hawley HA et al (1987) *Aedes albopictus* in North America: probable introduction in tires from northern Asia. Science 236. https://www.ncbi.nlm.nih.gov/pubmed/3576225. Accessed 17 July 2017
63. See Kek K et al (2014) Feeding host range of *Aedes albopictus* (Diptera: Culicidae) demonstrates its opportunistic host-seeking behavior in rural Singapore. J Med Entomol 51(4). https://www.ncbi.nlm.nih.gov/pubmed/25118424/. Accessed 24 July 2018
64. Marcondes C, de Fátima Freire de Melo Ximenes M (2016) Zika virus in Brazil and the danger of infestation by *Aedes* (*Stegomyia*) mosquitoes. Rev Soc Bras Med Trop 49(1). http://www.scielo.br/scielo.php?script=sci_arttext&pid=S0037-86822016000100004. Accessed 24 July 2018
65. See Richards SL et al (2006) Host-feeding patterns of *Aedes albopictus* (Diptera: Culicidae) in relation to availability of human and domestic animals in suburban landscapes of Central North Carolina. J Med Entomol 43(3). https://www.ncbi.nlm.nih.gov/pubmed/16739414. Accessed 24 July 2018
66. Ngoagouni C et al (2016) Invasion of *Aedes albopictus* (Diptera: Colcidae) into Central Africa: what consequences for emerging diseases. In: Haddow A (ed) Zika Virus. (SCIRP) Scientific Research Publishing, Wuhan, China; Kamgang B, Nchoutpouen E, Simard F, Paupy C (2012) Notes on the blood-feeding behavior of *Aedes albopictus* (Diptera: Culicidae) in Cameroon. Parasites & Vectors 5(57). https://www.ncbi.nlm.nih.gov/pubmed/22433236. Accessed 7 Aug 2018
67. Faraji A (2014) Comparative host feeding patterns of the Asian tiger mosquito, *Aedes albopictus*, in urban and suburban northeastern USA and implications for disease transmission. PLOS Negl Infect Dis 8(8). http://journals.plos.org/plosntds/article?id=10.1371/journal.pntd.0003037. Accessed 26 June 2017
68. Oxitec (2016) Florida keys project. Oxitec Website. http://www.oxitec.com/programmes/united-states/. Accessed 27 May 2017
69. Beck J (2016) What to know about *Aedes albopictus*, the other mosquito that carries Zika. The Atlantic, May 2. https://www.theatlantic.com/health/archive/2016/05/the-other-zika-mosquito-aedes-Albopictus-asian-tiger/480828/. Accessed 3 Feb 2017
70. Grard G, Caron M, Mambo IM, Nkoghe D et al (2014) Zika virus in Gabon (Central Africa)—2007: a new threat from *Aedes albopictus*? PLoS Negl Trop Dis. https://doi.org/10.1371/journal.pntd.0002681
71. Wong P-SJ et al (2013) *Aedes* (*Stegomyia*) *albopictus* (Skuse): a potential vector of Zika virus in Singapore. PLOS Negl Trop Dis 7(8). https://www.ncbi.nlm.nih.gov/pmc/articles/PMC3731215/. Accessed 14 May 2017
72. Smartt C et al (2017) Evidence of Zika virus RNA fragments in *Aedes albopictus* (Diptera: Culicidae) field-collected eggs from Camaçari, Bahia, Brazil. J Med Entomol. https://academic.oup.com/jme/article/3105868/Evidence-of-Zika-Virus-RNA-Fragments-in-Aedes. Accessed 14 May 2017
73. WHO (2016) One year into the Zika outbreak: how an obscure disease became a global health emergency. Emergencies, http://www.portal.pmnch.org/emergencies/zika-virus/articles/one-year-outbreak/en/. Accessed 2 Oct 2016
74. O'Neill N (2017) There's another mosquito carrying Zika virus. The New York Post, April 14. http://nypost.com/2017/04/14/theres-another-mosquito-carrying-zika-virus/. Accessed 26 May 2017
75. Gallagher G (2017) Researchers find Zika RNA in Brazilian *A. albopictus* mosquitoes. Healio Infectious Disease News, April 20. http://www.healio.com/infectious-disease/emerging-diseases/news/in-the-journals/%7B95c7d904-9021-472b-9c7c-3d1c01f7c140%7D/researchers-find-zika-rna-in-brazilian-a-Albopictus-mosquitoes. Accessed 2 June 2017
76. Smartt C et al (2017) Evidence of Zika virus RNA fragments in *Aedes albopictus* (Diptera: Culicidae) field-collected eggs from Camaçari, Bahia, Brazil. J Med Entomol. https://academic.oup.com/jme/article/3105868/Evidence-of-Zika-Virus-RNA-Fragments-in-Aedes. Accessed 19 July 2017

77. Buhagiar JA (2009) A second record of *Aedes* (*Stegomyia*) *albopictus* (Diptera: Culicidae) in Malta. Eur Mosquito Bull 27:65–67
78. *Aedes albopictus* (2016) European Centre for Disease Prevention and Control. http://ecdc.europa.eu/en/healthtopics/vectors/mosquitoes/Pages/aedes-Albopictus.aspx. Accessed 25 Oct 2016
79. Lafrance, Adrienne. (2016). The Sneak-Attack Mosquito. The Atlantic. April 25. https://www.theatlantic.com/science/archive/2016/04/aedes-aegypti/479619/. Accessed May 18, 2017.
80. dos Santos TP, Roiz D, de Abreu FVS, Luz SLB et al (2018) Potential of *Aedes albopictus* as a bridge vector for enzootic pathogens at the urban-forest interface in Brazil. Emerg Microbes Infect. https://doi.org/10.1038/s41426-018-0194-y
81. Vogel G (2016) Scientific sleuths hunt for Zika-carrying mosquitoes. Science, June 1. http://www.sciencemag.org/news/2016/06/scientific-sleuths-hunt-zika-carrying-mosquitoes. Accessed 9 June 2017; Lourenço-de-Oliveira R et al (2017) *Culex quinquefasciatus* mosquitoes do not support replication of Zika virus. J General Virol 99(2). https://www.ncbi.nlm.nih.gov/pubmed/29076805. Accessed 15 May 2019
82. Kay J (2017) Florida tests bacteria-infected mosquitoes to control Zika. York dispatch. http://www.yorkdispatch.com/story/news/2017/04/18/florida-tests-bacteria-infected-mosquitoes-control-zika/100622084/. Accessed 17 May 2017
83. Wong P-SJ et al (2013) *Aedes* (*Stegomyia*) *albopictus* (Skuse): a potential vector of Zika virus in Singapore. PLOS Negl Trop Dis 7(8). https://www.ncbi.nlm.nih.gov/pmc/articles/PMC3731215/. Accessed 19 July 2017
84. Fox M, Edwards E (2016) Zika virus is coming and we're not ready, U.S. experts say. NBC News, May 3. http://www.nbcnews.com/storyline/zika-virus-outbreak/zika-virus-coming-we-re-not-ready-u-s-experts-n567261. Accessed 14 April 2017
85. Gubler DJ (2002) Epidemic dengue/dengue hemorrhagic fever as a public health, social and economic problem in the 21st Century. Trends Microbiol, 10:100–103. https://doi.org/10.1016/S0966-842X(01)02288-0
86. Song B-H et al (2017) Zika virus: History, epidemiology, transmission, and clinical presentation. J Neuroimmunol 308. https://www.ncbi.nlm.nih.gov/pubmed/28285789. Accessed 19 July 2017
87. Bargielowski I, Blosser E, Lounnibos LP (2015) The effects of interspecific courtship on the mating success of *Aedes aegypti* and *Aedes albopictus* (Diptera: Culicidae) males. Ann Entomol Soc Am 108(4). https://www.ncbi.nlm.nih.gov/pubmed/27418696. Accessed 25 July 2018
88. Leahy MG, Craig GB (1967) Barriers to hybridization between *Aedes aegypti* and *Aedes albopictus*. Evolution 21(1). https://www.ncbi.nlm.nih.gov/pubmed/28556103. Accessed 25 July 2018
89. Harper JP, Paulson S (1994) Reproductive isolation between Florida strains of *Aedes aegypti* and *Aedes albopictus*. J Am Mosquito Control Assoc 10(1). https://www.ncbi.nlm.nih.gov/pubmed/8014633. Accessed 25 July 2018
90. Lee HL, Aramu M, Nazni WA, Selvi S, Vasan S (2009) No evidence for successful interspecific cross-mating of transgenic *Aedes aegypti* (L.) and wild type *Aedes albopictus* Skuse. Trop Biomed 26:312–319
91. Honório N et al (2018) Male origin determines satyrization potential of *Aedes aegypti* by invasive *Aedes albopictus*. Biol Invasions 20. https://link.springer.com/article/10.1007/s10530-017-1565-3. Accessed 8 July 2018
92. Marcela P et al (2015) Interspecific cross-mating between *Aedes aegypti* and *Aedes albopictus* laboratory strains: implications of population density on mating behaviors. J Am Mosquito Assoc 31(4). https://www.ncbi.nlm.nih.gov/pubmed/26675452. Accessed 8 July 2018
93. Leahy MG, Craig Jr GBB (1967) Barriers to hybridization between *Aedes aegypti* and *Aedes albopictus* (Diptera: Culicidae). Evolution 21. https://www.jstor.org/stable/2406739?seq=1#page_scan_tab_contents. Accessed 8 July 2018; Thomas V, Yap PL (1973) Hybridization between *Aedes aegypti* and *Aedes albopictus* in Malaysia. Southeast Asian J Trop Med Public Health 4. https://www.ncbi.nlm.nih.gov/pubmed/4749074. Accessed 8 July 2018

94. Harper JP, Paulson S (1994) Reproductive isolation between Florida strains of *Aedes aegypti* and *Aedes albopictus*. J Am Mosquito Control Assoc 10. https://www.ncbi.nlm.nih.gov/pubmed/8014633. Accessed 8 July 2018; Nazni W, Lee H, Dayang H, Azahari A (2009) Cross-mating between Malaysian strains of *Aedes aegypti* and *Aedes albopictus* in the laboratory. Southeast Asian J Trop Med Public Health 40. https://www.ncbi.nlm.nih.gov/pubmed/19323032. Accessed 8 July 2018
95. Nasci R, Hare SG, Willis FS (1989) Interspecific mating between Louisiana strains of *Aedes albopictus* and *Aedes aegypti* in the field and laboratory. J Am Mosquito Control Assoc 5(3). https://www.ncbi.nlm.nih.gov/pubmed/2584975. Accessed 8 July 2018
96. Webster R (2017) Comment. In: Atkins K (ed) First travel-related Zika case of 2017 reported. Florida Keys News, February 15. http://www.flkeysnews.com/news/local/environment/article132829289.html. Accessed 7 March 2017
97. Guo X et al (2016) Culex pipiens quinquefasciatus: a potential vector to transmit Zika virus. Emerg Microbes Infect 5:e102
98. Guedes DRD et al (2017) Zika virus replication in the mosquito *Culex quinquefasciatus* in Brazil. Emerg Microbes Infect 6:e69
99. Guo X-z et al (2016) *Culex pipiens quinquefasciatus*: a potential vector to transmit Zika virus. Emerg Microbes Infect. https://www.ncbi.nlm.nih.gov/pubmed/27599470. Accessed 14 May 2017
100. Ayres C (2016) Identification of Zika virus vectors and implications for control. The Lancet 16. http://thelancet.com/journals/laninf/article/PIIS1473-3099(16)00073-6/fulltext. Accessed 28 May 2017
101. Guedes DRD et al (2016) Zika virus replication in the mosquito *Culex quinquefasciatus* in Brazil. bioRxiv: The Preprint Server for Biology. http://biorxiv.org/content/early/2016/09/02/073197. Accessed 7 March 2017. https://doi.org/10.1101/073197 (not peer reviewed)
102. Amraoul F et al (2016). Culex mosquitoes are experimentally unable to transmit Zika virus. Eur Surveillance 21(15). http://www.eurosurveillance.org/ViewArticle.aspx?ArticleId=22573. Accessed 7 Feb 2017
103. Worth K (2016) A common U.S. mosquito may transmit Zika virus, study says. PBS Newshour, March 7. http://www.pbs.org/newshour/rundown/a-common-u-s-mosquito-may-transmit-zika-virus-study-says/. Accessed 30 June 2017
104. Hawaii Department of Health (2013) Mosquitoes. Bulletin 3. http://health.hawaii.gov/about/files/2013/06/VCB-bulletin_03_11.pdf. Accessed 15 May 2017
105. Alameda County Mosquito Abatement District (n. dat.) Mosquito biology—California Species. http://www.mosquitoes.org/education/california-species/. Accessed 15 May 2017
106. UF/IFAS (2016) Featured creatures. http://entnemdept.ufl.edu/creatures/aquatic/southern_house_mosquito.htm. Accessed 15 May 2017
107. Ayres C (2016) Identification of Zika virus vectors and implications for control. The Lancet 16.. http://thelancet.com/journals/laninf/article/PIIS1473-3099(16)00073-6/fulltext. Accessed 12 Jan 2017
108. Bajak A (2016) For the U.S., a more worrisome Zika vector? Undark, May 22. https://undark.org/2016/05/02/realistic-culex-mosquito-carry-zika/. Accessed 15 May 2017
109. Phillips D (2016) Zika is found in common Culex mosquitos, signaling a potentially larger risk. The Washington Post, July 21. https://www.washingtonpost.com/world/the_americas/zika-is-found-in-common-culex-mosquitos-signaling-a-potentially-larger-risk/2016/07/21/e1b37e0e-4f6f-11e6-bf27-405106836f96_story.html?utm_term=.ada3a37c283c. Accessed 28 May 2017
110. Adaja A et al (2016) Genetically Modified (GM) mosquito use to reduce mosquito transmitted disease in the US: a community opinion survey. PLOS. http://currents.plos.org/outbreaks/article/genetically-modified-mosquito-use-to-reduce-mosquito-transmitted-disease-in-the-us-opinion-survey/. Accessed 29 Sept 2016
111. Adalja A et al (2016) Genetically Modified (GM) mosquito use to reduce mosquito-transmitted disease in the US: a community opinion survey. PLOS Curr Outbreaks. https://doi.org/10.1371/currents.outbreaks.1c39ec05a743d41ee39391ed0f2ed8d3

112. Diamond A (2019) Why do mosquitoes exist? Why do elephants and donkeys represent the G.O.P. and the democrats? And more questions from our readers. Smithsonian.com, December. https://www.smithsonianmag.com/smithsonian-institution/mosquitos-exist-elephants-donkeys-used-represent-gop-democrats-180973517/. Accessed 21 May 2022
113. Bhattacharya S, Pal S, Acharyya D (2015) Are mosquitoes 'a necessary evil'? Int J Fauna Biol Stud, 124–129
114. Marten G (2007) Turtles. Biorational control of mosquitoes. Am Mosquito Control Assoc Bull (Floore TG, ed). Suppl to #2. http://gerrymarten.com/publicatons/pdfs/GM_Turtles.pdf. Accessed 21 May 2022
115. Kirchner J (2022) Meet the squad of mosquito-eating species. NWF Blog. August 16. https://blog.nwf.org/2020/08/meet-the-squad-of-mosquito-eating-species/. Accessed March 11, 2023
116. Rydell J, McNeill D, Eklof J (2002) Capture success of little brown bats (*Myotis lucifugus*) feeding on mosquitoes. J Zool 256:379–381
117. Wray AK et al (2018) Incidence and taxonomic richness of mosquitoes in the diets of little brown and big brown bats. J Mammal 99(3):668–674
118. Hoff S (2018) Study bolsters bats' reputation as mosquito devours. UW News, May 22. https://news.wisc.edu/study-bolsters-bats-reputation-as-mosquito-devourers/#:~:text=The%20study%20also%20found%20that,needs%20of%20the%20larger%20bats. Accessed 21 May 2022
119. McCay B (2016) Mosquitoes are deadly, so why not kill them all? Wall Street Journal, September 2. http://www.wsj.com/articles/mosquitoesaredeadlysowhynotkillthemall1472827158. Accessed 15 Sept 2016
120. Fang J (2010) A world without mosquitoes. Nature 466:432–434. http://www.nature.com/news/2010/100721/full/466432a.html. Accessed 4 Oct 2016
121. Oxitec Report to CVM FDA (2016) Environmental assessment for investigational use of *Aedes aegypti* OX513A, August 5. https://www.fda.gov/downloads/AnimalVeterinary/DevelopmentApprovalProcess/GeneticEngineering/GeneticallyEngineeredAnimals/UCM514698.pdf. Accessed 12 May 2017
122. Center for Food Safety (2016) This election, keys resident vote "NO" on GE mosquitoes. Santa Cruz IMC, November 9. https://www.indybay.org/newsitems/2016/11/09/18793252.php. Accessed 3 April 2017
123. Mizejewski D (2020) What purpose do mosquitoes serve. https://blog.nwf.org/2020/09/what-purpose-do-mosquitoes-serve. Accessed 21 May 2022
124. Peach D, Gries G (2016) Nectar thieves or invited pollinators? A case study of tansy flowers and common house mosquitoes. Arthropod-Plant Interact 10:497–506
125. Irwin RE, Brody AK (1999) Nectar-robbing bumble bees reduce the fitness of *Ipomopsis aggregata* (Polemoniaceae). Ecology 80:1703–1712
126. Center for Food Safety (2016) This election, keys residents vote "NO" on GE Mosquitoes. Santa Cruz IMC, November 9. https://www.indybay.org/newsitems/2016/11/09/18793252.php. Accessed 3 April 2017
127. Jaunzems M (N.dat.) Blunt-leaf Orchid (*Platanthera obtusata* (Bands ex Pursh) Lindley). USFS. https://www.fs.fed.us/wildflowers/plant-of-the-week/platanthera_obtusata.shtml. Accessed 21 May 2022
128. Gorham TR (1976) Orchid pollination by Aedes mosquitoes in Alaska. Am Midland Nat 95(1):208–210
129. Lahondere C, Vinauger C, Okubo R et al (2019) The olfactory basis of orchid pollination by mosquitoes. BioRxiv. https://doi.org/10.1101/643510. Accessed 21 May 2022
130. Brantjes NBM, Leemans J (1976) *Silene Otites* (Caryophllaveae) pollinated by nocturnal lepidoptera and mosquitoes. Acta Botanica Neellandica 25(4):281–295
131. Dexter JS (1913) Mosquitoes pollinating orchids. Science 37(962):867
132. Seale DB (1980) Influence of amphibian larvae on primary production, nutrient flux, and competition in a pond ecosystem. Ecology 61:1531–1550. https://doi.org/10.2307/1939059

133. Schafer M (2004) Mosquitoes as part of wetland biodiversity (summary). Faculty of Science and Technology, University of Uppsala, Uppsala. http://www.diva-portal.org/smash/get/diva2:165446/FULLTEXT01.pdf. Accessed 3 June 2022
134. Mokany A (2007) Impact of tadpoles and mosquito larvae on ephemeral pond structure and processes. Mar Freshw Res 58:436–444
135. Hanford JK, Webb CE, Hochuli DF (2019) Management of urban wetlands for conservation can reduce aquatic biodiversity and increase mosquito risk. J Appl Ecol. https://doi.org/10.1111/1365-2664.13576
136. Ostrowski J (2016) Frankenskeeters to the rescue? Modified mosquitoes touted to stop Zika. Palm Beach Post, September 23. http://www.mypalmbeachpost.com/news/news/local-govt-politics/frankenskeeters-to-the-rescue-modified-mosquitoes-/nsdkL/. Accessed 29 Sept 2016
137. Adler J (2016) A world without mosquitoes. The Smithsonian, June 38. http://www.smithsonianmag.com/innovation/kill-all-mosquitos-180959069/?no-ist. Accessed 2 Oct 2016
138. Musso D, Gubler DJ (2016) Zika virus. Clin Microbiol Rev 29(3):487–524
139. Hanley KA et al (2013) Fever versus fever: the role of host and vector susceptibility and interspecific competition in shaping the current and future distributions of the sylvatic cycles of dengue virus and yellow fever virus. Infect Genet Evol. https://www.ncbi.nlm.nih.gov/pubmed/23523817. Accessed 11 June 2019
140. Althouse BM et al (2016) Potential for Zika virus to establish a sylvatic transmission cycle in the Americas. PLOS Negl Trop Dis. https://journals.plos.org/plosntds/article?id=10.1371/journal.pntd.0005055. Accessed 15 May 2019
141. Thangamani S et al (2016) Vertical transmission of Zika virus in *Aedes aegypti* mosquitoes. Am J Trop Med Hygiene 95(5). http://www.ajtmh.org/content/journals/10.4269/ajtmh.16-0448. Accessed 4 June 2017
142. Ferguson NM (2016) Countering the Zika epidemic in Latin America. Science. https://doi.org/10.1126/science.aag0219
143. Verou R (2016) Zika virus, vectors, reservoirs, amplifying hosts, and their potential to spread worldwide: what we know and what we should investigate urgently. Int J Infect Dis 48. https://www.ncbi.nlm.nih.gov/pubmed/27208633. Accessed 26 July 2018
144. Heesterbeek H et al (2015) Modeling infectious disease dynamics in the complex landscape of global health. Science 347. http://science.sciencemag.org/content/347/6227/aaa4339. Accessed 27 June 2017
145. Silva GS et al (2018) Zika virus: report from the task force on tropical diseases by the world Federation of Societies of intensive and critical care medicine. J Crit Care 46. https://www.ncbi.nlm.nih.gov/pubmed/29779827. Accessed 26 July 2018
146. Ratna K (2017) The secret Life of Zika virus. Speaking Tiger Books, New Delhi
147. Fox M (2016) Zika virus birth defects may be 'tip of the iceberg', experts say. NBC News, May 1. http://www.nbcnews.com/storyline/zika-virus-outbreak/zika-virus-birth-defects-may-be-tip-iceberg-experts-say-n565631. Accessed 14 April 2017
148. Coyne CB, Lazear HM (2016) Zika virus—reigniting the TORCH. Nat Rev Microb 14. http://www.nature.com/nrmicro/journal/v14/n11/full/nrmicro.2016.125.html. Accessed 14 April 2017
149. Froeschl G et al (2017) Long-term kinetics of Zika virus RNA and antibodies in body fluids of a vasectomized traveller returning from Martinique: a case report. BMC Infect Dis. https://www.ncbi.nlm.nih.gov/pmc/articles/PMC5223480/. Accessed 14 May 2017
150. RoseWrites (2017) Red-whiskered bulbul: Zika's ideal reservoir host. InfoBarrel, January 27. http://www.infobarrel.com/Red-Whiskered_Bulbul_Zikas_Ideal_Reservoir_Host. Accessed 31 May 2017
151. RoseWrites (2017) An open letter to Dr. Margaret Chan, Director-General of WHO. InfoBarrel, February 5. http://www.infobarrel.com/An_Open_Letter_to_Dr_Margaret_Chan_Director-General_of_WHO. Accessed 7 March 2017
152. Malmlov A, Bantle C, Aboellail T, Wagner K et al (2019) Experimental Zika virus infection of Jamaican fruit bats (*Artibeus jamaicensis*) and possible entry of virus into brain via activated microglial cells. PLoS Negl Trop Dis 13(2):e0007071

153. Ding Q, Ploss A (2018). Factor important for ZIKA virus host species restriction. Science Daily, June 18. https://www.sciencedaily.com/releases/2018/06/180618222421.htm. Accessed 23 July 2018
154. Lessler J et al (2016) Assessing the global threat from Zika virus. Science 353(6300). http://science.sciencemag.org/content/early/2016/07/13/science.aaf8160. Accessed 19 May 2017
155. Cunningham A (2017) As Zika fades from public consciousness, scientists continue to pursue the virus. The Washington Post, December 30. https://www.washingtonpost.com/national/health-science/as-zika-fades-from-public-consciousness-scientists-continue-to-pursue-the-virus/2017/12/29/3fae96dc-e1d3-11e7-bbd0-9dfb2e37492a_story.html?utm_term=.4232eb7d5058. Accessed 10 July 2018
156. Saey T (2017) Zika virus 'spillback' into primates raises risk of future human outbreaks. Science News, February 8. https://www.sciencenews.org/article/zika-virus-spillback-primates-raises-risk-future-human-outbreaks. Accessed 31 May 2017
157. Matthews KRW, Herricks JR (2016) Mosquito-transmitted epidemics: Zika virus in the United States and Mexico. Policy brief no. 03.04.16. Rice University's Baker Institute for Public Policy, Houston, TX
158. Wetsman N (2017) The missing pieces: lack of Zika data from Africa complicates search for answers. Nat Med 23(8). https://doi.org/10.1038/nm0817-904
159. Ishtiaq F (2018) A call to introduce structured Zika surveillance in India. Trends Parasitol 34(2):92–95
160. Campos GS et al (2020) New challenge for Zika virus infection: human reservoirs? Viral Immunol 1–4. https://doi.org/10.1089/vim.2019.0187
161. Lazear Helen M et al (2016) A mouse model of Zika virus pathogenesis. Cell Host Microbe. http://www.cell.com/pb/assets/raw/journals/research/cell-host-microbe/online-now/chom1441_r.pdf. Accessed 31 May 2017
162. Mansuy JM et al (2016) Zika virus: high infectious viral load in semen, a new sexually transmitted pathogen? Lancet Infect Dis. http://www.thelancet.com/journals/laninf/article/PIIS1473-3099(16)00138-9/abstract. Accessed 17 July 2017
163. Nicastri E et al (2016) Persistent detection of Zika virus RNA in semen for six months after symptom onset in a traveller returning from Haiti to Italy, February 2016. Eurosurveillance 21(32). https://www.ncbi.nlm.nih.gov/pubmed/27541989. Accessed 26 May 2017; Mansuy JM et al (2016) Zika virus: high infectious viral load in semen, a new sexually transmitted pathogen? Lancet Infect Dis. http://www.thelancet.com/journals/laninf/article/PIIS1473-3099(16)00138-9/abstract. Accessed 26 May 2017; Musso D et al (2015) Potential sexual transmission of Zika virus. Emerg Infect Dis 21(2). https://wwwnc.cdc.gov/eid/article/21/2/pdfs/14-1363.pdf. Accessed 26 May 2017
164. Robbiani DF et al (2017) Recurrent potent human neutralizing antibodies to zika virus in Brazil and Mexico. Cell 169. https://www-sciencedirect-com.prox.lib.ncsu.edu/science/article/pii/S0092867417304750. Accessed 7 July 2018
165. Aid M et al (2017) Zika virus persistence in the central nervous system and lymph nodes of rhesus monkeys. Cell 169. https://www.ncbi.nlm.nih.gov/pubmed/28457610. Accessed 5 June 2017
166. Williams R (2017) Zika virus persists in the nervous system and elsewhere. The Scientist, April 27. http://www.the-scientist.com/?articles.view/articleNo/49296/title/Zika-Virus-Persists-in-the-Nervous-System-and-Elsewhere/. Accessed 1 July 2017
167. US Department of Health and Human Services, FDA, Center for Veterinary Medicine (2017) Guidance for industry—regulation of mosquito-related products: draft guidance. https://www.fda.gov/downloads/AnimalVeterinary/GuidanceComplianceEnforcement/GuidanceforIndustry/UCM533600.pdf. Accessed 18 April 2017
168. Molteni M (2017) When is a mosquito not an insect? When it's a pesticide. Wired. https://www.wired.com/story/oxitecs-genetically-modified-mosquitoes-are-now-the-epas-problem/. Accessed 24 July 2018
169. Servick K (2017) Proposed U.S. biotech rules raise industry hopes and anxieties. Science. http://www.sciencemag.org/news/2017/01/proposed-us-biotech-rules-raise-industry-hopes-and-anxieties. Accessed 31 May 2017

170. Goodhue D (2016) Legal action threatened against FDA over Oxitec decision. Florida Keys News, November 25. http://www.flkeysnews.com/news/local/article116965063.html. Accessed 18 April 2017
171. Molteni M (2017) U.S. EPA to review oxitec's genetically modified mosquito as pesticide. The Henry J Kaiser Family Foundation, October 18. https://www.kff.org/news-summary/u-s-epa-to-review-oxitecs-genetically-modified-mosquito-as-pesticide/. Accessed 24 July 2018; Molteni M (2017) When is a mosquito not an insect? When it's a pesticide. Wired. https://www.wired.com/story/oxitecs-genetically-modified-mosquitoes-are-now-the-epas-problem/. Accessed 24 July 2018
172. Eisen L et al (2009) Proactive vector control strategies and improved monitoring and evaluation practices for dengue prevention. J Med Entomol 46(6). https://www.ncbi.nlm.nih.gov/pubmed/19960667. Accessed 7 Aug 2018
173. Pandit PS (2018) Predicting wildlife reservoirs and global vulnerability to zoonotic flaviviruses. Nat Commun 9:5425. https://www.nature.com/articles/s41467-018-07896-2. Accessed 24 May 2022
174. Favoretto SR, Araujo DB, Duarte NFH et al (2019) Zika virus in peridomestic neotropical primates, northeast Brazil. EcoHealth 2019(16):61–69
175. Musso D, Ko AI, Baud D (2019) Zika virus infection—after the pandemic. N Engl J Med 381:1444–1457. https://doi.org/10.1056/NEJMra1808246

Chapter 11
Diagnoses, Treatments, Vaccines

ZIKV is challenging to diagnose. There are no standardized treatments for ZIKV. And while there was a rush to produce vaccines, the pandemic ebbed, and so did the motivation to build a vaccine. Even though the infection levels had ebbed considerably, experts were still concerned that ZIKV could re-emerge in one form or another.

What may have been learned from ZIKV would provide more information about arboviruses, flaviviruses, and maybe other infectious diseases. However, once there is no significant return on investment from marketing diagnostic tests, treatments, and even vaccines, the pharmaceutical vaccine industry closes the books on infection and moves on to something more profitable.

11.1 Research

Two sizeable epidemiological research studies: U.S. NIH and Brazil's Fiocruz (Fundação Oswaldo Cruz-Fiocruz), Zika in Infants and Pregnancy (ZIP), and ZikaAlliance were scheduled to take place across multiple sites in South America and the Caribbean.

Currently, the ZIP study is faced with dwindling numbers of ZIKV cases. As Anthony Fauci puts it, capturing enough data in an outbreak, where numbers of cases fluctuate from place to place and over time and, at times, dry up altogether, is always an issue. "That's just a risk you accept," he says. "Sometimes a study that you think would take two years will take four or five years." [1].

With the drop in new cases, the ZIKAlliance is also reconsidering its research protocols. The ZIKAlliance intended to capture what cases it can at its sites and is considering focusing resources on sites where ZIKV has been rarer, such as Bolivia, where future flare-ups might be more likely. "We will track cases where they are," Xavier de Lamballerie, ZikaAlliance led consortium Scientific Coordinator, says, "It's a race against the clock." [1].

D. M. Berube, *Pandemics and Resilience: Lessons we should have learned from Zika*, Risk, Systems and Decisions, https://doi.org/10.1007/978-3-031-25370-6_11

Of course, it is impossible to list all the research that would have been done that would not have happened. Their findings simply will go unreported. Along with it, so might advances in testing, treatment, and vaccines.

11.2 Diagnosis

Testing protocols were incredibly confusing when it came to the ZIKV. Testing implied slow, expensive access to the regimen and produced false results. Current diagnostic testing remains suboptimal for detecting congenital ZIKV infections and hampers the implementation of clinical management. Although the pandemic has subsided, sufficient information is lacking on ZIKV seroprevalence to evaluate the potential for future epidemics among the more than two billion people who live in regions at risk for ZIKV transmission [2].

Like the CDC guidance, the WHO recommended testing women reporting symptoms with additional ultrasounds to identify any fetal malformations if they could get an ultrasound.

WHO argued that testing asymptomatic women, even in areas of high ZIKV incidence, she failed on three of the Wilson and Junger screening criteria [3]: poorly understood natural history, lack of effective treatment, and low specificity of available diagnostic tests.

The design of standard-of-care and prospective study protocols must be sensitive to the cultural, religious, and medico-legal context. In some countries, a confirmed diagnosis of ZIKV in pregnancy could be of little benefit to the mother and could even lead to harm. In such cases, protocols that go beyond multiple uterine ultrasounds may not be feasible.

The diagnosis of ZIKV infection in Brazil relies on identifying the virus through RT-PCR during acute illness. The virus is detectable in blood during acute viremia and initial symptoms and is shed in the urine for 3 to 14 days. Because RT-PCR assays for ZIKV are usually unavailable, most cases of ZIKV infection in Brazil are diagnosed clinically, without laboratory confirmation [4].

Although ZIKV RNA detection provides conclusive evidence of an infection, a negative result does not rule out the diagnosis. A positive nucleic acid test shows the presence of ZIKV RNA but does not necessarily indicate the presence of an infectious virus [2].

The diagnosis of ZIKV infection in pregnant women and infants poses several challenges. ZIKV RNA is often detected transiently in an infected mother and fetus despite the observation of prolonged viremia during pregnancy. ZIKV RNA can be seen in amniotic fluid, but a negative result does not rule out the diagnosis [2].

In addition, serodiagnosis is complicated by false-positive results due to cross-reactivity in persons exposed to other flaviviruses [6].

These findings' overall public health impact highlights the need for a rapid but specific diagnostic test for blood banks worldwide to identify those infected, counsel pregnant women, or contemplate pregnancy [7].

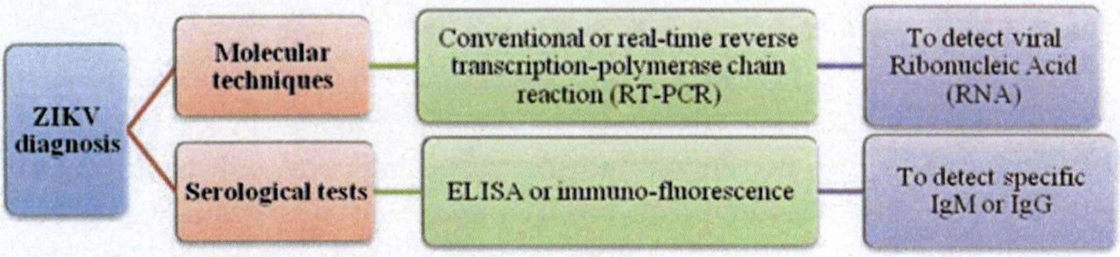

Fig. 11.1 Diagnostic techniques for ZIKV [5]

As of this date, there are no commercially licensed diagnostic tests to tell if someone had ZIKV in the recent past; the current test can only tell if someone has Zika [8].

Charrel et al. acknowledged a handful of commercial tests would be available soon [9]. For example, a team from Colorado State uses an existing technology called loop-mediated isothermal amplification, or LAMP. They claim a new low-cost but accurate way to identify whether a mosquito carries the ZIKV. The test could also be applied to humans to look for the ZIKV in a patient's blood, saliva, or bodily fluids. "The advantage," Dr. Joel Rovnak said, "comes in the cost per sample and the simplicity."

The standard way to test for the ZIKV involves machines that cost tens to hundreds of thousands of dollars and are based in medical laboratories. Rovnak said a LAMP test could be conducted in the field—providing health officials with a quick way to know whether mosquitoes in their area are carrying Zika. That makes the test ideal, he said, for being, "deployed in low-resource environments…."

Rovnak said a children's hospital in Nicaragua would soon begin using the LAMP tests to identify cases of ZIKV in its little patients and compare the results side-by-side with the standard diagnostic test. If the LAMP test holds up, it could be used in small clinics and doctors' offices worldwide that don't have access to the expensive machines needed for the standard test, Rovnak said [10]. Sure enough, by simply squishing Zika-infected mosquitoes, adding them to tubes containing LAMP reagents, and incubating the slurry at 63 °C, the team could robustly amplify the virus (correctly identifying 15 out of 15 infected insects and three out of three negative controls). The LAMP approach would need considerably more development before becoming a diagnostic tool, said Rovnak, but encouragingly it is as robust, accurate, and sensitive as RTPCR, yet reagents cost "about half." [11].

None of the assays developed for ZIKV have been validated in cases where patients show no disease symptoms, making such testing useless and unnecessarily distressing for patients showing what could be a false positive. "Until the risk for ZIKV can be further mitigated by vaccination or improved vector control, enhanced surveillance and diagnostic test validation must underpin the tools used in shared decision making with travelers to Zika-affected areas while incidence is decreasing yet unpredictable," Dr. Graciaa and colleagues stated [12].

11.2.1 Polymerase Chain Reaction (PCR)

The first blood or urine test, the polymerase chain reaction (PCR), is relatively straightforward but effective only if a person has the infection (only approximately up to five days after infection), which stays in the body for about two weeks. A negative result requires a patient to move on to an antibody test called IgM, which can show if someone has had the virus in the last 12 weeks. But if someone tests positive on the antibody test—and many do not—a more complicated antibody test typically follows, conducted only by the CDC or a CDC-approved laboratory. This test pinpoints whether the antibody is related to ZIKV, not dengue or chikungunya [13].

While PCR can determine the presence of ZIKV in blood samples, it is only reliable within the first week after someone has begun showing symptoms of the disease. This can become problematic for some women since the delays could complicate already distressing decisions about whether to terminate their pregnancies if they test positive; for example, it is important to underline that the State of Florida forbids abortions after 24 weeks [13].

The current ZIKV serologic diagnosis is mainly based on [14] IgM-capture ELISA, which takes over two days to complete under the best circumstances [15]. This test can find the antibodies starting five days after infection and can continue to detect them for up to 12 weeks.

Confirming inconclusive and positive IgM results requires further testing with the plaque reduction neutralization test (PRNT), which can be performed only at highly specialized reference laboratories and is also subject to false-positive results [2].

By July 2016, all state public health laboratories were able to test for the virus using IgM enzyme-linked immunosorbent assays (ELISA), real-time polymerase chain reaction tests (RT-PCR), and plaque reduction neutralization test (PRNT) assays [16].

If the test is sent from an outside laboratory, it can take five weeks to confirm results and deliver them because it is complicated, said a C.D.C. spokesman, Benjamin Haynes [13]. In addition, "tests are done and have been delivered, and sometimes there are bureaucratic reasons; they are in someone's computer or fax machine," says Lillian Rivera, Florida Health Department Administrator for Miami-Dade County.

When Florida began to offer to test, Florida handled the bulk of the tests itself but also had to start contracting some testing to private companies, LabCorp and Quest Diagnostics, but takes the bulk of the tests itself.

For pregnant women with the resources and opportunities to have a private test, private laboratories process these tests in seven to ten days, costing from $120 to $750. This underscores the concern expressed throughout this piece that a decisive socioeconomic or income gap is involved in the ZIKV crisis. It should be expected that the most significant impacts to be found in populations already shouldering primary survival burdens.

In Florida, pathologists also reported the backlog had grown tremendously, raising fears that the wait would increase considerably if locally acquired ZIKV cases spread to other counties in Florida. Dr. David M. Andrews, Medical Director of Pathology Laboratories at Jackson Memorial Hospital, said the hospital was waiting for the results of 800 to 900 specimens tested for ZIKV [13].

It certainly is undesirable that controversy around some tests has surfaced beyond the earlier delays. For example, officials at a Washington D.C. public health laboratory confirmed to ABC News that they are retesting hundreds of samples from people in the area for ZIKV over concerns about the accuracy of the initial test results [17]. Already, samples taken from three pregnant women, who tested negative for the virus, have tested positive for likely ZIKV infection [18].

A federal audit found testing at D.C.'s Public Health Lab did not comply with Clinical Laboratory Improvement Amendments (CLIA) requirements, the national regulatory standards for all clinical laboratory testing on humans, excluding trials and basic research. CLIA required an additional negative control be run, CDC's Department of Forensic Sciences said. "We notified the CDC immediately, and they took about a week to notify all the other public health labs," DFS Director Jennifer Smith said. The CDC confirmed it updated its procedures to the more than seventy-five laboratories that use its test for ZIKV and similar viruses. Smith said that even when following the revised guidelines, the current tests supplied by the CDC are inadequate [18].

In addition, The ZIKV serological assay is challenging because ZIKV antigens cross-react with other flavivirus amino acid sequences [19]. ZIKV diagnosis is challenging because it shares vectors, geographic distribution, and symptoms with dengue virus and chikungunya infection, [20] and the three illnesses are often misdiagnosed.

Given the risk for adverse pregnancy outcomes in women infected with ZIKV during pregnancy, it is essential to distinguish between the three viruses [21].

> One of the region's significant public health and biomedical challenges is represented by cocirculation and coinfection of different flaviviral and nonflaviviral arboviruses. Their differentiation as their cross-reactivity between flaviviruses implies immunological and diagnostic problems when used in serological tests. The cocirculation of ZIKV, chikungunya virus, and dengue virus, in addition to other emerging and remerging pathogens (such as Yellow Fever virus and Mayaro virus (Infection with Mayaro virus causes an acute, self-limited dengue-like illness of 3–5 days duration), Oropouche virus (sudden onset of high fever, headache, myalgia, rash, joint pain, and vomiting, lasts 3–6 days), among others), presents several challenges for clinical care and laboratory diagnosis in endemic areas. Patients infected with one or more of these viruses can present with similar clinical manifestations over a wide range of viremia, in some cases, suspecting that with coinfections, clinical symptoms of one of the infections can predominate while the other is asymptomatic. [22]

An EU team funded by the ZikaAlliance project warned laboratories that they have noticed "the recent ZIKV outbreak in Latin America substantially affects the dengue serology in routine diagnostic laboratories." They confirmed a high specificity of the dengue NS1 antigen. Still, they showed that in a cohort of ninety-three travelers with confirmed ZIKV infection, about one-third of ZIKV IgM and more than half of ZIKV IgG antibodies cross-reacted in the Euroimmun dengue virus (enzyme-linked immunosorbent assay) ELISA [23].

Serodiagnosis is hampered by the persistence of ZIKV IgM for up to 12 weeks after the onset of illness and the inability of PRNT to distinguish between recent and past infections [2].

In ZIKV infections, the virus persists in plasma or serum for only a relatively brief period. ZIKV can be detected in bodily fluids such as saliva, urine, or semen. This supports the goal of combining serological and RNA assays to provide clinically relevant information over a longer period, particularly useful in treating pregnant women [21].

11.2.2 Blood Tests

Researchers from The University of Texas Medical Branch at Galveston, in conjunction with the New York State Department of Health's Wadsworth Center, developed a detection test for ZIKV that is faster and more accurate than a currently available test. The new test can detect ZIKV in a tiny blood sample in less than four hours.

The new test is detailed in EBioMedicine. "The new diagnostic test was designed to more accurately detect ZIKV for a longer period after infection and reduce false positives due to cross-reactivity with other flaviviruses," said Pei Yong Shi, UTMB Professor in the department of biochemistry and molecular biology. "UTMB and the Wadsworth Center have jointly patented the technology. We anticipate the new test, called a microsphere immunofluorescence assay (MIA), will soon be approved for use in a clinical setting." [24].

A different team from Albany also worked on a multiplex microsphere immunoassay approach. It captures the diagnostic power of viral envelope protein (that elicits robust yet cross-reactive antibodies to other flaviviruses) and the differential power of viral non-structural proteins NS1 and NS5 (that induce more virus-type specific antibodies).

Moving forward, the multiplex capacity of MIA allows one to add more antigens to expand the diagnostic coverage of the assay. Since ZIKV, DENV, WNV, and Chikungunya virus may often cocirculate in the same geographic regions, it would be helpful to add antigens that could differentiate infections with these viruses. Compared with ELISA, another advantage of the MIA assay format is its high throughput and low cost [25].

The multiplex MIA achieves a rapid diagnosis (turnaround time < 4 h) and requires a small specimen volume (10 μl) in a single reaction. This serologic assay could be developed for the clinical diagnosis of ZIKV infection and monitoring immune responses in vaccine trials [25].

11.2.3 Urine Tests

Gourinat et al. [26] studied the diagnostic utility of urine in detecting the ZIKV through the urine sample still needed to be analyzed using RT-PCR (see above). Their six patient studies reported suitability.

They argued urine would be less invasive and easy to use, and RNA was detectable in urine at a higher load and with a longer duration than in serum. They admit confirmation involving a larger cohort is needed [26].

11.2.4 Saliva Tests

Preliminary data from Sabalza et al. [21] using saliva samples spiked with ZIKV showed that their LAMP diagnostic system detects ZIKV RNA in saliva. They used a benchtop isothermal amplification device and then adapted the assay to the microfluidic device, the Rheonix CARD1 cartridge. They demonstrated that a fully integrated RT-LAMP assay coupled with the reverse dot-blot (RDB) technique and processed by the Encompass Optimum workstation could detect ZIKV RNA spiked saliva samples [21].

Another team from the São Paulo Zika Virus Research Network (Rede Zika), led by Paolo Zanotto, announced a similar breakthrough. One of the team members, Edison Durigon, says he planned to distribute the recombinant protein free of charge, first to the centers that belong to the Zika Network and soon afterward to the public health system.

Nonetheless, using a saliva sample increased the rate of molecular detection of ZIKV at the acute phase of the disease. Still, it did not enlarge the window of detection of ZIKV RNA. ZIKV RNA detection can be negative in saliva while positive in blood, and a blood sample is required for other laboratory tests; saliva cannot replace blood samples [27].

11.2.5 Quick Tests

Zika, dengue, and chikungunya are three mosquito-borne viruses having overlapping transmission vectors. They cause diseases with similar symptoms in human patients but require different immediate management steps. Therefore, rapid (< one hour) discrimination of these three viruses in patient samples and trapped mosquitoes is needed.

A team led by Yaren from Firebird Biomolecular Sciences in Florida and GenePath Dx from India was designing a kit, complete with a visualization device, which is now available for point-of-sampling detection of Zika chikungunya, and dengue. The assay output is read in ca. 30 min by visualizing (human eye) three-color coded fluorescence signals. Assay in dried format allows it to be run in low-resource environments [28].

11.3 Treatments

Several compounds have shown activity against ZIKV in vitro, but none have been evaluated in clinical trials. Knowledge-driven drug repurposing, structure-based discovery, RNA interference, long non-coding RNAs, miRNAs, and peptide inhibitors may pave the way for discovering such novel agents.

Since regulatory agencies have approved no antiviral agents for treating ZIKV infection, the clinical management of acute ZIKV infection is supportive care [29]. Though many compounds against ZIKV have been discovered in the past few years, very few have been evaluated in clinical trials. Twenty anti-ZIKV drugs/agents in different classes have shown in vivo efficacy in animal models. Only one, BCX4430, has completed a Phase I clinical trial [2].

A PRC team led by Li [30] believes they have identified a molecular basis for the potential development of therapeutics against ZIKV.

> ZIKV infection stimulates a type I interferon (IFN) response in host cells, suppressing viral replication. Type I IFNs exert antiviral effects by inducing the expression of hundreds of IFN-stimulated genes (ISGs). To screen for antiviral ISGs that restricted ZIKV replication, we individually knocked out 21 ISGs in A549 lung cancer cells and identified PARP12 as a potent inhibitor of ZIKV replication. Our findings suggest that PARP12 mediated the ADP-ribosylation of NS1 and NS3, nonstructural viral proteins involved in viral replication and modulating host defense responses. This modification of NS1 and NS3 triggered their proteasome-mediated degradation. [30]

Some drugs were being studied for repurposing. Such as sofosbuvir and chloroquine and showed promising results in animal models. Though they were approved for clinical use with known safety profiles, there is lack of clinical trial data on their efficacy against ZIKV in humans, particularly in pregnant women [29].

Natural products such as hippeastrine hydrobromide, an active ingredient of the medicinal plant *Lycoris radiata*, emetine, extracted from the ipecac root, and quercetin-3-β-O-D-glucoside (Q3G), a natural derivative of quercetin have been discussed due to their record with treating vector-borne diseases, such as malaria. They may account for rich leads for antivirals [29].

Here are some examples. A team from Rockefeller University had presumably identified a ZIKV antibody called Z004. They report that 2004 was particularly effective at neutralizing Zika. "These antibodies could be beneficial shortly. One could envision, for example, administering Z004 to safely prevent ZIKV among pregnant women or others at risk of contracting the disease," said Davide F. Robbiani, Research Associate Professor and Colead [31]. "These antibodies could be beneficial shortly," said Robbiani. "One could envision, for example, administering Z004 to safely prevent ZIKV among pregnant women or others at risk of contracting the disease. The dengue one virus–a close relative of ZIKV and one of four types of dengue–has a ridge quite like the ZIKV. Z004 was also effective at neutralizing dengue one." [31].

Some Brazilian samples revealed evidence of prior dengue one infections, which could explain why specific volunteers' immune systems were more effective at neutralizing ZIKV. "It appears that, much like a vaccine, dengue one can prime the immune system to respond to Zika," Coauthor Margaret MacDonald said [32]. Their detailed maps revealed how the antibody pinches a ridge on the virus when it binds. While some efforts to develop a vaccine use all or most of the virus to stimulate the immune system, the researchers believe it could be safer to employ only a tiny fragment containing this ridge [31].

The drug hydroxychloroquine (from COVID fame) appears to reduce the viral load of ZIKV in the placenta and the brains of baby mice, which were normal sized at birth [33]. "We found that the ZIKV manipulates the garbage recycling system to its advantage. The ZIKV infection ramps up autophagy. So, when we use a drug that inhibits or suppresses this ramping up, we can block the virus from infecting the fetus," says Indira U. Mysorekar [34].

"It was not allowed to cross over into the placenta into the area where the blood flow, nutrient, and oxygen exchange is happening, so the babies were fine," said Dr. Mysorekar. Professor Mysorekar says what works in mice should also work in people. "This is going into human trials. The first round of human trials starts with this antibody," said Dr. Mysorekar. "We do not know the risk of long-term treatment with the drug [Hydroxychloroquine]. These tests must be done, which will be the next step as the findings leave the lab and head to the clinic," said Prof. Mysorekar [35].

11.4 Vaccines

ZIKV comprises a single strand of RNA encoding fourteen distinct proteins, but the immune system mainly responds to just two that protrude from the virus's outer surface. A cryo-EM examination by Sirohi et al. [36] of the ZIKV found a difference because of five residues in ZIKV relative to the dengue virus, reflecting a highly variable region of the E protein. The differences in E protein structure among ZIKV and other flaviviruses may govern cellular tropism and contribute to disease outcomes [36].

There are no approved vaccines against ZIKV infection. Until a vaccine is developed or effective treatment for ZIKV is discovered, healthcare providers must be aware, vigilant, and proactive to lessen the spread and impact of the implicated devastating birth defects (microcephaly) and other neurological disorders (e.g., Guillain–Barre syndrome) of this infection [37]. Women in high-risk areas are just recommended to delay their pregnancies since no available treatment can effectively prevent infections. The only other way to provide immunity is with a vaccine.

If children are infected before puberty, they presumably become immune and cannot pass ZIKV along to their children. And in parts of the world where ZIKV is endemic, people are more likely to have been exposed and become immune while young [38].

Due to its distinct pathogenesis, ZIKV presents unique challenges in developing and identifying antiviral agents that are safe, potent, and specific to prevent and treat infection in pregnant women [39].

- Must be pregnancy category A/B (do not impair the fetus or compromise the maternal/infant interface);
- Must inhibit replication devoid of toxicity across all relevant and permissive cell types; and
- Should be orally bioavailable for maintenance dosing.

There was a general concern that a vaccine for ZIKV might arrive too late to impact this outbreak substantially [40]. At one point, developing a safe and effective ZIKV vaccine was an urgent global health priority [41]. More than 20 vaccines [42] were under development, and one was predicted to become available as early as 2018 [43] (others predicted 2019) [33], which did not occur.

Report after report about a successful vaccine proliferated during the pandemic.

Virion Inc., a late clinical-stage company from San Diego, announced that it has collaborated with the Eunice Kennedy Shriver National Institute of Child Health and Human Development (NICHD) to identify new drug candidates for the treatment of the ZIKV and other significant virus infections. The new collaboration aims to capitalize on Viriom existing antiviral platform to provide an innovative solution to the ZIKV outbreaks and curb the global spread of the established and emerging viral diseases [44].

> The European Vaccine Initiative, Institut Pasteur Paris, Themis Bioscience, and the Commissariat à l'énergie atomique et aux énergies alternatives have joint forces and announce the start of the ZIKAVAX project. This consortium aims to develop a safe, fast-track, effective and affordable preventive vaccine against ZIKV infection based on specific know-how and technology at each partner.... ZIKAVAX is funded by an approximately €5 million grant from the EU's Horizon 2020 program, which invests €45 million in research to combat the outbreak of the ZIKV disease and other emerging infections transmitted by mosquitoes. Over the next 48 months the consortium will work to ultimately demonstrate the safety and immunogenicity of the recombinant measles ZIKV vaccine candidate in adults in phase I clinical trial.... The ZIKAVAX project addresses this urgent public health issue by proposing a vaccine based on a measles vaccine vector developed at Institut Pasteur by Dr. Frédéric Tangy. This vector has demonstrated proof of principle in humans and a preclinical track record of rapid adaptability and effectiveness for various pathogens. [45]

The Indian Bharat Biotech International company reported that it has two ZIKV vaccine candidates entering preclinical animal trials, which commenced in late February 2016 [46]. In a recent paper, Richner and Himansu et al. [47] reported that their mRNA vaccine for ZIKV managed to protect mice against infection. All the mice on placebo succumbed to the virus after 42 days, while nearly all of the mice who got Moderna Therapeutics's (RNA) vaccine survived [47]. Also, Pardi et al. [48] reported a single low-dose intradermal immunization with lipid nanoparticle-encapsulated nucleoside modified mRNA (mRNA-LNP) encoding the premembrane and envelope (prM-E) glycoproteins of a 2013 ZIKV outbreak strain elicited potent and durable neutralizing antibody responses in mice and non-human primates (rhesus macaques) [47]. Bharat Biotech's vaccines, now christened "Zikavac," are ready for preclinical trials; this makes these two vaccines head and shoulders ahead of the other international efforts, which are still literally efforts on the drawing board [49].

Another team of researchers at the University of Hawaii (UH) at Manoa, John A. Burns Medical School, claimed to have successfully developed a vaccine candidate for the ZIKV, showing that it effectively protects both mice and monkeys from the infection. They said their "vaccine candidate shows much promise particularly as it showed to require only two immunizations given three weeks apart and is a potentially safer alternative to other candidates already in clinical trials," said Axel Lehrer, Assistant Professor of Tropical Medicine, and Infectious Disease [50].

A different team from Queensland, Australia, reported in a blog (2017) that they had found a plant extract and the "…plaque assays found that the extract from this common native plant killed 100% of the ZIKV infection in cells," said Lead Researcher Trudi Collet from Queensland University of Technology (QUT) in Australia. "The research is in the initial stages, but we aim to ultimately synthesize the compounds in question and turn our attention to preclinical testing," Collet said [51]. She did not name the plant, presumably for commercial reasons, but said it was standard, and its compounds were found to kill 100% of the ZIKV infection in cells. QUT researchers were collaborating with Australian-based biotech company Health Focus Products Australia [52].

An entirely different approach seems to support inhibiting the action of LTRIN (a 15-kilodalton protein) from disrupting ZIKV transmission. LTRIN is found in the salivary glands of *Ae. aegypti* mosquitoes. It seems to have coevolved to moderate host immune responses and facilitate pathogen transmission. Lin et al. [53] have recently explained the underlying molecular mechanisms. They claimed to have identified a specific salivary protein from mosquito, LTRIN, that promoted the transmission of ZIKV by counteracting the LTβ R-mediated host immune response. This means a specific mosquito salivary protein can exacerbate the pathogenesis of ZIKV in mammalian hosts and that LTβR (lymphotoxin-β receptor) signaling has a critical role in infection with ZIKV. This study sheds light on the transmission process of ZIKV in nature [53].

> Our study demonstrated that inhibition of LTβR signaling was critical for promoting the transmission of ZIKV in Iftar–/–mice but not wild-type mice, which would suggest that the early immune responses mediated by LTβR signaling and those mediated by type I interferons have synergistic effects on counteracting infection with ZIKV. At the same time, ZIKV evolved to use LTRIN to escape the host immune response. [53]

Some efforts relied on the traditional method of growing or purifying a weakened or killed virus. A candidate reported by Abbink et al. was "[a] purified inactivated virus vaccine induced ZIKV-specific neutralizing antibodies and completely protected monkeys against ZIKV strains from Brazil and Puerto Rico." [54].

Another inactivated virus vaccine formulated with an adjuvant, a substance that enhances the immune response is under development by the Takeda Group and the Biomedical Advanced Research and Development Authority (BARDA) [55]. FDA granted fast-track designation to TAK-426, a purified, inactivated, alum-adjuvanted vaccine with the whole ZIKV. In November 2017, Takeda advanced the candidate into a 240-subject phase 1 study in the continental U.S. and Puerto Rico [56].

An experimental (inactivated) vaccine, called ZPIV, has already proved effective when designed to target a virus-like ZIKV, Japanese encephalitis. The vaccine has already been tested in monkeys, proving effective against Zika. ZPIV uses a strategy that has worked in many other vaccines: Scientists cripple the virus so it cannot cause disease, but that inactivated form still triggers an immune reaction [57]. Preclinical findings have been promising. Walter Reed Army Institute Research scientists are rushing to test the ZPIV vaccine, and they plan to start human testing at their clinic in Silver Spring before the end of 2017.

The U.S. Army has not decided whether to offer Sanofi an exclusive license for their ZIKV vaccine called Zika Virus Purified Inactivated Vaccine (ZPIV). According to the NGO Médecins Sans Frontières (MSF), it is worried that if given the exclusive license for the ZIKV vaccine, the pharmaceutical company will follow the same path and neglect countries with great need but fewer opportunities for profit, according to Judit Rius Sanjuan, MSF's U.S. Access Campaign Manager [58].

Nelson Michael from WRAIR and Dan Barouch from the Center for Virology and Vaccine Research at Harvard Medical School reported that the inactivated virus (formalin-inactivated viral particles) and a DNA-based vaccine—encoding for the viral premembrane and envelop proteins—both protected mice against Zika [58]. The researchers found that antibodies mainly mediated the protection afforded by the killed virus vaccine. When they transferred immunoglobulin G (IgG) from the plasma of vaccinated monkeys into naïve monkeys, the antibody recipients were also protected from infection. "That means antibodies and vaccines are plausible solutions," Crowe told The Scientist [59]. The researchers also reported that low antibody levels did not promote any detectable signs of antibody-dependent enhancement (ADE) of ZIKV infection [59].

Others employ newer technology, using DNA or RNA pieces, which allows a faster timeline [59].

> DNA and RNA vaccines have generated a lot of excitement, says Wilder-Smith, because they have the potential to be rapidly scaled up and deployed during an outbreak, should they prove to be efficacious. DNA vaccines work by creating a plasmid (a circle of DNA) that encodes the genes for the virus's envelope proteins. Once this plasmid makes its way into the target cell's nucleus, the cell's molecular machinery begins producing the viral proteins, inducing an immune response. "The body's cells become the factory for these vaccines," says Joseph Kim, chief executive of the Pennsylvania-based Inovio Pharmaceuticals, developing a DNA vaccine for Zika. RNA vaccines work in the same way, although the RNA can start producing viral proteins when it enters the cell without penetrating the nucleus. [60]

Reported breakthroughs predominantly involving DNA vaccines (a DNA vaccine is a type of vaccine that transfects a specific antigen-coding DNA sequence into the cells of an organism as a mechanism to induce an immune response) were announced regularly.

Part of the reason DNA vaccines are tricky, researchers say, is because the vaccine's DNA must reach the nucleus of a person's cells before it can begin instructing them to make ZIKV proteins. Not all vaccine is injected into the body and reaches the nucleus, lowering their effectiveness. To solve this problem, some companies are trying a similar method that uses RNA, a molecule in the body that is more flexible than DNA. Among the things it can do is carry out, on its own, the instructions contained in DNA. The vaccine would not need to reach a cell's nucleus to trigger the immune response [60].

For example, researchers from the Institut Pasteur and the Inserm Research Institute (French National Institute of Health and Medical Research) also claimed to have identified a protein target that, when overexpressed, can block ZIKV infections. It opens the possibilities for developing therapies to prevent ZIKV congenital disabilities by targeting the IFITM3 protein [61].

Interferon-induced transmembrane protein 3 is a protein that in humans is encoded by the IFITM3 gene. It plays a critical role in the immune system's defense against Swine Flu, where heightened levels of IFITM3 keep viral levels low, and the removal of IFITM3 allows the virus to multiply unchecked.

Monel et al. [62] demonstrated that IFITM3 is an essential component of the type I interferon response, conferring protection against ZIKV infection. Interferon-induced transmembrane proteins inhibit the replication and pathogenesis of a wide array of viruses, including flaviviruses, and confirmed that IFITM3 overexpression blocks the replication of ZIKV [61].

Another research team from Academia Sinica, Taipei, Taiwan, examined the effects of ZIKV infection on mouse testicular tissue. The virus damaged cells, impaired regular gene expression, damaged sperm, and infected sperm cells. They reported it has recently become apparent that semen acts directly on tissues in the female reproductive tract. Indeed, a substantial level of seminal pro-inflammatory cytokines and reactive oxygen species (ROS) may provoke inflammatory responses in the female reproductive tract and cause tissue injury. Consequently, these processes may create access to favorable conditions for seminal infectious ZIKV to establish an initial infection in the female reproductive tract before its systemic disease [63]. Notably, ZIKV caused increased testicular inflammation and oxidative stress, characterized by prominent levels of potentially damaging by-products of normal cellular processes known as reactive oxygen species. Next, the researchers treated infected mice with ebselen, an anti-inflammatory and antioxidant drug that can neutralize reactive oxygen species. They found that ebselen alleviated the observed testicular symptoms and prevented sexual transmission of ZIKV via sperm from infected male mice to uninfected female mice [63]. They reported Ebselen treatment significantly reduced ZIKV-induced testicular oxidative stress, leucocyte infiltration, and production of pro-inflammatory response. Furthermore, it improved testicular pathology and prevented the sexual transmission of ZIKV in a male-to-female mouse sperm transfer model [63].

Another team from Washington University School of Medicine in St. Louis, led by Dr. Indira Mysorekar from the Department of Obstetrics and Gynecology, was looking at ways to stop the spread of ZIKV from mother to child. Researchers infected pregnant mice with ZIKV and injected other mice with antibodies that blocked the virus [33].

"Carrying out a trial on pregnant women is always challenging. But given that the drug is already approved and ZIKV infections have such terrible consequences, there may be more chances of going forward," she said [33]. "We are excited because it's a prevalent drug, very cheap, available worldwide, and FDA-approved for pregnant women," Mysorekar said. "Often, scientists can work for decades in these studies, and they don't necessarily translate to patient care. It's very gratifying to repurpose this drug potentially." [33].

Dr. Sarah George is an infectious disease specialist from Saint Louis University. She is one of the several doctors chasing a vaccine for the ZIKV and is testing one. It is a two-dose shot that contains an inactivated form of the virus. In the study, more than 90% of volunteers showed an immune response to ZIKV. While ZIKV cases have dropped dramatically since that first outbreak, a vaccine could keep pregnant women and babies safe against future threats [64].

Another team, this time from UTMB Galveston, Dr. Peiyong Shi and Dr. Scott, were working on what some could be one of the most promising cures to the ZIKV epidemic. A vaccine they have created is effective in mice. The subsequent trial with larger animals is currently wrapping up. "They [the Brazilian government] are independently validating the vaccine simultaneously so we can speed up the advancement of the vaccine," Dr. Peiyong Shi said. This drug is more coveted than others because it is a single dose. "We can intervene quickly to slow down or stop the outbreak," Dr. Weaver said. Weaver said the vaccine starts working within a week and keeps people protected for many years. "Typically, vaccines like these that are live attenuating give protection for at least a decade, often longer," Dr. Weaver said. That's one form of protection women seems to be open to. UTMB Ob-Gyn Dr. Abbey Berenson said of the patients surveyed, most say they would get vaccinated [65].

Still another development involves research presented by Farshad Guirakhoe, Chief Scientific Officer of GeoVax, Inc, at the ASM Microbe 2017 meeting showed a new ZIKV vaccine that gives 100% protection in mice. "A single dose of GeoVax's NS1 vaccine candidate protected 100% of vaccinated animals," said Guirakhoo. In contrast, 80–90% of sham-immunized control animals died within 7–10 days. The NS1 proteins interfered with the mosquito's innate immune system and evolved to fight foreign pathogens, allowing the flavivirus to replicate and disseminate in mosquitoes. Therefore, said Guirakhoo, "a vaccine that can induce effective antibodies to NS1 and disable its function has the potential to reduce growth and transmission of ZIKV in its mosquito vector. This could enhance vaccine effectiveness in endemic areas by blocking the virus transmission from infected individuals to other community members." [66].

The following companies seemed to have been active in the ZIKV vaccine landscape: Bharat Biotech, Inovio Pharmaceuticals and GeneOne Life Sciences, Intrexon, Cerus, Sanofi-Pasteur, NewLink Genetics, Immunovaccine, Valneva, and GlaxoSmithKline [67].

Many of these companies were working on RNA vaccines. RNA vaccines work by introducing an mRNA sequence (the molecule which tells cells what to build) coded for a disease-specific antigen; once produced within the body, the immune system recognizes the antigen.

NIAID provided funding for some vaccine candidates including a DNA-based support a Phase II clinical trial of a DNA-based vaccine candidate developed by NIAID's Vaccine Research Center; Whole-Particle Inactivated—support preclinical and clinical evaluation of a Whole-Particle Inactivated vaccine developed in collaboration with WRAIR/BARDA; live-attenuated Zika Chimera—support product development including preclinical and Phase I/II clinical trials of a Live-Attenuated Zika Chimera vaccine developed in NIAID's Intramural laboratories in collaboration with the Butantan Institute, Brazil; Self-Amplifying mRNA—develop, manufacture, and conduct preclinical studies and Phase I clinical trials of a self-amplifying mRNA-based vaccine candidate developed by NIAID's Vaccine Research Center in collaboration with Glaxo Smith Kline; Vesicular Stomatitis virus—support further development, manufacture of pilot lots, preclinical and Phase I testing of a recombinant Vesicular Stomatitis virus vectored vaccine currently under development through a collaboration with Harvard University; and Virus-Like Particle and Monoclonal Antibody—support the discovery manufacture and preclinical and clinical evaluation of additional candidates such as virus-like particle and monoclonal antibody-based vaccines [68].

Despite the lack of published experimental data, NIAID announced that in collaboration with GlaxoSmithKline (GSK), a new ZIKV vaccine would be developed using a self-amplifying mRNA vaccine platform and enter clinical trials in late 2017 [69]. Messenger RNA (mRNA) vaccines teach cells how to make a protein that will trigger an immune response inside the body.

Sapparapu et al. (2016) [70] isolated a panel of human monoclonal antibodies from previously infected subjects with ZIKV. A subset of antibodies recognizes diverse epitopes on the envelope (E) protein and exhibits potent neutralizing activity. They evaluated the therapeutic efficacy of ZIKV-117 in pregnant and non-pregnant mice and found the monoclonal antibody treatment markedly reduced tissue pathology, placental and fetal infection, and mortality in mice. Thus, neutralizing human antibodies can protect against maternal–fetal transmission, illness, and disease and reveal essential determinants for structure-based rational vaccine design efforts [69]. They concluded that the most potent neutralizing antibodies exhibited a breadth of inhibitory activity against African, Asia, and the Americas strains. Even a single ZIKV-117 dose given five days after infection protected mice against lethal disease, a timeline like the most protective antibodies against other flaviviruses [70].

Developers in the United States, France, Brazil, India, and Austria worked on over twenty vaccine projects. WHO Assistant Director-General Marie-Paul Kieny said the feasibility of an "emergency-use" vaccine was being examined. "As it will be used to protect pregnant women and women of childbearing age, a vaccine must meet extremely high safety standards," Kieny said [71]. At one point, PAHO reported five potential vaccine candidates heading for clinical trials [72].

11.4.1 Vaccine Markets

According to Thomas Monath, CEO of the infectious disease division at NewLink Genetics Corp., ZIKV is the most significant opportunity for a new vaccine that has come along in my career. I have been in vaccines for 40 years [73]. Around 45 vaccine candidates are listed in the WHO's pipeline tracker. They range from traditional live and inactivated virus vaccines to more experimental DNA, Mrna, and protein subunit vaccines [60].

However, even though some candidates are now in Phase II trials, the transient nature of the outbreak is making wide-scale testing of the new vaccines complex, and some research programs have already been curtailed [60].

It turned out the trickiest thing about getting a ZIKV vaccine to market is that the outbreak just did not last long enough. By November 2016, the WHO ended the public health emergency, and the number of cases had fallen markedly since the outbreak's height. That is good news for people affected, but it makes testing the new vaccines much harder. "Because the rates are declining, it's difficult to do a Phase III trial to prove efficacy," says Mark Challberg, who oversees flavivirus research at NIAID. "There are not enough infections out there to get good statistical data in a reasonable amount of time." [60].

Conventional vaccines take years to develop and test. They often cost more than pharmaceutical companies are likely to recoup from sale in the primarily poor tropical countries where the diseases originate. The economics discourage many firms from producing them for new, emerging diseases. From a business perspective, the market may be small if public health authorities determine that a ZIKV vaccine need only be stockpiled for emergencies rather than administered routinely to the general population [73].

Who will benefit from a vaccine? Some argue that low-income families in Africa might bear the most significant burden of ZIKV despite ongoing coverage and prevailing wisdom. Guinea-Bissau has reported three microcephalic babies with clinical data suggesting ZIKV exposure to the WHO last year [43]. African strains may be even more deadly. According to one study by French virologists, a strain of ZIKV isolated in the Central African Republic was twice as toxic to brain stem cells as the variant circulating in Brazil [43].

11.4.2 Vaccine Developments for Pregnant Women

Testing vaccines in pregnant women is a general problem; the reluctance to pursue such studies has resulted in a shortage of information regarding drug safety and efficacy during pregnancy and is an obstacle to adequate prenatal care. Evaluating pregnant women must be a crucial component of new ZIKV interventions, and the development of innovative approaches to ethically conducting these studies may benefit maternal health more generally [43].

ZIKV infection in pregnant women can have devastating impacts on the developing fetus. Two vaccine platforms developed in 2017 promise to reduce fetal illness and ZIKV-induced fetal disease in pregnant animal models. These include live-attenuated and lipid-encapsulated modified mRNA-based approaches that produce neutralizing antibodies that prevent vertical transmission of ZIKV across the placental barrier, thus limiting fetal infection [74].

A passively administered, appropriately engineered human nmAb (neutralizing monoclonal antibodies) may be able to protect a fetus from ZIKV infection. nmAb therapy is likely one of the most promising interventions to prevent and treat ZIKV infection during pregnancy [75]. Researchers believe that this antibody combination will be safe enough to administer to pregnant women and likely cross the placenta, protecting both the pregnant woman and fetus from the virus [76].

New anti-ZIKV (ZIKV) vaccine platforms promise to limit congenital ZIKV syndrome effectively. They include lipid-encapsulated mRNA vaccine-based immunity established before pregnancy in mice is sufficient to protect against congenital infection and disease. Live-attenuated vaccines induce elevated levels of neutralizing antibodies, prevent infection and damage to the testis in males and abrogate in utero transmission and fetal disorder during pregnancy [77].

11.4.3 Vaccine Animal Trials

GLS-5700 was one of a few ZIKV vaccines that have shown promising results in animal models. Tebas et al. [78] found that the GLS-5700 vaccine generated a protective response against multiple ZIKV isolates, including the African lineage MR766 ZIKV strain in a neuronal-cell neutralization assay PR209 Caribbean sublineage of Asian ZIKV in a challenge model in IFNAR knockout mice [75].

Another team (Muthumáni et al., 2017) [79] researchers used genetic sequences of ZIKVes isolated between 1952 and 2015 to generate a novel DNA sequence called ZIKVprME. They demonstrated that following ZIKVprME vaccine injection, mice had significantly higher cytotoxic T cells (a type of immune cell involved in innate immunity). Postvaccine injection, mice had significantly higher levels of the ZIKV-specific IgG antibodies, and this level remained elevated long term. Then, they tested the vaccine in IFNAR-/- mice, a strain of mice highly susceptible to the ZIKV and succumb to the disease within 6–7 days. Again, the vaccine successfully elicited a robust immune response and neutralized the virus significantly. Moreover, 100% of the vaccinated IFNAR-/- mice survived, whereas most of the unvaccinated mice succumbed to the disease [78].

IFNARs are activated by type I interferons (IFNs), antiviral proteins released by the mother's immune system upon viral infection. Previous human research has suggested that the loss of IFNARs, and the subsequent pathway triggered by IFN signaling, is associated with greater susceptibility to ZIKV infection. Indeed, the IFNAR-deficient fetuses in this study showed higher viral loads. The Yale team of Iwasaki et al. (2018) [80] also discovered that the pathway triggered by IFNAR was dangerous to the fetus even in the absence of ZIKV infection, pointing to a possible role in other types of problems during pregnancy [79].

11.4.4 Vaccine Human Trials

Scientists at the National Institute of Allergy and Infectious Diseases (NIAID) in Bethesda, Md., make a vaccine consisting of the genes that encode these two proteins. In theory, once inside the body, a person's cells manufacture copies of the two ZIKV proteins, assemble them, and trigger an immune response [80]. The Vaccine Research Center (VRC) scientists state that the vaccine does not contain live infectious material, so it cannot cause active ZIKV infection in participants. It has entered phase 2/2b of clinical trial testing. The multisite phase trial is called VRC 705 and involves nearly 2500 participants from the United States, Puerto Rico, Brazil, Peru, Costa Rica, Panama, and Mexico. The study is expected to be completed by 2019, but initial results could be available at the end of 2017 [81].

A top candidate was developed by Inovio, GeneOne, and the Wistar Institute. The Wistar Institute, Penn Medicine, and Inovio Pharmaceuticals announced the receipt of FDA approval for the commencement of a first-in-human clinical trial to probe the safety and tolerability of a new synthetic DNA-encoded monoclonal antibody (DMAb) therapeutic technology to prevent ZIKV infection [82].

In DMAb therapeutic technology, the DMAbs are made inside people rather than at manufacturing plants as in other therapeutic antibodies. Patients are given the DNA instructions so that their bodies are equipped with the tools to generate their antibodies against pathogenic targets such as bacteria, virus-infected cells, and cancer cells [83].

It is a DNA-based vaccine called GLS-5700. Inovio Pharmaceuticals, Inc. and GeneOne are developing the Zika DNA vaccine, GLS-5700, with academic collaborators from the Wistar Institute, Philadelphia, Pennsylvania, and Université Laval in Quebec City, Canada. GLS-5700 is manufactured by VGXI, Inc., GeneOne's wholly owned subsidiary, located in The Woodlands, Texas. VGXI is the only facility with manufactured DNA vaccines for Ebola, MERS, and Zika [83]. VGMI is a contract manufacturer (CMO) of DNA plasmids.

Inovio announced they had introduced electroporation, whereby a mild electrical current follows the delivery of the vaccine. This opens cell membranes to let the DNA inside, avoiding the most significant complaint against DNA-based vaccines: "they are not easily absorbed." Inovio uses a pen-like instrument to deliver a mild electric pulse to open the cells so that the DNA can get inside before injecting the vaccine—electroporation. That increases the vaccine cost, though Kim says that can be driven down in mass production [84].

Inovio, based in Plymouth Meeting, Pennsylvania, is scheduled to start Phase I clinical trials. They received approval to begin human clinical trials on its ZIKV vaccine candidate, GLS-5700 [60]. In preclinical testing, this synthetic vaccine induced robust antibody and T cell responses in small and large animal models, demonstrating the product's potential to prevent infection from this harmful pathogen in humans [85].

Inovio initiated a clinical study of its preventive ZIKV vaccine (GLS-5700) in 160 subjects in Puerto Rico, where the ZIKV outbreak has been declared a public health emergency [86]. Inovio's ZIKV vaccine went from concept to first human dose in just seven months. Joseph Kim, Chief Executive of the Pennsylvania-based Inovio Pharmaceuticals, believes this is the fastest vaccine that has been put into practice in history. The company published the results of its Phase I trial of the vaccine in October 2017—in which 100% of participants produced antibodies to the virus after three doses of the vaccine [87].

GeneOne, headquartered in Seoul, South Korea, has received approval from the U.S. FDA and Health Canada's Health Products and Food Branch to conduct the clinical trial for GLS-5700. GeneOne Life Science, Inc. announced on July 16, 2016, the dosing of the first subjects in its multicenter Phase I trial to evaluate the Zika DNA vaccine (GLS-5700).

Both Inovio Pharmaceuticals and GeneOne Life Science said in late July 2016 that they tested a different DNA vaccine against ZIKV and tested it on volunteers in Miami [60].

11.4.5 Vaccine Concerns

On the vaccine front, there is some reason for optimism. After the infection clears, a person is thought to have lifelong protection against Zika, so lasting immunity appears possible. Geard et al. claim that immunity may be lifelong after comparisons with similar viruses, mainly dengue fever. An exemption may be challenged if the virus evolves. On the other hand, some immunity may open new risks, like dengue fever, where infection may produce immunity to a different strain for 10 to 20 years. Still, those infected after immunity laxes experience a more severe illness [88].

11.4.5.1 Cross-Reactivity

Recent research has shown that the ZIKV only has one serotype (dengue has four, maybe five, different serotypes), which means that even as the virus evolves into various strains, these strains should all be closely enough related that one vaccine should be able to protect against all of them [89].

Cross-reactivity with dengue is a particular concern, analysis indicates both cross-protection, and enhancement could shorten the time until epidemics can re-occur and might increase the chances of long-term endemicity [90].

Barbi [91] reported some cases of ZIKV have probably been misclassified because of the molecular similarity of the virus with dengue leading to serological cross-reactivity in serologic tests employing antibodies for laboratory diagnosis [92].

Regarding dengue, there is epidemiological evidence that a primary infection protects from re-infection with the same serotype but represents a risk factor for developing severe disease upon re-infection with a different serotype [91].

The close relationship between flaviviruses, such as ZIKV and dengue, also presents challenges. Flaviviruses can sometimes cross-react—an infection with one of the four strains of the dengue virus, for example, can make a subsequent infection with another strain more severe. The antibodies to one may not completely neutralize the other and allow it to enter cells more efficiently, and T-cells overreact, producing too many cytokines that cause dangerous inflammation. The result can be the more difficult dengue hemorrhagic fever. "There is some concern that ZIKV infections could worsen dengue infections, or vice-versa," says Monica McArthur, Infectious Disease Researcher at the University of Maryland. And since vaccines produce the same antibodies as an infection, they have the potential to cause the same cross-reactions [93].

A characteristic of specific flaviviruses is the disease-enhancing activity of cross-reactive antibodies elicited by the previous infection by heterologous viruses (see antibody-dependent enhancement or ADE below). ZIKV is structurally and antigenically closely related to other flaviviruses such as DENG, yellow fever (YF), Japanese encephalitis (JEV), and cross-reactive, and non-neutralizing antibodies have the potential to exacerbate secondary flavivirus infections through antibody-dependent enhancement (ADE) mechanisms [60].

ADE occurs cross-reactive mAbs generated against one flavivirus serotype may inadvertently be the cause of severe disease during a subsequent infection with other flavivirus serotypes [69].

Therefore, it is essential to consider the possibility of ADE in developing a ZIKV vaccine and select candidates unlikely to produce cross-reactivity with undesirable effects, particularly with DENG, where ADE can lead to a fatal outcome.

ADE can transform a secondary dengue infection into a potentially fatal hemorrhagic fever, and antidengue or anti-WNV (West Nile virus) sera can enhance ZIKV infection in vivo. Immune responses against ZIKV and dengue cross-react, raising the possibility that a vaccine for one of these viruses could result in a poor outcome after infection with another. Individuals with a diminished immune response against a ZIKV vaccine risk a more severe ZIKV (or dengue) infection. For this reason and others, it will take several years of testing before a ZIKV vaccine can safely be administered to women of childbearing age [94].

Because monoclonal antibodies are generally safe, they believe that this antibody cocktail might be appropriate for uninfected pregnant women; because the antibodies will likely cross the placenta, the researchers hope that administration during pregnancy may protect both the pregnant woman and the fetus from ZIKV [95].

Dr. Philip Russell, Former Director of the Walter Reed Army Institute of Research and Commander of the U.S. Army Medical Research and Development Command, as well as Founding President and Chairperson of the Sabin Vaccine, said that the fact that ZIKV is occurring in areas where dengue has been endemic hints at a potentially severe problem with ADE and ZIKV vaccine development. Dejnirattisai et al. [19] concur [96]. "The current epidemic of Zika, which is usually a mild disease, is made a lot worse in these populations," Russell said. "I think there's a major effect, but the studies haven't been done to sort that out." [97].

Dengue cross-reactivity has been debated hotly mainly since "ADE was not observed in vivo in patients who had been infected by dengue followed by a secondary infection by ZIKV." [98].

> According to a hypothesis called antibody-dependent enhancement (ADE), incomplete immunization appears to help the virus enter defense system cells, where it reproduces, increasing the number of copies of itself in the organism and intensifying the severity of the infection. Because dengue and ZIKV are both flaviviruses and genetically similar, it was believed that the partial immunization observed after dengue infection might also occur in Zika-infected individuals with prior dengue infection. This suspicion was strengthened in mid-2016 when research showed that antibodies against the dengue virus protect individuals against the ZIKV but do not neutralize it completely. In March 2017, U.S. researchers found partial immunization to explain the multiplication of ZIKV in a study using mice with weakened immune systems. [99]

Recently, Terzian et al. [99] claimed individuals infected with ZIKV who have dengue fever do not appear to become more severely ill than those with ZIKV who have never had dengue [100]. They claim their study is the first to show that prior dengue infection in humans infected by ZIKV does not necessarily lead to a worse illness. Previous research using only cells and rodents suggested that initial dengue infection would intensify ZIKV disease by facilitating virus replication of the virus. The study of what is true of cells cultured in vitro and laboratory mice does not necessarily apply to humans [100].

> In this study, we investigated a cohort of patients who had been previously exposed to dengue virus (DENV) via natural infection and had experienced secondary infection by DENV2 or ZIKV. We analyzed viral load and pro-inflammatory and anti-inflammatory cytokines levels and assessed possible correlations between viral load and cytokine levels in this very original and valuable cohort. The most significant finding of our study is that ADE was not observed in our in vivo investigation into patients who had been infected by DENV, followed by secondary infection by ZIKV. Moreover, none of the patients in this cohort exhibited a severe course of the disease, and all recovered after the recommended supportive therapy for infections caused by arboviruses. [100]

"These findings don't entirely rule out the possibility that ADE occurs, but they constitute important evidence that having dengue doesn't increase the severity of ZIKV disease," said Kalil, Study Coauthor. "Some people who have had dengue present with a milder infection when they contract Zika. If ADE caused by dengue led to microcephaly, we would have identified hundreds of cases in São José do Rio Preto and Ribeirão Preto. Still, we found none," Nogueira said [99].

Although developing a vaccine could provide a long-term solution to the current ZIKV epidemic, there remains an unmet clinical need to identify drugs that can limit or prevent the consequences of congenital infection. According to Vanderbilt Immunologist and Infectious Disease Expert James Crowe, the recent findings [100] that antibodies alone may be enough to prevent infection is good news because antibodies can travel across the placenta, meaning a vaccine or passive immunization administered to pregnant women might provide protection for developing fetuses [59].

A recent screen of a subset of Food and Drug Administration (FDA) approved compounds against ZIKV in hepatic cells identified several anticancer, antimicrobial, antiparasitic, and antifungal drugs with anti-ZIKV activity [59]. Over 30 FDA-approved molecules are endowed with anti-ZIKV activity [101].

For example, a team from Florida State and Johns Hopkins found three drugs being tested and in final-stage clinical trials used to treat liver disease and parasitic worms promising [102]. A Helsinki team [103] has studied obatoclax, saliphenylhalamide, and gemcitabine, claiming that more than three anti-influenza compounds effectively inhibit inhibiting this ZIKV infection in human cells [104].

Repurposing studies involve enfuvirtide, favipiravir, chloroquine, and nucleoside analogs. Also, teams at the University of Pennsylvania and Emory University work on small molecule compounds and antivirals. For example, Sofosbuvir, a drug in clinical use against the hepatitis C virus (HCV), is among the FDA-approved substances endowed with anti-ZIKV activity [103]. Ongoing studies will be required to determine whether azithromycin, daptomycin, sofosbuvir, and other inhibitors or combinations can reduce ZIKV infection in the critical cell types [102]. Kao et al. (2017) [105] believe that drugs already approved to treat hepatitis C and others in development to treat other flaviviruses offer prime candidates for screening [106].

In 2018, a drug meant to treat and cure hepatitis C reportedly could effectively treat people infected with Zika, especially in protecting the fetus of an infected pregnant woman. Researchers associated with the University of California San Diego and colleagues in Brazil found that in cellular and mice trials, the antiviral medication Sofosbuvir, brand name Sovaldi, was influential in repairing cells damaged by ZIKV and blocking the transmission of the virus to the fetus of a pregnant mouse. "This suggests that one, the drug was well-tolerated by the Zika-infected pregnant mice and two, more importantly, that it was able to arrest ZIKV replication in vivo and stop transmission from mother to fetus," Senior Author Dr. Alysson Muotri, Professor in the U.C. San Diego School of Medicine departments of Pediatrics and Cellular and Molecular Medicine, said in a statement [105].

According to Ferreira et al. [102], preliminary treatment with sofosbuvir doubled the percentage and time of survival of ZIKV-infected animals. Sofosbuvir also prevented the acute neuromotor impairment triggered by ZIKV. Sofosbuvir prevented hippocampal- and amygdala-dependent memory loss in the long-term behavioral analysis of ZIKV-associated sequelae. Results indicate that sofosbuvir inhibits ZIKV replication in vivo, consistent with the prospective necessity of antiviral drugs to treat ZIKV-infected individuals [107]. Moutri and Mesci [108] also reported sofosbuvir-mediated sparing of human neural cell types from ZIKV-mediated cell death in vitro. They reduced viral burden in vivo in animal models of chronic infection and vertical transmission, strengthening the growing evidence for sofosbuvir anti-ZIKV activity [102].

Ferriera's results indicate that sofosbuvir treatment at a pharmacologically relevant concentration inhibits ZIKV replication in vivo, reducing mortality and blocking behavioral sequelae in the short- and long-term analysis. These results are an essential proof of concept and consistent with a future need for antiviral drugs to treat ZIKV-infected individuals [108].

There is a crowdsourcing effort through IBM's World Community Grid underway. However, they warn that future preclinical and clinical studies will need to address the potential impact of cross-reactive antibodies against the dengue virus and other flaviviruses [102].

Cross-reaction and antibody-dependent enhancement (ADE) has been found in dengue cases. "When someone first becomes infected with dengue, they experience mild flulike symptoms while developing antibodies. But when the person becomes infected again, though antibodies will recognize the virus particles, they cannot neutralize the virus and instead help it enter the person's cells to cause a more severe form of the disease, known as an antibody-dependent enhancement." [109]. Antibody-dependent enhancement (ADE) of different dengue virus (DENV) serotypes has been shown to correlate with increased viremia and disease severity. For example, preliminary studies using the Nigeria 1968 and Cambodia 2010 strains of ZIKV revealed distinct ADE curves in vitro [110].

In other words, if a person is infected with one dengue serotype and then gets infected with another serotype, the antibodies from the previous infection then make the new serotype infect cells that it otherwise would not. Then the virus can reproduce in extremely high numbers in these cells leading to severe disease. In addition, other researchers, such as Stettler et al. [93, 111], have shown that preexisting immunity to ZIKV can enhance dengue virus disease severity in animals [93].

Given that DENV is also endemic in many regions currently battling ZIKV infections, the development of ZIKV type-specific mAbs, which do not cross-react with other flaviviruses, is desirable to avoid any potential antibody-dependent enhancement [38].

A team from Mount Sinai in New York discovered that administering DENV- or WNV-convalescent plasma into ZIKV-susceptible mice resulted in increased morbidity—including fever, viremia, and viral loads in the spinal cord and testes—and increased mortality. ADE may explain the severe disease manifestations associated with recent ZIKV outbreaks and highlight the need to exert great caution when designing flavivirus vaccines [94]. On one level, the authors argue that the results suggest that preexisting immunity to DENV may have contributed to the rapid spread of ZIKV in the Americas and possibly is associated with increased viremia and clinical symptoms, including microcephaly. These studies showed that in vitro, some DENV antibodies are cross-reactive to ZIKV and can enhance ZIKV infection at specific concentrations [111].

If these studies hold in the clinic, cocirculation of ZIKV and dengue could increase the severity of both viruses [112]. Wang et al. [94] believe they have found a potential solution. The isolation of thirteen specific human monoclonal antibodies has been reported from a single patient infected with ZIKV. Two isolated antibodies (Z23 and Z3L1) demonstrated potent ZIKV-specific neutralization in vitro without binding or neutralizing activity against strains 1 to 4 of dengue virus, the closest relative to ZIKV. These two antibodies provided post-exposure protection to mice in vivo. Structural studies revealed that Z23 and Z3L1 are bound to tertiary epitopes in envelope protein domains I, II, or III, indicating potential targets for ZIKV-specific therapy [38].

Another critical question is whether enhancement can also be driven by immunity to other related flaviviruses, such as WNV. More than three million people in the United States possess preexisting antibodies to WNV. This virus is also endemic throughout Europe, Africa, the Middle East, and Australia [94]. This was surprising to the medical and public health communities and impacted the use of preexisting flavivirus antibodies on Zika-induced diseases and ZIKV vaccine development [113].

More than three million people in the United States possess preexisting antibodies to WNV. This virus is also endemic throughout Europe, Africa, the Middle East, and Australia [111].

11.4.5.2 Public Reception

Vaccine acceptance and hesitancy are severe problems and may be the best case for vector-based management [113]. The more likely someone is to believe in the false association between the use of the MMR vaccine and autism, the less likely that person is to use a ZIKV vaccine. The Wakefield Effect is named after the physician who made the first association between the MMR vaccine and autism. The spillover effect between the purported linkage between the MMR vaccine and autism has been demonstrated with the HPV vaccine.

Public support for ZIKV vaccines has been studied intensely since the COVID pandemic. A University of Kansas Medical Center of one hundred pregnant women receiving routine prenatal care found nearly half of the 100 patients surveyed (48%) strongly agreed to get a hypothetical ZIKV vaccine while pregnant. Women indicating strong acceptance of a hypothetical ZIKV vaccine were also more likely to feel strongly about the importance of children being up to date on all their vaccinations (97% versus 83%) and getting recommended vaccinations during pregnancy (97% versus 79%) [114].

A 2018 study found that people's erroneous beliefs about an association between the measles, mumps, and rubella (MMR) vaccine and autism indicated people's lessened intention to get a ZIKV vaccine. "People's erroneous beliefs about an association between the measles, mumps, and rubella (MMR) vaccine and autism were predictors of people's lessened intention to get a ZIKV vaccine." [115]. "We found that the misbelief about the MMR vaccine's association with autism was more influential on whether to get vaccinated for ZIKV than even perceptions of ZIKV itself, which is worrisome, especially in light of the persistence of that misinformation." [116].

The study also found that people's perceptions of the severity of the ZIKV and their general belief in the power of science to solve problems increased their intention to get the vaccine. As seen in the concluding chapter, these findings explain vaccine hesitancy and offer suggestions for antivaccine communications.

People's intentions to use vaccines are influenced by their beliefs about the specific vaccine and the disease it prevents. In the absence of firm ideas about the ZIKV (ZIKV), individuals may base their intentions to vaccinate against it on beliefs about other vaccines, specifically the misbelief that MMR causes autism [117].

"When a new disease arises, people who lack understanding of the new threat may extrapolate from their knowledge of other diseases," said Yotam Ophir, who coauthored the study with APPC Director Kathleen Hall Jamieson. "We found that the misbelief about the MMR vaccine's association with autism was more influential on whether to get vaccinated for ZIKV than even perceptions of ZIKV itself, which is worrisome, especially in light of the persistence of that misinformation." [118].

Two of the key findings were (1) the more likely someone is to believe in the false association between the use of the MMR vaccine and autism, the less likely that person is to use a ZIKV vaccine; and (2) people who believe in the ability of science to overcome problems were more likely to intend to use a ZIKV vaccine [117]. Another finding is interesting: People who were engaged in behaviors to protect against ZIKV were less likely to intend to get the vaccination—which "may be the result of their confidence that their actions pre-empt the need to be vaccinated," researchers say [117].

However, new research indicated that it spilled over to the anti-ZIKV vaccine, a vaccine still not on the market.

> A study of 3300 individuals between August to September 2016, published in the ***Journal of Public Health***, found that people's erroneous beliefs about an association between the measles, mumps, and rubella (MMR) vaccine and autism were a predictor of people's lessened intention to get a ZIKV vaccine. The study also found that people's perceptions of the severity of the ZIKV and their general belief in the power of science to solve problems increased their intention to get the vaccine. "When a new disease arises, people who lack understanding of the new threat may extrapolate from their knowledge of other diseases," said Yotam Ophir, a Ph.D. candidate at Penn's Annenberg School for Communication who co-authored the study with APPC Director Kathleen Hall Jamieson. "We found that the misbelief about the MMR vaccine's association with autism was more influential on whether to get vaccinated for ZIKV than even perceptions of ZIKV itself, which is worrisome, especially in light of the persistence of that misinformation." [116]

A study on willingness to receive a ZIKV vaccine among 989 pregnant women in Malaysia found 94% willingness to be vaccinated and a greater desire to obtain a ZIKV vaccine if recommended by a physician, friend, or relative [117].

Vielot et al. [22] reported some concerning findings as well. Respondents reported similar willingness to receive a ZIKV vaccine across all levels of knowledge concerning ZIKV transmission and prevention in pregnancy, suggesting that knowledge of ZIKV is insufficient to generate interest in ZIKV vaccination. Given that ZIKV symptoms outside of pregnancy are generally mild and self-limited, increased awareness of ZIKV symptoms, transmission, and prevention methods may reduce vaccine interest, as current findings suggest among travelers outside U.S. states [119].

Risk perception in the Vielot et al. [22] study, indicated by the perceived likelihood of being bitten by a ZIKV-infected mosquito, was not associated with increased willingness to receive a ZIKV vaccine. However, respondents with some ZIKV risk perception were nearly twice as likely to express concern about ZIKV infection, suggesting that risk perception indirectly affects willingness to receive a ZIKV vaccine through the fears [120].

11.4.6 Vaccines Canceled

One step could significantly speed up the process: injecting healthy, non-pregnant volunteers with live ZIKV. While rare, that "human challenge trial" would give researchers a better understanding of how the virus causes disease and which prospective vaccines would most effectively prevent infection. Consequently, promising ZIKV vaccine candidates cannot be thoroughly tested in humans because of a decline in conditions. Bioethicists are conflicted over whether to allow testing in healthy people [120].

Fauci warned that the low level of cases would also affect field trials of experimental vaccines. One such problem, the NIH Vaccine Research Center 705 Phase II trial, began in March and aimed to enroll at least 2490 volunteers in seven countries in the Americas [121]. The unexpected development makes the disease harder to study and threatens to hamper trials of experimental vaccines that might protect pregnant women in future outbreaks.

$100 million vaccine trials were underway at 17 sites in nine countries in 2018. Now it faces an expected ironic challenge. Most of these vaccines seem dead in their tracks with the ebbing of the epidemic, and the statement by the WHO that ZIKV is no longer an international emergency.

Although the waning epidemic means fewer people are falling ill, the decrease in cases also makes it tougher to conduct late-phase clinical studies to prove vaccine efficacy; Sanofi cited that as one of the reasons behind its decision to drop out of an agreement to license a ZIKV vaccine candidate developed by the U.S. Army [1].

"[S]cientists from the University of Texas Branch have said it may be a decade until an FDA-approved treatment is available. That is a lot of time for the virus to spread and more than enough time to turn into a global threat instead of a regional one." [56].

Cases of ZIKV have plummeted to levels so low that most people vaccinated in the trial likely will never be exposed to the virus, making it impossible to tell whether the vaccine works [85]. Any evaluation of the safety and efficacy of ZIKV vaccine candidates is now faced with an increasingly small number of sites with sufficient ZIKV incidence [122].

Sanofi-Pasteur halted work on its vaccine licensed from Walter Reed in September 2017. Citing reduced funding from the U.S. Biomedical Advanced Research & Development Authority (BARDA), drugmaker Sanofi is shutting down the development of its vaccine for ZIKV. Last year, BARDA committed $43.2 million to support the manufacture and Phase II clinical study of Sanofi's ZIKV vaccine. Now, the government organization is limiting its contract with Sanofi to a ZIKV surveillance study, which could benefit all vaccines in the pipeline [123].

According to Susan Hills, MBBS, MTH, Medical Epidemiologist in the CDC's Division of Arboviral Diseases, at least nine ZIKV vaccine candidates are currently in development—although most are still in the preliminary stages. There are no approved vaccines available to prevent ZIKV infection. When one is available, "vaccination of women of reproductive age would be the priority," Hills said. Cassetti explained that several vaccines and therapies were "rapidly initiated" during the 2016 outbreak. Still, because the epidemic "waned rapidly," researchers could not evaluate the efficacy of candidates in field trials. "The epidemic came and went too fast" to conduct phase 3 tests, she said [124].

"One of the things we have learned in recent years with Zika, dengue, and West Nile is there are no easy solutions for these ecologically complex mosquito-borne arbovirus diseases" says David M. Morens, Senior Adviser to Director of the National Institute of Allergy & Infectious Diseases. The problem, Morens explains, that vaccines help control significant outbreaks. Still, viruses tend to flare up, peter out, and then pop up again in small, unpredictable clusters. In Puerto Rico, the hardest-hit U.S. territory, new cases of ZIKV dropped from nearly 35,000 in 2016 to less than five hundred this year. Although everyone is happy to see the virus abate, it leaves public health officials with a quandary: Who, if anyone, should be vaccinated? [125].

This has led the U.S. NIAID (National Institute of Allergy and Infectious Diseases) to consider deliberately infecting people with the ZIKV (called a human challenge trial). They have had difficulties getting people to participate because so many of them are testing positive for antibodies. In addition, an ethics committee convened by NIAID, and the Walter Reed Army Institute of Research called it premature and worried about inadvertent transmission between sexual partners. NIAID hoped to develop new protocols (enrolling only women to avoid semen transmission) and start infecting volunteers with a live but weakened ZIKV. "Regardless of whether the trial leads to an approved vaccine, Fauci has no regrets about launching it. "ZIKV was a very ominous threat just a couple of years ago, and there is certainly the possibility that it will come back." [124].

11.5 Conclusion

ZIKV vaccines would face the same problem as vaccines for chikungunya, West Nile, St. Louis encephalitis, and other arboviruses. Since epidemics appear sporadically and unpredictably, preemptively vaccinating large populations in anticipation of outbreaks may be prohibitively expensive and not cost effective.

Yet, vaccine stockpiling followed by rapid deployment may be too slow to counter sudden explosive epidemics [126].

The next chapter will review the vectors involved in infecting humans with the ZIKV and the concerns expressed by some experts that the ZIKV will remain in reservoirs (the reservoir of an infectious agent are the habitats in which the agent normally lives, grows, and multiplies. Reservoirs include humans, animals, and the environment) and will reappear in later manifestation.

References

1. Butler D (2017) Decline in Zika throws trials into doubt. Nature. 545:395–396 May 25
2. Musso D, Ko AI, Baud D (2019) Zika virus infection—after the pandemic. N Engl J Med 381:1444–1457. https://doi.org/10.1056/NEJMra1808246
3. Ades AE et al (2020) Researching Zika in pregnancy: lessons for global preparedness. Lancet Infect Dis 20:e61-68
4. Brasil P et al (2016) Zika virus infection in pregnant women in Rio de Janeiro. New Engl J Med 375:2321–2334. December 15. https://doi.org/10.1056/NEJMoa1602412. http://www.nejm.org/doi/full/10.1056/NEJMoa1602412. Accessed 16 Jan 2017
5. Kumari et al. (2016) Zika virus: an informative note. Critical review in pharmaceutical sciences. 5:1. p. 9. http://earthjournals.in/crps_167.pdf. eISSN 2319-1082
6. CDC (2017) Overview of Zika virus testing: guidance for U.S. laboratories testing for Zika virus infection. July 24. https://www.cdc.gov/zika/laboratories/lab-guidance.html
7. Chang C et al (2016) The Zika outbreak of the 21st century. J Autoimmun. 68:1–13. February 28 https://www.ncbi.nlm.nih.gov/pubmed/26925496. Accessed 27 Oct 2016
8. Beck J (2016) Zika is the most difficult emergency health response ever, CDC official says. The Atlantic. June 24. http://www.theatlantic.com/health/archive/2016/06/zika-is-the-most-difficult-emergency-health-response-ever-says-cdc-official/488579/. Accessed 16 Jan 2017
9. Charrel R et al (2016) Background review for diagnostic test development for Zika virus infection. Bull World Health Organ 94:574–584D. https://doi.org/10.2471/BLT.16.171207. http://www.who.int/bulletin/volumes/94/8/16-171207.pdf. Accessed 28 Mar 2017
10. Ingold J (2017) Colorado state university pioneers new way to identify Zika virus. The Denver post. May 3. http://www.denverpost.com/2017/05/03/colorado-state-university-zika-virus-identification/. Accessed 5 June 2017
11. Williams R (2017) Quick and cheap Zika detection. The scientist. May 3. http://www.the-scientist.com/?articles.view/articleNo/49333/title/Quick-and-Cheap-Zika-Detection/. Accessed 1 July 2017
12. Graciaa DS, Collins MH, Wu HM (2018) Zika in 2018: advising travelers amid changing incidences. Ann Intern Med. July 16. https://doi.org/10.7326/M18-1112
13. Alvarez L (2016) Pregnant women anxious as Florida's Zika test results take weeks, New York Times. September 12. http://www.nytimes.com/2016/09/13/us/zika-test-delays-florida-pregnant.html. Accessed 25 Oct 2016
14. Edwards SB. The Zika virus. ABDO Publishing, Minneapolis, MN
15. Wong SJ et al (2003) An immunoassay targeting nonstructural protein 5 to differentiate West Nile virus infection from dengue and St. Louis encephalitis virus infections, and form flavivirus vaccination. J Clin Microbiol 41. https://www.ncbi.nlm.nih.gov/pmc/articles/PMC193845/. Accessed 1 July 2017
16. CSTE (Council of State and Territorial Epidemiologists) (2018) 2018 Zika preparedness resources toolkit. CSTE, Atlanta, GA. https://cdn.ymaws.com/www.cste.org/resource/resmgr/zika/Zika_Virus_Preparedness_Reso.pdf. Accessed 8 Aug 2018
17. ABC News (2017) Hundreds of DC Zika virus tests to be reexamined after technical issues. ABC News. February 17. http://abcnews.go.com/Health/hundreds-dc-zika-virus-tests-examined-technical-issues/story?id=45562356. Accessed 17 July 2017
18. Segraves M (2017) 3 retest positive for Zika in DC; CDC updates testing procedures. NBC4. May 29. http://www.nbcwashington.com/news/health/3-Retest-Positive-for-Zika-in-DC-CDC-Updates-Testing-Procedures-421656883.html. Accessed 19 July 2017
19. Dejnirattisai W et al (2016) Dengue virus sero-cross-reactivity drives antibody-dependent enhancement of infection with zika virus. Nat Immunol 17(9). https://doi.org/10.1038/ni.3515 PMID: 27339099. https://www.nature.com/articles/ni.3515. Accessed 25 July 2018
20. Kelser EA (2016) Meet dengue's cousin, Zika. Microbes Infect. 18:3. March. https://www.sciencedirect.com/science/article/pii/S1286457915002592?via%3Dihub. Accessed 25 July 2018

21. Sabalza M et al (2018) Detection of Zika virus using reverse-transcription LAMP coupled with reverse dot blot analysis in saliva. PLOS One. 13:2. February 5. http://journals.plos.org/plosone/article?id=10.1371/journal.pone.0192398. Accessed 25 July 2018
22. Sanchez-Duque JA, Rodriguez-Morales AJ, Trujillo AM, Cardona-Ospina JA, Villamil-Gomez WE (2018) Cocirculation and coinfection associated to Zika virus in the Americas. In: Roderigues-Morales AJ (ed) Current topics in Zika. Intechopen, London. https://doi.org/10.5772/intechopen.77180
23. van Meer MPA et al (2017) Re-evaluation of routine dengue virus serology in travelers in the era of Zika virus emergence. J Clin Virol 92. July. https://www.ncbi.nlm.nih.gov/pubmed/28505571. Accessed 20 July 2017
24. Reyes R (2017) UTMB researcher is co-inventor of a faster and more accurate test for diagnosing Zika. Galveston.com. January 17. https://www.pinterest.com/pin/171347960803653883/. Accessed 30 May 2017
25. Wong SJ et al (2017) A multiplex microsphere immunoassay for Zika virus diagnosis. EBioMedicine. February 16. https://www.ncbi.nlm.nih.gov/pubmed/28094237. Accessed 1 July 2017
26. Gourinat A-C, O'Connor O, Calvez E, Goarant C, Dupont-Rouzeyol M (2015) Detection of Zika virus in urine. Emerg Infect Dis 21(1):84–86. January
27. Musso D, Roche C, Nhan T-X, Robin E et al (2015) Detection of Zika virus in saliva. J Clin Virol 68:53–55
28. Yaren Z et al (2017) Point of sampling detection of Zika virus within a multiplexed kit capable of detecting dengue and chikungunya. BMC Infect Dis. 17. April 20. https://www.ncbi.nlm.nih.gov/pmc/articles/PMC5399334/. Accessed 19 July 2017
29. Han Y, Mesplede T (2018) Investigational drugs for the treatment of Zika virus infection: a preclinical and clinical update. Expert Opin Investig Drugs https://doi.org/10.1080/13543784.2018.1548609
30. Li L et al (2018) PARP12 suppresses Zika virus infection through PARP-dependent degradation of NS1 and NS3 viral proteins. Sci Signal. 11:535. eaas9332. June 19. http://stke.sciencemag.org/content/11/535/eaas9332. Accessed 24 July 2018
31. FirstPost (2017) Zika antibody discovery may lead to new vaccine. FirstPost. May 9. http://www.firstpost.com/living/zika-antibody-discovery-may-lead-to-new-vaccine-3431012.html. Accessed 5 June 2017
32. Duffy S (2017) Discovery of Zika antibody spurs hope for vaccine. Courthouse News Serv. May 8. https://www.courthousenews.com/discovery-zika-antibody-spurs-hope-vaccine/. Accessed 5 June 2017
33. Bernhard B (2017) Researchers in St. Louis making headway against Zika virus. Stltoday.com. July 10. http://www.stltoday.com/lifestyles/health-med-fit/health/researchers-in-st-louis-making-headway-against-zika-virus/article_173e8152-421e-545b-a8d0-3a86a774eb3c.html. Accessed 21 July 2017
34. Prasad R (2017) Malaria drug shields foetus from Zika. The Hindu. July 17. http://www.thehindu.com/sci-tech/science/malaria-drug-shields-foetus-from-zika/article19297529.ece. Accessed 21 July 2017
35. McFadden M (2017) Doctors warn of a Zika virus resurgence. WNDU. April 28. http://www.wndu.com/content/news/Doctors-warn-of-a-Zika-virus-resurgence-420761043.html, Accessed 5 June 2017
36. Sirohi D et al (2016) The 3.8 Å resolution cryo-EM structure of Zika virus. Science 352:6284. April 22. http://science.sciencemag.org/content/early/2016/03/30/science.aaf5316. Accessed 19 July 2017
37. Pergolizzi J Jr et al (2020) The Zika virus: lurking behind the COVID-19 pandemic? J Clin Pharm Ther. October 22. https://doi.org/10.1111/jcpt.13310
38. Isern S (2017) Dengue virus antibodies may worsen a Zika infection. The conversation. January 4. http://theconversation.com/dengue-virus-antibodies-may-worsen-a-zika-infection-63980. Accessed 16 May 2017

39. Aliota MT et al (2017) Zika in the Americas, year 2: what have we learned? What gaps remain? A report from the global virus network. Antiviral Res 114. https://www.ncbi.nlm.nih.gov/pubmed/28595824. Accessed 15 May 2019
40. Geard N et al (2016) Is the end of Zika nigh? How populations develop immunity. The conversation. July 21. http://theconversation.com/is-the-end-of-zika-nigh-how-populations-develop-immunity-62816. Accessed 13 Sept 2016
41. Abbink P et al (2016) Protective efficacy of multiple vaccine platforms against Zika virus challenge in rhesus monkeys. Science https://doi.org/10.1126/science.aah6157
42. McNeil DG Jr. (2017) Houston braces for another brush with the peril of Zika. The New York times. July 17. https://www.nytimes.com/2017/07/17/health/zika-virus-houston-texas.html. Accessed 21 July 2017
43. Adams P, Nutt C (2016) A Zika vaccine, but for whom? The New York times. December 28. https://www.nytimes.com/2016/12/28/opinion/a-zika-vaccine-but-for-whom.html?_r=0. Accessed 10 Jan 2017
44. PR Newswire (2017) Virion announces collaboration with NIH on Zika and other major viral diseases. PR Newswire. January 6. http://www.prnewswire.com/news-releases/viriom-announces-collaboration-with-nih-on-zika-and-other-major-viral-diseases-300387005.html. Accessed 28 May 2017
45. European Vaccine Initiative (2016) The quest for an effective vaccine against Zika virus infection. December 14. http://cordis.europa.eu/news/rcn/136995_en.html. Accessed 14 April 2017
46. Chen H-L, Tang R-B (2016) Why Zika virus infection has become a public health concern? J Chin Med Assoc 79:174–178. http://www.sciencedirect.com/science/article/pii/S1726490116300065. Accessed 10 April 2017
47. Garde D (2017) Moderna breaks its publishing silence, but can it sell its Zika vaccine? STAT. February 17. https://www.statnews.com/2017/02/17/moderna-zika-vaccine/. Accessed 14 April 2017; Richner J et al (2017) Modified mRNA vaccines protect against Zika virus infection. Cell. 168:6. March 9. http://www.cell.com/cell/pdf/S0092-8674(17)30195-2.pdf. Accessed 14 April 2017
48. Pardi N et al (2017) Zika virus protection by a single low-dose nucleoside-modified mRNA vaccination. Nature. 543. February. https://www.nature.com/nature/journal/v543/n7644/full/nature21428.html. Accessed 28 May 2017
49. Laavanya VP et al (2017) A review on microcephaly associated with Zika fever in new born babies. Res J Pharm Technol 10:1. January. http://rjptonline.org/HTMLPaper.aspx?Journal=Research%20Journal%20of%20Pharmacy%20and%20Technology;PID=2017-10-1-68. Accessed 17 July 2017
50. Hackett DW (2018) University of Hawaii study identifies potential Zika vaccine. Zika News. December 5. https://www.zikanews.com/zika-vaccine-candidates-are-various-phases-human-clinical-trials. Accessed 2 June 2022
51. Mosquito Squad. (2017). Australian plant extract, kill Zika virus, and Zika. Blog. June 5. https://www.mosquitosquad.com/blog/archives/2017/06/05/june-5-2017-australian-plant-extract-can-kill-zika-virus-study/. Accessed 17 July 2017
52. Ferrier T (2016) Native aussie plant kills Zika virus. Australian associated Press. June 2. http://www.news.com.au/national/breaking-news/native-aussie-plant-kills-zika-virus/news-story/5ac6d4ab19ea34c8923c72b67ce9f9c0. Accessed 17 July 2017
53. Lin J et al (2018) Salivary factor LTRIN from Aedes aegypti facilitates the transmission of Zika virus by interfering with the lymphotoxin-β receptor. Nat Immunol 19. April. https://www.nature.com/articles/s41590-018-0063-9. Accessed 8 July 2018
54. Abbink P et al (2016) Protective efficacy of multiple vaccine platforms against Zika virus challenge in rhesus monkeys. Science. 353:6304. September 9. http://science.sciencemag.org/content/353/6304/1129. Accessed 17 May 17 2017

55. Sun KH (2017) The race to develop a vaccine: scientists inch closer to preventing Zika. The Washington post. January 12. https://www.washingtonpost.com/national/health-science/the-race-to-develop-a-zika-vaccine-researchers-move-forward-with-safety-testing/2017/01/12/a65db0b6-d383-11e6-a783-cd3fa950f2fd_story.html?utm_term=.fcc64aa8f21c. Accessed 2 June 2017
56. Liu A (2018) Takeda's Zika vaccine candidate wins FDA fast track status. FiercePharma. January 29. https://www.fiercepharma.com/vaccines/takeda-s-zika-vaccine-gets-fda-fast-track-though-virus-no-longer-emergency. Accessed 7 July 2018
57. Harris R (2016) Testing begins on an experimental Zika vaccine with inactivated virus. NPR New. November 7. http://www.npr.org/sections/health-shots/2016/11/07/501015866/testing-begins-on-an-experimental-zika-vaccine-with-inactivated-virus. Accessed 14 May 2017
58. Edwards S (2017) Zika vaccine could be delayed, unaffordable after US army grants exclusive rights to pharma company. DEVEX Newswire. January 27. https://www.devex.com/news/zika-vaccine-could-be-delayed-unaffordable-after-us-army-grants-exclusive-rights-to-pharma-company-89519. Accessed 14 April 2017
59. Keener A (2016) More Zika vaccines progress toward human trials. The Scientist. August 4. http://www.the-scientist.com/?articles.view/articleNo/46725/title/More-Zika-Vaccines-Progress-Toward-Human-Trials/. Accessed 17 May 2017
60. Owens B (2018) Zika vaccine development: two years on from the outbreak. Pharm J. February 1. https://www.pharmaceutical-journal.com/news-and-analysis/features/zika-vaccine-development-two-years-on-from-the-outbreak/20204340.article?firstPass=false. Accessed 25 July 2018
61. Rodriguez-Fernandez C (2017) French researchers discover a target to prevent Zika birth defects. Labiotech. May 8. http://labiotech.eu/research-zika-birth-defects/. Accessed 5 June 2017
62. Monel B et al (2017). Zika virus induces massive cytoplasmic vacuolization and paraptosis-like death in infected cells. EMBO J. June 15. 36:12. https://www.ncbi.nlm.nih.gov/pubmed/28473450. Accessed 17 July 2017
63. Simanjuntak Y et al (2018) Ebselen alleviates testicular pathology in mice with Zika virus infection and prevents its sexual transmission. PLOS Pathog 14(2):E1006854. http://journals.plos.org/plospathogens/article?id=10.1371/journal.ppat.1006854. Accessed 26 July 2018
64. Boutott C (2018) HealthWatch: vaccine for Zika virus. WeAreGreenBay. June 14. https://www.wearegreenbay.com/health-watch/healthwatch-vaccine-for-zika-virus/1224940177. Accessed 23 July 2018
65. Hernandez H (2017) Zika virus: new warning all families need to know. Click2Houston. May 8. http://www.click2houston.com/health/zika-new-warning-all-families-need-to-know. Accessed 5 June 2017
66. ASM Newsroom (2017) Research shows new Zika virus vaccine that offers 100% protection in mice. News-Medical Net. June 5. http://www.news-medical.net/news/20170605/Research-shows-new-Zika-virus-vaccine-that-offers-10025-protection-in-mice.aspx. Accessed 27 June 2017
67. Berkshire Hathaway (2017) Zika virus market 2017 new research study. Bus Wire. January 4. http://www.businesswire.com/news/home/20170104006265/en/Zika-Virus-Vaccines-Market-2017-Company-Profile. Accessed 29 May 2017; Newsmaker (2017) Zika virus market 2017 new research study. Newsmaker Rep Web. March 30. http://www.newsmaker.com.au/news/282148/zika-virus-market-2017-new-research-study#.WSxB-GjyuUk. Accessed 29 May 2017
68. Burwell S (2016) Zika supplemental funding spend plan. Department of health and human services. October 26. https://www.naccho.org/uploads/downloadable-resources/HHS-Zika-Spend-Plan-to-Congress.pdf. Accessed 27 Aug 2018
69. Lagunas-Rangel FA, Viveros-Sandoval ME, Reyes-Sandoval A (2017) Current trends in Zika vaccine development. Virus Eradication. July. https://www.ncbi.nlm.nih.gov/pmc/articles/PMC5518240/. Accessed 7 July 2018

70. Sapparapu G et al (2016) Neutralizing human antibodies prevent Zika virus replication and fetal disease in mice. Nature. 540. December 15. https://www.ncbi.nlm.nih.gov/pubmed/27819683. Accessed 2 June 2017; Hasan SS et al (2017) A human antibody against Zika virus crosslinks the E protein to prevent infection. Nat Commun. March 16. https://www.nature.com/articles/ncomms14722. Accessed 17 July 2017
71. Le Roux M (2016) WHO warns of potential for marked increase in Zika cases. The Straits Times. April. http://www.straitstimes.com/world/americas/who-warns-of-potential-for-marked-increase-in-zika-virus-cases. Accessed 19 May 2017
72. Carribean360 (2016) Zika's emergency phase gives way to a long-term public health challenge. October 12. http://www.caribbean360.com/news/zikas-emergency-phase-gives-way-long-term-public-health-challenge. Accessed 15 March 2017
73. McKay B, Loftus P (2016) America's next defense against Zika and other foreign invaders. Wall Street J. December 16. https://www.wsj.com/articles/americas-next-defense-against-zika-and-other-foreign-invaders-1481810402. Accessed 21 May 2017
74. Coyne CB, Lazear HM (2016) Zika virus—reigniting the TORCH. Nat Rev Microbiol. 14. http://www.nature.com/nrmicro/journal/v14/n11/full/nrmicro.2016.125.html. Accessed 14 April 2017
75. Diamond M, Coyne C (2017) Closing in on a Zika virus vaccine. Nat Rev: Immunol. December 4. https://www.nature.com/articles/nri.2017.132. Accessed 10 July 2018
76. Magnani DM et al (2017) Neutralizing human monoclonal antibodies prevent Zika virus infection in macaques. Sci Trans Med. 9. October 4. https://www.ncbi.nlm.nih.gov/pubmed/28978754. Accessed 10 July 2018
77. BioPharm International Editors (2018) MAbs against Zika show promise. BioPharm Int. January 12. http://www.biopharminternational.com/mabs-against-zika-show-promise. Accessed 27 June 2018
78. Tebas P, Roberts CC, Muthumani K, Reuschel EL et al (2021) Safety and Immunogenicity of an Anti-Zika virus DNA vaccine. N Engl J Med 385:e35. https://doi.org/10.1056/NEJMoa1708120
79. Shah H (2016) Novel Zika virus vaccine ZIKV-prME shows promising results in mice and monkeys. Med News Bull. December 14. https://www.medicalnewsbulletin.com/novel-zika-virus-vaccine-zikv-prme-shows-promising-results-mice-monkeys/. Accessed 2 June 2017; Muthumáni K et al (2017) In vivo protection against ZIKV infection and pathogenesis through passive antibody transfer and active immunization with a prMEnv DNA vaccine. NPKJ vaccines. https://www.nature.com/articles/npjvaccines201621. Accessed 2 June 2017
80. Offord C (2018) Maternal response to Zika damages mouse fetuses. Sci January 5. https://www.the-scientist.com/the-nutshell/maternal-response-to-zika-damages-mouse-fetuses-30448. Accessed 7 July 2018
81. Beil L (2016) Vaccines may offer defense against dengue, Zika and chikungunya. Sci News. June 15. https://www.sciencenews.org/article/vaccines-may-offer-defense-against-dengue-zika-and-chikungunya. Accessed 11 April 2017
82. Clemons N (2017) NIH begins phase 2 Zika vaccine trials. Infect Dis Advisor. April 21. http://www.infectiousdiseaseadvisor.com/zika-virus/zika-virus-vaccine-phase-2-trial/article/652016/. Accessed 5 May 2017
83. Clinical Trials Arena (2019) Company news. January 10. https://www.clinicaltrialsarena.com/news/trial-dmab-therapy-safety-zika-infection/. Accessed 5 June 2022
84. VGXI (2016) GeneOne life science doses first subjects in Zika vaccine trial. VGXI website. July 26. http://vgxii.com/geneone-life-science-doses-first-subjects-in-zika-vaccine-clinical-trial/. Accessed 29 May 2017
85. Bukspan D (2017) The race is on to stop a Zika virus epidemic in the US. CNBC. April 22. http://www.cnbc.com/2017/04/11/the-race-is-on-to-stop-a-zika-virus-epidemic-in-the-us.html. Accessed 5 May 2017

86. Inovio Pharmaceuticals (2016) Inovio pharmaceuticals and geneone life science receive approval for first-in-man Zika vaccine clinical trial. June 20. http://ir.inovio.com/news/news-releases/news-releases-details/2016/Inovio-Pharmaceuticals-and-GeneOne-Life-Science-Receive-Approval-for-First-in-Man-Zika-Vaccine-Clinical-Trial/default.aspx. Accessed 15 May 2017
87. Inovio Press Release (2016) Inovio launches Zika vaccine trial in midst of Puerto Rico epidemic to explore early signals of vaccine efficacy. Inovio Website. August 29. http://ir.inovio.com/news/news-releases/news-releases-details/2016/Inovio-Launches-Zika-Vaccine-Trial-in-Midst-of-Puerto-Rico-Epidemic-to-Explore-Early-Signals-of-Vaccine-Efficacy/default.aspx. Accessed 29 May 2017
88. Regalado A (2016) U.S. government starts test of Zika vaccine in humans. Technol Rev. August 2. https://www.technologyreview.com/s/602073/us-government-starts-test-of-zika-vaccine-in-humans/. Accessed 29 May 2017
89. Johnston I (2016) Zika strain that causes microcephaly found in Africa for the first time, WHO confirms. The Independent. May 20. http://www.independent.co.uk/life-style/health-and-families/health-news/zika-strain-that-causes-microcephaly-found-in-africa-for-the-first-time-who-confirms-a7039641.html. Accessed 15 Sept 2016
90. Beck J (2016) Three Zika vaccines found successful in monkeys. The Atlantic. Aug 4. https://www.theatlantic.com/health/archive/2016/08/three-kinds-of-vaccines-protect-monkeys-from-zika/494564/. Accessed 3 Feb 2017
91. Barbi L et al. (2018). Prevalence of Guillain-Barré syndrome among Zika virus infected cases: a systematic review and meta-analysis. Braz J Infect Dis 22:2. https://www.ncbi.nlm.nih.gov/pubmed/29545017. Accessed 15 May 2019
92. Ferguson NM (2016) Countering the Zika pandemic in Latin America. Science. July 14. https://doi.org/10.1126/science. aag0219
93. Stettler K et al (2016) Specificity, cross-reactivity, and function of antibodies elicited by Zika virus infection. Science 353:6301. August 19. http://science.sciencemag.org/content/353/6301/823. Accessed 17 May 2017
94. Wang Q et al (2016) Molecular determinants of human neutralizing antibodies isolated from a patient infected with Zika virus. Sci Transl Med 8. December 14. https://www.ncbi.nlm.nih.gov/pubmed/27974667. Accessed 30 June 2017
95. Magnani DM et al (2017). Neutralizing human monoclonal antibodies prevent Zika virus infection in macaques. Sci Trans Med. 9. October 4. https://www.ncbi.nlm.nih.gov/pubmed/28978754. Accessed 7 July 2018
96. NIH Media Advisory (2017) Monoclonal antibodies against Zika show promise in monkey study. Press Release. https://www.nih.gov/news-events/news-releases/monoclonal-antibodies-against-zika-show-promise-monkey-study. Accessed 4 May 2018
97. Dejnirattisai W et al (2016) Dengue virus sero-cross-reactivity drives antibody-dependent enhancement of infection with Zika virus. Nature. Immunology. https://www.ncbi.nlm.nih.gov/pmc/articles/PMC4994874/. Accessed 26 June 2017
98. Mercola J (2016) Zika: Brazil admits it's not the virus. Mercola.com. August 16. http://articles.mercola.com/sites/articles/archive/2016/08/16/birth-defects-brazil-not-zika-virus.aspx. Accessed 23 May 2017
99. Terzian ACB et al (2017) Viral load and cytokine response profile does not support antibody-dependent enhancement in dengue-primed Zika-infected patients. Clin Inf Dis. https://academic.oup.com/cid/article-abstract/doi/10.1093/cid/cix558/3872368/Viral-load-and-cytokine-response-profile-does-not. Accessed 21 July 2017
100. MedicalXpress (2017) Prior dengue infection does not increase Zika disease severity. MedicalXpress. July 17. https://medicalxpress.com/news/2017-07-prior-dengue-infection-zika-disease.html. Accessed 21 July 2017
101. Retallack H et al (2016) Zika virus cell tropism in the developing human brain and inhibition by azithromycin. PNAS. December 13. 113:50. http://www.pnas.org/content/113/50/14408.abstract. Accessed 11 April 2017

102. Ferreira AC et al (2017) Sofosbuvir protects Zika virus infected mice from mortality, preventing short- and long-term sequelae. Nature. 7:9409. August 25. https://www.ncbi.nlm.nih.gov/pubmed/29316963. Accessed 7 July 2018
103. Kuivanen S et al (2017) Obatoclax, saliphenylhalamide and gemcitabine inhibit Zika virus infection in vitro and differentially affect cellular signaling, transcription and metabolism. Antiviral Res. 139. March. http://www.sciencedirect.com/science/article/pii/S0166354216306040. Accessed 2 June 2017
104. McDaniels A (2016) Drug to treat parasites, liver disease shows promise for Zika. Baltimore Sun. August 29. http://www.baltimoresun.com/health/bs-md-zika-existing-drugs-20160829-story.html. Accessed 18 Sept 2016
105. Paddock C (2017) Zika virus: cure steps closer with protein-mapping study. Nat: Commun March 28. https://www.nature.com/articles/ncomms14762. Accessed 27 May 2017
106. Retallack H et al (2016) Zika virus cell tropism in the developing human brain and inhibition by azithromycin. PNAS. December 13. 113:50. http://www.pnas.org/content/113/50/14408.abstract. Accessed 29 May 2017; Onorati M et al (2016) Zika virus disrupts phospho-TBK1 localization and mitosis in human neuroepithelial stem cells and radial glia. Cell Rep 16. September 6. https://www.ncbi.nlm.nih.gov/pubmed/27568284. Accessed 18 July 2017
107. Kelly L (2018) Researchers find hepatitis C drug effective in treating Zika-infected individuals. The Washington times. January 28. https://www.washingtontimes.com/news/2018/jan/25/researchers-find-hepatitis-c-drug-effective-treati/. Accessed 7 July 2018
108. Mesci P, Muotri A et al (2018) Blocking Zika virus vertical transmission. Nat Sci Rep 8:1218. https://doi.org/10.1038/s41598-018-19526-4
109. Fratti K (2016) Are some people immune to Zika virus? There's no evidence yet. Romper. April 16. https://www.romper.com/p/are-some-people-immune-to-the-zika-virus-theres-no-evidence-yet-9014. Accessed 23 September 2016
110. Senthilingam M (2016) New dengue vaccine could lead to more cases, experts warn. CNN. September 1. http://www.cnn.com/2016/09/01/health/dengue-vaccine-increase-disease/index.html. Accessed 20 July 2017
111. Bandina S et al (2017) Enhancement of Zika virus pathogenesis by preexisting anti-flavivirus immunity. Science 356:175–180. April 14. http://science.sciencemag.org/content/early/2017/03/29/science.aal4365. Accessed 4 May 2017
112. Petersen LR et al (2013) Estimated cumulative incidence of West Nile virus infection in US adults, 1999–2010. Epidemiol Infect 141:3.591–595. May 28. http://search.proquest.com.prox.lib.ncsu.edu/docview/1324557733/fulltext/1E0121B99D3B467DPQ/1?accountid=12725. Accessed 4 May2017
113. Bandina S et al (2017) Enhancement of Zika virus pathogenesis by preexisting anti-flavivirus immunity. Science 356:175–180. April 14. http://science.sciencemag.org/content/early/2017/03/29/science.aal4365. Accessed 3 May 2017
114. Shepard DS et al (2004) Cost-effectiveness of a pediatric dengue vaccine. Vaccine 22. https://www.ncbi.nlm.nih.gov/pubmed/15003657. Accessed 10 Sept 10 2018
115. Alholm Z et al (2017) Pregnant women's acceptance of hypothetical Zika vaccine. Session 162. October 6. Maternal/Infant Immunization. Poster abstracts. OFID 2017:4 (Suppl 1). https://www.ncbi.nlm.nih.gov/pmc/articles/PMC5631272/. Accessed 18 July 2019
116. Rozansky M (2018) Who'll get the Zika vaccine? Depends on these factors. Futurity. March 16. URR
117. Annenberg Public Policy Center (2018) False beliefs about MMR vaccine found to influence acceptance of Zika vaccine. Commun Initiative Netw. May 15. http://www.comminit.com/global/content/does-scientific-breakthrough-increase-confidence-science-news-zika-vaccine-and-trust-sci. Accessed 10 July 2018
118. Ophir Y, Jamieson KH (2018) Intentions to use a novel Zika vaccine: the effects misbeliefs about the MMR vaccine and perceptions about Zika. J Public Health. March 15. https://doi.org/10.1093/pubmed/fdy042. Accessed 1 April 2018
119. Wong L, Alias H, Hassan J, Abu Bakar S (2017) Attitudes towards Zika screening and vaccination acceptability among pregnant women in Malaysia. Vaccine 35:43. October. https://www.ncbi.nlm.nih.gov/pubmed/28886944. Accessed 26 July 2018

120. Vielot NA et al (2018) United States travelers concern about Zika infection and willingness to receive a hypothetical Zika vaccine. Am J Trop Med Hyg 98:6. June. https://www.ncbi.nlm.nih.gov/pubmed/29692314. Accessed 26 July 2018
121. Baumgaertner E (2018) High-resolution snapshot of Zika virus reveals clues to fighting it. The New York times. June 26. https://www.nytimes.com/2018/06/26/health/zika-virus-image.html. Accessed 13 June 2019
122. Cohen J (2018) Steep drop in Zika cases undermines vaccine trial. Science 361(6407):1055–1056 September 14
123. O'Reilly KM et al (2018) Projecting the end of the Zika virus epidemic in Latin America: a modelling analysis. BMC Med 16:180. https://doi.org/10.1186/s12916-018-1158-1168
124. Jarvis L (2017) Sanofi ends Zika vaccine research. Chem Eng News 10. September 11
125. Children's National Hospital (2021) Zika 5 years later: Still much to learn as 'likely' future outbreak looms. January 21. HealioNews. https://www.healio.com/news/infectious-disease/20210114/zika-5-years-later-still-much-to-learn-as-likely-future-outbreak-looms. Accessed 25 May 2022
126. Fauci A, Morens D (2016) Zika virus in the Americas—yet another arbovirus threat. New Engl J Med 374(7):602–606. February 18. http://www.nejm.org/doi/full/10.1056/NEJMp1600297#t=article. Accessed 4 Sept 2016
127. Thomas K (2016) The race for a Zika vaccine. The New York times. November 19. https://www.nytimes.com/2016/11/20/business/testing-the-limits-of-biotech-in-the-race-for-a-zika-vaccine.html?_r=0. Accessed 4 June 2017
128. PLOS (2018) Antioxidant treatment prevents sexual transmission of Zika in mice promising drug also appears to alleviate testicular Zika symptoms. Sci Daily. February 15. https://www.sciencedaily.com/releases/2018/02/180215141843.htm. Accessed 7 July 2018

Chapter 12
Twentieth-Century Vector Control

Altogether preventing an individual from being exposed in one year has no impact on lifetime disease risk if a high-level risk of exposure resumes the following year—the only effect of such transient interventions is to postpone infection [1].

The issue of whether eradication of *Ae. aegypti* was feasible has been debated since at least the 1920s. In 1934, after successfully eradicating the mosquito from several cities in the northeast region, Brazil gave unofficial approval to a proposal for the eradication of *Ae. aegypti* within its borders, and in 1942, the Government of Brazil officially endorsed eradication. By 1962, 18 continental countries and many Caribbean Islands had been eradicated [2].

After 1962 interest waned, and countries became reinfected by countries that had not eradicated the mosquito, including Cuba, the U.S., Venezuela, and several Caribbean countries. PAHO reports over time, in most countries that had stopped the programs against *Ae. aegypti* lost political importance, and surveillance gradually declined so small reinfestations could no longer be detected. In addition, once a reinfestation was discovered, its response was slow. A combination of insufficient environmental sanitation, expansion in travel, insecticide resistance, and disinterest in establishing global solutions allowed *Ae. aegypti* to reestablish itself [2].

The risk is potentially more significant with the *Ae. albopictus* infestation in the Americas given its geographical location.

Today, mosquito prevention and control involve a careful mix of environmental spraying on the population level, mosquito repellents on the personal level, and alteration of human behavioral and ecological actions [3]. Today, there is a widespread perception that *Ae. aegypti* control "has failed" or the existing method will not reduce disease transmission. Therefore, existing approaches should be abandoned, and alternative strategies should be invested in and pursued. There is little reliable evidence from appropriately designed trials to conclude current control methods [4]. The following three chapters will examine more creative and sometimes controversial approaches.

There are three essential factors in transmitting mosquito-borne infectious diseases: the virus, the mosquitoes, and the humans [5]. Most mosquito control

D. M. Berube, *Pandemics and Resilience: Lessons we should have learned from Zika*, Risk, Systems and Decisions, https://doi.org/10.1007/978-3-031-25370-6_12

strategies involve a set of approaches. The number of mosquitoes biting humans and passing on diseases can be controlled and exposure to the mosquitoes can be reduced by eliminating where they breed. Many communities have adopted strategies like Houston, Texas' 3-Ds.

- **Drain**: Drain standing water around your home so mosquitos don't have a place to breed.
- **Dress**: Dress in long sleeves and pants outdoors to limit exposed skin.
- **DEET**: Apply EPA-approved insect repellent with the ingredient DEET [6].

The only metric available that verifies that a program is working is a decrease in disease incidence [7]. An exciting feature of some of the claims in vector control has been reductions in mosquito populations rather than decreases in disease incidence, which has been contentious. Given concerns regarding the numbers of species of mosquitoes that may be able to carry the virus and pass it onto humans, data on disease reductions rather than population reductions remain woefully wanting.

Chemical approaches seem to dominate community-level mosquito control planning. Short of carpet-bombing an area with insecticide, there are few effective ways of getting rid of urban *Aedes*. Even intensive house-to-house eradication campaigns often fail to root out the insect enough to make a dent in numbers [8]. In addition to chemicals, experts have looked at education efforts, biological methods (like the use of animals that eat mosquito larvae), and combinations of the three. Overall, they found that there is not much good evidence for any of these methods—which is not to say that they cannot work, just that they have not been well-studied [9].

As a result, according to Oxitec, even the best-funded and most organized public health authorities cannot successfully control *Ae. aegypti* in urban environments with more than 30% efficacy (up to 50% for the Florida Keys Mosquito Control District, world leaders in mosquito control). Still, even these levels may be insufficient to prevent disease outbreaks [10].

Little was done to contain the 2015–2016 outbreak in South America. The risk of endemic ZIKV depends on a host of variables, including the competence of mosquito control, the available funds to improve the management of mosquito populations, and the age distribution of those infected because the most severe outcomes of ZIKV infection were associated with pregnancy. After an initial post-invasion epidemic, the time until there is a risk of other epidemics will be driven by the replenishment of susceptible people through births and waning immunity (the latter seems unlikely based on evidence that other flaviviruses are provided lifelong immunity to the infecting strain) [11].

Suppose the virus can establish endemic or enzootic circulation (as has been suggested as occurring in Asia based on seroprevalence data). In that case, stable herd immunity may prevent future epidemics and the congenital Zika syndrome in Brazil [12].

The herd immunity models suggest the outbreak will die out on its own in three years. After that, further episodes may not be seen for at least a decade [13]. When a high proportion of a community is immune, diseases become hard to spread from person to person. This phenomenon is known as herd immunity. Herd immunity protects people indirectly by reducing their chances of contracting an infection.

By decreasing the number of people susceptible to infection, vaccination can starve an infectious disease outbreak like firebreaks can die a bushfire: reducing the fuel it needs to keep spreading. If the immune proportion is high enough, attacks can be prevented, and disease can even be eliminated locally [14].

Of course, any prediction based on herd immunity is subject to a relatively extensive list of assumptions. First, there are minimal changes in the genetic structure of the ZIKV, which did provoke some pushback on the use of transgenic mosquitoes. Second, the range for the mosquito vector does not increase from factors like climate change which might change the maximum altitudes at which the mosquitoes prosper; third, where populations of people select to live and play, and so on.

The war has been waged with shovels and insecticides, repellant and mosquito-eating fish, traps, nets, screens, and even rolled-up newspapers. Still, puddles fill up, insects evolve resistance, and predators are sated [15].

12.1 Direct Preventative Approaches

With insecticide resistance to existing public health insecticides increasing, control of adult *Aedes* will soon depend on a few insecticides with novel targets, some still in development. Meanwhile, there has been an increase in innovative strategies for deploying existing insecticides such as toxic sugar baits, insecticide-treated paper, window curtains or have access tubes, autodissemination traps, and mass deployment of lethal ovitraps, and combinations of repellents and attractants (push–pull) [4].

While there are over 3,000 species of mosquitoes, the approaches detailed below are mainly directed toward reducing a subspecies to a point where they can no longer support a human pandemic. No one below, by and large, advocates eliminating all mosquitoes from varied ecosystems. The approaches below are primarily associated with *Ae. aegypti* and maybe *albopictus,* the two primary mosquito species responsible for human infection by the ZIKV. There is an essential benefit to reducing these populations significantly, which has to do with all the other diseases traced back to these species. One of the strategies to reduce the viral load involves crashing problematic species.

Why has *aegypti* consumed nearly all the rhetorical space around vectors? Put simply, the *Ae. aegypti* is the most prominent carrier of a list of dangerous viral diseases such as dengue, chikungunya, yellow fever, and ZIKV.

12.1.1 Light Traps

Scientists use a light trap that captures more mosquitoes so mosquito control officials can save time and money in their spraying efforts. Light traps are generally used to track mosquito populations and are part of a more sophisticated approach using chemical insecticides. The traps help experts decide on their insecticidal strategy.

They have a negligible effect in reducing the population except near the traps. The light traps catch anything small enough to get sucked in when they get close to the mouth at the top of the web [16].

In addition, according to the American Mosquito Association, light traps are ineffective in most cases for the surveillance of *Ae. aegypti* and *Ae. albopictus* [17].

12.1.1.1 Biological Approaches

A meta-review assessing *Aedes* control measures was published in 2016. Thirteen systematic reviews investigating control measures' effects on entomological parameters or disease incidence were included. Biological controls seem to achieve a better reduction of entomological indices than chemical controls [18].

Professional staff implemented most biological approaches, and studies were of short duration and covered small areas. Unsurprisingly, biological control interventions are often weakened by failure to maintain populations of larvivorous organisms over the intervention.

Some have recommended that biological control interventions against (in this case) dengue vectors (nearly identical to Zika vectors) should be conducted alongside community-based educational programs that aim to train householders to use water containers appropriately to maintain water populations of larvivorous organisms. Biological control interventions should be locally adapted and consider cultural practices relating to water storage and the social acceptability of keeping living organisms in storage containers of drinking water [19].

Plants

Some plant species have been used as larvicides. Some species have essential oils (EOs) that possess acute contact, fumigant toxicity to insects, repellent activity, antifeedant activity, and development and growth inhibitory activity. The alternative use of EOs has two advantages over other natural pesticides. The first benefit is its low toxicity on mammals, birds, and fish. The second advantage is its great structural diversity as a pathway to a potential source of bioactive molecules [20].

For example, several species of the genus *Piper* have been investigated regarding their insecticidal activity. In India, species like *Piper longum, P. beetle,* and *P. cubeba* have shown insecticidal activity against mosquitoes and flies [20].

Elephant Mosquitoes

Toxorhynchites mosquitoes are known as elephant mosquitoes. Members of the mosquito genus *Toxorhynchites* are the largest of the world's mosquitoes (family

Culicidae), with wingspans exceeding 0.4 inches (12 mm). Unlike most mosquitoes, adult *Toxorhynchites* do not take blood meals. Instead, they fill up on protein while they are aquatic larvae—and, luckily, they prey upon the larvae of their blood-sucking mosquito relatives [21].

They are predaceous as larvae on immature mosquito species and often turn cannibalistic. They do not drink human blood; they only feed on plant nectar. *Toxo* females lay their eggs in the exact locations *Aedes* females do. *Toxo* mosquito larvae work by eating the larvae of *Aedes* mosquitoes. The *toxo* mosquito is known for eating other mosquito larvae and thus could help decrease the *Aedes* population by eating its larvae. Its larva lifecycle is 30 days; during that time, it can devour up to four hundred *Aedes* larvae.

The *toxo* mosquito is known for eating other mosquito larvae and thus could help decrease the *Aedes* population by eating its larvae. The critical thing about *toxo* mosquitoes is that they feed on plant nectar, not blood, and thus will not bite humans [22]. Once they reach adulthood, elephant mosquitoes, both males and females, feed only on flower nectar [21].

Toxo is another non-chemical approach that could be adopted to rid an area of mosquitoes. And it has already been tested in this country. Three years ago, *toxorhynchites* mosquitoes were released into the Subang Jaya area to curb mosquitoes from breeding.

Although *Toxorhynchites* alone are unlikely to reduce pest or vector species below operational thresholds, they can be a valuable management tool in areas where containers and tree holes contribute substantially to the standing crop of mosquitoes [23].

There are two advantages to this approach. First, a *Toxorhynchites* adult mosquito can disperse and lay eggs in areas most likely to escape insecticide treatment. Second, this biological control method can result in favorable press coverage and often allows a mosquito control district to stress ecologically sound practices while controlling mosquitoes [24].

Disadvantages generally include high labor costs of mass rearing and distribution and the eggs' short "shelf life". On the downside, elephant mosquitoes take a lot of time to grow and are expensive to produce [25].

Biological control using predatory mosquitoes seldom results in complete control of the targeted population [23]. While no one seems to advantage from a species control program exclusively built on *Toxorhynchites*, some speculate it might be part of a more comprehensive program. The Subang Jaya Municipal Council, Selangor, Malaysia, reported a significant decrease in dengue cases following a tested release of *Toxo* mosquitoes in 2011–2013 [26].

Texas Harris County's mosquito control program is one of the more robust in the country. It includes Houston and a swath of Texas, where travel to Latin America is daily. *Aegypti* even more so. Some mosquitoes in the thick of summer will pick up ZIKV and spread it. "It's just a matter of when not if", says Umair Shah, the executive director of Harris County Public Health and Environmental Services [27]. A Harris County, Texas lab is working on the elephant mosquito option [28].

Trials have been undertaken in American Samos, Fiji, Manilla, and Panama. Still, some failed when the *Toxo* could not get a toehold in the ecosystem, and some simply were unable to control populations sufficiently [29].

Copepods, Turtles, and Fish

Copepods are widely used biological agents. They are crustaceans in water storage that eat mosquito larvae. Two species, *M. thermocyclopoides,* and *M. aspericornis,* have proven effective against dengue vectors. Copepods can survive up to six months in containers; they are often lost when water is removed, containers are cleaned, or (copepod) food is limited. Reintroduction may be necessary for sustainable control. A large-scale vector control program using copepods and clean-up campaigns in Viet Nam successfully controlled dengue transmission for several years and, in this region at least, appeared to be sustainable [30].

Copepods (*Mesocyclops spp.*) were effective for vector control in five community studies in Vietnam (in three rural communes in the central Vietnam provinces of Quang Nam, Quang Ngai, and Khanh Hoa), including long-term control of larval and adult *Ae. aegypti* and dengue incidence [31]. *Ae. aegypti* were reduced by approximately 90% by year 1, 92.3 – 98.6% by year 2, and *Ae. aegypti* immature forms had been eliminated from two of three communes by June 2003 [32]. Before implementing this vector control program, clinical dengue infections were eliminated in three rural communes in central Vietnam with a high incidence of this disease [32].

This approach demands training and follow-up with communities using copepods to control larvae. One reason is that the water in many containers was exhausted during the dry season. On refilling, particularly in the preliminary stages of the project, collaborators did not have sufficient experience in breeding and maintaining *Mesocyclops* populations for public use. Householders did not know how to reintroduce *Mesocyclops* each time they refilled a water container. By the end of the project, *Mesocyclops* populations were successfully maintained in these containers [32].

However, this success was not replicated in studies conducted elsewhere. The authors attributed Vietnam's success to community participation and environmental and biological factors [18].

According to the AMCA, these smaller aquatic predators (e.g., predacious copepods) may control mosquito larvae that develop in containers; however, source reduction is the optimal control strategy for these species of mosquitoes [17].

Other species, larvivorous fish, have been used to reduce mosquito populations. The most promising fish species known to have larvivorous potential belong to *Poeciliidae*, *Cyprinidae*, *Cyprinodontidae,* and *Cichlidae.* Unfortunately, mass production is required, but fish farmers do not culture them in their fisheries due to their low food value.

The mosquitofish, *Gambusia affinis*, is also used as a larvicide in permanent water bodies such as cisterns, abandoned pools, and ornamental ponds [33]. According to the AMCA, larger aquatic predators such as *Gambusia spp* may control mosquito larvae in permanent or semipermanent bodies of water but will not fully control adult mosquitoes [17].

This strategy targets the larvae rather than the adult mosquitoes and is potentially safer for humans as it does not involve insecticides. However, the method is only suitable for use where mosquito vector breeding is well defined and where transmission is seasonal. The use of larvivorous fish has been introduced in several countries of the Eastern Mediterranean Region, and its potential and implementation methods have been tested in two countries [34].

However, there is insufficient research to support the use of larvivorous fish, and their effect on native fish and non-target species is not fully understood [35].

12.1.2 Chemical Approaches

Because chemical control is part of an integrated control effort to combat the vector, its continuous use increases the selection of resistant individuals to those molecules. Furthermore, these same molecules can be toxic to other animal species or may contaminate the soil [36].

Numerous factors have eroded the effectiveness of overreliance on mass pesticide spraying, including increases in pesticide resistance, an awareness of the detrimental side effects of pesticide use, decreased government funding for public health services, and a push from the global health community to move toward horizontal programs integrating education and community participation [38].

Larvicides and insecticides have traditionally been a primary method of mosquito control. However, in some cases, insecticide use for vector control has sparked concern among populations desiring a less "toxic" approach. "Think about how people control mosquitoes now", Ravi Durvasula, a Medicine and Infectious Diseases professor at the University of New Mexico School of Medicine, said. "Dumping pesticides over hundreds of square miles … driving through the city with canisters on motorbikes, spraying willy-nilly. That's an incredibly toxic approach [39].

In addition, chemical approaches to mosquito control may be costly for resource-poor countries, and they may be unwilling or unable to invest in this approach.

Some insecticides, like dichlorodiphenyltrichloroethane (commonly known as DDT)–which are effective against mosquitoes but can be toxic to humans and animals in specific contexts–have diminished in use. Additionally, in recent years, *Aedes* mosquitoes in certain regions have become increasingly resistant to some of the other insecticides, rendering chemical intervention less effective [40]. Also, space spraying and larviciding require trained personnel to deliver potentially toxic insecticides using specialized equipment and dedicated resources, such as spraying equipment, insecticides, transport, and are dependent on vertical municipality-driven programs [41].

Also, one must remember that many stakeholders, especially the public, are suspicious and concerned about the heavy use of insecticides, mainly in the aerial application. While other stakeholders may be concerned about releasing genetically modified mosquitoes to reduce vectors, as will be seen later, pushback from another group of stakeholders affects decision-making over chemical approaches.

For example, during an outbreak of West Nile Virus in Orange County, California (2014–2015), during which eight died, antipesticide activists pressured the County Mosquito and Vector Control District Public Health Board to reject aerial spraying for health reasons as well as the impact on species such as butterflies and bees [42].

Another example, this was predicated on concern about the spread of the ZIKV across the south after a New York Times report on September 1, 2016. It seems local officials targeted a 15-square-mile area near Charleston with naled, a pesticide that has been in use in the United States for more than 50 years. The death toll exceeded two million bees.

Federal officials have said the chemical can harm honeybees while posing brief risks to aquatic invertebrates and terrestrial wildlife. According to Dr. Dennis van Engelsdorp, a bee researcher at the University of Maryland, "I think this shows how easy it is to forget best practices when faced with emerging issues—and killing bees is probably not the biggest impact", said. "If you're killing honeybees, you're killing a lot of other non-honeybee pollinators, too, and those populations could take a long time to recover" [43].

Typical mosquito control includes spraying mosquitos outside at dusk and dawn. However, this strategy is not likely to work well for preventing ZIKV (as well as chikungunya and dengue). The mosquito that transmits ZIKV, the *Aedes* mosquito, has different behaviors than the more well-known *Culex* mosquitos that live outdoors and are primarily active at dusk and dawn [44].

Aliota et al. reported on insecticides and larval source reduction. These strategies, insecticides and larval source reduction, have not prevented the invasion of this virus into new locales and have not been adequate to control the virus upon arrival [46].

Authors of a study [45] debunking pyriproxyfen as a cause for microcephaly in Pernambuco commented in their article that it should be noted that the mosquito vector control strategies during the past three decades, mainly based on chemical insecticides and larvicides, have proven ineffective, citing Regis et al [45]. and de Silva Augusto et al [45]

However, the indiscriminate use of synthetic insecticides has led to the emergence of resistant strains of mosquitoes and an uncontrolled increase in the mosquito population. Another neglected source of resistance involves farmers who use crop protection insecticides like deltamethrin. Previous researchers have reported that exposure of mosquitoes to crop protection insecticides could also result in insecticide resistance development [45].

The advent of potent, long-lasting chemical insecticides in 1940 was promising. Still, overdependence on this tool for vector control resulted in the evolution of resistance in natural populations, hindering its effectiveness in the long term. Today, the dissemination of insecticide resistance throughout vector populations is much faster than the rate of development of new insecticides. In addition, cross-resistance, based on the activation of general detoxifying mechanisms in the vector, can shorten the lifespan of alternative insecticides or even prevent their implementation [48]. The growing prevalence of insecticide-resistant mosquitoes, environmental contamination, and destruction of non-target organisms have resulted in criticism of chemical-based methods [49].

No amount of insecticide, larvicide, and fogging will appreciably reduce the mosquitoes as it is impossible to reach all available breeding sites. Esu and colleagues stated, "Based on a comprehensive search of available peer-reviewed literature, the effectiveness of peridomestic space spraying in reducing dengue transmission has not been conclusively demonstrated" [50].

> Bouzid's 2016 meta-analysis concluded:
>
> Chemical control, commonly used, is not associated with sustainable reductions in mosquito populations over time. Indeed, by contributing to a false sense of security, chemical control may reduce the effectiveness of educational interventions to encourage local people to remove mosquito breeding sites. [18]

Since the 1980s, vector control efforts have been cut back or hampered by a lack of effective pesticides, allowing the Aedes species to return to North and South America and the Caribbean. The CDC's revised vector surveillance estimates show that *Ae. aegypti* is present over a wider geographic area in the U.S. than previously thought, as far North as New York and Northern California [51]. The Brazilian Association of Collective Health has called to stop chemical products against Ae. aegypti, especially in household water reservoirs, and prioritize sanitary measures [52].

There are two categories of chemical approaches for controlling mosquitoes as an infection disease vector: one involves larvicides and the other adulticides, associated with insecticides. One kills the larva and the other the adult mosquito.

12.1.2.1 Larvicides

Larvicides used in Florida and Key West include Bacillus thuringiensis israelensis (Bti), Bacillus sphaericus (Bs), methoprene, temephos, and Spinosad, or oil dispersants such as Kontrol or CocoBear.

Miami-Dade County's aerial spraying includes BTI insecticide (Bacillus thuringiensis israelensis, a naturally occurring bacterium). This bacterium produces toxic proteins leading to high mortality among larvae after ingestion and targets mosquito larvae instead of adult mosquitoes [53]. Bti is an organic larvicide. It is a group of bacteria that helps eradicate mosquitoes, fungus gnats, and blackflies but has nearly zero impact on other organisms [54]. The endotoxin produced by the entomopathogenic bacteria Bacillus thuringiensis var. israelensis (Bti) has high larvicidal activity in mosquitoes and is non-toxic to important beneficial organisms. Various formulations (i.e., wettable powders, granules, and briquettes) are effective, and newer slow-release long-lasting formulations may reduce the need for frequent reapplication [30].

In California, VectoBac WDG is used. It is presumably environmentally friendly, is approved for application on organic crops, and has no effect on people or pets at the amounts used for mosquito control, Jill Oviatt, the Coachella Valley Mosquito and

Vector Control District's public information manager, said. The aerial application was in response to the presence of Ae. aegypti mosquitoes in Indio [55].

Methoprene is an insect growth regulator that inhibits mosquito larvae from developing into viable adults. Spinosad causes excitation of the mosquito's nervous system leading to paralysis and death.

These products are rotated to avoid prolonged exposure of mosquito larvae to a particular larvicide's mode of action to reduce the likelihood of resistance.

Standard treatment of larval *Ae. aegypti* is Bti if the larvae are first through the third instar [33]. (An instar from the Latin "form", "likeness" is a developmental stage of arthropods, such as insects, between each molt (ecdysis) until sexual maturity is reached).

Pyriproxyfen is an insect juvenile hormone analog, active against many arthropods, and has been used for agricultural pest control for about 15 years. It is highly effective against mosquitoes and can prevent the emergence of *Ae. aegypti* at concentrations as low as one ppb or less, while highly high concentrations do not inhibit oviposition. New pyriproxyfen formulations can retain efficacy for six months, reducing the need for frequent reapplication, and recent studies indicate its high effectiveness [30].

Pyriproxyfen is effective at inhibiting adult *Ae. aegypti* emergence at concentrations of less than or equal to one part per billion can be applied in various formulations (e.g., sticks, granules) and is cost competitive. It is already in veterinary and agricultural use. Up to five months, it remains effective longer than *Bacillus thuringiensis israelensis*, *methoprene*, or *temephos* and is less toxic. Adult mosquitoes exposed to pyriproxyfen have decreased fecundity. Importantly, contaminated adults can disseminate lethal doses from treated to untreated sites [7].

Orac over at the Respectful Insolence blog points out that "humans do not make or use sesquiterpenoid hormones (aka juvenile insect hormones), which is what pyriproxyfen targets", and over the years, a great deal of research has been conducted on the pesticide's physicochemical properties, toxicology, and safe levels [57].

In Brazil, two different larvicides are employed, the organophosphate *temephos* and the chitin synthesis inhibitor *diflubenzuron*, mainly in localities with confirmed resistance to the organophosphate [52]. Temephos is an organophosphate larvicide used for the control of *Ae. aegypti* larvae [33].

A recent study by Arosteguí et al [56]. failed to demonstrate that several factors could explain a significant protective association between temephos and entomological indices. These include ecological adaptability of the vector, the resistance of *Aedes* to the larvicide, operational deficiencies of the vector control program, or a decrease in preventive actions by households resulting from a false sense of protection fostered by the centralized government program using chemical agents. Whatever the explanation, the implication is that temephos affords less protection under normal field conditions than expected from its efficacy under experimental conditions [56]. The occurrence of temephos resistance due to the intensive use of organophosphates for larval control has led to the gradual replacement of these insecticide families by Bti and insect growth regulators [4].

Despite all the strategies implemented in Brazil by the National Plan for Dengue Control (PNCD), investigations of *Ae. aegypti*, populations in sixty-seven cities revealed selective resistance to temephos in 1999 and 2000; this resistance was mainly found in Northeastern and Southeastern Brazil. In a study conducted in two towns in Pernambuco state in the northeast part of the country, the designated procedures from the PNCD were followed but yielded no positive suppression of the vector [37].

Although official data point out that 92% of urban households in Brazil were connected to public water in 2010, there are nearly four million unserved households. Intermittent water supply forces the population to store water for everyday consumption, favoring mosquito breeding. Rural Brazil has a more significant challenge, with only 28% of rural households connected to public water supplies [52]. Some communities have reported successes.

For example, with modest support from the Environmental Health Unit of the Ministry of Health, Fijian communities significantly reduced *Ae. aegypti* breeding sites. Before the intervention, 51% of tires and 21% of drums contained *Ae. aegypti*. The percentage of primary positive containers for *Ae. albopictus* also reduced significantly from 33 to 5% for tires and 42% to 8% for drums. The number of container habitats for *Ae. aegypti* decreased during the nine months with active community participation [39].

The AMCA recommends that the direct application of larvicides and pesticides be considered part of a comprehensive program to control container-inhabiting mosquitoes [14]. Also, larvicides are impractical since *Aedes* lay their eggs in small containers [60].

One Mosquito Control System (MCS) known as the Kit Bebas Denggi is a simple but effective and sustainable method; the MCS aims to eliminate mosquitoes in their larval stage using a non-toxic biolarvicide called Mousticide. This biolarvicide is sprinkled into an Aedes larvae ovitrap (A lot) filled with water to attract mosquitoes. The biolarvicide stops protein digestion in mosquito larvae, causing metabolic starvation and death [61].

According to the AMCA, larval source reduction is the most effective vector control to prevent mosquito production. Larval source management (LSM) involves removing, modifying, treating, and monitoring aquatic habitats to reduce mosquito propagation and human-vector contact. Interventions for LSM range from simple—draining marine sites or treating them with larvicidal chemicals and removing water-holding containers capable of producing mosquitoes—to complex, such as implementing Rotational Impoundment Management or Open Marsh Water Management techniques [17].

12.1.2.2 Insecticides (Adulticides)

According to the AMCA, ultra-low volume (ULV) space sprays are the only effective means of rapidly reducing transmission risk during arboviral disease outbreaks and effectively reducing populations of adult container-inhabiting Aedes in peridomestic environments, even when applied at night [17]. Globally, mosquito chemical control utilizes broad-spectrum insecticides, including organochlorines, carbamates, organophosphates, and pyrethroids which have known off-target effects on honey bees, native pollinators, and other beneficial insects [62].

However, opponents of pesticides have used legal suits and other mechanisms to limit or even prevent mosquito control spraying efforts. This opposition can seriously affect an unprepared program's ability to respond quickly to mosquito-borne disease during emergencies [63].

ULV (ultra-low volume) applications are often believed to be ineffective in controlling diurnally active urban mosquitoes, such as Ae. aegypti and *Ae. albopictus*, potentially due to structural obstacles that protect gravid or engorged females resting during nighttime ULV applications. However, some evidence suggests that such applications may reduce adult mosquito populations. There is growing evidence that containers inhabiting *Aedes* in peridomestic environments may be active even at night. ULV applications within urban and suburban habitats may penetrate habitats previously believed to be inaccessible [17].

The WHO recommends limiting spatial spraying to emergencies to prevent a developing epidemic or halt one already in process. Outdoor space spraying alone has not been proven effective in controlling outbreaks. WHO recommends including outdoor space spraying in an integrated vector management plan involving using larvicides, reducing breeding sites, and personal protection measures to mitigate human-vector contact [64].

Although insecticides are effective in many contexts, the financial cost of their application can be prohibitively high, their widespread application logistically challenging in both very urban and remote areas, and their efficacy unstable owing to the evolution of resistance in their target insects. Despite the successes, the ongoing case burden demonstrates that insecticides are not sufficient to bring mosquito vector diseases under control as they are currently being deployed [65].

Local governments commonly practice outdoor spatial spraying to prevent dengue and other Aedes-borne viruses. However, it has been publicly criticized for its potential to reduce community-level actions such as clearance of mosquito habitat due to a resulting ªfalse sense of security. According to Reyes-Castro et al [64], "respondents who reported outdoor space spraying (the OSS group) as a dengue prevention strategy exhibited lower frequencies of cleaning and trash disposal, greater use of chemical pest controls around the house and had a higher number of positive containers with *Ae. aegypti* larvae/pupae around their homes. These findings suggest that residents who cite outdoor space spraying to prevent dengue fever may be less likely to remove potential breeding sites that provide immature *Ae. aegypti* habitat may produce more vectors in their yards than residents who report that outdoor space spraying prevents dengue [64].

The U.S. CDC has written that "adulticiding, applying chemicals to kill adult mosquitoes by ground or aerial applications, is usually the least efficient mosquito control technique [66].

12.1.2.3 Resistance, Effectiveness, and Toxicity

Two final elements deserve repetition, and the other should be underlined. Resistance is a grave issue, especially for the species of mosquitoes involved in the dissemination of the ZIKV. A second element needs to be heavily emphasized; chemical approaches involve poisons, and while though there are a lot of pushbacks over environmentally safe reduction strategies in the following few chapters, especially over genetically designed mosquitoes, there is a shortage of discussion of the costs and benefits associated with the chemical and poison approaches versus the genetic engineering approaches.

An exciting part of this debate that has received much too little coverage has been genetics' role in controlling vectors without introducing polluting pesticides. Chemical spraying is hardly sustainable, and some experts claim it provides the public with an unjustified sense of security. Populations drop for a few days but soon return. Many even suggest it is mostly fear management and has little to do with health. The labor-intensive approach exposes pest control employees and whole communities to high exposure to insecticides.

Quincy Perkins, a Key Haven, Florida resident and supporter of Oxitec, is deeply concerned about the amount of insecticide sprayed on the island. He hopes Oxitec mosquitoes could help reduce that. "The big thing for me is that I'm sick of pesticides. It's mind-blowing that there's not more outrage about them". The district sprays thousands of pounds of six insecticides with varying toxicity degrees yearly [92].

On Resistance

Insecticide resistance has evolved in many *Ae. aegypti* mosquito populations worldwide, and there is evidence that it has compromised the success of control interventions. The levels of resistance in *Ae. albopictus* is relatively low at present compared to *Ae. aegypti*, possibly due to the reduced exposure of this more exophilic species to insecticides, particularly those targeting the adult stage. However, due to the expansion of *Ae. albopictus* populations into areas where insecticides are used intensively (adulticides or selection pressure by agriculture in its new breeding sites), it is highly likely that insecticide resistance will eventually negatively impact the ability to control this vector shortly [67].

Resistance to insecticides has spread in most mosquito vectors through selection—generally, widespread elevated levels of insecticide resistance in South American *Ae. aegypti* impacts the efficacy of conventional insecticides in the prevention of arbovirus transmission. Vector control, mainly by insecticides, plays a crucial role in disease prevention. Still, the use of the same chemicals for more than 40 years and the dissemination of mosquitoes by trade and environmental changes resulted in the spread of insecticide resistance [4].

Repeatedly confronted with insecticidal molecules, mosquito populations have developed resistance to insecticides, making vector control more complex, [68] and this is particularly true for *Ae. aegypti* [69]. Results from bioassays to determine the presence of insecticide resistance of *Ae. aegypti* shows that the mosquitoes collected from thirty-eight locations in twenty-three of the seventy-eight municipalities in Puerto Rico are either resistant or partially resistant to many insecticides tested. Resistance has been observed in permethrin and malathion, two commonly used insecticides in Puerto Rico [70].

In Argentina, vector "control" is carried out using pyrethroids, a little less toxic but banned in Europe because of their effect on people [71]. Similarly, the use of pyrethroids for space spraying and existing cross-resistance mechanisms previously selected by DDT led to the rapid rise of pyrethroid resistance [4]. Generally, pyrethrins or pyrethroids are not used to control *Aedes* because they have developed a resistance to them.

[I]insecticide resistance alleles generally increase quickly in natural mosquito populations exposed to intense insecticide application. The point is that the overuse of insecticides, even for a brief period, poses a formidable selection pressure on natural mosquito populations, quickly selecting those insects with resistance alleles. The rapid increase of insecticide resistance alleles in *Ae. aegypti* field populations emphasize the need for developing alternative strategies [70].

Resistance to all four classes of insecticides (carbamates, organochlorines, organophosphates, and pyrethroids) has developed in *Ae. aegypti*, and there is mounting evidence that this resistance is compromising the success of control interventions [67].

> Since 2010, 60 countries have reported resistance to at least one class of insecticide, with forty-nine countries reporting resistance to two or more types. However, this is likely an underestimate of the prevalence of resistance since many countries do not routinely monitor local insecticide resistance locally.... Insecticide resistance is broadly categorized into two groups: metabolic and target-site. The former occurs when resistant mosquitoes develop enzymes that more rapidly detoxify pesticides, preventing the active ingredient from reaching its physiologic target. The latter is observed when a mutation alters the pesticide target on the mosquito.... Behavioral resistance may also occur. For example, when resting surfaces are treated with pesticides, some mosquitoes in the target population may never land on them. This difference in exposure alters the survival rates of the next mosquito generation and may increase the frequency of any genetic factors contributing to the avoidance behavior. If this is true, progressively fewer mosquitoes will be killed by the pesticide over time [17].

In the past, control directed at adult mosquitoes has been limited to using ultra-low-volume (ULV) application of insecticides, usually by vehicle-mounted apparatus. There is controversy regarding the efficacy of this type of control, with several studies indicating that its effect is rarely, if ever, significant [30].

On Effectiveness

And while everyone seems to be worried about what kinds of terrible things the "mutant DNA" from these autocidal mosquitoes (Oxitec's) will do, few seem concerned that they are constantly being coated in pesticides instead. The pesticides being used are the safest available. However, ecologists have raised concerns about the effects on non-target wildlife, including potential negative impacts on the beautiful aquatic habitats that draw tourists [72].

Sustainable vector control programs can save valuable local emergency response resources. Vector control programs are relatively inexpensive, costing approximately $3.67 per person. This figure pales compared to the costs associated with the emergency use of contractors, equipment, and pesticides. For example, the cost associated with a West Nile virus outbreak in Louisiana for eight months from 2002-to 2003 was $20.1 million. It included $9.2 million for public health response, $4.4 million for medical costs, and $6.5 million for non-medical expenses [63].

The insecticides applied with spray have more impact on the human population than on mosquitoes [73]. And it is gross overkill. "[S]praying affects all kinds of animals, rather than targeting the *Ae. aegypti* mosquito accounts for less than 1 percent of all mosquitoes in the Southern U.S., according to Oxitec" [74].

Control of homes can be achieved by fumigating indoors, but these actions are often hampered by limited access to houses and resources. They are killing the adult female *Ae. aegypti* using traditional fogging with insecticide aerosols is generally ineffective unless implemented within homes or other places exposed to virus infection. These vectors tend to stay indoors, where they are unlikely to penetrate. Indoor spraying, including residual insecticides with repellant activity, can be more effective but expensive and labor-intensive [75]. Adulticides are most effective when used in the home. This is no simple task. When insecticide is applied indoors by health professionals or others, residents should leave the house during the treatment and keep it shut for 20 min after fumigation to guarantee that the mosquitoes die. In addition, when insecticide is applied indoors by health professionals or others, care should be taken to ensure that kitchen utensils, food, and water for human and animal consumption are adequately covered or kept in enclosed spaces [76].

Aircraft-delivered and truck-mounted ultra-low volume spraying have limited efficacy against Ae. aegypti because the vapor frequently does not penetrate buildings where adult mosquitoes rest, although it is often used to symbolize governmental action during emergencies [74]. Fumigation in airplanes or trucks usually has limited effectiveness against Ae. aegypti because the steam does not penetrate inside buildings where adult mosquitoes are at rest, although it is often used in emergencies as "a visible symbol of government action" [73].

Because *Ae. aegypti* likes to fly low, it is too low to spray with pesticides from helicopters or trucks, unlike the high-flying mosquito species that transmit West Nile [27]. Aerial pesticide spraying has not eradicated the ZIKV in Miami Beach, adding fuel to critics who have warned that buildings are too tall and ocean breezes too strong for that method to work there, claimed critics [77].

Numerous studies have demonstrated that vehicle-mounted space spraying is ineffective when closed structures provide refugia for *Ae. aegypti*. Despite these research findings, vehicle-mounted space spraying remains a staple vector control method in many dengue-endemic countries. This is mainly due to a perceived lack of more effective alternatives [78].

Spraying is both expensive and ineffective. Michael Doyle, the director of Monroe County's Mosquito Control District, reported that it takes a $16,000,000 budget to eradicate half the live mosquitoes [80]. University of Texas Dr. Robert Test recently said in the American Journal of Tropical Medicine and Hygiene that, as with many related viruses, including dengue and yellow fever, Zika could be transmitted from female mosquitoes to their offspring. The virus can be passed along to mosquito offspring, making Zika harder to control. "Pesticides in the Keys, one of the best mosquito control programs globally, has about a 50 percent success rate with *Ae. aegypti*, which are rapidly becoming immune to some products", officials in the Keys said [81]. In addition, "Spraying affects adults, but it does not usually kill the immature forms—the eggs and larvae. Spraying will reduce transmission, but it may not eliminate the virus", Tesh said [82].

On Toxicity

Another few hundred pages could be spent examining the toxicological footprint of chemical approaches to controlling mosquito populations. The following is merely a snapshot from the literature.

Mosquito populations are typically treated using an arsenal of adulticides, including organophosphates like naled and malathion and pyrethroids like permethrin and sumithrin. According to the Citizens Environmental Research Center [79], malathion can cause acute and long-term neurological health problems and has been linked to congenital disabilities in a variety of animals; permethrin is more acutely toxic to children than to adults and has been reported to contaminate ground and surface water, while sumithrin seems to have harmful effects on the central nervous system and may be estrogenic and antiprogestagenic [79].

These pesticides have been linked to adverse effects, including neurotoxicity, cancer, and reproductive dysfunction. Further, adulticiding is the least effective method for reducing mosquito populations, resulting in pesticide drift affecting human health and non-target organisms like honeybees. Visit the Pesticide Induced Disease Database (PIDD) for more pesticide-related diseases [83].

Puerto Rico's Governor Alejandro Garcia Padilla announced on July 22, 2016, that he would not authorize aerial spraying with the insecticide naled to fight an increase in ZIKV cases, as U.S. health officials have urged. Concerns were raised during several recent protests about its potentially harmful effects on people and wildlife [84]. According to the Guardian, Governor Padilla sent it back when the CDC sent a shipment of naled to Puerto Rico to combat the ZIKV [85].

Brazil fumigates against adult *Aedes* using Malathion, a carcinogenic organophosphate compound according to WHO. Paraguay acquired thousands of tons of chlorpyrifos to kill mosquitoes. Purportedly chlorpyrifos affects the developing brain of fetuses and newborns.

The go-to insecticide currently used is Dibrom, and its main ingredient is naled. Dibrom has been used in the United States over the past 50 years, said Timothy J. Donnelly, vice president, chief administrative officer, and general counsel of AMVAC Chemical Corp., which makes the insecticide". It is regularly sprayed in twelve states on over fifteen million acres yearly", Donnelly said in August 2016 [53].

The EPA has determined evidence of non-carcinogenicity in humans for naled per se (i.e., naled is a Group E chemical). Dichlorvos (DDVP), a metabolite of naled, has been classified as a Group C (possible human) carcinogen [86]. Annual domestic use of naled is approximately one million pounds of the active ingredient, with about 70% used in mosquito control and about 30% in agriculture [86]. The chemical kills mosquitoes on contact. Sprayers produce fine tiny droplets to stay airborne and intercept mosquitoes in flight. Naled spray droplets remain airborne for an extended period, and the chemical begins to break down once exposed to sunlight or water [53].

For example, In research presented at the Pediatric Academic Societies 2016 Meeting, aerial pesticide exposure was linked to an increased risk of developmental delays and autism spectrum disorder among children. The study compared children living in zip codes where aerial pesticide spraying was used each summer to combat mosquitoes that carry the eastern equine encephalitis virus with children residing in non-aerial-spraying zip codes. Children exposed to aerial pesticide spraying were about 25 percent more likely to be diagnosed with autism or have a documented developmental delay than those living in areas that used other pesticide application methods (such as manual spreading of granules) [87].

Naled is an organophosphate approved by the EPA since 1959 and has been implicated in the deaths of twenty-five children in India, where restrictions are lax. The deaths due to overexposure occurred in 2013. Like most organophosphates, it is a nerve agent. Organophosphates break down the communication between nerves and muscles, leading to suffocation [86]. Naled can cause cholinesterase inhibition in humans; it can overstimulate the nervous system, causing nausea, dizziness, confusion, high exposures (e.g., accidents or major spills), respiratory paralysis, and death [88].

A recent Chinese study by University of Michigan authors, who say it is the first to examine real-world exposure to naled outside workplace accidents or lab experiments, used cord blood from 237 mothers who gave birth to healthy babies at a hospital in Southeast China between 2008 and 2011. At six weeks, the babies displayed no problems. But at nine months, the babies suffered from slight issues with coordination, movement, and other motor functions. Lead author Monica Silver said it was impossible to determine how the Chinese mothers ended up with naled in their blood, although she suspected it was used on crops or mosquito spraying [85].

The team used cord blood collected between 2008 and 2011 by coauthor Betsy Lozoff for another study examining iron deficiency and brain development. They found several pesticides but focused on five that occurred in traceable levels. Researchers used a standard motor skill test to measure reflexes, body control, movement, and hand and eye coordination problems. As exposure to naled increased, they found deficits also rose [89].

EPA has assessed the risks of naled and reached an Interim Reregistration Eligibility Decision (IRED) for this organophosphate (OP) pesticide. Provided that risk mitigation measures are adopted, naled fits into its own "risk cup"—its individual, aggregate risks are within acceptable levels [86]. The U.S. Environmental Protection

Agency says when applied in ultra-low volumes, naled does not harm people—even toddlers playing in or eating grass in an area sprayed [60].

The "No Spray Coalition" has expressed concern about its long-term effectiveness. For example, researchers from the New York Department of Health showed that 11 years of naled spraying was "successful in achieving short-term reductions in mosquito abundance, but populations of the disease-carrying mosquito of concern "increased 15-fold over the 11 years of spraying [66]. Secondly, they are seriously concerned about its long-term effect on human health. Naled's breakdown product is dichlorvos (another organophosphate insecticide). Dichlorvos interferes with prenatal brain development. In laboratory animals, exposure for just three days during pregnancy reduced brain size by 15 percent. It also causes cancer, according to the International Agency for Research on Carcinogens. In laboratory tests, it caused leukemia and pancreatic cancer. Two independent studies have shown that children exposed to household "no-pest" strips containing dichlorvos have a higher incidence of brain cancer than unexposed children [66].

The group "Hormones Matter" has much more to say. If the sole goal is to kill insects, Naled is a satisfactory solution. But the goal should not be to kill insects (even ones as nasty as mosquitoes). It should be to prevent microcephaly and other brain development problems in fetuses. Naled is one of many organophosphate pesticides, and any chemical that changes levels of thyroid hormones is an endocrine disrupter [90]. Thousands, if not millions, of people, have been exposed to Naled because of the recent spraying to "prevent Zika". Exposing significantly more fetuses to endocrine-disrupting brain development and inhibiting pesticides from preventing a few cases of ZIKV-induced microcephaly is INSANE [90].

Though the CDC and county say naled, the main pesticide being used, is safe, the European Union has banned it. EU regulators say naled poses an "unacceptable risk" to human health [77]. Naled degrades to dichlorvos, a toxic chemical, in the naled. The US EPA has classified dichlorvos as a Group B2, a probable human carcinogen [91]. In addition, Bryn Phillips, a specialist in the University of California, Davis environmental toxicology department, said his research indicates that the insecticide breaks down into different chemical versions of naled. One of these versions, dichlorvos, is toxic to aquatic species at the "low end of the food chain", like aquatic insects and frog larvae—"basically fish food", Phillips said [53].

Like Rachel Carson before him, Bill Irwin, a Florida Keys Community College biology professor, felt that using altered insects was more environmentally friendly than pesticides. Irwin feared increased spraying to guard against Zika would harm the treasured habitat he wanted to protect. Some insecticides kill more than just mosquitoes; South Carolina recently started spraying for adult *Ae. aegypti* and residents there said the spraying was followed by an apocalyptic die-off of other insects, including bees [92]. Millions of bees died after heavy insecticide spraying in South Carolina the following week [93]. Spraying in Summerville, South Carolina, associated with four travel-related cases of Zika in Dorchester County, recently led to allegations of bee death. Since the variety of mosquito that carry ZIKV is the most active at predawn and sunset, all spraying must occur. Bees can start foraging at dawn. "In the summertime, the bees are already out at dawn, when aerial spraying is recommended. So, spraying in the morning is the worst thing they can do for bees", says Mississippi State entomologist Jeffrey Harris [53]. Keepers are advised to cover their bees during spraying operations.

Chemical spraying may make people complacent. "People think, 'I don't need to worry about removing breeding sites because, hey, it's been sprayed, so everything's going to be fine'", says Paul Hunter, a professor of health protection at East Anglia [94].

For decades, the drawbacks of heavy pesticide use have been observed. Some of the proposed alternatives to reducing the population of mosquitoes that can infect humans with the ZIKV are also alternatives to spraying neighborhoods with often hazardous chemical pesticides.

No amount of insecticide, larvicide, and fogging will appreciably reduce the mosquitoes as it is impossible to reach all the available breeding sites. "Insecticides can only suppress the population of *Ae. aegypti* by 30 percent to 50 percent at best, which is not enough, said Dr. Derric Nimmo, Oxitec's head scientist" [95]. With salt marsh mosquitoes, you can kill 95 percent on a good night", said Michael Doyle, the director of the district, from 2011 until he resigned on Sept. 1. "With *Ae. aegypti*, you're lucky if it's 50 percent". The only programs that have successfully combated *Ae. aegypti* have been militaristic: Door-to-door campaigns punished people for not getting rid of standing water [94].

According to Andy Stone, Naled and similar pesticides are far more dangerous to beneficial environmental creatures than reducing the *Ae. aegypti* mosquito populations. This mosquito strain is not native to the U.S., and there are plenty of other mosquitoes [96].

12.1.3 Some Unique Approaches

The first approach has had very mixed results, and some in the Wolbachia and even the gene drive fields are rediscovering it, while the second is much too new even to evaluate (and mostly speculative) but exciting.

12.1.3.1 Irradiation Approaches

One population and vector control approach is "the sterile male". While the "sterile insect technique" (SIT) may include sterile females, most research seems to be on sterile males. For example, sterile male methods aim to suppress target populations. Modified sterile males are released to mate with wild females; the modification results in the death of some or all the offspring of such mating. If sufficient sterile males are removed for an adequate time, the target population will be suppressed and eliminated [97].

There have been some successful SIT programs in different areas since the 1970s, e.g., for eradication or suppression of Mediterranean fruit fly, screwworm in Central

America, tsetse fly from Zanzibar, melon fly from Okinawa, and medfly from Mexico [98].

"Think of it as a method of birth control. We produce sterile male mosquitos using radiation that sterilizes the sperm in the male mosquito", says Rosemary Lees, a medical entomologist with the IAEA. "When we release many these males, we flood a region with sterile males so that the wild females are more likely to mate with them". Since female mosquitos usually only mate once, mating with infertile males would stop the further reproduction of Aedes mosquitos [102].

Radiation-induced sterilization is one of many sterile insect techniques often associated with the acronym SIT. SIT has been employed over the last 50 years using radiation-based sterilization. This technique is not new. Nevertheless, it is " being developed to stop Zika. Presumably, the controlled mass release of male mosquitoes sterilized by low doses of radiation crashes a population. When sterile male mates, the female's eggs do not survive. When the sterile males outnumber the fertile males in a natural environment, the mosquito population dies out" [100].

David Hoel, an assistant director of Lee County Mosquito Control in Florida, talked about Lee County's plans to release mosquitoes sterilized through gamma radiation starting in 2019. He said a colony of locally collected bugs is kept on the property of Lee County Mosquito Control, where they feed on chickens and mate. The X-ray machine has not arrived yet, but once it does, male mosquito eggs from the colony will be put in a pupal separator, then run through the X-ray machine to be sterilized. It's the gamma rays that fix the bugs [101].

Insect sterilization using radiation is not new, but this type of sterilization often damages the insect's fitness. It was used on the screwworm fly and some fruit flies. Using radiation to cause an insect's genes to mutate randomly, scientists made problematic species like the screwworm unable to produce viable offspring. By 1982, screwworm was eradicated from the US. Sterile insects released in the wild will depress populations. SIT uses ionizing radiation to sterilize mass-reared target insects and cast them into nature. These sterilized insects mate with wild insects but do not produce offspring [99].

In partnership with the Food and Agriculture Organization (FAO), the IAEA has committed over two million Euros to help fight Zika in Latin America and the Caribbean, where close to four million people could be infected this year. Thanks to nuclear technology, science may be able to stop the spread of what could soon be a global disease epidemic [102].

Even the WHO suggests that more research is needed to establish the public health value of this approach. Irradiation has generated mixed results. Kristen Brown put it bluntly. The elevated levels of radiation required to guarantee sterility can turn male insects into sad, unattractive mates for wild females—and if they do not mate, the technique does little to cull their numbers [99]. It seems mosquitoes are too fragile to blast with radiation rays and still be capable of mating in the wild, forcing scientists to turn to other techniques [99].

12.1.3.2 Nanotechnology

Emerging insecticide resistance in disease vectors is of great public health concern. The discovery of new targets and novel strategies for insecticidal interventions to control vector-borne diseases is a public health imperative.

Nanotechnology offers enormous potential for molecular genetic investigations and the delivery of effector molecules to control disease vectors. Nanotechnology involves works on the nanoscale (one nanometer is one-billionth of a meter) and has become facilitated by a new generation of tools built over the last three to four decades [105].

Novel approaches to deliver effector molecules and compounds to improve vector or pest control would also be of immense value. Nanotechnology offers exciting potential in both areas. Researchers have developed a hydrogel nanoparticles (NPs) toolbox with the biodistribution and tissue tropism characteristics for gene structure/function studies and delivery of lethal vector cargoes to adult mosquitoes. Improved nanoparticle (NP) delivery of dsRNA to induce RNAi to silence and functionally characterize genes and cause insect mortality offers exciting new potential for research and insect vector and pest management [107].

Paquette et al [104] demonstrated the potential use of and preferred physical characteristics of hydrogel NPs for delivery of cargoes (e.g., dsRNA) to silence mosquito genes for functional analyses and mosquito control [106].

> Positively charged particles are more efficiently internalized in vector cells, but negatively charged NPs were detected abundantly in tissues in the proboscis, regardless of the challenge mode. Cells in the labella of the proboscis of mosquitoes frequently contained or were associated with significant accumulations of NPs [104].

RNA interference (RNAi) is a promising strategy to suppress the expression of disease-relevant genes and induce post-transcriptional gene silencing. RNA interference (RNAi) refers to the process of exogenous double-stranded RNA (dsRNA) silencing the complementary endogenous messenger RNA. RNAi has been widely used in entomological research for functional genomics in various insects.

Its potential for RNAi-based pest control has been increasingly emphasized mainly because of its high specificity [108]. Although current research has proven RNAi has immense potential to be used in insect pest management, a better understanding of dsRNA uptake mechanisms will facilitate the development of effective RNAi-based technologies for insect pest management [109].

Bioengineered silver nanocomplexes exhibited maximum mosquito larvicidal activity (100%) against the *An. stephensi* and *Ae. aegypti*. In addition, the silver nanohybrids showed enhanced antibacterial activity toward the tested microbial pathogens. Results show that the biomolecules coated silver nanoparticles were a promising and potential target larvicidal nanodrug that can be used for protecting malaria, Chikungunya, Zika, and dengue fever [109].

Barry Beaty from the University of Colorado Denver's School of Public Health has described using nanoparticles as molecular mosquitocides. He discussed the efficacy and targets of different shapes and sizes of nanoparticles and the search for an environmental uptake and delivery system [111].

A team from Periyar University led by Jinu et al [109] accepted the challenge by using silver nanoparticles. The ability to deliver effector molecules through oral or contact challenges for gene structure–function studies would be greatly valued. It would preclude confounding effects of injection on gene regulation (e.g., induction of innate immune genes by penetration of the cuticle) and minimize mortality in experimental insects due to injection [107].

12.2 Indirect Preventative Approaches

In general, the quality of the evidence was low to extremely low for most vector interventions. The effectiveness of any control strategy is setting-dependent [18]. And University of Texas' Scott Weaver says current mosquito control methods do not work well against *Ae. aegypti*, live remarkably close to people [112]. Mosquitoes like *Ae. aegypti* breed in and around houses, so draining wetlands would not be influential. Similarly, bed nets will not be effective against the mosquitoes that bite during the day [114] though both LLIN (long-lasting insecticide-treated nets) window curtains and water-container covers can be effective [30].

However, there are challenges relating to the distribution of insecticide-treated bed nets and the maintenance of their effectiveness. There is evidence that mosquito behavior is shifting from indoor to outdoor biting or night to dawn biting in areas where these nets are used [114]. For example, Miroux et al [113] demonstrated a switch in malaria vectors' biting behavior after implementing LLIN at universal coverage. "Results showed that compared with the baseline survey years after implementing LLIN at the community level, *Ae. funestus* bit later during the night (almost at dawn) and frequently outdoors [113].

"The best eradication strategy is to go inside people's houses and spray residual insecticides on the walls and their closets in dark places where mosquitoes like to rest, which is extremely labor-intensive". The CDC says the mosquitoes have developed resistance to insecticides (see above).

"*Ae. aegypti* is an exceedingly difficult mosquito to control and eliminate. It will require an extremely aggressive and concerted effort", Anthony Fauci, the director of the National Institute for Allergy and Infectious Diseases, said [112].

12.2.1 Individual Scale Efforts

Many protective measures can be taken against mosquito bites, which include proper use of house screens, air conditioning, or sleeping under a bed net, appropriate disposal of household debris, and effective use of insect killers, especially in dark and humid places as mosquitoes will dwell there and limit outdoor activities [115].

Among the best preventive measures against ZIKV are house screens, air-conditioning, and removal of the yard and household debris and containers that provide mosquito-breeding sites, luxuries often unavailable to impoverished residents of crowded urban locales where such epidemics hit hardest [116]. In Mexico, improving housing by using long-lasting insecticide-treated house screens and targeted larviciding resulted in significant reductions in *Ae. aegypti* infestations [117]. Insecticide-treated window curtains and insecticide-treated domestic water container covers can reduce dengue vector densities to low levels [114]. However, screens and air conditioning are considered luxuries in many neighborhoods in South America.

Bed nets are less effective because of *Ae. aegypti* bites daily. There is evidence that mosquito behavior is shifting from indoor to outdoor biting or night to dawn biting in areas where these nets are used [114]. *Ae. aegypti* exhibit predominantly indoor resting and blood-feeding behavior, and barriers to access would be expected to impact this species [41].

ZIKV infection can be prevented through mosquito repellents, protection with long-sleeved clothing and trousers, and wearing clothes impregnated with permethrin [119].

Some educational campaigns have sprouted. According to Bouzid's meta-analysis, educational campaigns are essential to reduce breeding sites and interrupt disease transmission [18]. For example, the Guam Department of Public Health and Social Services announced it would be launching a Zika campaign early in 2017. The campaign aims to educate health providers and the community on recognizing and preventing mosquito-borne illnesses, including Zika. Guam has a tire-shredding program and is increasing its lab [120]. In Brazil, *Ae. aegypti* control is increasingly based on community participation campaigns that seek to fortify the importance of mechanical control through the adequate coverage/elimination of potential breeding sites [58].

The primary means of educating the public involve various educational pamphlets. What constitutes public education and outreach matters. Barlett-Healy et al [118] reported the brochures were ineffective in encouraging middle and high-income homeowners to take the recommended action. Instead of relying on these passive forms of education, future educational efforts will depend on more active community-based approaches to teaching since the most effective education campaigns are ones where the community has ownership of the program. These will involve community peer educators to empower residents to reduce mosquito habitats [119].

The AMCA warns against this form of "passive education".

Passive education (distribution of educational materials) is not highly effective in engaging the public in control efforts. In one study, six communities were randomly selected to receive 1 of 3 strategies: 1) education and mosquito control, 2) education only, and 3) no education or mosquito control. The education program included a 5-day elementary school curriculum in the spring and three door-to-door distributions of educational brochures. The number of mosquito-larval container habitats was counted in fifty randomly selected homes per study area before and after each educational event. Although there were reductions in container habitats in sites receiving education, they were not significantly different from the control. These results suggest that conventional passive public education is insufficient to motivate residents to reduce backyard mosquito-larval habitats [17].

People are eradicating *Ae. aegypti,* and it mainly involves boots and trucks. Compliance officers' empty containers of water where the larvae develop into mosquitoes, and communities use insecticides (see above). Policing sitting water where mosquitoes breed is incredibly expensive. Anna Maria Barry-Jester described the daily activities of 56-year-old Billy Ryan, an employee of the Monroe County Mosquito District. He visits auto-body shops and boatyards, overturning containers of standing water. He climbs into boats and empties plastic buckets, pipes, and other bits of debris. "You think a spray truck is gonna get in here?" he muttered. "No way, you have to do this by hand".

Draining water collected in containers around the home also helps, but this approach demands high community compliance, and no one believes a campaign built upon eliminating stand water is sufficient. The *Ae. aegypti* is a container breeder. Larvae have been found in many artificial containers, such as discarded plastic cups and bottle caps, plates under potted plants, birdbaths, vases in cemeteries, and pet water bowls [121]. These mosquitoes will lay eggs in the stagnant water collected in birdbaths or old tires, sure, but also in the few droplets gathered in a discarded candy wrapper, on the damp surface of a fallen leaf, or in a practically empty soda can, [122] even in toilets [123].

Finally, people who eliminate mosquito habitats are still vulnerable to neighbors who do not. Participation is often high only during epidemics. No examples were found where education campaigns had a lasting influence on behavior. Although "bottom-up" policy is attractive, it is unrealistic to expect it to work without strong "top-down" leadership and support [7].

Unfortunately, half measures will not slow the virus. *Ae. aegypti* lay eggs that can cling to a dry surface until the rain they need arrives. If it is still too cool for the baby mosquitoes, they can stay in the larval phase for many more months if they have enough water to remain submerged [122]. It is not enough merely to empty the rows of conch shells, for instance, that decorate outdoor restaurants in Miami. "You have to scrub each one out", Mr. Joseph Conlon, an American Mosquito Control Association advisor, said [124].

These methods are expensive and require enormous human resources: Thirty-five full-time inspectors must cover the entire Keys, which stretch 110 miles from end to end, although *Ae. aegypti* accounts for less than 1 percent of the mosquitoes in the Keys; the district says it spends about $1 million of its $10 million budget battling the species each year [92].

In general, successful mechanical control is labor-intensive, costly, and time-consuming. Some containers are bound to be overlooked regardless of strenuous efforts and scrutiny [58].

12.2.1.1 Repellants

WHO recommends covering the skin with clothing as much as possible and using insect repellents as effective measures to protect against bites from mosquitoes that transmit viruses such as chikungunya, dengue, yellow fever, and Zika [100].

They report: Effective repellents contain DEET (diethyltoluamide), or IR 3535 (3[NbutylNacetyl], amino propionic acid ethyl ester) or KBR3023 (also called Icaridin or Picaridin). These are simply the most common active biological ingredients in repellents.

Some people have negative attitudes about mosquito repellent. Survey research indicates that some people do not like how traditional mosquito repellents smell or feel on the skin; others have concerns about product safety [63].

DEET is not without its detractors. For example, DEET is a neurotoxin implicated in illnesses among some veterans of the Persian Gulf War (Abdel-Rahman, Shetty and Abou—Donia [125]). Lab animals exposed to the average human equivalent dose of DEET (N, NDiethyl tolbutamide) performed much worse in neurobehavioral tasks (such as those involving coordinated muscle movement).

> Subchronic dermal application of DEET and permethrin to adult rats, alone or in combination, leads to diffuse neuronal cell death in the cerebral cortex, the hippocampal formation, and the cerebellum. Collectively, the above alterations can lead to many physiological, pharmacological, and behavioral abnormalities, particularly motor deficits and learning and memory dysfunction.... These alterations can lead to physiological and behavioral abnormalities, particularly motor deficits and learning and memory dysfunction. The above alterations are likely the contributory factors for neurobehavioral abnormalities observed earlier in adult rats following exposure to DEET and permethrin, alone or in combination. Thus, it is likely that subchronic exposure to DEET and permethrin experienced by service personnel during the Persian Gulf War has played an essential role in developing illnesses in some veterans after the Gulf War [125].

Former New York attorney general Eric Schneiderman has warned parents about marketing worthless products. He called some of the manufacturers of these products shameless. He specifically called out products based on B vitamins and "essential oils", along with those that emit ultrasound energy to purportedly deter mosquitoes (like the iGuard 2.0 Ultrasonic Insect Pest Repellant and STAR Ultrasonic Pest Repeller). He focuses his ire on Wildheart Outdoors Natural Mosquito Repellent Bracelet and Kenza High-Quality Zika Mosquito Repellent Smiley Patch. These approaches are not supported by evidence or approved by the U.S. Centers for Disease Control and Prevention [127].

Schneiderman also urged people to use mosquito repellants that are effective and are registered by the U.S. Environmental Protection Agency: DEET, picaridin (also known as KBR 3023, icaridin, or bayrepel), IR3535, oil of lemon eucalyptus, and paramenthanediol [127].

12.2.1.2 Isoxazolines

The researchers found a class of drugs called isoxazolines, sold in veterinary products such as fluralaner (Bravecto) and afoxolaner (NexGard), to protect pets from fleas and ticks, also kills species of disease-carrying mosquitoes that feed on human blood [128].

> The Calibr and TropIQ scientists and their collaborators tested two of the drugs, fluralaner, and afoxolaner. They found they also kill disease-carrying mosquitoes and sand flies that feed on human blood infused with the insecticides. The drugs were also effective against insect strains resistant to typical insecticides. Based on existing data from studies of the medicines in animals, the researchers estimated that a single human dose of the drugs would convey an insecticide effect against mosquitoes and sand flies lasting 50-90 days [128].

Based on safety studies of isoxazoline use in animals, the drugs can be safe if repurposed for human use. The research team is planning to evaluate the efficacy of the medicines in humans and anticipates these studies will take around two years.

A research team (Miglianico et al [126]) determined via experimental studies on mosquitoes and computational modeling that giving isoxazoline drugs to less than a third of the population in areas prone to seasonal outbreaks of insect-borne diseases could prevent up to 97 percent of all cases of infection [127].

They concluded: that "even a 30% population coverage could substantially reduce the clinical incidence of malaria or ZIKV fever. In conclusion, repurposing isoxazoline compounds promises to develop a single-dose vector control drug based on a novel mode of action compared with commonly used insecticides, with activity against a broad range of relevant disease vectors [126].

"Insect-borne infectious diseases remain primary causes of severe illnesses and fatalities worldwide, and novel approaches to preventing outbreaks of these diseases are critically needed", said Peter Schultz, Ph.D., chief executive officer of Calibr and Scripps Research. "Our findings suggest that isoxazolines might effectively control outbreaks of diseases carried by mosquitoes and other insects in regions with limited medical infrastructure" [128].

12.2.2 Ecosystem Efforts

Ecosystem management interventions such as insecticide-treated materials such as window/indoor net curtains in houses and the targeted treatment of productive breeding sites have shown potential for integrated dengue vector control in many geographical contexts [117].

The combination of long-lasting insecticidal screens fitted to external windows and doors and targeted treatment of the most productive *Ae. aegypti* breeding sites can significantly impact dengue vector populations and sustain that impact for up to 24 months [117].

Controlling productive containers year-round, such as water tanks and metal drums, has a long-term effect on vector density, immatures, and adults [117].

12.2.2.1 Traps

There are many types of traps. These traps fall into four categories: (1) backpack aspirators, (2) sticky ovitraps, (3) nets, and (4) even some high-tech traps. Traps are used to monitor mosquito populations and trap and kill. Several field tests using insecticidal gravid traps have shown a lack of consistent results on their impact on Ae. aegypti populations. One factor that works against the efficacy of gravid traps and ovitraps is the presence of naturally occurring containers that function as refuges for the reproductive population instead of the trap's sink effect [129].

Ovitraps are simple, inexpensive devices consisting of a small cup that holds water, often mixed with a failure, and provide a substrate on which gravid mosquitoes may lay their eggs. Ovitraps are useful for Aedes because they tend to oviposit in artificial containers. These devices have been used extensively for conducting surveillance for invasive Ae. aegypti and *Ae. albopictus* [17].

The AMCA suggests lethal oviposition cups and BGS (BG-Sentinel™) traps should be used together to monitor both sexes and all physiologic stages of *Aedes*. BGS traps, gravid Aedes Trap (GAT), and CDC-autocidal gravid ovitrap (CDC-AGO) are the most widely used [117].

> Similarly, Ae. aegypti and *Ae. albopictus* do not fly far from larval developmental sites. BGS traps baited with the BG-Lure have reduced population abundance and human biting rates compared with no intervention. Recent studies in the United States utilizing Mosquito Magnets and human-scented and octanol lures have shown that these traps may outperform BGS traps for capturing *Ae. albopictus* up to 6-fold. Cost and labor are significant issues in using BGS traps for control because trap density and maintenance requirements are high [17].

Microsoft is evaluating a clever trap to isolate and capture *Aedes* in Texas. *Aegypti* mosquitoes, known Zika carriers, for study by entomologists to give them a jump on predicting outbreaks. Most conventional mosquito traps capture all comers—moths, flies, other mosquito varieties—leaving a pile of specimens for entomologists to sort through. The Microsoft machines differentiate insects by measuring a feature unique to each species: the shadows cast by their beating wings. In Texas, ten mosquito traps made by Microsoft are operating in Harris County, which includes the city of Houston. The Houston tests, begun last summer, showed the traps could detect *Ae. aegypti* and other medically necessary mosquitoes with 85% accuracy, Microsoft engineer Ethan Jackson said. The traps are prototypes now. But Microsoft's Jackson said the company eventually hopes to sell them for a few hundred dollars each, roughly conventional traps. The goal is to spur wide adoption, particularly in developing countries, to detect potential epidemics before they start [130].

The AMCA warns that caution is warranted when considering egg surveillance methods based on conflicting results between eggs and adult populations of Aedes mosquitoes. AMCA reports although oviposition cups are valuable for determining the presence and absence of Aedes vectors, they are not always dependable for adult population estimation [17].

> Oviposition cups have several potential limitations. First, the data generated must be interpreted cautiously because oviposition cups compete with natural larval habitats, presenting a problem, particularly after source reduction campaigns. Second, microscopy may be needed to count eggs, especially if debris is present on the oviposition surfaces accurately. Third, trained personnel must hatch, rear, and identify species [17].

Aspirator devices, such as sweepers, suction traps, and hand-held battery-operated flashlight aspirators, may collect resting mosquitoes on either natural resting harborage or artificial structures. The CDC-Backpack Aspirator has been widely used for indoor collections of certain domestic mosquito species, including Aedes; however, it has several limitations, including weight and cost. As an alternative, a less expensive, battery-powered, relatively light aspirator, the ProkoPack, has been developed that efficiently collects adult mosquitoes [17].

The second subset of traps involves those that can be placed around the home. In principle, ovitraps could kill adult mosquitoes if the ovistrip was treated with insecticide or destroy progeny using fine nylon netting to trap the larvae. In Brazil, lethal ovitraps with deltamethrin-treated ovistrips killed 89% of Ae. aegypti adults produced more than 99% larval mortality during one-month field trials. The advantages of lethal ovitraps for controlling Aedes vectors include their simplicity, specificity, and effectiveness against container breeders like *Ae. aegypti* and potential for integration with other chemical or biological control methodologies [30].

The fertility of *Ae. aegypti* populations can be reduced by using autocidal oviposition cups that prevent the development of mosquitoes inside the trap by mechanical means or larvicides, as well as by releasing sterile, transgenic, and para-transgenic mosquitoes. This approach relies on creating a "constellation" of "sinkholes" that will crash the mosquito population, Roberto Barrera, head of the CDC's entomology and ecology activities and one of the trap's inventors, explained [131].

Sticky AGO traps are effective at eliminating gravid *Ae. aegypti* females and function as a population sink both reproductive adults and egg stages. Autocidal gravid ovitraps (AGO trap) reduced the *Ae. aegypti* population and prevented mosquito outbreaks in Southern Puerto Rico. Barrera et al [129] claimed the population of *Ae. aegypti* were significantly reduced by 79% [131].

Third, there is netting. At dawn in Harris County, Texas, workers hoist nearly invisible nets in local parks to capture birds for blood sampling. (West Nile and several other dangerous viruses are found in birds, although Zika is not.) More than four hundred mosquito traps are scattered around the county. Some are baited with dry ice, emitting carbon dioxide, the element of human breath that draws mosquitoes. Some traps exude the lactic acid-ammonia mix of human sweat, while others use water to attract egg-laying females [132].

A screen prevents the mosquito from reaching the water, and the insects get caught in sticky, non-toxic resin inside the opening. The traps cost about $11 each, and although they have been used to monitor mosquito populations in the past, they are just now being considered eradication tools [131].

Finally, there are advanced "tech" traps. Some traps attract females to lay their eggs, and the CDC reportedly likes one involving water and hay [60]. The BG trap is an attractant trap that requires electricity and costs US$100–US$300 per unit (http://www.biogents.com/en/index.html), which will limit its use in resource-strapped environments [7]. Survival and fertility can be simultaneously reduced by capturing gravid female Ae. aegypti with sticky gravid traps. The presence of three to four AGO control traps per home in 81% of the houses prevented outbreaks of Ae. aegypti, which would be expected after rains. Mosquito captures in BG-Sentinel, and

AGO traps were significantly and positively correlated, showing that AGO traps are helpful and inexpensive mosquito surveillance devices. Barrara et al. found the use of AGO (autocidal gravid ovitrap) to manage *Ae. aegypti* populations are compatible with other control means such as source reduction, larviciding, adulticiding, sterile insect techniques, induced cytoplasmic incompatibility, and dominant lethal gene systems [133]. An autocidal gravid ovitrap has been shown to reduce *Ae. aegypti* populations under field conditions in two isolated urban areas of Puerto Rico by 53–70 percent using 3–4 traps per home in 81 percent of houses [75].

And unlike the mainland, Puerto Rico rejected the widespread use of insecticides like naled to fight the menace, fearing their toll on people and the environment. And that's where Barrera's team is stepping in. In early February 2017, the CDC worked with the local government to turn this mid-sized city into a mosquito graveyard deadly enough to kill the ZIKV. They are making Caguas a test case for chemical-free Zika reduction [131].

The problem involving the use of traps is that it requires community participation. Barrera et al [133] described some challenges they confronted when researching the efficacy of these traps.

The use of autocidal gravid ovitraps requires ample participation in the community. For example, most houses (81–85%) in each intervention area have had three traps in their yards since December 2011 in La Margarita and since February 2013 in Villodas, a locality in Guayama. Permission to enter properties for trap servicing must be requested every two months. However, many residents have authorized personnel to enter their yards for this purpose, even in their absence. Furthermore, to reach the coverage required for servicing most of the traps, fieldwork schedules have been adjusted to include weekends and after working hours to increase the likelihood that technicians would find residents at home.

12.2.2.2 Codes and Laws

Here are two examples. In November, a Green candidate in Mosquito Control District 1 elections announced that he endorses using code enforcement to combat Zika-carrying mosquitoes. Oliver Kofoid, not a fan of the Oxitec approach, said inspectors would check properties for standing water, where they populate. Those who maintain their properties would get a tax break, saving the district money on spraying [134].

Singapore adopted a more autocratic approach. The National Environment Agency launched its annual Mozzie Wipeout Campaign 3 months earlier this year, instigated by the Zika scare, and found and destroyed 13,000 breeding spots in homes. They also clamped down on households where mosquito breeding spots were detected. It imposed fines on those breeding mosquitoes at home, not just those who lived in dengue clusters [135].

12.2.2.3 Clean-Up Campaigns (TIP and Toss)

Probably, the most widespread practices to suppress dengue vector populations are clean-up campaigns, typically community-driven and in tandem with education and health promotional campaigns, as well as numerous additional approaches. When clean-up campaigns were evaluated, they were one element within multiple interventions or continued to be promoted as background across all the arms within a study [41].

Thus, source reduction or clean-up campaigns were applied in twenty studies (Bowman et al [41]) but were associated with interventions ranging from fogging or water container covers targeting adult mosquitoes to larviciding and copepods for control of immatures. It is impossible to dissect their specific contribution to reducing vector populations or their impact on dengue transmission [41].

A study in two New Jersey counties examined education with mosquito control against education only and no education or mosquito control. Although Bartlett-Healy et al [118] saw reductions in container habitats in sites receiving education, they were not significantly different from the power. Their results suggest that traditional passive means of public education, often considered the gold standard for mosquito control programs, are insufficient to motivate residents to reduce backyard mosquito-larval habitats. However, specific container habitats, such as birdbaths, tarpaulins, and plant pot receptacles, are challenging to eliminate and can continue as potential oviposition sites for mosquitoes [118].

Although this drop suggests these efforts might have been practical, noticing the same reduction in containers per house in sites that did not receive education, resulting in areas that were not statistically different. It is possible that the presence of mosquito personnel counting containers in all places could have indirectly motivated residents to conduct source reduction behavior in their backyards. Mosquito personnel were instructed not to educate homeowners in the control sites. This may have raised concerns with residents, who might be worried about being cited or fined due to producing mosquitoes in their backyard [118].

The AMCA warns that for container-inhabiting Aedes, given many potential larval sites and many of these containers are located on private property, the direct application may have limited success and is labor-intensive and time-consuming, requiring public education efforts and close cooperation with the community [17].

Removal of conspicuous open containers may "push" *Ae. albopictus* females to oviposit in cryptic habitats; therefore, it is critical to locate and assess all potential container sources, including those that may be more difficult to identify, access, and treat with larvicides [17].

This approach involves community action. Affected communities, empowered through education and advocacy, can mobilize and mount effective control operations relatively independently via horizontal or community-based efforts [R]eductions in potential larval development sites can be achieved with householders and communities taking responsibility, supported by education and social mobilization [41].

If conducted comprehensively, source reduction is the most effective control method against container inhabiting Aedes species. However, this method is operationally challenging to implement and sustain. Container removal programs and so-called "tip-and-toss" techniques (overturning containers holding water) effectively eliminate habitat and may be combined with natural larvicide treatments. Given many potential container sources and circumstances where many of these containers are situated on private property, this approach may have limited success while being labor-intensive and time-consuming, requiring public education efforts and close cooperation with the community [17].

12.3 Conclusion

Vector control has been an important feature in the scholarship on mosquito control for many decades. Without a doubt, malaria, and dengue more than any other infectious diseases involving mosquitoes have been the driving force in these discussions. When ZIKV broke onto the scene, the epidemic provided fuel for the arguments about new vector control approaches. In the next three chapters, two approaches are examined that nearly simultaneously with the ZIKV epidemic in Latin America. One approach involves a bacterium and the other genetic engineering. The last of three chapters reviews much of the speculation associated with gene drive technologies.

References

1. Ferguson NM (2018) Challenges and opportunities in controlling mosquito-borne infections. Nature 559:490–497, 26 Jul. https://doi.org/10.1038/s41586-018-0318-5
2. PAHO (1997) The feasibility of eradicating Aedes aegypti in the Americas. Revista Panamericana de Salud Pública 1(1):68–72
3. CSTE (Council of State and Territorial Epidemiologists) (2018) 2018 Zika preparedness resources toolkit. Atlanta, GA: CSTE. Retrieved from https://cdn.ymaws.com/www.cste.org/resource/resmgr/zika/Zika_Virus_Preparedness_Reso.pdf. Accessed on 8 Aug 2018
4. Corbel V et al (2017). International workshop on insecticide resistance in vectors of arboviruses, December 2016, Rio de Janeiro, Brazil. Parasites Vectors 10:278, 2 Jun. https://doi.org/10.1186/s13071-017-2224-3. Accessed on 7 Aug 2018
5. Tepedino K cited by Ramos N (2016) Brazil GM mosquitoes to breed out diseases. In: MENAFN.com, 31 Oct. Retrieved from https://www.yahoo.com/news/brazil-mutant-mosquitoes-breed-diseases-015740667.html. Accessed on 22 Dec 2016
6. McDonald S (2018) Houston health department urges mosquito prevention. In: Patch, 15 Mar. Retrieved from https://patch.com/texas/houstonheights/houston-health-department-urges-mosquito-prevention. Accessed on 7 Jul 2018
7. Morrison AC et al (2008) Defining challenges and proposing solutions for control of the virus vector aedes aegypti. PLOS Med 5(3):e68, Mar. https://doi.org/10.1371/journal.pmed.0050068. Accessed on 26 May 2017
8. Yong Ed (2016) The answer to zika may be more mosquitos. In: The Atlantic, 27 Oct. Retrieved from https://www.theatlantic.com/science/archive/2016/10/zika-fighting-mosquitoes-take-off-in-south-america/505523/. Accessed on 2 Jul 2017

9. Beck J (2016) Pesticides aren't the best way to fight Zika-carrying mosquitoes. In: The Atlantic, 9 Dec. Retrieved from https://www.theatlantic.com/health/archive/2016/12/pesticides-arent-the-best-way-to-fight-zika-carrying-mosquitoes/510167/. Accessed on 3 Feb 2017
10. Oxitec (2016) Florida keys project. In: Oxitec Website. Retrieved from http://www.oxitec.com/programmes/united-states/. Accessed on 27 May 2017
11. Lessler J et al (2016) Assessing the global threat from Zika virus. Science 353(6300):1–11, 14 Jul. Retrieved from http://science.sciencemag.org/content/early/2016/07/13/science.aaf8160. Accessed on 19 May 2017
12. Costa, Federico et al. (2016). Emergence of congenital zika syndrome. Ann Intern Med 164(10):689–691, 24 Feb. https://doi.org/10.7326/M16-0332. Retrieved from http://annals.org/aim/fullarticle/2498549/emergence-congenital-zika-syndrome-viewpoint-from-front-lines. Accessed on 27 Jun 2018
13. Geard N et al (2016) Is the end of Zika nigh? How populations develop immunity. The Conversation, 21 Jul. Retrieved from http://theconversation.com/is-the-end-of-zika-nigh-how-populations-develop-immunity-62816. Accessed on 13 Sep 2016
14. Geard N, Wood J, McVernon J (2015) Explainer: what is herd immunity? In: The Conversation.com, 14 Dec. Retrieved from https://theconversation.com/explainer-what-is-herd-immunity-52377. Accessed on 25 Sep 2016
15. Adler J (2016) A world without mosquitoes. Smithson 38, Jun. Retrieved from http://www.smithsonianmag.com/innovation/kill-all-mosquitos-180959069/?no-ist. Accessed October 2, 2016.
16. Li C-X, Smith ML, Fulcher A, Kaufman PE, Zhao T-Y, Xue R-D (2015) Field evaluation of three new mosquito light traps against two standard light traps to collect mosquitoes (diptera: culicidae) and non-target insects in northeast Florida. Fla Entomol, Jun 2015. Retrieved from https://www.sciencedaily.com/releases/2015/06/150609092805.htm. Accessed on 28 Feb 2022
17. American Mosquito Control Association (2017) Best practices for integrated mosquito management: a focused update, Jan. Retrieved from https://www.researchgate.net/publication/315924484_Best_Practices_for_Integrated_Mosquito_Management_A_Focused_Update. Accessed on 27 Aug 2018
18. Bouzid M et al (2016) Public health interventions for aedes control in the time of zikavirus—meta-review on effectiveness of vector control strategies. PLOS Negl Trop Dis, 7 Dec. https://doi.org/10.1371/journal.pntd.0005176. Accessed on 9 Mar 2017
19. Erlanger TE, Keiser J, Utzinger J (2008) Effect of dengue vector control interventions on entomological parameters in developing countries: a systematic review and meta-analysis. Med Vet Entomol 22(3):203–221. Retrieved from https://www.ncbi.nlm.nih.gov/pubmed/18816269. Accessed on 7 Aug 2018
20. Marques AM et al (2017) Larvicidal activity of ottonia anisum metabolites against aedes aegypti: a potential natural alternative source for mosquito vector control in Brazil. J Vector Borne Dis 54:61–68, March. Retrieved from https://www.ncbi.nlm.nih.gov/pubmed/28352047. Accessed on 17 Jul 2017
21. Eye on Nature (2013) An elegant mosquito. In: Eye on Nature, 10 Feb. Retrieved from https://eyeonnature.wordpress.com/2013/02/. Accessed on 24 Oct 2016
22. Yeoh O (2017) SAVVY: mosquitoes begone! In: New Straight Times, 19 Nov. Retrieved from https://www.nst.com.my/lifestyle/sunday-vibes/2017/11/304995/savvy-mosquitoes-begone. Accessed on 10 Jul 2018
23. Goettle BJ, Adler PH (1999). Elephant (or tree hole) predatory mosquito. Retrieved from http://www.dnr.sc.gov/cwcs/pdf/Predmosquit.pdf. Accessed on 14 Oct 2016
24. South Carolina Department of Natural Resources (2006) Elephant (or tree hole) predatory mosquito. Retrieved from http://www.dnr.sc.gov/cwcs/pdf/Predmosquit.pdf. Accessed on 24 Oct 2016
25. Interlandi J (2016) Fighting zika with genetically modified mosquitoes. In: Consumer Reports, 14 Sep. Retrieved from http://www.consumerreports.org/conditions-treatments/fighting-zika-with-genetically-modified-mosquitoes/. Accessed on 23 Sep 2016

26. El Sen T (2016) Usage of elephant mosquitoes shows positive results in reducing aedes. In: Astro Awani Network. Retrieved from http://english.astroawani.com/malaysia-news/usage-elephant-mosquitoes-shows-positive-results-reducing-aedes-31234. Accessed on 25 Sep 2016
27. Zhang S (2015) Ride with the mosquito hunters protecting the US against zika. In: Wired, 8 Feb. Retrieved from https://www.wired.com/2016/02/mosquito-control-story/. Accessed on 24 Oct 2016
28. Green J (2019). The nature of an assassin: harnessing deadly killers to fight pests. In: Beyond Bones, 15 Aug. Retrieved from https://blog.hmns.org/2019/08/the-nature-of-an-assassin-harnessing-deadly-killers-to-fight-pests/. Accessed on 25 Jun 2022
29. Flores C, Villalobos-Cerrud D, Borace J, Fabrega L et al (2021) Epidemiological aspects of maternal and congenital toxoplasmosis in panama. Pathogens 10(6):764, June. https://doi.org/10.3390/pathogens10060764
30. McCall PJ, Kittayapong P (2006) Control of dengue vectors: tools and strategies. In: Report of the Scientific Working Group on Dengue. Geneva: TDR/World Health Organization. Retrieved from http://www.who.int/tdr/publications/documents/swg_dengue_2.pdf. Accessed on 20 Jul 2017
31. Lazaro A, Han WW, Manrique-Saide P, George L, Velayudhan R, Toledo J et al (2015) Community effectiveness of copepods for dengue vector control: systematic review. Trop Med Int Health 20(6):685–706
32. Vu SN et al (2005) Elimination of dengue by community programs using mesocyclops (copepoda) against aedes aegypti in central Vietnam. Am J Trop Med Hyg 72(1):67–73. Retrieved from https://www.ncbi.nlm.nih.gov/pubmed/15728869. Accessed on 9 Jun 2017
33. Oxitec Report to CVM FDA (2016) Environmental assessment for investigational use of aedes aegypti OX513A. In: Center for Veterinary Medicine United States Food and Drug Administration Department of Health and Human Services, 5 Aug. Retrieved from https://www.fda.gov/downloads/AnimalVeterinary/DevelopmentApprovalProcess/GeneticEngineering/GeneticallyEngineeredAnimals/UCM514698.pdf. Accessed on 12 May 2017
34. WHO (2003) Use of fish for mosquito control. WHO: Regional Office for the Eastern Mediterranean, Cairo. Retrieved from http://applications.emro.who.int/dsaf/dsa205.pdf?ua=1. Accessed on 2 Oct 2016
35. Calder JAM (2017) Zika virus in the Americas: is it time to revisit mosquito elimination? J Environ Health, Sep 80(2):26—27
36. Walshe DP, Garner P, Abdel-Hameed Adeel AA, Pyke GH, Burkot T (2013) Larvivorous fish for preventing malaria trans-mission. Cochrane Database Syst Rev 12:CD008090
37. Araujo HRC, Carvalho D, Ioshino R, Costa-da-Silva A, Capurro M (2015) Aedes aegypti control strategies in Brazil: incorporation of new technologies to overcome the persistence of dengue epidemics. Insects 6:576–594
38. Kyle JL, Harris E (2008) Global spread and persistence of dengue. Annu Rev Microbiol 62:71–92. Retrieved from https://www.ncbi.nlm.nih.gov/pubmed/18429680. Accessed on 18 May 2017
39. Lafrance A (2016) Genetically modified mosquitoes: what could possibly go wrong? In: The Atlantic, 26 Apr. Retrieved from https://www.theatlantic.com/technology/archive/2016/04/genetically-modified-mosquitoes-zika/479793/. Accessed on 18 May 2017
40. Adalja A et al (2016) Genetically modified (GM) mosquito use to reduce mosquito-transmitted disease in the US: a community opinion survey. PLOS Curr Outbreaks, 25 May. Retrieved from https://doi.org/10.1371/currents.outbreaks.1c39ec05a743d41ee39391ed0f2ed8d3
41. Bowman LR, Donegan S, McCall PJ (2016) Is dengue vector control deficient in effectiveness or evidence?: Systematic review and meta-analysis. PLOS Negl Trop Dis 10(3):e0004551. Retrieved from https://www.ncbi.nlm.nih.gov/pubmed/26986468. Accessed on 7 Aug 2018
42. Chandler J (2016) Aerial spraying for mosquitoes rejected for O.C. – at least for now. In: Orange County Register, 22 Apr. Retrieved from http://www.ocregister.com/2016/04/22/aerial-spraying-for-mosquitoes-rejected-for-oc-at-least-for-now/. Accessed on 14 May 2017

43. Blinder A (2016) Aimed at zika mosquitoes, spray kills millions of honeybees. In: New York Times, 1 Sep. Retrieved from https://www.nytimes.com/2016/09/02/us/south-carolina-pesticide-kills-bees.html. Accessed on 24 May 2022
44. Matthews KRW, Herricks JR (2016) Mosquito-transmitted epidemics: zika virus in the United States and Mexico. Policy brief no. 03.04.16. Houston, Texas: Rice University's Baker Institute for Public Policy
45. de Fatima P Militão de Albuquerque M et al (2016) Pyriproxyfen and the microcephaly epidemic in Brazil-an ecological approach to explore the hypothesis of their association. Mem Inst Oswaldo Cruz 111(12):774–776. Rio de Janeiro, Dec. Retrieved from http://www.scielo.br/scielo.php?script=sci_arttext&pid=S0074-02762016001200774. Accessed on 18 Jul 2018; Regis LN et al (2014) Characterization of the spatial and temporal dynamics of the dengue vector population established in urban areas of Fernando de Noronha, a Brazilian oceanic island. Acta Trop 137:80–7, Sep. Retrieved from https://www.ncbi.nlm.nih.gov/pubmed/24832009. Accessed on 30 Jul 2018; de Silva Augusto LG et al (2016) Aedes aegypti control in Brazil. Lancet 387(10023):P1052–1053, 12 Mar. Retrieved from https://www.thelancet.com/pdfs/journals/lancet/PIIS0140-6736(16)00626-7.pdf. Accessed on 8 Aug 2018
46. Aliota M et al (2016) The wMel strain of wolbachia reduces transmission of zika virus by aedes aegypti. Nat Sci Rep 6:28792. https://doi.org/10.1038/srep28792. Retrieved from http://www.nature.com/articles/srep28792. Accessed on 16 Jan 2017
47. Ukpai OM, Ekedo CM (2018) Insecticide susceptibility status of culex quinquefasciatus [Diptera: Culicidae] in Umudike, Ikwuano LGA Abia State, Nigeria. Int J Mosq Res 6(1):114-118
48. Maciel-de-Freitas R et al (2012) Why do we need alternative tools to control mosquito-borne diseases in Latin America? Mem Inst Oswaldo Cruz 107(6):828–829, Sep. Retrieved from http://www.scielo.br/scielo.php?script=sci_arttext&pid=S0074-02762012000600021. Accessed on 7 Aug 2018
49. Christodoulou M (2011) Biological vector control of mosquito-borne diseases. Lancet 11(2):84–5, 11 Feb. Retrieved from http://www.thelancet.com/journals/laninf/article/PIIS1473-3099(11)70017-2/abstract. Accessed on 10 Apr 2017
50. Esu E, Lenhart A, Smith L, Horstick O (2010) Effectiveness of peridomestic space spraying with insecticide on dengue transmission; systematic review. Trop Med Int Health 15(5):619–31
51. Adalja et al (2016) Genetically modified (GM) mosquito use to reduce mosquito transmitted disease in the US: a community opinion survey. PLOS, 25 May. Retrieved from http://currents.plos.org/outbreaks/article/genetically-modified-mosquito-use-to-reduce-mosquito-transmitted-disease-in-the-us-opinion-survey/. Accessed on 29 Sep 2016
52. da Silva Augusto LG et al (2016) Aedes aegypti control in Brazil. Lancet 387(10023):1052–1053, 12 Mar. Retrieved from https://www.ncbi.nlm.nih.gov/pubmed/26944024. Accessed on 7 Aug 2018
53. Sanchez R, Goldschmidt D, Scutti S (2016) Zika spraying in Miami: what you need to know. In: CNN.com, 9 Sep. Retrieved from http://edition.cnn.com/2016/09/08/health/florida-zika-spraying-concerns/. Accessed on 29 Sep 2016
54. Rose W (2016) Safe mosquito eradication that works: using coffee, Bti, rubbing alcohol, and a cat. In: InfoBarrel, 29 Oct. Retrieved from http://www.infobarrel.com/Safe_Mosquito_Eradication_That_Works__Using_Coffee_Bti_Rubbing_Alcohol_and_a_Cat. Accessed on 20 Jul 2017
55. Reyes J (2017) Vector control will conduct helicopter spraying in Indio due to aedes aegypti mosquitoes. In: KESQ, 6 Feb. Retrieved from http://www.kesq.com/news/vector-control-to-conduct-helicopter-spraying-in-indio-for-mosquitoes/315431616. Accessed on 30 May 2017
56. Arosteguí J et al (2017) Beyond efficacy in water containers: temephos and household entomological indices in six studies between 2005 and 2013 in Managua, Nicaragua. BMC Public Health 17(Suppl 1):434. Retrieved from https://www.ncbi.nlm.nih.gov/pmc/articles/PMC5506593/. Accessed on 6 Aug 2018

57. BEC Crew (2016) Report says pesticide is to blame for microcephaly outbreak - not zika. In: Science alert, 16 Feb. Retrieved from https://www.sciencealert.com/argentinian-report-says-monsanto-linked-pesticide-is-to-blame-for-microcephaly-outbreak-not-zika. Accessed on 5 Jun 2017
58. Maciel-de-Freitas R et al (2014) Undesirable consequences of insecticide resistance following aedes aegypti control activities due to a dengue outbreak. PLOS One, 27 Mar. https://doi.org/10.1371/journal.pone.0092424. Accessed on 10 May 2017
59. Raju AK (2003) Community mobilization in aedes aegypti control programme by source reduction in peri-urban district of Lautoka, Viti Levu, Fiji Islands. Dengue J 27:149–155. Retrieved from http://apps.who.int/iris/bitstream/handle/10665/163791/dbv27p149.pdf;jsessionid=FC58F0B1A6A1CE84C3FA8EA91A6D7A83?sequence=1. Accessed on 9 Aug 2018
60. Fox M (2016) Sprays, traps, and GM Bugs: a look at our tools to fight zika. In: NBC News, 9 Sep. Retrieved from http://www.nbcnews.com/storyline/zika-virus-outbreak/here-s-look-our-tools-fight-zika-n645766. Accessed on 25 Sep 2016
61. Kamaruddin N, Pentai L (2017) Kill aedes at the breeding stage. In: The Star Online, 12 Apr. Retrieved from http://www.thestar.com.my/opinion/letters/2017/04/12/kill-aedes-at-the-breeding-stage/. Accessed on 17 May 2017
62. Parker C, Bernaola L, Lee BW, Elmquist D et al (2019) Entomology in the 21st century: tackling insect invasions, promoting advancements in technology, and using effective science communication—2018 student debates. J Insect Sci 19(4):1–11. https://doi.org/10.1093/jisesa/iez069
63. ASTHO (2015) Before the swarm: guidelines for the emergency management of vector-borne disease outbreaks. Retrieved from http://www.astho.org/Programs/Environmental-Health/Natural-Environment/Before-the-Swarm/. Accessed on 28 Aug 2018
64. Reyes-Castro PA et al (2017) Outdoor spatial spraying against dengue: a false sense of security among inhabitants of Hermosillo, Mexico. PLOS: Negl Trop Dis 11(5):e0005611, 17 May. https://doi.org/10.1371/journal.pntd.0005611. Accessed on 11 Aug 2018
65. McGraw EA, O'Neill SL (2013) Beyond insecticides: new thinking on an ancient problem. Nat Rev Microbiol 11(3):181–193, Mar. Retrieved from https://www.ncbi.nlm.nih.gov/pubmed/23411863. Accessed on 9 Aug 2018
66. NALED insecticide fact sheet (2016) Np spray. Retrieved from http://nospray.org/naled-insecticide-fact-sheet/. Accessed on 15 May 2017
67. Vontas J, Kioulos E, Pavlidi N, Morou E et al (2012) Insecticide resistance in the major dengue vectors aedes albopictus and aedes aegypti. Pestic Biochem Physiol 104:126–131
68. Amraoul F et al (2016) Culex mosquitoes are experimentally unable to transmit zika virus. Euro Surveill 21(35):30333, 1 Sep. Retrieved from http://www.eurosurveillance.org/ViewArticle.aspx?ArticleId=22573. Accessed on 7 Feb 2017
69. Adelman ZN, Tu Z (2016) Control of mosquito-borne infectious diseases: sex and gene drive. Trends Parasitol 32(3):219–229, Mar. Retrieved from https://doi.org/10.1016/j.pt.2015.12.003. Accessed on 7 Apr 2017
70. CDC (2016) Insecticide resistance. In: CDC, 15 Nov. Retrieved from https://www.cdc.gov/zika/vector/insecticide-resistance.html. Accessed on 3 Apr 2017
71. Reduas.com (2016) Report from physicians in the crop sprayed town regarding dengue-zika microcephaly and massive spraying with chemical poisons. Retrieved from http://reduas.com.ar/report-from-physicians-in-the-crop-sprayed-town-regarding-dengue-zika-microcephaly-and-massive-spraying-with-chemical-poisons/. Accessed on 22 Sep 2016
72. Wilcox C (2015) Journalists vector GM fears as FDA considers Oxitec's Keys mosquito plan. In: Discovery, 27 Jan. Retrieved from http://blogs.discovermagazine.com/science-sushi/2015/01/27/journalists-vector-gm-fears-fda-considers-oxitecs-keys-mosquito-plan/#.WVkgJYjys2w. Accessed on 1 Jul 2017
73. Troncoso A (2016) Zika threatens to become a huge worldwide pandemic. Asian Pac J Trop Biomed 6(6):520–527. Retrieved from http://www.sciencedirect.com/science/article/pii/S2221169116302921. Accessed on 4 Jun 2017

74. Martin N (2017) Genetically modified mosquitoes could kill their own kind, cut West Nile, zika risks. In: Dallas News, 10 Apr. Retrieved from https://www.dallasnews.com/news/dallas-county/2017/04/10/genetically-modified-mosquitoes-coming-houston-dallas-county-officials-hope-next. Accessed on 21 May 2017
75. Weaver SC et al (2016) Zika virus: history, emergence, biology, and prospects for control. Antiviral Res 130:69–80. Retrieved from https://www.ncbi.nlm.nih.gov/pubmed/26996139. Accessed on 30 Jun 2017
76. PAHO (2016) Zika virus infection step-by-step guide to risk communication and community engagement. Retrieved from http://iris.paho.org/xmlui/handle/123456789/18599 (Nov). Accessed on 28 May 2017
77. Iannelli J (2016) Miami Beach asks FDA for emergency permission to release anti-zika GMO mosquitoes. In: Miami New Times, 1 Nov. Retrieved from http://www.miaminewtimes.com/news/miami-beach-asks-fda-for-emergency-permission-to-release-anti-zika-gmo-mosquitoes-8893201. Accessed on 15 May 2017
78. Eisen L et al (2009) Proactive vector control strategies and improved monitoring and evaluation practices for dengue prevention. J Med Entomol 46(6):1245–1255, Nov. Retrieved from https://www.ncbi.nlm.nih.gov/pubmed/19960667. Accessed on 7 Aug 2018
79. CCE & CERI (2002) The health effects of pesticides used for mosquito control (Aug). Farmingdale, NY: Citizens Campaign for the Environment and Citizens Environmental Research Institute
80. d'Albissin A, Girard A (2016) Oxitec: lord of the mosquitoes. In: The Blue Paper: Key West, 19 Jun. Retrieved from http://thebluepaper.com/lord-of-the-mosquitoes/. Accessed on 5 Sep 2016
81. Alvarez L (2016) In Florida Keys, some worry 'science and government' more than zika. In: New York Times, 29 Aug. Retrieved from http://www.nytimes.com/2016/08/25/us/zika-florida-keys-mosquitoes.html?_r=0. Accessed on 18 Sep 2016
82. Steenhuysen J (2016) Adult mosquitoes can pass zika to their offspring: U.S. study. In: Reuters, 29 Aug. Retrieved from http://www.reuters.com/article/us-health-zika-mosquitoes-idUSKCN1142CD. Accessed on 5 Sep 2016
83. ABC News (2017) Beyond pesticides daily news blog. In: Beyond Pesticides, 1 May. Retrieved from http://beyondpesticides.org/dailynewsblog/2017/05/wolbachiamosquitoesreleasednewfightzikamosquitobornediseases/. Accessed on 26 Jun 2017
84. Coto D (2016) Puerto Rico rejects insecticide to fight zika amid protests. In: AP NEWS, 22 Jul. Retrieved from https://apnews.com/3e93dacd616f4b1087eed9d5103e4f8f. Accessed on 24 May 2022
85. Schlanger Z (2017) Swarmed with mosquitoes after Harvey, Texas calls in the US Air Force. In: Quartz, 13 Sep. Retrieved from https://qz.com/1075935/swarmed-with-mosquitoes-after-harvey-texas-calls-in-the-us-air-force/. Accessed on 10 Jul 2018
86. EPA (2006) Reregistration eligibility decision for naled. Retrieved from https://archive.epa.gov/pesticides/reregistration/web/html/status.html. Accessed on 9 Jun 2017
87. Mercola J (2016) Zika: Brazil admits it's not the virus. In: Mercola.com, 16 Aug. Retrieved from http://articles.mercola.com/sites/articles/archive/2016/08/16/birth-defects-brazil-not-zika-virus.aspx. Accessed on 23 May 2017
88. Alliot D (2016) Should GM mosquitoes be used to fight zika. In: Newsmax, 16 Sep. Retrieved from http://www.newsmax.com/Health-News/zika-Aedes-aegypti-oxitec-naled/2016/09/16/id/748717/. Accessed on 18 Sep 2016
89. Staletovich J (2017) Study links pesticide used to fight zika in Florida to health impacts in Chinese babies. In: Miami Herald, 8 Jun. Retrieved from http://www.miamiherald.com/news/local/environment/article154960599.html. Accessed on 27 Jun 2017
90. Bloomquist L (2016) Pesticides sprayed to prevent zika may cause neural defects. In: Hormones Matter, 7 Sep. Retrieved from https://www.hormonesmatter.com/zika-naled-microcephaly-dangers/. Accessed on 15 May 2017
91. Dichlorvos (2000) Health effects notebook. Retrieved from https://www.epa.gov/haps/health-effects-notebook-hazardous-air-pollutants. Accessed on 29 Sep 2016

92. Barry-Jester AM (2016). Small island, big experiment: how a tiny Florida community could influence the way we fight zika around the world. In: FiveThirtyEight. Retrieved from https://fivethirtyeight.com/features/zika-mosquito-florida-vote/. Accessed on 17 Jan 2017
93. Brown KV (2016) Genetically modified mosquitoes could wipe out the world's most deadly viruses. If we let them. In: Fusion.net, 19 Sep. Retrieved from http://fusion.net/story/347298/oxitec-genetically-modified-mosquitoes/. Accessed on 25 Sep 2016
94. Beck. J. (2016). Mosquitoes can pass zika to their offspring. In: The Atlantic, 30 Aug. Retrieved from https://www.theatlantic.com/health/archive/2016/08/mosquitoes-can-pass-zika-to-their-offspring/497960/. Accessed on 3 Feb 2017
95. Bukspan D (2017) The race is on to stop a zika virus epidemic in the US. In: CNBC, 22 Apr. Retrieved from http://www.cnbc.com/2017/04/11/the-race-is-on-to-stop-a-zika-virus-epidemic-in-the-us.html. Accessed on 5 May 2017
96. Stone J (2016) Zika is on the ballot in florida and science-based decision-making. In: Forbes, 8 Nov. Retrieved from https://www.forbes.com/sites/judystone/2016/11/08/zika-and-science-based-decisions-are-on-the-ballot-in-florida/#54033543cf7e. Accessed on 2 Jun 2017
97. Alphey, Luke. (2014). Genetic Control of Mosquitoes. Annu Rev Entomol 59:205–224. Retrieved from http://www.oxitec.com/genetic-control-of-mosquitoes-luke-alphey/. Accessed on 7 Apr 2017
98. Macer DRJ (2013) UNDP/World Bank/WHO. Ethical, legal, and social issues of genetically modified disease vectors in public health. Geneva, Switzerland: WHO. Retrieved from http://www.who.int/tdr/publications/tdr-research-publications/seb_topic1/en/. Accessed on 8 Aug 2018
99. Brown KV (2017) The EPA just approved lab-grown mosquitoes to fight disease. In: GIZMODO. Retrieved from https://gizmodo.com/the-epa-just-approved-lab-grown-mosquitoes-to-fight-dis-1820231191. Accessed on 23 Jul 2018
100. WHO (2016) Dispelling rumours around zika and complications. In: WHO. Retrieved from http://www.who.int/emergencies/zika-virus/articles/rumours/en/. Accessed on 24 Sep 2016
101. Atkins K (2018) How best to kill skeeters? Experts pitch their plans. In: Florida Keys News, 23 Feb. Retrieved from http://www.flkeysnews.com/news/local/article201838619.html. Accessed on 18 Jul 2018
102. CNA (2016) The nuclear connection to combatting the zika virus. In: CNA website, July. Retrieved from https://cna.ca/news/nuclear-connection-combating-zika-virus-3/. Accessed on 10 Apr 2017
103. Brown KV (2016) If genetically modified mosquitoes freaked you out, you won't like what's coming. In: GIZMODO, 13 Dec. Retrieved from http://gizmodo.com/if-genetically-modified-mosquitoes-freaked-you-out-you-1790021060. Accessed on 15 Mar 2017
104. Paquette CCH et al (2015). Biodistribution and trafficking of hydrogel nanoparticles in adult mosquitoes. PLOS Negl Trop Dis 9(5):e0003745, 21 May. https://doi.org/10.1371/journal.pntd.0003745. Accessed on 19 Jul 2017
105. Berube D (2006) Nano-hype: the truth behind the nanotechnology buzz. Prometheus Books, Amherst, NY
106. Stone J (2017) How Brazil's zika epidemic women's everyday plight: watch. In: Forbes, 15 Jul. Retrieved from https://www.forbes.com/forbes/welcome/?toURL=https://www.forbes.com/sites/judystone/2017/07/15/how-brazils-zika-epidemic-highlights-womens-every-day-plight-human-rights-watch/&refURL=https://www.google.com/&referrer=https://www.google.com/. Accessed on 21 Jul 2017
107. Paquette CH et al (2015) Biodistribution and trafficking of hydrogel nanoparticles in adult mosquitoes. PLOS Negl Trop Dis 9(5):e0003745, 15 May. Retrieved from https://doi.org/10.1371/journal.pntd.0003745. Accessed on 21 Jul 2017
108. Yu Na et al (2013) Delivery of dsRNA for RNAi in insects: an overview and future directions. Insect Sci 20(1):4–14. https://doi.org/10.1111/j.1744-7917.2012.01534.x/abstract. Accessed on 21 Jul 2017
109. Jinu U, Rajakumaran S et al (2018) Potential larvicidal activity of silver nanohybrids synthesized using leaf extracts of cleistanthus collinus (Roxb.) Benth. ex Hook.f. and Strychnos

nux-vomica L. nux-vomica against dengue, chikungunya and zika vectors. Physiol Mol Plant Pathol 101:163–171
110. Zhu KY (2013) RNA interference: a powerful tool in entomological research and a novel approach for insect pest management. Insect Sci 20(1):1–119. https://doi.org/10.1111/1744-7917.12006/abstract. Accessed on 21 Jul 2017
111. Povelones M, Christophides GK (2013) Meeting report of the mosquito kolymbari Meeting 2013. Pathog Glob Health 107(8):393–399. Retrieved from https://www.ncbi.nlm.nih.gov/pmc/articles/PMC4073524/. Accessed on 17 Jul 2017
112. Fox M, Edwards E (2016) Zika virus is coming and we're not ready, U.S. experts say. In: NBC News, 3 May. Retrieved from http://www.nbcnews.com/storyline/zika-virus-outbreak/zika-virus-coming-we-re-not-ready-u-s-experts-n567261. Accessed on 14 Apr 2017
113. Miroux N et al (2012) Changes in anopheles funestus biting behavior following universal coverage of long-lasting insecticidal nets in Benin. J Infect Dis 206(10):1622–1629, 15 Nov. Retrieved from https://www.ncbi.nlm.nih.gov/pubmed/22966127. Accessed on 9 Aug 2018
114. McGraw EA, O'Neill SL (2013) Beyond insecticides: new thinking on an ancient problem. Nat Rev Microbiol 11(3):181–193, Mar. Retrieved from http://www.nature.com/nrmicro/journal/v11/n3/full/nrmicro2968.html. Accessed on 22 May 2017; Russell TL et al (2011). Increased proportions of outdoor feeding among residual malaria vector populations following increased use of insecticide-treated nets in rural Tanzania. Malar J 10:80, 9 Apr. https://doi.org/10.1186/1475-2875-10-80. Accessed on 9 Aug 2018
115. Aziz H et al (2017) Zika virus: global health challenge, threat, and current situation. J Med Virol 89(6):943–951. Retrieved from https://doi.org/10.1002/jmv.24731/abstract. Accessed on 26 Jun 2017
116. Fauci AS, Morens DM (2016) Zika virus in the Americas – yet another arbovirus threat. N Engl J Med 374(7):602–606, 18 Feb. Retrieved from https://doi.org/10.1056/NEJMp1600297#t=article. Accessed on 4 Sep 2016
117. Che-Mnedoza A et al (2015) Long-lasting insecticide-treated house screens and targeted treatment of productive breeding-sites for dengue vector control in Acapulco, Mexico. Trans R Soc Trop Med Hyg 109(2):106–115, Feb. Retrieved from https://www.ncbi.nlm.nih.gov/pubmed/25604761. Accessed on 7 Aug 2018
118. Bartlett-Healy K et al (2011) Source reduction behavior as an independent measurement of the impact of a public health education campaign in an integrated vector management program for the Asian tiger mosquito. Int J Environ Res Public Health 8(5):1358–1367, May. Retrieved from https://www.ncbi.nlm.nih.gov/pubmed/21655124. Accessed on 27 Aug 2018
119. Silva GS et al (2018) Zika virus: report from the task force on tropical diseases by the world federation of societies of intensive and critical care medicine. J Crit Care 46:106–109, Aug. Retrieved from https://www.ncbi.nlm.nih.gov/pubmed/29779827. Accessed on 26 Jul 2018
120. Baza I (2016) Zika virus educational campaign starting in 2017. In: KUAM News, 14 Dec. Retrieved from http://www.kuam.com/story/34047132/2016/12/Wednesday/zika-virus-educational-campaign-starting-in-2017. Accessed on 3 Feb 2017
121. WHO (2016) Mosquito control: can it stop zika at source? Emergencies. In: WHO, 17 Feb. Retrieved from http://www.who.int/emergencies/zika-virus/articles/mosquito-control/en/. Accessed on 2 Oct 2016
122. Lafrance A (2016) The sneak-attack mosquito. In: The Atlantic, 25 Apr. Retrieved from https://www.theatlantic.com/science/archive/2016/04/aedes-aegypti/479619/. Accessed on 18 May 2017
123. Edwards SB (2016) The zika virus. Minneapolis, MN: ABDO Publishing
124. McNeill Jr. DG (2016) Predict zika's spread? It's hard enough to count the cases. In: New York Times, 19 Sep. Retrieved from https://www.nytimes.com/2016/09/20/health/zika-spread-predictions.html. Accessed on 15 May 2017
125. Abdel-Rahman A, Shetty AK, Abou-Donia MB (2001) Subchronic dermal application of N,N-Diethyl m-Toluamide (DEET) and permethrin to adult rats, alone or in combination, causes diffuse neuronal cell death and cytoskeletal abnormalities in the cerebral cortex and the hippocampus, and purkinje neuron loss in the cerebellum. Exp Neurol 172(1):153–171. Retrieved from https://www.ncbi.nlm.nih.gov/pubmed/11681848/. Accessed on 17 Jul 2017

126. Miglianico M et al (2018) Repurposing isoxazoline veterinary drugs for control of vector-borne human diseases. Proc Natl Acad Sci USA 115(29):E6920–E6926, 2 Jul. Retrieved from http://www.pnas.org/content/early/2018/06/26/1801338115. Accessed on 7 Aug 2018
127. Hamblin J (2016) Parents are being sold zika protection that doesn't work. In: The Atlantic, 3 Aug. Retrieved from https://www.theatlantic.com/health/archive/2016/08/zika-products-that-dont-work/494363/. Accessed on 14 May 2017
128. Scripps Research Institute (2018) A well-known animal health drug could stop outbreaks of malaria and Zika virus: veterinary flea and tick drugs kill mosquitoes and other insects that carry diseases between humans. In: ScienceDaily, 2 Jul. Retrieved from https://www.sciencedaily.com/releases/2018/07/180702154731.htm. Accessed on 25 Jul 2018
129. Barrera R, Amador M, Acevedo V, Hemme R, Felix G (2014) Sustained, area-wide control of aedes aegypti using CDC autocidal gravid ovitraps. Am J Trop Med Hyg 91(6):1269–1276
130. Reuters (2016) Here's how microsoft & other tech firms have waged a war against disease-carrying mosquitoes. In: ScoopWhoop News. Retrieved from https://www.scoopwhoop.com/battle-against-zika-virus-google-and-microsoft-have-come-together-to-fight-it-off/#.xdkc5md04. Accessed on 21 Jul 2017
131. Wyss J (2017) How one city in Puerto Rico is fighting Zika — without using chemicals. In: Miami Herald, 3 Feb. Retrieved from http://www.miamiherald.com/news/nation-world/world/americas/article130514329.html. Accessed on 1 Jul 2017
132. McNeil Jr. DG (2017) Houston braces for another brush with the peril of zika. In: The New York Times, 17 Jul. Retrieved from https://www.nytimes.com/2017/07/17/health/zika-virus-houston-texas.html. Accessed on 21 Jul 2017
133. Barrera R et al (2014) Use of the CDC autocidal gravid ovitrap to control and prevent outbreaks of aedes aegypti (diptera: culicidae). J Med Entomol 51(1):145–154, 1 Dec. Retrieved from https://www.ncbi.nlm.nih.gov/pubmed/24605464. Accessed on 5 Jun 2017
134. Editorial (2016) Our recommendations in the Nov. 8 election. In: Florida Keys News, 3 Oct. Retrieved from http://www.flkeysnews.com/opinion/editorials/article105619031.html. Accessed on 4 Oct 2016
135. Lai L (2016) Dengue epidemic fears wiped out. In: The Straits Ties, 29 Dec. Retrieved from http://www.straitstimes.com/singapore/dengue-epidemic-fears-wiped-out. Accessed on 19 May 2017

Chapter 13
Wolbachia

This is the first of two chapters examining a specific approach to vector management, both associated with the *Ae. aegypti* mosquito and the ZIKV pandemic. Traditional and some mildly fantastic methods were discussed above. Now, it is time to look at the two approaches receiving the most technical and widespread media attention and driving much of the literature on vector management.

Vector management is essential due to the many other benefits, not only including the reduction of other viral infections coming from these mosquitoes but also potential reductions in the costs imposed on all levels of government and professional services provided to those who have become infected and long-term care for those who suffer from adverse pregnancies and some of the most debilitating effects on born offspring, such as microcephaly and developmental disorders. In addition, many of the vector management approaches undertaken in the past have involved the use of some questionably safe chemicals.

There are two models. One is called pay-for-service, and they are becoming popular. However, while such initiatives may provide service for those areas willing to pay the fees, the environmental justice implications of such a model raise concerns that low-income areas would not receive the same treatment as wealthier communities. Furthermore, pay-for-service models do not consider the short-lived nature of adulticide application or that adult mosquitoes may migrate in from untreated, nonpaying areas [1]. The other approach involves the species' long-term extinction, hoping another reservoir for the virus does not fill the space after the species-cide. If so, then this approach is not recurring. Once the species is debilitated, there is little reason to waste it again. Finally, both models involve a lot of stakeholders from localities, counties, or other territorial subdivisions that provide vector control as a public service.

In addition, it just happens that traditional methods do not healthily control the mosquito in question. "ZIKV and other diseases spread by [the *Ae. aegypti* mosquitoes] are not controllable with current technologies," said Thomas Frieden of the CDC. "We will see this become endemic in the hemisphere" [2] without some new way to control these vectors.

D. M. Berube, *Pandemics and Resilience: Lessons we should have learned from Zika*, Risk, Systems and Decisions, https://doi.org/10.1007/978-3-031-25370-6_13

Enter Wolbachia. The first thing to realize in studying this approach is there are many different strains of Wolbachia. The many strains of Wolbachia can do some amazing things to their insect hosts: changing their sex, killing their offspring, and possibly even creating new species [3]. Only three Wolbachia strains from the many that exist in nature have been introduced and characterized in the *Ae. aegypti* mosquito [4] debate.

Some recent reports have shown that exposure of larvae to higher rearing temperatures can significantly affect the density of Wolbachia in adults, leading to reduced rates of maternal transmission. As *Ae. aegypti* larvae are often found in bodies of water experiencing daytime heating, temperature susceptibility could potentially limit the invasive capacity of a Wolbachia strain and its ability to inhibit virus transmission [4].

As such, some strains are more effective than others in their ability to exterminate or substantially curtail the population of a specific mosquito species population. For example, a team from the University of Glasgow's MRC Centre for Virus Research learned some Wolbachia strains might not be effective at blocking the transmission, or they are not inherited efficiently by the mosquitoes in tropical countries.

Ant et al. [4] undertook the study [5].

> To compare the effectiveness of two new Wolbachia strains, wAu and wAlbA, the researchers introduced them, along with two previously tested ones, wMel and wAlbB, to *Ae. aegypti* mosquitoes. After feeding on dengue or ZIKV-infected blood, the mosquitoes infected with wAu had lower levels of viral RNA—which carries genetic information—in their bodies than the mosquitoes infected with the other strains. wAu also showed high rates of inheritance, including under high-temperature conditions. [6]

Some early research from Walker et al. [7] established the working premise. The wMelPop-CLA Wolbachia strain reduces the lifespan of adult *Ae. aegypti* mosquitoes in stably transinfected lines; several Wolbachia infections, including wMelPop-CLA, can also directly influence the susceptibility of insects to infection with a range of insect and human pathogens, and induce the reproductive phenotype cytoplasmic incompatibility with minimal apparent fitness costs and high maternal transmission, providing optimal phenotypic effects for invasion, and block transmission of dengue serotype 2 (DENV-2) in *Ae. aegypti*, forming the basis of a practical approach to dengue suppression [6].

The strategy remains challenging [4]. However, wMelPop imposes high costs on various traits, including reduced longevity, fecundity, and egg survival in quiescence. These adverse fitness effects have made the introduction of wMelPop into wild host populations problematic; despite unidirectional solid cytoplasmic incompatibility, recent field trials in Vietnam and Australia failed to achieve population replacement using this strain [7].

Another strain has proved more effective. Scientists from the University of Glasgow's MRC Centre for Virus Research (CVR) concluded that the "wAu" strain is more effective at blocking virus transmission than those currently being used". Professor Steven Sinkins said: "The Wolbachia transmission blocking strategy shows great promise for controlling mosquito-borne viruses and is now being deployed on a

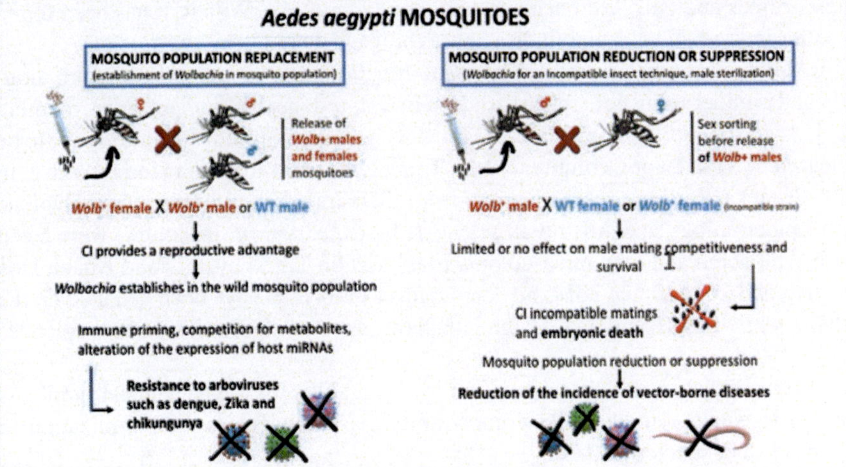

Fig. 3 Exploiting *Wolbachia* for vector control. Left panel: mosquito population replacement approach, in which *Wolbachia*-infected female and male mosquitoes are released; through CI, this strategy allows the spread of *Wolbachia* in the natural population. The presence of *Wolbachia* provides a fitness advantages (determined by CI) and can reduce the arbovirus transmission. Right panel: mosquito population reduction or suppression strategy. This approach involves the release of *Wolbachia*-infected males into an area; when these mosquitoes mate with wild *Wolbachia*-negative females (or female mosquitoes harbouring an incompatible strain of *Wolbachia*), a strong reduction in the rate of egg hatching is observed (CI incompatible matings). Thus, repeated releases of *Wolbachia*-infected males result in reduction or suppression of mosquito populations. *CI* cytoplasmic incompatibility, *WT* wild-type mosquitoes, *Wolb+ Wolbachia*-infected mosquitoes

Fig. 13.1 Wolbachia approach Manoj et al. (2021). Wolbachia: endosymbiont of onchocercid nematodes and their vectors. Parasites & Vectors. 14:245

large scale in several tropical countries" [4]. "Our results with the wAu strain showed the effective transmission blocking for all the viruses we tested, and it provides an exciting new option to explore for disease control programs." [4, 8]. The scientists say their research can be 100% effective when blocking the transmission (Fig. 13.1).

A simulation model by Ruang-areerate and Kittayapong [9] confirmed population replacement by transinfected *Ae. aegypti* is possible over time. The establishment of Wolbachia double infections in *Ae. aegypti* by direct adult microinjection and the demonstration of CI expression in this new host suggest that Wolbachia could be experimentally transferred into vector species and used as a gene-driving system to manipulate vector populations genetically [9]. Studies showed that the parasite could spread to the point of "fixation" over 10 to 20 weeks and remain in the population for at least five years [10].

In addition, in this case, the virulence of Wolbachia is believed to be due to an inability of this strain, *Ae. aegypti*, to control its replication in the adult nervous and muscle tissues resulting in an approximate halving of the adult lifespan of the fly [11]. This makes the female mosquitoes less viable as disease vectors.

The endosymbiotic bacterium Wolbachia is present in an estimated 20–40% of all known terrestrial insect species [12]. While lower than the previous estimate (66% high), it still points to a surprisingly high number of arthropods harboring the bacteria [13]. Many bacteria find homes on the surface of organisms and even more inside their orifices and guts. Wolbachia is a bit more aggressive. With it, it has been found to enter individual cells and finds a home for itself there [14].

wMel is currently the favored infection for Wolbachia interventions internationally and is undergoing field release trials in Brazil, Indonesia, Vietnam, and Colombia [15]. According to Vietnam's Health Ministry, Wolbachia-laced mosquitos were to be released in Vinh Luong commune, Nha Trang City, central Khanh Hoa Province, in 12–18 weeks. Earlier, a research group from Vietnam's National Institute of Hygiene and Epidemiology and Australian scientists bred *Ae. aegypti* mosquitos were laced with Wolbachia and then released on a trial basis on Tri Nguyen Island Khanh Hoa in 2013 and 2014. Since 2014, no dengue fever outbreaks have been detected on the island, while Nha Trang and Khanh Hoa have reported significant disease episodes [16].

Wolbachia, a maternally transmitted microorganism of the Rickettsial family, is known to cause cytoplasmic incompatibility, parthenogenesis, or feminization in various insect species [17].

Shorten life spans:

One Wolbachia-induced trait is the shortening of the adult insect lifespan; this trait is uniquely associated with a particular Wolbachia strain, wMelPop. Shortening the lifespan of mosquito vectors could theoretically reduce the transmission of several viruses and parasites because of the importance of the extrinsic incubation period (EIP) to disease dynamics [18].

After a female mosquito ingests an infectious blood meal, parasites, or arboviruses, such as dengue, penetrate the mosquito's midgut and replicate in various tissues before infecting the salivary glands, where they are transmitted to a new host during subsequent blood-feeding. This period from pathogen ingestion to potential infectivity is termed the extrinsic incubation period (EIP) and lasts ~ 2 weeks for both dengue and malaria [19].

Extrinsic incubation periods (EIPs) can vary from 7 to 30 days, depending on the disease transmission system. As such, most disease transmission is done by older arthropods. Wolbachia infections that shorten adult lifespan can potentially reduce disease transmission by preferentially removing that segment of the vector population responsible for most disease transmission [20].

Wolbachia infection reduces the lifespan of adult female mosquitoes by half. Mathematical modeling of vectorial capacity shows that even small shifts in average vector lifespan can significantly impact the transmission dynamics of a disease [20]. It has been proposed that life-shortening Wolbachia strains, such as wMelPop, might be used to skew mosquito population age structure toward younger individuals, thereby reducing pathogen transmission without eradicating the mosquito population [19].

Introducing the life-shortening Wolbachia strain into a population of insect disease vectors, like *Ae. aegypti*, could reduce disease transmission [19]. Dutra et al.'s results indicate that Wolbachia infection can significantly reduce mosquitoes' capacity to the harbor and transmit various medically necessary pathogens, including the dengue and chikungunya viruses, which extend to ZIKV [21].

Cytoplasmic incompatibility (CI): A term regularly seen in these debates is cytoplasmic incompatibility, a phenomenon that results in sperm and eggs being unable to form viable offspring.

CI, the most common manipulation in insects, occurs when Wolbachia-infected male insects mate with Wolbachia-free female insects and produce non-viable offspring [22].

When these male mosquitoes mate with wild females who do not carry the same strain of Wolbachia, the resulting fertilized eggs do not hatch because the paternal chromosomes do not form properly [23].

Wolbachia modifies the sperm of infected males, which results in the generation of non-viable progeny when mated to uninfected females. Infected females, in contrast, "rescue" this sperm modification, producing viable progeny and resulting in a relative fitness advantage that can drive and maintain Wolbachia at high population infection frequencies [4]. Put simply, sperm and eggs are unable to form viable offspring.

Advocates are using sterility for the Wolbachia bacteria approach. Sterility can also be created by artificial infection with various strains of Wolbachia, a diverse group of intracellular bacteria. The disease affects insect sperm to prevent successful reproduction between infected males and uninfected females [18]. The key seems to be to coordinate the vector strategy with the incubation period of the species.

When these male mosquitoes mate with wild females who do not carry the same strain of Wolbachia, the resulting fertilized eggs do not hatch because the paternal chromosomes do not form properly [23].

The bacteria are maternally transmitted, like mitochondria, passed from the mother to her offspring. Infected males are useless to the maternally inherited Wolbachia for propagation; they produce modified sperm that produce viable zygotes only with eggs from infected females. Wolbachia, in effect, conducts a sterile male attack on uninfected females [24].

Inhibit replication of RNA viruses. Wolbachia works by preventing RNA arboviruses, like dengue, Chikungunya, and ZIKV, from replicating in the mosquito [25]. The Wolbachia "immunizes" the mosquito against the viruses they eat with their blood meal. Furthermore, Wolbachia infections protect mosquitoes against RNA viruses by competing for fatty acids needed for viral replication [26].

These bacteria are of particular interest in controlling arboviral diseases as they are known to inhibit the replication of RNA viruses in insects. Infections of Wolbachia from *Drosophila melanogaster* and *Ae. albopictus* were recently introduced experimentally into *Ae. aegypti* and were found to suppress the transmission of dengue, ZIKV, chikungunya, yellow fever, and West Nile viruses [15].

Parthenogenesis. The bacteria can cause parthenogenesis in the wasp: Unfertilized eggs become diploid females instead of the usual haploid males—the bacteria feminized genetic males into functional females in the woodlouse, Armadillidium Vulgare [17].

Wolbachia gets rid of the males entirely. Infected females produce unfertilized eggs that do one of two things: copy their chromosomes without dividing or undergo a normal cell division followed by the fusion of the two nuclei. In either case, the animal winds up with two sets of chromosomes and develops as a standard, if infected, female [14].

In some isopods, the bacteria head for the organ that produces male sex hormones and destroys it, ensuring the embryo develops as a female. In insects, Wolbachia seems to manipulate the sex determination pathway more directly. It is not clear what the bacteria do. Still, if the insects are given antibiotics partway through development, females will develop normally, but males that started out developing as females will be awkward between the two sexes [14].

The genomes of several strains of Wolbachia have been sequenced, only to find that there isn't a lot unique about them. The genomes in one to two million base pairs tend to be minor but are not as small as some dedicated parasites. They also seem to have an unusual amount of junk compared to most bacteria, with about 15% of the genome repeated elements like transposons and viruses. They also have a high number of duplicated genes compared to other bacteria. All these would suggest that the genome is in a regular state of flux and has plenty of raw materials to evolve new functions [14].

Wolbachia-infected mosquitoes have one clear advantage over genetically modified ones. The genetic tweaks created in the lab do not remain in the wild because they cause the modified mosquitoes' offspring to die. While that assuages fears about genetic modifications spreading into nature, any country using GM mosquitoes must keep buying fresh supplies from Oxitec to keep the *Ae. aegypti* population under control. By contrast, Wolbachia-infected mosquitoes keep passing the parasite on to the next generation [27].

Unfortunately, none of these technologies has demonstrated evidence that they can be implemented at a scale beyond small pilots. Regardless, Oxitec's mosquitoes are on a suicide mission. Wolbachia mosquitoes are missionaries designed not to wipe out wild mosquito populations but to transform them [10].

13.1 Two Companies

Wolbachia is a gram-negative obligate intracellular bacterium. It is a non-toxic bacterium found throughout the world. No mosquitoes are killed in this approach. Instead, the males are primarily sterile; hence, it becomes impossible to sustain a viable population. The Wolbachia strains being used have been successfully transferred between insect species. It has been shown to induce strong cytoplasmic incompatibility (CI) and shorten adult lifespan in its new host [18].

There have been two leading companies in the world of bacteria-based mosquito control.

13.1.1 Eliminate Dengue (World Mosquito Program)

The **Eliminate Dengue** approach involves *reducing the transmissibility of the virus among mosquitoes*. This team of researchers was interested in their well-stated goal when the ZIKV became prominent in 2015 and 2016. They have a head start on some of the other approaches that started only when ZIKV went pandemic. They are taking their approach and targeting ZIKV in over one hundred countries [28].

Eliminate Dengue has run trials in Colombia, Brazil, Vietnam, Indonesia, and Australia over the past five years [29]. Researchers in Australia use Wolbachia to reduce virus replication inside the mosquito. The **Eliminate Dengue Program**, a not-for-profit international collaboration led by Australia's Monash University, uses the bacteria Wolbachia to stop disease-causing viruses from replicating inside the mosquito and being transmitted between people [30].

Associated studies have shown that the dengue virus transmission among *Ae. aegypti* mosquitoes has been reduced by 66 to 75% when mosquitoes are infected with Wolbachia [31]. Ferguson et al. argue that wMelPop infection confers numerous phenotypic traits on *Ae. aegypti*, including refractoriness to DENV infection, reduced life span, reduced viability of desiccated eggs, and reduced blood-feeding success [31].

Researchers from Sun Yat-sen University in Guangzhou, China, and Michigan State University in East Lansing began field trials of Wolbachia-infected mosquitoes last year on Shazai Island in Guangzhou. In March 2016, the tests expanded to Dadao Island, also in Guangzhou. The researchers are releasing 1.5 million male Ae. albopictus per week, with plans to increase that to five million per week by August [23].

An extensive risk analysis by the Australian Commonwealth Scientific and Industrial Research Organization considered the forthcoming release as having "negligible risk," adds O'Neill. After its researchers completed additional risk analysis, the project finally received regulatory approval from the Australian Pesticides and Veterinary Medicines Authority [26].

This program is supported by the Bill and Melinda Gates Foundation and the biomedical research charity Wellcome Trust. In October, Wellcome Trust committed to investing $18 million if the **Eliminate Dengue Project** can prove effective in Latin America [30].

Partnering with the United Nations' Food and Agriculture Organization, the Vienna-based International Atomic Energy Agency (IAEA), and Sun Yat-Sen University and Guangzhou Wolbaki Biotech, both in China, **Eliminate Dengue Project** conducted the 2-year trial on two islands in Guangzhou's Pearl River.

The investigators, led by Zhiyong Xi, Ph.D., a microbiology and molecular genetics professor at Michigan State University, mass-reared male mosquitoes in a controlled environment and irradiated them to cause sterilization. In this trial, the investigators nearly eliminated the local population of the Asian tiger mosquito (*Aedes* albopictus), a prolific disease spreader. After subjecting the mosquitoes to the sterile insect technique (SIT), the team applied the incompatible insect technique (IIT).

One of the benefits of using this combination SIT/IIT technique is that Wolbachia on its own partly sterilizes male mosquitoes, meaning they can receive a lower dose of radiation during SIT than would be needed if only SIT were being used. This lower dose of radiation keeps the males attractive enough for females to mate with, which is not always the case when a total dose of radiation is used. And using IIT on its carries the risk that stray females in the mass-reared population would be infected with Wolbachia along with the males which-because the genders would no longer be cytoplasmically incompatible in offspring should they mate [23].

According to Zheng et al. [32], combining incompatible and sterile insect techniques (IIT–SIT) enabled the near elimination of field populations of the world's most invasive mosquito species, *Aedes* albopictus, in the Chinese field trial. The combined IIT–SIT approach is environmentally friendly and cost-effective, enabling vector control in complex and inaccessible urban habitats in which implementation of standard vector control is difficult 27, 28, as released males actively seek wild females and allows the release of much higher numbers of male mosquitoes in comparison to IIT alone, while simultaneously protecting against accidental female release. Area-wide application of this approach will necessitate the development of novel technologies, especially in scaling-up production capacity and enabling the efficient mass release of mosquitoes [32].

This report, however, is not without its detractors. According to Moretti and Calvitti [33], in promoting the combination technique as the most promising of the available techniques, Zheng et al. [32] stated that lines of *Ae. albopictus* mosquitoes that have been "cured of their native double wAlbA/wAlbB infection" and transinfected with the wPip Wolbachia strain were "either unsuitable for IIT—as they are pathogenic or do not inhibit arboviruses—or their appropriateness has not been fully determined." This statement does not reflect the current state of research in the field of IIT.

They wrote: "Overall, we express doubts about the value of combining incompatible and sterile insect techniques if the irradiation dose typical of SIT alone is needed and if the mating competitiveness of the released males, based on the authors' estimate, is almost halved compared to wild-types1. Furthermore, despite the high irradiation dose, some fertile HC eggs were collected in the field after the releases1, demonstrating that irradiation does not entirely mitigate the risk of successfully reproducing released females." [33].

13.1.2 Mosquito Mate

The top player in the U.S. bacterial approach to vector control seems to be **MosquitoMate**. Their approach is different from **Eliminate Dengue**, involving using Wolbachia *to reduce the number of mosquitoes by releasing sterile males.*

In late August 2016, the US Environmental Protection Agency approved **MosquitoMate**'s bacteriological approach involving Wolbachia and *Aedes albopictus*. Since this is a biological solution, the FDA is not involved. Nonetheless, the process can be demanding.

In 2017, the EPA approved **MosquitoMate**'s lab-reared mosquitoes infected with the bacterium Wolbachia pipientis to be released into the wilds of 20 states and Washington, DC. [34]. The approval lasts five years and is approved for Washington, D.C., California, Connecticut, Delaware, Illinois, Indiana, Kentucky, Massachusetts, Maine, Maryland, Missouri, New Hampshire, New Jersey, Nevada, New York, Ohio, Pennsylvania, Rhode Island, Tennessee, Vermont, and West Virginia [35].

It is not just people that **MosquitoMate**'s founder, Stephen Dobson, must convince. The EPA has never dealt with a "pesticide" like this before. "We kind of blew their minds," says Dobson, who has been working with the agency on the application since 2008. "This has been a lengthy process, a lot of emails, a lot of flights, and many stacks of paper going back and forth." [36].

MosquitoMate Inc. collaborating with researchers at the University of Kentucky claims that if female *Aedes* mate with a male with Wolbachia, her eggs will not hatch. They have recently received approval and have begun trials in Clovis, California. The plan in Clovis was to release 400,000 male *Ae. aegypti* mosquitoes over ten weeks and 120 acres of the suburban landscape [36].

Verily, the life sciences arm of Google umbrella company Alphabet is bugging out. To the tune of twenty million machine-reared, bacteria-infected mosquitoes that it is about to release into Fresno. The "Debug Project" is a collaborative venture with MosquitoMate and Fresno County's Consolidated Mosquito Abatement District [37].

Verily says it has built a robot that can raise a million mosquitoes a week and has used it to produce infertile male insects. The company has started releasing the first batches of twenty million sterilized mosquitoes in Fresno County, California. Verily's high-tech approach will allow for the release of one million mosquitoes a week, 25 times more than the Kentucky company was able to do before [38]. This has been made possible by Verily's automated mass-rearing and sex-sorting processes. Additionally, software algorithms and on-the-ground release devices will allow the distribution of the sterile male mosquitoes evenly throughout Fresno's mosquito season. It is believed these advancements could have a meaningful impact on what is traditionally a very labor-intensive process and could reduce the number of biting *Ae. aegypti* in Fresno County [39].

For five years, the EPA has given commercial approval for **MosquitoMates**' *Aedes* albopictus ZAP mosquitoes. Dobson says it also has an experimental permit for the company's *Ae. aegypti* mosquitoes. The company plans to sell the ZAP mosquitoes commercially to individuals or businesses like hotels in spring 2018 when mosquito season starts to kick off [39].

"This study will be the largest U.S. release of sterile male mosquitoes treated with Wolbachia, a naturally occurring bacterium, and will take place over 20 weeks in two neighborhoods, each approximately 300 acres in size," wrote Verily in a blog post [39]. Crawford et al. [39] reported "at peak mosquito season, the number of female mosquitoes was 95.5% lower (95% CI, 93.6–96.9) in release areas compared to non-release areas, with the most geographically isolated neighborhood reaching a 99% reduction." Bouyer et al. (2022) were less impressed and demanded that a competitiveness index be applied to the data. Crawford et al. (2022) responded "measuring impact on adult Ae. aegypti female density is the most direct measure of efficacy and impact, and we focused our efforts accordingly." [39].

Verily's separation technology may give MosquitoMate a leg up on the competition. At MosquitoMate's labs in Lexington, immature mosquitoes are forced through a sieve-like mechanism that separates the smaller males from the females. These mosquitoes are then hand sorted to weed out stray females that slip through. "That's done using eyeballs," said Stephen Dobson, MosquitoMate's chief executive. Enter Verily. The company is automating mosquito sorting with robots to make it faster and more affordable. Company officials declined to be interviewed. But on its website, Verily says it combines sensors, algorithms, and "novel engineering" to speed the process [40].

Unlike Oxitec's approach, they injected an egg and raised generations of mosquitoes that carried Wolbachia.

13.2 Claims

Two approaches, two companies, and two sets of claims. What follows has been carefully located such that the claims being made are associated with the approaches taken and the companies involved.

13.2.1 World Mosquito Program (Formerly Eliminate Dengue)

When the infected males are released into the wild, they mate with females, making them sterile. As part of a one-two punch, the researchers also will seed Wolbachia-infected females to establish a viral-resistant population. Once the resistant population grows to a specific density, males carrying a second Wolbachia strain are released to suppress the people further. By reducing mosquito density and mosquitoes' ability to transmit viruses, the project, when fully deployed, expects to arrest disease transmission immediately, Xi Zhiyong, director of Sun Yat-sen University-Michigan State University Joint Center of Vector Control for Tropical Diseases, said [41].

The **World Mosquito Program**'s research has shown that when introduced into the *Aedes* aegypti mosquito, Wolbachia can help reduce the transmission of these viruses to people. This vital discovery can transform the fight against life-threatening mosquito-borne diseases. On October 13, 2018, the **World Mosquito Program**, with approval from Fiji's Ministry of Health and Medical Services, released Wolbachia-carrying mosquitoes in Suva [42]. Dip Chand, its chief health inspector, reported the project had been fully funded by the Australian and New Zealand governments and would involve Fiji, Kiribati, and Vanuatu [42].

Eliminate Dengue says the cost of eliminating the viruses could be as low as $1 a person after a campaign releasing about a million mosquitoes in a year over several kilometers [43].

13.2.2 Mosquitomate

Dutra et al. report that Wolbachia-carrying mosquitoes are highly resistant to ZIKV and display reduced virus prevalence and intensity. Saliva from Wolbachia-carrying mosquitoes did not contain infectious virus in the saliva, suggesting the possibility of blocking ZIKV transmission. Their data indicate that the use of Wolbachia-harboring mosquitoes could represent an effective mechanism to reduce ZIKV transmission and should be included as part of ZIKV control strategies [44].

> They observed the wMel Wolbachia infection in *Ae. aegypti* significantly inhibited ZIKV infection in mosquito abdomens, and it reduced disseminated infection in heads and thoraces and ZIKV prevalence in mosquito saliva. Our results suggest that saliva from wMel-infected mosquitoes did not contain the infectious virus. This inhibition occurred for two ZIKV isolates that circulated in Brazil during the 2015 epidemic and for mosquitoes with a wild-type genetic background, suggesting that wMel could significantly reduce ZIKV transmission in field populations of *Ae. aegypti*, which would likely reduce the frequency of ZIKV-associated pathology in humans. [44]

Since Wolbachia is transmitted vertically via the egg, female-biased reproductive manipulations can drive Wolbachia infections to high frequencies in wild populations. Results from semifield and field trials in Australia have shown that Wolbachia can be persistently established in wild *Ae. aegypti* populations [45].

13.3 State of the Wolbachia Option

Mosquitoes could evolve resistance against a particular strain of Wolbachia, reducing its densities or restricting its tissue distribution. There is precedence for exactly this occurring in *Drosophila simulans* in response to transinfection with Wolbachia wMelPop, and it is also possible that pathogens themselves will evolve a means to evade Wolbachia-based blocking [18].

Wolbachia is a maternally inherited intracellular bacterium in numerous insect species worldwide, including mosquitoes, butterflies, beetles, ants, and bees. Interestingly, *Ae. aegypti* has no native Wolbachia symbionts. Aliota et al.'s findings optimism that Wolbachia biocontrol represents a significant new development in the fight against arbovirus transmission. Indeed, because dengue, chikungunya, Yellow

Fever, and Zika viruses cocirculate in many parts of the tropics, Wolbachia biocontrol could potentially be used as a multivalent strategy against all four of these *Ae. aegypti*-transmitted viruses [46].

The successful transfer of a life-shortening strain of the inherited bacterial symbiont, Wolbachia, into the primary mosquito vector of dengue, *Ae. aegypti* halved adult life span under laboratory conditions. The association is stable, and the Wolbachia strain is maternally inherited at high frequency. It can induce complete cytoplasmic incompatibility, facilitating its invasion into natural field populations and its persistence over time. McMeniman et al.'s data suggest that targeting mosquito age with inherited Wolbachia infections may be a viable strategy to reduce the transmission of pathogens such as dengue viruses [19].

According to Araujo et al. [47], field trials have already been conducted in four sites in Australia, one in Vietnam, and two in Indonesia; trials in Brazil have been underway since September 2014 using *Ae. aegypti* infected with the wMel Wolbachia strain. Laboratory and field assays performed in Brazil showed that the wMel infection had no detrimental fitness effects in Brazilian *Ae. aegypti* mosquito populations and would theoretically be able to successfully invade the mosquito populations in distinct urban landscapes.

13.4 Public Support

Neither the **World Mosquito Program** nor **MosquitoMate** has met much resistance to its technology, unlike Oxitec, presumably because their approaches involve naturally occurring bacteria, not engineered genes. **World Mosquito Program** reports more than 90% acceptance of its work in communities where field trials have been launched. In Colombia, they created a "Wolbachia Warriors" program, where children raise the mosquitoes and release them [30].

In Townsville, Australia, the community participated in their release through Mosquito Release Containers (MRCs). A key feature of using MRCs for mosquito releases is the possibility of mobilizing the community to undertake the deployment instead of employing program staff. Community-based waivers were conducted in three ways; school programs where students were given MRC kits to take home, direct community releases where MRC kits were given to householders who had signed up to participate through community engagement activities, and finally, by having large employers within the city distribute MRCs to staff who were willing to participate [46].

The Wolbachia approach has a conservative price tag (deployment cost should be reduced to less than US$1 per person). In contrast to most other interventions, the charges should not be ongoing since once Wolbachia is introduced, it is expected to maintain itself in populations. This suggests that the use of Wolbachia for arbovirus control, as described in this study, can be an extremely cost-effective intervention compared with existing methods and many other proposed interventions that feature the release of modified mosquitoes [46].

This led the Florida Keys Mosquito Control District and the privately owned, Kentucky-based company **MosquitoMate** to deploy field trials in April 2017. The results were examined.

> Results from field trials in Florida and California demonstrated reduced A. aegypti populations in treated areas compared to areas where PPF-treated males were not released. These results indicate that the release of PPF-dusted A. aegypti males can impact A. aegypti populations as measured by both reduced larval survival and lower numbers of adult female by Brelsfoard et al. [48] A. aegypti. [46]

Community support will be critical as the EPA considers **MosquitoMate**'s applications to use Wolbachia-infected mosquitoes more extensively. The application has received ten comments, at least three from people who confused the Wolbachia mosquitoes with genetically modified insects [36].

On November 3, the EPA registered **MosquitoMate**'s Asian Tiger mosquito (*Ae. albopictus*) as a new biopesticide, with a five-year license to sell in twenty different states [49]. Lab-reared mosquitoes will deliver the bacterium to wild mosquito populations. The decision—which the EPA has not formally announced—allows the company based in Lexington, Kentucky, to release the bacteria-infected mosquitoes in 20 US states and Washington DC [36].

Infecting mosquitoes with natural bacteria like Wolbachia do not require multiple costly releases of genetically engineered mosquitoes, warns Dana Perl, Senior Campaigner for Friends of the Earth [50].

Wolbachia mosquitoes could have a more significant advantage over genetically modified ones (see below) because genetic tweaks kill just one generation. At the same time, Wolbachia is passed on to the next generation via infected females [51].

It is essential to point out that extensive public engagement will be required before the release of Wolbachia-infected mosquitoes can be scaled up for use in other areas [50].

13.5 Controversies

The Wolbachia approach is much less controversial than the Oxitec approach (see below). In Florida, some opponents to genetic modification technology actively lobby for **MosquitoMate**'s Wolbachia approach [52]. On February 14, 2017, the Mosquito Control Board met for a Keys Wolbachia trial workshop [21].

13.6 Test Locations and Trials

In terms of field trials, the Wolbachia option has been deployed before. It has a history of deployment in north Queensland, Australia. While these have provided a basic template for release, this environment differs substantially from the large urban centers in Southeast Asia and Latin America, where a cluster-randomized trial would probably be done [53]. Researchers have recently used the bacteria, known as Wolbachia, in cracks in places including Australia and Brazil. But those efforts were small, reaching areas with tens of thousands of residents [54].

13.6.1 World Mosquito Program (Formerly Eliminate Dengue)

Outside of the United States, the first Wolbachia-infected mosquitoes approved by the Australian Pesticides and Veterinary Medicine Authority (APVMA) were released into the suburbs of Cairns, Australia (Yorkeys Knob [614 houses] and Gordonvale [668 homes]), in 2011 and quickly spread into the wild, replacing the disease-carrying population with the new, disease-free one, and replicating quickly into subsequent generations [54].

"It worked better than expected," says Eliminate Dengue's Scott O'Neill. By May, Wolbachia was sitting happily in 80% of the Gordonvale mosquitoes and 90% in Yorkeys Knob. The dengue-proof insects had almost totally replaced the native ones in just four months [29]. According to the research group Eliminate Dengue, testing so far shows the helpful bacteria remain. Additional field trials are underway in Indonesia, Vietnam, Colombia, and Brazil [54]. Over the next few years, they will release their insects in three to-be-confirmed locations in Brazil, Colombia, and the Asia–Pacific. Two of those will be home to around 2.5 million people. "We're talking close to entire cities," says O'Neill [55].

The results of the Cairns experiment offer one of the many approaches that may be effective in vector control. While Hoffman et al. [56] investigated the role and function of Wolbachia-infected mosquitoes in controlling dengue, the application to ZIKV control is evident.

Because wMel and other Wolbachia strains inhibit dengue virus transmission in the laboratory, this technology could potentially be used for area-wide suppression of dengue transmission. It has been demonstrated that it is possible to produce Wolbachia-infected mosquito populations that can function as "nursery" areas for future human-assisted collection and further dispersal of Wolbachia-infected mosquitoes without rearing other mosquitoes in an insectary. This should provide a strategy for sustainable dengue control at a low cost, with a relatively simple deployment system suitable for implementation in developing countries [56].

Under the banner of Eliminate Dengue—the nonprofit organization that O'Neill and his colleagues founded—they established teams and pilot studies in Indonesia, Brazil, Colombia, and Vietnam. The organization's focus was initially trained on dengue. But after ZIKV took the world by surprise earlier this year, Eliminate Dengue researchers quickly showed that Wolbachia also curtails the spread of that disease—perhaps even more potently than it does for dengue [55].

And in the island of Nha Trang, Vietnam—the site of one of the pilot projects—there were just a couple of dengue cases last year, compared to the dozens the island used to see annually. That is encouraging but also anecdotal. To get better evidence, Eliminate Dengue is now running a randomized controlled trial in Yogyakarta, Indonesia, releasing their mosquitoes over some areas and not others to see if dengue cases fall in the former. "That will provide the gold-standard evidence that epidemiologists would like to see," O'Neill says [55].

"Traditional methods of insecticide spraying are expensive, difficult to do, and often not very effective," said Scott O'Neill, dean of science at Monash University and head of the global Eliminate Dengue initiative, which has released Wolbachia-infected mosquitoes in Australia, Vietnam, Brazil, and Indonesia in a similar program to Xi Zhiyong's. October 2014 marked its first citywide pilot in Townsville, Queensland [57]. Four million Wolbachia mosquitoes over sixty-six square kilometers were released.

In the 44 months after the releases began, local authorities recorded just four locally acquired dengue cases, compared with fifty-four locally caught cases over the forty-four preceding months. (During the same period after the release, fifty-one imported cases were reported) [58].

Dr. Soumya Swaminathan, Director-General of the Indian Council of Medical Research, "We are about to sign a memorandum of understanding next month with Monash University for vector control using Wolbachia-infected Ae. aegypti mosquitoes." He added it was a low-cost solution because it is self-sustaining, unlike to Oxitec approach [59].

After almost a decade of research, O'Neill's team managed to rear mosquitoes carrying Wolbachia in the laboratory. The first Wolbachia-infected *Aedes* was released in 2011 in Cairns, a small coastal city in Queensland, Australia. O'Neill's team did a similar experiment in Townsville, 346 km to the south, in October 2014, and other members of the research consortium have done identical field trials in Colombia, Indonesia, and Viet Nam [60].

O'Neill et al. [46], in their study of the Townsville experiment, demonstrated that the wMel strain of Wolbachia can be deployed effectively across large geographic areas at low cost; that once the intervention is deployed, it is stable and self-sufficient; and that communities are accepting of the release of mosquitoes and are willing to participate in deployments when effectively engaged [61].

"Based on those trials, the company believes its strategy would result in a self-sustaining, Wolbachia-infected *Aedes* population—i.e., the mosquitoes would not have to be redeployed indefinitely, making them a potentially more cost-effective option". [52]. Large-scale trials are planned in Brazil and Colombia. Financial backers include the Gates Foundation [62].

In December, Radio New Zealand [63] reported Eliminate Dengue would release Wolbachia-laced mosquitoes on the island of New Caledonia. According to the Scientist, Indonesia, and Australia, too, are expanding the number of sites where researchers are releasing bacteria-bearing mosquitoes [64].

The Gates Foundation and Wellcome Trust provide $18 million in funding to EDP to launch large-scale Wolbachia deployment programs in Latin America. Along with support from USAID, the U.K. government, and $3.7 million from the Brazilian Ministry of Health, EDP will begin its Wolbachia interventions in large urban settings across Bello, Antioquia, and parts of Rio de Janeiro [65]. The scope of these Latin tests could allow Eliminate Dengue to leap ahead of for-profit efforts by Oxitec [43].

In 2017, Eliminate Dengue announced its intention to expand from field trials to pilot sites in Nha Trang, Vietnam, this coming March, a few months after the organization unveiled plans to increase the release of bacteria-treated mosquitoes in Colombia and Brazil. "The release of mosquitos with Wolbachia has been implemented on a trial basis in the city's Tri Nguyen island and has proved effective," according to Xinhua, "no dengue fever outbreaks have been reported." [27].

Brazil is becoming a proving ground for tailored mosquitoes. About six hundred kilometers to the east, in the coastal cities of Niterói and Rio de Janeiro, another lab strain of mosquitoes is on the wing. Bred by a nonprofit organization called Eliminate Dengue, this one is infected with a bacterium called Wolbachia pipientis that protects it from dengue, ZIKV, and a third virus named chikungunya [10].

Wolbachia-treated mosquitoes have not been wholly effective. However, it is not yet clear if Wolbachia will be an effective method eventually. Brazil's Eliminate Dengue program launched the first trial of Wolbachia mosquitoes in 2014. Early results were good after artificially infected mosquitoes were put into the wild population, some 65% *Ae. aegypti* caught the parasite. But after that, the people of Wolbachia-infested mosquitoes shrank. Wolbachia weakens mosquitoes so much that they become susceptible to insecticides, which untreated Ae. aegypti can avoid [27].

Eliminate Dengue already has an efficacy study underway in Yogyakarta, Indonesia, tracking disease rates in twenty-four areas of about 14,500 people each, half of which will receive Wolbachia mosquitoes. A network of clinics across the city has been testing continuously for dengue diseases for over two years. An even more extensive study is being planned in Vietnam. Oxitec is enlisting independent experts to design a trial tentatively slated for 2018 [10].

In Colombia, Eliminate Dengue developed a Wolbachia Warrior program, where schoolchildren aged 8 to 11 can raise their mosquitoes and do their releases [66].

13.6.2 Mosquitomate

MosquitoMate's first trials focused on *Aedes* albopictus, another mosquito species that can spread diseases to people. (It can transmit ZIKV, although not as efficiently as *Ae. aegypti.*) But it has also started trials targeting *Ae. aegypti*, and mosquito control officials in the Florida Keys approved a problem with MosquitoMate mosquitoes beginning in March 2017 [27]. MosquitoMate also listed Key Largo and Stock Island as potential test sites in Florida. However, they will need the approval of the Florida Department of Agriculture and Consumer Services and the Mosquito Control Districts. In a letter to State Surgeon General Celeste Philip detailing local plans for combating ZIKV, Miami-Dade Mayor Carlos Gimenez said mosquito control officials had contacted the University of Kentucky's Department of Entomology to explore the possibility of using its Wolbachia-infected mosquitoes [67]. Roughly 100,000 male *Ae. aegypti* mosquitoes with Wolbachia would be released every week into three one-acre trial areas along with twenty locations on a ten-acre site on Stock Island and Key Largo 2 times a week for the next 12 weeks [68].

With the advent of the ZIKV, MosquitoMate has shifted gears from focusing on Asian Tiger mosquitoes (*Aedes* albopictus) to the species *Ae. aegypti* prefers feeding during dusk and dawn. Trials using albopictus mosquitoes have been conducted in three states since 2013, while a problem of aegypti mosquitoes occurred last year in California [69].

The Lower Keys trial is MosquitoMate's second U.S. test with *Ae. aegypti* mosquitoes, after a similar problem in Clovis, California, in 2016. Stock Island is about 130 miles southwest of Miami, where *Ae. aegypti* mosquitoes were blamed for spreading the ZIKV last year [70].

According to the Florida Keys Mosquito Control District, 20,000 male *Ae. aegypti* mosquitoes were released on Stock Island on April 18 for a field trial that will last 12 weeks. They will be removed twice a week for twelve weeks. The mosquitoes, which do not bite, have been manually infected with Wolbachia's naturally occurring bacteria [54].

By mid-October 2016, **MosquitoMate Inc.**, working through the University of Kentucky, received tentative approval from the Florida Department of Agriculture and Consumer Services to reduce the population of *Ae. aegypti* mosquitoes in the Florida Keys use mosquitoes infected with the natural bacteria Wolbachia [71].

Beth Ranson with the Florida Keys Mosquito Control District confirms that the agency started releasing Wolbachia bacteria-infected males on February 8, 2018, in mid-April in South Miami's Brewer Park. The plan was six months, and six million male mosquitoes were released in batches of about 20,000. The Florida Department of Health is funding the $4.1 million program with a grant from the Centers for Disease Control [72].

The Wolbachia approach is seen as more natural and generated little controversy. According to commission officers, only three residents and a reporter attended the Mosquito Control district's meeting to discuss the MosquitoMate trial [73].

MosquitoMate has tested Wolbachia in Ae. albopictus mosquitoes over the past three years. The approach has reduced the wild mosquito population by more than 70% in those areas, says Stephen Dobson, an entomologist at the University Of Kentucky who founded the company [23].

13.7 Disposition: Successes and Concerns

For ZIKV, there is some speculative evidence on unintended implications from Wolbachia.

Before getting to them, there is another source of discontent: a Canadian blogger called RoseWrites, who publishes her blogs on InfoBarrel. She claims she has a background in nursing and was a frontline worker during the SARS outbreak. She sells ZIKV-related products on Zazzle, the profits from which she claims goes to ZIKV research. While an outsider, her comments are often footnoted in the peer-reviewed literature. She published a letter to the Pattaya Mail (Thailand) editor and remade some of these claims.

She claims, there are considerable risks to Wolbachia-based mosquito control. One concern presumably from the WHO acknowledged that *Culex* is also a vector (researchers from Brazil, Canada, and China confirmed this). *Culex* is prevalent in Florida and a likely vector of ZIKV. The problem is that Wolbachia in *Culex* enhances West Nile virus and Malaria (and probably does with ZIKV too). On another level, Dr. Anthony James, University of California Irvine, said in response to Dr. Thomas Scott, University of California Davis: "These genetic tools might not be the best strategies for ZIKV given that at this point there seem to be multiple vectors not only at the species but also at the population level. The current genetic technologies would not appropriately apply to such complex systems." [74].

On the other hand, the consensus seems to be more in line with Popovici et al. "Considering the wide distribution of Wolbachia and no indications that it is impacting negatively on infected insect populations, we would contend that the consequences of any harm arising from low probability transfer events would be negligible." [75].

13.7.1 General Feasibility

Previously, it was difficult to check if *Ae. aegypti* mosquitoes were infected with Wolbachia: Mosquitoes show no visual symptoms, and current diagnostic tests are costly and hard to read.

In a new PLOS Neglected Tropical Diseases study, Sanchita Bhadra of the University of Texas in Austin and colleagues describe their new diagnostic tool that will aid research "in the field." Their new device uses a cellphone camera and application to identify the mosquito's species and whether it carries Wolbachia. The researchers combined fluorescent probes for *Ae. aegypti* DNA and Wolbachia DNA with a LAMP assay, a simple, famous DNA detection test. When the target DNA is present, the probes bind to it and produce what the researchers deem a "yes/no" fluorescence which is detectable by a cellphone camera [76].

They say they can test a mosquito's DNA using a smartphone camera, a small 3D-printed box, and a chemical test. They could use their 97% accurate LAMP OSD device to replace the complicated DNA removal of insects and expensive lab testing. And they can see whether a mosquito-control strategy called Wolbachia works by revealing if the mosquito has come into contact with it. The team is now working on a similar test to determine whether mosquitoes carry viruses like ZIKV, dengue, and yellow fever [77].

13.7.2 Separating Males from Females

Suppressing the mosquito population of an entire city is likely to require the weekly production of millions of these mosquitoes. To reach that level, **MosquitoMate** must efficiently find a way to separate male mosquitoes from females. The company's technicians now separate them both by hand and mechanically, Dobson, founder of **MosquitoMate**, says. The scientists use mechanical sorters to separate males from females, based on size differences at the pupal stage, with more than 99% efficiency, says Zhiyong Xi, a medical entomologist and microbiologist at Michigan State University, who leads the project. They expose the remaining mosquitoes to X-ray radiation at a dose that sterilizes any remaining females but is too low to affect the males [78].

13.7.3 Heat Stress

[M]osquito larvae experience extreme temperatures in their natural habitat. While the thermal limits of *Ae. aegypti* are generally well-understood; research has not assessed Wolbachia's reproductive effects in *Ae. aegypti* at the elevated temperatures they can experience in the field. Ulrich and others [76] recently demonstrated the density of wMel in *Ae. aegypti* decreased sharply when larvae experienced cycling temperatures of 28.5 °C diurnally to 37.5 °C during development. Ross et al. showed *Ae.*'s wMel and wMelPop-CLA infections for the first time. *Aegypti* exhibit incomplete cytoplasmic incompatibility when immature stages experience cyclical temperatures of 26–37 °C (79–99 °F) during development. It is also shown that these infections are not transmitted to the next generation when infected mosquitoes experience these conditions over their entire lifecycle. Ulrich et al. concluded the spread of Wolbachia in wild *Ae. aegypti* populations and any consequent protection from dengue and ZIKV viruses might be limited in ecosystems that experience periods of extreme heat. Still, Wolbachia levels recover partially after temperatures return to normal [78].

This suggests that the reproductive effects of Wolbachia could also be altered if infected larvae develop under similar conditions in the field [15]. Ross et al. added, "[m]aximum daily temperatures of larval mosquito habitats in nature can reach or exceed the maximum temperature tested in our study." This should be a careful consideration for additional research in this area [15].

In early January (2017), reports began to surface that some strains of Wolbachia were sensitive to elevated temperatures. High temperatures have been known for some time to hurt Wolbachia. In other arthropods, high temperatures can reduce the density of Wolbachia in its host, weaken the reproductive effects induced by Wolbachia and even eliminate Wolbachia [15].

Specifically, Ross found that the wMel Wolbachia strain, currently used in many field trials worldwide, survived significantly lower numbers as the temperature rose.

But other strains of Wolbachia were tested, and some proved resistant to these higher temperatures [15].

Results also indicate that the release of mosquitoes infected by other strains, such as wMelPop, may need to be ongoing. This does not represent an autonomous intervention, as would be the case for different Wolbachia strains like wMel that have been shown to persist after a relatively small introduction of infected mosquitoes successfully [79].

"Although it was alarming to see that Wolbachia is vulnerable to high temperature, it is promising that other strains are more robust," said lead author Perran Ross, a graduate student at the University of Melbourne, in a press release. "These strains are also effective at blocking viruses." [64].

Ross et al. [15] published a corroborating study claiming "the wMel strain of Wolbachia which is currently being used for dengue and ZIKV control in several countries may have reduced effectiveness at invading populations when mosquitoes experience heat stress. Since mosquito larvae experience extreme temperatures in their natural habitat, these results have implications for current and future releases of Wolbachia-infected mosquitoes.". They note that "the ability of wMel and wMelPop-CLA-infected *Ae. aegypti* to invade and persist in natural populations will be adversely affected by heat and the [m]aximum daily temperatures of larval mosquito habitats in nature can reach or exceed the maximum temperature tested in this study [79].

Ross et al. [15] observed that when all life stages were maintained at 26–37 °C, the wMel and wMelPop-CLA infections were not transmitted to the next generation. The wAlbB infection also exhibited some maternal transmission leakage despite maintaining high densities and complete cytoplasmic incompatibility when only larvae were exposed. This suggests that both the duration of exposure and the maximum temperature reached will affect Wolbachia density [64].

Ross et al.'s heat tolerance study of three strains of Wolbachia, namely wMel, wAlbB, and wMel-Pop-CLA, validates Ulrich and others. They found that only the wAlbB strain was heat tolerant and able to transmit from mother to offspring with high reliability. Malaysia's Institute for Medical Research (IMR) will be using the wAlbB strain, and it was reported that Singapore is using a similar strain [80].

13.7.4 Under-Regulated

The Cartagena Protocol—a United Nations safety regulation for the transfer, handling, and use of genetically modified organisms, signed by 170 countries—does not apply to Wolbachia-infected mosquitoes because the bacteria are considered non-transgenic. Therefore, the release of insects hosting Wolbachia was not subject to these regulations. As far as it is known, no country has limitations specifically about Wolbachia-infected insect release or mitigation strategies to deal with unexpected results. Even if a nation enacts such legislation, it will not extend to other countries, whereas Wolbachia-infected insects, at least in theory, can easily cross political borders [79].

13.7.5 Environmental Complications

Under laboratory conditions, it may be possible, with limited success, to infuse infection with parasites in vector(s) since the life-shortening Wolbachia strains do not seem to occur in mosquitoes naturally. Also, it may be possible to manipulate and create an imbalance artificially in the host-parasite relationship under laboratory conditions. However, it may not be possible to reproduce the same in the field where the environment governs the balance and is beyond human control [81].

However, environmental conditions such as temperature, nutrition, and pathogen infection are known to modulate Wolbachia densities in other insects. Given the importance of bacterial density in determining Wolbachia's reproductive effects (cytoplasmic incompatibility and maternal transmission fidelity), fitness costs, and viral blocking effects, work is needed to determine if environmental impacts modulate densities in experimental infections of *Ae. aegypti* [15].

Ambient temperature and relative humidity are some of the most important environmental variables that determine the duration of development of parasites/pathogens in vectors (the "extrinsic incubation period"). Therefore, the life-shortening effect of mosquitoes by Wolbachia alone would not be sufficient to interrupt the transmission of infection. With the increases in temperature that have been widely experienced in many areas, viral pathogens such as dengue could easily reach their infective titer in a shorter duration in the field by simple multiplication ("propagative" nature) [81].

"With climate change projections of increasing temperatures, we may see mosquitoes with dengue migrating from Queensland further down to southern Australia, particularly if another invasive species that transmit diseases, *Ae. albopictus* enters mainland Australia. Also, we expect higher temperatures to become more common," University of Melbourne Professor Ary Hoffmann said [82].

While **MosquitoMate** is undergoing testing in the USA, the **World Mosquito Program** can be found in more countries.

Some concern has been expressed about the consumption of Wolbachia-infected *Ae. aegypti* mosquitoes by some predators. Research does not seem to bear this out. Hurst et al. indicated that wMelPop infected and uninfected *Ae. aegypti* larvae and adults were equally susceptible to predation by all six tested predators. In addition to evaluating potential fitness costs to the infected host, Hurst could not demonstrate the horizontal transfer of wMelPop via consumption of infected *Ae. aegypti* larvae to six naturally occurring predator species; cyclopoid copepods, fish, predatory Toxorhynchites mosquito larvae, and a salticid jumping spider. These results and the ecology of Wolbachia and mosquito predators show the horizontal transfer of wMelPop from *Ae. aegypti* into naturally occurring predators is not cause for concern [83].

Although Wolbachia typically undergoes vertical transmission through its host population's maternal line, there is compelling evidence from molecular phylogenies that extensive horizontal (intertaxon) transmission must have occurred. Some of the best candidate vectors for the horizontal transmission of Wolbachia are insect parasitoids, which comprise around 25% of all insect species and attack arthropods from an enormous range of taxa [84].

Heath et al. [84] argued Wolbachia evolved between 80 and 100 million years ago, while the evolution of an arthropod occurred at least two hundred million years ago. "As Wolbachia isotypes are present in a wide variety of arthropods, this observation leads inevitably to the conclusion that Wolbachia have undergone horizontal transmission—and the life history and virtual ubiquity of insect parasitoids in terrestrial ecosystems implicate them as vectors of Wolbachia.

Popovici et al. [75] argue that horizontal transfer is an infrequent event. A common phenomenon is now being seen; that is, 65% of insect species infected is a product of contact between them and Wolbachia over millions of years [83].

The only recorded natural horizontal transfers have been between different parasitoids superinfecting the same insect host. However, this seems to be an unlikely explanation for a general mechanism. Furthermore, parasitoids are not significant parasites of mosquitoes. The scientific community's consensus is that natural horizontal transfer events are infrequent. The wide distribution of Wolbachia among insects is explained by the many millions of years that Wolbachia is believed to have been associated with insects, which in turn has allowed time for numerous rare events to accumulate [83].

Phil Lounibos, a professor at the Florida Medical Entomology Laboratory at the University of Florida, said Eliminate Dengue's approach had been proven effective in Australia and Indonesia—not necessarily as a mosquito control program, but as a mosquito "replacement program" that produced generations of Wolbachia-infected mosquitoes incapable of transmitting viruses to people. Of the two approaches to using Wolbachia in mosquitoes, Lounibos said he prefers the Eliminate Dengue model. "It's ecologically sounder," he said, adding that both methods may help reduce the spread of disease. Still, only one will require repeated releases of sterile male mosquitoes, which could mean long-term expense for local taxpayers [75].

The Wolbachia option could be troubling. While the Oxitec possibility (below) can be raised on horse blood, does the Wolbachia mosquito require human blood? Talk about an opportunity to spread human pathogens! It seems there should be more questions being asked about Wolbachia mosquitoes [85]! The downside is that the release of even a single female mosquito infected with Wolbachia could "potentially lead to the alien bacteria spreading in the target population," says a June 2013 Pathogens and Global Health report [86].

Also, there are similar concerns about the Wolbachia approach still functioning as a GMO. Wolbachia is known to transfer some of its genes to its insect hosts. The authors highlight the irony that while the artificial transfer of the entire Wolbachia genome in the lab is considered "natural," one or a few Wolbachia genes had been artificially transferred to the mosquito. It would have been considered a GMO!! [85]

The Wolbachia trial uses naturally occurring bacterium that is already found in insects. So, for that reason, many people in Florida believe this is a more natural alternative to genetically modified trials [87].

Elsewhere in the world, a Chinese factory is breeding twenty million male mosquitoes a week and releasing them into nature to copulate with wild female mosquitoes. Its aim—culling the population and eradicate disease. The world's most prominent "mosquito factory," with a total area of 3500 m^2 (38,000 sq. ft.), and four workshops that each can breed five million mosquitoes a week, has been established in southern China's Guangzhou City [51]. Millions of Wolbachia-infected *Ae. albopictus* mosquitoes are released every week. The mosquitoes are released on Shazai Island, a village thirty-seven miles (60 km) from the factory. A 300-m (984 feet) bridge separates the town of 1900 people from the mainland. According to Xi, the isolated location keeps incoming mosquito populations at bay: Mosquitoes can fly 50–75 m (164–246 feet) at most [88].

A $1 million grant was recently awarded to Zhiyong Xi, MSU associate professor of microbiology and molecular genetics, to build a mosquito factory in Yucatan, Mexico. The laboratory will be modeled after a facility in Guangzhou, China, a center Xi leads in partnership with Sun Yat-sen University. Xi will coordinate Yucatan efforts with multiple Mexican government agencies, the governor of Yucatan, and Universidad Autonoma de Yucatan to breed and conduct field tests in the region [41].

This approach involves using the Wolbachia pipientis bacterium. The eggs need to be injected just once, and they pass down the bacteria from mother to offspring. Sun Yat-sen University and Michigan State have a Joint Center of Vector Control for Tropical Diseases.

In Guangzhou, southern China, "Some residents have even asked to get mosquitoes from us to release in their own home," said Xi Zhiyong of Michigan State University, who heads the project [49]. Reportedly, the mosquito population on Guangzhou island had dropped by half by June that year, Xi told the Beijing News (link in Chinese) [51].

[F]ewer concerns are typically raised about the Wolbachia mosquitoes. The bacterial approach also has had an easier regulatory route in the US: The mosquitoes are considered "biopesticides" by the EPA and do not need FDA approval to be released, unlike the Oxitec mosquitoes (see below) [29].

The U.S. EPA approved the experimental use of the UK's Wolbachia-infected *Ae. aegypti* mosquitoes this year for limited tests in Monroe County, which encompasses the Florida Keys, and in California, where officials in Fresno County have released thousands of non-biting males dusted with the bacteria into the wild, where they produced offspring that never matured [30].

Ae. aegypti mosquitoes infected with Wolbachia bacteria are currently being released for arbovirus suppression worldwide. Their potential to invade populations and persist will depend on interactions with environmental conditions, particularly as larvae are often exposed to fluctuating and extreme temperatures in the field [89].

For example, the wMel-Pop strain used in Thailand and China, while a good dengue blocker, was found to cause high mortality among adult *Aedes* mosquitoes [80].

13.7.6 Genetic Diversity

Aedes mosquitoes breed in clean water, prefer human hosts, and are eaten by birds. Once infected by Wolbachia, they will introduce a new pathogen in birds, fish, other species, and water. And surprisingly, RoseWrites claims Wolbachia can survive for at least a week in a dead host [90]. However, While Wolbachia obtained in this manner can be maintained in a cell-free medium and retain viability for at least a week, no replication is apparent. Nonetheless, these results indicate that Wolbachia can survive at least for a limited time outside host cells. However, the fact that Wolbachia cannot synthesize many essential amino acids is likely a significant factor limiting the extent of extracellular survival [91].

However, few have focused on the probability of Wolbachia strains being transferred to other insects and the potential environmental and economic impacts of the host shift. Wolbachia strains (including wMel-like strains) can be shared horizontally among distantly related arthropods in a short evolutionary time. Moreover, some parasites can carry Wolbachia strains to other species [92].

Although mosquitoes deliberately infected with Wolbachia could reduce the need for insecticide use, the consequences of Wolbachia host shift to native species are, for now, unpredictable. Arthropods present complex and poorly understood ecological relationships and alterations in reproductive parameters of non-target species can generate ecological disturbances [92].

Wolbachia was associated with different mitochondria, suggesting multiple horizontal transmission events and paternal transmission leakage of mitochondrial and Wolbachia. Since Wolbachia is coinherited with mitochondria, natural selection acting over the bacterium will also affect mitochondria. Depending on the infection context, this hitchhiking effect may increase or decrease mitochondrial genetic diversity [90].

The release of insects hosting Wolbachia strains should be more carefully considered. Further studies of the potential impact of these bacteria on biodiversity should be undertaken before this strategy can be widely used [92].

13.7.7 Irreversibility

Wolbachia bacteria often infect insects' reproductive tissues, where they can improve the chances that their invertebrate hosts will reproduce—sometimes eliminating the need for mating. It helps the bacteria pass through eggs from one generation to the next. This approach also means that any dangerous side effects could be cast in tiny reproductive stones for generations after the infected specimens are released [93].

Penn State entomologist Jason Rasgon tested Wolbachia's effect on *Culex* (*Cx. tarsalis*) mosquitoes' ability to spread West Nile Virus. They reported the need for caution in moving forward with *Wolbachia*-based mosquito control efforts when they discovered the human pathogen had been enhanced with a Wolbachia infection [93].

Rasgon's team took a laboratory strain of mosquitoes captured initially in California's Yolo County. They knocked young female mosquitoes unconscious using carbon dioxide and injected some of them with Wolbachia and others with a control substance. A week later, they were fed cow blood infected with West Nile. Two weeks after the feeding, mosquitoes were killed and tested for the disease. The results of the experiments came as a disappointment. "We had to repeat it a couple of times before believing the result," Rasgon said. "We were expecting this to block West Nile." Instead, mosquitoes treated with the wAlbB strain of Wolbachia were more likely to carry the virus that the researchers were trying to protect them from [93].

13.7.8 Bacterial Pathogen

Humans have been exposed to Wolbachia for thousands of years. Wolbachia is extremely common in the environment, naturally infecting many insect species, including pests of stored food products and insects that bite humans, such as nuisance mosquitoes like *Aedes* albopictus and Culex quinquefasciatus. There is no evidence showing neither that residues of Wolbachia-infected insects in food products are harmful to humans nor are any adverse effects reported from people being bitten by insects containing Wolbachia [75].

Over prolonged periods, human volunteers exposed to thousands of bites from Wolbachia-infected mosquitoes have antibody responses to mosquito saliva but have not developed IgG antibodies specific to Wolbachia [75].

13.7.9 West Nile Virus, River Blindness, Elephantiasis

Dodson et al. [94] evaluated the effects of Wolbachia (wAlbB strain) on infection, dissemination, and transmission of West Nile virus (WNV) in the naturally uninfected mosquito *Culex tarsalis*, which is a crucial WNV vector in North America. Wolbachia did not inhibit WNV in this mosquito. Instead, the WNV infection rate was significantly higher in Wolbachia-infected mosquitoes compared to controls. This is the first observation of Wolbachia-induced enhancement of a human pathogen in mosquitoes, suggesting that caution should be applied before releasing Wolbachia-infected insects as part of a vector-borne disease control program. The authors did notice: It should be noted that these experiments were performed with mosquitoes transiently infected in the somatic tissues with Wolbachia rather than a stable maternally inherited infection. It remains to be seen whether a normal wAlbB condition will similarly enhance WNV.

Though Wolbachia does not infect vertebrates, it can impact human health. The nematode that causes onchocerciasis (river blindness) in tropical countries can only lay eggs and reproduce in humans thanks to the Wolbachia bacteria. Wolbachia, not the nematode itself, triggers the immune response that damages the eye [94].

Filarial nematodes cause some of the most debilitating diseases in tropical medicine. Recent studies, however, have implicated the parasites' endosymbiotic Wolbachia bacteria, rather than the nematode, as the cause of inflammatory-mediated filarial disease. Wolbachia is the principal cause of acute inflammatory filarial disease. Accumulated exposure to acute episodes of inflammation may also underlie the development of chronic filarial pathology. Filariasis causes some of the most debilitating global diseases, including elephantiasis and river blindness [95].

13.7.10 Human Male Sterility

The lab-grown versions are infected with a disease that prevents natural mosquito populations from breeding. But some activists fear the disease could transfer to humans, ultimately making all human males sterile. Despite claims that it is safe for humans, there are also some concerns that it could affect other arthropods, such as spiders that prey on mosquitoes, increasing populations eventually, or even people, rendering humans unable to breed. Perhaps some world regions are so overpopulated that even sure governments are willing to fund something likely to cause men to become sterile (Wolbachia) and are the unspoken cofactor amplifying ZIKV." [96].

Canadian blogger Rose Webster is behind a campaign to stop the releases. She has spent months speaking to scientists and fears that it may be possible for Wolbachia to infect humans and prevent individuals from breeding. She is also convinced that Wolbachia releases have helped spread the ZIKV. She said: "Wolbachia is responsible for the most widespread pandemics in the animal kingdom" [96].

13.8 Conclusion

Wolbachia is a clever approach to crashing the *Ae. aegypti* population. However, it has significant competition from a small UK company called Oxitec. Oxitec has grown immensely since this author began following it half a decade ago, and their story follows.

References

1. ASTHO (2015) Before the swarm: guidelines for the emergency management of vector-borne disease outbreaks. http://www.astho.org/Programs/Environmental-Health/Natural-Environment/Before-the-Swarm/. Accessed 28 Aug 2018
2. Nikolau L (2016) Bacteria-injected mosquitoes: the latest in the battle against Zika. Humanosphere. October 27. http://www.humanosphere.org/science/2016/10/bacteria-injected-mosquitoes-the-latest-in-the-battle-against-zika/. Accessed 26 May 2017

3. Timmer J (2011) Meet wolbachia: the male-killing, gender-bending, gonad-eating bacteria. Ars Technica. October 24. https://arstechnica.com/science/2011/10/meet-wolbachia-the-male-killing-gender-bending-gonad-chomping-bacteria/. Accessed 27 June 2017
4. Ant T et al (2018) The Wolbachia strain wAu provides highly efficient virus transmission blocking in *Aedes* aegypti. PLoS Pathog 14(1):2018. https://doi.org/10.1371/journal.ppat.1006815.AccessedJune27
5. Ant TH, Herd CS, Geoghegan V, Hokkmann AA, Sinkins SP (2018) The wolbachia strain wAu provides highly efficient virus transmission blocking in Aedes aegypti. PLoS Pathog. January 25. https://doi.org/10.1371/journal.ppat.1006815
6. Channel Asia News (2018) New wolbachia strain more effective at blocking dengue and Zika transmission: study. Channel Asia News. January 26. https://www.channelnewsasia.com/news/health/new-wolbachia-strain-more-effective-at-blocking-dengue-and-zika-9895686. Accessed 27 June 2018
7. Walker T et al (2011) The wMel wolbachia strain blocks dengue and invades caged *Aedes* aegypti populations. Nature 476. August 25. https://www.nature.com/nature/journal/v476/n7361/full/nature10355.html. Accessed 30 June 2017
8. Burki T (2020) Wolbachia, a bacterium fighting on our side. Lancet Infect Dis 20(6):662–663. https://doi.org/10.1016/S1473-3099(20)30384-4. PMID: 32473140; PMCID: PMC7255110.
9. Ruang-areerate T, Kittayapong P (2006) Wolbachia transinfection in *Aedes* aegypti: a potential gene driver of dengue vectors. Proc Natl Acad Sci 103(33). August 15. https://www.ncbi.nlm.nih.gov/pmc/articles/PMC1567913/. Accessed 31 May 2017
10. Servick K (2016) Brazil will release billions of lab-grown mosquitoes to combat infectious disease. Will it work? Science. October 13. http://www.sciencemag.org/news/2016/10/brazil-will-release-billions-lab-grown-mosquitoes-combat-infectious-disease-will-it. Accessed 31 May 2017
11. Brownstein JS, Hetta AZ, O'Neill SL (2003) The potential of virulent wolbachia to modulate disease transmission by insects. J Invertebr Physiol 84. http://www.sciencedirect.com/science/article/pii/S002220110300082X. Accessed 16 March 2017
12. Müller MJ et al (2012) Wolbachia pipientis is associated with different mitochondrial haplotypes in natural populations of Drosophila willistoni. J Invertebr Pathol 109. https://www.ncbi.nlm.nih.gov/pubmed/21945051. Accessed 17 July 2017; Zug R, Hammerstein P (2012) Still a host of hosts for wolbachia: analysis of recent data suggests that 40% of terrestrial arthropod species are infected. PLOSOne 7:6. http://journals.plos.org/plosone/article?id=10.1371/journal.pone.0038544. Accessed 14 April 2017
13. Zug R, Hammerstein P (2012) Still a host of hosts for wolbachia: analysis of recent data suggests that 40% of terrestrial arthropod species are infected. PLOSOne 7:6. http://journals.plos.org/plosone/article?id=10.1371/journal.pone.0038544. Accessed 14 April 2017
14. Timmer J (2011) Meet wolbachia: the male-killing, gender-bending, gonad eating bacteria. Ars Technica. October 24. https://arstechnica.com/science/2011/10/meet-wolbachia-the-male-killing-gender-bending-gonad-chomping-bacteria/. Accessed 27 June 2017
15. Ross PA et al (2017) Wolbachia infections in *Aedes* aegypti differ markedly in their response to cyclical heat stress. PLOS Pathog. January 5. http://journals.plos.org/plospathogens/article?id=10.1371/journal.ppat.1006006. Accessed 31 May 2017
16. XinhuaNet (2018) Vietnam to release anti-dengue mosquitos. XinhuaNet. January 15. http://www.xinhuanet.com/english/2018-01/15/c_136896537.htm. Accessed 10 July 2018
17. Min K-T, Benzer S (1997) Wolbachia, normally a symbiont of Drosophila, can be virulent, causing degeneration and early death. Proc Natl Acad Sci 94. September. https://www.ncbi.nlm.nih.gov/pmc/articles/PMC23488/. Accessed 23 May 2017
18. McGraw E, O'Neill S (2013) Beyond insecticides: new thinking on an ancient problem. Nat Microbiol 11. March. http://www.nature.com/nrmicro/journal/v11/n3/full/nrmicro2968.html. Accessed 22 May 2017
19. McMeniman C et al (2009) Stable introduction of a life-shortening wolbachia infection into the mosquito *Aedes* aegypti. Science 323. January 2. http://science.sciencemag.org/content/323/5910/141. Accessed 22 May 2017

20. Sinkins SS, O'Neill SL (2000) Wolbachia as a vehicle to modify insect populations. In: Handler AM, James AA (eds) Insect transgenes: methods and applications. CRC Press, Boca Raton, FL, pp 271–288
21. Dutra H et al (2016) Wolbachia blocks currently circulating Zika virus isolates in Brazilian *Aedes* aegypti mosquitoes. Cell Host Microbe 19. June 8. https://www.ncbi.nlm.nih.gov/pubmed/27156023. Accessed 14 April 2017
22. Lambrechts L et al (2015) Assessing the epidemiological effect of wolbachia for dengue control. Lancet Infect Dis 15. June 4. http://thelancet.com/journals/laninf/article/PIIS1473-3099(15)00091-2/fulltext. Accessed 19 May 2017
23. Waltz E (2016) Infected mosquitoes could fight Zika. Nature 533. May 26. https://www.nature.com/polopoly_fs/1.19967!/menu/main/topColumns/topLeftColumn/pdf/533450a.pdf?origin=ppub. Accessed 30 June 2017
24. Alphey L (2014) Genetic control of mosquitoes. Ann Rev Entomol 59:205–224. http://www.oxitec.com/genetic-control-of-mosquitoes-luke-alphey/. Accessed 7 April 2017
25. Stone J (2016) Smart science confirms wolbachia's value in fighting Zika as well as dengue. Forbes. May 4. https://www.forbes.com/sites/judystone/2016/05/04/smart-science-confirms-wolbachias-value-in-fighting-zika-as-well-as-dengue/. Accessed 2 June 2017
26. Christodoulou M (2011) Biological vector control of mosquito-borne diseases. The Lancet. February 11. http://www.thelancet.com/journals/laninf/article/PIIS1473-3099(11)70017-2/abstract. Accessed 10 April 2017
27. Rathi A (2016) Brazil is fighting its biggest epidemics with weaponized mosquitoes. Quartz Media. January 20. https://qz.com/598432/brazils-fighting-its-biggest-epidemics-with-weaponized-mosquitoes/. Accessed 29 May 2017
28. Parnell S (2017) Scott O'Neill: bacteria key to battling dengue fever. The Australian. January 18. http://www.theaustralian.com.au/news/australians-australian-of-the-year/scott-oneill-bacteria-key-to-battling-dengue-fever/news-story/928c1bec86a3ec7ff16e0cd2e15f9e26. Accessed 28 May 2017
29. Joseph A (2016) Rio to fight Zika with massive release of bacteria-infected mosquitoes. Sci Am STAT. October 26. https://www.scientificamerican.com/article/rio-to-fight-zika-with-massive-release-of-bacteria-infected-mosquitoes/. Accessed 17 May 2017
30. Chang D, Driscoll A (2016) Miami-dade considers new weapon in Zika fight: disease-fighting bacteria. Miami Herald. November 11. http://www.miamiherald.com/news/healthcare/article114267723.html. Accessed 22 Dec 2016
31. Ferguson N et al (2015) Modeling the impact on virus transmission of wolbachia-mediated blocking of dengue virus infection of *Aedes* aegypti. Sci Trans Med 7:279. March 18. http://stm.sciencemag.org/content/7/279/279ra37. Accessed 3 April 2017
32. Zheng X, Zhang D et al (2019) Incompatible and sterile insect techniques combined eliminate mosquitoes. Nature 572:56–61
33. Moretti R, Calvitti M (2021) Issues with combining incompatible and sterile insect techniques. Nature 590. February 3. E1-E2. https://www.nature.com/articles/s41586-020-03164-w. Accessed 23 May 2022
34. Brown K (2017) The EPA just approved lab-grown mosquitoes to fight disease. GIZMODO. https://gizmodo.com/the-epa-just-approved-lab-grown-mosquitoes-to-fight-dis-1820231191. Accessed 23 July 2018
35. Pereira S (2017) EPA approves killer mosquitoes to stop harmful diseases like Zika. Newsweek. November 8. https://www.newsweek.com/epa-approves-killer-mosquitoes-stop-harmful-diseases-zika-705484. Accessed 23 July 2018
36. Zhang S (2016) A California city is fending off Zika by releasing 40,000 mosquitoes every week. Wired. August 4. https://www.wired.com/2016/08/california-city-fending-off-zika-releasing-40000-mosquitoes-every-week/. Accessed 25 Sept 2016
37. Mukherjee S (2017) Why google's verily is unleashing 20 million bacteria-infected mosquitoes in Fresno. Fortune. July 17. http://fortune.com/2017/07/14/google-verily-zika-mosquitoes/. Accessed 21 July 2017

38. Mulin E (2017) Verily robot will raise 20 million sterile mosquitoes for release in California. MIT Technol Rev. July 14. https://www.technologyreview.com/s/608280/alphabet-has-built-a-robot-that-is-releasing-millions-of-sterile-mosquitoes-in-california/. Accessed 21 July 2017
39. Crawford J (2017) Debug Fresno, our first U.S. field study. Verily blog. July 14. https://blog.verily.com/2017/07/debug-fresno-our-first-us-field-study.html. Accessed 21 July 2017
40. Reuters (2016) Here's how microsoft & other tech firms have waged a war against disease-carrying mosquitoes. ScoopWhoop News. https://www.scoopwhoop.com/battle-against-zika-virus-google-and-microsoft-have-come-together-to-fight-it-off/#.xdkc5md04. Accessed 21 July 2017
41. MSU Today (2017) MSU lands $1 million USAID grant to fight Zika. MSU Today. January 5. http://msutoday.msu.edu/news/2017/msu-lands-1-million-usaid-grant-to-fight-zika/. Accessed 26 May 2017
42. Naigulevu F (2018) Wolbachia-carrying mosquitoes to be released in Suva today. Fiji Times. October 11. https://www.fijitimes.com/wolbachia-carrying-mosquitoes-to-be-released-in-suva-today Accessed 23 May 2022
43. Regalado A (2016) Are altered mosquitoes a public health project, or a business? October 27. Technol Rev. https://www.technologyreview.com/s/602720/are-altered-mosquitoes-a-public-health-project-or-a-business/. Accessed 29 May 2017
44. Dutra H et al (2016) Wolbachia blocks currently circulating Zika virus isolates in Brazilian *Aedes* aegypti mosquitoes. Cell Host Microbe 19. June 8. https://www.ncbi.nlm.nih.gov/pubmed/27156023. Accessed 12 April 2017
45. Lambrechts L et al (2015) Assessing the epidemiological effect of wolbachia for dengue control. Lancet Infect Dis 15. June 4. http://thelancet.com/journals/laninf/article/PIIS1473-3099(15)00091-2/fulltext. Accessed 19 May 2017; see Hoffmann AA et al (2011) Successful establishment of *Wolbachia* in *Aedes* populations to suppress dengue transmission. Nature 476:7361. August 25. https://doi.org/10.1038/nature10356; Walker T et al (2011) The wMel wolbachia strain blocks dengue and invades caged *Aedes* aegypti populations Nature 476:7361. https://doi.org/10.1038/nature10355
46. O'Neill SL et al (2018) Scaled deployment of wolbachia to protect the community from *Aedes* transmitted arboviruses. Gates Open Res 2:36. August 16. https://gatesopenresearch.org/articles/2-36/v1. Accessed 10 Sept 2018
47. Araujo HRC, Carvalho D, Ioshino R, Costa-da-Silva A, Capurro M (2015) *Aedes* aegypti control strategies in Brazil: incorporation of new technologies to overcome the persistence of dengue epidemics. Insects 6:576–594
48. Brelsfoard CL, Mains JW, Mulligan S, Cornel A, Holeman J et al (2019) Aedes aegypti males as vehicles for insecticide delivery. Insects 10:230. https://doi.org/10.3390/insects10080230
49. Graham K (2017) Lab-grown mosquitoes to combat disease carrying mosquitoes. Digit J. November 9. http://www.digitaljournal.com/tech-and-science/science/startup-given-epa-approval-to-release-lab-grown-mosquitoes/article/507242. Accessed 10 July 2018
50. Center for Food Safety (2016) This election, key residents vote "NO" on GE mosquitoes. Santa Cruz IMC. November 9. https://www.indybay.org/newsitems/2016/11/09/18793252.php. Accessed 3 April 2017
51. Huang Z (2016) A Chinese "mosquito factory" releases 20 million of the little buggers into the wild every week. Quartz. March 16. https://qz.com/640394/a-chinese-mosquito-factory-releases-20-million-of-the-little-buggers-into-the-wild-every-week/. Accessed 15 May 2017
52. Interlandi J (2016) Fighting Zika with genetically modified mosquitoes. Consum Rep. September 14. http://www.consumerreports.org/conditions-treatments/fighting-zika-with-genetically-modified-mosquitoes/. Accessed 23 Sept 2016
53. Atkins K (2017) First travel-related Zika case of 2017 reported. Florida Keys News. February 15. http://www.flkeysnews.com/news/local/environment/article132829289.html. Accessed 7 March 2017
54. Goldschmidt D (2017) Florida releases experimental mosquitoes to fight Zika. CNN. April 20. http://www.cnn.com/2017/04/20/health/florida-mosquito-wolbachia-trial-zika/. Accessed 19 May 2017

55. Yong E (2016) How to beat dengue and Zika: add a microbe to mosquitoes. The Atlantic. August 8. https://www.theatlantic.com/science/archive/2016/08/how-to-beat-dengue-and-zika-add-a-microbe-to-mosquitoes/494036/. Accessed 2 July 2017
56. Hoffman AA et al (2011) Successful establishment of Wolbachia in *Aedes* populations to suppress dengue transmission. Nature 476. August 25. https://www.nature.com/nature/journal/v476/n7361/full/nature10356.html. Accessed 17 July 2017
57. Branigan T (2015) Sterile mosquitoes released in China to fight dengue fever. The Guardian. May 24. https://www.theguardian.com/world/2015/may/24/sterile-mosquitoes-released-in-china-to-fight-dengue-fever. Accessed 8 March 2017
58. Callaway E (2018) Dengue rates plummet in Australian city after release of modified mosquitoes. Citing SL O'Neill et al. Gates Open Res 2:36. August 8
59. Prasad R (2017) GM mosquito trials to control dengue, chikungunya launched. The Hindu. January 25. http://www.thehindu.com/sci-tech/health/GM-mosquito-trials-to-control-dengue-chikungunya-launched/article17093840.ece. Accessed 14 May 2017
60. WHO (2016) Promising new tools to fight *Aedes* mosquitoes. Bull World Health Organ 94:562–563. http://www.who.int/bulletin/volumes/94/8/16-020816/en/. Accessed 2 Oct 2016
61. Collier R (2009) Revised WHO pandemic scale requires higher incidence of disease for most alert levels. Can Med Assoc J 180:12. June 9. https://www.ncbi.nlm.nih.gov/pmc/articles/PMC2691422/. Accessed 18 Sept 2018
62. McCay B (2016) Mosquitoes are deadly, so why not kill them all? Wall Street J. September 2. http://www.wsj.com/articles/mosquitoesaredeadlysowhynotkillthemall1472827158. Accessed 15 Sept 2016
63. Radio New Zealand (2016) New Caledonia will fight disease-carrying mosquitoes with a bacteria. Radio New Zealand. http://www.radionz.co.nz/international/pacific-news/321346/new-caledonia-will-fight-disease-carrying-mosquitoes-with-a-bacteria. Accessed 14 May 2017
64. Grens K (2017) Bacteria-treated mosquitoes released in more locations. The Scientist. January 17. http://www.the-scientist.com/?articles.view/articleNo/48030/title/Bacteria-Treated-Mosquitoes-Released-in-More-Locations/. Accessed 14 May 2017
65. Moses S-L (2016) Philanthropy versus mosquitoes: the funders giving big money to fight a tiny insect. Inside Philanthropy. December 6. https://www.insidephilanthropy.com/home/2016/12/6/attacking-killer-mosquitoes-these-funders-are-throwing-big-money-to-fight-a-tiny-insect. Accessed 26 May 2017
66. Yong E (2016) The answer to Zika may be more mosquitos. The Atlantic. October 27. https://www.theatlantic.com/science/archive/2016/10/zika-fighting-mosquitoes-take-off-in-south-america/505523/. Accessed 2 July 2017
67. AFP (2016) Mutant mosquitoes to breed out diseases. October 30. http://www.newvision.co.ug/new_vision/news/1438897/mutant-mosquitoes-breed-diseases. Accessed 22 Dec 2016
68. Klingener N (2017) Keys starts trial of new mosquito fighting method: bacteria. WLRN. April 18. http://wlrn.org/post/keys-starts-trial-new-mosquito-fighting-method-bacteria. Accessed 17 May 2017
69. Cara E (2017) Scientists are turning mosquitoes against each other in Florida. AOL News. April 20. https://www.aol.com/article/news/2017/04/20/scientists-are-turning-mosquitoes-against-each-other-in-florida/22047814/. Accessed 5 May 2017
70. Kay J (2017) Florida tests bacteria-infected mosquitoes to control Zika. York dispatch. April 18.http://www.yorkdispatch.com/story/news/2017/04/18/florida-tests-bacteria-infected-mosquitoes-control-zika/100622084/. Accessed 17 May 2017
71. Atkins K (2016) Wolbachia-infected bugs approved for March trial. Florida Keys News. October 19. http://www.flkeysnews.com/news/local/article109137307.html. Accessed 25 Oct 2016
72. CBS Miami (2018) The war against Zika: fighting mosquitoes with mosquitoes. February 10. https://miami.cbslocal.com/2018/02/08/zika-fighting-mosquitoes-with-mosquitoes/. Accessed 10 July 2018

73. Shastri D (2017) Mosquito battle gets political. J Sentinel. October 5. https://projects.jsonline.com/news/2017/10/5/mosquito-battle-gets-political.html. Accessed 25 July 2018
74. Webster R (2016) Letter to the editor: huge risks to wolbachia-based mosquito control. November 18. http://www.pattayamail.com/mailbag/huge-risks-wolbachia-based-mosquito-control-155486. Accessed 30 June 2017
75. Popovici J et al (2010) Assessing key safety concerns of a wolbachia-based strategy to control dengue transmission by *Aedes* mosquitoes. Memórias do Instituto Oswaldo Cruz 105:8. https://www.ncbi.nlm.nih.gov/pubmed/21225190. Accessed 15 May 2017
76. Keating A (2018) New app identifies mosquito species and wolbachia. Eureka Alert. August 30. https://www.eurekalert.org/news-releases/671992. Accessed 23 May 2022; Bhadra S, Riedel TE, Saldaña MA, Hegde S, Pederson N et al (2018) Direct nucleic acid analysis of mosquitoes for high fidelity species identification and detection of Wolbachia using a cellphone. PLOS Neglected Trop Dis 12(8):e0006671. https://doi.org/10.1371/journal.pntd.0006671
77. Blanchard S (2018) Zika-carrying mosquitoes can be tested with smartphones to see if they are likely to carry deadly disease. The Daily Mail. September 26. https://www.dailymail.co.uk/health/article-6209335/Zika-carrying-mosquitoes-tested-smartphones.html. Accessed 11 June 2019
78. Ulrich JN et al (2016) Heat sensitivity of wMel wolbachia during *Aedes* aegypti development. PLoS Neglected Trop Dis 10(7). http://journals.plos.org/plosntds/article?id=10.1371/journal.pntd.0004873. Accessed 31 May 2017
79. Ross PA et al (2017) Wolbachia Infections in *Aedes* aegypti differ markedly in their response to cyclical heat stress. PLOS Pathogens. January 5. http://journals.plos.org/plospathogens/article?id=10.1371/journal.ppat.1006006. Accessed 19 July 2017
80. Datuk F (2017) IMR to use wolbachia strain that is heat tolerant. The Star Online. March 31. http://www.thestar.com.my/opinion/letters/2017/03/31/imr-to-use-wolbachia-strain-that-is-heat-tolerant/. Accessed 17 May 2017
81. Shanmugavelu S, Purushothaman J (2012) What ails wolbachia transinfection to control disease vectors? Trends Parasitol 28(1). January. https://www.ncbi.nlm.nih.gov/pubmed/22079163. Accessed 2 June 2017
82. Medical Xpress (2017) Bacteria deployed to destroy mosquito borne dengue can't take the heat. Medical Xpress. January 6. https://medicalxpress.com/news/2017-01-bacteria-deployed-mosquito-borne-dengue.html. Accessed 22 May 2017
83. Hurst T (2012) Impacts of wolbachia infection on predator prey relationships: evaluating survival and horizontal transfer between wMelPop infected *Aedes* aegypti and its predators. J Med Entomol 49(3). https://www.ncbi.nlm.nih.gov/pubmed/22679870. Accessed 15 May 2017
84. Health BD et al (1999) Horizontal transfer of wolbachia between phylogenetically distant insect species by a naturally occurring mechanism. Curr Biol. March 15. https://www.ncbi.nlm.nih.gov/pubmed/10209097. Accessed 21 July 2017
85. Morphd (2016) Comment: Florida keys may determine whether GMO mosquitoes will be employed globally to fight Zika. FiveThirtyEight. October 19. https://www.geneticliteracyproject.org/2016/10/19/florida-keys-may-determine-whether-gmo-mosquitoes-will-employed-globally-fight-zika/. Accessed 16 Jan 2017
86. Hotez P (2016) Zika is coming. The New York times. April 8. https://www.nytimes.com/2016/04/09/opinion/zika-is-coming.html?_r=0. Accessed 15 May 2017
87. Allen G (2016) Florida keys approves trial of genetically modified mosquitoes to fight Zika. NPRNow. http://www.npr.org/sections/health-shots/2016/11/20/502717253/florida-keys-approves-trial-of-genetically-modified-mosquitoes-to-fight-zika. Accessed 16 Jan 2017
88. Jozuka E (2016) Inside China's 'mosquito factory' fighting Zika and dengue. CNN. December 28. http://www.cnn.com/2016/12/28/health/china-mosquito-factory-zika-dengue/. Accessed 17 May 2017
89. Ross PA et al (2017) Wolbachia infections in *Aedes* aegypti differ markedly in their response to cyclical heat stress. PLOS Pathog. January 5. http://journals.plos.org/plospathogens/article?id=10.1371/journal.ppat.1006006. Accessed 14 May 2017

90. RoseWrites (2016) Wolbachia infected mosquitoes might reduce dengue, enhance Zika, and cause a million souls to become sterile. InfoBarrel. November 25. http://www.infobarrel.com/Wolbachia-Infected_Mosquitoes_Might_Reduce_Dengue_Enhance_Zika_and_Cause_a_Million_Souls_to_Become_Sterile. Accessed 31 May 2017
91. Pietri JE, DeBruhl H, Sullivan W (2016) The rich somatic life of wolbachia. Microbiol Open. May 28. https://doi.org/10.1002/mbo3.390
92. Loreto E, Wallau G (2016) Risks of wolbachia mosquito control. Science 351:6279. March 18. http://science.sciencemag.org/content/351/6279/1273.2. Accessed 17 July 2017
93. Upton J (2014) Tackling west nile with bacteria may worsen the disease. Pac Stan. July 11. https://psmag.com/social-justice/west-nile-treatment-problem-85708#.p88je2cmi. Accessed 15 May 2017
94. Dodson B et al (2014) Wolbachia enhances west nile virus (WNV) infection in the mosquito culex tarsalis. PLOS Neglected Trop Dis 8:7. July 10. http://journals.plos.org/plosntds/article?id=10.1371/journal.pntd.0002965. Accessed 26 June 2017
95. Taylor MJ (2003) Wolbachia in the inflammatory pathogenesis of human filariasis. Ann New York Acad Sci 990. https://www.ncbi.nlm.nih.gov/pubmed/12860672. Accessed 31 May 2017
96. Astin J (2017) Shock claim: lab created super-mosquitos released into wild could 'make all men infertile'. Express. November 15. https://www.express.co.uk/news/science/880050/Lab-designed-mosquito-released-wild-USA-Wolbachia-Asian-Tiger-Mosquito. Accessed 23 July 2018
97. ITV News (2018) Scientists make breakthrough in fight against Zika. January 26. ITV Report. http://www.itv.com/news/2018-01-26/scientists-make-breakthrough-in-fight-against-zika/. Accessed 10 July 2018
98. Brownstein JS, Hetta AZ, O'Neill SL (2003) The potential of virulent wolbachia to modulate disease transmission by insects. J Invertebr Physiol 84:24–29. http://www.sciencedirect.com/science/article/pii/S002220110300082X. Accessed 16 March 2017
99. Ferguson N et al (2015) Modeling the impact on virus transmission of wolbachia-mediated blocking of dengue virus infection of *Aedes* aegypti. Sci Trans Med 7:279. March 18. http://stm.sciencemag.org/content/7/279/279ra37. Accessed 14 April 2017
100. Saloman L (2019) Combination birth control shows promise in wiping out mosquito population. ContagionLive. August 13. Combination birth control shows promise in wiping out mosquito population (contagionlive.com). Accessed 23 May 2022
101. Sifferlin A (2017) There's a new way to wipe out mosquitoes in the U.S. time. November 16. http://time.com/5025250/mosquito-control-mosquitomate/. Accessed 10 July 2018
102. Crawford JE, Clarke DW, Criswell V, Desnoyer M, Cornel et al (2020) Efficient production of male wolbachia-infected Aedes aegypti mosquitoes enables large-scale suppression of wild populations. Nat Biotechnol 38:482–492. April 6; Bouyer J, Maiga H, Vreyen JB (2022) Assessing the efficiency of Verily's automated process for production and release of male wolbachia-infected mosquitoes. Nat Biotechnol. May 26. https://doi.org/10.1038/s41587-022-01324-z; Crawford JE, Hopkins KC, Buchman A, Zha T et al (2022) Reply to: assessing the efficiency of Verily's automated process for production and release of male wolbachia-infected mosquitoes. Nat Biotechnol. May 26. https://doi.org/10.1038/s41587-022-01325-y
103. Savike J (2018) Plans to increase wolbachia carrying mosquitoes. Fiji Times. October 13. The Fiji times » plans to increase wolbachia carrying mosquitoes. Accessed 23 May 2022
104. Aliota M et al (2016) The wMel strain of wolbachia reduces transmission of Zika virus by *Aedes* aegypti. Nat: Sci Rep 6:28792. https://doi.org/10.1038/srep28792. http://www.nature.com/articles/srep28792. Accessed 16 Jan 2017
105. Waltz E (2017) US government approves 'killer' mosquitoes to fight disease. Nature. November 6. https://www.nature.com/news/us-government-approves-killer-mosquitoes-to-fight-disease-1.22959. Accessed 10 July 2018
106. Ulrich JN et al (2016) Heat sensitivity of wMel wolbachia during *Aedes* aegypti development. PLoS Neglected Trop Dis. 10(7). http://journals.plos.org/plosntds/article?id=10.1371/journal.pntd.0004873. Accessed 20 July 2017

107. Ross PA et al (2017) Wolbachia Infections in *Aedes* aegypti differ markedly in their response to cyclical heat stress. PLOS Pathog. January 5. http://journals.plos.org/plospathogens/article?id=10.1371/journal.ppat.1006006. Accessed 7 Feb 2017
108. Nguyen TH et al (2016) Field evaluation of the establishment potential of wmelpop wolbachia in Australia and Vietnam for dengue control. Parasites Vectors. October 28. https://parasitesandvectors.biomedcentral.com/articles/10.1186/s13071-015-1174-x. Accessed 26 May 2017
109. RoseWrites (2016) Why we need to investigate wolbachia infected mosquito releases. InfoBarrel. December 31. http://www.infobarrel.com/Why_We_Need_to_Investigate_Wolbachia-Infected_Mosquito_Releases. Accessed 31 May 2017
110. Upton J (2014) Tackling west nile with bacteria may worsen the disease. Pacific Stan. July 11. https://psmag.com/social-justice/west-nile-treatment-problem-85708. Accessed 9 June 2017
111. Sanders R (2002) Bizarre parasite that kills male insects and disrupts insect sex lives is not all bad: it can make sterile fruit flies fertile again. Berkeley Press Release. July 3. http://www.berkeley.edu/news/media/releases/2002/07/03_paras.html. Accessed 30 May 2017

Chapter 14
Oxitec

The alternatives to genetically engineered (GE) mosquitoes are more significant exposure to mosquitoes and the health implications of ZIKV, and other dangerous vector-borne diseases, or those from greater mass spraying of dangerous insecticides. The good news is that even though opponents are loud, most Americans hate mosquitoes [1]. The ZIKV issue may be a bellwether for the entire genetic engineering movement in the U.S. How it is handled may have significance beyond the case of ZIKV since their approach has been justified in cases of diseases like dengue malaria.

A fascinating comment surfaced in October 2016. "It appears ZIKV is a well-timed "New" malady to try to scare us to genetically modified organism (GMO) mosquitoes" [2]. Whether this is true or not is irrelevant. For too many opponents to genetic engineering, ZIKV and the other well-known arboviruses are being used by scientists in genetics to impose upon the public genetic engineering as a solution to some of these most pressing problems. The highly suspicious nature of many opponents to genetic engineering has framed their understanding of population crash as a vector control for some of the heinous diseases brought by the bite of the mosquito as a trick to try to sneak genetic engineering into the playbook by which some other thorny issues are resolved.

The mosquito genocide is beginning. Millions of genetically modified versions of the useless vampire insects have been released in Brazil and elsewhere. If all goes according to plan, the mosquitoes will have a huge sex party and begin to kill off all their natural counterparts [3].

Recent technological advances have made disease control strategies using GM vectors more plausible. Selecting an appropriate field site for research with GM mosquitoes may be one of the research process's most complex and significant aspects.

Among the critical considerations of the process is the need to address ethical, legal, and cultural (ESC) issues. When ZIKV broke, there were no guidelines had been developed for this complicated and sensitive process [4]. Recent WHO guidelines have greatly improved conditions for researchers in the 2020s [5].

D. M. Berube, *Pandemics and Resilience: Lessons we should have learned from Zika*, Risk, Systems and Decisions, https://doi.org/10.1007/978-3-031-25370-6_14

Like it or not, the concept of genetically modified organisms or GMOs carries a stigma for some, which has not been helped by years of horror movies about scientists' creations running amuck; "when people think of GMOs or GM products, even insects, they often think of an organism that has been designed to persist in the environment," Oxitec's former product development manager Derric Nimmo says [6]. Dr. Nimmo moved on to another organization a few years ago.

Site selection is an essential consideration in research with GM insects. It is the first formal step in establishing long-term research collaborations between researchers from high-income countries and collaborators from the host endemic sites, primarily in low- and middle-income countries [4]. Early during a pandemic, site selection if done poorly could implicate other opportunities associated with genetic engineering.

While genetic modification was a beacon for the future's glittering possibility at one point in history, today, "genetic" and "modified" can scarcely be uttered without causing a fuss [7]. This reaction may be more strongly associated with the United States than anywhere else. A 2016 PEW survey asked 4726 people how they felt about gene editing (next chapter) and other human-enhancing technologies. More than 60% said that they were "very" or "somewhat" worried about technologies that could make people brighter, healthier, and more robust, including gene editing to reduce the risk of disease in babies, brain chips for cognitive enhancement, and synthetic blood transfusions to improve speed, strength, and stamina [8].

Interestingly, the more minor or impermanent the change, the easier people found it to swallow. Only 28% of adults, for example, said a synthetic blood transfusion would be an appropriate use of technology if it produced enhancements "far above that of any human known to date." But a more significant 47% said synthetic blood products would be okay if the magnitude of change were smaller. More than a third of respondents said implants that could enhance cognitive ability would be more acceptable if they could be turned on and off. More than half said brain chips would be less suitable if the results were irreversible [8].

During ZIKV outbreak, none of the GE methods for vector control have been deployed in the field long enough to test their long-term stability empirically [9]. Oxitec claims in 2022 to have released "more than 1 billion Friendly™ males with no evidence of ill effects, etc. And it is now fully deregulated (i.e., approved for biosafety) by Brazilian national regulators—it is self-limiting in the environment, the science shows it is safe, and it is proven to suppress target *Aedes*" [10].

The looming question for these projects remains simple: Do these disease-fighting mosquitoes reduce disease? "The goal here is not to kill mosquitoes. It is to prevent people from getting infected, sick, and dying," says Thomas Scott, an epidemiologist and insect ecologist at the University of California, Davis. So far, there is no proof that either approach does that, and this remains a challenge for the industry. It may seem intuitive that fewer infectious mosquitoes mean fewer infections, but just a few *Ae. aegypti* may be enough to transmit disease through a susceptible population, Scott says [11].

This is a long chapter, but this is a big issue. After following this company for over half a decade, Oxitec qualifies as the shaker and mover in vector control of the *Ae. aegypti* mosquito. The Oxitec website remains an important source for information.

In addition, it has attracted a lot of pushbacks that may be especially useful when beginning to design messages for the public and the regulators. Maintaining objectivity in this discursive examination has been complex and challenging, and I hope you are forgiving.

14.1 Introducing Oxitec

Oxitec was founded in 2002 as a start-up from Oxford University and has changed dramatically since then. Below there are the names of many former employees who at the time of the Brazilian outbreak were active in the company. The first time their names are mentioned they are identified as former. There are two different Oxitec GE mosquitoes. At the time of the Brazilian outbreak, it was OX513A. In 2019 that approach was retired and OX5034 rose in prominence.

Oxitec claims many of the concerns about OX513A are now historical artifacts. Since this chapter is not a press release for the company, extensive effort has been made to keep this distinction prevalent and the tenses accurate. While ZIKV was a pandemic the GE mosquito was OX513A, today there is OX5034 and the distinctions between them are described below.

Oxitec is one of a few successful organizations applying genetic engineering for the control of insect pests. Oxitec has developed a mosquito control program that adapts the sterile insect technique (SIT). This methodology has successfully managed several insect species in different countries over the last 50 years using radiation-based sterilization. In the 1950s, the sterile insect technique was used to eliminate the screwworm, and the responsible entomologists were awarded the World Food Prize in 1992 [12].

Absent a vaccine against ZIKV, Oxitec's technology or one like it could be the most effective tool in fighting ZIKV. GE mosquitoes could go a long way toward fighting some of the world's deadliest viruses. The bigger question is whether we will let them [13].

While **Oxitec's approach is not a gene drive approach**, trying to convince the opponents and the public of the difference might be one of the most challenging scientific arguments for the public to digest.

Also, Oxitec is not trying to eliminate ***all*** mosquitoes. "There are over 3000 species of mosquitoes and discussed here are about one, two or every three. Oxitec is getting rid of one mosquito species from a localized population to stop it from transmitting pathogens to humans," says the University of California, San Diego molecular biologist Omar Akbari to Donavyn Coffey at Scientific American.

"And this mosquito species—*Ae. aegypti*—is invasive and does not have an important purpose in this environment." Science writer Theresa Machemer and most others think there "will not be any negative impact from removing the species from the environment" [14].

Oxitec's original approach involved engineering male mosquitoes to die after procreation as they are genetically programmed to do. Since starting this book, it is noteworthy that Oxitec has generated its 2.0. Oxitec is already piloting its second-generation *Ae. aegypti* mosquito in Brazil after securing regulatory approval for field releases there. It is reviewed below.

In general, male mosquitoes do not bite humans or animals and therefore cannot transmit vector viruses or other saliva constituents [15]. When the modified OX513A, male mosquitoes were released into the wild, they mated with wild females. The females passed on the death gene to their eggs, which died before they could breed or spread the disease. [16]. The male mosquito's offspring would carry an inbuilt flaw that will cause them to pass quickly. Oxitec bred the so-called almost whimsically Friendly™ *Ae. aegypti.* The OX513A approach involved the modification of two genes to produce: conditional lethality and fluorescence.

This was Oxitec's first entry into the vector control field involving ZIKV involving a self-limiting strategy. In its self-limiting processes, the modification would disappear from the target population unless replenished by the periodic release of additionally modified insects (Fig. 14.1).

In contrast, autonomous strategies use a change intended to persist indefinitely and perhaps even spread within the initial target population or other populations. The gene drive systems (following chapter) are designed to apply within the target population and are almost invariably self-sustaining [17].

The sterile male methods are self-limiting as the lethal or sterile factor disappears rapidly from the target population, maintained only by periodic release of additionally modified males.

OX513A male adults required three resources: (a) access to plant sugars for food, (b) mates, and (c) resting sites. Female adults needed the same three resources and two more (e) a bloodmeal and (d) oviposition sites to lay eggs. The females of their offspring would die before adulthood, while the surviving males could mate again with wild females. The self-limiting gene could survive for up to ten generations, after which no genetically modified mosquitoes remain [18].

OX513A Friendly™ *Aedes*' offspring also inherited a fluorescent marker that allowed tracking and monitoring at a level never achieved, making the effectiveness assessment more accurate throughout the whole Friendly™ *Aedes* deployment program [19]. Oxitec could collect larvae from the area of release and use the color marker to see how many Oxitec mosquitos there are and how many are wild pest mosquitos. Using this color marker, they could adjust the number of Oxitec mosquitos needed to release and monitor their mating success [20]. Unlike other approaches, Friendly™ Aedes mosquitoes died with their offspring and did not persist in the environment or leave any ecological footprint [19]. The OX513A Oxitec approach reduced the population by introducing males that were unable to survive without tetracycline.

This approach was retired. The OX5034 version (see below) induces female-specific mortality in all female carriers and has been demonstrated as being highly effective in suppressing *Ae. aegypti* in densely populated urban areas (see press releases for more details, in Oxitec news page on website; peer-reviewed paper publication pending).

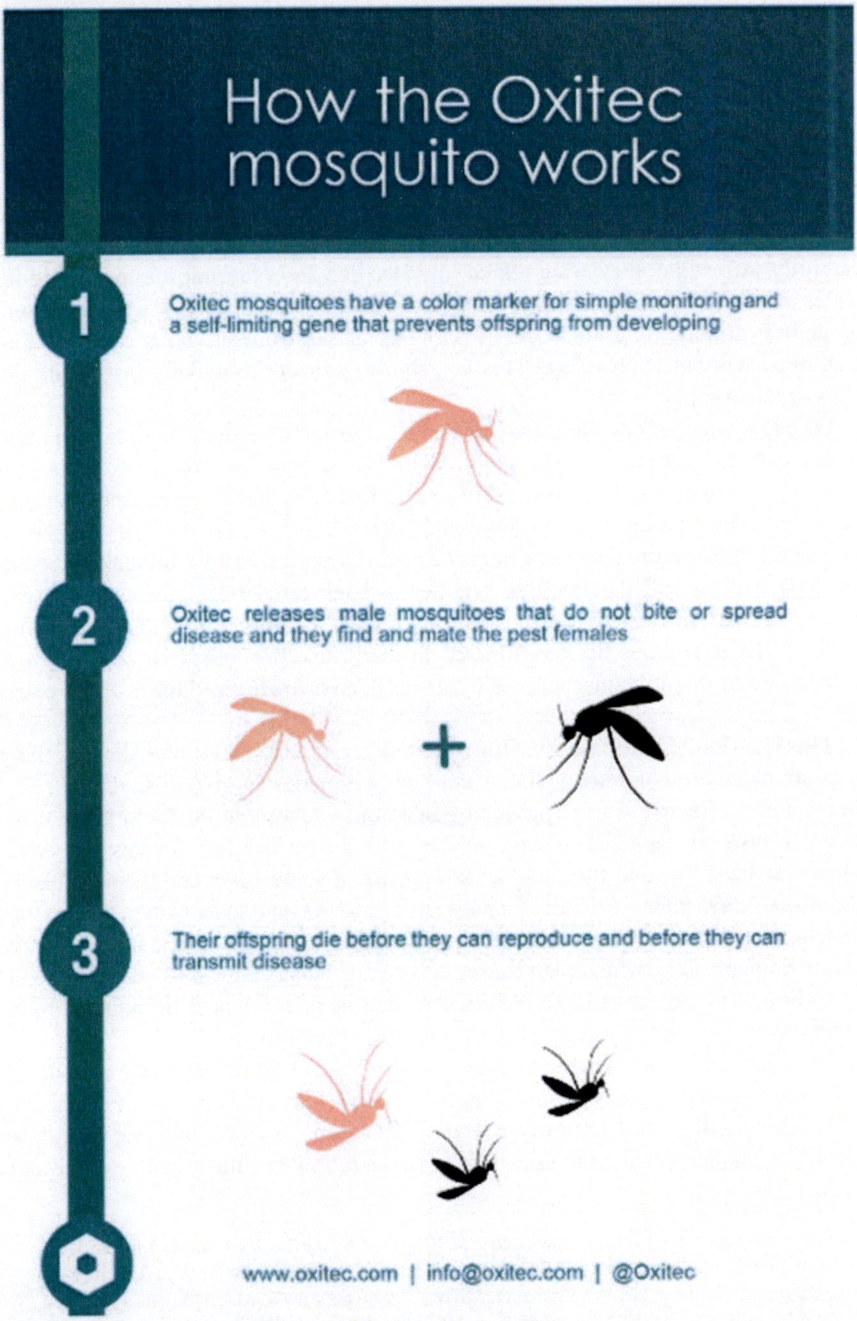

Fig. 14.1 How the Oxitec mosquito works. Image courtesy Oxitec

With the original approach involving OX513A, the genetically modified males bred with normal females, and most offspring died before becoming adults. This approach made the larvae dependent on the presence of a tetracycline antibiotic; in its absence, only 3–4% of the larvae survive as adults. This suppressed the population, including female mosquitoes, which are the ones that bite and infect.

The tetracycline diluted in water was used only during the larval period in the rearing process to turn off the self-limited gene to allow the development of males from eggs to adults. Once they develop into pupae, the mosquitoes are washed carefully. After that, they were reared in water like any other mosquito. The adult males were released. This conditional lethality trait or "self-limiting" trait prevented progeny from inheriting the #OX513 rDNA construct from surviving to functional adulthood without tetracycline. This is a similar concept to making insects sterile with irradiation [21].

OX513A mosquitoes were reared in the presence of tetracycline as a dietary supplement, as such the expression of tTAV (a cellular protein) is repressed, allowing normal cell function and survival. tTAV in quantities that cause cell malfunction and more than 95% mortality before adulthood [22].

The OX513A approach that was introduced during the ZIKV outbreak involved the repeated controlled release of genetically engineered (GE) male *Ae. aegypti* mosquitoes (strain OX513A) expressing conditional lethality traits and a fluorescent marker. The strain was first constructed in 2002, and a publication about it in a peer-reviewed scientific journal in 2007. It has been characterized for over ten years [23].

This is no longer the case with the updated version OX5034. With the OX5034, eggs are placed in non-tetracycline water and only males survive to adulthood. So, this is a cost-effective way of producing male-only cohorts in the factory, but now they can also be deployed as eggs in the field. Just-add-water devices are used, which can be shipped to the end-user who pours in some water and the eggs hatch and—a few days later—Friendly™ males emerge to find and mate with pest females. Developmental failure occurs when the cells cannot make the proteins they require to function normally, which then causes cell death. This is known as transcriptional squelching [21]. Because this sterile insect technique does not push through multiple generations, some saw this approach as much less disruptive.

Enter Randal J. Kirk. Intrexon acquired Oxitec, the Oxford University spinout, in 2015. An industry analysis published in May 2016 by Spotlight Research characterized Kirk's holdings as "an intricate web of microcap, zero revenue, free cash flow negative companies that seem [allegedly] to exist to inflate [Intrexon's] revenue and profitability" [24].

> Kirk, who has homes in San Francisco and Virginia but spends most of his time in West Palm Beach, Fla., sold his drug distribution company, General Injectables & Vaccines, in 1998 for $65 million. He then started New River Pharmaceuticals, which the Shire acquired for $2.6 billion in 2007, followed by Clinical Data, a genetic-testing company acquired by Forest Laboratories for $1.2 billion in 2011.
>
> Around 2004, he invested $300 million into Intrexon, which a molecular geneticist named Thomas Reed founded in 1998 as the human genome was being sequenced, assembling a

library of standardized DNA components— "hot rod parts," as Kirk calls them—that could be used to make designer genes. Kirk took the company public in 2013.

Before acquiring Oxitec, Intrexon bought AquaBounty Technologies, which genetically modifies salmon to grow faster, and Okanagan Specialty Fruits, which makes apples that don't turn brown. [24]

Intrexon (then and former owner of Oxitec) is based in the US. It has been around for more than 17 years and commercialized effectively zero meaningful commercial products using its technology.

It has experienced its own controversies. Nobody can explain exactly what it does not its employees, not sell-side analysts, and not the people who own the stock. The company can only generate revenue through related party transactions and selling stock. The fair value of this company is minuscule compared to its current $4.5 billion market capitalization [25]. However, genetically engineered mosquitoes, high-protein insect fish feed, and non-browning apples could cause Intrexon's stock to rise [26]. Analysts predict Oxitec's mosquito could bring up to "$400 million in annual sales for its parent company, Intrexon" [27].

To add more drama to the mix, it was reported that the Rosen Law Firm was investigating Intrexon Corporation on allegations of issuing misleading business information to the investing public. The speculation involved the effect of this information on depressing the stock price before significant contractual opportunities come from the Oxitec initiative. Current investors saw the stock price drop by 25%, making it available for wily investors to buy cheap and reap significant profits.

Seeking Alpha released an anonymous report, and shares of Intrexon fell nearly 30% to $27.09 (April 21) based on the information and only gained back 7% (April 22, 2016). Intrexon said the "materially false and misleading" report appears to be part of a hedge fund campaign to manipulate the company's stock and damage its reputation of the company. It smelled like a pump and dump scheme where short-sellers plant rumors to drive down the price.

Some parts of the reports raised detailed concerns regarding the background of Intrexon's former CEO Randall Kirk (see above), a lawyer by trade, not a scientist. And suggested it is unclear if Intrexon produces anything. It has acquired companies that produce products and has partnerships [28]. Overall, the investment community found the report suspicious.

In December 2016, the Independent Republic reported: "The stock has a 1-month performance of 22.2% and is 0.66% year to date as of the recent close. About 118.35 million shares outstanding, which made its market cap $3.54 billion" [29].

In 2020, Oxitec was acquired by Third Security, a venture capital company in Radford, Virginia. Third Security is led by Intrexon Executive Chairman, Randal J. Kirk former Intrexon CEO and Third Security purchased Oxitec from Intrexon for $53 million in cash plus the contingent right to receive certain additional amounts that Third Security may earn from these assets after closing. In addition, Third Security has agreed to purchase from the company $35 million of shares of Intrexon's common stock.

Oxitec has received some notable support from The Bill & Melinda Gates Foundation which joined forces with Oxitec to breed an antimalaria GE mosquito. Oxitec received a sizable investment of over $4 million from the Bill and Melinda Gate Foundation to a lab creating mosquitoes that self-destruct by producing offspring that die before adulthood [30].

The Gates Foundation also supported antimalaria efforts involving the *Aedes* mosquito. Still, those efforts focus on a different strategy that uses Wolbachia bacteria (see previous chapter) to reduce the bugs' ability to transmit the disease. Gates supports a dengue initiative involving a team from Cal-Irvine led by Anthony James, and it did fund Oxitec for about $5 million [31].

In April 2021, Britain's Wellcome Trust, one of the world's largest charitable health foundations, gave the company $6.8 million to develop ways to release the *Ae. aegypti* mosquitoes more widely, including a large two-year area-wide project in Brazil, where regulators have already approved their routine use [32].

Finally, it should be noted that despite the story that follows regarding the 2016 petition to release GE mosquitoes in Key Haven, Florida that in April 2021, it was announced that Oxitec would release approximately 150,000 mosquitoes across six locations in Florida [33]. A continuation of the Florida Keys Mosquito Control District (FKMCD)-Oxitec Mosquito Project, Oxitec and FKMCD announced that a new phase of the project ("Pilot D") was initiated on or after July 7, 2022. This phase of the project examined single-point releases of Oxitec's male mosquitoes.

For the first part of this pilot project in 2021, Oxitec released under five million male, non-biting mosquitoes. The maximum number of non-biting male Oxitec mosquitoes released during the entire 2022 project is expected to be fewer than seven million [34].

Oxitec Ltd. had planned to build its first centralized Friendly™ *Aedes* egg production unit, which could generate one billion mosquito eggs per week, in Oxfordshire, UK, to help meet the increasing the global demand for *Ae. aegypti* control programs to reduce wild populations worldwide [35]. In 2018, the plan was for the GM male mosquitoes to be reared in Marathon, Nimmo told the Keynoter nearly five years ago. Today, mosquito eggs are shipped from Oxitec's lab in the U.K.—no colony rearing happens in Florida (e.g., no blood-feeding) [36]. Focus has shifted considerably to Brazil and Oxitec's Campinas facilities since the ZIKV outbreak.

Even the blood used for this process was collected using a sterile apparatus and processed aseptically from a closed herd of healthy animals permanently housed in the UK, under regular veterinarian supervision, which is screened for the virus, bacteria, and other pathogens to minimize the potential for contamination of the final defibrinated blood product [37].

This story focuses on an earlier attempt to affect population control of the *Ae. aegypti* mosquito was responsible for the spread of ZIKV during the Brazilian outbreak from 2015 to 2017.

Oxitec faced some stopgaps in Florida, but they have made an impressive treatment footprint elsewhere and eventually made it back to Florida in 2022.

Their original approach (OX513A) dates back to the original pandemic in Brazil and involved a self-limiting gene causing offspring to die and a marker gene to track them when released. OX513A is a bi-sex RIDL (release of insects with dominant lethality) strain. Males are released to mate with wild females. The progeny dies as late larvae. Continued releases reduce the target population sufficiently to impact the spread of the disease. The OX513A has demonstrated stable performance over sixty generations.

OX5034 is not OX513A. Consequently, the controversy that precipitated at the outbreak of Zika was not about OX5034. However, the purpose of this work is not about the technology per se but more about the action and reaction of communities and regulators, the cognitive consequences of the approach. Still, some of the claims and counterclaims in the debate over effectiveness and safety involving OX513A are mostly moot. The motivations for these debates remain relevant.

14.1.1 Claims of Success

Starting by examining the claims made by both sides in this debate as it occurred during the outbreak in 2015 and 2016. The focus of this work is the failure to learn the lessons from the past so a little time travel back to the pandemic is needed to frame the debate accurately. However, at the end of the chapter, the present will be revisited when the current state of the art in this technology is examined.

According to the **Oxitec Website**, their original argument was: "We have introduced a gene into the mosquitos so that their cells do not function as they should, and this effect is specific to the mosquito (also known as a 'self-limiting gene because it makes mosquito reproduction a dead-end—the offspring do not survive to adulthood). As a result, over 95% of Oxitec mosquitos die before becoming adults and do not live to breed further, reducing the population. *Ae. aegypti* does not produce offspring with other species, so the control is species-specific, and the genes do not spread. The released mosquitos and their offspring die, so the genes do not persist in the environment. The color marker and self-limiting genes are non-toxic and non-allergenic, so if Oxitec mosquitos or predators bit animals ate Oxitec mosquitos, it would be the same as getting bitten by or eating wild ones" [20].

Oxitec has conducted five open field trials between 2011 and 2014 of the self-limiting mosquitoes in Brazil, Panama, and the Cayman Islands in conjunction with independent collaborators. Each test saw a 90% reduction of the wild *Ae. aegypti* population within six months [38]. Oxitec claimed since 2009, their approach has suppressed *Ae. aegypti* mosquito population by 80–90% with no reports of adverse effects.

Oxitec mosquitoes would need to be released three times a week at locations some two hundred feet apart since mosquitoes fly only three hundred feet in any direction [39]. Oxitec's Nimmo states, "If you release enough of our males, over a long enough period, you can get the population to crash" [40].

Intrexon's former Head of Health Sector, Samuel Broder, offers "a new paradigm of species vector control resulting in dramatic reductions of dangerous mosquitoes without persistence or harm to the ecosystem" [41].

Oxitec not only wants to gain commercial approval for its mosquitoes but in the U.S., Brazil, India, and other places where it has projects in the works. This would make it readily available to any city that wants self-destructing mosquitoes to fight a local mosquito problem and hopes to court consumers.

Oxitec's former CEO Haydn Parry said it might sell its genetically modified mosquitoes at the garden store, competing with citronella candles to prevent bites in yards. "I can send you eggs, and you can protect your biosphere." In Parry's imagining, dried-out mosquito eggs would be sold alongside the ladybugs you can purchase to ward off aphids, ready to be hatched in under an hour in just a little water. But first, the company needs the world to be comfortable releasing GMOs that fly [42].

Since this approach dampens the population rather than changing the genome of the wild mosquito population itself, Oxitec would need to repeat the release of these modified mosquitoes, which brings along the US FDA. An online petition against the release garnered over 150,000 signatures, and opponents swarmed community meetings carrying "No Consent" signs. Oxitec has received more than 1200 public comments on the FDA's environmental impact assessment of mosquitoes [43].

14.1.2 Testing and Approval

Their investigational field trial aimed to evaluate the mating ability of released OX513A mosquitoes with local wild-type *Ae. aegypti* females, to assess the survival of the resultant progeny to estimate mortality related to the inheritance of the OX513A rDNA construct and determine the efficacy of sustained releases of OX513A mosquitoes in the suppression of a local population of *Ae. aegypti* in the defined release area [15]. They were initially held in highly controlled environments. The likelihood of escape, survival, and establishment of OX513A would be highly improbable due to a combination of physical, geophysical, geographic, and biological measures during egg production, transport, local rearing, and release. There are three levels of containment.

1. Geographic containment is provided by the siting of the egg production unit in the U.K., which is beyond the isothermal range of the mosquito (i.e., it is too cold for *Ae. aegypti* to survive outside the climate-controlled environment of the laboratory).
2. Geophysical containment is provided by the island location of the release site (original Key Haven site), where the area is predominantly surrounded by ocean, and the mosquito in any life stage cannot survive due to the high salinity of the waters.
3. Biological containment is afforded by introducing the conditional lethality trait into the OX513A *Ae. aegypti* line, where on mating with the local females of the same species, more than 95% of the progeny will die in the absence of tetracycline, leading to the overall reduction in the population of *Ae. aegypti* at a given site.

The trials were anticipated to be short, and any unanticipated adverse effects were unlikely to be widespread or persistent in the environment. Most importantly, the status of the domain was restored when releases were stopped (i.e., the released mosquitoes all die, and the setting reverts to the pretrial rate) [15].

Finally, data were to be collected via traps. Adult traps directly capture adult *Ae. aegypti*. As *Ae. aegypti* is a mosquito that lives near humans, traps were to be located predominantly by domestic dwellings, although other sites (e.g., garages, commercial buildings) could be included [15].

"After consideration, several countries approved limited field trials as part of an incremental testing and scale-up process. Public perception has generally been positive, though it is too soon to determine long-term trends beyond the case instant.

Public response to innovative technologies, including genetic control, may vary considerably depending on social, political, technical, epidemiological, presentational, and cultural factors. The genetic aspect is only one and may change over time, warned Oxitec's Luke Alphey" [17].

Countries seemed to be lining up. It was reported that the Netherlands was considering releasing Oxitec's genetically modified mosquitoes to fight dengue, chikungunya, and ZIKV in Saba, a Dutch Caribbean Island. A report published on July 7, 2017, by the National Institute of Public Health and the Environment (RIVM) concluded that releasing the mosquitoes would not pose any risks to human health and the environment [44].

France's High Council for Biotechnology (HCB) also backed the use, with caution, of genetically modified mosquitoes in France. Oxitec expects the report published by the HCB will help shape the policy to regulate the genetically engineered insects in France and eventually allow testing of the technology in the French Caribbean [44].

14.1.2.1 Oxitec Field Tests in Latin and South America

Beginning with a review of the field tests in Latin America, the hotspot for ZIKV during the epidemic. "In Brazil, the National Biosafety Technical Commission (CTNBio) gave its technical approval in April 2014 after deeming the release of these modified mosquitoes safe and risk-free [45].

Seven years after releasing the world's first genetically modified (GM) mosquito, Oxitec chose Brazil as the site of a significant scale-up. It was moving from small-scale pilot projects like the CECAP/Eldorado to planned releases covering tens of thousands of people [11].

Brazil's Piracicaba's CECAP/Eldorado district became the world's first municipality to partner directly with Oxitec and, in April 2015, started releasing its self-limiting mosquitoes whose offspring do not survive. Results: Wild mosquito larvae reportedly decreased by 82%, and dengue fever cases dropped by 91% in its first year of use [46]. This deal protecting a city of 300,000 people would generate up to $3 million in annual revenue [47].

The Municipality of Santiago de Cali, Colombia, and Oxitec Ltd. announced on April 10 a memorandum of understanding to deploy Friendly™ *Aedes* in the Comuna 16 region to help protect over 104,000 residents. Friendly™ *Aedes* is an innovative approach to fighting the *Ae. aegypti* mosquito transmits ZIKV, dengue, chikungunya, Mayaro, and yellow fever [48].

Oxitec Ltd. announced that Oxitec do Brasil signed a contract on July 11, 2017, to launch the Friendly (™) *Aedes* Project in Juiz de Fora, Brazil. Juiz de Fora became the second Brazilian municipality and the first city in Minas Gerais state to deploy innovative Friendly (™) *Aedes* in the fight against dangerous *Ae. aegypti* mosquitoes—the primary vector of dengue, ZIKV, chikungunya, and yellow fever. "Expansion of Oxitec's bio-based solution into a second state within Brazil is important as we seek to significantly lower the population of the primary vector of many dangerous arboviruses—*Ae. Aegypti*—across the entire country. The transmission of diseases has particularly impacted Minas Gerais through mosquitoes. We are enthusiastic about tackling this task head-on and helping protect its residents," said Dr. Mark Carnegie-Brown, Chief Executive Officer of Oxitec [49].

In Panama, the National Biosafety Commission and Ministries of Agricultural Development and Commerce and Industry approved in 2014 to conduct trials with these mosquitoes [45]. Similar trials have occurred across Brazil, Malaysia, and the Cayman Islands with 90% suppression. Oxitec has released more than 150 million GM-engineered mosquitoes in at least five field tests since 2011 in Brazil, Panama, and the Cayman Islands. According to Oxitec firm, five field tests conducted between 2011 and 2014—in Panama and the Cayman Islands, and the Northeastern Brazilian state of Bahia—showed the population of wild *Ae. aegypti* insects dropped by 90% after the mutant mosquitoes were released [50].

Brazil

Based on adult trap data, a 13-week field test in Itaberaba, a suburb of Juazeiro, was 95% successful [51]. There, mosquito larvae were reduced by 82% in 2017, which led to a massive 91% drop in dengue fever cases. Released in a neighborhood of Piracicaba, the company has achieved an 81% reduction in mosquito larvae in the second year. In addition, a new trial in a second, central neighborhood of Piracicaba has shown a 78% mosquito reduction in only six months [52].

Oxitec opened its new Friendly™ *Aedes* mosquito production facility in Piracicaba, Brazil. Touted as the "first and biggest factor" of genetically modified mosquitoes [53], the new 5000 m^2 (53,000 square foot) facility can produce sixty million Friendly™ *Aedes* per week. Oxitec plans to utilize a portion of the new facility's significant production capability to support its ongoing deployment program in Piracicaba's downtown area and CECAP/Eldorado district to suppress *Ae. aegypti* [54].

"The scalability of our biological solution took a meaningful step forward with the inauguration of this new world-class facility in Brazil," according to Lt. General (Ret.)

Thomas Bostick, Ph.D., senior vice president & head of Intrexon's Environment Sector [38].

Oxitec's facility in Piracicaba can protect 300,000 people from breeding up to three times the current output of China's largest mosquito factory [55]. The company is also setting its sights on the lucrative U.S. market [56].

The city expanded the Friendly™ *Aedes* project on September 6, 2016, to an additional thirteen neighborhoods by the end of the year, covering an area home to about 60,000 residents. This should be the largest deployment area ever for a genetically modified animal [56]. Oxitec do Brasil's former director Glenn Slade said the new facility took only five months to create and could quickly be reproduced elsewhere. He said the laboratory would increase Oxitec's production capacity 30-fold in the South American country [57].

The project was a partnership between Moscamed and the University of São Paulo (USP), and it has the support of the Ministry of Health (MOH), the Secretary of State of Bahia Health (SESAB), and the City of Juazeiro. Based on adult trap data, Carvalho et al. reported a 95% reduction in mosquito population in Juazeiro, Bahia, Brazil [58].

Brazil's Piracicaba's Epidemiologic Surveillance Service released data showing the incidence of dengue had decreased by 91% to just 12 cases in the CECAP/Eldorado district where Friendly™ *Aedes* mosquitoes were released, compared to a 52% reduction in the rest of the city during the same 12-month period [59]. During the first year of releases in Piracicaba's CECAP/Eldorado neighborhood, Friendly™ *Aedes* suppressed the wild larvae population of *Ae. aegypti* by 82% in the treated area versus the non-treated zone. In the second year, an 81% reduction was achieved with 59% fewer Friendly™ *Aedes* mosquitoes released than the year prior. In the São Judas neighborhood, where Friendly™ *Aedes* was first deployed in mid-2016, wild larvae decreased 78% compared to the non-treated area within just six months [19]. This is evidence of the strength and durability of Oxitec's solution [49]. Oxitec seems to have demonstrated that its approach can sustain mosquito reduction long-term.

Since the OX513A approach suppressed populations locally and did not extinguish them, it had to be repeated. Brazil's National Biosafety Committee (CTNBio) approved its commercial use and received a sales permit from Brazil's Anvisa (Agência Nacional de Vigilância Sanitária) health authorities [60]. The mayor of Piracicaba from signed a four-year, $1.1 million deal with Oxitec. In its first wave, the company would release ten million factory-bred mosquitos weekly into this city of 360,000 people [61]. Importantly, a survey conducted by the CW7 Market Research Institute in mid-2016 showed that 98% of Piracicaba's citizens supported innovative tools to fight dengue, ZIKV, and chikungunya, and 88% support the use of Friendly™ *Aedes* [54].

There seems to be substantial validation. Carvalho et al. (2015) demonstrated effective control of a wild population of *Ae. aegypti* by sustained releases of OX513A male *Ae. aegypti*. The *Ae. aegypti* population was diminished by 95% based on adult trap data and 78% based on ovitrap. Carvalho et al. conclude that sustained release of OX513A males was an effective and widely practical method for suppressing the key dengue vector *Ae. aegypti.* The observed level of suppression would likely be sufficient to prevent dengue epidemics in the locality tested and other areas with the similar or lower transmission [62].

The "Friendly *Aedes*" project began in April 2014, a year after an epidemic of dengue fever that caused more than 1.5 million cases in Brazil. After ten months of testing in two small neighborhoods, the number of dengue cases among 5600 residents fell from 133 in a year to only one [63].

Cayman Islands

Oxitec released millions of genetically altered male insects in a selected area of West Bay, where the population of *Ae. aegypti* has now fallen by almost 90% compared to nearby areas where the GM insects have not been released. The *Ae. aegypti* population in the project area is now just 12% of the numbers found in West Bay's comparative non-treatment study area [64].

In mid-January 2017, the number of *Ae. aegypti* eggs collected in traps in the treatment area was 88% less than in the nearby non-treatment area, while the fluorescent larvae averaged 94% in samples during the last two months. "Both these statistics showed that the program was working as anticipated," said Bill Petrie, the director of the Mosquito Research and Control Unit (MRCU) working in partnership with Oxitec [64].

Oxitec reported positive results from its test in the Cayman Islands. In the Cayman Islands, Oxitec declared roughly a 96% reduction in the mosquito population in the tiny 0.16 km^2 release area [11].

Surveys indicated that 69% of Grand Cayman support using Oxitec's Friendly™ *Aedes* [48]. Their success has led to a proposal by the Cayman Mosquito Research and Control Unit for import permits to deploy Oxitec on Carmen Brac. Also, this proposal contemplated the construction of new mobile labs converted from shipping containers for hatching the larvae [65].

However, reports of success in the Cayman Islands have been somewhat controversial. Before moving forward, keep in mind that GeneWatch has been one of the primary critics of the genetic engineering of mosquitoes to effectuate a population crash. They insist: There remains no evidence that the GM mosquito releases reduce the risk of dengue, ZIKV, or chikungunya transmission [66].

GeneWatch reported that latest information showed that the releases have been ineffective, and large numbers of biting female GM mosquitoes have been released. Dr. Helen Wallace, the director of GeneWatch UK, "Oxitec's GM technology fails in the field and poses unnecessary risks. Islanders' money should not be thrown away on an approach which has not been successful" [66]. GeneWatch claimed Oxitec's positive effects occurred during the dry season when both the numbers were low, and spraying occurred. In addition, controversy surrounds their use of egg traps to determine how practical the approach is

> [E]gg traps do not measure the number of adult females who bite and cause disease. In the annual report, for the first time, adult female numbers are reported, measured by the egg traps, and are preceded by increases (spikes) in adult female numbers associated with the releases: (1) spikes in adult females which follow the GM releases might be caused by the accidental releases of GM females; (2) the project allowed up to 1,000 biting female GM mosquitoes to be released per week, but in practice up to 9,000 biting females were released per week due to a problem with sorting males and females; and (3) once scaled up, this could lead to up to 180,000 GM biting females being released per week if the sorting problem is not solved, or 20,000 GM biting females per week even if it is [66].

Dr. John Norris of the Florida Keys Mosquito Control Board told Key Haven commissioners he had visited with officials in the Cayman Islands, where Oxitec had started previous releases in 2016. He wanted data about antibiotic resistance seen in people living at the trial site. His concern: "These insects are designed to get into people's houses and cause the extinction of whatever *Aedes* live there, but the bacteria they leave behind are left to breed because it has no death chain," he said. According to data provided by officials, there has been a 62% suppression rate of the virus-carrying mosquitoes in the Caymans [67]. However, that report presumably contained inaccurate information that overstated the program's success, according to recently released internal emails between government and MRCU officials.

> Emails state that the government or the MRCU did not formulate the report. Still, by Oxitec itself, which stood to gain from the report's positive results – around that time, the company was aiming to close an $8 million deal that would have expanded the project to the entire island. When an MRCU official brought this to his superiors, he was told that the public could not be said that a private sector company formulated the ostensibly public report.... Meanwhile, it's not clear what the program's actual suppression rate was in the West Bay target area, nor is it clear the scope of the current MRCU-Oxitec suppression project. [68]

As concerns mounted, in 2018, the Cayman government announced that the planned national roll-out of the bio-engineered mosquitoes was shelved. While the West Bay pilot program would continue, the MRCU would be looking at other options.

The counterclaim was that budget restrictions had stopped the planned national release. Still, the GeneWatch report suggested issues over the efficacy and the unfulfilled claims by Oxitec may be the real reason [69].

Reports corroborate. The decision appears to have been driven by a mix of budget issues and concerns that more data are needed to assess the effectiveness of the method of suppressing local populations of the disease-spreading *Ae. aegypti* mosquito. There is no evidence of safety or public health concerns on either side, and the primary issue seems to be value for money [70].

Mr. Kerrie Cox, a lawyer who made the Freedom of Information Act (FOIA) request on behalf of his client, U.S. nonprofit GeneWatch, said the emails showed the technology had been overhyped. "Although the actual test results from the West Bay 'pilot deployment' of GMMs [genetically modified mosquitoes] have been redacted from the FOIA disclosure, they did not achieve anything like the rates of *Ae. aegypti* suppression was repeatedly suggested as achievable by former MRCU Director Bill Petrie," said Mr. Cox [70].

In a letter to Jennifer Ahearn, the head civil servant in the ministry responsible for the MRCU, Richard Adey, Oxitec's regional manager for the Caribbean, expressed concern that the national rollout was not proceeding. He said Oxitec was "surprised" that the decision to "expand our technique throughout Grand Cayman appears to have been revised." In exchanges between Mr. Adey and Ms. Nancy Barnard, an acting director of the MRCU, both acknowledge that the GM mosquitoes have yet to meet "suppression targets," though this is partly attributed to an unspecified "setback" unconnected to the efficacy of the technique. Ms. Banard said the decision not to go ahead with a national rollout was "partially dictated by budget and partially to allow MRCU scientists to assess a new integrated vector management approach with Oxitec in the same area of West Bay as in 2018" [70].

India

The Gangabishan Bhikulal Investment and Trading Ltd (GBIT) and Oxitec established a partnership in 2011 and have conducted successful laboratory-based studies demonstrating the compatibility of Oxitec's mosquitoes with the local population of wild *Ae. aegypti* population [71].

In January 2017, Oxitec and GBIT announced the launch of outdoor caged trials in Dawalwadi, India, evaluating Oxitec's genetically engineered ability mosquitoes to suppress the local *Ae. aegypti* population. If successful, open field trials were to follow [72]. The trials began on January 23, 2017. Laboratory-based studies have already been carried out in India since 2012 by GBIT and Oxitec, and these studies have demonstrated the compatibility of Aedes aegypti mosquitoes.

Laboratory-based studies had already been carried out in India since 2012 by GBIT and Oxitec, and these studies have demonstrated the compatibility of *Ae. aegypti* mosquitoes. "The efficiency in killing offspring was over 99%, and male mosquitoes imported from the U.K. were able to mate with locally available wild female mosquitoes, and the longevity of imported mosquitoes was the same as the wild ones," says Dr. Shaibal Dasgupta, Project Leader, GBIT, Delhi [73]. Following these cage trials, GBIT and Oxitec had plans to conduct open field trials of the Friendly™ mosquitoes, pending approval from the Indian regulatory authorities [74].

Malaysia

On December 21, 2010, 6000 genetically modified (GM) mosquitoes were released in an uninhabited forest in Malaysia. The purpose of the deliberate release was a limited "marked release and recapture" (MRR) experiment, a standard ecological method in entomology, to evaluate under field conditions the flight distance and longevity of the sterile male *Aedes* aegypti strain OX513A strain [75].

After extensive studies and regulatory scrutiny, Lacroix et al. (2012) reported the field release of engineered mosquitoes was safe and successful in Malaysia. The engineered strain showed similar field longevity to an unmodified counterpart, though dispersal was reduced relative to the unmodified strain in this setting. As with previous field releases, the transgene disappeared rapidly from the environment post-release and was not detected more than a few hundred meters beyond the release area. Consistent with the prior risk assessment, no features were revealed that suggested any adverse effect on human health or the environment. The transgene seems to have a little negative impact on lifespan; the apparent reduction in dispersal is not of a magnitude to prohibit operational use [76].

The release received mixed responses as with any other GM technology. Due to the complexity of this process, there is not a precise mechanism or guideline for effective communication and outreach, especially for health-related projects. This is further complicated by negative perceptions in some quarters over the use of GM technology [75]. Some of the reportage on this experiment examined outreach and public engagement, which seems to plague most public efforts to employ GM technology, no less so for GM mosquitoes (see Chap. 15).

An aside at this point, as the scientific community debates the public engagement strategies for similar GM releases, diseases, including dengue, chikungunya and ZIKV continue to rise with a heavy toll on morbidity, mortality, and healthcare budgets. Meanwhile, the wild female *Aedes* aegypti continues to breed offspring, surviving and evading conventional interventions for vector control.

14.1.2.2 Oxitec Field Tests in the USA?

There have been some calls for testing in the U.S. [77]. For example, Harris County officials in Texas submitted a similar application to Florida's and was awaiting approval to begin planning a trial. "We're not sure if we get it if we get it in time for this year, but we'll look for that possibility next year," said Elizabeth Perez, communications director for Harris County Public Health [78]. Harris County sought federal approval to conduct a pilot test of the genetically engineered mosquitoes in a neighborhood in the Houston area. It would be the first urban area in the United States to use the technology. Jenkins and Dallas County Health Director Zachary Thompson met last year with Oxitec, and both were impressed. They wanted Dallas to be next in line for a free trial because they have seen what the *Ae. aegypti* can do [79].

Controversy ensued. One of Jenkins' colleagues, Commissioner Elba Garcia, a practicing dentist, hates the idea of conducting a trial in Dallas County. She worries that the engineered mosquitoes have not been tested enough yet and that there could be unintended consequences—perhaps decades later—on humans, animals, and the ecosystem. "He wants us to be a guinea pig?" she said of Jenkins. "You're asking me if I will put 2.3 million people as a testing tube?" [79].

The mosquitoes that carry the ZIKV and other mosquito-borne illnesses showed up in thirty-eight new counties in the U.S. in 2016—and thirty-four of those counties are in Texas. Now Oxitec is ready to bring trials to the U.S.—despite receiving a less than warm welcome from some Texans. After an April story in The Dallas Morning News, Dallas County Health Director Zachary Thompson said he was "inundated with emails" (the emails followed the same format, taken from a website, GMO-Free USA)—he estimated about 50—protesting the use of genetically modified mosquitoes. Dallas County Commissioner Elba Garcia, a practicing dentist, said she did not want a trial in Dallas County, worrying that the engineered mosquitoes have not been tested enough and could be unintended consequences for humans and the animal's ecosystem [78].

14.1.3 Oxitec and the FDA'S Center for Veterinary Medicine

When Oxitec began its venture to produce its genetically engineered sterile male mosquito (OX315A), the U.S. Food and Drug Administration (FDA) was responsible for regulating the genetic material introduced into the Oxitec mosquito as a "new animal drug"—like how it governs flea medicines and analgesics for dogs and cats. The rationale is that introducing DNA into the genome of the mosquitoes is analogous to dosing them with a drug. Thus, to be marketed, the genetic material, like other "drugs," must be shown to be safe and effective for the animal.

That presents a bizarre and possibly insoluble regulatory conundrum because to grant commercial approval to the Oxitec product; the FDA needed to employ logic that only a bureaucrat could love: Regulators would somehow have to conclude that the genetic material that causes a male mosquito to self-destruct after producing defective offspring is safe and effective for the mosquito.

Consequently, the FDA could find itself tied up in legal knots if, as is likely, its ultimate approval of the insect was to be challenged in court by environmentalists or antigenetic-engineering activists [80].

The FDA resisted issuing an Emergency Use Authorization (EUA) because the statutory language permits the emergency use only of human drugs but not animal drugs and the agency classified the Oxitec mosquito as an "animal drug" as such [the EUA] is not applicable [81]. Others find this reasoning abhorrent.

The FDA's legal interpretation may be incorrect, and some feel it fails the common-sense test. Section 564(a)(1) allows the use of an unapproved "drug" in an emergency, and the FD&CA defines the term "drug" as encompassing both human and animal drugs. Section 564(a)(1) says that "notwithstanding any provision of this Act," the Secretary may authorize an unapproved product even if not approved for humans. The statute requires that four statutory criteria be met—(1) the presence of a severe or life-threatening disease, (2) a "maybe effective" standard for effectiveness, (3) a risk–benefit analysis, and (4) the absence of alternatives. All of these apply [82].

Nonetheless and with due diligence, the FDA opened the decision-making process to public comment to offer informed general consent. The FDA's public comment period closed on May 13, 2016, and they reviewed public remarks, all of which were examined for this chapter.

In addition, research published by Cinnamon Bloss in the Journal of the American Medical Association analyzes the 2624 public comments provided to the FDA during its review. She found that 75% of the comments opposed the trial. Of those, 49%

cited concerns about ecological safety, 61% discussed human health, 68% genetically modified organisms, and 30% mistrusted the government or industry. Bloss cautioned that the comments were from the people most invested in the decision but said the number who expressed distrust of government was significant [77]. As such, it was difficult to isolate the motive for the remarks.

The FDA's Center for Veterinary Medicine had to review the Investigation New Animal Drug (INAD) application from Oxitec. It concluded there seemed to be no significant health issue (the introduced genes are non-toxic and non-allergenic, and the *Ae. aegypti* OX513A only mate within species).

After votes were logged in Key Haven and Monroe County, the FDA released the proposed rule (Draft Guidance for Industry 236), open to public comment through February 19, 2017.

Corroboratively, even though the WHO admitted the technology has demonstrated an ability to reduce the population of mosquitoes in small-scale field trials, that there was insufficient data yet on the epidemiological impact [83]. Regardless, import and contained testing approvals were given by Brazil, Cayman Islands, France, India, Malaysia, Singapore, Thailand, the USA, and Vietnam.

The FDA likewise said that its finalization of the EA (environmental assessment) and FONSI (finding of no significant impact) does mean that Oxitec's mutant mosquitoes are approved for commercial use. However, the agency said that the biotech company must ensure all local, state, and federal requirements are met before conducting the proposed trial. It is also up to the company and the Florida Keys Mosquito Control District to decide when to start the field trial [83].

The FDA approved the Florida experiment on August 5, 2016. Still, the Mosquito Control Board would not authorize release under the two non-binding referenda (one for Key Haven and the other for Monroe County) on November 8.

The countywide ballot poll asks, "Are you in favor of the Florida Keys Mosquito Control District conducting an effectiveness trial in Monroe County, Florida, using genetically modified mosquitoes to suppress an invasive mosquito that carries mosquito-borne diseases?" [84]. Some residents were very outspoken and many of the residents were not backing down in Key Haven, where imposing houses mix with the original working-class variety and cars park next to boats in the driveways [85].

During this time frame, Oxitec had a facility in the UK that produced eggs. Planning for FDA approval, Oxitec said they could set up breeding facilities or "mosquito factories" quickly anywhere in the U.S., as they have already done overseas. So, we ship them somewhere like Miami, and we have a facility that produces male mosquitoes," Oxitec's Nimmo said [86].

The Mosquito Control Board contended to postpone final approval until a review of the FDA Draft Environmental Assessment or the Final Environmental Assessment [87]. The role played by the FDA was under public assault as well. "The issue comes down to the fact that the FDA was doing something kind of on the skirt (not experienced) of what it has experience doing," said Lower Keys physician John Norris about the assessment and the change in regulation [88].

This led to a peculiar situation. "We're in this crazy situation where the FDA has said you can move with your field trial, but we're going to this public referendum in the general election," Oxitec Chief Scientific Officer Dr. Simon Warner said. "Which has set precedence. That does not normally happen. When the FDA says go, you can normally go. People do not vote whether a new drug or medicine can go to field trials or not?" [89].

Unsurprisingly, the role played by the FDA was redefined in 2017. The phrase "articles (other than food) intended to affect the structure, or any function of the body of man or other animals" in the Food, Drug and Cosmetic Acts (FD&C) drug definition [21 U.S.C. 321 (g)(1)(C)] does not include articles intended to function as pesticides by preventing, destroying, repelling, or mitigating mosquitoes for population control purposes. FDA believes that this interpretation is consistent with congressional intent and provides a rational approach for dividing responsibilities between FDA and the Environmental Protection Agency (EPA) in regulating mosquito-related products [90].

If the "guidance" was adopted, genetically modified bugs like those produced by British biotech company Oxitec be regulated by the EPA, not the FDA [88]. This occurred with "the issuance of final guidance #236 and Oxitec Ltd's genetically engineered mosquito, with its proposed claim to control the population of wild-type *Ae. aegypti* mosquitoes fell under the EPA's regulatory authority, and all related regulatory questions should be directed to the EPA" [91].

Doing due diligence, the EPA collected public comments for a second round by Oxitec to test in Florida. They follow in the section on disposition.

The company applied for and waited for an experimental use permit from the EPA to be allowed to release the GMO bugs in twelve yet-to-be-determined sites [92].

Should approval be granted, and Oxitec receives federal funding, the location for the trials in the Keys would be a joint decision between Oxitec and the Florida Keys Mosquito Control District [93].

Beth Ranson, public education and information officer with the Florida Keys Mosquito Control District, whose board of elected commissioners approved Oxitec's proposal, said the sites would likely be chosen among areas where voters were more receptive to the 2016 referendum [92].

In a second referendum for the entire county, 31 of 33 precincts approved an "effectiveness trial in Monroe County. In Ocean Reef, 84% are in favor. The Middle Keys ranged from a 56% approval at Kirk of the Keys precinct to 72% approval at Key Colony Beach City Hall precinct. Summerland Key and parts of Big Pine Key and Sugarloaf were also overwhelmingly in favor at near 67%" [93].

It seemed a bit odd that a decision would rest based on a public vote. For example, Oxitec's Nimmo said the commenters might not represent the Keys population.

"The EPA stated that, although public support or opposition may help guide important policy, agencies make determinations for a proposed action based on sound reasoning and scientific evidence rather than majority votes," Nimmo said [92].

The EPA is not basing its final decision on what people write in the Federal Register, even if they are. Of course, this would not quell the general controversy and discord.

14.1.3.1 Controversies

The release of the Oxitec mosquitos in experimental test trials in Key Haven suffered from the NIMBY effects [94].

Not In My Backyard (NIMBY) is a phenomenon generally associated with enormous energy, chemical, and waste product facilities. A community reads the risk signature of locating some facility very negatively primarily because the location is nearby. The primary variable in determining the public's risk signature is proximity rather than objective scientific and engineering data. NIMBY has become associated with environmental racism whereby those with means can protest and litigate the location of a facility, for example, while those without the means find the facility located in their neighborhoods. Socially influential persons are generally more effective at preventing trials they perceive to be risky in their area or, conversely, attracting social resources and away from weaker persons in the community [95]. A lot of the opposition to NIMBY is voiced by a small minority of the community stakeholders. Many of the same variables discussed in the academic literature on NIMBY appeared in the early test location debates.

This situation was bit different because there was "no facility" per se. In this case, there would be an experimental release and the location was not chosen because of its socioeconomic credentials. As mentioned earlier, the location was selected because of its unique location.

However, opponents to the Key Haven tests did not have a history of ZIKV infection and found the location of the experiments highly intrusive and potentially problematic. It did not seem important to the residents that the area had been selected because it was ideally experimentally rather than an immediate response to a crisis.

What follows is a distillation of the issues that surfaced and were expressed in Key Haven. In public meetings, on local radio, and online, opponents hijacked the conversation about mosquitoes and ZIKV in the Keys. They called Oxitec's tactics unethical and underhanded. They called the company's science untested, unproven, and unsafe. Above all, they were worried about unintended consequences—their not-so-affectionate name for the Oxitec mosquitoes: Frankenflies [24].

14.1.4 Disposition: Successes and Concerns

Noting this is not Oxitec's only genetically engineered insect. Fifteen million Oxitec pink bollworm moths have been released in the US. GE pink bollworm have been released in the US in Arizona by the USDA since 2006 as part of their SIT program to control this significant cotton pest. In addition, Oxitec's GE diamondback moths (DBM) completed field cage studies. DBMs are the world's worst insect pest of brassica vegetable crops such as cabbage, canola, broccoli, cauliflower, and kale—costing farmers $45 billion annually worldwide. Animal and Plant Health Inspection Service (APHIS) approved field trials of diamondback moths in November 2014 after issuing a Finding of No Significant Impact (FONSI). This was considered a remarkable success because the Oxitec GE moths could reduce growers' dependence on insecticides [96]. Jack Bobo, the former senior vice president and chief communication officer for Oxitec parent company Intrexon, said the GE moths "could provide a new tool" for growers to control the moths and reduce the need to use insecticides "that have limited efficacy today" [96].

In addition, before examining the complaints alleged against the OX513A mosquito release in Key Haven, it is essential to highlight and continue to emphasize Oxitec has introduced a second-generation mosquito, OX5034.

> The eggs of the 2nd Generation mosquito produce ONLY males. Male mosquitoes don't bite and are incapable of spreading diseases such as dengue, ZIKV, chikungunya, and Yellow fever. Oxitec's *Ae. aegypti* carry a self-limiting gene that prevents female offspring from surviving, allowing for male-only production. The previously proposed method required adult mosquitoes to be carefully sorted by gender. With the 1st Generation of the Oxitec mosquito released, males mated with a female in the wild, and all male and female offspring would die. The 2nd Generation just targets the females. Male progenies also survive, carrying the self-limiting gene to half its male-only offspring that are male-only … and so on. The 2nd Generation technology allows Oxitec to reduce the number of times the egg box must be refilled, reduces the total population, and Oxitec says, when releases stop, the 2nd Generation mosquito dies out in the wild after a few generations. And finally, Oxitec said the 2 Generation mosquitoes it plans to release do not encounter tetracycline at any stage, either as eggs or as adults. [97]

What follows is a summary of the debates over OX513A and not OX5034. However, the EPA comment solicitation examined below was on OX513A and the commentary collected now constitutes a historical artifact. They remain included because the public reaction to GE mosquitoes is not necessarily built on sound science but rather on risk perception. Motivations for opposition are often built on incomplete information and misinformation is often spread by those with their own agendas of self-interest or penchant for contrarianism.

In general, the scientific literature is clear on the efficacy of OX513A. Scientists who are not affiliated with either side of the battle, like Tom Miller, a retired entomology professor at the University of California Riverside, vouched for the technology. The U.S. is not the first place Oxitec has tested this technology: The company has run trials in the Cayman Islands, Panama, and Brazil over the past decade, claiming to have reduced the *Aedes* aegypti population in the places where they've been used by up to 96% [98].

Oxitec got more than its fair share of pushback. This led some opponents such as Barry Wray, the executive director of the Florida Keys Environmental Coalition to repeat: "They have not passed muster as being proven safe. When this all started, everybody was intrigued. But when you peel this onion, you find some layers are not as pretty as you want" [16].

Others may never be convinced. For example, Dr. Helen Wallace, the director of GeneWatch UK, stated in September. 30, 2016: "Current permits for releases should now be revoked until regulators recognize the downsides of Oxitec's technology and the need to consider all the impacts on the ecosystem. The consequences of mass releases of GE mosquitoes could be harmful if other disease-carrying mosquito species move in. Risk assessments in Brazil, the Cayman Islands, and the U.S. must be revised" [99].

All of this led to years of "what-about-isms" with Oxitec and its promoters playing the carnival game of Whack-A-Mole. While some of the concerns expressed were seemingly legitimate, many were suppositions with no or truly little data. While stakeholders who understand this behavior, who know little about the science and engineering involved, come from academics and research scientists.

With all honesty, it is necessary to notice that the claims against the Oxitec mosquitoes are like those against any bioengineered product, especially GM food. After reading the literature, it becomes more apparent that much of the criticism had little to do with ZIKV and mosquitoes and more with anything bioengineered.

As opportunities for public statements were made available when Oxitec submitted their OX513A mosquito for regulatory approval, some availed themselves of the options. A review of all their complaints and the associated responses are found below.

14.1.4.1 Oxitec can not be Trusted

Complaint: There were some ancillary concerns about transparency, given the agreed role Oxitec may play in general publicity associated with the experiment. There were a few complaints about public outreach. For example, Oxitec has been allowed by the state government of Maharashtra to organize the trials. A source in the UK warned that after regular lab tests are completed, the insect would be released for open trials initially in three states in India. One of the sites chosen was in the Jalna district of Maharashtra. Two villages were identified, and the trials were set to begin below the media's radar and the public. Therefore, the district's locals were unaware of such trials, the implications of genetically engineered insects, and the danger they can pose to the region's biodiversity [100].

Oxitec might have made a better effort at eliciting consent from residents in Malaysia and the Cayman Islands. In both cases, millions of GE mosquitoes were released, and the public learned of it a month later [101]. For example, De Mier said she realized that as successful as Oxitec's method was, it had never faced severe public scrutiny before coming to the US.

"I saw what they did in Brazil," she said. "They brought a truck around with a loudspeaker and made a song." "God sent you the mosquito to heal you" [24].

According to one resident in the Key Haven test area, "This is Jurassic science," said David Berthier, a Key West resident and one of the critics of the experiment. "People distrust this because there is so much corporate spin" [85].

Because FDA approval was filed with the FDA's Center for Veterinary Medicine, and the experiment is evaluated as an animal drug test, no human consent is required. While Oxitec was correct in explaining their approach affects mosquito populations, the experiment released higher concentrations in high human population neighborhoods. The district must submit all publications to review by Oxitec. If Oxitec objects, the issue goes to a steering group of three Oxitec employees and at least two

Mosquito Control District staffers. While unanimity is not required for all informational releases, a local group named "Never Again" argued this transfers a leadership role to a private company [102].

Fact Check: Nimmo claimed the public can count on Oxitec to be transparent in Florida. "We have been very open about everything," Nimmo said. "We publish all of our results, all of the technology—how it works—and have been very public about how much it will be costing." Oxitec does use vehicles with loudspeakers in Brazil, but it also distributes more scientifically based information [24].

> For example, Oxitec reports that Oxitec insects (mosquitos and other species) have been tested by independent laboratories, including the Institute Pasteur in Paris, the Gorgas Memorial Institute in Panama, the Institute for Medical Research in Malaysia, the University of Colorado (USA) and the United States University of Health Sciences (USUHS). The Center for Medical and Veterinary Entomology (CMAVE) at Gainesville, FL, has also tested our mosquitos in outdoor cage trials.
>
> Outdoor trials have been carried out by the Mosquito Research and Control Unit Cayman, the Gorgas Memorial Institute in Panama, the Institute for Medical Research in Malaysia, Moscamed, and the University of São Paulo in Brazil. In each outdoor suppression trial, the wild population of *Ae. aegypti* in the areas has been reduced by over 90%. The United States Department of Agriculture (USDA), the Animal and Plant Health Inspection Service (APHIS), and the Center for Plant Health Science Technology (CPHST) have tested the same Oxitec approach in another insect species. A thorough examination of the risks and benefits has been conducted under the National Environmental Protection Act (NEPA), which concluded that this technology was 'environmentally preferable' to alternatives. [20]

14.1.4.2 Oxitec Works Without Peer Review

Complaint: Critics, especially the late Mila de Mier, claimed Oxitec operated without peer review.

Fact Check: Oxitec has provided the OX513A strain and technical support and nearly 100 peer-reviewed publications on both its strains as of 2022. These external evaluators are the Mosquito Research Control Unit (MRCU) in Cayman, the Institute of Medical Research (IMR) in Malaysia, Moscamed, and University of Sao Paulo in Brazil, and the Gorgas Memorial Institute in Panama. In each case, the institutes above can publish or publicize their results which they have done through a combination of publications, conference presentations, or other media [20].

The Pasteur Institute in Paris and researchers at the University of Colorado replicated Oxitec's findings. In addition, Chris Creese emailed links to eleven articles to Mier reviewing the company's modified mosquito. Nature Biotechnology, the Journal of Applied Ecology, and Proceedings of the National Academy of Sciences are some periodicals to which Oxitec has submitted articles for peer review.

14.1.4.3 Oxitec's Successes Have Been Overstated

Complaint: In Panama, despite triumphant statements by Oxitec, its government partner, the Gorgas Institute, called the trial's results "inconclusive": $620,000 was spent on reducing the number of mosquitoes in a neighborhood of only two hundred families.

In the 2010 trial in the Cayman Islands, Oxitec released three times more mosquitoes than projected. Twenty-five transgenic males were released for every wild male mosquito to achieve an 80% reduction in the mosquito population.

According to a 2017 Interim Report from GeneWatch, Oxitec did not manage population suppression until the dry season (early 2017), when the number of mosquitoes had already dropped significantly. The Mosquito Research and Control Unit (MRCU) also had the first spray as well to get the observed effect [103]. GeneWatch continued. The MRCU annual report reveals problems with releasing adult female GM mosquitoes (which bite and may transmit disease, discussed further below) and spikes in the adult female mosquito population due to the releases, both of which may pose risks that have not been fully considered.

Also, Oxitec has been criticized for using egg traps to claim its GM mosquito releases have suppressed the local mosquito population because egg traps do not measure the number of adult females which bite and cause disease [104]. Female GM mosquito larvae, produced by matings between released GM males and wild females, might have unexpectedly high survival rates. Alternatively, more wild female mosquitoes might have flown in from surrounding areas to mate with the released males [103]. When the roll-out is scaled up, this could lead to up to 180,000 GM biting females being released per week if the sorting problem is not solved, or 20,000 GM biting females per week even if the sorting criterion is met [104].

Danilo Carvalho at the University of São Paulo in Brazil helped analyze the data. He says that "the numbers are more like 60–70% reduction, not ninety, and called into question Oxitec's methods and said their analysis was below scientific standards" [103]. In Northeast Brazil, the overlapping of the release of Oxitec transgenic mosquitoes with a ZIKV epidemic is a source of much controversy as to the safety of the Oxitec protocol.

The WHO complained there was not sufficient data yet on the epidemiological impact [105]. None of the trials done by Oxitec thus far have been designed to measure the effect on disease [45]. Further, there is no evidence that releases of GM mosquitoes can reduce the incidence of harm to health caused by ZIKV, dengue, or chikungunya [103]. In other words, is an 80-percent reduction in the mosquito population enough to stop the spread of ZIKV? The short answer is: We don't know [106].

Note well: These complaints were filed against the OX513A GE mosquito and not the OX5034 second-generation mosquito developed by Oxitec and replacing the OX513A mosquito since late 2019 (more on this throughout the following).

Fact Check: "There was no scientific data linking mosquito numbers to disease," former CEO Parry said. "And [our trials] are not designed to measure disease. But no mosquito means no disease." Others warned, however, that current releases are too small to demonstrate any effect on disease, said Mark Benedict, a malaria research advisor and entomologist at the University of Perugia in Italy. But randomized controlled trials are costly, so he adds that it may not surprise that there have been no demonstrations of the effects of this technology on disease outcomes [45].

14.1.4.4 Oxitec Mosquitoes are Expensive

Complaint: The Florida Keys Mosquito Control District (FKMCD) spends about $1 million annually to control the A. aegypti species alone [45]. While the price gets cheaper as the *Ae. aegypti* population decreases and fewer Oxitec mosquitoes need to be released, the treatments are not a short-term prospect: To ensure *Ae. aegypti* does not come back; the company continues releasing its mosquitoes on an open-ended basis [24].

Grayson Brown from the University of Kentucky's Public Health Entomology Lab said the going rate for the Frankenskeeters is ten cents. The mosquitoes must be released by the thousands, in large enough quantities that the lab-bred males outnumber wild males seven to one. "That blows out most mosquito control budgets," Brown says [16]. The need to release mosquitoes frequently over extended periods could make it financially taxing [107].

"Since *Ae. aegypti* eggs can lie dormant for over six months and hatch at any time, the Oxitec solution, as with any vector control solution, may require repeat releases to maintain the suppression of populations of *Ae. aegypti*," Oxitec's former senior scientist Nimmo wrote [108].

The Piracicaba expansion will cost the city roughly $1.1 million over two years—some $10 per person in the treated area—about half of which will come out of the existing mosquito control budget [11].

Not every community in Brazil will likely be able to afford mosquitoes, which raises new questions, says Fred Gould, an insect geneticist at North Carolina State University in Raleigh. "There's a whole issue of justice: Who gets these mosquitoes and who doesn't? Who gets ZIKV, and who doesn't?" [11].

"Mosquitoes do not fly more than 500 feet, so you would have to release multiple iterations of modified mosquitoes over a vast geographic area to get 'scale,'" Anthony Fauci of the National Institute of Allergy and Infectious Diseases at the U.S. NIH said. The need to release mosquitoes frequently over lengthy periods could make it financially taxing. Malaysia, where Oxitec conducted trials, decided not to go ahead with the project after tests, citing cost [107]. Brown reported that negotiations were underway to release more mosquitoes in Panama, but the price is also a factor. In Brazil, the pilot project would cost about $50,000, which the city of Piracicaba split with Oxitec. Piracicaba anticipates paying about $2.7 million years when it scales up the number of mosquitoes released [7].

Fact Check: Oxitec's Nimmo disagrees. Oxitec could effectively work in the Florida Keys for no more than the same amount the county spends now on pesticides—$1.1 million for an area of 25,000 residents or about $40 per capita [79].

Brazilian officials working on a test with Oxitec have cited a cost of $0.50 a year per person to protect them from bites using Oxitec's mosquito releases [109].

14.1.4.5 Oxitec Mosquitoes will Develop Resistance to the Killing Gene

Complaint: "For example, of key concern is how good Oxitec's strain is at transmitting viruses compared with wild mosquitoes—its so-called vector competence. "So far, studies of such changes in a post-release population are missing." Insect geneticist Max Scott of North Carolina State University in Raleigh said it was a "theoretical possibility" that wild mosquitoes could become immune, but that should not stop the tests" [110].

Food and Water Watch expressed concern stemming from lab-bred GE mosquitos "can evolve resistance to the lethal gene" [99].

Fact Check: Oxitec's modified mosquitos reared in the production facility require the antidote to stay alive, so there is no selection pressure to develop resistance to the self-limiting gene that has been introduced. As a result, Oxitec has not observed any sign of resistance developing in our mosquitos, despite rearing them in large numbers over more than 150 generations. It is worth noting that the Oxitec mosquitoes are the same as wild ones, except they cannot reproduce effectively [20].

14.1.4.6 Tiger Mosquitoes (Albopictus) will Fill the Niche

Complaint: If *Ae. aegypti* has eradicated another species—*Ae. albopictus* (the Asian tiger mosquito)—could move in. "We have seen elsewhere where these two species' distribution is distributed when there is a vacant niche *Ae. albopictus* will move in, and it is a competent vector of the same viruses," says Phil Lounibos, an entomologist who studies mosquito–human disease transmission at the University of Florida [111].

The Center for Food Safety agrees and insists on eliminating *Ae. aegypti*, the yellow fever mosquito engineered by Oxitec, may open space for *Ae. albopictus*, the tiger mosquito which carries the same diseases [112]. University of Campinas ecology professor Jose Ferraz worries about the unintended effects of eliminating *Ae. aegypti* is only replaced by another type of mosquito known as *Ae. albopictus*. He said that could lead to another simply replacing one vector for diseases.

But a memo written by a member of Brazil's biological regulatory commission in 2014 proposes another theory. "Scientific studies show that until the eighteenth and nineteenth centuries, *Ae. albopictus* was the species that frequently bit people in the daytime in Asian cities," the report reads. "It has the aggressiveness and potential to occupy that niche again" [56].

In addition, documents before the Cayman court suggest Oxitec's use of GE mosquito species *Ae. aegypti* may be increasing the number of an equally, possibly more virulent disease-carrying variety, the *Ae. albopictus* species, according to four environmental and food safety groups, including the International Center for Technology Assessment, GeneWatch UK, Food and Water Watch, and Friends of the Earth.

Wenonah Hauter of Food and Water Watch warned that "to hire a company to sell a technology to reduce one mosquito species, so then they can also sell a technology to deal with the species that replace it... is not worth the effort, expense, and potential risk for communities in the US to start down this path" [99]. In Malaysia, where Oxitec conducted trials of its mosquitoes, the country decided not to go ahead with the project after trials, citing cost [113].

Fact Check: An indirect effect that may occur is that the ecological niche *Ae. aegypti* will be vacated, and other mosquito species could move into the vacated place. This is not an intrinsic consequence of using the rDNA construct in the OX513A strain. The same would be expected with other mosquito control measures, as all control methods for mosquitoes aim to reduce or eliminate the mosquito from an area significantly [15].

This is such a severe issue that Oxitec is preparing for it. The company's 2014 application to the Cayman Islands Department of Environment states, "Should *Ae. albopictus* begin to occupy the *Ae. aegypti* niche upon reduction in their numbers, a concurrent operation will reduce the numbers of *Ae. albopictus*" [99].

In response, some Oxitec employees say their program in Piracicaba has not increased the albopictus mosquitoes' population. And in any event, the company says the *albopictus* feeds primarily off animal blood, so it does not pose as much of a threat to humans [56]. Overall, it is less efficient in spreading diseases to humans. In addition, genetic modification technologies are being developed that could target *Ae. albopictus* as well [114]. *Ae. albopictus* has shown competence in ZIKV dissemination in laboratory studies but has never been implicated in ZIKV epidemiology outside of Africa [115].

A 2015 study by Gorman et al. (from Oxitec Limited and the Gorgas Memorial Institute for Human Health) reported no significant change in either *Ae. albopictus* prevalence or spread observed throughout the study. Therefore, there was no evidence to support the theory that *Ae. albopictus* fills vacant niches when the local population of *Ae. aegypti* is primarily removed from the environment [22].

Though the Gorman study was short-term and may not have allowed enough time for species substitution, the authors made the admission and underscored their conclusions.

It may be considered unlikely that the exploitation of vacated niches results from the removal of *Ae. aegypti* would occur throughout this relatively short study. Assessing the potential for interactions with coexisting species will only be possible as longer-term datasets are available at various geographical scales. Nonetheless, with a requirement to advance such pioneering studies stepwise, it was evident that significant suppression of *Ae. aegypti* by >82% for 84 days did not increase *Ae. albopictus* at these sites [22].

14.1.4.7 Female Aegypti do not Like to Mate with the Oxitec Aegypti Males

Complaint: Regarding the success of their program in Brazil, a group of doctors claimed: "Currently, in Brazil, nearly 15 million GM mosquitoes have been released, and the failure is complete. Where field tests were carried out, less than 15% of larvae were transgenic; that is to say wild females do not accept the English mosquito from Oxitec" [116].

Fact Check: First, in mosquitoes, mating is highly species-specific. In addition, *Ae. aegypti* does not form part of a species complex (i.e., a group of insects of similar form that are often indistinguishable at the species level), and matings with closely related mosquito species do not produce viable offspring [21].

Second, there is plenty of data. This is not an issue. There is lab data: A study showed *Aedes* cross-species insemination in the field. *Ae. aegypti* and *Ae. albopictus,* but these interspecific matings encounter many barriers. Only low frequencies of this type of mating appear to occur (a single *Ae. albopictus* was found to have *Ae. aegypti* sperm in this study, and *Ae. albopictus* inseminated three *Ae. aegypti* females), but no viable progeny resulted [117]. Also, if the OX513A male were equally attractive to the female as a wild-type male, mating competitiveness would be equal to 0.5. The OX513A strain performed successfully against all the wild-type strains tested regardless of the genetic background, as none of the mating competitiveness estimates differ significantly from 0.5 [15].

And there are field data: "We have already begun to see evidence of males mating with wild females," Bill Petrie, Head of Grand Cayman's mosquito program, told Kristin Brown. "It's too early to see a reduction in the population, but it is encouraging" [7].

14.1.4.8 Females will be Plentiful and Still Bite

Complaint: Oxitec is convinced the population will crash. They explain. The male bugs live a brief time (3–6 days) and are designed to be self-limiting leaving no trace of themselves in the environment once they have mated with natural females and died. Oxitec has admitted that a tiny percentage of female GM insects are released, but they are also sterile, so again, self-limiting. The offspring will be infertile even if they mate with a male natural mosquito. On occasion, some larvae are viable for a brief period, but even if the tiny fraction that survives for a while were able to mate, the prodigy would still die very quickly [118].

Regarding the claim made by physicians in the Crop-Sprayed Town that the unintentional spread of the disease is based on the assumption female mosquitoes "sting" when pregnant and generating eggs after being fertilized by a male; presumably, when millions of male mosquitoes are released, there will be many more fertilized females looking to suck the blood of mammals, thus increasing the spreading of the disease from infected people to healthy people [116].

Food and Water Watch believes bites from female mosquitos could cause allergies [99].

Only 0.03% of the engineered mosquitoes released by Oxitec will be females (it is hard to separate millions of mosquitoes by sex) [111]. In 2017, GeneWatch claimed the accidental release of many female GM mosquitoes (which may bite and transmit disease) [103].

Oxitec released virtually all males. Oxitec sorts the genetically altered males from the females as pupae and has a greater than 99.9% sorting efficiency. But inevitably, there will be some females released, but this is negligible compared to the number of wild females taken out of the environment through this approach.

For instance, in the recent trial in Panama, the wild *Ae. aegypti* population was reduced by over 90%. During the Panama trial, the number of female Oxitec mosquitos released was less than one per person per year. More than ninety million Oxitec mosquitos have been released worldwide in different countries, and there have been no reports of adverse impacts [20].

Fact Check: Only female *Ae. aegypti* mosquitoes bite humans and suck their blood, transmitting diseases. So, the engineered protein in the males would never be passed to humans.

Any females generated in the production facility will be disposed of by incineration by an external contractor. To date, sorting successes of 99.93% were reported in the Cayman trials by Harris (2012) [119]. and 99.98% in the Brazilian trials by Cavalho (2014) [120].

Even if some OX513A were to escape the containment, they would not live longer than their short lifespan. The introduced lethality trait and the dependence on the presence of tetracycline for survival will prevent establishment in the environment. It is doubtful that OX513A will survive longer than their short lifespan or disperse beyond the proposed trial site, and therefore, the likelihood that OX513A *Ae. aegypti* will survive and distribute negligibly [15].

Important note: With the OX5034 only males survive, hence this concern is currently moot.

14.1.4.9 Biting Females Could Transfer an Engineered Gene into Humans

Complaint: Regarding the claim that the protein from the engineered gene could make its way into humans from the females, modified mosquitoes released would include females due to imperfections in the company's extraction process. Jeff Smith of IRT said these modified females could contain undetected allergens and create additional problems by biting humans [114]. Oxitec's documents reveal that it cannot prevent up to 1% of GE female mosquitos from being released [99].

[T]here's simply no way that even if a GM mosquito bites you, any DNA from the mosquito would alter your DNA in any form or otherwise cause you harm. That is simply not how biology works [121].

Suppose a person was bitten by an OX513A female inadvertently released at the trial site or by a female OX513A progeny that survived. In that case, the immunological response to these bites in humans and other animals is not expected to differ from the immunological reaction to bites by wild-type *Ae. aegypti* mosquitoes. Inadvertently released female OX513A mosquitoes would not be likely to transmit any disease because (1) released mosquitoes would be maintained in conditions and with procedures that prevent contamination with the virus, and (2) dengue virus takes a long time to develop in a mosquito to the point when it can be transmitted so that shorter-lived females such as the OX513A females are less likely to pass on diseases [21].

Again, see above and again moot. Only males are surviving in their exposure to the OX5034 strain.

Fact Check: It is doubtful that the rDNA construct could be transferred to humans or other animals through biting. The rDNA construct is stably integrated into the mosquito genome and cannot remobilize due to altered ITR sequences even when treated with appropriate transposases [21].

> Oxitec injects the DNA they want to be expressed into mosquito eggs alongside other things designed to help insert the genome. Then, they heat shock them, and even after that, they readily admit that their transformation success rate is low (about 1%). And that's how unlikely it is when the DNA is directly injected into the cell. Think for a second about how a mosquito bites—its proboscis is far too large to inject anything into that tiny a target. And even if a mosquito were to inject its DNA into one of your cells when it bit (and somehow not cause the cell to explode from the volume), the odds that the tiny piece of synthetic DNA, out of all the genes in mosquito's genome, would be incorporated into that cell are unfathomably low, and even then, it wouldn't matter: blood cells are final products of our bodies. They don't split to produce new cells. So if you impossibly got the modified mosquito DNA injected into a blood cell, and it incorporated into the genes of that cell, which is indescribably unlikely, the cell would do one of two things: either the DNA would land somewhere where it isn't expressed and be completely ignored for the remainder of the cell's lifespan (less than a month, for a white blood cell), or it would cause the cell to stop working correctly, and the cell would explode, trigger a self-destructive cascade, or be terminated on the spot by immune cells. The foreign DNA would be chopped up into nucleotides, never seen again. There's no chance that you'll be transformed or mutated. None. [121]

The officials who conducted the risk assessment for the release of GM mosquitoes in Brazil said Oxitec has completed "a thorough study" and "rightly concluded that protein fails to exhibit allergenic potential." They said, "though there is a small probability of an individual being repeatedly bitten by female GM mosquitoes, the protein is not allergenic, and the damage is null" [121].

> It's not easy to make a genetically modified organism. Cells don't just "pick up" DNA willy-nilly and insert it into their genomes for fun. If they did, we all would be hodgepodges of the genes from everything around us. All things alive contain DNA, which means all our

> foods, from corn to chicken, are chockfull of genes—yet we manage to eat them every day, and our genomes remain intact. When you cut your finger chopping onions, you introduce onion DNA directly into your bloodstream—yet no one would be concerned about potential mutagenic effects. You don't have to worry about how much DNA you breathe in on pollen grains or whether your yogurt will force yeast genes into your stomach cells because enzymes readily chop up DNA that enters your body. It does not infect your cells—end of the story. [121]

So far, the firm has said that it has released more than seventy million mutant mosquitoes without receiving reports of any side effects from mosquito bites [122].

14.1.4.10 Risks to Other Species

Complaint: Attorneys for Key Haven and concerned citizens noted that the FDA did not complete adequate testing on the potential impact the release would have on people, the local environment, and area species concerned threatened and endangered.

Fact Check: Most importantly, as a non-native species, the *Aegypti* mosquito has not been in the ecosystem for sufficient time to develop an essential ecosystem function [15].

Some experts say Oxitec's technology has a smaller ecological footprint than spraying pesticides, which can harm beneficial insect populations [123].

In 2010, a release occurred in Malaysia to counter yellow fever. In its application to test, Oxitec reported. The exchange of genetic materials between different insect species in the natural environment rarely happens as insects exchange gametes internally and have complex mating behaviors and structures to prevent interspecies mating [124]. For example, the report that interspecies mating conducted between the closely related species, *Ae. aegypti* and *Ae. albopictus* resulted in no fertile hybrids [124].

CVM reported the only threatened species in Key Haven's proposed site is the Stock Island Tree Snail. "Except for the Stock Island Tree Snail, there is no habitat overlap of OX513A mosquitoes with threatened or endangered species as *Ae. aegypti* is a mosquito strongly associated with human habitats. The trial will not likely adversely affect the Stock Island Tree Snail as it does not propose removing or modifying its habitat" [21].

In addition, in October of 2000, the Stock Island tree snail was relocated to public and private property throughout the Florida Keys. The remaining populations are currently being monitored and tended to by the US Fish and Wildlife Service [15].

This led the CVM to report: "None of the critical habitats of the identified species overlap with *Aedes* peridomestic/domestic habitat. *Aegypti*, meaning that released OX513A mosquitoes would not occupy the same habitat as these threatened and endangered species." In addition, CVM added, "tTAV and DsRed2 proteins in the mosquitoes lack any toxic potential and, therefore, do not pose any significant risks to non-target animals, including endangered species" [21].

Oxitec also reported data from experiences with Elephant mosquitoes. "OX513A larvae reared both on and off tetracycline were fed to *Toxorhynchites* larvae as 100% of their diet. The experiment was conducted over six successive generations of *Toxorhynchites,* and results showed no difference between *Toxorhynchites* provided on OX513A and the wild-type control." [124]

"In addition, due to the conditional lethality trait of the OX513A strain, the progenies of this strain will die, limiting or eliminating the possibility of the gene being persistent in the environment or transferring to other organisms. They concluded that since no toxic elements have been incorporated into the OX513A, potential hazard arising from the dead insects persisting in the environment is highly unlikely." [124]

The scientific literature review determined that no sequences in the construct were directly or indirectly likely to be toxic, allergenic, or pathogenic to humans, animals, or the environment [15]. The proteins of the introduced genes (color marker and self-limiting gene) are non-toxic and non-allergenic. Animals that eat the OX513A *Ae. aegypti* mosquitoes will be exposed to nutritional elements–protein, fat, carbohydrate, and others–as they would from eating any mosquito. Still, they cannot take up genes through this route [15].

"They don't churn the soil, replenish anything, or do anything useful in the environment," Oxitec's Nimmo said. "Removing this mosquito restores the environment because it's not meant to be here. It lives in a very artificial environment around our homes, transmitting diseases" [79].

14.1.4.11 Antibiotic-Resistant Bacteria will be Released

Complaint: Some local doctors in Florida raised another concern: Oxitec's mosquitoes may promote the spread of antibiotic-resistant bacteria. Dr. John Norris expressed concern over spreading antibiotic-resistant bacteria. Norris wanted to learn more about the bacteria the insects have on the surface of their bodies, claiming the tetracycline could result in that bacterium becoming resistant [125]. "Dr. John Norris submitted a petition to the Mosquito Control District, signed by 19 colleagues, wanting to know more about the types of bacteria the insects have on the surface of their bodies." Doctors are concerned that these bacteria may already have developed antibiotic resistance in the lab before being spread by mosquitoes in the environment [126]. Norris and thirty other doctors filed a petition with the district wanting a sample of the mosquitoes to be independently tested.

It is doubtful that antimicrobial-resistant bacteria, even in the larval or pupal stages, would be present in adult OX513A mosquitoes because their gut bacteria are lost during mosquito metamorphosis from larvae to adults. It is doubtful that any antimicrobial resistance in bacteria in the rearing water would arise. This trait would be transferred to other bacteria that could cause food or water-borne diseases due to the short duration of the mosquito lifecycle and the trial in general [21].

Fact Check: Nimmo, a former project manager from Oxitec, responded negatively, claiming that the likelihood of adverse effects is extremely low, and that the FDA would not allow it [127].

"The problem of antibiotic-resistant S. aureus is a significant health challenge, and we take it very seriously. So does the Food and Drug Administration. The FDA-led review team, which also included experts from the Centers for Disease Control and Prevention and Environmental Protection Agency, looked specifically into this issue and concluded that 'the likelihood of the adverse effects associated with the development of antimicrobial resistance is extremely low and the risk is negligible,'" Nimmo said in a letter to Norris [128].

First, there is not an injection per se. Larvae are reared in water with tetracycline. Second, mosquito skin (known as a cuticle) is an inhospitable environment for bacteria: It is smooth and dry, unlike human skin, which harbors many different bacterial species. Regarding internal bacteria that may have been exposed to tetracycline, they are doubtful to be continued into the adults. When larvae become pupae, they expel their gut bacteria, with few bacteria found in newly emerged adults [129].

14.1.4.12 Tetracycline in the Environment will Counteract the Kill Gene

Complaint: Regarding the claim that GM mosquitoes could pick up tetracycline in water and soil, especially pet food, and may negate their short lifespans as procreating parents, about 3% may survive, as was found in past trials. That percentage could increase if tetracycline—a common antibiotic—is present in the environment. Thus, the introduction of the modified mosquitoes could permanently alter the gene pool over time, Smith from IRT said [114]. With its massive animal husbandry industry, there is tetracycline in every puddle in Brazil. Lots more mosquitos are released, and there are a lot more diseases. They think the combination in Brazil created the crisis it was supposed to eliminate [130].

Fact Check: Oxitec reported that "its mosquitoes are raised in water with very high concentrations of tetracycline, which cannot occur outside the lab environment" [56]. The Pasteur Institute was unimpressed by the likelihood that mosquitoes might survive exposure to tetracycline already in the environment. This misunderstanding is from misreading an Oxitec document submitted for their environmental assessment. Sustained access to tetracycline from a water bowl involves an unlikely scenario whereby water is not changed for weeks. The mosquitoes have sustained access (when they fly, the self-limited gene becomes active, and they die). CVM added: There are no commercial farms, aquaculture facilities, or hospitals near the proposed release site that have the potential to provide sufficient levels of tetracycline residues [21].

The amount of tetracycline used on the mosquitoes is "meager," Oxitec's Nimmo says, trivial next to what would be found on a typical pig farm [24].

In addition, Agwuh and MacGowan (2006) used concentrations approximately ten times higher than the highest dose found in the literature in human blood. The results showed no increased survival of the OX513A mosquito female offspring if they were to take a blood meal from a human that has recently received a therapeutic dose of tetracycline [131].

Theoretically, they could reproduce in that water, but offspring moving away from the source would still die. Monitoring throughout the trial, the color marker means that this would be easily identifiable. The area could be treated with other mosquito control methods such as insecticides before recommencing with Oxitec release [20].

In the second-generation technology (OX5034), only female breeders are treated with tetracycline in the lab in the U.K. [132]. Then, the male-only eggs can be shipped globally for distribution. Oxitec said that the eggs and the adult male mosquitoes that hatch from them are not exposed to tetracycline.

14.1.4.13 Tetracycline as Waste

Complaint: The use of tetracycline is a concern for Norris and a group of Lower Keys doctors because bacteria may be riding on the backs of the mosquitoes since bacteria develop resistance to antibiotics after exposure to too much or too little. The surviving bacteria can become even more powerful and resistant to medicine down the road. Dr. John Norris told Mosquito Control District commissioners it is not the release the doctors are worried about—they want the mosquitoes to be swabbed for bacteria [36].

Fact Check: Oxitec uses concentrations of tetracycline approximately ten times higher than the highest dose found in humans treated with tetracycline and five times higher than the highest dose found in the blood of animals treated with tetracycline [21]. In addition, where the mosquitoes would be reared would produce less tetracycline waste per week than a single person ingests per day to treat acne.

> Nathan Rose, the Oxitec scientist, agreed that tetracycline is commonly used as a genetic tool switch in lab scenarios. But he said antibiotics are also used heavily in Florida for agricultural citrus health and medicinal purposes. Furthermore, he said that the amount of tetracycline used to treat the female "breeders" for a trial in the US is minuscule. "It would be about 5 grams, comparable to a teaspoon of sugar, or the equivalent of two courses (10 days each) of antibiotics prescribed to a human to fight an infection," he said. "In comparison, in Florida, the agricultural use of antibiotics, including tetracycline, is 88 million times higher." [133]

The amount of tetracycline antidote given to mosquito larvae so they can survive to adulthood for controlled release is insignificant compared to usage for pets, farming, and human therapeutic use. Just considering use by people for context–the amount of tetracycline in wastewater at the mosquito rearing facility in Florida would be 140 times lower than the average amount of tetracycline entering wastewater in the Florida Keys through human therapeutic use alone [20].

The levels of tetracycline in the wastewater are expected to be relatively low and would be quickly degraded or adsorbed in the environment [21].

Again, with OX5034 only female breeders are treated with tetracycline in the lab in the U.K. [132]. Then, the male-only eggs can be shipped globally for distribution. Oxitec said that the eggs and the adult male mosquitoes that hatch from them are not exposed to tetracycline.

14.1.4.14 Piggybacking Transposon

Complaint: Concerns have been expressed regarding whether some transposon (a chromosomal segment that can undergo transposition) might take the opportunity to jump into the *Ae. aegypti* population.

Oxitec's line has been: [o]nly female *Ae. aegypti* mosquitoes bite humans and suck their blood, transmitting diseases. So, the engineered protein in the males would never be passed to humans [79].

"FDA determined that the OX513 rDNA construct is stably integrated into the OX513A mosquito genome and completely refractory to remobilization, even when deliberately re-exposed to the *piggyBac* transposase used for insertion into the mosquito genome." [21]

"Therefore, it is doubtful that the OX513 rDNA construct could be transferred to humans or other animals and pose a higher risk to them than expected from any mosquito bite." [37]

Oxitec's Simon Warner preempted this in the Brazilian risk assessment Oxitec submitted when they applied for a permit. "The OX513A mosquito has now been through more than one hundred generations, and there has never been an instance where the self-limiting gene has behaved as if it were a jumping gene. In human terms, one hundred generations are equivalent to a period from the early AD years to the present day" [134].

14.1.4.15 ZIKV will Evolve

Complaint: Dr. Jan Medlock from Oregon State [135] found risks associated with a transgenic mosquito triggering an evolutionary response in dengue, producing an even more virulent strain of dengue. Still, he would not admit a parallel mutation with Oxitec's approach, and ZIKV was likely [101]. Could the transgenic mosquito have caused the "evolutionary response" described in Dr. Medlock's mathematic model? Could the transgenic mosquitoes have helped the ZIKV mutate from the benign headache and skin rash to the nightmarish head-shrinking virus it is today? [101]. Food and Water Watch are also concerned that "mosquitoes' disease could become more dangerous" [99].

Fact Check: "The likelihood that the production and release of OX513A mosquitoes would lead to the development of antimicrobial-resistant prokaryotes is extremely low. This is partly because resistant bacteria, even if present in the larval or pupal stages, would be highly unlikely in adult OX513A mosquitoes because gut bacteria are lost during mosquito metamorphosis from larvae to adults" [21].

14.1.4.16 The Gene Pool will be Polluted

Complaint: One critic claims that no more than 0.2% or up to sixty-two biting females per person in Key Haven (anticipated release site in Monroe County and five miles northeast of Key West) will be released, and about 5% of males released will not have the lethal gene. They are crippled and live shorter lifespans than wild mosquitoes [136].

Fact Check: The Frankenstein mosquitoes Oxitec created do not moan or accidentally drown little girls in ponds. But when they breed with good old-fashioned mosquitoes, the resulting larvae die. Release enough into the wild, and voila: *Ae. aegypti* is toast. But even if someone somehow were bitten by a modified *Ae. aegypti*, it would not matter. No genetically modified DNA would enter the bloodstream, the company said [137].

14.1.4.17 It will Release Nasty Pathogens

Complaint: Another dramatic Reddit post claimed that the 3–4% of the larvae that survive to adulthood without tetracycline are free to go on and reproduce. The fitness cost of staying under these circumstances may make the surviving mosquitoes more vulnerable to specific pathogens, which may be passed on to humans [138]. This claim is accompanied by a nasty set of responses on Reddit.

Fact Check: Mosquito saliva proteins introduced via mosquito bites elicit an allergic response in some humans. OX513A saliva is not expected to differ from that of wild-type *Ae. aegypti* in its overall composition irrespective of its genome's OX513 rDNA construct. Therefore, regardless of the current allergic status to mosquito saliva proteins, OX513A saliva is expected to have no additional impact on the response of humans to mosquito bites [21]. In the end, Oxitec reports: "The OX513A mosquitoes were susceptible to all the chemicals tested, except DDT (this chemical is not in use anymore). The current standard adulticides effectively control OX513A strains when used at standard concentration" [124].

The next set of controversies was associated with selecting the original test location: Key Haven, Florida.

14.1.5 Test Location: Key Haven, Florida (Raccoon Key)

Key Haven had been selected as a location to test Oxitec's mosquito because of its overall geography. Key Haven has 450 homes and intersects by an extensive canal network, like a little Venice. Single-family homes comprise 41% of the housing types in Stock Island (SI) and Key Haven (KH) communities, with 64% of those single-family homes located in Key Haven. The island is divided into thirds by these canals, and it was this unusual landscape that made Key Haven an ideal place to conduct the

Oxitec trial; the company needs a research area that can be split into three defined but contiguous sections: a test zone where the mosquitoes would be released, a buffer zone, and a control zone that would be free of modified mosquitoes. According to the Mosquito Control District, the island already has plenty of *Ae. aegypti* [139].

If the trial had proceeded as planned, the mosquitoes would have been released in Key Haven in an experimental zone separated from a control zone by a buffer area. (The mosquitoes only fly a few hundred yards in their lives.) Experts then compare the mosquito populations in the experimental and control zones to see if the trial reduced the insects' numbers [140].

A 2013 survey of Key Haven residents (Monroe County) conducted by local mosquito control staff concluded only about 4–6% of the population surveyed in the potential release area was opposed or thought the release was unsafe [101]. In 2013, 59% of residents in Key Haven, the neighborhood where the trial was planned, supported Oxitec's project, with only 9% opposed. A few years later, 58% were in opposition [107]. What happened?

A few years later, in May 2015, a UPMC Center for Health Security team canvassed the area by mail (22% responded), and 58% of those who responded were opposed or strongly opposed to the trials. In comparison, 36 (42%) were either neutral or "support" or "strongly support" GM mosquito use for vector control in their community [141].

Reasons included the possible harmful effects of this intervention, specific worries about GM mosquitoes' human and animal health impacts, and environmental concerns about potential adverse effects on the ecosystem [142].

This survey also reported that when residents were asked directly to what extent they support using GM mosquitoes for control in their community. Of the eighty-six who responded, 50 (58%) said they either "oppose" or "strongly oppose" GM mosquito use.

According to the UPMC, based on the mean score of each control method, residents were most supportive (1 is most preferred and 5 is the least preferred) of the process of "draining standing water on private property to reduce mosquito breeding" (mean = 1.98), followed by treating standing water with larvicides (mean = 2.49), spraying pesticides at ground level (mean = 3.01), spraying insecticides in the air (mean = 3.14). Residents were least supportive of using "GM sterile male mosquitoes to reduce the mosquito population (mean = 4.14)" [142].

Residents were more likely to oppose GM mosquito use if they had a low perception of the potential risks posed by diseases like dengue and chikungunya. They were less concerned about the need to control mosquitoes in general. These findings suggest a need for novel approaches to risk communication, including educational efforts surrounding mosquito control and reciprocal dialogue between residents and public health officials [142]. Effective risk communication can decrease the danger of a potential vector-borne disease outbreak. For example, depending on the species, new mosquitoes may appear 5–10 days after a water-related natural disaster, and increased mosquito activity may continue for several weeks after that. Even without disease-carrying mosquitoes, large numbers of biting nuisance mosquitoes can seriously hamper power restoration activities, impede recovery efforts, and pose significant public health hazards (see chapter seventeen) [143].

The question about GM mosquito use was followed up with two questions to understand why they either supported or did not support this control option. Residents were asked to check all reasons that applied. For those who supported or were neutral on GM mosquito use, the survey provided four possible reasons for supporting this method. Of these reasons for supporting GM mosquito use, respondents chose one reason most often: "GM mosquitoes could reduce the need to use pesticides/larvicides for mosquito control" ($n = 24$). Other reasons, including a reduction in pesticide-resistant mosquitoes ($n = 16$), concern about the mosquito-transmitted disease ($n = 15$), and removal of mosquito nuisance ($n = 14$), were chosen less often [142].

The selection of Key Haven was not based on an outbreak of ZIKV per se. Instead, it was selected as a high-quality location to study the population crash of the mosquito. Separating these two rationalizations seemed to evade those purporting to perform the study and the people of Key Haven who saw themselves as lab rats.

"We will not be laboratory mice," said Jitka Olsak, who moved to Key Haven ten years ago from the Czech Republic. "Nature takes care of its things" [85].

They're using us as guinea pigs," said Gina Fox, who has lived in Key Haven for 15 years. She calls it "Jurassic" science that she does not want to be a part of. "I think they're lying. I think they're just using us," said Fox [40].

"Whenever you try to alter mother nature, mother nature responds. It's not always a positive thing. It could be worse than you initially thought," said Gregory O'Flynn, another Key Haven resident [40].

Oxitec proposed a public release in Key Haven, Florida, and anticipated starting in 2015, but that time passed. Key Haven is a community of about 1000 people. Surrounded by water, it is easier to isolate the area during testing.

Fort Haven has not reported ZIKV cases, so immediacy is more minor an issue there, and this is where some of the problems may lie.

A comment by Sloan Bashinsky posted on the Blue Paper website claims to be providing direct information from Steve Smith, a Monroe County Mosquito Control Board member. "There aren't many *Ae. aegypti* mosquitoes on Key Haven or Stock Island. *Aedes*, likes to hang out in Key West because of its extensive tree canopy. During the day, *Aedes* hides under the trees' leaves, which are tough to reach with aerial insecticide spray released from Mosquito Control's helicopters. So Key West is where any genetically altered mosquitoes should be released [144].

"We've never had a mosquito problem here. Why would I want genetically altered mosquitoes?" explained 80-year-old Mary Murray, who moved to the Keys 20 years ago [145].

14.1.5.1 Vocal Opponents: Russo, Wray and De Mier

One of Oxitec's former chief scientists, Derric Nimmo, told this story. "When I visited the mosquito control board, an hour's drive north of Key West in Marathon, a dozen or so protesters campaigned outside with signs such as "We are not your experiment." One of the protesters followed me inside. She sat next to me as the mosquito board chairperson talked on camera with the local news. For every fact he offered the camera, she hissed in my ear that it was "hearsay" or "lies" [7].

Let's meet three of them: Russo, Wray, and de Mier.

Ed Russo was the chairperson of the Florida Keys Environmental Coalition. This group surfaced during the DeepWater Horizon oil spill in 2010. Then Presidential candidate Donald Trump was the first donor to the coalition. Russo's group was among those incensed that the MCD appeared to be fast-tracking the Oxitec experiment by having it overseen by the FDA, which would treat the GMO technology as an animal drug rather than as a biopesticide, which might have required a complete environmental impact statement (EIS) from the EPA. "If you even want to take down one tree in a wetland, you need an EIS," Russo says. "And these clowns don't want to do an EIS? And we're considered antiscience?" [24].

At a particularly heated community meeting with the MCD board in 2012, Russo asked questions about Oxitec's protocols and whether the MCD was prepared for problems. Russo said the board had no answers for him that day and no answers over the next two months. Then the MCD announced that it had decided to move the venue for the experiment after consulting with the FDA. Instead of Key West, the proposed trial would be conducted in Key Haven, a small community of 144 homes on a neighboring island. "That's when all the flags went up, and the sirens wailed," Russo said. "Would you let your family participate in a scientific experiment without your informed consent in writing? If you are a prisoner in an institution in the U.S., you are given that right" [24].

Another member of the Coalition was Barry Wray. He served as the group's executive director. Reporting on him paints a smarmy character. "Allowing the release would require us to submit to blind flying hypodermic needles. It would be an experiment we could not get out of" [146].

Oxitec's Nimmo fought fears and fiction by going door-to-door in Key West to explain Oxitec's science. The company placed an ad for canvassers to help with the spread. But some dissenters, like Wray, Nimmo said, will never be swayed by science or fact. "If people don't want to listen," he told me, "There's nothing I can do" [7].

The Florida Keys Environmental Coalition takes no official position on the Oxitec-ZIKV conspiracy theory (see Dorothy below). But Barry Wray, the group's executive director, was willing to keep the conversation going. "We're witnessing the results of the microcephaly question gradually evolve," he said. Wray was also happy to give oxygen to another roundly denied conspiracy theory—that Oxitec won over the local government in a less-than-above board way. "There are some more nefarious things that have occurred right now, and I'm not at liberty to talk about those right now," he said. Then presumably he smiled. "I would say that if you don't consider the spectrum of things you've experienced over your life, you're not thinking broadly enough" [24].

Mila de Mier was a real estate agent in the Keys and a single parent with three boys. Her biggest concerns were the lack of long-term research on the mosquitoes, the fact that not all the offspring die immediately, and concerns that the mosquitoes could permanently affect the local environment [147]. She described the testing protocols as "a human rights issue" [27].

She was at all meetings, and other trial critics have mentioned the issue multiple times at the MCB meetings. She can be found all over the Internet, standing near or behind "No Consent" signs. She claimed she wanted to "protect my kids and community from becoming guinea pigs/ lab rats against our will."

> If this proceeds, it will effectively make my kids and the residents of Key Haven unwilling human "guinea pigs" in a for-profit experiment that lasts 22 months and continually releases a GMMS for an estimated twenty-two million genetically modified mosquitoes (GMMs). Thousands of the released GMMs will be biting females, and even more females will be born and survive to adulthood. Less than a mile from the testing site is the only hospital (within a fifty-mile radius), an elementary school, the only community college, a senior citizen center, and a golf course. [148]

De Mier expressed concern over Oxitec's environmental assessment and emerged as a grass-roots leader of the resistance. She claimed that as successful as Oxitec's method was, it had never faced severe public scrutiny before coming to the US.

"I saw what they did in Brazil," she says. "They brought a truck around with a loudspeaker and made a song. 'God sent you the mosquito to heal you.' That was the public engagement." (Oxitec does use vehicles with loudspeakers in Brazil, but it also distributes more scientifically based information.) [24].

She wrote, "The environmental assessment was written by Oxitec, which means we are one step closer to becoming non-consenting "Guinea Pigs" in a for-profit genetic experiment. There is no health oversight component (FDA, CDC, NIH, Dept of Health) for the people of Key Haven during the experiment" [148]. She saw herself as a soldier in a war against genetic technology.

"This is an island with the Atlantic Ocean and the Gulf on one side," she said. "The winds are coming in all kinds of directions. The mosquitoes are not going to stay there." In 2012, she said, she decided to act and has become the face of those opposed to the trial. In 2015, she launched a Change.org petition that garnered national coverage and more than 168,000 signatures, more than double the 75,000 people who live in the Florida Keys [147]. At the MCB meetings, she and other trial critics have discussed the issue multiple times.

After the referendum in Key Haven, she said she planned on personally shepherding lawsuits against the trial should it go forward elsewhere in Monroe County [149]. She passed away in a drowning accident in 2018 at the Cambria Suites Hotel in the Washington DC Convention Center [150].

Members of Wray's group have traveled to Washington D.C. to petition Congress. And they have tapped into wider online communities of anti-GMO activists and conspiracy theorists for support, circulating rumors of ZIKV cover-up scandals, underhanded corporate controversy, and general government mistrust.

14.1.5.2 Some Proponents

Everyone associated with the Key Haven experiment was not opposed to it. "The opponents have very little information, and they are led by a few people who are non-science based," Phil Goodman, chairperson of the FKMCD Board of Commissioners, said. "We have tried to explain the real answers to them. They are not interested in the truth" [85].

"This is a tool we have to fight a disease that we could be using today if we weren't trying to get the public on board," Phil Goodman, the chairperson of the mosquito control board, told me. "We are in the business of public health, not public opinion" [107].

"We've got to stop that ZIKV from being a flash fire coming through us to the rest of the country," Clay Greager, a Key West entrepreneur and author, said with a sense of urgency. "And if they say it can't happen, they have no idea." According to an ABC News report, when Greager looked up the petition online, he found insufficient scientific evidence to dissuade him from supporting the trial. "I grew up through tuberculosis. I grew up through measles and mumps," Greager said while sitting on his back porch. "I remember the first heart transplant. And all these things that no longer exist, [it] was all by the medical profession and scientists taking this harm away from us."

"We are the front door," Greager said of what experts believe is the inevitable invasion of ZIKV into the United States. "That way is Miami. We are their front door, and if we looked at it that way, I would say I want to close that door" [147].

Phil Goodman, the FKMCD chairperson, said: in September 2016, "[e]ighty percent of Florida lawmakers have signed a petition to the FDA for permission to use the GM mosquitoes in their districts" [114]. They claimed that delaying its use poses an **unnecessary health risk** [151].

Some residents were apprehensive about going public with their support for experimental release in Key Haven, "afraid of being harassed by the opposition." The debate was getting vicious [152]. One Key Haven resident, Quincy Perkins, has spoken publicly about the research. He has received petty insults on Facebook and claims a recent online threat directed at his wife and daughter prompted him to call the police [152].

14.1.5.3 NGOS and Nonprofits

A Cayman Island testing release received positive results after the controlled project in West Bay. It did not take long for an American nonprofit to interfere with the anticipated release across the Caymans. The Institute for Responsible Technology (IRT), is financing the efforts, also backed by some local supporters, raising more opposition and at least stall, if not stop, the broader project that aims to reduce the invasive Oxitec mosquito in Cayman drastically [118].

The IRT was started by self-published author Jeffrey Smith and is headquartered in Fairfield, Iowa. The IRT is an anti-GMO website that claims policy successes across the planet [153]. Smith has self-published two books about the apocalyptic dangers of GM crops and foods and a self-produced documentary narrated by the wife of Dr. Oz. He goes on Dr. Oz and other fringe science shows. And he travels worldwide, addressing rabidly anti-GMO audiences with the enthusiasm of a religious fanatic. Perhaps his sentiment is related to his membership in the Maharishi religious cult that seems to make opposition to innovation in agricultural biotechnology one of its central tenets [154].

Several campaign groups lobbied against releasing the Oxitec insects in Key Haven, Florida, including the Center for Food Safety, Friends of the Earth, Foundation Earth, the International Center for Technology Assessment, the Florida Keys Environmental Coalition, and Food & Water Watch.

Dana Perls, the senior food and technology campaigner with the environmental group Friends of the Earth, commented: "This victory protects local communities from reckless experiments. The FDA should never let people and ecosystems be treated as laboratories. We need long-term and sustainable solutions to prevent mosquito breeding grounds" [155].

The public is genuinely concerned the playing field may not be level. For example, Citizens for Safe Science, the Key West-based group against releasing the mosquitoes, had $2860 in contributions to the committee as of October 14. Chairwoman Dina Schoneck said the group has paid for radio ads to "get the word out that there is opposition" to the release of the mosquitoes.

The opposition may not be the only team using the reach of a public interest group. Goodhue reported that Oxitec formed a political action committee called the Florida Keys Safety Alliance and hired a Tallahassee public relations firm, Vancour Jones, to handle PAC communication [156].

The PAC posted a Craigslist ad on August 26, 2016, seeking door-to-door canvassers to speak with voters [157] responding with a counterpetition to the 100,000 signatures online petition collected by Never Again, a citizen's group, against the experiment in Key Haven [127].

The Florida Keys Safety Alliance reportedly had a $100,050 campaign account. Two donations of $50,000 came from Intrexon, and the alliance issued a press release in August 2016 that it "has launched an education awareness campaign to reach the residents in Monroe County who will be voting on a non-binding referendum related to the use of genetically modified mosquitoes to suppress the invasive *Ae. aegypti* mosquito" [158].

The robocalls have been reported on the Never Again Facebook site to worsen matters. "In a recording of one of the calls, a female robotic voice is heard in a short statement cut off at the beginning asking, 'Monroe County Florida using genetically modified mosquitoes to suppress an invasive mosquito that carries mosquito-borne diseases? Press one for no or three for yes'" [156]. Oxitec and Vancore Jones disavow any knowledge of the robocalls through Vancour Jones admitted living operator call polling in summer 2016 [156].

Some organizations representing broader anti-GE interests have been especially virulent in their condemnations. Richard Levine and the Genetic Literacy Project provide an exciting exercise in contrasting apoplectic when he explains how anti-GMO forces, while in arms over sterile GMO mosquitoes, had little to say about the release of irradiated sterile screwworm flies which decimates their population by the mid-1960s. Screwworm flies are flesh-eating blowflies.

Levine correctly assesses public concerns about crashing or exterminating a species population. Whether screwworm or *Ae. aegypti* mosquito, generally it is the purposeful extinction of a species being discussed. Levine does not grasp the nearly uniquely personalized public response to genetic engineering.

Although the screwworm maggots pose terrible consequences for wildlife, they are far less problematic than the *Ae. aegypti* mosquito. Through its ability to transmit viruses such as ZIKV, dengue, and yellow fever, the insect has caused the deaths of millions of people and has even altered human history [159].

14.1.6 Controversy Across Florida

Outside of Key Haven, there was public support. A phone survey by the University of Pennsylvania's Annenberg Public Policy Center between August 18 and 22, 2016, found that 60% of Floridians support the release of GE mosquitoes [160].

Pinellas County in Florida was much less concerned. Pinellas County claimed there was no reason to wait. Pinellas County's first local ZIKV case was announced on August 23rd, more than three weeks after the first local transmission was confirmed in Miami-Dade County, where the only active transmission zones in the US—in Wynwood and Miami Beach—have been identified [161]. Some state representatives came together to demand an emergency use exception. When the FDA told them rules prohibit them from authorizing emergency use, the Pinellas team approached Oxitec directly [162].

In another bipartisan letter sent in late August by House Representative Chris Sprowls and the elected leadership of Pinellas County, Florida, Department of Health and Human Services HHS Secretary Burwell, the representatives of the nearly one million residents stated they wanted to ensure the county has every tool at its disposal to combat a potential spread of the ZIKV. They also requested that an Emergency Use Authorization (EUA) be issued to allow the county to utilize Oxitec's Friendly™ *Aedes* mosquitoes [46].

By early September 2006, the debate over using GE mosquitoes accelerated. For example, 61 Republican and Democratic Representatives from Florida's General Assembly wrote a letter to federal officials urging them to release genetically modified mosquitoes, including an EAU [163]. The letter expressed concern over the delay, arguing that such a delay presents an unnecessary health risk to the people of Florida, resulting in a loss of life [46].

"I think we need to get started sooner than later to end, as quickly as possible, the ZIKV outbreak," said Rep. Heather Fitzenhagen, (R) District 78. "I wanted to try to engage any methods that we can" [163].

There has been some discussion about testing the Oxitec mosquitoes in another county if necessary. In another letter, Rep. Matt Caldwell (R) District 79 would like the FDA to allow Oxitec to choose another Florida county to use as a test site [163].

While many outlets report strong Floridian opposition to Oxitec's mosquitoes, that was not the case outside Key Haven and a few highly outspoken individuals. For example, a recent article in the New American begins with the sentence: "Florida residents are abuzz with opposition to US health regulators' approval of releasing genetically engineered mosquitoes to combat the insect-transmitted ZIKV" [164]. At the same time, "abuzz" is an excellent word in this context, which is invalid.

Most Floridians favor GE mosquitoes to fight the spread of ZIKV and are significantly more likely to approve of it than people who live outside Florida, the latest Annenberg Science Knowledge (ASK) survey has found. The survey found that 40% of Floridians "strongly favor" the release of the mosquitoes, which are modified so that when they breed, their offspring die early, reducing the population. Twenty percent of the Floridians surveyed "somewhat favor" the use of GM mosquitoes, 11% "somewhat oppose" it, and 19% "strongly oppose" it [165].

14.1.7 Controversy Across the U.S.

Plans for an across-the-board US arrival of the GE modified mosquitoes had met with hardened obstruction from anti-GMO associations, which went entryway to entryway to work up resistance four years prior in the Florida Keys when preliminaries were first proposed [166].

A team from Purdue University led by Nicole Widmar of the Department of Agricultural Economics reported the results of an online survey of 964 US residents (February 1–12, 2016): 84% were aware of the ZIKV outbreak, and 81% were aware of the potential for microcephaly when pregnant women contract the virus, and 78% supported the introduction of genetically modified mosquitoes to fight the ZIKV [167].

These are remarkable findings given the apprehension expressed over genetically modified foods and organisms. Widmar said: "We can certainly say that what we've discovered is startling, and we're pleased that the US public has demonstrated a willingness to be open to all the tools we've got in fighting this outbreak" [167].

The research involved a nationally representative sample (gender, income level, age, and residence). The questions on the introduction of genetically modified mosquitoes used in the Purdue study did not differentiate from the gene drive approaches that carry on from generation to generation and the population crash approach where no genetic material is carried into future generations.

Furthermore, the Purdue sample may not reflect the genuine public interest surrounding ZIKV control. Too many respondents see themselves as unaffected by a mosquito-borne disease like ZIKV. In addition, the general sample is affected by the "optimism" bias. "Bad things happen more to other people than to me." Whether releasing genetically modified mosquitoes constitutes a "bad act" or not is an issue here; instead, public samples seem to think problems more often affect others.

While the Purdue study surveyed a much broader population, it may indicate potential changes in public support for GM mosquito use that may emerge as communities face the threat of ZIKV transmission in their neighborhoods [168].

Whether to release genetically modified mosquitoes does involves the public indirectly via the decisions reached by the regulators. In 2016, the decision on whether to release the OX513A mosquitoes into Key Haven rested with the Food and Drug Administration's Center for Veterinary Medicine. In contrast, the decision to use MoquitoMate's Wolbachia-infected rests with the Environmental protection Agency and both these agencies did their jobs. The delays occurred after the Mosquito Control Boards' decisions determined how to move forward or not.

The next level of the complaint rests with Congress's inability to fund efforts to prepare for ZIKV when it arrives on the mainland in full force and aids Puerto Rico, which is experiencing a significant outbreak. Florida, Louisiana, and Texas are three highly vulnerable states for mosquito-carry disease outbreaks. In time, ZIKV may head northward, but in the short term, the front lines are in Puerto Rico and, to a much lesser extent, Florida.

It may be a matter of time when the travel-related cases that lead to sexually transmitted circumstances will be matched with infections from local mosquitoes. "No local mosquitoes have transmitted ZIKV in the Dallas area, but officials say it becomes more likely that local mosquitoes will contract the disease when people carry it here from other countries" [79].

The CDC predicted that ZIKV would infect 25% of Puerto Rico's population by the end of 2016. That is nearly a million people. And as of September 30, the administration warned that the current funds available to fight ZIKV will be used up [169].

"If the federal government follows its normal bureaucratic processes, it might take years for Florida to access this technology. Such a delay presents an unnecessary health risk to the people of our state. Red tape is never an acceptable justification for the loss of human life," the Florida lawmakers wrote [170]. While these lawmakers indicated concern over the use of genetically engineered mosquitoes, they might have well addressed the overall budget for the NIH and the CDC to fight ZIKV.

14.1.8 Federal Government to the Rescue

In February 2016, the Obama Administration called for $1.9 billion in emergency funding for ZIKV preparedness. While waiting on Congress to act, Obama redirected $589 million from unexpended Ebola funding. Fauci and others warned that money dedicated to research malaria, tuberculosis, and a universal flu vaccine would be redirected unless emergency funding was authorized [171].

In June, the House of Representatives offered $1.1 billion in ZIKV funding, but the money was not new funding. It mainly was repurposed funding from other programs, notably Obamacare and Planned Parenthood, and efforts to clean up after the Ebola epidemic that killed 11,000 people in West Africa in 2014–15. According to some critics, while Congress allocated $1.1 billion to combat ZIKV at home, it cut $109.5 million previously dedicated to improving laboratory capacity in parts of West Africa devastated by Ebola to do it, funding that might have helped to define and address the threats of ZIKV in Africa. The African Development Banks raised only $2 million for ZIKV surveillance in Africa [172].

Finally, on September 29, the Senate passed a bill that authorized $1.1 billion to fight ZIKV. The bill includes continuing funding for the federal government through early December and assistance to Flint to address its water crisis [173]. The House passed the ZIKV spending bill on Wednesday, September 28th. Obama signed the bill into law on Thursday, September 29. The law includes $15 million targeted for states with local transmissions, the only state with local transmissions was Florida. It also includes $60 million specifically for territories like Puerto Rico. Puerto Rico had the highest number of infected American citizens with ZIKV. About $400 million would go toward mosquito control, and a similar amount would be for the ZIKV vaccine and better testing research [174].

The Zika Response and Preparedness Act, 2017 (H.R. 5325) provided $933 million in emergency supplemental funding to the Department of HHS for ZIKV preparedness and response activities. The Act allocated $394 million to the Centers for Disease Control and Prevention (CDC), $152 million for the National Institutes of Health, and $387 million to the Public Health and Social Services Emergency Fund (PHSSEF). The Public Health and Social Services Emergency Fund appropriation was directed to support preparedness and response activities within the Assistant Secretary for Preparedness and Response (ASPR), the Health Research and Service Administration (HRSA), for oversight activities, and the Centers for Medicare and Medicaid Services (CMS) [175].

In contrast, many nonprofits were profoundly disappointed in the inadequate funding levels proposed in H.R. 5243, the Zika Response Appropriations Act. This initiative failed to provide appropriate resources to address needed public health efforts to protect pregnant women and infants from the ZIKV. The bill provided less than one-third of its funding to the CDC for the public health efforts desperately needed, such as laboratory capacity, vector control, and public education. It also appeared to restrict funding for the National Institutes of Health only to engaging in vaccine development, which would prevent its use for time-critical needs like cohort studies to determine the risk of and susceptibility to congenital disabilities caused by ZIKV. This funding was only available until the end of September placing unwarranted and counterproductive constraints on its use [176].

Funding to health departments often takes the form of Public Health Emergency Preparedness (PHEP) funding, and the 2016 appropriations bill shifted funding from traditional PHEP funds into ZIKV-related activities.

A group known as the National Association of County and City Health Officials (NACCHO) complained. The reprogramming of these FY 2016 PHEP funds reduced state and local preparedness funding by approximately 7% beginning July 1, with less than 1% in some jurisdictions to up to 10% in others. For public health agencies around the country, this meant a reduced capacity to manage concurrent "everyday" threats like other major disease outbreaks, severe weather events, national security events, active shooters, and other mass casualty incidents [177].

Underfunding ZIKV prevention efforts came with a heavy human and economic toll. In addition to experiencing the pain of families with infants born with dreadful preventable congenital disabilities, the US can expect approximately $10 million in lifetime care costs for each affected baby. Most of those costs will likely be borne by the government through Medicaid programs [176].

More than half of local health departments rely solely on federal funding for emergency preparedness. State and local public health preparedness programs have been reduced by more than a third since peak levels in 2003. Additionally, more than half of hospital and health system emergency preparedness funding has been cut [178].

NACCHO studies revealed that anticipated PHEP spending cuts would decrease the local health department's ability to plan and respond to emergencies. The studies also indicate local health departments are estimated to receive an 8.5% reduction in PHEP grant funding. Half of the local health departments that responded to NACCHO's study expected PHEP funding cuts to have some or significant impact on their local health department's jurisdiction's ZIKV preparedness and response and other emergency efforts. Many respondents expressed concern that the redistribution of funds threatens the sustainability of preparedness programs and sets a dangerous precedent [179].

> "When you weaken the local public health infrastructure, you weaken a community's ability to respond to emerging threats, natural disasters, or any emergency. Local health departments are rightfully concerned because their ability to respond after an emergency is directly related to their capacity and preparedness before the emergency," said LaMar Hasbrouck, MD, MPH, NACCHO's executive director. Local health departments reported that public health preparedness capabilities most negatively impacted by PHEP funding reprogramming are community preparedness, volunteer management, and medical countermeasure dispensing. They also said that funding cuts would hamper pre-event readiness, the availability of supplies, and staffing levels. Nearly half of local health departments reported a decrease in staffing capacity due to cuts, with local health departments predicting the high possibility of staffing cuts, hiring freezes, layoffs, and the reassignment/reduction of staffing duties. [179]

According to NACCHO, this cut will result in a loss of skilled public health professionals, reduce first responders' training and exercising, defer upgrading critical laboratory and interoperable communications equipment, and cancel plans to replenish stockpiles with lifesaving medical countermeasures, among other anticipated impacts.

Ironically, most states and territories felt this funding reduction would considerably affect their overall ZIKV preparedness and response efforts. Restoration of these funds is paramount [177].

In addition, funding for ZIKV by 2017 had run its course. NACCHO and others found this commitment lacking. According to STAT News, some who have been working to combat the outbreak worry that it is not enough: In fact, some experts thought it was merely half what the US would need to stop ZIKV's spread [180]. In addition, the measure would also weaken a decades-old pesticide restrictions to combat mosquitoes, a move Democrats oppose [169].

The blame belongs on both sides of the aisle. While Democrats charged that Republicans had booby-trapped the legislation by adding provisions that would restrict the role of Planned Parenthood, the women's health organization, and similar clinics in providing contraceptive services related to fighting ZIKV, which can be transmitted sexually. The Democrats also said that Republicans had inserted a provision cutting $540 million in financing from the Affordable Care Act, President Obama's signature healthcare law, and that they had stripped a House provision that would ban the flying of the Confederate battle flag in federal cemeteries (the flag provision was removed in a later version of the bill) [181].

Maybe, the most controversial of all was the decision by the Republican House to cut funding for Planned Parenthood. And it failed to pass because Democrats would not go along with the move against Planned Parenthood. Even Democrat Bill Nelson from Florida voted against it.

Even with adequate and responsible funding, states must confront the federal bureaucratic red tape. For example, in Florida, the agency distributing that money through a research grant would be the Florida Department of Health's Biomedical Research Advisory Council or BRAC. BRAC Chair Daniel Armstrong says his council has added another area to consider the long-term effects on kids and adults. That includes cognitive impairment in adults to the more common ZIKV congenital disability known as microcephaly. "We know this grant mechanism's funding period will be three years," Armstrong added. "There is an emphasis on soliciting the best science proposals that can also move most quickly to helping us identify the vaccines, the testing methods, and the associated impacts of this disease" [174].

"Still, for women who have been infected with ZIKV and are carrying fetuses with severe, life-threatening congenital disabilities, less access to reproductive health care—including abortion—is deeply concerning" [180].

14.1.9 Evans and Powell Article

Oxitec became the focus of a final controversy. Evans et al. [182] claimed some offspring of the GM mosquitoes released in Brazil, the vanguard remediation location for Oxitec, survived and produced offspring that made it to sexual maturity. According to the study, some survivors were viable hybrid mosquitoes that are "very likely" to be "more robust" than the original mosquito population. This conclusion is based on a genetic theory called "hybrid vigor". The study's text suggested that the hybrid mosquitoes could be more resistant to insecticides or better at spreading infectious viruses, such as dengue and ZIKV. The suggestion spurred alarming headlines and played to the fears of anti-GMO advocates [183].

Barry Wray from the Florida Keys Environmental Coalition mentioned above framed the report as "another verification of the unquantified risk associated with these genetically modified mosquitoes," he said. "A new version based on the same lethal gene technology represents the same problems, with just a new marketing discussion wrapped around it" [184].

The Evans team had studied Jacobina in Bahia, Brazil. It followed the release of 450,000 Oxitec's first-generation OX315A males bred from wild mosquitoes from Cuba and Mexico into the Pedra Branca, Catuaba, and Incoop neighborhoods. The following quotation comes from the discussion section of the original article.

> Our data clearly show that the release of the OX513A has led to a significant transfer of its genome (introgression) into the natural Jacobina population of *Ae. aegypti*. The degree of introgression is not trivial. Depending on the sample and criterion used to define unambiguous introgression, about 10% to 60% of all individuals have some OX513A genome. [182]

Initially, the authors suggested genetic mixing could have made the mosquito population "more robust"—more resistant to insecticides, for example, or more likely to transmit disease—triggered anti-GM news reports, a backlash from some scientists, and strong pushback from Oxitec [185].

"That's frustrating because that's the exact opposite of what we've done with the Oxitec mosquito," Nathan Rose, the head of scientific and regulatory affairs for Oxitec, said. He said Oxitec has purposely made the mosquito sensitive to the most common insecticides. Rose said the most overlooked part of the paper deals with the trial's success in Brazil—that the test achieved 85% suppression of mosquitoes in the wild. "Not only was it not creating a "super mosquito," it was reducing the local wild mosquito population that is the real threat because they are the ones that spread disease," he said [133].

The company had a lot at stake. The article came out near when Oxitec had submitted a new generation of its GM mosquitoes for U.S. regulatory review and hoped to conduct its first U.S. field trials the following year.

As such, the article was subject to heavy criticism from Oxitec. Oxitec and other critics said that such a suggestion is incendiary conjecture, and that the study's text is flawed. For specific complaints from Oxitec, visit their website [186].

A small number of such hybrids were expected, Oxitec and other researchers noted. These hybrid mosquitoes contained chunks of genetic material from the strain of mosquito used to make OX513A—a cross of *Ae. aegypti* strains found in Mexico and Cuba. But the hybrids did not contain the transgene and thus were not gene drive mosquitoes per se. More importantly, there was no evidence in the paper or elsewhere that hybrid mosquitoes were more dangerous than the original mosquitoes [183].

"The gene transfer referenced in the title does not refer to the persistence of Oxitec's self-limiting and marker genes," said Nathan Rose. "Yes, some of the background genes were present 27 months after the trial ended, but they decreased in time until they completely disappeared. That was unsurprising to us. We knew that about 3–5% of the first-generation offspring would survive, and we published that in our first paper in 2007 and shared that information with regulators. The study authors confirmed that with good data" [133].

In an emailed statement to Ars Technica in September, an Oxitec spokesperson called the study "an unqualified research article with misleading, speculative, and unsubstantiated claims and statements about Oxitec's mosquito technology." Oxitec said it was working with the publisher of Scientific Reports to "remove or substantially correct this article" [183]. Others joined them.

> Scientists responded quickly and sharply to this speculation. Michigan State University plant geneticist Chad Niederhuth tweeted, "Some of the claims made by this paper are completely unfounded, and the title is downright irresponsible." According to Jacob Rasgon, "While there may be a potentially interesting population genetics story here, there is no epidemiological significance." On the paper's claim that the hybridization would result in "a more robust population than the pre-release population," Rasgon pointed out (in all caps) that "they present absolutely no evidence or data supporting this statement." [166]

The editors received a response to the concerns from the corresponding author and sought further advice from expert peer reviewers regarding both the issues raised and the response received. The reviewers confirmed that the scientific problems are valid and should be addressed [187]. These concerns are detailed in an appendix [187].

The AAAS reported that one coauthor of the paper contends it overstates the potential risks of the GM insects, and several coauthors have reportedly requested that it be retracted [188]. According to Ars Technica, six researchers are reportedly calling for the retraction of their study [183]. One of the authors Margareth Capurro from the Brazilian side of the team, publicly objected to the suggestion that Oxitec's hybrids resulted in a more robust population than the prerelease population, telling Revista Questão de Ciência, a publication of the Brazilian nonprofit Science Question Institute, that no data were supporting the idea of "hybrid vigor." She also said she had not seen the last version of the text before its publication [188].

Nature published an editorial expression of concern. Yale's study includes "numerous false, speculative, and unsubstantiated claims and statements about Oxitec's mosquito technology [189].

14.1.10 Back to Florida

Supporters say it is a new way to rid the area of annoying, disease-spreading mosquitoes. Furious opponents describe the biotech company as strongarming their community into serving as a petri dish for a poorly vetted gene-hacking experiment. "I find this criminal, that we are being bullied into this experiment," said Islamorada resident Meagan Hull at a heated town council meeting in March. "I find it criminal that we are being subjected to this terrorism by our own Florida Keys Mosquito Control Board" [190].

Oxitec revealed its "second-generation" mosquito technology in September 2019. In both the first and second versions of the technology, Oxitec takes a biologically modified mosquito and uses tetracycline to keep it alive to adulthood so that it can mate. A basic layman's explanation of the procedure is that when the tetracycline applications stop (when the mosquitoes are released into the wild), the mosquitoes and offspring die.

Oxitec finally did get their opportunity to lab test their GE mosquito in Florida. On May 1, 2020, the federal EPA approved field tests for Oxitec's biologically modified mosquitoes. Under a two-year Experimental Use Permit (EUP), Oxitec has been granted permission to release over one billion genetically modified mosquitoes across 6600 acres in Monroe County, Florida, and Harris County, Texas [191].

The EPA determined the company's information "to be adequate to support a finding of no unreasonable adverse effects to man and the environment during the proposed EUP." They also got permission from an advisory board of the CDC and the Florida Department of Agriculture and Consumer Services [192].

And finally, on August 18, 2020, the Florida Keys Mosquito Control District approved the agreement, and the trial began in the summer of 2020 in the Keys. The permit included Harris County, Texas (Houston area) in 2021 [132].

Oxitec hopes to demonstrate through field trials that their latest GM mosquito strain can reduce local populations of *Ae. aegypti*. These mosquito species transmit dengue, yellow fever, chikungunya, and the ZIKV. Oxitec says the mosquitoes, all males who do not bite humans, will then breed with wild females who bite. But they will pass on the gene, a hereditary OX5034 payload that prevents any female offspring from reaching adulthood. The theory is that the more the gene-hacked mosquitoes and their descendants reproduce, the fewer biting female mosquitoes will be in the area [190].

Some scientists wanted to hit pause on Oxitec's Florida trial, to find what they say is a fairer process in deciding to release the mosquitoes. Others want to see more apparent proof that this technology is even necessary, claiming that the company has only released its most positive data to the public and has kept other critical data, including whether the mosquitoes curb disease transmission, private [189].

"Before the GE mosquito is allowed or even considered for release, there should be caged trials," Perls said. "This is essentially a new pesticide. So, there must be a caged trial in a Florida-like environment. At a time in history when biodiversity is the most devastated it has ever been, during a global public health crisis, it is necessary to emphasize scientific evidence."

Kenneth Labbe, the EPA spokesperson, defended the lack of caged trials, saying that skipping over them was a deliberate aspect of the project rather than the careless oversight that the experiment's opponents say it is. "Based upon a thorough review of all information available to the agency, EPA has concluded that neither the mosquitoes nor the trial will present unreasonable adverse effects to humans or the environment," Labbe told Futurism [190].

There are safeguards. "In the unlikely event Oxitec finds genetically modified female offspring, they are required to cease releases immediately, apply conventional pesticides targeting the adult and larval mosquito stages and continue monitoring until no female OX5034 mosquitoes are found for two consecutive generations," EPA spokesperson Kenneth Labbe said [190]. The EPA also prohibits Oxitec from releasing mosquitoes within 500 m of anywhere tetracycline is being used—several times the distance a typical *Aedes* mosquito will travel in its lifetime [190].

A few points for the opponents: EPA did not require that water within the release site be tested for traces of the compound, and male mosquitoes are perfectly capable of spreading disease on their own, even if they don't bite humans.

Male mosquitoes can infect or be infected by females when they mate, *Ae. aegypti,* according to a study from 2017. According to Robitski, releasing millions at a time might inadvertently spread dangerous diseases in the area and make it increasingly more likely that any surviving wild females are carrying a dangerous virus in their blood [190].

Opponents have also expressed concerns that the EPA has not imposed strong enough measures to monitor Oxitec's experiment and prevent unintended downstream damage to local ecosystems. They call for specific criteria, including an open-source registry for genetically modified organisms that reproduce in the wild, to help scientists better understand what is happening [193].

But alas, approval has been given, and, on balance, the proponents' arguments succeeded. Phil Goodman, the chairperson of the FKMCD, an independently elected commission carrying out mosquito control within Monroe County, says that many of those discredit Oxitec's evidence do not understand the technology. "They're fearmongering," he says. "They have minimal credibility here in the Florida Keys as far as I'm concerned," he adds.

Opponents are correct: No genetically engineered mosquito had been trialed in the United States. However, the country previously allowed tests of a genetically engineered diamondback moth in New York and an engineered pink bollworm in Arizona, both developed by Oxitec [194].

Like before, there was a 30-day public forum with 31,174 comments opposing the release and fifty-six in support [189]. Nonetheless, on Tuesday, June 16, 2020, it was announced that the Florida department of agriculture and consumer services had given the green light to a plan to release 144,000 mosquitoes in the Florida Keys, the string of picturesque islands that extend from the southern tip of the state, beginning in the 2020 summer [195].

Oxitec's GE male mosquitoes for pest control started growing from tiny eggs in toaster-sized, hexagonal boxes on private suburban properties in late April.

On May 12, 2021, experiment monitors confirmed that males had matured enough to start flying off on their own to court American female mosquitoes. This short-term Florida experiment marks the first outdoor test in the United States of a strain of GE male mosquitoes as a highly targeted pest control strategy [196].

Aedes aegypti makes up about 4% of the mosquito population in the Keys, a chain of tropical islands off the southern tip of Florida. But it is responsible for practically all mosquito-borne diseases transmitted to humans in the region, according to the FKMCD, which works closely with Oxitec on the project [194].

Oxitec employees have taken precautions against vandalism by placing mosquito boxes on private, fenced-in properties and not disclosing their precise locations to the public, fearing public efforts to disrupt the test.

Guy Reeves, a genetic researcher at the Max Planck Institute for Evolutionary Biology in Germany, stresses that he does not think the company's approach is unsafe. "If the population in the Florida Keys becomes so sensitized to this issue—that they can no longer cooperate—that's good for the mosquitoes, not good for the people," he adds [189].

For a relatively complete timeline of the Oxitec regulatory hurdles and opposition from Florida residents, see Waltz (2021) [194].

Nevertheless, they reported meaningful results recently in a YouTube online seminar (https://www.youtube.com/watch?v=zDfVcvnNQIY). The company has not yet published data from the experiment, but representatives said during an online seminar earlier this month that the results were promising [197].

In February 2022, FKMCD and Oxitec announced the continuation of their collaboration into 2022. On March 7, 2022, the EPA approved an extension of the project. State-level approval was granted from the Florida Department of Agriculture & Consumer Services (FDACS) on May 4, 2022. The 2022 project launched the week of May 9 [198]. In 2022, Oxitec got the green light from U.S. regulators to release over two million genetically modified mosquitoes in Florida and California [199].

In late April 2021, six blue-and-white hexagonal boxes were placed by Oxitec on the properties of six volunteers. Water was added, and the modified eggs activated and hatched. About 12,000 Oxitec's male mosquitoes flew out of the boxes each week for the next 12 weeks [14].

14.2 State of the Oxitec Option and Oxitec 2.0

Oxitec announced in May 2018 that it was adding to its Friendly™ *Ae. aegypti* mosquito line. "OX5034" began open field trials on May 23, 2018. **OX5034 is the next generation of Oxitec's non-biting Friendly™ *Aedes* mosquitoes**, designed to reduce disease-spreading *Ae. aegypti* mosquito populations. It offers significant advances from the first-generation mosquito, known as "OX513A" [200].

> OX5034 is a second-generation mosquito control vector related to the approved mosquito OX513A. As with the first-generation mosquito, males mate with wild females upon release into the wild OX5034, and all biting female offspring die. One of the new benefits is that male offspring survive, allowing additional mating cycles that reduce the pest population. This multi-generation function of OX5034 is time-limited as, in subsequent generations, fewer and fewer males pass on their self-limiting genes. This amplifies the impact of a release program while still ensuring that Oxitec self-limiting male mosquitoes will not persist in the environment, as they will disappear from the environment ten generations after releases stop. [200]

This first field trial of this second-generation Friendly™ *Aedes* mosquito took place in Indaiatuba, a municipality in São Paulo and part of the metropolitan region of Campinas, Brazil. The trial was designed to generate data for an application to Brazil's regulatory agency, the National Technical Commission for Biosecurity (CTNBio), to secure approval for commercial use of the new Friendly™ mosquitoes throughout the country [200]. The CTNBio is in charge of assessing direct biologic risks resulting from releasing a GMO into the environment.

"This trial was an excellent first demonstration of this new strain's future potential for new vector control approaches in urban settings most prone to dengue outbreaks. We achieved elevated levels of suppression even with lower release rates of our mosquitoes in relatively small sites in densely populated urban areas, which were subject to pressure from mosquitoes migrating from neighboring communities not treated with Friendly™ mosquitoes," said Natalia Ferreira, Oxitec do Brasil Country Director [201]. The CTNBio reports that *Ae. aegypti* pose no additional risks to the environment, human beings, and animals [202].

It seems Oxitec has received approving nods from France and the Netherlands. This may open the doors to testing this new technology in the French Caribbean. As for the Netherlands, they were reportedly considering releasing Oxitec's genetically modified mosquitoes to fight dengue, chikungunya, and ZIKV in Saba, a Dutch Caribbean Island [203].

14.3 Conclusion

Oxitec seems to have occupied an essential niche in the vector control industry. Their positive test results and a competitive trend in contracts suggest that their approach may come to dominate community and government efforts to control mosquito infestation. While it seems reasonable to assume that there might be some correlative effect on the disease outbreaks, it is simply waiting to be seen.

There's not much time to rest on their laurels, given recent developments in gene drive technology, which takes genetic engineering to a new level (see next chapter). Burkina Faso, Mali, and Uganda—are building the groundwork to eventually let loose "gene drive" mosquitoes, which would contain a mutation that would significantly and quickly reduce the mosquito population [204].

References

1. Neeley J (2016) If you want to stop Zika, start promoting GMOs. The Federalist. August 29. http://thefederalist.com/2016/08/29/zika/. Accessed 4 Sept 2016
2. roddcarlson (2016) Comment: a look inside key west's battle to prevent release of GMO mosquitos. October 6. Zero hedge. http://www.zerohedge.com/news/2016-10-06/look-inside-key-wests-battle-prevent-release-gmo-mosquitos. Accessed 27 Oct 2016
3. Jones R (2016) Scientists hope to eradicate disease with massive mosquito orgy. GIZMODO. October 30. https://www.google.com/search?q=Scientists+Hope+to+Eradicate+Disease+With+Massive+Mosquito+Orgy&oq=Scientists+Hope+to+Eradicate+Disease+With+Massive+Mosquito+Orgy&aqs=chrome..69i57.767j0j4&sourceid=chrome&ie=UTF-8. Accessed 17 May 2017
4. Lavery JV, Harrington LC, Scott TW (2008) Ethical, social, and cultural considerations for site selection for research with genetically modified mosquitoes. Am J Trop Med Hyg 79:3. September. https://www.ncbi.nlm.nih.gov/pubmed/18784220. Accessed 8 Aug 2018

5. WHO (2021) Guidance framework for testing of genetically modified mosquitoes, 2nd edn. May 19. https://www.who.int/publications/i/item/978924002523. Accessed 10 Aug 2022
6. Bukspan D (2017) The race is on to stop a Zika virus epidemic in the US. CNBC. April 22. http://www.cnbc.com/2017/04/11/the-race-is-on-to-stop-a-zika-virus-epidemic-in-the-us.html. Accessed 5 May 2017
7. Brown, K (2016) Genetically modified mosquitoes could wipe out the world's most deadly viruses. If we let them. Fusion.net. September 19. http://fusion.net/story/347298/oxitec-genetically-modified-mosquitoes/. Accessed 5 May 2017
8. Brown, K (2016) Americans are terrified genetic enhancements will turn the rich into sci-fi supermen. October 31. Fusion.net. http://fusion.net/americans-are-terrified-genetic-enhancements-will-turn-1793861544. Accessed 4 May 2017
9. McGraw E, O'Neill S (2013) Beyond insecticides: new thinking on an ancient problem. Nature Microbiol 11. March. http://www.nature.com/nrmicro/journal/v11/n3/full/nrmicro2968.html. Accessed 22 May 2017
10. Morrison N (2022) Personal communication. August 10, 2022
11. Servick K (2016) Brazil will release billions of lab-grown mosquitoes to combat infectious disease. Will it work? Science. October 13. http://www.sciencemag.org/news/2016/10/brazil-will-release-billions-lab-grown-mosquitoes-combat-infectious-disease-will-it. Accessed 31 May 2017
12. Adler J (2016) A world without mosquitoes. The Smithsonian. June. 38. http://www.smithsonianmag.com/innovation/kill-all-mosquitos-180959069/?no-ist. Accessed 2 Oct 2016
13. Brown K (2016) Genetically modified mosquitoes could wipe out the world's most deadly viruses. If we let them. Fusion.net. September 19. http://fusion.net/story/347298/oxitec-genetically-modified-mosquitoes/. Accessed 25 Sept 2016
14. Machemer T (2021) Genetically modified mosquitoes take flight to fight invasive species in Florida. Smithsonian Magazine. May 17. https://www.smithsonianmag.com/smart-news/genetically-modified-mosquitoes-take-flight-fight-invasive-species-florida-180977748/. Accessed 31 May 2022
15. Oxitec Report to CVM FDA (2016) Environmental Assessment for Investigational Use of Aedes aegypti OX513A. August 5. https://www.fda.gov/downloads/AnimalVeterinary/DevelopmentApprovalProcess/GeneticEngineering/GeneticallyEngineeredAnimals/UCM514698.pdf. Accessed 27 May 2017
16. Ostrowski J (2016) Frankenskeeters to the rescue? Modified mosquitoes touted to stop Zika. Palm Beach Post. September 23. http://www.mypalmbeachpost.com/news/news/local-govt-politics/frankenskeeters-to-the-rescue-modified-mosquitoes-/nsdkL/. Accessed 29 Sept 2016
17. Alphey L (2014) Genetic control of mosquitoes. Annu Rev Entomol 59:205–224. http://www.oxitec.com/genetic-control-of-mosquitoes-luke-alphey/. Accessed 7 Apr 2017
18. Fernandez C (2018) Gates foundation & oxitec fight malaria with genetically modified mosquitoes. LABIOTEC. June 20. https://labiotech.eu/gates-foundation-oxitec-malaria-mosquito/. Accessed 24 July 2018
19. Warren M (2017) Oxitec's friendly™ Aedes achieves 81% suppression of wild Aedes aegypti in CECAP/Eldorado, Piracicaba, in Second Year of Project. Cision PR Newswire. March 30. http://www.prnewswire.com/news-releases/oxitecs-friendly-aedes-achieves-81-suppression-of-wild-aedes-aegypti-in-cecapeldorado-piracicaba-in-second-year-of-project-617620493.html. Accessed 30 June 2017
20. Oxitec (2016) Florida keys project. Oxitec Website. http://www.oxitec.com/programmes/united-states/. Accessed 27 May 2017
21. Oxitec Report to CVM FDA (2016) Environmental assessment for investigational use of Aedes aegypti OX513A. August 5. https://www.fda.gov/downloads/AnimalVeterinary/DevelopmentApprovalProcess/GeneticEngineering/GeneticallyEngineeredAnimals/UCM514698.pdf. Accessed 12 May 2017
22. Gorman K et al (2015) Short-term suppression of Aedes aegypti using genetic control does not facilitate Aedes albopictus. Pest Manage Sci. October 16. https://www.ncbi.nlm.nih.gov/pubmed/26374668. Accessed 18 Apr 2017

23. Phuc HK, Andreasen MH, Burton RS, Vass C, Epton MJ, Pape G, Fu G, Condon KC, Scaife S, Donnelly CA et al (2007) Late-acting dominant lethal genetic systems and mosquito control. BMC Biol 5:1–11
24. Kolker R (2016) Florida's feud over Zika-fighting GMO mosquitoes. Bloomberg Businessweek. October 6. http://www.bloomberg.com/features/2016-zika-gmo-mosquitos/. Accessed 25 Oct 2016
25. Seeking Alpha (2016) Intrexon: Zika virus hype is nonsensical. Seeking Alpha. April 21. http://seekingalpha.com/article/3966892-intrexon-zika-virus-hype-nonsensical?page=2. Accessed 29 Sept 2016
26. Chatsko M (2016) 3 Reasons Intrexon corp. Stock could rise. The Motley Fool. November 15. https://www.fool.com/investing/2016/11/15/3-reasons-intrexon-corp-stock-could-rise.aspx. Accessed 3 Apr 2017
27. McCauley L (2016) Florida keys residents resist controversial GMO mosquito trial. Common Dreams. August 15. https://www.commondreams.org/news/2016/08/15/florida-keys-residents-resist-controversial-gmo-mosquito-trial. Accessed 21 May 2017
28. English Carleton (2016) Intrexon bruised by blind short-seller report. The Street: Real Money. April 25. http://realmoney.thestreet.com/articles/04/25/2016/intrexon-bruised-blind-short-seller-report. Accessed 29 Sept 2016
29. Howard I (2016) Two biotechnology names are hot: PharmAthene, Inc. (PIP), Intrexon Corporation (XON). The Independent Republic. December 6. http://theindependentrepublic.com/2017/03/14/two-biotechnology-names-are-hot-biopharmx-corporation-bpmx-regulus-therapeutics-inc-rgls/. Accessed 15 May 2017
30. de Graff M (2018) Bill Gates donates $4 million to creating self-destructive mosquitoes in desperate bid to eradicate malaria 'within a generation'. The Daily Mail. June 22. http://www.dailymail.co.uk/health/article-5874509/Bill-Gates-donates-4-million-creating-self-destructive-mosquitoes.html. Accessed 6 July 2018
31. Gates Foundation (2005) Genetic strategies for control of dengue virus transmission. https://gcgh.grandchallenges.org/grant/genetic-strategies-control-dengue-virus-transmission. Accessed 28 May 2022; Boyle A (2018) Gates foundation teams with oxitec on new breed of malaria-blocking mosquito. GeekWire, June 19. https://www.geekwire.com/science/. Accessed 28 May 2022
32. Hotz RL (2021) Genetically altered mosquitoes target deadly dengue fever and Zika. Wall Street J. May 31. https://www.wsj.com/articles/genetically-altered-mosquitoes-target-deadly-dengue-fever-and-zika-11622476316. Accessed 31 May 2022
33. Dapcevich M (2021) Bill Gates releasing genetically modified mosquitoes in Florida? Here's the Whole Story. Snopes. June 3. https://www.snopes.com/fact-check/bill-gates-release-gmo-mosquitoes/. Accessed 26 June 2022
34. Oxitec (2022) The Florida keys mosquito control district & oxitec announce launch of next phase of ground-breaking project. Oxitec Website. https://www.oxitec.com/en/news/the-florida-keys-mosquito-control-district-amp-oxitec-announce-launch-of-next-phase-of-ground-breaking-project. Accessed 10 Aug 2022
35. Business Facilities (2017) Oxitec to build mosquito egg production unit in UK. September 19. https://businessfacilities.com/2017/09/oxitec-to-build-mosquito-egg-production-unit-in-uk/. Accessed 23 July 2018
36. Atkins K (2018) How best to kill skeeters? Experts pitch their plans. Florida Keys News. February 23. http://www.flkeysnews.com/news/local/article201838619.html. Accessed 18 July 2018
37. Oxitec Report to CVM FDA (2016) For public comment: preliminary finding of no significant impact. March. https://www.fda.gov/downloads/animalveterinary/developmentapprovalprocess/geneticengineering/geneticallyengineeredanimals/ucm487379.pdf. Accessed 12 May 2017
38. Walter K (2016) Firm to open mosquito facility in Brazil. RandDMagazine. October 31. https://www.rdmag.com/article/2016/10/firm-open-mosquito-facility-brazil. Accessed 30 June 2017

39. Levy J (2016) West Bay seniors get lowdown on Zika. Cayman Compass. September 12. https://www.caymancompass.com/2016/09/12/west-bay-seniors-get-lowdown-on-zika/. Accessed 15 Sept 2016
40. Polansky R (2016) GMO mosquitoes to fight Zika opposed by Fla. Keys residents. NBC2 News. http://www.nbc-2.com/story/32901966/gmo-mosquitoes-to-fight-zika-opposed-by-fla-keys-residents#.V96x3CgrKUk. August 30. Accessed 18 Sept 2016
41. Ferrari E (2016) Expansion of oxitec's vector control solution in Brazil attacking source of Zika virus and dengue fever after positive program results. MarketWatch. January 19. http://www.marketwatch.com/story/expansion-of-oxitecs-vector-control-solution-in-brazil-attacking-source-of-zika-virus-and-dengue-fever-after-positive-program-results-2016-01-19-720317. Accessed 4 Sept 2016
42. Brown K (2016) This company wants to sell you genetically modified mosquitoes for your backyard. Fusion.net. September 21. http://fusion.net/story/349250/gmo-mosquitoes-for-all/. Accessed 25 Sept 2016
43. Zhang S (2016) A California City is fending off Zika by releasing 40,000 mosquitoes every week. Wired. August 4. https://www.wired.com/2016/08/california-city-fending-off-zika-releasing-40000-mosquitoes-every-week/. Accessed 25 Sept 2016
44. Labiotech (2017) France and the Netherlands deem Oxitec's GM mosquitoes safe. Labiotech.com. July 17. http://labiotech.eu/genetically-modified-mosquitoes-oxitec-netherlands/. Accessed 21 July 2017
45. Chakradhar S (2015) Buzzkill: regulatory uncertainty plagues rollout of genetically modified mosquitoes. Nature Med 21(5):416–418. http://www.nature.com/nm/journal/v21/n5/full/nm0515-416.html. Accessed 3 Apr 2017
46. StreetInsider.com (2016) Intrexon (XON) announces Florida request for emergency authorization to combat mosquitoes. StreetInsider.com. September 7. http://www.streetinsider.com/Corporate+News/Intrexon+(XON)+Announces+Florida+Request+for+Emergency+Authorization+to+Combat+Mosquitoes/12009728.html. Accessed 25 Sept 2016
47. Chatsko M (2017) Where will Intrexon be in 10 years? Motley fool. April 18. https://www.fool.com/investing/2017/04/18/where-will-intrexon-be-in-10-years.aspx. Accessed 5 May 2017
48. Oxitec (2017) The municipality of Santiago de Cali, Colombia, and Oxitec Ltd. Announce memorandum to deploy friendly™ Aedes. Oxitec website. April 10. http://www.oxitec.com/municipality-santiago-de-cali-colombia-oxitec-ltd-announce-memorandum-deploy-friendly-aedes/. Accessed 27 May 2017
49. PR Newswire (2017) Oxitec launching friendly (TM) Aedes project in Juiz de Fora, Brazil. Market Watch. July 17. http://www.marketwatch.com/story/oxitec-launching-friendlytm-aedes-project-in-juiz-de-fora-brazil-2017-07-11-10203219. Accessed 21 July 2017
50. Costa Rica Star (2016) Brazil ready to release mutant mosquitos to wipe out Zika carriers. Costa Rica Star News. November 1. http://news.co.cr/brazil-ready-release-mutant-mosquitos-wipe-zika-carriers/52632/. Accessed 3 Apr 2017
51. Fang J (2015) Genetically modified mosquitoes released in Brazil. IFLSCIENCE! July 7. http://www.iflscience.com/plants-and-animals/dengue-fighting-mosquitoes-are-suppressing-wild-populations-brazil/. Accessed 4 Sept 2016
52. Fernandez C (2017) New results show GM mosquitoes keep dengue and Zika at Bay in Brazil. LABIOTECH. March 4. https://labiotech.eu/oxitec-dengue-zika-brazil/. Accessed 24 July 2018
53. Walker P (2016) Scientists prepare to unleash millions of mosquitoes to have sex with and kill their cousins. The Independent. October 31. http://www.independent.co.uk/news/science/mosquito-sex-experiment-brazil-millions-to-mate-kill-cousins-zika-dengue-a7388871.html. Accessed 30 June 2017
54. Wang B (2016) Factory produces 60 million modified mosquitos per week will reduce breeding of Zika mosquitos. October 31. New Big Future. http://www.nextbigfuture.com/2016/10/factory-produces-60-million-modified.html. Accessed 27 May 2017
55. DiStasio C (2016) Brazil unleashes millions of genetically modified mosquitoes to combat Zika. Inhabitat.com. October 31. http://inhabitat.com/brazilian-scientists-unleash-millions-of-genetically-modified-mosquitoes-to-combat-zika/. Access 11 Apr 2017

56. Rueda M (2016) Welcome to the Brazilian town that's experimenting with mutant mosquitoes. Fusion.net. September. http://fusion.net/story/348992/welcome-to-the-brazilian-town-thats-experimenting-with-mutant-mosquitoes/. Accessed 25 Sept 2016
57. Whitaker P (2016) Intrexon says Brazil mosquito laboratory could protect millions. Reuters. October 26. http://www.reuters.com/article/health-zika-oxitec-idUSL1N1CW2BP. Accessed 2 July 2017
58. Carvalho D et al (2016) Suppression of a field population of *Aedes aegypti* in Brazil by sustained released of transgenic male mosquitoes. PLOS Negl Trop Dis 9(7):e0003864. https://doi.org/10.1371/journal.pntd.0003684
59. PR Newswire (Oxitec Press Release) (2016) Oxitec opens large scale mosquito production facility in Brazil. PR Newswire. October 26. http://www.prnewswire.com/news-releases/oxitec-opens-large-scale-mosquito-production-facility-in-brazil-300351373.html. Accessed 28 May 2017
60. Ramos N (2016) Brazil GM mosquitoes to breed out diseases. MENAFN.com. October 31. https://www.yahoo.com/news/brazil-mutant-mosquitoes-breed-diseases-015740667.html. Accessed 22 Dec 2016
61. AFP (2016) Mutant mosquitoes to breed out diseases. October 30. http://www.newvision.co.ug/new_vision/news/1438897/mutant-mosquitoes-breed-diseases. Accessed 22 Dec 2016
62. Carvalho D et al (2015) Suppression of a field population of Aedes aegypti in Brazil by sustained release of transgenic male mosquitoes. PLoS Negl Trop Dis 9(7):e0003864. https://doi.org/10.1371/journal.pntd.0003864
63. Ferreira FD (2016) Inside the mosquito factory that could stop dengue and Zika. MIT Technol Rev. February 16. https://www.technologyreview.com/s/600821/inside-the-mosquito-factory-that-could-stop-dengue-and-zika/. Accessed 24 July 2018
64. Cayman News Now (2017) Genetic modification project in the Cayman Islands cuts mosquitoes by over 80 percent. Cayman News Now. January 28. http://www.caribbeannewsnow.com/headline-Genetic-modification-project-in-the-Cayman-Islands-cuts-mosquitoes-by-over-80-percent-33345.html. Accessed 15 Mar 2017
65. Whittaker J (2017) Plan hatched to release GM mosquitoes across Cayman Islands. Cayman Compass. March 28. https://www.caymancompass.com/2017/03/28/plan-hatched-to-release-gm-mosquitoes-across-cayman-islands/. Accessed 29 June 2017
66. Caribbean News Now (2018) New documents show genetically modified mosquitoes released in the Cayman Islands are ineffective and risky. Caribbean News Now. September 13. https://wp.caribbeannewsnow.com/2017/09/13/new-documents-show-genetically-modified-mosquitoes-released-cayman-islands-ineffective-risky/. Accessed 23 July 2018
67. Atkins K (2018) Oxitec hopeful for GM mosquito release this year. Florida Keys News. January 20. http://www.flkeysnews.com/news/local/article195745094.html. Accessed 23 July 2018
68. Silva K (2018) Emails reveal GM mosquito program impact was overstated. Cayman Compass. May 14. https://www.caymancompass.com/2018/05/14/emails-reveal-gm-mosquito-program-impact-was-overstated/. Accessed 25 July 2018
69. Cayman News Service (2018) GM mozzie trial in WB may have failed. Cayman News service. May 15. https://caymannewsservice.com/2018/05/genewatch-gm-mosquitoes-failed/. Accessed 23 July 2018
70. Whittaker J (2018) Government backs away from genetically modified mosquito rollout. Cayman Compass. February 14. https://www.caymancompass.com/2018/02/14/government-backs-away-from-genetically-modified-mosquito-rollout/. Accessed 23 July 2018
71. PR Newswire (2017) Oxitec and GBIT announce launch of friendly™ Aedes project in India. PR Newswire. January 23. http://www.prnewswire.com/news-releases/oxitec-and-gbit-announce-launch-of-friendly-aedes-project-in-india-611503055.html. Accessed 28 May 2017
72. House D (2017) Project launched in India to evaluate Oxitec's engineered mosquitoes. Seeking Alpha. January 23. https://seekingalpha.com/news/3236607-project-launched-india-evaluate-oxitecs-engineered-mosquitoes. Accessed 15 May 2017

73. Prasad R (2017) GM mosquito trials to control dengue, chikungunya launched. The Hindu. January 25. http://www.thehindu.com/sci-tech/health/GM-mosquito-trials-to-control-dengue-chikungunya-launched/article17093840.ece. Accessed 14 May 2017
74. Kannan S (2020) Deep dive: Florida's GM mosquito experiment aims to rewrite rules of vector-borne diseases. India Today. August 26. https://www.indiatoday.in/news-analysis/story/florida-gm-mosquito-experiment-aims-to-rewrite-rules-of-vector-borne-diseases-1715063-2020-08-26. Accessed 26 June 2022
75. Subramaniam TS et al (2012) Genetically modified mosquito: the Malaysian public engagement experience. Biotechnol J. November 5. https://doi.org/10.1002/biot.201200282
76. Lacroix R et al (2012) Open field release of genetically engineered sterile male Aedes aegypti in Malaysia. PLoS ONE 7(8):e42771. https://doi.org/10.1371/journal.pone.0042771
77. Shastri D (2017) Mosquito battle gets political genetic engineering plan raises fears of 'Jurassic Park' invasion. Milwaukee J Sentinel Online. October 5. https://www.infowars.com/mosquito-battle-gets-political-genetic-engineering-plan-raises-fears-of-jurassic-park-invasion/. Accessed 10 July 2018
78. Caldera C (2017) Genetically engineered mosquitoes could wipe out Zika, but some in Dallas County oppose local trials. Dallas News. July 14. https://www.dallasnews.com/news/zika-virus/2017/07/14/genetically-engineered-mosquitoes-wipe-zika-dallas-county-oppose-local-trials. Accessed 21 July 2017
79. Martin N (2017) Genetically modified mosquitoes could kill their own kind, cut West Nile, Zika risks. Dallas News. April 10. https://www.dallasnews.com/news/dallas-county/2017/04/10/genetically-modified-mosquitoes-coming-houston-dallas-county-officials-hope-next. Accessed 21 May 2017
80. Miller H (2016) Why are the feds blocking technologies to control the mosquitoes that spread Zika Virus? Forbes. September 7. http://www.forbes.com/sites/henrymiller/2016/09/07/the-feds-are-failing-to-use-available-tools-to-control-the-vector-for-zika-virus/2/#754b018d12ea. Accessed 25 Sept 2016
81. Kaiser Family Foundation News (2016) FDA classification prevents emergency authorization for potential Zika prevention technology. October 31. http://kff.org/news-summary/fda-classification-prevents-emergency-authorization-for-potential-zika-prevention-technology/. Accessed 23 May 2017
82. Miller H, Kershen D (2016) FDA's hostility blocks Zika-prevention technology. The Hill. October 28. http://thehill.com/blogs/congress-blog/healthcare/303290-fdas-hostility-blocks-zika-prevention-technology. Accessed 23 May 2017
83. Lee R (2016) FDA approves releasing Oxitec's genetically modified mosquitoes in Florida to fight Zika virus. TechTimes. August 6. http://www.techtimes.com/articles/172793/20160806/fda-approves-releasing-oxitecs-genetically-modified-mosquitoes-in-florida-to-fight-zika-virus.htm. Accessed 15 Sept 2016
84. Duncan C (2016) GM mosquitoes on track in West Bay trial. Cayman Compass. October 23. https://www.caymancompass.com/2016/10/23/gm-mosquitoes-on-track-in-west-bay-trial/. Accessed 25 Oct 2016
85. Alvarez L (2016) In Florida Keys, some fear 'science and government' more than Zika. New York Times. August 29. http://www.nytimes.com/2016/08/25/us/zika-florida-keys-mosquitoes.html?_r=0. Accessed 18 Sept 2016
86. Randles T (2016) Researchers look to genetics to combat mosquito problem. WMC Action news 5 (Memphis). September 26. http://www.wsmv.com/story/33248347/researchers-look-to-genetics-to-combat-mosquito-problem. Accessed 29 Sept 2016
87. Latest information on the Proposed Oxitec GM Mosquito Project (2016) http://keysmosquito.org/latest-gm-information/. Accessed 4 Sept 2016
88. Atkins K (2017) FDA wants public comment on mosquito regulation. Florida Keys News. February 1. http://www.flkeysnews.com/news/local/article130015924.html. Accessed 7 Feb 2017
89. Aboraya A (2016) Genetically modified mosquito talks draws protester in Orlando. WMFE. September 30. http://www.wmfe.org/gmo-mosquito-talk-draws-protester-in-orlando/64851. Accessed 4 Oct 2016

90. FDA (2017) Clarification of FDA and EPA jurisdiction over mosquito-related products guidance for industry. #236. October. https://www.fda.gov/AnimalVeterinary/NewsEvents/CVMUpdates/ucm578420.htm. Accessed 10 July 2018
91. House D (2017) EPA now the regulator for Oxitec's engineered mosquitoes; parent Intrexon up 1%. Seeking Alpha. October 4. https://seekingalpha.com/news/3299379-epa-now-regulator-oxitecs-engineered-mosquitoes-parent-intrexon-1-percent. Accessed 10 July 2018
92. Goodhue D (2018) This firm would like to release GMO mosquitoes in the keys. The EPA wants your thoughts. Miami Herald. May 8. https://www.miamiherald.com/latest-news/article210711864.html. Accessed 24 July 2018
93. Matthis S (2018) Mosquito: oxitec is courting the keys again. Florida Keys Weekly. February 20. https://keysweekly.com/42/mosquito-oxitec-is-courting-the-keys-again/. Accessed 24 July 2018
94. Foster D, Warren J (2021) The NIBMY problem. J Theoret Polit October 24. 3(1):145–172
95. Macer D (2013) UNDP/World Bank/WHO. Ethical, legal, and social issues of genetically modified disease vectors in public health. WHO, Geneva, Switzerland. http://www.who.int/tdr/publications/tdr-research-publications/seb_topic1/en/. Accessed 8 Aug 2018
96. Davies S (2017) APHIS gives tentative thumbs-up to release of GE moths. AgriPulse. April 19. https://www.agri-pulse.com/articles/9175-aphis-gives-tentative-thumbs-up-to-release-of-ge-moths. Accessed 3 May 2017
97. Matthis S (2019) Oxitec reveals new technology up for EPA consideration. Florida Keys Weekly. September 11. https://keysweekly.com/42/oxitec-reveals-new-technology-up-for-epa-consideration/. Accessed 31 May 2022
98. Costly D (2020) Genetically engineered mosquitoes could soon be unleashed in the U.S. OneZero Medium. May 6. https://onezero.medium.com/genetically-engineered-mosquitoes-could-soon-be-unleashed-in-the-u-s-61896cc647b4. Accessed 28 May 2022
99. Spounias J (2016) Plan to release genetically modified mosquitoes sparks serious concerns among health activists. American Free Press. October 21. http://americanfreepress.net/plan-to-release-genetically-modified-mosquitoes-sparks-serious-concerns-among-health-activists/. Accessed 25 Oct 2016
100. The Guardian (2018) GM mosquito trials a health risk in India. The Sunday Guardian. January 7. https://www.sundayguardianlive.com/news/12314-gm-mosquito-trials-health-risk-india. Accessed 10 July 2018
101. d'Albissin A, Girard A (2016) Oxitec: lord of the mosquitoes. The Blue Paper: Key West. June 19. http://thebluepaper.com/lord-of-the-mosquitoes/. Accessed 5 Sept 2016
102. Goodhue D (2016) Agreement places district staffers under Oxitec's control. Florida Key News. August 25. http://www.flkeysnews.com/news/local/article97912907.html. Accessed 4 Sept 2016
103. GeneWatch (2017) Oxitec's genetically modified mosquitoes: failing in the field? GeneWatch UK. September. http://www.genewatch.org/uploads/f03c6d66a9b354535738483c1c3d49e4/Oxitec_GWbrief_Sep17_fin.pdf. Accessed 10 July 2018
104. GeneWatch (2017) New documents show genetically modified mosquitoes released in the Cayman Islands are ineffective and risky. Caribbean News Now. September 13. https://wp.caribbeannewsnow.com/2017/09/13/new-documents-show-genetically-modified-mosquitoes-released-cayman-islands-ineffective-risky/. Accessed 10 July 2018
105. Lee R (2016) FDA approves releasing oxitec's genetically modified mosquitoes in Florida to fight Zika Virus. Tech Times. August 6. http://www.techtimes.com/articles/172793/20160806/fda-approves-releasing-oxitecs-genetically-modified-mosquitoes-in-florida-to-fight-zika-virus.htm. Accessed 19 May
106. Lafrance A (2016) Genetically modified mosquitoes: what could possibly go wrong? The Atlantic. April 26. https://www.theatlantic.com/technology/archive/2016/04/genetically-modified-mosquitoes-zika/479793/. Accessed 18 May 2017
107. Brown K (2016) Genetically modified mosquitoes could wipe out the world's most deadly viruses. If we let them. Fusion.net. September 19. http://fusion.net/story/347298/oxitec-genetically-modified-mosquitoes/. Accessed 26 Oct 2016

108. Beck J (2016) Mosquitoes can pass Zika to their offspring. The Atlantic. August 30. https://www.theatlantic.com/health/archive/2016/08/mosquitoes-can-pass-zika-to-their-offspring/497960/. Accessed 3 Feb 2017
109. Regalado A (2016) Are altered mosquitoes a public health project, or a business? October 27. Technol Rev. https://www.technologyreview.com/s/602720/are-altered-mosquitoes-a-public-health-project-or-a-business/. Accessed 29 May 2017
110. Astin J (2017) SHOCK CLAIM: lab created super-mosquitos released into wild could 'make all men infertile'. Express. November 15. https://www.express.co.uk/news/science/880050/Lab-designed-mosquito-released-wild-USA-Wolbachia-Asian-Tiger-Mosquito. Accessed 23 July 2018
111. Walker M (2016) Florida votes to release millions of Zika-fighting mosquitos. Wired. November 10. https://www.wired.com/2016/11/florida-votes-release-millions-zika-fighting-mosquitos/. Accessed 30 June 2017
112. Center for Food Safety (2016) This election, keys residents vote "NO" on GE mosquitoes. Santa Cruz IMC. November 9. https://www.indybay.org/newsitems/2016/11/09/18793252.php. Accessed 3 Apr 2017
113. Brown K (2016) If genetically modified mosquitoes freaked you out, you won't like what's coming. GIZMODO. December 13. http://gizmodo.com/if-genetically-modified-mosquitoes-freaked-you-out-you-1790021060. Accessed 15 Mar 2017
114. Carroll M (2016) Florida officials want to enlist frankenskeeters' in Zika fight. AMI Newswire. September 13. https://aminewswire.com/stories/511009888-florida-officials-want-to-enlist-frankenskeeters-in-zika-fight. Accessed 24 Sept 2016
115. Charrel R et al (2016) Background review for diagnostic test development for Zika virus infection. Bull World Health Organ 94:574–584D. https://doi.org/10.2471/BLT.16.171207. http://www.who.int/bulletin/volumes/94/8/16-171207.pdf. Accessed 28 Mar 2017
116. Reduas.com (2016) Report from physicians in the crop sprayed town regarding Dengue-Zika microcephaly and massive spraying with chemical poisons. http://reduas.com.ar/report-from-physicians-in-the-crop-sprayed-town-regarding-dengue-zika-microcephaly-and-massive-spraying-with-chemical-poisons/. Accessed 22 Sept 2016
117. Tripet F et al (2011) Competitive reduction by satyrization? Evidence for interspecific mating in nature and asymmetric reproductive competition between invasive mosquito vectors. Am J Trop Med Hyg August. 85(2). https://www.ncbi.nlm.nih.gov/pubmed/21813845. Accessed 27 May 2017
118. Cayman News (2017) US non-profit backs local an-GM mosquito campaign. Cayman News. February 5. https://caymannewsservice.com/2017/02/us-non-profit-backs-local-anti-gm-mosquito-campaign/. Accessed 28 Mar 2017
119. Harris AF et al (2012) Successful suppression of a field mosquito population by sustained release of engineered male mosquitoes. Nature Biotechnol 30. September 10. http://www.nature.com/nbt/journal/v30/n9/full/nbt.2350.html. Accessed 27 May 2017
120. Carvalho DO et al (2014) Mass production of genetically modified Aedes aegypti for field releases in Brazil. J Visualized Exp 83. https://www.ncbi.nlm.nih.gov/pmc/articles/PMC4063546/. Accessed 27 May 2017
121. Wilcox C (2015) Journalists vector GM fears as FDA considers Oxitec's keys mosquito plan. Discovery. January 27. http://blogs.discovermagazine.com/science-sushi/2015/01/27/journalists-vector-gm-fears-fda-considers-oxitecs-keys-mosquito-plan/#.WVkgJYjys2w. Accessed 1 July 2017
122. St. Fleur N (2015) The genetically modified mosquito bite. The Atlantic. January 27. https://www.theatlantic.com/health/archive/2015/01/Genetically-Modified-Mosquitoes-May-Be-Released-in-Florida-Keys/384859/. Accessed 2 June 2017
123. de la Garza A (2021) Genetically modified mosquitoes have come to the U.S. will they work? Time 2030. May 9. https://time.com/6047051/genetically-modified-mosquitoes/. Accessed 28 May 2022

124. Steinbrecher R (2010) Application for approval for limited mark-release-recapture of Aedes aegypti (L.) wild type and OX513a strains, NBB REF NO: NRE(S)609-2/1/3. FACT SHEET—National Biosafety Board, Malaysia. September. http://www.econexus.info/publication/release-gm-mosquito-aedes-aegypti-ox513a. Accessed 21 May 2017
125. Goodhue D (2016) Oxitec's US owner forms PAC to promote GMO mosquitoes before vote taken in Keys. Miami Herald. August 30. http://www.miamiherald.com/news/local/community/florida-keys/article98953257.html. Accessed 4 Sept 2016
126. Terrell R (2016) Floridians oppose FDA-approved genetically modified mosquitoes. The New American. September 6. http://www.thenewamerican.com/usnews/health-care/item/23997-floridians-oppose-fda-approved-genetically-modified-mosquitoes. Accessed 25 Sept 2016
127. Goodhue D (2016) Pro-GM mosquito PAC to pass out petitions. Florida Key News. September 2. http://www.flkeysnews.com/news/local/article99504542.html. Accessed 5 Sept 2016
128. Unger S (2016) Oxitec: drug resistant bacteria not a threat. KeysNews.com. October 17. http://keysnews.com/node/78398. Accessed 9 June 2017
129. Nimmo D (2016) Oxitec has faith in its science. Florida Keys News. September 29. http://www.flkeysnews.com/opinion/letters-to-the-editor/article104862331.html. Accessed 2 Oct 2016
130. Sandhillexit (2016) Comment: a look inside key West's battle to prevent release of GMO mosquitos. October 6. Zero Hedge. http://www.zerohedge.com/news/2016-10-06/look-inside-key-wests-battle-prevent-release-gmo-mosquitos. Accessed 27 Oct 2016
131. Agwuh KN, MacGowan A (2006) Pharmacokinetics and pharmacodynamics of the tetracyclines including glycylcyclines. J Antimicrob Chemother 58(2). August. https://www.ncbi.nlm.nih.gov/pubmed/16816396. Accessed 27 May 2017
132. Matthis S (2020) Feds approve Oxitec mosquito trial. Florida Keys Weekly. May 7. https://keysweekly.com/42/feds-approve-oxitec-mosquito-trial/. Accessed 28 May 2022
133. Matthis S (2019) Locals respond to Oxitec's latest proposal for biologically modified mosquitoes. Florida Keys Weekly. October 4. https://keysweekly.com/42/locals-respond-to-oxitecs-latest-proposal-for-biologically-modified-mosquitoes/. Accessed 31 May 2022
134. Risk Assessment, Record ID 4556 (2014) Biosafety clearinghouse. August 29. https://bch.cbd.int/database/record.shtml?documentid=105831. Accessed 5 Sept 2016
135. Medlock J, Luz PM, Struchiner CJ, Galvani AP (2009) The impact of transgenic mosquitoes on dengue virulence to humans and mosquitoes. Am Naturalist 174:4. October. https://doi.org/10.1086/605403. Accessed 5 Sept 2016
136. Bethune D (2016) Comments. In Goodhue D (ed) Pro-GM mosquito PAC to pass out petition. Florida keys reported. September 2. http://www.flkeysnews.com/news/local/article99504542.html. Accessed 4 Sept 2016
137. Moyer JW (2015) Genetically modified killer mosquitoes may attack Florida Keys—to target other killer mosquitoes. The Washington Post. January 26. https://www.washingtonpost.com/news/morning-mix/wp/2015/01/26/gentically-modified-killer-mosquitoes-may-attack-florida-keys-to-target-other-killer-mosquitoes/?utm_term=.e20fb1c072ac. Accessed 26 May 2017
138. Genetically modified mosquitoes released in Brazil in 2015 linked to the current Zika epidemic? (2016). https://www.reddit.com/r/conspiracy/comments/42mhii/genetically_modified_mosquitoes_released_in/. Accessed 4 Sept 2016
139. Barry-Jester AM (2016) Small Island, big experiment: how a tiny Florida community could influence the way we fight Zika around the world. FiveThirtyEight. https://fivethirtyeight.com/features/zika-mosquito-florida-vote/. Accessed 17 Jan 2017
140. Joseph A (2016) Florida Keys voters split on genetically modified mosquito trial. STAT. November 8. https://www.statnews.com/2016/11/08/florida-keys-voters-split-on-genetically-modified-mosquitoes/. Access 17 May 2017
141. Klingener N (2016) Majority of Key Haven survey respondents oppose GMO mosquito test. Health News Florida. June 8. http://health.wusf.usf.edu/post/majority-key-haven-survey-respondents-oppose-gmo-mosquito-test#stream/0. Accessed 5 Sept 2016

142. Adaja A et al (2016) Genetically modified (GM) mosquito use to reduce mosquito transmitted disease in the US: a community opinion survey. PLOS. May 25. http://currents.plos.org/outbreaks/article/genetically-modified-mosquito-use-to-reduce-mosquito-transmitted-disease-in-the-us-opinion-survey/. Accessed 29 Sept 2016
143. ASTHO (2015) Before the swarm: guidelines for the emergency management of vector-borne disease outbreaks. http://www.astho.org/Programs/Environmental-Health/Natural-Environment/Before-the-Swarm/. Accessed 28 Aug 2018
144. Bashinsky S (2016) Comments. In: d'Albissin A, Girard A (2016) Oxitec: lord of the mosquitoes. The Blue Paper: Key West. June 19. http://thebluepaper.com/lord-of-the-mosquitoes/. Accessed 5 Sept 2016
145. Brown K (2016) Genetically modified mosquitoes could wipe out the world's most deadly viruses. If we let them. Fusion.net. September 19. http://fusion.net/story/347298/oxitec-genetically-modified-mosquitoes/. Accessed 5 May 2016
146. Danley-Greiner K (2017) FDA backs down from releasing genetically modified mosquitoes after lawsuit threatened. Florida Record. January 8. http://flarecord.com/stories/511066542-fda-backs-down-from-releasing-genetically-modified-mosquitoes-after-lawsuit-threatened. Accessed 11 Apr 2017
147. Mohney G, Quart J (2016) Fighting Zika in the US: the battle over GMO mosquitoes. ABC News. http://abcnews.go.com/Health/deepdive/fighting-zika-40277607/. Accessed 25 Oct 2016
148. de Mier M (2016) Urgent I need your help only until tomorrow to tell the FDA reject the GMO Mosquitoes. Change.org. https://www.change.org/p/say-no-to-genetically-modified-mosquitoes-release-in-the-florida-keys/u/16560317. Accessed 25 Oct 2016
149. Glenza J (2016) Genetically modified mosquitoes could be released in Florida Keys by spring. The Guardian. November 26. https://www.theguardian.com/us-news/2016/nov/26/zika-virus-genetically-modified-mosquitoes-florida. Accessed 18 Apr 2017
150. Filosa G (2018) Anti-GMO mosquito activist dies in a swimming pool. Miami Herald. April 11. https://www.miamiherald.com/news/local/community/florida-keys/article208567304.html. Accessed 1 July 2022
151. Brown K (2016) Florida voters just approved the release of mutant mosquitoes to combat Zika. Fusion.net. November 8. http://fusion.net/florida-voters-just-approved-the-release-of-mutant-mosq-1793863574. Accessed 5 May 2017
152. morphd (2016) Comment: Florida Keys may determine whether GMO mosquitoes will be employed globally to fight Zika. FiveThirtyEight. October 19. https://www.geneticliteracyproject.org/2016/10/19/florida-keys-may-determine-whether-gmo-mosquitoes-will-employed-globally-fight-zika/. Accessed 16 Jan 2017
153. IRT website (DATE). http://responsibletechnology.org/about/. Accessed 28 Mar 2017
154. Entine J (2015) Jeffrey Smith: 'I know nothing about GMOs but that doesn't stop me from promoting junk science'. Genetic Literacy Project. April 6. https://www.geneticliteracyproject.org/2015/04/06/jeffrey-smith-i-know-nothing-about-gmos-but-that-doesnt-stop-me-from-promoting-junk-science/. Accessed 28 Mar 2017
155. Sandle T (2016) Decision taken not to release GM mosquitoes in Florida. Digit J. December 17. http://www.digitaljournal.com/news/environment/decision-taken-not-to-release-gm-mosquitoes-in-florida/article/481921. Accessed 31 May 2017
156. Atkins K (2016) Another option for Zika? Florida Key News. September 14. http://www.flkeysnews.com/news/local/environment/article101750562.html. Accessed 22 Sept 2016
157. Goodhue D (2016) Pro-GM mosquito PAC to pass out petition. Florida Key News, September 2. http://www.flkeysnews.com/news/local/article99504542.html. Accessed 4 Sept 2016
158. Atkins K (2016) Key West group fires back at Oxitec. Florida Key News. October 15. http://www.flkeysnews.com/news/local/article108470607.html. Accessed 25 Oct 2016
159. Levine R (2016) Environmental activist hypocrisy? 'No' to GMO mosquito but 'yes' to irradiated screwworm fly. Genetic Literacy Project. December 11. https://geneticliteracyproject.org/2016/12/11/environmental-activist-hypocrisy-no-gmo-mosquito-yes-irradiated-screwworm-fly/. Accessed 19 May 2017

160. Klingener N (2016) Survey finds most Floridians support GMO mosquitoes. Health News Florida. August 28. http://health.wusf.usf.edu/post/survey-finds-most-floridians-support-gmo-mosquitoes#stream/0. Accessed 5 Sept 2016
161. Staletovich J (2016) Unleash GMO mosquitoes against Zika under emergency rule, Gulf Coast leaders urge. Miami Herald. August 26. http://www.miamiherald.com/news/local/environment/article98185027.html. Accessed 5 Sept 2016
162. Editorial (2016) Let Pinellas use GMO mosquito to fight Zika. Tampa Bay Times, September 1. http://www.tampabay.com/opinion/editorials/editorial-let-pinellas-use-gmo-mosquito-to-fight-zika/2291413. Accessed 4 Sept 2016
163. Polansky R (2016) Fla. Reps. urging feds to use GMO mosquitoes in Zika fight. NBC2.com WBBH News. September 15. http://www.nbc-2.com/story/33106339/fla-reps-urging-feds-to-use-gmo-mosquitoes-in-zika-fight#.V96xsygrKUk. Accessed 18 Sept 2016
164. Terrell R (2016). Floridians oppose FDA-approved genetically modified mosquitoes. The New American. September 6. http://www.thenewamerican.com/usnews/health-care/item/23997-floridians-oppose-fda-approved-genetically-modified-mosquitoes. Accessed 25 Sept 2016
165. Annenberg Public Policy Center (2016) Most in Florida favor use of genetically modified mosquitoes to fight Zika. Annenberg Public Policy Center. August 26. http://www.annenbergpublicpolicycenter.org/mostinfloridafavoruseofgmmosquitoestofightzika/. Accessed 25 Sept 2016
166. Porterfeild A (2019) Oxitec's GMO mosquitoes spread their genes in the wild? Separating science from hype after controversial new study. Genetic literacy project. September 23. https://geneticliteracyproject.org/2019/09/23/oxitecs-gmo-mosquitoes-spread-their-genes-in-the-wild-separating-science-from-hype-following-release-of-controversial-new-study/. Accessed 28 May 2022
167. Agriculture News (2016) Survey: public supports use of GMO mosquitoes to fight Zika virus. Agriculture News, Purdue University. February 22. https://www.purdue.edu/newsroom/releases/2016/Q1/survey-public-supports-use-of-gmo-mosquitoes-to-fight-zika-virus.html. Accessed 4 Oct 2016
168. Adalja A et al (2016) Genetically Modified (GM) mosquito use to reduce mosquito-transmitted disease in the US: a community opinion survey. PLOS Curr Outbreaks. May 25. https://doi.org/10.1371/currents.outbreaks.1c39ec05a743d41ee39391ed0f2ed8d3
169. Newton-Small J (2016) Here's why Zika funding is stalled in Congress. Time. September 6. http://time.com/4481394/zika-funding-congress-stalled/. Accessed 25 Sept 2016
170. Kennedy J (2016) Florida House leaders ask feds to approve use of genetically altered mosquitoes in Zika fight. Palm Beach Post. September 7. http://postonpolitics.blog.palmbeachpost.com/2016/09/07/florida-house-leaders-ask-feds-to-approve-use-of-genetically-altered-mosquitoes-in-zika-fight/. Accessed 18 Sept 2016
171. Gardner T, Mason J (2016) U.S. officials warn Zika 'scarier' than initially thought. Reuters: Aerospace & Defense. April 11. http://www.reuters.com/article/us-health-zika-whitehouse-idUSKCN0X825A. Accessed 23 Sept 2016
172. Adams P, Nutt C (2016) A Zika vaccine, but for whom? The New York Times. December 28. https://www.nytimes.com/2016/12/28/opinion/a-zika-vaccine-but-for-whom.html?_r=0. Accessed 10 Jan 2017
173. CNBC (2016) Senate approves $1.1 billion to fight Zika virus. September 28. http://www.cnbc.com/2016/09/28/us-senate-poised-to-move-forward-on-bill-averting-government-shutdown.html. Accessed 2 Oct 2016
174. Cordner S (2016) Questions remain as to how, when federal Zika funds will be distributed to Florida. WUSF. http://news.wfsu.org/post/questions-remain-how-when-federal-zika-funds-will-be-distributed-florida, Accessed 4 Oct 2016
175. Burwell S (2016) Zika supplemental funding spend plan. Department of Health and Human Services. October 26. https://www.naccho.org/uploads/downloadable-resources/HHS-Zika-Spend-Plan-to-Congress.pdf. Accessed 27 Aug 2018
176. NACCHO (2016) Joint letter opposing house Zika response appropriations act. May 18. https://www.naccho.org/uploads/downloadable-resources/05-18-16-House-Zika-bill-coalition-ltr.pdf. Accessed 28 Aug 2018

177. NACCHO (2016) Joint letter to congress requesting emergency Zika funding. May 10. https://www.naccho.org/uploads/downloadable-resources/Congressional-Zika-Letter_Senate-Leadership.pdf. Accessed 28 Aug 2018
178. NACCHO (2017) Local health departments respond to Zika. https://www.naccho.org/uploads/downloadable-resources/flyer_advocacy_zika.pdf. Accessed 28 Aug 2018
179. NACCHO (2016) One in two local health departments concerned about the impact of looking cuts in their Zika response efforts. Press Release. May 16. https://www.naccho.org/uploads/downloadable-resources/Zika-Study-2016-Press-Release-Final.pdf. Accessed 28 Aug 2018
180. Norman A (2016) Romper. What will happen with Zika virus research under president trump? Things don't look good. Romper. November 17. https://www.romper.com/p/what-will-happen-with-zika-virus-research-under-president-trump-things-dont-look-good-22895. Accessed 26 May 2017
181. Herszenhorn D (2016) Zika bill is blocked by senate democrats upset over provisions. The New York Times. June 28. http://www.nytimes.com/2016/06/29/us/politics/congress-zika-funding.html?_r=0. Accessed 25 Sept 2016
182. Evans BR et al (2019) Transgenic Aedes aegypti mosquitoes transfer genes into a natural population. Nature. September 10. 9:13047. https://doi.org/10.1038/s41598-019-49660-6
183. Mole B (2019) Study claimed a GMO trial went horrifically wrong. The study's authors disagree. Ars Technica. October 2. https://arstechnica.com/science/2019/10/study-claimed-a-gmo-trial-went-horrifically-wrong-the-studys-authors-disagree/. Accessed 28 May 2022
184. Klingener N (2019) GMO mosquitoes proposed again for keys—as new study finds they can interbreed with wild insects. WLRN. September 12. https://www.wlrn.org/news/2019-09-11/gmo-mosquitoes-proposed-again-for-keys-as-new-study-finds-they-can-interbreed-with-wild-insects. Accessed 28 May 2022
185. Servick K (2019) GM mosquito study draws fire. Science 365(6459). https://doi.org/10.1126/science.365.6459.1234. Accessed 28 May 2022
186. Oxitec (2019) Oxitec responds to article entitled 'Transgenic Aedes Aegypti mosquitoes transfer genes into a natural population. September 18. https://www.oxitec.com/en/news/oxitec-response-scientific-reports-article. Accessed 28 May 2022
187. Editors (2020) Editorial expression of concern: transgenic Aedes aegypti mosquitoes transfer genes into a natural population. Nature Res Sci Rep 10(5524) March 24. Editorial Expression of Concern: Transgenic Aedes aegypti Mosquitoes Transfer Genes into a Natural Population | Scientific Reports (nature.com). Accessed 13 May 2022
188. Servick K (2019) Dissent splits authors of provocative transgenic mosquito study. Science. October 1. https://www.science.org/content/article/dissent-splits-authors-provocative-transgenic-mosquito-study. Accessed 28 May 2022
189. White T (2021) First GMO mosquitoes to be released in the Florida Keys. UNDARK. April 12. https://research.ncsu.edu/ges/2021/04/first-gmo-mosquitoes-to-be-released-in-the-florida-keys-undark/. Accessed 13 May 2022
190. Robitski D (2021) A biotech company is releasing 500 million gene-hacked mosquitoes in Florida—and residents are enraged. Futurism. https://futurism.com/biotech-unleashes-gene-hacked-mosquitoes-florida. Accessed 31 May 2022
191. Kofler N, Kuzma J (2020) Before genetically modified mosquitoes are released, we need a better EPA. The Boston Globe. June 22. https://apps.bostonglobe.com/opinion/graphics/2021/06/meet-globe-opinion/. Accessed 28 May 2022
192. Kipnis V (2021) A billion lab-grown mosquitoes are being released and people are frisking out. Vice News. April 29. https://www.vice.com/en/article/pkbd8v/genetically-modified-oxitec-mosquitos-florida-texas. Accessed 28 May 2022
193. Robitski D (2020) Scientists fight plan to release gene-hacked mosquitoes in TX, FL. Futurism. June 5. https://futurism.com/the-byte/scientists-fight-genetically-modified-mosquitoes-tx-fl. Accessed 31 May 2022
194. Waltz E (2021) First genetically modified mosquitoes released in the United States. Nature. May 3. https://www.nature.com/articles/d41586-021-01186-6. Accessed 13 May 2022

195. Millman O (2020) Plan to release genetically modified mosquitoes in Florida gets go-ahead. The Guardian. June 17. https://www.theguardian.com/environment/2020/jun/17/genetically-modified-mosquitoes-florida-texas. Accessed 13 May 2022
196. Milius S (2021) The U.S.'s first open-air genetically modified mosquitoes have taken flight. Science News. May 14. https://www.sciencenews.org/article/mosquito-genetically-modified-us-florida-keys-pest-control-zika-dengue#:~:text=After%20long%20debate%2C%20Oxitec%20pits,approach%20to%20the%20pests'%20control. Accessed 13 May 2022
197. Kuta S (2022) First U.S. Open-air test of genetically modified mosquitoes deemed a success. The Smithsonian. April 21. https://www.smithsonianmag.com/smart-news/first-us-open-air-test-of-genetically-modified-mosquitoes-deemed-a-success-180979960/. Accessed 13 May 2022
198. FKMCD Oxitec Mosquito Project (2022) Keeping the Florida keys safe. https://www.keysmosquitoproject.com/. Accessed 28 May 2022
199. Fitzsimons T (2022) EPA OKs plan to release 2.4 million more genetically modified mosquitoes. NBC NEWS. March 11. https://www.nbcnews.com/news/us-news/epa-oks-plan-release-24-million-genetically-modified-mosquitoes-rcna19738. Accessed 13 May 2022
200. Oxitec (2018) Oxitec launches field trial in Brazil for next generation addition to friendly™ mosquitoes platform. PRNewswire. May 24. https://www.oxitec.com/oxitec-launches-field-trial-in-brazil-for-next-generation-addition-to-friendly-mosquitoes-platform/. Accessed 25 July 2018
201. Oxitec (2019) Oxitec successfully completes first field deployment of 2nd generation friendly™ Aedes aegypti technology. June 3. https://www.oxitec.com/en/news/oxitec-successfully-completes-first-field-deployment-of-2nd-generation-friendly-aedes-aegypti-technology. Accessed 28 May 2022
202. Araujo HRC, Carvalho D, Ioshino R, Costa-da-Silva A, Capurro M (2015) Aedes aegypti control strategies in Brazil: incorporation of new technologies to overcome the persistence of dengue epidemics. Insects 6:576–594
203. Glandorf DCM (2017) Technical evaluation of a potential release of OX513A Aedes aegypti mosquitoes on the island of Saba. National Institute for Public Health and the Environment, The Netherlands. https://biotechnologie.rivm.nl/document/technical-evaluation-potential-release-ox513a-aedes-aegypti-mosquitos-island-saba. Accessed 28 May 2022
204. Swetlitz I (2018) For the first time, researchers will release genetically engineered mosquitoes in Africa. STAT. September 5. https://www.statnews.com/2018/09/05/release-genetically-engineered-mosquitoes-africa/. Accessed 19 June 2019

Chapter 15
Gene Drives

A "gene drive" is a version of gene editing—a newer, more precise way to change a DNA (or RNA) sequence; in this case, combining a guide RNA with an enzyme that can make a gene drive takes this to another level, making sure that a new or altered genetic sequence has a greater than 50 percent chance of being inherited. This can be done in several ways, some of which already exist in nature and some no different from "traditional" gene editing.

Scientists think gene drive organisms could help solve many problems, including wiping out other insect-borne diseases such as ZIKV. Mosquito-borne infectious diseases such as dengue, chikungunya, and now ZIKV pose a public health threat to the US, particularly Florida, the Gulf Coast states, and Hawaii. Recent autochthonous transmission of dengue and chikungunya in Florida, the recent dengue outbreak in Hawaii, and the potential for future local spread of ZIKV in the US have led to the consideration of novel approaches to mosquito management.

One novel approach, releasing sterile genetically modified mosquitoes, has been proposed as a possible intervention. A trial release of GM mosquitoes is being considered in one Florida community [1]. However, Oxitec has so far steered away from producing gene drives [2].

The politics of genetic engineering may leave the public at odds with the science because of the GM aspects and the fact that more mosquitoes must be released to be effective. The permanent presence of a mosquito with novel traits is an inherently difficult topic with which to deal, mainly due to the unforeseen future risks [3].

Genetic strategies for vector control are usually divided into two steps. The first step consists of population suppression, containment, or eradication that aims to reduce or even eliminate specific insect species by developing genes that are either (conditionally) lethal or capable of making the insects sterile. The second step involves population transformation or replacement. The aim is not to eliminate the vector but to create a substitution responsible for introducing an effector gene to reduce or block disease transmission into the wild population [4].

D. M. Berube, *Pandemics and Resilience: Lessons we should have learned from Zika*, Risk, Systems and Decisions, https://doi.org/10.1007/978-3-031-25370-6_15

15.1 Synthetic Biology

Synthetic biology is in its infancy, and the first commercial applications will likely appear incremental to traditional genetic modification. Depending on the pace of development, it might be expected to see fully commercialized outputs from synthetic biologists in total production within the next ten years [5].

One of the first examples of the application of synthetic biology was the production of the antimalarial therapy artemisinin in yeast by introducing additional genes encoding the biosynthesis of artemisinic acid from natural fatty acid precursors [19].

The scientific community is largely undisturbed by the idea of removing the A. aegypti mosquito from the face of the earth. "The disappearance of a few species, while a pity, does not bring a whole ecosystem crashing down," evolutionary biologist Olivia Judson has written." [6].

Species-ide: The engineering of extinction: There is nothing sinister about extinction; species go extinct all the time. The disappearance of a few species, while a pity, does not bring a whole ecosystem crashing down: There is not a wasteland every time a species vanishes. Removing one species sometimes causes shifts in the populations of other species, but different needs do not mean worse [7].

Mosquitoes are highly sexually dimorphic. There is a clear separation between harmless, nectar-feeding males and deadly blood-feeding females for the significant vectors of malaria and dengue. The identification of molecular switches and genetic programs that control the decision made in the early developing mosquito embryo to proceed as male or female may be used to improve existing sterile insect strategies for managing mosquito-borne disease agents [8].

Synthetic biologists design gene drives—artificial selfish genetic elements with a strong propensity to spread through target populations or species. Gene drive refers simply to the ability of a gene to be inherited more frequently than Mendelian genetics would dictate, thus, increasing in frequency, perhaps even to fixation [8].

Synthetic biology is described or treated as the application of engineering principles to genetic modification or as a generic set of tools, technologies, and approaches (essentially services) for achieving biotechnology objectives or as simply a synonym for biotechnology with no meaningful difference between the two [19].

Industry research estimated that equity funding to private synthetic biology companies topped $1 billion in 2016, helping drive market forecasts to an estimate of close to $40 billion by 2020 [19].

Gene drive systems for population modification of vector mosquitoes have been proposed for nearly half a century [9].

Because gene drives could rapidly propagate novel DNA through an entire population in the wild, they could be used, proponents say, to eradicate marauders. They might make mosquitoes resistant to the microbes that cause malaria or dengue fever or even block the gene that makes locusts swarm, saving millions of tons of crops yearly. But gene drives might also doom important species to extinction, change the course of evolution, and perhaps be used to create bioweapons [10].

> Scientists could alter mosquito genetics to spread a fatal flaw through the entire population, reducing overall numbers; they could modify mosquitoes to produce more male offspring than female offspring, reducing the number of mosquito bites; or they could equip mosquitoes with genes to help them fend off malaria, reducing transmission of the disease within mosquito populations and thus to humans, too. [11]

"Gene drives have enormous potential for the control of populations of insect vectors and pests," mosquito researchers Tony Nolan and Andrea Crisanti wrote this week in The Scientist. "They are species-specific, self-sustaining, and have the potential to be long-term and cost-effective." [11].

15.1.1 CRISPR

What is CRISPR? Of course, it's more than one thing. Generally, it is simply a segment of DNA containing short repetitions of base sequences, involved in the defense mechanisms of prokaryotic organisms (is a single-celled organism that lacks a nucleus and another membrane-bound organelle) to viruses. CRISPR is also a genetic engineering tool that uses a CRISPR sequence of DNA and its associated protein to edit the base pairs of a gene. Clustered Regularly Interspaced Short Palindromic Repeats (CRISPR) is often followed by a suffix, such as Cas is a term used to denote the enzymes that cut and edit DNA in the CRISPR method (Cas9). So, artificial CRISPR-Cas systems are also known simply as "CRISPR." Several new CRISPR inhibitors have been discovered, including one for CRISPR/Cas12a, and it is becoming a popular alternative to Cas9 mostly because it is a more compact protein than Cas9 and, therefore, may be easier to get into humans. DARPA funded the Cas12a initiative [12].

Even if Oxitec is chased away by the naysayers, which seems very unlikely, the idea of undermining *Ae aegypti* at a genetic level will persist. More-sophisticated gene-editing techniques, such as CRISPR/Cas9, have been developed, and new businesses will emerge to take advantage of them. This involves the computerization of biological engineering.

Jason Rasgon, a professor of entomology and disease epidemiology at Penn State, says that he is seen older genetic techniques that were inefficient, random, and slowly paved the way for new methods like CRISPR, which enables efficient, rapid, and precise editing of mosquito genomes, and ReMOT Control which eliminates the requirement for injecting materials into mosquito embryos. These innovative technologies make GM mosquitoes for disease control not a question of "if" but rather a question of "where" and "when" [13].

Mosquitoes modified with gene drive systems are being proposed as new tools to complement current practices aimed at reducing or preventing the transmission of vector-borne diseases such as malaria [14].

Gene drive systems based on CRISPR/CRISPR associated protein can potentially spread beneficial traits through interbreeding populations of malaria mosquitoes.

However, the characteristics of this technology have raised concerns that necessitate careful consideration of the product development pathway [14].

15.1.2 CRISPR and the DOD

A sidebar. This technology has not escaped the interest of the military and its powerful research arm, Defense Advanced Research Projects Agency (DARPA) [15]. There is a rich body of literature in this area. What follows is a brief introduction to the military gene drive connection.

> Announced by DARPA in November 2016, the Insect Allies program was reportedly backed by more than $27 million in awarded research contracts. In July 2017, the first of three consortia announced that they had been awarded a contract from DARPA to develop systems for insect dispersion of genetically modified viruses. These are contracts for completing a 4-year work plan culminating in large-scale greenhouse demonstrations of the fully functional insect-dispersed HEGAA (horizontal environmental, genetic alteration agents) approach.

In 2017, DARPA announced a $65 million research program called "Safe Genes" to develop countermeasures and prophylactic treatments against unwanted gene editing. This concept is covered below when examining some precautions discussed to make gene drive modified organisms like mosquitoes less ominous.

Teams from Berkeley and UC San Francisco have received defense money to help discover new CRISP inhibitors. The legendary Dr. Jennifer Doudna leads the Berkeley team [16] (Nobel-prize-winning biochemist), while Dr. Joseph Bondy-Denomy leads the UCSF team [17].

As scientists march closer to using CRISPR gene editing to treat diseases, the U.S. Department of Defense is prepping for the possibility of more nefarious use of CRISPR: its weaponization to harm humans, animals, or crops [18].

"It could be a national security interest to ensure that CRISPR goes forward safely," Bondy-Denomy says. He adds that the Cas12a inhibitors are just one small part of many ongoing projects funded by DARPA. "Weaponized gene editing might sound a little bit like science fiction, but in the short term, it is obvious that having safe gene editing makes sense." [12].

15.2 Gene Drives Are Here

Gene drives for mosquitos have recently been designed to cause little or no disadvantage to offspring receiving a copy of the gene drive from only one parent but to cause sterility in females who receive the gene drive from both parents. This design is intended to allow the gene drive to spread rapidly through populations until it accumulates. At this point, the population numbers will crash, with little opportunity for mosquitos to escape this fate through natural selection. The aim is to deploy this system in the environment to rapidly reduce the mosquito population to below the threshold level that supports the spread of diseases like malaria and dengue fever and, therefore, to reduce or even eliminate these diseases massively [19].

For a while, there was much litigation over patents and ownership. However, on September 10, 2018, a U.S. federal court ruled in favor of the Broad Institute of MIT and Harvard, agreeing with a lower court that Broad's patents for CRISPR/Cas9 gene editing did not interfere with a patent application from CRISPR's other inventors at the University of California, Berkeley, and the University of Vienna.

15.3 Gene Drives and Malaria

The primary driver at this time may not even be the mosquitoes associated with ZIKV but instead *Anopheles gambiae*, the malaria-carrying mosquito. Malaria is an awful disease. In 2020, there were 241 million cases worldwide, and the estimated number of deaths stood at 627,000. The WHO African Region carries a disproportionately high global malaria burden. In 2020, the region was home to 95% of malaria cases and 96% of malaria deaths. Children under five accounted for about 80% of all malaria deaths in the Region [20]. In addition, projects involving gene drive-CRISPR solutions to malaria have attracted investments from the Bill and Melinda Gates Foundation to get regulatory approvals in place by 2024 and the first gene drive mosquitoes ready for use by 2026 [21].

The approach involves the release of mosquitoes equipped with "gene, a technology that overrides nature's genetic rules to give every baby mosquito a particular trait that normally only half would acquire. Once such an insect gets out into the wild, it will move indiscriminately and spread its modified feature without respect for political borders [22].

Geneticists at Imperial College London are designing the gene drive mosquitoes. Specifically, they are studying two ways to disrupt the reproductive system: reduce the number of female babies or stop the mosquitoes from having offspring. To make the population predominantly male, Austin Burt, Target Malaria's primary investigator, and collaborators study an "X shredder"—a gene that destroys the X chromosome in sperm, making all offspring males. Alongside that, they are looking at reducing the number of mosquitoes of both sexes by creating genes that make them sterile [22].

Burkina Faso, Mali, and Uganda are building the groundwork to eventually let loose "gene drive" mosquitoes, which would contain a mutation that would significantly and quickly reduce the mosquito population [23]. Researchers in Mali and Uganda are working under the "Target Malaria" project, propelled by $75 million from the Bill and Melinda Gates Foundation and support from research laboratories in England and Italy. It may be six years before the gene drive mosquitoes are released in Burkina Faso, but scientists are already working around the clock to prepare the community for their release [22].

It will take years to reach the point that scientists will be ready to test the gene drive mosquitoes in the wild. In the meantime, they are facing the challenge of winning over residents who might be wary of these new creatures. No living thing—mammal, insect, or plant—with a gene drive has ever been set free. But if all goes as planned, it might happen here, in a remote village of about a thousand people, where the residents do not even have a word for "gene".

15.4 Gene Drives and Mosquitoes

The gates have been flung open. The race is on. What follows is hardly a complete listing of ongoing research involving gene drive and CRISPR technologies as they might be employed to reduce mosquito-vectored infectious diseases (Fig. 15.1).

15.4.1 A ZIKV-Resistant Mosquito

A team from the University of California, San Diego, collaborated with investigators in Australia and Taiwan and used genomic technology to generate a mosquito resistant to ZIKV. To do that, investigators took Aedes aegypti mosquitoes at the embryonic stage. They injected them with an anti-ZIKV gene and another gene expressing red eyes to differentiate the modified ones. The engineered mosquitoes displayed lower viral infection, dissemination, and transmission levels. "Our results demonstrate that engineered mosquitoes express a polycistronic cluster of small synthetic RNAs designed to target the ZIKV genome. As a result, homozygous mosquitoes were refractory to ZIKV infection and therefore could not transmit the virus," the investigators wrote [24].

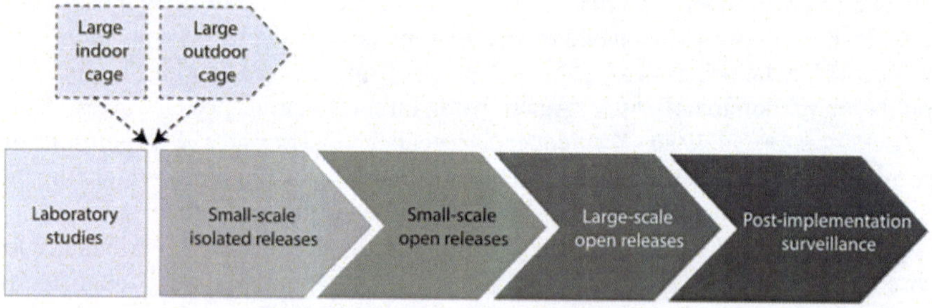

Fig. 15.1 Pathway to deployment of gene drive mosquitoes. James et al. [14]

15.4.2 Underdeveloped and Weakened Females

Optimal species-specific body size is essential for other characteristics, such as metabolism, habitat, life history, and extinction risk. Control of body size is exceptionally complex and provided by numerous intrinsic and environmental cues. Attaining optimal body size and nutritional status is critical for mosquitoes to become reproductively competent and effective disease vectors. Ling and Raikhel (2018) have identified the fat-body-specific serotonin signaling involved in regulating body size and metabolism in mosquitoes [25].

CRISPR-Cas9 disruption of the fat-body-specific serotonin receptor Aa5HT2B impairs body growth and lipid accumulation in Aedes aegypti mosquitoes, the vectors of dengue fever, yellow fever, and ZIKV. Aa5HT2B controls insulin-like peptides. Using the CRISPRCas9 approach, the differential roles of insulin-like peptides in controlling body size and metabolism have been uncovered. There has also been a link between blood-feeding and the fat-body-specific serotonin signaling [25].

15.4.3 Sterilized Males

UCSD also reported on a new precision-guided sterile insect technique, or pgSIT alters genes linked to male fertility—creating sterile offspring—and female flight in Aedes aegypti. UCSD's Omar Akbari, cited earlier, uses CRISPR to inactivate a fertility gene in Aedes aegypti to sterilize future generations of females [26].

Leveraging advancements in CRISPR-based genetic engineering, researchers at the University of California San Diego have created "a new scalable genetic control system that uses a CRISPR-based approach to engineer deployable mosquitoes that can suppress populations," said UC San Diego Biological Sciences Professor Omar Akbari. "Males don't transmit diseases, so the idea is that as you release more and more sterile males, you can suppress the population without relying on harmful chemicals and insecticides." [27].

15.4.4 Hermaphrodites

For the first time, researchers have begun large-scale releases of the engineered insects into a high-security laboratory in Terni, Italy. The lab was specially built to evaluate the modified insects as close to a natural environment as possible without the risk of releasing them into the wild. There are deep concerns regarding unforeseen effects on the environment.

Scientists have tried to keep genetically engineered organisms from spreading their mutations to prevent unforeseen environmental effects. But in this case, researchers want the modification to apply. So, they engineered mosquitoes with a gene drive. Usually, traits are passed to only half of all offspring. With the gene drive, nearly all the progenies inherit the modification [28].

Researchers created mosquitoes using the powerful new gene-editing technique known as CRISPR. Ruth Mueller, an entomologist who runs the lab, likens it to a "molecular scissor which can cut at a specific site in the DNA." The cut altered a gene known as "doublesex," which is involved in the sexual development of the mosquitoes. "The females become a bit more male," Mueller says. "A kind of hermaphrodite. "While genetically female, the transformed insects have mouths that resemble male mosquito mouths. That means they cannot bite and cannot spread the malaria parasite. In addition, the insects' reproductive organs are deformed, so they cannot lay eggs [28]. Her lab is working on Anopheles gambiae, the main species of mosquito that spreads malaria.

15.4.5 Sexual Biasing

The discovery of the first male-determining factor in mosquitoes, combined with the gene-editing capabilities of the CRISPRCas9 system, could be used to bias mosquito populations from deadly, bloodsucking females toward harmless, nectar-feeding males, thus helping to prevent the spread of mosquito-borne diseases [29].

In female embryos expressing this so-called M factor, a sex-determination gene called Nix triggered the development of external and internal male genitalia. Nix was required and sufficient to initiate male development in Aedes aegypti, a primary carrier for dengue, yellow fever, ZIKV, and chikungunya viruses. "This discovery sets the stage for future efforts to leverage the CRISPRCas9 system to drive maleness genes such as Nix into mosquito populations, thereby converting females into males or simply killing females," Tu says [30].

Austin Burt and fellow Imperial College London biologist Andrea Crisanti focus on disrupting reproduction. Burt said. Using an enzyme that targets some two hundred sites on the X chromosomes in mosquito eggs, the team has shreds X chromosomes so thoroughly "that it's too much for the cell to repair," Burt said. The result: Eggs carry only the Y chromosome, which makes sons, and no X, which makes daughters. So far, they have gotten 95 percent of sons using an editing tool other than CRISPR [10].

15.4.6 Hardening Eggshells [11]

Jun Isoe and Roger Miesfield, a team from the University of Arizona, believe they have isolated the gene responsible for hardening (tanning) the eggshells laid by disease-carrying mosquitoes.

Shortly after the eggs are deposited in a moist area such as a flowerpot or the edge of a pond, the eggshell hardens through a process called "tanning," in which the eggshell turns from white to brown during the maturation process. Because the drought-resistant larvae will hatch from the hardened mosquito eggs in the presence of water during the monsoon season in areas such as Tucson, the complete formation of the eggshell within a few hours of egg-laying is essential to mosquito reproduction [31].

> To discover mosquito genes that are uniquely required for mosquito eggshell synthesis in blood-fed female Aedes aegypti mosquitoes, we used a computer-based approach to identify genes that have evolved to be unique to mosquitoes and are not found in closely related insects such as fruit flies and honeybees, nor animals such as ourselves. Among the roughly one hundred mosquito-specific genes we disrupted in blood-fed *Ae. aegypti*, we found one we call Eggshell Organizing Factor 1 (EOF1) required to complete the eggshell synthesis. We discovered that 100% of the eggs laid by mosquitoes lacking the EOF1 protein were missing a tanned eggshell, and none of the larvae survived. The lethal effect of an EOF1 deficiency was partly because the eggs did not complete the tanning process required for eggshell maturation. [31]

Mansfield and Iso claim: This does not mean that mosquitoes should be eliminated from the ecosystem, as that could have unknown consequences. Instead, mosquito populations should be selectively reduced by decreasing reproductive rates at specific times of the year, such as the rainy season [31].

15.4.7 Invisibility Cloak

For the first time, scientists have used the gene-editing tool CRISPR-Cas9 to render humans effectively invisible in the eyes of *Ae. aegypti* mosquitoes, which use dark visual cues to hunt, according to a paper published in the journal *Current Biology*. By eliminating two of that mosquito's light-sensing receptors, the researchers knocked out its ability to visually target hosts.

When female Aedes aegypti sense CO_2 from human breath, their attention to other human-derived stimuli, including visual cues, dramatically increases. Using CRISPR, Zhan et al. knock out two of the five opsins in the eyes and show that this eliminates attraction to visual targets after CO_2 stimulation, whereas other vision-dependent behaviors remain [32].

The new paper could inform future strategies to control mosquito populations. If female mosquitoes could not see hosts, they would have more difficulty finding the blood required for their eggs to develop. "The population would crash," Dr. Montell said. The researchers have yet to expose the double mutants to hosts. When they do, Dr. Thakre is curious to know precisely how impaired vision affects the ability of mosquitoes to feed on blood, given the insects' many other senses. "The thing you want to control is a mosquito bite," Dr. Thakre said [33].

15.5 Gene Drives Debates

"This is a technology where we don't know where it will end. We need to stop this right where it is," says Nimmo Bassey, director of the Health of Mother Earth Foundation in Nigeria. "This is an experimental technology which could have devastating impacts," says Dana Perls of Friends of the Earth, an environmental group part of an international coalition fighting this new generation of modified organisms [28].

Over a decade ago, Xi et al. wrote that "a gene drive vehicle is a key component of vector population replacement strategies, providing a mechanism for the autonomous spread of desired transgenes into the targeted population. Compared with strategies that rely on inundating releases and Mendelian inheritance, gene drive strategies would require relatively small "seedings" of transgenic individuals into a field population. Perhaps more important than increased cost-efficacy, gene drive strategies can facilitate population replacement with transgenic individuals with lower fitness relative to the natural population." [34].

As the pioneering synthetic biologist Jack Newman said to Kristen Brown, "What stands between us and addressing one of the world's biggest public health issues (mosquitoes and infectious diseases) is not science. It is how we talk about science." [35].

15.5.1 Reservations on CRISPR

Many articles discuss drawbacks to CRISPR; there are many more that hail its discovery and its endless list of applications. It is not the purpose of this chapter to review them substantially. If you want to learn about CRISPR, there are online courses and many excellent articles and books on the subject. Instead, this chapter reviews a few problems anticipated by some critics that are specific to its use in engineering mosquitoes for vector control.

Some have argued CRISPR may produce unwanted off-target mutations that then are passed on to another generation. They warn that strategies for reducing genome-wide off-target effects are imperfect, possessing only partial and unproven efficacies and other limitations [36].

Using Clustered Regularly Interspaced Short Palindromic Repeats (CRISPR) and others have triggered a backlash from environmental activist groups—non-governmental organizations (NGOs) that utilize fear for their political ends [37].

15.5.2 Fears and Reservations

Negative representations of GMOs tap into people's fears by stressing that genetic modification can contaminate the environment and the food people eat. The fear of contamination is powerful, universal, and challenging to control. Fear impedes people's ability to reason and affects their risk perception [38].

Gene drives have triggered a backlash from environmental activist groups—non-governmental organizations (NGOs) that some suggest utilizing fear for their political ends [37]. On the other hand, many critics are concerned about releasing gene drive modified mosquitoes.

Editing pernicious genes to make a disease-causing mosquito or a pathogen-carrying rodent less harmful sounds like an appealing idea. But there are earnest questions about the ethics of engineering a wild species and potential environmental consequences that might change ecosystem dynamics or spread well beyond the targeted location. The risks inherent in gene drives designed to spread indefinitely to all populations are likely to be greater than with those that are designed to have limits to spread. Concerns with unlimited gene drive approaches have resulted in researchers considering genetic manipulations that would make non-target populations immune to these gene drives [39].

15.5.3 *Public Understanding of Gene Drives*

There seems to be widespread public support for releasing genetically modified mosquitoes, less so for a gene drive modified mosquito. Recall the difference: A "gene drive" is a version of gene editing—a newer, more precise way to change a DNA (or RNA) sequence; in this case, combining a guide RNA with an enzyme that can make a gene drive takes this to another level, making sure that a new or altered genetic sequence has a greater than 50 percent chance of being inherited. Genetically modified mosquitoes like the ones developed by Oxitec do not drive the genes into the next generation. That gene modification makes the offspring less viable to a point where the population crashes.

While there is a difference, one ponders whether the public understands the difference or would react differently if one class of GM mosquitoes were released rather than the other.

A nationwide survey published in February 2007 by the University of Pennsylvania's Annenberg Public Policy Center found that 35% of respondents believed that GM mosquitoes cause the spread of ZIKV. There is speculation on how much this opinion might impact local support for GM mosquitoes [40].

A little less than a decade later and little more than half of U.S. adults (53%) favor having scientists release GM mosquitoes to minimize the spread of the ZIKV, according to a (2016) survey also by the Annenberg Public Policy Center (APPC) of the University of Pennsylvania. Twenty-nine percent oppose it [41].

> Two years later, in 2018, Annenberg reported on yet another of their surveys.
>
> The survey's most widely accepted use of genetic intervention in animals involves mosquitoes. Seven-in-ten Americans (70%) believe that genetically engineering mosquitoes to prevent their reproduction and, therefore, the spread of some mosquito-borne diseases

would be an appropriate use of technology, while about three-in-ten (29%) see the use of genetic engineering for this purpose as taking technology too far [42].

At a deeper granular level, about three in ten of those who said genetic engineering of mosquitoes would be taking technology too far explained that humankind would be disrupting nature (23%) or interfering with God's plan (8%). Some 24% of those with objections to reducing mosquitoes' fertility through genetic engineering to reduce mosquito-borne illnesses raised concerns about the possible impact on the ecosystem.

A year later, in 2019, there was a gene drive survey reported by Jones et al. on U.S. adult attitudes toward agricultural gene drives. When informed about potential risks, benefits, and two previously researched applications, respondents' support/opposition depends heavily (+22%/−19%) on whether the spread of drives can be limited, non-native versus native species are targeted (+12%/−9%), or the industry replaces versus suppresses target species (±2%) [2].

In their discussion, they observed that while seekers of non-GMO food are relatively less supportive of pursuing gene drives, roughly half of this subpopulation still supports the pursuit of controllable applications to non-native species. This result suggests that a sizable portion of this subpopulation does not uniformly transfer their revealed preferences regarding GMO food to other genetic engineering technologies [2].

Partnerships with supportive governments, local collaborators, and a willing public are crucial to they are establishing field-based testing in environmentally relevant areas (essentially, where releases might occur) [43]. Modeling suggests that a powerful nuclease-based gene drive system could become found in the wild with the accidental release of just a few individuals during such testing. However, the outcome may depend highly on local landscapes, vector population density, and dispersal characteristics [44]. This means securing permission for field-based testing of highly active (CRISPR) gene drive constructs may be challenging in epidemiologically relevant (dengue or malaria-endemic) areas without guarantees that the introduced gene could be removed from the study area if needed [8].

The anxiety here is that of the unknown. About three-quarters of adults thought the technology would be used before the health effects were fully understood. Unsurprisingly, among those who were already aware of gene-editing technology, a higher percentage (57%) said they were inclined to give it a whirl [45].

In a paper published in Science Advances, Zack Brown, an assistant professor of resource economics at NC State, and his colleagues surveyed 1000 American adults on their opinions of gene drives. Instead of opposition, they found support for the technology, with a few caveats: 61% of respondents supported gene drives that suppressed populations of non-native pests, while 57% supported drive drives that would replace non-native pests. Nearly 50% supported uncontrolled gene drives that eliminated native pest species (25% did not have a strong opinion) [46].

The science community has retorted that most people embrace biotechnology when they recognize it benefits them directly.

When Malaysia was evaluating the Introduction of the Oxitec mosquito (not developed by gene drive), Dr. Ricarda Steinbrecher from Oxford sent this comment on field trials to the Director-General of the Biosafety Ministry. Her complaint was given the importance of missing data, experiments, investigation, and knowledge for ensuring the "protection of human, plant and animal health, the environment and biological diversity" I would suggest that it is too early for any open field releases. This is particularly the case since RIDL (release of insects carrying a dominant lethal) is not 100% dependable, and thus LMO mosquitoes will escape via surviving progeny [47].

Like many others, her concern is steeped in what is not known.

15.6 Conclusion

Gene drives represent a classic case of "post-normal science," for which purely technical expertise is not enough to address a scientific issue's technological, social, ethical, and legal complexities. Unlike "normal" scientific problems, risk assessment can be based primarily on scientific inputs; post-normal science must rely on many perspectives when assessing risks and benefits [39].

Having told the ZIKV story, it is time to move to the last four chapters dedicated to outreach and communication and building resilience into pandemic/epidemic management strategies so there will be better preparation for ZIKV return or a whole new virus.

References

1. Adalja A et al (2016) Genetically modified (GM) mosquito use to reduce mosquito-transmitted disease in the US: a community opinion survey. PLOS Curr Outbreaks. May 25. https://doi.org/10.1371/currents.outbreaks.1c39ec05a743d41ee39391ed0f2ed8d3
2. Jones MS, Delborne JA, Elsensohn J, Mitchell PD, Brown Z (2019) Does the U.S. public support using gene drives in agriculture? And what do they want to know? Sci Adv 5:eaau8462. September 11. https://www.science.org/. https://doi.org/10.1126/sciadv.aau8462. Accessed 15 June 2022
3. Sikka V et al (2016) The emergence of Zika virus as a global health security threat: a review and a consensus statement of the INDUSEM Joint working Group (JWG). J Glob Infect Dis 8:1. https://pubmed.ncbi.nlm.nih.gov/27013839/. Accessed 13 June 2022
4. Araujo HRC, Carvalho D, Ioshino R, Costa-da-Silva A, Capurro M (2015) Aedes aegypti control strategies in Brazil: incorporation of New technologies to overcome the Persistence of Dengue Epidemics. Insects 6:576–594
5. (Society of) Lloyd's (2009) Lloyd's emerging risks team report: synthetic biology influencing development. July. https://www.lloyds.com/news-and-risk-insight/risk-reports/library/technology/synthetic-biology. Accessed 4 Sept 2018
6. Kolker R (2016) Florida's feud over Zika-fighting GMO mosquitoes. Bloomberg Businessweek. October 6. http://www.bloomberg.com/features/2016-zika-gmo-mosquitos/. Accessed 25 Oct 2016
7. Judson O (2003) A Bug's death. The New York Times. September 25. http://www.nytimes.com/2003/09/25/opinion/a-bug-s-death.html. Accessed 5 June 2017

8. Adelman Z, Tu Z (2016) Control of mosquito-borne infectious diseases: sex and gene drive. Trends Parasitol 32(3):2017. https://doi.org/10.1016/j.pt.2015.12.003.AccessedApril7
9. Curtis CF (1968) Possible use of translocations to fix desirable genes in insect pest populations. Nature 218(5139). April 27. https://www.nature.com/nature/journal/v218/n5139/abs/218368a0.html. Accessed 14 May 2017
10. Begley S (2015) Gene drive gives scientists power to hijack evolution. STAT. November 15. https://www.statnews.com/2015/11/17/gene-drive-hijack-evolution/. Accessed 98 Mar 2017
11. Brown K (2017) This controversial genetic engineering technology could eliminate malaria. GIZMODO. January 14. http://gizmodo.com/controversial-genetic-engineering-technology-could-elim-1790727517. Accessed 14 Mar 2017
12. Cross R (2018) New CRISPR inhibitors found. Chem Eng News. September 17. https://cen.acs.org/biological-chemistry/biotechnology/New-CRISPR-inhibitors-found-help/96/web/2018/09. Accessed 15 June 2022
13. Rasgon J (2018) Genetically modified mosquitoes may be best weapon for curbing disease transmission. The Conversation. August 20. https://theconversation.com/genetically-modified-mosquitoes-may-be-best-weapon-for-curbing-disease-transmission-100719. Accessed 13 June 2022
14. James S et al (2018) Pathway to deployment of gene drive mosquitoes as a potential biocontrol tool for elimination of Malaria in Sub-Saharan Africa: recommendations of a scientific working group. Am J Trop Med Hyg 98(6). https://www.ncbi.nlm.nih.gov/pubmed/29882508. Accessed 10 Sept 2018
15. Reeves RG, Voeneky S, Caetano-Anollés D, Beck F, Boëte C (2018) Agricultural research, or a new bioweapon system? Science 362:35–37; Eureka Alert! (2017) BTI receives DARPA "Insect Allies" Award. EurekAlert! www.eurekalert.org/pub_releases/2017-07/btibrd072717. Accessed 15 June 2022
16. Watters KE, Fellmann C, Bai HB, Ren SM, Doudna JA (2018) Systematic discovery of natural CRISPR-Cas12a inhibitors. Science 362(6411). https://doi.org/10.1126/science.aau5138
17. Marion ND, Zhang JY, Borges AL, Sousa AA, Leon LM et al (2018). Discovery of widespread type I and type V CRISPR-Cas inhibitors. Science 362(6411). https://doi.org/10.1126/science.aau5174
18. Cross R (2018) Synthetic biology could enable bioweapons development. Chem Eng News. June 19. https://cen.acs.org/biological-chemistry/synthetic-biology/Synthetic-biology-enable-bioweapons-development/96/i26. Accessed 15 June 2022
19. Polizzi KM, Stanbrough L, Heap JT (2018) A new lease of life: Understanding the risks of synthetic biology. An emerging risks report published by Lloyd's of London. https://www.lloyds.com/news-and-risk-insight/news/lloyds-news/2018/07/a-new-lease-of-life. Accessed 4 Sept 2018
20. WHO (2022) Malaria. April 6. https://www.who.int/news-room/fact-sheets/detail/malaria. Accessed 13 June 2022
21. Gates B (2019) Test-tube mosquitoes might help us beat malaria. Gates Notes. April 15. https://www.gatesnotes.com/Health/Test-tube-mosquitoes-might-help-us-beat-malaria. Accessed 13 June 2022
22. Swetlitz I (2017) In a remote West African village, a revolutionary genetic experiment is on its way—if residents agree to it. STAT. March 14. https://targetmalaria.org/stat-news-in-a-remote-west-african-village-a-revolutionary-genetic-experiment-is-on-its-way-if-residents-agree-to-it-march-2017/. Accessed 2 June 2022
23. Swetlitz, Ike. (2018). For the first time, researchers will release genetically engineered mosquitoes in Africa. STAT. September 5. https://www.statnews.com/2018/09/05/release-genetically-engineered-mosquitoes-africa/. Accessed June 19, 2019.
24. Ward, Alexandra. (2019). Tackling Zika transmission at the source with genetically engineered resistant mosquitoes. Contagion Live. February 14. https://www.contagionlive.com/view/tackling-zika-transmission-at-the-source-with-genetically-engineered-resistant-mosquitoes. Accessed 13 June 2022

25. Ling L, Raikhel AS (2018) Serotonin signaling regulates insulin-like peptides for growth, reproduction, and metabolism in the disease vector Aedes aegypti. PNAS 115(42):E9822–E9831. https://doi.org/10.1073/pnas.1808243115
26. McCay B (2016) Mosquitoes are deadly, so why not kill them all? Wall Street J. September 2. http://www.wsj.com/articles/mosquitoesaredeadlysowhynotkillthemall1472827158. Accessed 15 Sept 2016
27. Aguilera M (2021) New technology designed to genetically control disease-spreading mosquitoes. UC San Diego News. September 10. https://ucsdnews.ucsd.edu/pressrelease/new-technology-designed-to-genetically-control-disease-spreading-mosquitoes. Accessed 13 June 2022
28. Stein R (2019) Scientists release controversial genetically modified mosquitoes in high-security lab. NPR. February 20. https://www.npr.org/sections/goatsandsoda/2019/02/20/693735499. Accessed 13 June 2022
29. Cell Press (2016) Can CRISPR help edit out female mosquitoes? Sci News. February 17. https://www.sciencedaily.com/releases/2016/02/160217125535.htm. Accessed 2 June 2017
30. Cell Press (2016) Can CRISPR help edit out female mosquitoes? Sci News. February 17. https://www.sciencedaily.com/releases/2016/02/160217125535.htm. Accessed 2 June 2017; and Adelman Z, Tu Z (2016) Control of mosquito-borne infectious diseases: sex and gene drive. Trends Parasitol 32:3. March. https://doi.org/10.1016/j.pt.2015.12.003. Accessed 7 Apr 2017
31. Miesfeld R, Isoe J (2017) Mosquito eggs without eggshells disrupt the ability to reproduce. January 31. Tucson.com. http://tucson.com/news/local/mosquito-eggs-without-eggshells-disrupt-the-ability-to-reproduce/article_7fb6d4fa-f853-5f1a-ab88-47701e9a9403.html. Accessed 23 May 2017
32. Zhan Y, San Alberto DA, Rusch C, Riffell J, Montell C (2021) Elimination of vision-guided target attraction in Aedes aegypti using CRISPR. Curr Biol 31:1–8. September 27
33. Imbler S (2021) What if you could become invisible to mosquitoes? The New York Times. August 17. https://www.nytimes.com/2021/08/17/science/crispr-mosquito-vision.html. Accessed 13 June 2022
34. Xi Z, Khoo C, Dobson S (2005) Wolbachia Establishment and Invasion in an Aedes aegypti laboratory population. Science 310(5746). October 14. https://www.ncbi.nlm.nih.gov/pubmed/16224027. Accessed 2 Jul 2017
35. Brown K (2016) Genetically modified mosquitoes could wipe out the world's most deadly viruses. If we let them. Fusion.net. September 19. http://fusion.net/story/347298/oxitec-genetically-modified-mosquitoes/. Accessed 26 Oct 2016
36. Kleinstiver BP, Pattanayak V, Prew MS, Tsai SQ et al (2016) High-fidelity CRISPR-Cas9 nucleases with no detectable genome-wide off-target effects. Nature 28(529):490–495. January
37. Porterfield A (2019) Viewpoint: public supports CRISPR, gene drives to battle infectious disease, plant pests—despite activist opposition. Genetic Literacy Project Newsletter. September 17. https://geneticliteracyproject.org/2019/09/17/viewpoint-public-supports-crispr-gene-drives-to-battle-infectious-disease-plant-pests-despite-activist-opposition/. Accessed 13 June 2022
38. Parker C, Bernaola L, Lee BW, Elmquist D et al (2019) Entomology in the 21st century: tackling Insect Invasions, promoting advancements in technology, and using effective science communication—2018 student debates. J Insect Sci 19(4):1–11. https://doi.org/10.1093/jisesa/iez069
39. Brossard D, Belluck P, Gould F, Wirz CD (2018) Promises and perils of gene drives: navigating the communication of complex, post-normal science. PNAS 116(16):7692–7697. April 16
40. Adalja A et al (2016) Genetically modified (GM) mosquito use to reduce mosquito-transmitted disease in the US: a community opinion survey. PLOS Curr Outbreaks. May 25. https://doi.org/10.1371/currents.outbreaks.1c39ec05a743d41ee39391ed0f2ed8d3; and Harvey C (2016) A shocking one-third of Americans believe this Zika conspiracy theory. Washington Post. February 23, 2016. https://www.washingtonpost.com/news/energy-environment/wp/2016/02/23/a-shocking-one-third-ofamericans-believe-this-zika-conspiracy-theory/. Accessed 14 Apr 2016

41. Annenberg Press Release (2016) Just over half of U.S. public favors using GM mosquitoes to fight Zika. https://www.annenbergpublicpolicycenter.org/just-over-half-of-u-s-public-favors-using-gm-mosquitoes-to-fight-zika/. Accessed 15 June 2022
42. Funk C, Hefferon M (2018) Most Americans accept genetic engineering of animals that benefits human health, but many oppose other uses, pew research center. August 16. https://www.pewresearch.org/science/2018/08/16/most-americans-accept-genetic-engineering-of-animals-that-benefits-human-health-but-many-oppose-other-uses/. Accessed 15 June 2022
43. Brown DM et al (2014) Criteria for identifying and evaluating candidate sites for open-field trials of genetically engineered mosquitoes. Vector Borne Zoonotic Dis 14:291–299
44. North A et al (2013) Modelling the spatial spread of a homing endonuclease gene in a mosquito population. J Appl Ecol 50:1216–1225
45. Brown K (2016) Americans are terrified genetic enhancements will turn the rich into sci-fi supermen. October 31. Fusion.net. http://fusion.net/americans-are-terrified-genetic-enhancements-will-turn-1793861544. Accessed 4 May 2017
46. Porterfield A (2019) Viewpoint: public supports CRISPR, gene drives to battle infectious disease, plant pests—despite activist opposition. Genetic Literacy Project Newsletter
47. Steenbrecker R (2010) NBB ref no: NRE(S)609—2/1/3 application for approval for limited mark—release—recapture of Aedes aegypti (L.) Wild Type and OX513A Strains. EcoNexus. September 3. http://www.econexus.info/publication/release-gm-mosquito-aedes-aegypti-ox513a. Accessed 2 June 2017

Chapter 16
Travel and Pregnancy Warnings

In 2017, the CDC reported 39,350 reported disease cases in the United States and territories, with 3947 in pregnant women [1].

The most difficult challenge about messaging is staying on the issues. Even the most experienced communicator will find herself swept toward a collateral issue. This happened when the CDC decided to issue a travel warning.

At first, it seemed unimportant until a deeper reading of this warning appeared to dump the entire issue set into the laps of women, primarily young women of color. The notices asked them to use repellants and cover-up. Still, even more importantly, it called upon women to delay their pregnancies if they or their partners could not stay isolated from areas where ZIKV was epidemic. If they were pregnant, they should stay away, and if they were thinking about becoming pregnant, they should stay away from the outbreak. This was an onerous burden for many young women, as discussed below.

ZIKV was an unprecedented and complex response for the CDC. During the height of the pandemic, and approximately 1900 CDC staff members supported the ZIKV response [2]. They provided funding for vector surveillance and control, supported laboratory capacity and equipment improvements, and supported surveillance, epidemiology, and public health investigations, especially with its Zika Pregnancy Registry. The U.S. Zika Pregnancy and Infant Registry (USZPIR) is a collaborative and innovative system to learn about ZIKV infection during pregnancy and after birth. Information from the USZPIR is used to make recommendations for healthcare providers caring for families affected by ZIKV and understand the spectrum of outcomes among infants with possible congenital ZIKV exposure.

CDC's ArboNET is a comprehensive public health surveillance data management system for ZIKV and other arthropod-borne viruses [3]. ArboNET is the national arboviral surveillance system managed by CDC and state health departments. In addition to human disease, ArboNET maintains data on arboviral infections among presumptive viremic blood donors, veterinary disease cases, mosquitoes, dead birds, and sentinel animals. Initially designed in 2000 for West Nile Virus, in 2003, it was expanded to include all other nationally notifiable domestic arboviruses.

D. M. Berube, *Pandemics and Resilience: Lessons we should have learned from Zika*, Risk, Systems and Decisions, https://doi.org/10.1007/978-3-031-25370-6_16

ArboNET is a passive electronic system managed by the CDC. Arboviral surveillance data are reported to ArboNET by fifty states and three local or territorial health departments (New York City, the District of Columbia, and Puerto Rico). Reporting jurisdictions receive data from public health and commercial reference laboratories, blood collection agencies, and healthcare providers [4].

Many comments express dissatisfaction with data transmission mechanisms, particularly its inability to accept National Electronic Disease Surveillance System (NEDSS) compliant messages. The National Electronic Disease Surveillance System (NEDSS) Base System (NBS) is a CDC-developed integrated information system that helps local, state, and territorial public health departments manage reportable disease data and send notifiable disease data to CDC. NBS provides a tool to support the public health investigation workflow and process, analyze, and share disease-related health information. NBS also provides reporting jurisdictions with a NEDSS-compatible information system to transfer epidemiologic, laboratory, and clinical data efficiently and securely over the Internet. Despite the expanding scope of the ArboNET system, national funding to state and local health departments to support arboviral surveillance and testing has been drastically reduced in recent years [4].

The U.S. CDC and have not been consistent in its recommendations. Understandably, as more is known, it must change its posted warnings. No one wants to sacrifice the good for the perfect. Waiting until the strategies and protocols are finalized may result in heightened infections, illnesses, and even deaths.

While the expert community understands conservative commentary as responsible behavior, varying the warnings, especially in instances where the gravity of the warning is increasing, leaves the public questioning the competence of the agencies at a point when trust is paramount.

Warmer weather also raises the possibility of local ZIKV transmission in the continental United States, federal health officials push state and local governments to devise plans to protect citizens, and in the case of the ZIKV, pregnant women from infection increased access to contraception and ensured better coordination of mosquito bites control districts [5]. The CDC also says the *Ae. aegypti* species of mosquito linked to ZIKV is present in at least part of thirty states, including areas in southeastern Indiana along the Ohio River [6]. The potential breadth of the warning zone is substantial.

The $1.1 billion in ZIKV funding that Congress passed in 2016 ran out in September. And in 2017, the Trump administration sought to cut the CDC budget by $1.2 billion, to what the agency had 20 years ago [7]. This budget includes funding a congenital disabilities surveillance program intended to monitor affected babies' development and connect them to health services. "That surveillance is critical for knowing what is going on," said Dr. Oscar Alleyne of the National Association of County and City Health Officials. "Otherwise, we're flying blind" [8].

Regarding policy recommendations, the CDC recommends that jurisdictions develop ZIKV action plans to guide preparedness and response activities through a phased, risk-based continuum. The continuum includes support for mosquito season preparedness and graduated measures in response to the detection of confirmed local mosquito-borne transmission and multiperson local mosquito-borne transmission if present. Mosquito season varies by jurisdiction but is typical during the summer months. However, year-round local transmission of the ZIKV may be possible in warmer locations. Jurisdictions with competent vectors should assess risk and institute vector control activities as indicated [9].

16.1 CDC Travel Warning

Important and reputable professional societies, including the CDC, issued guidelines for self-protection, screening, and management for pregnant patients with ZIKV exposure.

- People with Zika, dengue, or chikungunya symptoms should visit a health center.
- Avoid allowing standing water in outdoor containers (flowerpots, bottles, and containers that collect water) so they do not become mosquito breeding sites.
- Cover domestic water tanks so that mosquitoes cannot get in.
- Avoid accumulating garbage: Put it in closed plastic bags and keep it in closed containers.
- Unblock drains that could accumulate standing water.
- Use screens and mosquito nets in windows and doors to reduce contact between mosquitoes and people. To prevent mosquito bites, it is recommended that people who live in areas where there are cases of the disease, as well as travelers and, especially, pregnant women, should:
  - Cover exposed skin with long-sleeved shirts, trousers, and hats.
  - Use repellents recommended by the health authorities (and apply them as indicated on the label).
  - Sleep under mosquito nets.

Approximately half a million pregnant women are estimated to travel to the United States annually from the thirty-two (February 18, 2016) ZIKV-affected countries and U.S. territories with active ZIKV transmission. On January 15, 2016, the CDC issued a Level Two Alert. This is a travel alert. It reads: "Pregnant women should for now avoid traveling to Latin American regions experiencing the most ZIKV infections."

The rationales came from the number of confirmed cases of ZIKV infection reported among women who had traveled to one or more of the following nine areas with ongoing local transmission of ZIKV: American Samoa, Brazil, El Salvador, Guatemala, Haiti, Honduras, Mexico, Puerto Rico, and Samoa.

The CDC recommended that couples trying to get pregnant should wait six months if a man gets infected and has symptoms and wait eight weeks if a woman has symptoms. Women in Florida regularly were exposed to a warning that those who live in or frequently travel to Miami-Dade County diagnosed with Zika should wait at least eight weeks after symptoms start before trying to get pregnant. Men who live in or frequently travel to Miami-Dade County diagnosed with Zika should wait at least six months after symptoms start before trying to get their partner pregnant.

The CDC adds: Because the full clinical spectrum of congenital ZIKV infection is unknown, all infants born to women with laboratory evidence of possible recent ZIKV infection during pregnancy should receive postnatal neuroimaging and ZIKV testing to a comprehensive newborn physical exam and hearing screen. Identification and follow-up care of infants born to women with laboratory evidence of possible recent ZIKV infection during pregnancy and infants with possible congenital ZIKV infection can ensure that appropriate clinical services are available.

On transmission, the CDC offers the following guidance.

- Zika can be passed through sex from a person who has Zika to their sex partners. Women who have traveled to areas with active Zika transmission should wait at least eight weeks before trying to conceive, and men who have traveled should wait six months, regardless of symptoms. Use a condom if you engage in sexual activity eight weeks after returning from a Zika-affected area.
- If you are a man who has traveled and developed symptoms, please use a condom if you engage in sexual activity for six months after you return.
- Sex includes vaginal, anal, and oral sex and sharing sex toys.
- Zika can be passed through sex, even if the person does not have symptoms at the time.
- It can be passed from a person with Zika before their symptoms start, while they have symptoms, and after their symptoms end.
- Though not well-documented, the virus may also be passed by a person who carries the virus but never develops symptoms.

The CDC offered this warning for infants with laboratory evidence of congenital ZIKV infection but without findings consistent with congenital Zika syndrome.

- Perform auditory brainstem response within one month of birth.
- Consider repeat auditory brainstem response at age 4–6 months or perform behavioral diagnostic testing at nine months.
- Refer to audiology for any abnormal findings or parental or provider concerns.

Infants with laboratory evidence of ZIKV infection but without apparent abnormalities at birth are recommended to have additional monitoring until further information is available regarding outcomes because some neurologic sequelae of congenital ZIKV infection (e.g., seizures, cognitive impairment, and vision and hearing abnormalities) might be subtle or have delayed onset.

CDC also adds these recommendations for people who may have been infected

- Persons with suspected Zika infections should stay indoors and avoid mosquito bites for the first days of illness to prevent local transmission.
- All persons who have traveled to Zika-affected areas should avoid mosquitos for three weeks after their return to prevent local transmission, and
- Persons who have traveled to Zika-affected areas should not donate blood for 30 days upon return.

The CDC does not stop there but added to their website. "There is limited data about the persistence of ZIKV RNA in body fluids, and the risk for adverse pregnancy outcomes associated with maternal ZIKV infection around conception is unknown. Given this information, some couples in which one or both partners have had a possible ZIKV exposure might choose to wait longer or shorter than the recommended period to conceive, depending on individual circumstances (e.g., age, fertility, details of exposure) and risk tolerance."

"ZIKV hasn't gone away," said CDC acting director Dr. Anne Schuchat. "We can't afford to be complacent." The consequences of reduced tourist numbers for the millions of people in the Caribbean whose livelihoods depend on the sector can be severe, contributing to the adverse economic and social burden on Caribbean economies. As travel increases, the number of ZIKV cases among travelers will increase, and the potential negative impact on Caribbean tourism-dependent economies will likely escalate. As such, some locations still promoted travel during the pandemic.

Travel was contentious during the pandemic mainly because countermessaging occurred as well. An example comes from the Caribbean. The Caribbean Tourism Organization and the Caribbean Hotel and Tourism Association released the following to travelers (as well as providing some advice to hoteliers): In conclusion, a message should be sent out to travelers:

> Coming to the Caribbean, "Don't let the mosquitos ruin your travel"! While there is no vaccine to prevent contracting ZIKV or medicine to treat ZIKV, travelers can protect themselves against ZIKV and other mosquito-borne infections like chikungunya and dengue by avoiding mosquito bites. We thus urge travelers to: Use insect repellent; Stay and sleep in screened /air conditioning rooms (hotels/lodgings) and wear long-sleeved shirts and long pants. [10]

In mid-December 2018, a new ZIKV virus Travel Alert was issued by CDC for some regions of India. Public health officials have reported an unusual increase in confirmed ZIKV cases in Rajasthan and surrounding states in India [11]. Surprisingly, scientists at the Indian Council of Medical Research have shown that the ZIKV strains in India are less virulent than in Brazil and are not associated with microcephaly [12].

During the outbreak in India, the Indian Council of Medical Research made a case that ZIKV strains in India were less virulent than those found in Brazil. India's Secretary of their Department of Health Research petitioned the U.S. CDC to withdraw or modify its ZIKV travel alert [13].

The Alert was removed from the CDC site in mid-March 2019. It turned out that the strain in India was the African strain. This thinking was primarily based on a single study and seemed highly motivated by concerns over the effects the warnings in India may have been having on their travel industry. According to a study published by Yuan et al., the Indian public health authorities were misled into thinking that a single mutation can predict the risk of ZIKV-associated microcephaly [14].

Indian officials also used these data to ask the CDC to withdraw or modify its ZIKV travel alert to India because the strain lacks the "gene responsible for causing microcephaly" [12].

Yuan et al. evaluated different ZIKV mutations that arose between 2008 and 2013 and found that the S139N mutation enhanced neurovirulence in neonatal mice and in vitro viral replication in the human neuron. Critically, however, Yuan et al. did not assess whether the mutation enhances the ability of ZIKV to cross the placental barrier, the most crucial factor for congenital disease. Several studies indicate that ZIKV strains without the S139N mutation can still cross the placental barrier in animals and damage neurons. A microcephaly case recently diagnosed in Thailand was associated with a ZIKV strain that did not have the S139N mutation [15].

16.2 CDC/Who Birth Control Advice

While the primary audience for CDC warnings is the American population, many of its leads are followed by others. In addition, the WHO has acted by the CDC and sometimes in addition to the CDC.

Dr. Rebecca Gomperts of Women on the Web said that many regional governments have bungled their response and put the burden entirely on women. "The only calls that have gone out from health ministers are 'Don't get pregnant,' which is an unrealistic demand if contraception is not available for the poorest. Of course, men are responsible for having sex and getting women pregnant, but the reality is that men refuse to take this responsibility seriously," Gomperts said. "So, women are the ones who get pregnant, and they are the ones who are called upon to prevent getting pregnant." [16].

"We don't know how to prevent ZIKV, but we do know how to prevent pregnancy," says Christopher Zahn, the vice president of practice at The American College of Obstetricians and Gynecologists (ACOG). Birth control cannot keep someone from getting ZIKV, but ZIKV in adults is not that bad [17].

"If effective contraception is provided and people use it, it's theoretically 100% effective in preventing those ultimate outcomes," Zahn says. "The best way to avoid a potentially [affected] fetus is by avoiding pregnancy." [17].

Although a lack of financial resources or information can create barriers to accessing services, the causal relationship between access to health services and poverty also runs in the other direction. When health care is needed but is delayed or not obtained, people's health worsens, leading to lost income and higher healthcare costs, contributing to poverty [18].

Since poverty is one of the drivers associated with zoonotic infectious diseases, like ZIKV and microcephaly, advocating birth control as a solution without providing the health services support to provide contraceptives to the poor, let alone abortions seems unconscionable.

These recommendations provided a paradox, as greater than 50% of pregnancies in Colombia are unplanned. El Salvador has one of the highest rates of adolescent and teenage pregnancy in the region, with girls between the ages of 10 and 18 representing approximately one-third of all pregnancies. In addition, sexual violence is prevalent in both countries [19].

With the recognition that girls and women become infected with the ZIKV. At the same time, pregnant are at risk of developing such poor obstetrical outcomes as spontaneous abortion, stillbirth, infants with microcephaly, and other forms of fetal malformation; even if they exhibit no symptoms during gestation, the demand for abortion has increased [19].

In addition, microcephaly does not typically appear until the late second or early third trimester, when a woman is often past the legal window for termination of pregnancy should she desire it [20].

According to the CDC guidance, microcephaly can typically be detected on ultrasound between 18 and 20 weeks of gestation at the earliest. Still, it more often becomes apparent on an ultrasound during the late second and early third trimester, somewhere between 24 and 28 weeks gestation. It may be too late to decide to keep the fetus by that point.

Nielsen-Saines was involved in a study in Brazil where researchers followed women who had been infected in pregnancy. "Even after the brain is formed, some problems happen," she said. "There's atrophy, there's damage, even later in pregnancy. So just because someone has reached the second trimester, it does not mean that there could not be a problem if they were to contract ZIKV." That damage can take several weeks to become apparent in prenatal testing. When some women learn what ZIKV has done to the fetuses they carry, their access to legal abortion could be blocked in most states [21].

In the U.S., according to NYU's Zika Briefing Report, 62% supported the availability of federally financed abortion services. U.S. residents who identify as Democrats are nearly seven times as likely as those who identify as Republicans to support access to federally funded abortion services for pregnant women infected with the virus [22].

According to the Guttmacher Institute, sexual and reproductive health research and policy organization, fourteen states already ban abortions after 20 weeks except in cases of life or health endangerment to the woman [23]. Another eight states ban abortion at 24 weeks, and in another two states, 45% of all pregnancies in the United States are unintended [17]. Regarding abortion specifically, many states already ban abortions at 24 weeks and later. For women infected with ZIKV, this would significantly limit their options considering they may not know whether they are carrying a fetus with CZS until sometime in that 20 to 24-week window [23]. According to the CDC, fifteen babies in the U.S. have congenital disabilities. Seven babies with ZIKV-related congenital disabilities have been miscarried or aborted. Those numbers will likely grow as affected women progress in their pregnancies and deliver their babies [25].

One pregnant woman's ultrasounds showed average results until she was 29 weeks pregnant, according to a February 2016 report in The New England Journal of Medicine [26]. An ultrasound performed at 32 weeks found the fetus had microcephaly and other serious problems. The woman, who was infected while working in Brazil, returned to her home country of Slovenia for pregnancy care, where she chose to end the pregnancy at 32 weeks. "There are only a handful of providers in the U.S. who perform abortions late in pregnancy," Elizabeth Nash, senior state issues associate at the Guttmacher Institute, said [26].

According to the Guttmacher Institute, which researches reproductive health, just 1.2% of the one million abortions performed in the U.S. each year occur after the 21st week of pregnancy. A normal pregnancy lasts 40 weeks. Only six states—Georgia, Louisiana, South Carolina, Texas, Utah, and West Virginia—allow abortions in late pregnancy because of "lethal fetal anomalies," said Nash. Maryland also provides abortion in late pregnancy due to a fetal genetic anomaly. None of those exceptions are likely to apply to microcephaly or congenital ZIKV syndrome; as Nash said, the spectrum of ZIKV-related congenital disabilities has become known. "Women may have to cross several states to access second-trimester abortion services," Nash said [25].

Some have blinded themselves to the data on the link between ZIKV and microcephaly. NECSI questioned the linkage through most evidence indicating a severe relationship (see Chaps. 6 and 7). In response, prolifers have used the minority literature to advance a political agenda. Take, for example, the following taken from *The Federalist*.

> Mothers are being told that an endemic infection will cripple their babies and that they will be unable to care for them. They are being pressured to abort, and these countries are being pushed to expand abortion laws to facilitate later-term abortions. Information from pro-life and neutral organizations is far more encouraging and shows mothers with ZIKV infections will always have a healthy baby. ZIKV is not a reason to increase abortion access and pressure women into aborting their babies. [27]

Abortion opponents say ZIKV changes nothing. "Abortion does not cure ZIKV. It kills people with ZIKV," said Arina Grossu of the Family Research Council, advocating for conservative causes. "Should we kill people because of their special needs? That's inhumane and goes against the fundamental need of society to protect the people who are the most vulnerable." [25].

In March (2016), Indiana Gov. Mike Pence (and Trump's Vice President) signed a bill entirely blocking women from seeking an abortion "solely" because the fetus had a fetal abnormality—like microcephaly or Down syndrome [28].

As the threat of homegrown ZIKV spreads, Americans are getting more realistic about abortion regulations. A poll conducted by the Harvard School of Public Health and STAT last month found that while 61% of Americans oppose abortions after 24 weeks, a majority would support late-term abortions in the case of microcephaly—a condition in which a baby is born with an underdeveloped brain and skull. Microcephaly, found in infants whose mothers have been infected by ZIKV, is only detectable after 24 weeks of pregnancy [28].

A new STAT Harvard poll found that Americans' strong aversion to late-term abortions drops precipitously if a developing fetus is born with severe damage from the ZIKV. It showed that 59% of respondents thought women should have the right to end a pregnancy after 24 weeks of gestation if testing showed there was a severe possibility the fetus had microcephaly caused by the mother's ZIKV infection. Most Americans oppose late-term abortions. In a separate poll conducted a week earlier by STAT and Harvard, only 23% said they favored allowing a woman to obtain an abortion after 24 weeks—when the question did not raise the possibility of microcephaly [21].

Among Democrats, support for late-term abortion increased from 34 to 72% when there was a substantial likelihood of ZIKV-induced severe congenital disabilities. The STAT Harvard T. H. Chan School of Public Health poll found. But even among respondents who identified themselves as Republicans, support for abortion after 24 weeks was surprisingly high, with 48% saying it should be allowed if ZIKV-induced congenital disabilities were likely—compared with just 12% who felt that way about late-term abortion in general [21].

CDC issued a travel alert on January 15, 2016, advising pregnant women to consider postponing travel to areas with active transmission of ZIKV. Half a million pregnant women are estimated to travel to the United States annually from the 32 (as of February 18, 2016) ZIKV-affected countries and U.S. territories with active transmission of ZIKV (personal communication, Bradley Nelson, February 23, 2016) [30].

Puerto Rico, American Samoa, New Caledonia, and Saint Barthelemy were removed from the countries and locales where ZIKV was spreading after the transmission was deemed to have stopped there, and Brazil declared its ZIKV health emergency in mid-May and saw a status change [21].

16.3 Special Amplifiers

Newspapers and magazines cover events that have made their way onto the public agenda. This is true for ZIKV-related events. They were influenced by source competition (are other outlets covering a news item), journalist's training (news items may strip the knowledge capacity of the journalist to report effectively), "newsworthiness" (human interest factors and the amount of policy activity), news momentum (once some outlets cease to cover a news item others soon follow) and the organization of news beats and media outlets (those most knowledgeable [scientists] are least trained and often simply too busy to interface with media).

These variables affect why and how some risks are highlighted at times while others are entirely overlooked [32]. In addition, these same variables explain why the press and even TV news are ill-adapted for sustaining high-level coverage of long-term threats [32].

A study involving women of reproductive age in Florida found patients who desired or were unsure about wanting pregnancy in the next year ($n = 15$) and those who did not choose pregnancy ($n = 44$) received most of their information about the ZIKV from the media rather than their doctors [33].

16.3.1 Summer Olympics

The Summer Olympics scare is fascinating for a few reasons. Burattini and colleagues estimated that the resulting risk for ZIKV infection for tourists visiting Rio during the Carnival week festivities in February and the three weeks of the Olympic Games in August is 3.6/100,000 and 1.8/1,000,000 tourists, respectively [34]. According to this study, there will be less than one case of infection among the 500,000 tourists planning to come to the Olympics.

On the other hand, reports are done such as the one posted on August 9th, 2016; Martin Rogers of USA TODAY Sports wrote: "On Saturday, the influx of mosquitoes was extraordinary athletes were swatting mosquitoes away from their faces as they stood atop the podium [35] and an NPR post published August 26th, 2016, clearly shows mosquitoes hovering around South Korea's bronze medalist Ki Bo Bae during the medal ceremony for archery at the Rio Olympic Games" [36].

Only about 6% of the cases of ZIKV had been reported in the Southeast Region of Brazil, where Rio is found [37]. Others report athletes and spectators are likely to spend their time in places purged of mosquito breeding sites and sprayed heavily with insecticide. *Ae. aegypti* are poor flyers (two hundred meters), so game participants are unlikely to encounter mosquitoes unless they travel far away from "official" sites [38].

Accordingly, 1 in 31,250 would contract ZIKV at the 2016 Rio Olympics. The results of this study have been challenged because the authors failed to consider ZIKV as a sexually transmitted disease, and Brazil leads the world in sex trafficking. The Olympic village is anecdotally notorious for flings, and the sample excluded the effect visitors may have on native Brazilians [39]. Still, others are worried about infected visitors returning to tropical countries with the right mix of *Ae. aegypti* mosquitoes and overladen slums could establish local viral transmission and new outbreaks of microcephalic, brain-damaged children disabled for life [40].

A second study agreed. Scientists from Hospital São Paulo/Paulista School of Medicine of São Paulo and scientists from four other countries published in the Cambridge Journals, England, point out that the risk of tourists contracting the ZIKV during the Olympic and Paralympic Games is 1.8 cases per one million people [41].

Of course, the warnings of Canadian epidemiology professor from the University of Ottawa, Dr. Amir Attaran's should be heeded: "Brazil's outbreak is hypothesized to have started with just one infected carrier traveling in 2013 from French Polynesia to the Americas. Not 500 000, but just 329 travelers entered Brazil from Oceania's small islands that year—suggesting that even a few travelers can damage if they return to the right setting for the disease" [40].

> ...[T]he absolute risk of any traveler from these countries being infected with ZIKV during the games is minuscule: There were 36,923,504 passenger journeys to the United States from all ZIKV-affected countries and U.S. territories in 2015 and 38,798 journeys from Rio de Janeiro during August 2015; thus, the proportion of estimated U.S. travel from Rio de Janeiro for the Games, relative to that of all ZIKV-affected countries is 0.11%. [42]

Grills et al. [43] have assessed which countries are susceptible to ongoing ZIKV transmission resulting from an introduction by a single traveler to the games. CDC estimated that nineteen countries currently not reporting ZIKV outbreaks have the environmental conditions and population susceptibility to sustain mosquito-borne transmission of ZIKV if a case were imported from infection at the Games. Four countries—Chad, Djibouti, Eritrea, and Yemen—were at risk because "they do not have a substantial number of travelers to any country with local ZIKV transmission, except for anticipated travel to the Games" [42].

Of course, there are climate and weather considerations. "Contrary to the Northern Hemisphere, it is winter in Brazil, when historically and epidemiologically the rates of diseases transmitted by the mosquito are declining and reach their lowest rate. This period is in August and September when the Olympic and Paralympic Games will be held in Brazil." [41]. While this might be a godsend for the Olympic Games, Latin America is still left with an epidemic when the guests leave.

The fallout from the ZIKV crisis has some significance. For example, two top athletes—from golfer Vijay Singh to cyclist Tejay van Garderen—refuse to attend the Summer Olympic Games because of fears of the virus [45].

16.3.2 Undocumented Immigrants

Latin America's ZIKV is the latest undocumented immigrant to hit these shores. People from Central and South America, ground zero for ZIKV and other infectious diseases including tuberculosis, dengue, Chagas, chikungunya, and schistosomiasis (a disease caused by parasitic worms), make up 15% of the illegal-immigrant population in the U.S. [46]. In addition to the word "jihad," the terms measles, polio, diphtheria, tuberculosis, malaria, scabies, dengue, and now "Zika." must be reintroduced into the common vocabulary [47].

What happened before the spread of previously thought annihilated viruses (measles and polio)? The massive surge of undocumented immigrants, including "unaccompanied minors," at the southern border and their transportation to the country's interior by the federal government [47]. The assumption: illicit aliens and unrestrained immigration brought ZIKV to the USA.

The primary source for this claim was WorldNetDaily (WND). "Given that WND fearmongers about vaccines and is also home to various representatives of the far-right, antivaxxer Association of American Physicians and Surgeons—who also like to fearmonger about filthy, disease-ridden immigrants—it's no surprise Corsi is joining a long, dubious line of quack medical information." [48].

"In the wake of the WHO's decision to declare the ZIKV outbreak in Brazil an international health emergency, a glance at available evidence suggests open borders contribute to the vulnerability of the United States to the virus" [49]. This is not the first-time immigrants have been blamed for infectious diseases.

According to the World News Daily, in 2014, the dengue hemorrhagic fever mosquito surfaced in San Diego and Los Angeles. It is suspected that the disease-bearing mosquitoes were brought in the clothing and baggage of the "unaccompanied minors" [50].

This argument is bizarre due to its tie-in to explaining why people will believe incomprehensible and tortured explanations for the world rather than investing belief into the most expected cause.

> Illegal immigrants drain our economy, peril to our national security, and drag on our souls. They may also be hazardous to our health, thanks to lax U.S. immigration laws acting as incubators for diseases once foreign to North America—like the untreatable ZIKV now affecting dozens of Americans. Yet illegals-friendly federal health officials allude American travelers are transporting the bacteria instead of unlawful immigrants coming here unchecked from originating ZIKV regions in South and Central America and the Caribbean. It is no coincidence that the countries Centers for Disease Control fingers as ZIKV hotbeds—among them Brazil, Guatemala, Honduras, Haiti, and El Salvador—are experiencing a mass exodus to the U.S. Or that the highest number of undocumented immigrants live in Texas, New Jersey, Florida, and Illinois where the virus is confirmed. Contagion and illegal immigration are a marriage made in hell. Disturbing El Salvador remains a leading exporter of unlawful immigrants, preparing for another border surge. It, and neighboring Guatemala and Honduras, are nerve centers for infectious diseases, and smugglers rum-running locals and others to America, increasing the certainty that illegals—not Americans—are ZIKV carriers". [51]

This argument made a list for this reason and because it is quintessentially conspiratorial and false news.

16.4 Warnings

Warnings of all sorts have been problematic in determining behaviors. According to scholars in communication, asking the public to protect themselves and others by changing their behavior is complex. In his Protection Motivation Theory, Rogers' thesis is that behavioral modification for a person's well-being and those about her that means something to her demands that the behavior change is feasible [52]. The term from Witte's Extended Parallel Process Model is often used is efficacy, and it generally describes the means the receiver of the message has to behave in a way consistent with the news [53]. Finally, Cummings added the concept of secondary risk, whereby the behavioral change cannot produce a secondary risk that might weigh enough against the primary risk to mitigate the behavioral change [54]. These theories argue that behavioral change depends on efficacy and the net assessment of the behavior.

16.4.1 Travel Restrictions

Infectious disease specialists have critiqued travel restrictions. These include the propagation of stigma associated with an emerging infectious disease, interference with the rights of affected individuals, and impacts on the movement of healthcare supplies necessary to respond to the disease.

During the 2014 Ebola virus disease outbreak, the WHO discouraged the use of travel bans because of their potential: to create a false sense that the disease was being controlled; decrease the number of healthcare workers volunteering in affected countries; and negatively impact the economy and exacerbate humanitarian hardship secondary to reductions in essential trade, including food, fuel, and healthcare equipment [55].

The case with ZIKV is different. These travel restrictions involved mostly optional travel involving visiting sunbaked holiday locations and, in this unusual case, the Summer Olympics in Rio (see above). They were optional, though, of course, there were the traditional international travelers visiting for familial and professional reasons, but a lot of the likely travelers were traveling optionally.

In addition, there was never a point when flight restrictions cut off an area as during COVID-19 (Fig. 16.1).

16.4.2 Travel and Pregnancy Warning Poster

Unfortunately, women with the least access to contraceptive methods are those of lower socioeconomic status. These women also are at a greater risk of acquiring ZIKV due to increased exposure to mosquitos and are least equipped to deal with the potential consequences of microcephaly [56].

A recent Twitter study by Valente et al. [57] made two critical observations.

First, during the beginning of the ZIKV emergency, traditional media were responsible for originating much of the content related to the sexual transmission of ZIKV shared on social media like Twitter. While a significant reproductive health crisis was taking place in Latin America, content created by traditional media that circulated on social media, in this case, Twitter, was technical and impersonal.

Second, on the one hand, the predominance of traditional media-originated content may be less conducive to spreading misinformation and rumors, thus contributing to disseminating technically accurate information. On the other hand, this predominance may propagate epidemiological/clinical discourses that omit the voices and concerns of the people most affected by public health crises.

Prioritizing epidemiological/clinical aspects of epidemics may have a depoliticizing effect and contribute to overlooking socioeconomic determinants of the ZIKV epidemic and issues related to reproductive justice [53].

Several countries, including El Salvador and Colombia, have recommended that women avoid getting pregnant soon to circumvent the risk of transmission—although this might not be realistic.

Fig. 16.1 A typical tourism-related poster, Honolulu, HI. Taken by author

16.5 Effectiveness of the Warnings

In health communication theory, there is the concept of efficacy (see above). Regardless of how well-designed the health message if the audience does see themselves as having efficacy in fulfilling the recommendation found in the message, then the message will fail.

16.5.1 Travel

Travel warnings have always been contentious for many reasons, especially for Americans who dislike being told what to do. Americans have a sense of independence which exaggerates the interests of the individuals against those of all others. It is some twisted form of exceptionalism. The best example of this was the unwillingness of many Americans to refuse to wear masks during the COVID-19 pandemic.

Data on public understanding and information seeking were mixed. For example, a survey of 828 people nationwide conducted by Lori Pennington-Gray for the University of Florida's Tourism Crisis Management Initiative, or TCMI, shows more than 70% of potential visitors are concerned with the mosquito-borne ZIKV in Florida. Of those who expressed concern, less than 10% have changed their travel plans. The study also shows that 45% had medium to elevated knowledge of the ZIKV because of coverage on social media (36%) and television news coverage (27%) [58].

An online survey of 300 U.S. citizens also by TCMI found global concerns about ZIKV are not stopping Americans from making international travel plans. Still, many who plan to go abroad say they want more information about the virus from the CDC and destination resources.

Of those with international travel plans, more than 90% said they would keep them, and 44.3% will take extra precautions to protect themselves from ZIKV. In addition, more than 70% believe they should use EPA-registered insect repellents to protect themselves. In comparison, less than 55% believe wearing permethrin-treated clothing is an effective way to stay safe [59].

Fluent, a digital marketing company, released a poll of 3312 Americans and found that nearly half of Americans say they would avoid planning (44%) or cancel (43%) their trips to areas affected by the ZIKV. However, the study also found that frequent travelers and those with higher household incomes would be less likely to alter their plans. Florida (35%) and California (25%) are the most likely replacement destinations for those who want to avoid ZIKV-impacted areas. Only 5% would choose a currently unaffected Caribbean Island as an alternative [60]. About half (49%) of Americans are worried about the impact the ZIKV will have on world health, while somewhat fewer (37%) are concerned about contracting the virus.

16.5.2 Pregnancy Recommendation

> The psychological toll of reproductive threats cannot be underestimated. Increased rates of anxiety, depression, and stress were documented in women in Brazil during the ZIKV epidemic and the current COVID-19 pandemic. Maternal ZIKV infection in the Americas was associated with intimate partner violence and abandonment, and studies show similar impacts in the U. S. during the COVID-19 pandemic lockdown. These psychological impacts, exclusive of viral infection, are known to increase perinatal complications such as preeclampsia, depression, increased nausea and vomiting during pregnancy, preterm labor, low birth weight, and low Apgar scores. [61]

16.5.3 United States

There are dozens of surveys involving pregnant women to gauge their likelihood to take precautions, avoid pregnancy, or engage in safe sex during the ZIKV epidemic. This study [62] involved nine hundred travel-associated ZIKV cases in New York City. It found an overall poor understanding of ZIKV symptoms and complications, transmission modes, and current recommended prevention guidelines, and very importantly, reported travel history to ZIKV endemic regions is not significantly associated with ZIKV knowledge [62].

When asked about the risks pregnant women with ZIKV face, 41% identified that they might be sick, and even fewer, only 34%, determined that they may be at risk for a miscarriage. Only 10% recognized that she might be at risk from illegal and unsafe termination of pregnancy. Forty percent of persons do not think that the risk of ZIKV applies to them [62].

Nearly three-quarters of participants, 74%, correctly indicated that condoms should be used during intercourse if the male partner in a couple travels to a ZIKV endemic region and develops symptoms. However, in this situation, only 35% indicated that the couple should also refrain from intercourse for at least six months, keeping with current CDC recommendations [62].

> Even though over half of all participants could identify that ZIKV is sexually transmitted, only a small percentage of the participants identified condom use as a precaution to take if one is pregnant and traveling to a ZIKV endemic region. More participants identified condom use as a method to prevent ZIKV. Still, when asked what they would do in specific scenarios to prevent ZIKV transmission, condom use was not as commonly evoked. This discrepancy in behaviors and attitudes surrounding sexual and reproductive health in our participants illustrates a gap in knowledge and practice regarding ZIKV prevention; specifically, that condom use, an underused method to prevent transmission, can directly prevent what participants find most concerning about ZIKV—the birth of children with congenital anomalies. [62]

16.5.4 Brazil

The ZIKV outbreak in Brazil was associated with a decrease of more than 100,000 births between September 2015 and December 2016, which may have resulted from the postponement of pregnancies and increased clandestine abortions [63].

16.5.5 Puerto Rico

The data from Puerto Rico were not promising. Even though their healthcare providers had counseled more than 90% of their sample to avoid pregnancy and over 93% reported feeling or very worried about contracting ZIKV during their pregnancy, and 92% about the possibility of microcephaly or other congenital disabilities. Nevertheless, more than 20% of survey respondents reported not receiving testing for ZIKV infection in their first or second trimester [64].

The same 2017 D'Angelo et al. study reported that one in five (19.9%) women abstained from sex during the entire pregnancy, one quarter of whom did so specifically to avoid ZIKV infection [64].

Among sexually active women, less than a quarter (22.7%) reported using condoms consistently throughout pregnancy. Altogether, approximately one-third (38.5%) of respondents reported using at least one measure to prevent sexual transmission of ZIKV during pregnancy (abstinence or consistent condom use). Common reasons for not using condoms consistently were not thinking her partner had ZIKV (37.4%), not thinking that condom use was necessary during pregnancy (31.8%), and not wanting to use condoms (20.1%) [64].

Another study, this time qualitative involving focus groups by August et al. (2020), discovered a minority of the female participants indicated that they would consider postponing pregnancy because of potential ZIKV transmission. In contrast, no men attributed any contraception decision-making to concerns about adverse pregnancy and birth outcomes because of ZIKV infection [65].

Data showed that there was community awareness regarding ZIKV in Puerto Rico. However, it was not a motivating factor in contraception decision-making; economic factors were the significant drivers [65].

Some were unmotivated, and some of the literature traced those concerns to conspiracy theories circulating on how ZIKV was a depopulant.

Many participants demonstrated skepticism around ZIKV messaging. Some commented that they believed that the ZIKV outbreak was a conspiracy to increase mosquito repellent sales. In contrast, others felt it was a way for the government to influence reproductive behaviors in Puerto Rico (e.g., decreasing pregnancy/birth rates) [65].

Common reasons for not using condoms consistently were not thinking her partner had ZIKV (37.4%), not thinking that condom use was necessary during pregnancy (31.8%), and not wanting to use condoms (20.1%) [64].

16.6 Abortion

Complicating the issue is that microcephaly may not be detected until after 20 weeks of pregnancy, the threshold when abortion is illegal in many countries.

16.6.1 Testing and Abortions: In Florida

This is an interesting problem. ZIKV testing was offered in Florida to women through the Florida Department of Health. Former Governor Rick Scott promoted this program but had not anticipated the long waits in processing these tests. The pregnant women who availed themselves of this offer were indigent. Dr. Christine Curry, an obstetrician at the University of Miami Health System, said the governor's proposal was significant because it gave low-income pregnant women or uninsured women a chance to be evaluated. Still, she said the state had not been able to cope with the numbers rushing to be tested [66].

The Health Department said more than 6649 people had been tested for ZIKV statewide. So far, eighty-six pregnant women in the state have tested positive. 771 Florida residents have tested positive for the virus, and most were infected while traveling abroad [66]. This has produced a backlog in processing results. As a result, hundreds of pregnant women in Miami have waited for results for weeks. Doctors recommend that pregnant women receive a ZIKV test every trimester. Obstetricians in Miami-Dade County say that many of their patients will soon need to be retested without knowing the initial results [66].

Some delays in obtaining test results relate to the number of tests some women must take to get a definitive answer. That can require several tests, and, for some, a final test called the PRNT that the CDC must conduct [66].

16.6.2 Abortions: Latin America

Over half the women in Brazil are avoiding pregnancy due to fears about the deadly ZIKV that reveals a study by Diniz et al. (2017). The findings showed that over half (56%) of the women reported that they had avoided or tried to avoid pregnancy because of the ZIKV epidemic. The results, published online in the Journal of Family Planning and Reproductive Health Care, suggest an urgent need to reconsider abortion criminalization and improve reproductive health policies to ensure women access safe and effective contraceptives. "As indicated by the high proportion of women who avoided pregnancy because of ZIKV, the Brazilian government must place reproductive health concerns at the center of its response, including reviewing its continued criminalization of abortion," the study said [68].

Restrictive abortion laws in many countries in Latin America leave women who may wish to discontinue their pregnancies with little access to safe, legal termination and leave women exposed to the risks of unsafe procedures. In most Latin American countries, abortion is illegal or highly restricted, leaving pregnant women with few options. A 2016 study between two groups of countries (those with autochthonous ZIKV transmission and legally bound abortion and those without transmission) found groups with ZIKV transmission showed statistically significant increases of 36–108% over baseline in requests for abortion through Women on the Web (WoW). This nonprofit organization provides access to abortion medications after the PAHO announcement alerts [66]. This study may underestimate the effect of the advisories on demand for abortion since many women may have used an unsafe method, accessed misoprostol from local pharmacies or the black market, or visited local underground providers [69]. The Guttmacher Institute has estimated that there were 4.4 million abortions in Latin America in 2008 and that 95% of these were unsafe.

This part of the debate occurs on two levels. The first is made in the very words of a Latin American woman searching for pregnancy health resources.

> In a supplement to the 2018 Aiken study, the researchers have published some of the emails to Women on the Web, revealing their fears. "I need to do an abortion because of the elevated risk of infection with ZIKV here … Please help me. My economic situation is tough," said one woman in Brazil. Another in Colombia wrote: "Here ZIKV is a major problem and the health authorities do not help with it … I have no resources currently and want to ask for your help because fear overwhelms me. What if the baby is born sick?" An email from a woman in Venezuela said: "We are going through a dire situation for the economic and humanitarian crisis unleashed by ZIKV. There are no treatments, contraceptives, or pills to abort. I want to terminate my pregnancy, but I cannot". [71]

Women in impoverished areas whose pregnancy is more advanced have few options. "There is no support for those people," said Aiken. "They cannot pay for ultrasound scans. They do not have many choices about what they can do if there is microcephaly (a small head and brain damage). They have no access to anyone who can help them [71].

Abortion in Latin America is tenuous at best. Laws against abortion vary by country in Latin America. More than 97% of women in Latin America and the Caribbean live in countries where access to abortion is either restricted or banned altogether. As a result, unsafe abortion is widespread. It causes 10% of all maternal deaths in the region. About 760,000 women in the region are treated annually for complications from unsafe abortion before the ZIKV pandemic. Penalties can be extreme—up to 10 years in prison for mothers having abortions in Paraguay and Honduras. In El Salvador, several single mothers have been imprisoned for having miscarriages during their pregnancy; women convicted of having an abortion face imprisonment of up to 50 years [19].

Most of the following comes from the Kaiser Foundation. It was the complete source of abortion data in Latin America.

Of the twenty-one countries in Latin America and the Caribbean with confirmed ZIKV transmission as of April 14, the contraceptive prevalence rate (CPR) ranges from just 37.8% in Haiti to 79.5% in Nicaragua; for modern, more effective methods, the range is 33.6% in Haiti to 75.7% in Costa Rica. Most countries in the region have CPR for modern methods below 70%, including countries with CPR at 50% or below.

It is important to note that these CPR estimates are national averages that mask significant inequities within counties, particularly across geographic and income lines. Typically, poorer women and women in rural areas have less access to contraceptives and are likely to be most at risk for ZIKV infection. In addition, such averages do not consider the contraceptive method mix available to women, which in some cases may be limited [72].

El Salvador, the Dominican Republic, and Nicaragua outlaw abortion in all cases, including rape and incest. Six provide abortion only to save the women's life. Very few, such as Colombia, allow abortions when a fetus displays signs of a severe deformity—a narrow exception that abortion rights advocates argue should apply to microcephaly because microcephaly is not detectable until quite late in pregnancy (if at all), when many countries, even those with less restrictive abortion laws, would not permit abortion [72].

In eight of the nine countries with severe restrictions, the CPR for modern methods is below 70%; three countries have a CPR of 50% or below. Five of the nine reported stockouts of contraceptives, meaning that contraceptive access may be more problematic for women in these countries.

Unfortunately, Indigenous women frequently reside in poor, rural, and medically underserved areas and have little access to modern medical care and family planning education and interventions. Unsurprisingly, unsafe induced abortion is practiced among Indigenous women, contributing to the high maternal morbidity and mortality rate among these populations. It has been estimated that 18% of maternal mortality in Ecuador, 16% in Peru, and 28% in Colombia resulted from complications due to unsafe abortions [16].

According to official data, unsafe abortion is Brazil's fourth leading cause of maternal mortality. Since 2005, 911 women have died from unsafe abortion, including 69 in 2015 and 48 in 2016. Approximately 17% of the abortion-related deaths between 2011 and 2015 were adolescent girls and young women between 10 and 19 years old [73].

New data show that demand for abortions has soared among women in countries hit by the ZIKV, who fear having a baby with severe congenital disabilities. In unprecedented numbers, women in Latin America are accessing the website Women on the Web, which has a long history of helping those in countries where abortion is illegal to obtain pills that will terminate an early pregnancy.

In Brazil, Venezuela, and Ecuador, the requests for help have doubled, while they have risen by a third in other Latin American countries [71]. In an interview with The Washington Post, Gomperts said the number of Brazilian women contacting Women on the Web had nearly tripled, climbing from 100 during the first week of December (before the ZIKV outbreak became public) to 285 during the first week of February [16].

In more than a thousand emails to Women on Web, a Canada-based group that provides advice and medication for women wanting an abortion in countries where it is banned, the women beg for the pills that are prohibited by law in their respective countries of Brazil, Colombia, Venezuela, Peru, or El Salvador [16].

Aiken et al. [69] analyzed requests for abortion through Women on the Web between January 2010 and March 2016 in 19 Latin American countries. They compared the number of submissions made before and after PAHO issued the ZIKV alert. The number of requests in Brazil doubled compared with what was expected. One thousand two hundred ten women requested abortions from November 2015 to March 2016, while only 581.7 proposals were expected (relative change 108.0%, $P <$ 0.001). Abortion requests in Colombia increased by nearly 38.7%, with 141 requests compared with the 101.7 expected from previous trends (relative change of 38.7%, $P < 0.001$) [69].

While the actual global burden of unsafe abortion-related mortality remains unknown, employing the newest figures for international maternal mortality, the WHO estimates that in 2008 approximately 13% of maternal mortality worldwide, or 47,000 deaths was due to unsafe abortion [76]. In countries where abortion is restricted or illegal, women often seek abortion-related services outside the formal medical system. In such settings, due to social and cultural stigma, and fear of legal consequences, women are often reluctant to seek medical services in the event of complications or reveal to family members the underlying cause of the difficulties. Because of legal consequences for patients and providers alike, clinicians who provide abortion-related services may be reluctant to report abortion-related deaths [76].

Although Gomperts stressed that considering an abortion for any reason, especially in a country that does not allow it, is always a terrifying decision for a woman, ZIKV had made it even more unbearable [16].

Most countries hit hardest by the virus, including El Salvador and Colombia, have stuck by their government's harsh anti-abortion laws, including their limited dispersal of birth control. Brazil moved to tighten restrictions on abortions—promising incarceration to any women thought to have aborted their child due to microcephaly [28].

Only Guyana and French Guiana of the affected regions permit abortion without exception. Some countries are stringent, such as El Salvador, where women are jailed for having spontaneous miscarriages [77]. "In El Salvador, the recommendation to postpone pregnancy is offensive to women and even more ridiculous in the context of strict abortion laws and elevated levels of sexual violence against girls and women," Monica Roa, the vice president for strategy at Women's Link Worldwide, a women's rights group, told Reuters [78].

Abortion is prohibited in Brazil (except for a few situations, such as rape, anencephaly, or the mother's risk of death). Still, it is commonly performed illegally, and 16.4% of women reported having had > 1 abortion [79].

Recent research by Abigail Aiken from the University of Texas and others [75] indicated that in countries in Latin America where abortion is illegal and public advisories on ZIKV had been issues, there were statistically significant increases of

36–108% in requests for abortions through Women on Web (WoW). The authors claim this finding is underestimated since women may have used an unsafe method, got misoprostol from a local pharmacy or on the illegal market, or visited underground abortion providers.

"Accurate data on pregnant women's choices in Latin America is hard to obtain. If anything, our approach may underestimate the impact of health warnings on requests for abortion, as many women may have used an unsafe method or visited local underground providers," Aiken said. Aiken spoke of "a much, much wider problem with women who do not have access [to the internet] and live in impoverished rural areas and are in very a crisis and will be driven to less safe methods of illegal and underground abortion. We think we are looking at the tip of the iceberg" [75].

However, many women living in Latin America find it hard to protect themselves against the virus spread by mosquitoes and do not have the option of termination in an unwanted pregnancy. It is challenging for those in poorer communities, living in difficult conditions where mosquitoes readily breed [71].

Dr. Gomperts shared this letter. "I dare to write you because I'm a resident in Colombia, and here the ZIKV is a major problem, although the health authorities haven't recognized it," she wrote. "I want to ask for help because I am afraid that my baby will be born sick. I already have two girls and work long and hard as a single parent to provide for them. Life in Bogotá is difficult enough without being in charge of a sick child, especially with Colombia's precarious health system." Other women said they felt the ZIKV closing in on them like a mosquito in a closed room and thought it was only a matter of time until they—and their unborn children—were affected by the virus [16].

Women's reproductive rights have taken on a new focus in Brazil following the onset of the ZIKV outbreak, which authorities strongly suspect is linked to congenital disabilities. Brazilian law currently only allows abortion in cases of rape or specific dire health threats, which presently do not include the ZIKV [77].

"The poorest women are suffering from this crisis," Gomperts said. "It is not women in the upper-class neighborhoods who can protect themselves from mosquito bites. And that is the most frustrating part. We know that many of these women do not have access to the Internet. So, the women we read are only a few affected by this crisis." To make matters worse, Brazilian customs officials have been blocking Women on Web's pills from reaching their intended destination for several years. Gomperts said she is still waiting to see if ZIKV changes its mind. "We hope the people working in the customs have some empathy and humanity in them and that in the wake of ZIKV, they might just decide to suspend the confiscation of the medicines so that women can at least have safe abortions in the wake of this public health crisis," she said [16].

Many pregnant women said they had tested positive for ZIKV but could not travel or obtain pills to get an abortion. "I contracted ZIKV and cannot leave the country!" wrote one woman who asked to be sent abortion pills. Another woman said she was able to get Misoprostol in the underground economy but was unsure how to take the abortifacient [16].

The response of many Latin American governments to ZIKV infection during pregnancy was to recommend that women avoid or postpone their pregnancies. These recommendations were not possible for many women at risk in the affected countries, especially those uneducated or impoverished. As a result of the ZIKV pandemic, there has been an increased demand for abortion in many of the affected countries that carry the heightened risk for additional maternal morbidity and mortality because of the clandestine and illegal nature of the procedure [19].

In Latin America, the ZIKV pandemic has disproportionately affected women in the reproductive-age group and especially the most vulnerable members of society—those girls and younger women who live in conditions of poverty. It has brought renewed attention to the multifaceted human rights problems that, although predating the onset of the pandemic in 2015, have significantly been worsened by the spread of ZIKV infection. These include ethnic and socioeconomic health disparities, access to reproductive health education, sexual and reproductive rights restrictions, inadequate access to water and sanitation, and stigmatization and criminalization of women seeking to terminate their pregnancies [19].

Human Rights Watch analyzed these human rights problems through the lens of the ZIKV outbreak. Research has found gaps in the Brazilian authorities' response that have harmful impacts on women and girls and leave the general population vulnerable to continued outbreaks of serious mosquito-borne illnesses [81].

16.7 Post-epidemic Warnings

The CDC noted no areas with local mosquito-borne ZIKV transmission in the continental U. S. Still, ZIKV transmission continues in many places, with nearly 100 countries and territories still having an active risk of ZIKV. "For this reason, the CDC continues to urge pregnant women not to travel to areas with risk of ZIKV and recommends that men and women who travel to an area with risk of ZIKV wait before trying to conceive," the agency said in a news release [82].

The CDC now recommends that men with possible ZIKV exposure who are planning to conceive wait for at least three months after symptom onset (if symptomatic) or their last potential ZIKV exposure (if asymptomatic) before engaging in unprotected sex. Additionally, the agency now recommends that for couples who are not trying to conceive, men can consider using condoms or abstaining from sex for at least three months after symptom onset (if symptomatic) or their last possible ZIKV exposure (if asymptomatic) to minimize their risk for sexual transmission of ZIKV [82].

16.8 Conclusion

The CDC has embraced policies of reactive funding, and once a pandemic is off the radar, its budget is cut. "The program was never intended to be a long-term, high-level maintenance," Roger Nasci of the CDC says, "but to provide the core capacity to the states, with the hope and the assumption that they would pick up the components that were required for their particular jurisdictions" [83]. However, labs do not work that way, and funding cuts diminish capacity. And once budget cuts are undertaken, these labs are threatened.

The ZIKV epidemic was yet another lost opportunity to increase culturally sensitive family planning services since the costs of the ZIKV for maternal and perinatal health demanded a broad spectrum of health interventions [83].

The pandemic increased positive awareness about critical social justice issues, such as stigma toward and isolation of families of infants with congenital ZIKV syndrome, the reproductive rights of women, and access to safe abortion and contraception in Latin America [84].

References

1. Colli G (2017) Zika virus threat not over yet. My Statesman. January 27. http://www.mystatesman.com/news/national/zika-virus-threat-not-over-yet/MVfL2BT4fw2dpTmDsgNSgI/. Accessed 3 Apr 2017
2. Burwell S (2016) Zika supplemental funding spend plan. Department of Health and Human Services. October 26. https://www.naccho.org/uploads/downloadable-resources/HHS-Zika-Spend-Plan-to-Congress.pdf. Accessed 27 Aug 2018
3. CSTE (Council of State and Territorial Epidemiologists) (2018) 2018 Zika preparedness resources toolkit. CSTE, Atlanta, GA. https://cdn.ymaws.com/www.cste.org/resource/resmgr/zika/Zika_Virus_Preparedness_Reso.pdf. Accessed 8 Aug 2018
4. Lindsey NB et al (2012) State health department perceived utility of and satisfaction with ArboNET, the U.S. National Arboviral Surveillance System. Pub Health Rep 127:4. July/August. https://www.ncbi.nlm.nih.gov/pubmed/22753981. Accessed 8 Aug 2018
5. Chante C (2017) CDC urges states to prepare for Zika. The Philippine Star Philstar.com. February 5. http://www.philstar.com/opinion/2017/02/05/1669247/cdc-urges-states-prepare-zika. Accessed 3 Apr 2017
6. Kokomo Tribune (2017) Keep pests under control. The Daily Herald (Columbia). April 24. http://www.columbiadailyherald.com/lifestyle/20170424/home-life-editorial-keep-pests-under-control. Accessed 17 May 2017
7. McNeil DG (2017) Houston braces for another brush with the Peril of Zika. The New York Times. July 17. https://www.nytimes.com/2017/07/17/health/zika-virus-houston-texas.html. Accessed 21 Jul 2017
8. Neergard L (2017) Zika virus: The threat continues. News Tribune. July 6. http://www.newstrib.com/free/zika-virus-the-threat-continues/article_72fd02a2-6290-11e7-ab20-a39d99f70d7b.html. Accessed 21 Jul 2017
9. Cohen ERM et al (2008) Public engagement on global health challenges. BMC Pub Health 8:168. May 20. https://bmcpublichealth.biomedcentral.com/articles/https://doi.org/10.1186/1471-2458-8-168. Accessed 4 Sept 2018
10. Indar L (n.dat.) Zika and Tourism. Caribbean Public Health Agency. https://carpha.org/Portals/0/Documents/Zika%20Virus_Tourism.pdf. Accessed 8 June 2022

11. Hackett DW (2018) Pregnant women should not visit Rajasthan, India: Level 2 Travel Alert for Zika issued for India by the CDC. PrecisionVaccinations.com. December 14. https://www.precisionvaccinations.com/level-2-travel-alert-zika-issued-india-cdc. Accessed 13 June 2019
12. Ghosh A (2018) Not so alarming, revise Zika alert: India to CDC. The Indian Express. December 29. https://indianexpress.com/article/india/not-so-alarming-revise-zika-alert-india-to-cdc-5514459/. Accessed 21 May 2022
13. Hackett DW (2019) India Asks CDC to Withdraw Zika Travel Alert. Precision Vaccinations. January 4. https://www.vaxbeforetravel.com/zika-virus-india-reported-not-have-gene-responsible-causing-microcephaly. Accessed 8 June 2022
14. Yuan L et al (2017) A single mutation in the prM protein of Zika virus contributes to fetal microcephaly. Science 358:933–936
15. Grubaugh ND, Ishtiaq F, Setoh YX, Ko AI (2019) Misperceived risks of Zika-related microcephaly in India. Trends Microbiol 27(5). https://doi.org/10.1016/j.tim.2019.02.004
16. Miller M (2016) Zika virus: pregnant women begging online for abortion pills in countries where terminations are illegal. The Independent. February 18. https://www.independent.co.uk/life-style/health-and-families/health-news/zika-virus-pregnant-women-begging-online-for-abortion-pills-in-countries-where-terminations-are-illegal-a6880896.html. Accessed 24 May 2022
17. Beck J (2016) The importance of contraception to the Zika Fight. The Atlantic. July 1. https://www.theatlantic.com/health/archive/2016/07/the-importance-of-contraception-to-the-zika-fight/489767/. Accessed 3 Feb 2017
18. Peters DH et al (2008) Poverty and access to health care in developing countries. Ann New York Acad Sci 1136:161–172
19. Schwartz D (2017) Pregnant and out of options: the quest for abortion in Latin America Due to the Zika Virus Pandemic. Family Planning. Ed. Zouhair O. Amarin. IntechOpen, London
20. Aliota MT et al (2017) Zika in the Americas, year 2: what have we learned? What gaps remain? A report from the Global Virus Network. Antivir Res 114. https://www.ncbi.nlm.nih.gov/pubmed/28595824. Accessed 15 May 2019
21. Branswell H (2016) Most Americans favor late-term abortion if Zika harms fetus, STAT-Harvard poll finds. STAT. August 5. https://www.google.com/search?q=Most+Americans+favor+lateterm+abortion+if+Zika+harms+fetus%2C+STATHarvard+poll+finds&oq=Most+Americans+favor+lateterm+abortion+if+Zika+harms+fetus%2C+STATHarvard+poll+finds&aqs=chrome..69i57.791j0j4&sourceid=chrome&ie=UTF-8. Accessed 5 June 2017
22. Abramson D, Pitch-Loeb R (2016) U.S. Public's Perception of Zika risk: awareness, Knowledge, and Receptivity to Public Health Interventions. NYU Zika Briefing Report #1. https://www.nyu-pir2.org/research. Accessed 19 June 2019
23. Uffalussy J (2016) What Donald Trump's win could mean for women with Zika. Fusion. November 11. http://fusion.kinja.com/what-donald-trump-s-win-could-mean-for-women-with-zika-1793863700. Accessed 4 June 2017
24. Norman A (2016) Romper. What will happen with Zika virus research under president Trump? Things don't look good. Romper. November 17. https://www.romper.com/p/what-will-happen-with-zika-virus-research-under-president-trump-things-dont-look-good-22895. Accessed 26 May 2017
25. Szabo L (2016) Zika outbreak could reignite abortion debate. USA Today. August 5. https://www.usatoday.com/story/news/2016/08/05/zika-outbreak-could-reignite-abortion-debate/87961918/. Accessed 2 June 2017
26. Szabo L (2016) Zika outbreak could reignite abortion debate. USA Today. August 5. https://www.usatoday.com/story/news/2016/08/05/zika-outbreak-could-reignite-abortion-debate/87961918/. Accessed June 2, 2017, and Mlakar J et al (2016) Zika virus associated with microcephaly. The New England J Med 34(10):951–958. http://www.nejm.org/doi/full/https://doi.org/10.1056/NEJMoa1600651. Accessed 14 May 2017
27. Scheer H (2016) New studies reinforce that Zika is no excuse to sterilize women and kill children. The federalist. September 21. http://thefederalist.com/2016/09/21/new-studies-reinforce-zika-no-excuse-sterilize-women-kill-children/. Accessed 31 May 2017

28. Zelinski A (2016) Threat of Zika is changing Americans' minds about late-term abortion. Think Progress. August 5. https://thinkprogress.org/the-zika-virus-unlikely-silver-lining-ced498b7cae3. Accessed 2 Jul 2017
29. Harvard School of Public Health (2016) Zika virus and the election season. August. https://cdn1.sph.harvard.edu/wp-content/uploads/sites/94/2016/08/STAT-Harvard-Poll-August-2016-Zika.pdf. Accessed 17 Jul 2017
30. Meaney-Delman D, Hills SL, Williams C, Galang RR et al (2016) Zika virus infection among U.S. pregnant travelers—August 2015–February 2016. Morbid Mortal Week Rep 65(8):211–214
31. Branswell H (2017) Puerto Rico declares its outbreak of Zika virus is over. STAT. June 5. https://www.statnews.com/2017/06/05/puerto-rico-zika-outbreak/. Accessed 26 June 2017
32. Kitzinger J, Reilly J (1997) The rise and fall of risk reporting: media coverage of human genetic research, 'false memory syndrome' and 'mad cow disease.' Europ J Commun 12:3. September 1. http://journals.sagepub.com/doi/https://doi.org/10.1177/0267323197012003002. Accessed 18 Sept 2018
33. Billero V et al (2017) Knowledge and perceptions of Zika virus among reproductive-aged women in Miami, Florida, vol 129, issue 5. May. https://www.ncbi.nlm.nih.gov/pubmed/29111579. Accessed 13 June 2019
34. Burattini MN, Coutinho FA, Lopez LF et al (2016) Potential exposure to Zika virus for foreign tourists during the 2016 Carnival and Olympic Games in Rio de Janeiro Brazil. Epidemiol Infect 144:1904–1906
35. Rogers M (2016) An olympic venue where mosquitoes live, but there's little buzz about Zika. USA TODAY Sports. September 8. https://www.usatoday.com/story/sports/olympics/rio-2016/2016/08/08/olympic-venue-where-mosquitoes-live-but-theres-little-buzz-zika/88430868/. Accessed 20 Jul 2017
36. Doucleff M (2016) Guess how many Zika cases showed up at the olympics? NPR Goats and Soda. August 26. http://www.npr.org/sections/goatsandsoda/2016/08/26/491416709/guess-how-many-zika-cases-showed-up-at-the-olympics. Accessed 20 Jul 2017
37. Ministério da Saúde Brazil (2016) Ministério da Saúde confirma 1.656 casos de microcefalia. http://portalsaude.saude.gov.br/index.php/cidadao/principal/agencia-saude/24437-ministerio-dasaude-confirma-1-656-casos-de-microcefalia. Accessed and translated 23 Sept 2016
38. Editorial (2016) Zika virus at the games: is it safe. Lancet Infectious Diseases. May 10. http://www.thelancet.com/pdfs/journals/laninf/PIIS1473-3099(16)30069-X.pdf. Accessed 24 Sept 2016
39. Jaffe E (2016) What the debate over Zika at the Olympics ignores. Forbes. August 17. http://www.forbes.com/sites/sciencebiz/2016/08/17/what-the-debate-over-zika-at-the-olympics-ignores/#37aaf1087206. Accessed 18 Sept 2016
40. Attaran A (2016) Zika virus and the 2016 olympic games. Lancet Infect Dis 15. September. http://www.thelancet.com/pdfs/journals/laninf/PIIS1473-3099(16)30230-4.pdf. Accessed 24 Sept 2016
41. Mendes A (2016) Number of cases of Zika reported in Brazil drops by 87%. Brazilian Health Ministry-Latest News. June 10. http://combateAedes.saude.gov.br/en/latest-news/804-number-of-cases-of-zika-reported-in-brazil-drops-by-87. Accessed 18 Sept 2016
42. Grills A et al (2016) Projected Zika virus importation and subsequent ongoing transmission after travel to the 2016 Olympic and Paralympic Games—country-specific assessment. Morbid Mortal Week Rep. July 13. https://www.ncbi.nlm.nih.gov/pubmed/27442184. Accessed 27 June 2017
43. Grills A et al (2016). Projected Zika virus importation and subsequent ongoing transmission after travel to the 2016 Olympic and Paralympic Games—country-specific assessment. Morbid Mortal Week Rep. July 13. https://www.ncbi.nlm.nih.gov/pubmed/27442184. Accessed 21 May 2017
44. McConnell J (2016) Zika virus and the 2016 Olympic Games—Editors' reply. The Lancet. July 22. http://www.thelancet.com/journals/laninf/article/PIIS1473-3099(16)30266-3/fulltext?cc=y. Accessed 21 May 2017

45. Carless W (2016) Brazil now has doubts that Zika alone causes birth defects. PRI. August 2. https://www.pri.org/stories/2016-08-02/brazil-now-has-doubts-zika-alone-causes-birth-defects. Accessed 27 June 2017
46. Malkin M (2016) Zika virus and illegal immigration: the connection. Nat Rev. February 3. http://www.nationalreview.com/article/430713/zika-virus-illegal-immigration-connection. Accessed 17 Jul 2017
47. Bruce T (2016) When foreigners bring disease across the border. Washington Times. January 25. http://www.washingtontimes.com/news/2016/jan/25/tammy-bruce-when-foreigners-bring-disease-across-t/. Accessed 21 Jul 2017
48. Terry K (2016) WND's Corsi Quick to Baselessly Blame Zika virus on illegal immigrants. February 5. Conwebblog. http://conwebwatch.tripod.com/blog/index.blog?entry_id=2358218. Accessed 17 Jul 2017
49. Corsi J (2016) Zika virus join list of diseases brought be illegals. February 1. WND. http://www.wnd.com/2016/02/zika-virus-joins-list-of-diseases-brought-by-illegals/. Accessed 17 Jul 2017
50. Klein A (2014) Deadly Central American mosquito found in San Diego. WND Weekly. October 29. http://www.wnd.com/2014/10/deadly-central-american-mosquito-found-in-san-diego/. Accessed 17 Jul 2017
51. Abruzzo S (2016) Illegals, not American travelers, may be Mosquito-transmitted epidemics: Zika virus in the United States and Mexico bringing Zika to our shores. Brooklyn Daily. February 5. http://www.brooklyndaily.com/stories/2016/6/all-britview-zika-virus-2016-02-05-bd.html. Accessed 17 Jul 2017
52. Rogers RW (1975) A protection motivation theory of fear appeals and attitude change. J Psychol 91(1):93–114. https://doi.org/10.1080/00223980.1975.9915803.PMID28136248
53. Witte K, Meyer G, Martell D (2001) Effective health risk messages: a step-by-step guide. Sage, Newbury Park, CA
54. Cummings C, Rosenthal S, Wong EY (2021) Secondary risk theory: validation of a novel model of protection motivation. Risk Anal 204–220
55. Errett NA, Sauer L, Rutkow L (2020) An integrative review of the limited evidence on international travel bans as an emerging infectious disease disaster control measure. J Emerg Manage 18(1):7–14
56. Matthews KRW, Herricks JR (2016) Mosquito-transmitted epidemics: Zika virus in the United States and Mexico. Policy brief no. 03.04.16. Rice University's Baker Institute for Public Policy, Houston, Texas
57. Valente PK, Morin C, Roy M, Mercier A, Atlani-Duault (2020) Sexual transmission of Zika virus on Twitter: a depoliticized Epidemic. Glob Pub Health 15(11):1689–1701
58. UF News (2016) Visitors concerned about Zika but still plan to travel to Florida, UF study shows. August 11. https://news.ufl.edu/articles/2016/08/visitors-concerned-about-zika-but-still-plan-to-travel-to-florida-uf-study-shows.html. Accessed 8 June 2022
59. Pennington-Gray L (2016) Zika doesn't deter Americans from traveling abroad, study shows. Sci Daily. February 16. https://www.sciencedaily.com/releases/2016/02/160216140359.htm. Accessed 8 June 2022
60. Entertainment Close-up (2016) Fluent poll: Zika virus outbreak affects travel plans of half of Americans. Entertainment Close-Up
61. McBroom K (2021) A comparison of Zika virus and COVID-19: clinical overview and public health messaging. J Midwifery Women's Health 66:334–342
62. Samuel G, DiBartolo-Cordovano R, Taj I, Merriam A et al (2018) A survey of the knowledge, attitudes, and practices on Zika virus in New York City 18(98). https://doi.org/10.1186/s12889-017-4991-3
63. Musso D, Ko AI, Baud D (2019) Zika virus infection—after the pandemic. N Engl J Med 381:1444–1457. https://doi.org/10.1056/NEJMra1808246
64. D'Angelo D et al (2017) Measures taken to prevent Zika virus infection during pregnancy—Puerto Rico, 2016. Morbid Mortal Week Rep 66(22):574–578
65. August E et al (2020) Community understanding of contraception during the Zika virus outbreak in Puerto Rico. Health Prom Prac 21(1):133–141

66. Alvarez L (2016) Pregnant women anxious as Florida's Zika test results take weeks. New York Times. http://www.nytimes.com/2016/09/13/us/zika-test-delays-florida-pregnant.html. Accessed 25 Oct 2016
67. Alvarez L (2016) Florida gets help to deal with backlog of Zika tests. New York Times. September 14. http://www.nytimes.com/2016/09/15/us/florida-gets-help-to-deal-with-backlog-of-zika-tests.html?_r=0. Accessed 25 Oct 2016
68. Reuters (2016) Over half the population of women in Brazil shunning pregnancy for Zika fear. Econ Times Reuters. December 23. http://economictimes.indiatimes.com/magazines/panache/over-half-the-population-of-women-in-brazil-shunning-pregnancy-for-zika-fear/articleshow/56140264.cms. Accessed 30 May 2017
69. Aiken AAA (2016) Requests for abortion in Latin America related to concern about Zika Virus Exposure. New England J Med 365:4. July 28. https://doi.org/10.1056/NEJMc1605389. Accessed 11 June 2019
70. Aiken, Abigail A. A. (2016). Requests for abortion in Latin America related to concern about Zika virus exposure. New England J Med 365:4. July 28. https://doi.org/10.1056/NEJMc1605389. Accessed 11 June 2019
71. Boseley S (2016) Abortion demand soars in countries hit by Zika outbreak, study finds. The Guardian. June 23. https://www.theguardian.com/world/2016/jun/22/abortion-information-zika-virus-birth-defects-latin-america-study. Accessed 13 June 2019
72. Kaiser Family Foundation (2016) Zika virus: the challenge for women. February 1. https://www.kff.org/global-health-policy/perspective/zika-virus-the-challenge-for-women/. Accessed 24 May 2022
73. Ministry of Health of Vrazil (2017) "Painel de Monitoramento da Mortalidade Materna," Coordenação-Geral de Informação e Análise Epidemiológica, http://svs.aids.gov.br/dashboard/mortalidade/materna.show.mtw. See Human Rights Watch (2017). Neglected and Unprotected
74. The Impact of the Zika Outbreak on Women and Girls in Northeastern Brazil. July 12. https://www.hrw.org/report/2017/07/13/neglected-and-unprotected/impact-zika-outbreak-women-and-girls-northeastern. Accessed 24 May 2022
75. Aiken ARA, Scott JG, Gomperts R et al (2022) Requests for abortion in Latin America related to concern about Zika virus exposure. New England J Med 375:396–398. https://doi.org/10.1056/NEJMc1605389. Accessed 24 May 2022
76. Gerdts C, Vohra D, Ahern J (2013) Unsafe abortion-related mortality: a systematic review of the existing methods. PLoS ONE 8(1):e53346. https://doi.org/10.1371/journal.pone.0053346. https://search.proquest.com/docview/1289067218?pq-origsite=summon. Accessed 13 June 2019
77. Rose J (2016) Are you immune to Zika after you get it? This may be the only good news about the virus. Romper.com. April 13. https://www.romper.com/p/are-you-immune-to-zika-after-you-get-it-this-may-be-the-only-good-news-about-the-virus-8852. Accessed 25 Sept 2016
78. McNeil D (2016) Growing support among experts for Zika advice to delay pregnancy. The New York Times. February 5. https://www.nytimes.com/2016/02/09/health/zika-virus-women-pregnancy.html?_r=0. Accessed 22 May 2017
79. Paploski IAD et al (2016) Time lags between exanthematous illness attributed to Zika virus, Guillain-Barré syndrome, and microcephaly, Salvador, Brazil. Emerg Infect Dis 22:8. August. https://wwwnc.cdc.gov/eid/article/22/8/16-0496_article. Accessed 19 Jul 2017
80. Aiken ARA et al (2016) Requests for abortion in Latin America related to concern about Zika virus exposure. New England J Med 375:4. July 28. https://doi.org/10.1056/NEJMc1605389#t=article. Accessed September 24, 2016.
81. Human Rights Watch (2017) Neglected and unprotected the impact of the Zika outbreak on women and girls in Northeastern Brazil. July 12. https://www.hrw.org/report/2017/07/13/neglected-and-unprotected/impact-zika-outbreak-women-and-girls-northeastern. Accessed 24 May 2022
82. AAFP (2018) Children exposed to Zika in utero need long-term monitoring. August 17. https://www.aafp.org/news/health-of-the-public/20180817mmwr-zika.html. Accessed 14 May 2019

83. Lemoult C (2011) Mosquito research feels bite of budget cuts. NPR. August 25. https://www.npr.org/2011/08/25/139928697/mosquito-research-feels-bite-of-budget-cuts. Accessed 8 Aug 2018
84. Linde-Arias AR, Roura M, Siqueira E (2020) Solidarity, vulnerability, and mistrust: how context, information and government affect the lives of women in times of Zika. BMC Infect Dis 20:263. https://bmcinfectdis.biomedcentral.com/articles/. https://doi.org/10.1186/s12879-020-04987-8. Accessed 24 May 2022

Chapter 17
Communicating Pandemic Risks

The medical ethicist at NYU's Langone Medical Center, Arthur Caplan, wrote: "It would also appear that the ringing of bells by health professionals, including those who went as far as to suggest postponing or moving the 2016 Olympic Games in Rio de Janeiro, was unnecessarily alarmist." [1]. Nonetheless, ZIKV remains a concern, primarily due to the world's collective unpreparedness.

Leslie Lobel, an Israeli physician who worked with the US military and the Uganda Virus Research Institute to find a vaccine for Ebola, believes public panic over epidemics can cause more damage than the diseases themselves [2]. Lobel suggests the WHO declared an emergency with ZIKV this time because governments were nervous after failing to deal adequately with Ebola. Similarly, he said the U.S. has suddenly announced almost $2 billion in ZIKV research funding. It is worried the disease will head up through Mexico once the Northern Hemisphere summer arrives. He believes the money could be spent more effectively to deal with future, more severe challenges. "There is a lot that might be done that is not being done in a corrective way to control all diseases, not just ZIKV. This is just the tip of the iceberg. There are going to be a lot more." [2].

Florida House Speaker Richard Corcoran said in a prepared statement early August 29: "The outbreak of the ZIKV, coupled with the inability of current measures to stop the spread, clearly demonstrate that time is of the essence if we are to beat back the spread of this disease." [3].

Welcome to the world of "pandemic pornography," no less sensual and provocative than photos and videos of barely clothed people engaged in suggestive and, in some cases, actual sex acts. "Pandemic porn" is about titillation.

The subject of this chapter is why the media and government amplification stations (see Chap. 2) employed this approach to communicating outbreak-related health risks. Please keep in mind that the ZIKV epidemic is a current phenomenon, and it is ongoing in one form or another, so what appears here is what is best known at this time instant. Think of this article as a photograph or a painting taken at a time using what material was best at hand.

D. M. Berube, *Pandemics and Resilience: Lessons we should have learned from Zika*, Risk, Systems and Decisions, https://doi.org/10.1007/978-3-031-25370-6_17

The primary caution of this chapter remains that ZIKV is a serious health problem, and many amplification stations have legitimate and sometimes selfless reasons for portraying the outbreak as they did. According to Chicago's Director of public health in 1918, "It is our job to keep people from fear. Worry kills more than the disease." [4]. As mentioned earlier, using fear to change behavior has many drawbacks, not the least of which is what happens when the fear dissipates because what was causing the fear begins to lift.

According to Margaret Chan, Director-General of WHO, poor access to family planning services was one. The dismantling of national programs for mosquito control was another [5]. A deep concern was: "Simply educating the public about the potential risks and dangers of ZIKV infection may be insufficient to mobilize the people in the event of significant outbreaks." [6]. An innovative approach may be necessary, especially when one message involves delaying pregnancy. She wrote.

> Rather, willingness to delay pregnancy is driven by two factors: a sense of being at personal risk for ZIKV and living in the Gulf Coast states. All other factors being equal, U.S. residents who believe they are at risk for the ZIKV are one and a half times as likely to delay pregnancy as those who do not think they are at elevated risk. Residents of the Gulf Coast states are 1.6 times as likely to delay pregnancy (regardless of their sense of danger or other factors tested). [6]

The key to risk communication in the case of ZIKV is not overly complicated. Of course, that assumes someone is willing to take the time and effort to search out the experts who have worked in pandemic communication [7].

What can be done to help people who have traveled to those areas by reminding them that they can use insect repellent on their exposed skin for three weeks? Can this be done without referring to a community and the fundamental respect there should be for others? It is just treating people how you want to be treated, which means thinking through what is being done that might be disrespectful and communicating that information. Keep in mind that communication errors leave messages stained. Once stained, a message is much more difficult to rehabilitate than it would have been had the message been well designed in the first place.

The process of communicating the risk of the infectious disease involves many steps.

First, there is the step of **awareness** (the broadest appreciation of the existence of the disease). For example, the American public awareness of ZIKV increased from 74% in March 2016 to 95% in August, rising linearly [6]. The second step involves **knowledge**. For example, in the case of ZIKV, this encompassed a respondent's understanding that the ZIKV could (1) cause congenital disabilities, (2) be expressed as an asymptomatic infection, and (3) be sexually transmitted [6].

> According to the NYU report: although the proportion of the U.S. population or the subset of women of child-bearing age aware of ZIKV increased from April to July, more specific knowledge about the virus remained low and did not change over time. [6]

	APRIL-MAY 2016	JULY-AUGUST 2016
% of U.S. population **aware** of Zika	77.9	84.8
% of women of child-bearing age **aware** of Zika	75.7	84.6
% of U.S. population **knowledgeable** of Zika	38.7	38.2
% of women of child-bearing age **knowledgeable** of Zika	41.8	45.0

Figure SEQ Figure_5-1._Transmission_routes. * ARABIC 17-1 Awareness data from NYU study. Abramson, David and Pitch-Loeb, Rachael. (2016). U.S. Public's Perception of Zika Risk: Awareness, Knowledge, and Receptivity to Public Health Interventions. NYU Zika Briefing Report #1. https://www.nyu-pir2.org/research. Accessed June 19, 2019

The same NYU survey reported: Among the overall U.S. public, the wealthiest, the most educated, and Republicans are most likely to be aware. Women and adults with higher education are also more likely to know about the virus. Democrats are more knowledgeable than Independents or Republicans. Only political ideology separates the women regarding knowledge: Women who identify as Democrats and Independents are approximately twice as likely to be knowledgeable about the ZIKV as are Republican women (46% vs. 24%) [6]. How can knowledge about a health issue be so affected by someone's political party affiliation? Please keep in mind that the relationship may be irrelevant and may only mean that the reasons someone belongs to a political party might include the same variables as to why someone is unaware of a health event like ZIKV.

Other than by personal experience, messaging health risks generally are communicated from four primary sources: (1) conventional media, such as broadcast, print, or online news; (2) social media, friends, and family; (3) one's doctor; or (4) government (**confidence in government** is essential to increase receptivity to indoor spraying, and communication campaigns should demonstrate the knowledge and ability of governmental actors overall, and especially regarding the ZIKV) [6]. The bold highlight is intentional and might explain why it matters to be a Republican.

NYU reported: Those who list their primary source of information about ZIKV as conventional media are four and a half times as likely to be aware of ZIKV as are those who rely upon social media, friends, and family as their primary source of information and those who list government as their primary source of information (and this may be at any level, from federal to state to local) are more knowledgeable than those who report other sources of information [6].

As discussed below, source issues, while important, pale compared to messaging problems. The messages involved an array of subjects, individual to broad policy subjects: on the personal level, travel choices, emptying anything outdoors that collected water, indoor spraying, wearable insecticide, delaying pregnancy, and testing. In terms of community decision-making, traditional approaches like larvicide and insecticide use, habitat reduction for the relevant mosquito species, use of aggressive species that can reduce mosquito populations, especially the release of traditionally biologically engineered mosquitoes, and "new" gene drive approach to frustrate population growth and efforts toward extinction of the mosquito vector altogether.

Keeping the public informed is a critical public health function during an emergency response situation. Effective risk communication campaigns are successful because they provide accurate, straightforward, and timely information, reducing general anxiety and giving people concrete steps to protect themselves. Health departments should continually communicate with physicians and other healthcare providers to ensure that all medical professionals who may be involved in diagnosing vector-borne disease patients are apprised of what the current vectors and diseases of concern are, including what environmental conditions may facilitate the spread of vector-borne disease, and what public health resources are available [8].

"You may not want to wear insect repellent for three weeks but try it and see if just tomorrow you can put it on three weeks you have been traveling someplace where ZIKV, in a ZIKV affected area. See if you could just put it on tomorrow. If you can do it tomorrow, maybe you can do it the next day." [9].

The future of ZIKV remains unpredictable, but its recent spread confirms that ZIKV is following the path of chikungunya and dengue. However, the severe disease associated with ZIKV in French Polynesia and Brazil suggests that this virus will become a powerful global public health problem [10] for some years.

The rate of congenital disabilities found in confirmed ZIKV cases is more than thirty times higher than that of similar congenital disabilities in the United States before the ZIKV outbreak. The CDC report also revealed that many physicians are not carefully tracking pregnancies threatened by ZIKV. About one-third (35%) of infants with possible ZIKV infection during pregnancy were not tested for the virus at birth, and only 25% received neuroimaging after delivery to check for possible defects [11].

Although it is essential to stress ZIKV prevention, these communications must avoid creating undue anxiety and aim to provide measured recommendations based on ZIKV transmission probabilities in travel destinations and pregnancy intentions. Targeted communication between public health officials or providers about the risks of ZIKV and its sequelae could generate interest in a prophylactic ZIKV vaccine to prevent harmful long-term effects on infected individuals and their infants [12].

Risk communicators will need to consider highlighting various aspects of their messages—whether increasing knowledge of transmission routes, conveying the actual risks posed by multiple vectors, or promoting the trustworthiness of government or public health organizations—depending upon the intervention they wish to advance [6].

17.1 Pandemic Communication

One of the biggest challenges of public health is getting the correct information into the hands of the population that it would serve. With so many messages and means to deliver them, cutting through the noise [13] to deliver timely, relevant information on a health topic becomes fierce competition [14].

The next influenza pandemic could be from a mild or virulent virus. The single most important weapon against the disease will be a vaccine. The second most important will be communication. History has shown that governments must communicate well—both themselves and the public—cut vaccine production time, minimize economic and social disruption, and deliver health care and food [4].

Beyond its impact on morbidity and mortality, pandemics are interesting communication phenomena, mostly events that have been understudied. In 2016, the Global Health Risk Framework for the Future Commission called for reinforcing national public health capabilities and infrastructure as the foundation of a country's health system and the first line of defense against potential pandemics; second, strengthening international leadership and coordination for preparedness and response; and third, accelerating research and development in the infectious disease arena [15].

As a data-driven social scientist, my interest lies in trends. There is evidence that social science theory may play an essential role in the success of community-based programs. Research suggests that multifaceted interventions are more effective than single interventions because a wider variety of barriers to change can be addressed simultaneously.

Unfortunately, evidence that community-based dengue control programs alone and in combination with other control activities can enhance the effectiveness of dengue control programs is weak [16]. However, a literature review further suggests that this is due to lacking a formalized assessment plan and fine-tuned granular data concatenation more than any other factor [16].

Human health involves motivations that err in favor of false positives. No one wants to be responsible for instigating calm, only to discover later that an outbreak has been devastating. As such, amplifying pandemic risks has become an essential part of the rhetoric of health bureaucrats, technical and popular media, and even some non-governmental organizations in the public health fields.

In addition, a "potential" pandemic is being experienced in a communication setting that has become digitally dominant. So, not only are the people who have traditionally been given the responsibility to communicate health risks to those engaging in hyperbole, but a whole new college of health communicators on the Internet are doing the same, sometimes creating data rather than taking the time and effort to find out what is happening.

ZIKV may represent a trigger event that helps raise the visibility of a critical issue in science communication: conflation. When confronted with new phenomena, the public often uses simple heuristics to understand what is before them. These heuristics enable shortcuts in reasoning because that is the easiest way to make sense of something complicated. Susan Fiske [17] described people as cognitive misers, basically lazy. We expend only as much energy as we feel may be appropriate. When confronting a "new" and often "confounding" phenomenon, there is a desire to simplify it.

Communication scholars have examined how the public makes sense of things for decades. Petty and Cacioppo [18] labeled their mode of describing the public's thoughts as the Elaboration Likelihood Model (ELM). They argued that the quantity and quality of cognition dedicated to an issue reflect a set of variables, such as salience and exigence. Some cases tend to provoke high cognitive efforts, known as elaborations. Other problems do not. Elevated levels are reflected in a concept coded as a central route of reasoning. Issues that do not tend to provoke high cognitive effort are elaborated along a peripheral way. The peripheral course involves more incidental variables and is less phenomenologically essential to the issue.

Retired NYU Professor of Psychology Shelly Chaiken [19] offered a similar perspective regarding information processing. She has been attributed with the heuristic-systematic model, a model that significantly impacted the study of persuasion. People reason systematically under conditions (structured and logical and based on good reasons). Other times there may be a heuristically derived reasoning typified by mental shortcuts. Nearly, one hundred heuristics or mental shortcuts individuals have populated the literature used to make sense of things.

Finally, Princeton Professor Daniel Kahneman, best known as the first social scientist to have won the 2002 Nobel Prize in economics, with his late colleague and Stanford Professor Amos Tversky, Kahneman described two general types of thinking: System 1 and System 2 [20]. They argued there are two modes of thought: "System 1" is fast (heuristic), instinctive, and emotional; "System 2" is slower (systematic), more deliberative, and more logical.

These economics, psychology, and communication scholars are making somewhat similar claims. Under certain circumstances, people will seriously evaluate some issues because of intrinsic properties associated with the problem. In other cases, they elect to make the process less demanding and choose mental shortcuts.

It is unfair to assume certain people are systematic and another heuristic set. Everyone uses mental shortcuts, some of which are more reasonable than others at arriving at defensible conclusions. Experts, as well and inexperts (a term used throughout to describe naïve audiences) have any of the same mental shortcuts and a host of different ones.

In addition, some issues do not justify the mental effort involved in systematic thinking. If every recognized issue was carefully and systematically analyzed, there would be a rampant experience of what Kahneman has called "paralysis by analysis." The OED defines it as the inability to respond effectively to a situation due to an overanalytical approach or an excess of available information.

Like Goldilocks's dilemma, a healthy person making the most defendable claims and drawing reasonable conclusions must find the sweet spot, the "just right" balance. Failure to do so leads to frustration, hesitation, and debilitation. While the trespassing yellow-haired girl decided it was suitable for her and she was soundly situated to allow her to run into the forest to escape the angry homeowners, too many people remain lost in the woods or become the dinner for those who would mislead them.

The following story is typical of the general issue confounding science communication: Too many people conflate all versions of a thing under its most threatening risk profile. Science history is replete with examples of hysteria predicated on information failure. Some professional opinion managers know how to tap into certain biases, and heuristics do so not only to sell products no one needs but also to convince people something is much worse than it is.

An outbreak or an epidemic is an occurrence of any disease, sometimes infectious. An epidemic is simply a widespread infectious disease at a specific time. At the same time, a pandemic is an epidemic that has spread through populations across a large region. Technically, there was fear of an outbreak of ZIKV in Rio during the Olympics; there was a pandemic in Latin America, including Brazil, Colombia, and Puerto Rico.

The importance of moving quickly to fight new threats like the ZIKV is understood. But those working in global health also worry about losing focus on such age-old deaths every day from these three diseases killers as HIV/AIDS, malaria, and tuberculosis. Do not forget that more than 11,300 people died from Ebola during the 2014–16 outbreak in West Africa [21].

Increasing ease of global travel, continued encroachment of human populations into wildlife-occupied areas, climate change, and increasing microbial evolution and antimicrobial drug resistance rates have increased the likelihood that wildlife pathogens will be introduced into novel regions or naive populations and spill over into human populations [22].

Does labeling a disease outbreak event as a pandemic communicate something different from the absence of this "brand?" Government officials function much like the media in reporting the risk footprints or signatures of events such as an outbreak of infectious diseases. In the parlance, amplification occurs for many varied reasons. Some amplification is non-sensible, whence someone posits an erroneous opinion as a fact and sometimes goes further and adds false details. Some am-amplification occurs due to assumed or actual self-interest being involved. If you are involved with a government agency, event amplification into a crisis brings your agency attention to staff and budgetary increases. The same can be said of individuals whose careers place them in positions whereby amplification serves their professional interests, individually or collectively, as a non-governmental organization, even a nonprofit foundation. Demand attention and financial support are powerful incentives to amplify risk events.

Take the case of SARS. While it can be deadly for a small number of those infected, it can be controlled and cured given the proper treatment. It may not have been thought that was possible based on how the events were reported in China, Hong Kong, Japan, and Southeast Asia.

Still, the world would be much healthier if the same sense of urgency to stop Ebola and ZIKV could be applied to HIV/AIDS, TB, malaria, pneumonia, and diarrhea—scourges that have been battled for decades [21].

Rank ordering potentially pandemic events in terms of actual lethality does not correspond to the coverage they receive. The U.S. keeps records on fifty-two different infectious diseases. However, at times they are grossly hyperbolized. The problem for readers and viewers of media coverage of infectious disease events is reckless reporting. While it may be arguable true that with increasingly convenient and affordable air transportation, there is an increasing likelihood of transmission from an infected individual that is the exception rather than the rule.

17.1.1 Rhetoric of Warfighting

Rhetoric is more than language designed to have a persuasive or impressive effect on its audience but is often regarded as lacking in sincerity or meaningful content. Rhetoric is the art of persuasion, along with grammar and logic, one of the three ancient arts of discourse. Some of the techniques studied are tropes. Tropes are motifs and themes.

Traditionally, scholars and communication practitioners have studied how outbreaks and pandemics are portrayed as battles and wars with a rich use of warfighting metaphors.

"The fight against these diseases is like a war," George Dimopoulos, Entomologist from Johns Hopkins, says. "You can't win it with one weapon." [23]. Half of the global population is at risk of a disease transmitted by *Ae. aegypti*. Again, the health catastrophe in Latin America shows that human population is the greatest threat to public health. [24]

This simile instills worry, trepidation, and panic.

17.1.2 Panic

Panics can be costly. In 2002, several Israeli doctors analyzed "epidemic" and "outbreak" in medical dictionaries, epidemiology texts, and other medical and legal literature [25]. They recommended that "outbreak" be used to identify more limited types of epidemics but found that the terms were often used interchangeably. They concluded that the "interpretation of the term epidemic might vary according to the context in which it is used." [26].

In common parlance, an epidemic is more significant than the everyday occurrence of a health-related event in a specific region during a particular period. The word is commonly referred to as infectious and non-infectious diseases and physical conditions like obesity. There is no set minimum number of cases to justify using the term, and the period could range from several hours to several years.

The region could be a community or a country, but the word "pandemic" comes into play when borders are crossed [26]. The pandemic threat is terrifying as there is a lack of control, which tempts people into aimless activity. Health experts use "pandemic" to describe an epidemic of an infectious disease that has spread across a large region, such as a continent or even the entire world [26].

The decision to stockpile antivirals and influenza vaccines to control avian flu (2005–2006) and swine flu (2009) costs significant money. Both epidemic threats were mostly iatrogenic pandemics of panic, which, while neither event caused much human suffering, the global plans to control them were essentially a waste of money.

In addition, the modern disease expert knows a lot about the disease in question but does not necessarily know much about public health, health economics, health policy, or public policy, which are much more about priority setting and resource allocation between competing priorities. Disease experts are often biased as specialists and are increasingly part of health industrial networks. The importance of the drug industry is to make a profit [27].

On one level, money spent on stockpiling antivirals with hypothetical effectiveness against a hypothetical pandemic is not available for health care, education, or any other crucial human need thought to be underfunded.

On another level, it is essential to consider the ethical impact of events like stockpiling. For example, the scramble for neuraminidase inhibitors shows the effects of patent laws. The manufacturer has the monopoly, and the rational monopolist sets the price based on "what the market will bear." Rich countries are stockpiling the drugs at an average of $5 a person, representing less than 1% of their annual healthcare budgets. But $5 is higher than the yearly healthcare budget for three hundred million poor Africans. The Asian nations where the avian flu causes most deaths and where the risks of recombination of avian and human flu are highest cannot afford stockpiles of neuraminidase inhibitors [28].

Disease experts are necessarily fatally biased. It is not reasonable that they bear the entire responsibility for decisions related to their disease. Over the past years, this has been convincingly shown with the two iatrogenic pandemics of influenza panic. Of course, the advice of disease experts is valuable and indeed crucial. Still, this advice should be evaluated against the available evidence, balanced by other stakeholder views, and checked by the transparent evaluation of costs and values. At last, the final evidence-based policy advice should be drafted by independent scientists trained in the evaluation and priority setting [27].

The National Institute for Health and Clinical Excellence (NICE) uses independent, general scientists from the UK's epidemiology and public health, health economics, and medical ethics. In a transparent and collaborative process, the teams review the best available evidence and involve all stakeholders, including administrators, healthcare professionals, patients, carers, industry officials, and academic disease experts. Costs are explicitly and transparently evaluated and compared against the expected value of a specific policy. NICE introduces checks and balances necessary to safeguard cost-effective health care. NICE is held "accountable for reasonableness" by the public in a continuous debate [27].

Even in a pandemic of iatrogenic panic, a pandemic needs global health governance and an institution with an international mandate in health to intervene [28].

17.2 The Communicators

There are many health communicators. By and large, they tend to be associated with government entities. My concern is not to list them all.

17.2.1 US Regulators

Who is responsible for what is essential on so many levels. Why generally good people allow evil to surface is often jurisdictional. As bureaucracies become more diversified, they often become less responsive to the problems of the public. It is less about intent and more based on how humans are simply not designed to work in a bureaucracy. Organizations within bureaucracies develop a self-sustaining philosophy where their people and budgets define their importance.

The defining characteristic for the roll call was whether the unit functioned as a component of the ZIKV epidemic. Before starting, it is essential to highlight that trust in the U.S. CDC and the FDA remains high.

Using two national samples of adolescents ($N = 1125$) and adults ($N = 5014$), demographic factors were studied, with a focus on vulnerable groups, as correlates of awareness of and trust in the CDC, FDA, and the federal government. From nine different weighted and adjusted logistic regression models, there were high levels of awareness found of the existence of the FDA and CDC (ranging from 55.7% for adolescents' understanding of the CDC to 94.3% for adults' awareness of the FDA) and moderate levels of trust (ranging from a low of 41.8% for adults' trust in the federal government and a high of 78.8% for adolescents' trust of the FDA) [29].

17.2.1.1 The Centers for Disease Control and Prevention

The CDC provides up-to-date guidance, resources, and training for providers evaluating and caring for patients with ZIKV. CDC has a surveillance system for collecting data on ZIKV virus cases. Prevent the spread of ZIKV by controlling mosquitoes in and around your home and community.

CDC maintains an impressive web presence with a plethora of resources and has influenced both in the U.S. and abroad. In the U.S., it is one of the first places governments and the public turn to for their health information, but that has been changing.

Trust in the FDA and CDC is higher than in the federal government, and in the same 2015 poll, 51% and 71% of the U.S. population reported that they view the FDA and CDC favorably, respectively [30].

Even after COVID, Kalichman et al. (2017) reported that the government was trusted significantly less than all three sources of COVID-19 information. Finally, trust in the CDC was more significant than in state health departments [31]. However, the CDC did suffer losses of trust during COVID. Overall, the surveys show a statistically significant decrease in trust in the CDC. These surveys show that the CDC will need some perception rehabilitation, particularly among those who reported intending to vote for someone other than President Biden or not voting at all (who had low levels of initial trust that declined even further). Although it is known that the Black community has had low trust in the CDC, it is now similarly low across all the groups [32].

The CDC engages in many trust-building activities and will need to do so in the future. For example, during the ZIKV epidemic, they launched a multimedia campaign that included public service announcements and print and digital materials. They held community engagement events in Puerto Rico, the U.S. Virgin Islands, and American Samoa. The campaign aims to educate pregnant women and communities about ZIKV prevention and provide them with the steps they can take to protect themselves from ZIKV infection, mainly by preventing mosquito bites and avoiding potential sexual transmission [33].

The next chapter on engagement will offer some suggestions.

17.2.1.2 The Food and Drug Administration

> The US Food and Drug Administration (FDA) issued updated guidance for the industry to reduce the risk of transfusion-transmission of ZIKV in August 2016. These recommendations call for blood collection establishments in all states and U.S. territories to screen individual units of donated whole blood and blood components with a ZIKV screening test authorized by the FDA under an investigational new drug (IND) application or with a licensed test when available. Alternatively, an FDA-approved pathogen-reduction device may be used for plasma and certain platelet products. FDA has also issued guidance for reducing the risk of ZIKV transmission by human cell and tissue products. The Organ Procurement and Transplantation Network (OPTN) of the Health Resources and Services Administration (HRSA) has developed information on the ZIKV for organ transplant establishments and organ procurement organizations. [34]

And that does not include its role in Oxitec's testing regime (see Chap. 13). Oxitec found itself paralleling the experience that the genetically modified salmon producer Aqua Bounty Technologies faced. However, the data surrounding Aqua Advantage salmon's safety were first submitted to the FDA in 1996. Still, the agency has not decided whether the salmon can be commercially sold [35].

17.2.1.3 The Environmental Protection Agency

"[T]he FDA and EPA committed to considering mechanisms that would enable EPA to regulate certain mosquito-related products under the Federal Insecticide, Fungicide, and Rodenticide Act (FIFRA) when the developer claims they are intended to control mosquito population levels, and the FDA to regulate them under the FD&C Act when the developer makes other claims, such as a disease prevention claim." [36].

Its website offers information on repellants and pesticides. The agency plays a formative role in approving vector management activities involving larvicides, adulticides, and other approaches, including experimental trials (Chaps. 12 and 13) under its powers to grant experimental use permits (EUP).

Fundamentally, EPA ensures that state and local mosquito control departments have access to practical mosquito control tools that they can use without posing an unreasonable risk to human health and the environment.

17.2.1.4 US Department of State

In support of the Global Health Security, Agenda championed by President Obama, the Combating Zika and Future Threats Grand Challenge invested $30 million in groundbreaking innovations and interventions that enhance the ability to prevent, detect, and respond in both the short and long term by sourcing innovations that mitigate the spread and impact of ZIKV and improve the ability to combat future infectious disease outbreaks [37].

17.2.1.5 US Department of Health and Human Services

Dr. Fred Soper from the National Institutes of Health presented a paper at the American Public Health Associated in 1961.

> The failure of the U.S. to participate in the eradication of *Ae. aegypti* is to be deplored because the presence of this mosquito here constitutes a potential source of reinfestation to those countries which, under constant threat of infection of their cities with the virus of jungle yellow fever, have eradicated *Ae. aegypti*. Probably no less deplorable is the example set by this country is failing to join in the first official regional eradication program, which depends on the participation of all nations. [38]

The eradication of the *Ae. aegypti* mosquito in the Western Hemisphere is not new. Since 1947, countries have engaged in programs to protect themselves from Yellow Fever. This mosquito-borne disease came to public attention during the construction of the Panama Canal Program in the Americas to control urban yellow fever. In the 1970s, however, the American program was disbanded, and *Ae. aegypti* re-invaded most countries in the region [39].

The United States focused on its concerns, syphilis, tuberculosis, and poliomyelitis outbreaks. The principal delinquent in the international effort to eradicate *Ae. aegypti* was and is the United States [38].

Much like the decision by President Trump to back away from the Paris Accords, the United States has neglected its responsibilities. The children with microcephaly and congenital disabilities born in the United States from ZIKV rest partly on the thresholds of the White House and the U.S. Congress.

It remains unsurprising that the US effort in combatting ZIKV has been tepid. The virus is expected to ramp up in the U.S. as the weather gets warmer. Nine months after that, Lawrence Gostin, Professor of global health law at Georgetown, predicted: "There will be another congressional hearing, and you'll have poor mothers with their microcephalic babies testifying before Congress, and the public will ask, 'How did you let this happen?'" [40].

In 2016, HHS committed $561 million to the CDC, Food and Drug Administration, Centers for Medicare and Medicaid Services, and the Office of the Assistant Secretary for Preparedness and Response for efforts to combat ZIKV, former HHS Secretary Sylvia Burwell said.

"Approximately $40 million went to twenty-three community health centers and two public health nonprofits in the U.S. to fight the mosquito borne virus, which can also be sexually transmitted. Health centers in the three territories will receive approximately $39 million, which will be used throughout a three-year period to fund ZIKV related preventive and primary health care services. Two nonprofits were awarded a combined $1 million, which will be used to provide technical assistance and training to health centers in developing methods to prevent the virus and manage ZIKV related health care needs." [41].

Given limited investments in public health infrastructure, such surge funding for emergent threats like ZIKV is necessary. National Association of County and City Health Officials (NACCHO) cited the need for Congress to sufficiently fund the core infectious disease program at CDC to help avert these situations in the future. This program includes the vector-borne diseases program that provides state and local health departments resources to detect, control, and prevent bacteria and viruses transmitted by mosquitoes [42].

Some states in the USA are committing resources to protect the public from ZIKV. "The Texas Health and Human Services Commission (HHSC) announced it plans to bring back a Medicaid mosquito repellent program, beginning May 1, 2017. This time, the State will expand the benefit to include eligible boys aged fourteen and up, plus men and women ages 45–55. By having men, officials hope to reduce the risk of sexual transmission to women.... Texas Medicaid, a state and federal cooperative, serves primarily low-income families, children, pregnant women, people aged sixty-five and older, plus other needy people in the Lone Star State who might otherwise go without medical care." [43].

17.2.1.6 National Institutes of Health

U.S. health officials have said that a widespread outbreak is unlikely, given their experience with other mosquito-borne viruses like dengue, limited to small caseloads in Florida and Texas. "We still feel that that is the very likely scenario," Dr. Anthony Fauci, Director of the National Institute of Allergy and Infectious Diseases (NIAID) at the National Institutes of Health, said on April 7, 2016. But he cautioned that "ZIKV seems to surprise us a bit," leaving the possibility that it could be more widespread than dengue [40].

NIAID's response to the ZIKV epidemic provided support for the development, preclinical testing, and clinical evaluation of at least five vaccine candidates; funded the development of diagnostics, therapeutics, and vector control strategies; sustained the Zika in Infants and Pregnancy (ZIP) study begun in FY 2016 and support basic discovery research critical to understand ZIKV and its complications better and inform the development of new interventions [44].

ZIP involved improving the understanding of the health effects of ZIKV infection in pregnant women and infants by following 10,000 pregnant women for the duration of their pregnancies and their infants at several intervals for at least one year after birth [44].

$129 million was provided to support vaccine development and clinical trials. NIAID will continue to support the critical effect of vaccines, including the discovery and development of new and existing candidates, manufacturing activities, preclinical testing, the establishment of clinical trials sites in endemic regions, and the conduct of clinical trials to evaluate the safety, immunogenicity, and efficacy of vaccine candidates. NIAID advanced five lead candidates and claimed it was willing to support discovering and developing additional potential vaccine platforms and candidates, such as virus-like particles and other virus-vectored vaccines [44].

17.2.2 *International Entities*

An initial observation given the global nature of pandemics: One would expect more prominence associated with international organizations than reported in the literature.

During the same period, the US CDC launched a campaign for the continental United States that had to message travelers, including ads placed in United Airlines and American Airlines in-flight magazines. Ads ran in both 2016 and 2017 [45].

Risk communicators are part of a pandemic management complex. Many researchers have discovered that communicating about pandemics is incredibly challenging. A rational scientific approach to pandemic management is insufficient. A more nuanced sociopolitical blend of science, culture, and public perceptions offers a more substantial basis for public health policy [46].

Pandemic infections have been feared throughout history [47]. There is thus a strong consensus that governments should adopt persistent public health and other protective measures to prevent pandemics from developing and manage risks should a pandemic strike. However, there is much less consensus about how risk should be framed (is it a scientific, social, or political question?), what forms of governance should be deployed (e.g., quarantine or public education?), what disciplinary perspectives should be favored (medicine, law, politics, social sciences), and how that management is perceived or accepted by a culturally diverse public [46].

Eliminating mosquito breeding sites remains the most critical strategy for preventing and controlling ZIKV (dengue and chikungunya) infection. Therefore, communication plans for the response to ZIKV should include intersectoral action and community engagement to modify behaviors, encourage sustained practices to eliminate breeding sites and control the mosquito, and inform and educate target audiences about the steps they can take to prevent ZIKV transmission [48].

One of the most significant challenges confronting risk communicators has to do with sites of infection. Mosquitoes infiltrate the home and the backyard on a humid weekend evening. A worthwhile long-term goal for mosquito control is to convince people that transmission mainly occurs in the home environment [24].

There are many risk communicators involved in ZIKV. Authors include a recommendation in their articles, both scientific and popular ones. For example, Chen and Tang suggest preventing further mosquito bites by using insect repellant, wearing long-sleeved shirts and long pants, and treating clothing with permethrin to avoid passing on your infection to a second, may be uninfected mosquito and passing on the illness. They also warn insect repellent, by and large, should not be used on babies younger than two months of age [49].

The non-pharmacological treatment methods include (i) get plenty of rest; (ii) drink fluids to prevent dehydration; (iii) take medicine such as acetaminophen to relieve fever and pain; (iv) wear long-sleeved shirts and long pants; (v) use insect repellents containing DEET, picaridin, oil of lemon eucalyptus (OLE), or IR3535. Always use as directed; (vi) insect repellents containing DEET (according to Jim Fredricks of the National Pest Management Association, up to 20%) [50], picaridin, and IR3535 are safe for pregnant and nursing women and children older than two months when used according to the product label. Oil of lemon eucalyptus products should not be used on children under three years of age; (vii) if using both sunscreen and insect repellent, apply the sunscreen first and then the repellent; (viii) use permethrin-treated clothing and gear (such as boots, pants, socks, and tents); (ix) use bed nets as necessary; (x) stay and sleep in screened-in or air-conditioned rooms [51].

There are also many different communication channels available. PAHO recommends media briefings, social networking, micro-websites with FAQs, radio and television broadcasts, printer materials, interviews with trusted experts, simple fact sheets, telephone hotlines, and updates [52].

After noticing an increase in ZIKV test requests during the summer of 2016, the NYC Department of Health and Mental Hygiene assumed the additional messaging might be justified. They developed a survey of pregnant women with follow-up interviews. They reported a third of the respondent was unaware of the government advisory (avoid travel to areas with active ZIKV transmission). Half did not know there was active ZIKV transmission in places they traveled. More than one-third did not know they were pregnant during travel [53]. While the results cannot be generalized for many reasons, it does suggest that warnings alone may not be sufficient to communicate to vulnerable populations.

The situation in Brazil was disheartening. With over 174,000 Brazilians infected with ZIKV, a team undertook a national mixed-methods survey in June 2016 of a nationally representative sample of 2000 Brazilian women between 18 and 39 years. They reported that 27% of women responded that they had not tried to avoid pregnancy because of the ZIKV epidemic [54]. They found no significant difference among the main religious groups. However, women of color were more likely to report pregnancy than white women, which may reflect the disproportionate impact of ZIKV among the most vulnerable racial groups [54].

The purpose of reporting this finding is to raise some clamor over the effectiveness of the "avoid pregnancy" warning in the public health messaging. Three out of ten women were not taking notice seriously.

17.2.2.1 World Health Organization

The WHO has issued only three international health emergencies. Emergencies announced by the WHO are so scarce that bureaucrats have named them Public Health Emergency of International Concern (PHEICs). In 2009, it was swine flu. In 2013, it was Ebola in West Africa. In 2014, it was polio (this time an announcement of its near extinction) [55]. Since the onset of ZIKV, the WHO has issued many PHEICs. This work examines a snapshot of WHO activity and efforts since the mid-twenty-first-century teens deserve further criticism elsewhere.

The WHO uses a pandemic scale for influenza (see above). Any country that suspects or has verified such an event should urgently consult with WHO so that the situation can be jointly assessed, and a decision made by the affected government if implementing a rapid pandemic containment operation is warranted. Phase 4 indicates a significant increase in the pandemic risk but does not necessarily mean that a pandemic is a forgone conclusion [26].

There is a lot of discussion about whether the WHO may have overreacted. After the 2014 Ebola outbreak, Doctors Without Borders and others criticized the WHO for its slow response, responsible for 11,300 deaths. It has been suggested that the lessons learned involved getting ahead of this outbreak by declaring an early Public Health Emergency. The WHO reserves the PHEIC designation for "extraordinary" events that "constitute a public health risk to other states through the international spread of disease" such that they require "a coordinated international response." [56].

The World Health Organization (WHO) declared ZIKV a Public Health Emergency of International Concern on February 1, 2016, after 10,000 ZIKV infections were reported in Puerto Rico [57] and another 4,000 in other US territories [58]. By November 2016, the ZIKV no longer posed a global public health emergency [59].

"In November 2016, the WHO changed its classification for ZIKV from a "public health emergency" to a "long-term commitment." Some experts voiced their concern that the decision would dampen enthusiasm for vaccine development and, consequently, a lack of preparedness if the disease spreads again. Some have also warned that the reactive nature of vaccine R&D could shift developers' attention every time a new outbreak emerges, drawing comparisons with how ZIKV replaced Ebola on the international community's agenda [60].

In September 2017, Sano Sanofi decided to opt-out of developing the U.S. Army vaccine [61]. This is a case worth examining as infections wane, and there are fewer opportunities to test a new vaccine and fewer economic incentives to research vaccines. Organizations like the WHO declared the end to an emergency; they inadvertently dampened the vaccine activities to allow re-emergence to be dealt with promptly and efficiently.

The declaration has been withdrawn indeed does not mean it is not a risk anymore," said Andre du Plessis, who codirects the Zika Congenital Virus Program at the Children's National Medical Center. "Dropping our guard may be unwise." [62]. WHO spokesperson Daniel Epstein said: "We are (now) trying to convince people that international concern's global public health emergency is over. It does not mean that the problem is less serious (Fig. 17.1).

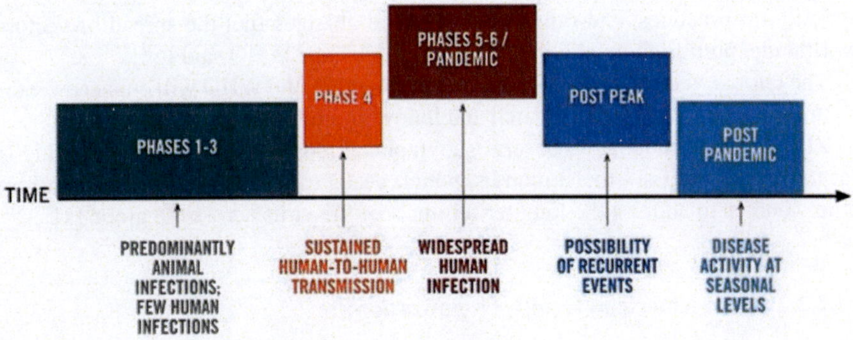

Fig. 17.1 WHO pandemic influenza phases (WHO 2009 Pandemic Influenza Preparedness and Response: A WHO Guidance Document. https://www.ncbi.nlm.nih.gov/books/NBK143061/. Accessed July 2, 2022)

The reality is the problem is more serious because it is now endemic, setting into a large group of countries. This virus is likely to stay around." [63].

Dr. Anthony S. Fauci, Director of the National Institute for Allergy and Infectious Diseases, funding efforts to find a ZIKV vaccine, suggested that it was premature to lift the state of emergency since summer is just beginning in the Southern Hemisphere. Other experts, like Dr. Fauci, were more critical. The WHO decision is "unwise," said Dr. Lawrence O. Gostin, Georgetown University's O'Neill Institute for National and Global Health Law Director [64].

[O]ther experts worried that the WHO's declaration might slow the international response to an epidemic still spreading and lull people at risk into thinking they were safe. Even if the outbreak no longer meets the technical definition of an emergency under 2005 international health regulations, there is an important psychological component to declaring an emergency [64].

But some experts worry that the change could result in less funding for ZIKV research. "I think WHO's decision is unwise," Lawrence Gostin, Global Health Law Expert at Georgetown University, told Reuters. "Although ZIKV's spread has waned, it still holds the potential for an explosive epidemic. If it were to reemerge in the Americas or jump to another part of the world, it would significantly threaten a new generation of children born with disabilities such as microcephaly." [65].

Peter Salama, Executive Director of WHO's health emergencies program, explained. "By placing it as a longer-term program of work, we are saying ZIKV is here to stay, and WHO's response is here to stay." [66]. That does not mean concern over ZIKV is over, but now that a link between ZIKV and microcephaly has been established, it is viewed as a long-term problem requiring constant attention [67].

The Global Vector Control Response (GVCR) 2017–2030 outlines several goals and activities to reduce mortality from vector-borne diseases by at least 75% and incidence by at least 60% by 2030 and prevent epidemics worldwide. The WHO said approximately 80% of the world's population is at risk for these diseases. "Rapid unplanned urbanization, massive increases in international travel and trade, altered agricultural practices, and other environmental changes fuel the spread of vectors worldwide, putting more and more people at risk," the WHO said [68].

The Global Vector Control Response developed by the WHO will enhance vector control capacity and foster research and innovation in this field [69].

ZIKV is now endemic in several geographical locations. Emergency funding is set to run out after six to 12 months, which needs to pivot to allocate funds from other sources to address the long-term impact of the virus across the globe [1].

17.2.2.2 Pan American Health Organization

PAHO is the Pan American Health Organization. On December 1, 2015, it issued a ZIKV epidemiological alert [70]. PAHO took the lead in informing the world about ZIKV on its website and published the first travel advisory.

PAHO reports that ZIKV transmission had occurred in over forty-eight countries or territories in the Americas, with 177,614 confirmed cases of locally acquired ZIKV in the region, over half a million suspected cases, and 2525 confirmed congenital syndrome-associated cases with ZIKV infections as of December 29, 2016 [67]. According to PAHO's latest update, five countries in the Americas have reported sexually transmitted ZIKV cases.

Globally, the risk assessment of ZIKV has not changed, and the virus continues to spread geographically to areas where competent vectors are present, PAHO said [71]. PAHO said experts now consider ZIKV a long-term public health challenge, following the declaration by WHO's Emergency Committee on Zika that the epidemic's emergency phase was over [71].

PAHO helps countries in the Americas by implementing country-specific vector control communication strategies to engage individuals and communities in mosquito control campaigns, launching a regional campaign against mosquitoes called Mosquito Awareness Week.

PAHO recommends the following for more effective risk communication about ZIKV. It may be essential to note that public health providers will be competing with rumors, false truths, and media amplification.

- Communicate information about ZIKV promptly, addressing the population's public health concerns and need for information about the potential complications of this disease. Prepare and evaluate the key messages.
- Segment audiences to emphasize the risks of ZIKV to the more vulnerable groups at higher risk—in this case, women of reproductive age, pregnant women, and health workers.
- Continue promoting individual behavioral changes, social mobilization, and community engagement to control the vector and eliminate breeding sites in dwellings.
- Keep the public fully informed (draft a transparency policy) about the risk of ZIKV infection, state what is known, and describe the efforts of health institutions and

the international community to respond to this health emergency, as well as the research and other action being taken to learn more about this new disease.

- Maintain the credibility of health institutions and public confidence by disseminating accurate, evidence-based information. The timely, transparent dissemination of accurate and accessible evidence-based information about ZIKV infection gives the public confidence in the action taken by the health authorities.
- Set up a system for monitoring public speculation and conjecture to dispel rumors and refute inaccurate information and misconceptions as quickly as possible.
- Quickly respond to the public's concerns and specific information needs, partners and allies, healthcare providers, and the public health community.
- Adopt a consistent and uniform government (national) approach to strategic and operational communications. Include partners and allies from the non-governmental sector (NGOs, private enterprises, and the community).
- Set up a system for ensuring the consistency of the messages conveyed by personnel from the national government, hospitals, and the rest of the local health authorities.
- Establish mechanisms to monitor communication effectiveness and methods for understanding the public's attitudes and motivations [52].

Educating communities and empowering women on how to prevent ZIKV transmission is the focus of a new collaborative effort by the Centers for Disease Control and Prevention (CDC), the Pan American Health Organization (PAHO), and the CDC Foundation, aimed mainly at pregnant women in U.S. territories and the Americas. The Bill & Melinda Gates Foundation supports these efforts, including a comprehensive health campaign on ZIKV prevention, surveys on risk perception and knowledge gaps in the Americas, and community engagement on mosquito control, primarily to protect pregnant women from ZIKV. [72]

PAHO, with support from partners, will also conduct Knowledge, Attitudes, and Practice (KAP) surveys in Latin America and the Caribbean to measure risk perceptions and address the knowledge gaps related to ZIKA. The countries' risk communication strategies will also be strengthened and operationalized using a PAHO-designed model and training [72].

17.2.2.3 Private Concerns

There are hundreds of examples. Here is one. The National Association of Chain Drug Stores (NACDS) Foundation announced that it would be rolling out a radio and online ZIKV prevention campaign on May 9, 2016.

The initiative aims to encourage women to speak to physicians and pharmacists about protecting themselves and their unborn children from the effects of ZIKV, including microcephaly in children or pregnancy loss [73]. Its work was noted in Puerto Rico.

"I appreciate this meaningful work to help the people of Puerto Rico address a public health situation facing this generation and generations to come," Puerto Rico

Governor Luis Fortuño said. "At times like this, the needs are great, and raising awareness is among those critical needs. I applaud the philanthropic and private entities playing a vital role in this emerging public health threat, like the NACDS Foundation's campaign that will help get the word out about prevention." [73].

17.2.2.4 ZIKAlliance

The ZIKAlliance was funded through the European Union's Horizon 2020 Program. The project involves fifty-two partners from eighteen countries. It began on October 1, 2016, and is still working. Coordinated out of Paris, France by the Institut National de la Sante et de la Recherche Medicale (INSERM). The EU grant was just under twelve million Euros.

> ZIKAlliance is a multidisciplinary project with a global "One Health" approach, built: on a multi-centric network of clinical cohorts in the Caribbean, Central & South America; research sites in countries where the virus has been or is currently circulating (Africa, Asia, Polynesia) or at risk for emergence (Reunion Island); a strong network of European and Brazilian clinical & basic research institutions; and multiple interfaces with other scientific and public health programs.
>
> ZIKAlliance will address three key objectives relating to (i) the impact of ZIKV (ZIKV) infection during pregnancy and short & medium-term effects on newborns, (ii) associated natural history of ZIKV infection in humans and their environment in the context of other circulating arboviruses and (iii) building the overall capacity for preparedness research for future epidemic threats in Latin America & the Caribbean. The project will utilize large standardized clinical cohorts of pregnant women and febrile patients in Latin America and the Caribbean, where the virus is circulating, expanding a preexisting network established by the IDAMS EU project. I will also benefit from solid expertise in basic and environmental sciences, with access to fieldwork and sophisticated technological infrastructures to characterize virus replication and physiopathology mechanisms.
>
> To meet its three key objectives, the scientific project has been organized into nine work packages, with WP2/3 dedicated to clinical research (cohorts, clinical biology, epidemiology & modeling), WP3/4 to basic research (virology & antivirals, pathophysiology & animal models), WP5/6 to environmental research (animal reservoirs, vectors & vector control), WP7/8 to social sciences & communication, and WP9 to management. The broad consortium set-up allows gathering the necessary expertise for a fundamentally interdisciplinary approach and operating in various countries with contrasting ZIKV epidemiological status. [74]

Their support for research remains impressive. For example, in mid-January 2017, the ZIKAlliance consortium made the news by revealing that the Gas6-AXL pathway mediated the ability of the ZIKV to infect glial cells in developing brains. They found that Aravive Biologics' Anti-AXL Candidate could play an antiviral role in addition to its previously reported anticancer activity [75].

This approach has been heavily disputed, however. Wells et al. [76] indicated that AXL is not essential for ZIKV infection of human neural progenitor cells (NPCs) and suggest that therapeutic inhibition of AXL alone would likely not be sufficient to mitigate ZIKV pathogenesis in developing brain tissue. Other studies have recently questioned the role of AXL in ZIKV infection of neural cell types. These findings challenge the utility of AXL inhibitors for preventing congenital disabilities after infection and suggest that further studies of viral attachment factors in NPCs are needed [76].

Nevertheless, ZIKAlliance has supported some important research. An important transfusion study was done in Martinique in 2016, evidencing ZIKV in blood supplies [77].

17.3 The Themes

Public health budgets are highly vulnerable to cuts. They wax and wane depending on events and ideological considerations. For example, Joseph McCormick, Regional Dean of the University of Texas School of Public Health, said severe budget cuts to medical research proposed by the Trump administration were worrying. "More resources will have to come, but it's not clear it will come from this administration," he said. "That part of it's bleak." [78].

Federal funding for ZIKV has mostly run out, with its future unclear. And new cases are already popping up in Florida. State lawmakers are considering a budget that includes money for research and additional state epidemiologists. But federal money that was supposed to last five years will likely run out this summer.

A recent analysis from the U.S. CDC found the risk to pregnant women is much greater than researchers initially thought. Its key finding: Last year, one in ten pregnant women with confirmed ZIKV infections in the United States had a baby or fetus with severe congenital disabilities [79].

For example, the House of Representatives passed the Zika Vector Control Act on May 24, 2016. The bill uses the threat of ZIKV as a cover for rolling back crucial EPA regulations that protect bodies of water from pesticides. "The reality is that the majority has been pushing this legislation for years under whatever name is convenient at the time," Rep. Raul Ruiz (D-CA) said during the floor debate. "This bill has nothing to do with combating ZIKV." Vector control agencies already have the authority to apply pesticides in emergencies to prevent the spread of infectious diseases without applying for a permit. If the bill becomes law, users can discharge pesticides into bodies of water without first applying for a license with the EPA, and users will not have to tell the EPA if their pesticides end up in bodies of water [80].

17.3.1 Poor Persons of Color Disease?

ZIKV ran havoc among the poorest areas of Brazil and Latin America. In the WHO warning, the phrase "Avoid visiting impoverished or overcrowded areas" was used.

When asked why "impoverished or overcrowded" areas were singled out, the *Ae. aegypti* mosquito that carries the ZIKV likes to breed in standing water that collects in discarded tires and flowerpots, says Aylward of WHO. "And it is in these poorer or impoverished areas where you are more likely to find still water or standing water that has been collected. So, again it is just looking at what factors drive the higher probability of getting bitten by mosquitoes." He believes the probability will be higher in those poorer areas [81].

Most of the population groups affected by ZIKV are of lower socioeconomic status. That access and inequality differ markedly across the rural–urban divide; these scores indicate a clear need to prioritize differentiated levels of access and support.

This is true in the U.S. as well. For example, "[m]ore than 1.3 million people live in the Rio Grande Valley, many in deprived neighborhoods known as colonias, where conditions are ripe for mosquitoes to breed: sprawling settlements limit the effectiveness of spraying, standing water is common, and many houses lack window screens or air-conditioning.... [M]any people do not have the equipment to cut their grass, which could hide breeding pools, and that the streets lack proper drainage." [78].

17.3.2 *Women's Disease?*

AIDS is/was a gay disease! When the press and health officials called the outbreak of an infectious illness later to be called AIDS, they selected a population for blame and responsibility. Was something happening other than reporting a specific class of victims, or was their choice of framing being HIV positive or having full-blown AIDS, blaming them, and isolating them as the cause? Were they to blame for this outbreak and those outside the class blameless?

"Women should avoid pregnancy" has been the mantra of choice from government health officials in their advice to the public. This is not the best advice to give to those most vulnerable individuals. The ZIKV outbreak has revealed the "conspicuous invisibility of women" in outbreak response.

With that came a second, gender-based war against microcephaly, a congenital disability associated with the ZIKV. In this battle, the burden of responsibility was put on women, who were expected to adopt preventive measures and avoid pregnancy [82].

As for the gender dimension, in Brazil, as elsewhere in South America, women have limited access to contraception, abortion is illegal, and rates of sexual violence are high. In the poorer northeastern states, there are also much higher levels of young pregnancy, much lower levels of education, and far fewer women with jobs than in the rest of the country. The so-called war on Zika fails to address these social factors. The most vulnerable groups who suffer from ZIKV impact have been historically excluded from fair wages and decent living conditions. They are not free to make their own decisions about sexual health. They do not have access to decent quality public education or health services. They do not enjoy freedom from violence [82].

U.N. High Commissioner for Human Rights, Zeid Ra'ad Al Hussein, and the WHO reinforced the importance of women's human rights in national responses to the ZIKV outbreak. Nevertheless, ZIKV gave way to a rollout of official declarations from ministries of health across Latin America telling women not to get pregnant. These echoed narrowly conceived and often ineffective public health campaigns of an earlier era concerning HIV/AIDS and abstinence in Africa [83].

Many critics warn a directive has been unleashed that has all the nuttiness of a vicious diatribe from a hate-filled patriarch. *Don't get pregnant*. In countries where contraception and abortion are banned, that translates as *Don't have sex.* Is this the best science has to offer? Is this a death wish in the springtime of life? How is this different from the medieval misogyny women live with every day? *Hide them, veil them, gag them, burn them, crush them in the egg* [70].

These "new" diseases put an additional strain on women, who carry the burden of the fear of the conditions during their pregnancies and often care for the potentially sick children. Women and children are disproportionately affected by the current ZIKV outbreak. Health systems need to be ready to respond to their specific health needs and rights by listening to their concerns, ensuring their autonomy, and involving them in the measures that affect them [84].

Five Latin American and Caribbean countries have advised pregnancy delays for different periods. Officials in Ecuador recently did the same. Colombia's Health Minister, Alejandro Gaviria, urged women to wait six to eight months. Jamaica's Health Ministry suggested waiting up to a year. And El Salvador asked women to wait until 2018 [85].

In Brazil, Dr. Claudio Maierovitch, Head of surveillance for the Health Ministry, suggested in December that women in the most ZIKV-infested areas postpone having children indefinitely [85]. Diniz indicated that 56% of the participating women in Brazil in the survey sample had tried to avoid getting pregnant since the epidemic began because they feared the consequences of ZIKV. This trend was more prevalent among women in the Northeast of Brazil (66%) than in women in the South (46%), which the Diniz et al. say potentially reflects the fact the epidemic has hit northern regions harder. Another concern is that [w]omen seem afraid of getting pregnant and try to avoid it as much as possible. There is concern that this will negatively impact their mental health," Lead Author Debora Diniz told IBTimes UK [86].

One direct impact on women is the restriction on sexual and reproductive rights. Women still lack access to education about reproduction and contraceptives, resulting in unwanted pregnancies. Then, despite Brazil proclaiming a right to healthcare, poor women (especially) are often unable to access adequate prenatal care and testing, let alone abortion, which is illegal [87].

Advocates for women mocked it as unrealistic, disconnected from the complex lives of women in a part of the world where contraception can be hard to obtain an abortion is often illegal [85]. Clandestine abortions in Brazil remain too necessary, resulting in needless deaths—the fourth leading cause of maternal death, per HRW. Since 2005, about 17% of these abortion-related deaths were in young girls and women only 10–19 years old [87]. Many single-parent families have been affected, mostly headed by women.

In ZIKV-affected countries, there are also high rates of sexual violence, elusive contraception, teen pregnancies, and lack of sexual education. According to a study published by the Guttmacher Institute in 2014, 56% of pregnancies in Latin America and the Caribbean are unintended because of a lack of access to contraceptives or gender violence [83]. "It is good to say women avoid pregnancy, but how do they do it? Northeastern Brazil lacks basic healthcare facilities, and most women do not receive proper gynecologic follow-ups. Contraceptives are not widely available and often ineffective," Aline Philibert, Researcher at Paris Descartes University on ZIKV and sexual violence, told IBTimes UK [88].

However, in the areas most affected by ZIKV—the north and northeastern regions of Brazil—women are less likely to have access to contraceptives, less likely to be using a contraceptive method that works, and less likely to have access to necessary medical care. The women in these regions are among the poorest in the country, with unmet needs for water, sanitation, and education and denied sexual and reproductive rights. As with other pandemics, there is the potential for the impact of the ZIKV to fall most heavily on the most disadvantaged members of society [89].

> According to the Guttmacher Institute, a research organization, many married women in some affected countries practice "modern methods" of birth control. In Colombia, the figure is 73%; in Brazil and the Dominican Republic, 70%; in El Salvador and Paraguay, 61%; in Ecuador, and 58%. However, the rates are lower in other countries, and the challenges will be more significant. In Guatemala, it is 34%; in Bolivia, 32%; in Haiti, and a mere 24%. [85]

Currently, women of color from Latin America have been suffering the most. But shifting responsibility for the disease to women's behavior isolates it from other socioeconomic factors that influence its transmissions, such as sanitation or environmental issues [83].

They have limited access to family planning and abortion services. Aborting a microcephalic child is unlikely because abortion services are unavailable, and where available, abortion would be the last term and illegal abortion. They live in neighborhoods that are often cluttered with detritus that allow water to gather in which mosquitoes and homes without windows or screens make them highly vulnerable to mosquito intrusion into the house.

In addition, there has also been increasing recognition of the disproportionate impact of neglected tropical diseases (NTDs) on the health of girls and women [90]. This relationship has been based on their long-standing adverse effects on maternal and child health and human productivity and labor.

Human Rights Watch states, "More than one-third of Brazil's population lacks access to a continuous water supply." So, women store water in containers that might become breeding grounds for the mosquitoes. Poor sanitation leaves standing water and sewage as breeding grounds for the mozzies [87].

Education and access to contraceptives are critically important. Beatriz Galli, Senior Latin America Policy advisor for Ipas, a reproductive rights NGO, explained that an underlying problem was focusing on the mosquito and fumigation rather than reproductive risks and prevention. As appears universally true, poor, disadvantaged women are disproportionately affected. In Brazil, it is by getting infected with ZIKV and having babies with severe neurologic defects and then not being able to access care for their children. She concludes, "These are not families that have a voice." [87].

In recent months, this point has been noted about ZIKV by UN High Commissioner for Human Rights, who has said that upholding women's human rights is an essential element of an effective response to the outbreak. The United Nations Population Fund has also pointed out that sexual and reproductive health services must be included in response to ZIKV. Yet to date, the focus seems to be on advising women on preventive measures such as practicing safe sex, avoiding pregnancy, and avoiding mosquito bites [89].

Statistically, women's experience in the ZIKV outbreak in South America has differed from men's. Women have been, and continue to be, disproportionately affected by both attacks. The dramatic drop in primary healthcare services during the Ebola outbreak heavily affects women and children. Furthermore, as other complex emergencies have shown, both episodes have illustrated that women are more likely to experience social and economic deprivation and limited access to resources [89].

There is a strong association and a strong correlation between sex and incidence. Overall, females had a 75% higher reported incidence rate of ZIKV disease than males (rate ratio, 1.75; 95% confidence interval [CI], 1.71–1.79); the speed was exceptionally high among women 20–49 years of age. This difference was also observed in the Yap Island (Micronesia) epidemic. It could be due to greater exposure to the intra-domiciliary mosquito vector, more severe symptoms among women in this age group, active healthcare-seeking behavior by females, or enhanced reporting by health workers, given the risk of infection during pregnancy [91].

The prevalence of ZIKV infection in countries where the virus is endemic is unknown, but seroprevalence surveys suggest less than 1% to more than 50%. Currently, there is a lack of adequate information on the percentage of reproductive-age women immune before pregnancy. Conversely, the age of susceptible pregnant women and the rate of women infected during pregnancy in these settings where the virus is endemic [92].

Recent experience with COVID communication in the USA and abroad made a robust case for improved communication among stakeholders during a pandemic.

17.4 Approaches

Communication studies have welcomed a few different theories to help explain why some risks seem more critical than others, even though the morbidity and mortality footprint does not match the public's rankings. For example, it is common knowledge the public has safety reservations about flying and prefers driving. The public perceives driving on the road as less safe than flying. The literature suggests that airplane crash events received much more media coverage than recurring automobile accidents [93]. Gigerenzer empirically established this phenomenon in 2004 [94]. His research identified a substantial increase in U.S. traffic fatalities after the September 11 terrorist attacks due to a substitution of driving for flying induced by fear of dread risks. Eight years later, he validated his findings [95].

17.4.1 Social Amplification of Risk Framework

Kasperson argued years ago that news items are often amplified through what he and his colleagues called "amplification stations." Three of these stations included government bureaucrats, mass media, and professional information brokers (a term that provides for non-governmental organizations and public health nonprofits). Over the last two decades, the mass media has been joined by digital media: Mass media participating within the digital world and social media platforms where an entirely new population of journalists have entered debates over social and political issues.

Once accepting that there is no objective standard to indicate when risk amplification occurs, actors are likely to correct for other actors' apparent risk amplifications and attenuation instead of simply correcting their own risk beliefs. Beliefs: This can have an intensely polarizing effect on risk beliefs and can produce residual worry and loss of demand for associated products and services after a crisis has passed [96].

As mentioned in the introduction, the ZIKV crisis has been hyperbolized. Indeed, hyperbole is a dominant trope in all communication over pandemics. The news outlets latched onto this hot item and bombarded the public with information that is sometimes legitimate, sometimes misleading, and sometimes some. The Internet age has allowed rapid dissemination of such information, causing fast reactions or responses by national and international agencies, regulatory bodies, and professional organizations [97].

Zoonosis outbreaks are also likely to produce the secondary economic and social effects associated with ideas of risk amplification. Amplification is what one actor thinks another is doing, rather than an objective statement about the social world. Busby and Duckett (2012) offer this explanation.

Zoonotic diseases of the kind seen in recently highly publicized events provide an excellent context for this idea of social risk amplification as an attribution. Risks to people are bound up with risks to animals; risks to animals are divided out among wild animals, animals kept as pets, and animals being reared for food; risks, as seen by producers, are interwoven with risks as seen by agricultural regulators and risks as seen by consumers are incorporated with risks as seen by food regulators. Thus, a wide range of social actors has a stake in how zoonoses are responded to and have a range of beliefs and interests that are sometimes in conflict and sometimes held in common [98].

Once a spiral of social amplification occurs, then we become hyperbolic. The amplifications associate with and then transform more social problems generating outrage [99].

Many exaggerated and unfounded reports about medical discoveries and treatment regimens provide misinformation and false promises that lead to false hopes, expanding medical tourism, online predatory marketing, and other unscrupulous practices. Furthermore, fake medical journals that provide hype, exaggerations, and disinformation are growing [100].

17.4.2 Branding

Branding is a separate way to assign meaning to artifacts and associated ideas. Branding is a framing-like effect that has been used primarily by opponents and amplified by media. It is mainly labeling. Sometimes the labels are positive and sometimes negative. GMO-free is a negative label. It has dominated branding discussions over processed foods [101].

Many genetic modifications bring mutant bugs and Frankenstein's monster to mind. "There is a hysterical response by many people, and there is even sometimes a religious response. Some people feel like that step on the toes." [102].

A recent Miami Herald article has referred to the mosquitoes in the proposed Oxitec experiment as "Franken-skeeters." [103]. AMI Newswire Reporter Michael Carroll also used "Frankenskeeters" to refer to Oxitec's GM mosquitoes [104]. Another concerned reader of the Key West Blue Paper claimed that the 3% of GE mosquitoes that remain viable after release amounts to 30,000 viable Frankenskeeter reproducers per million [105]. In late August, the Key West Blue Paper referred to Oxitec as a sorcerer's apprentice and included "a Frankenstein mosquito" among their wares [106].

Frankenskeeters are a brand.

17.4.3 Framing

To the opposition, successfully blocking the intervention in Key Haven is proof of what they already suspected: Residents do not want these experiments taking place in their neighborhoods. Empowered by the referendum results, the opposition uses this information to bolster their legitimacy and derail the project entirely. They attempt to control how the arguments are framed to control the public discourse [107].

Framing is a media theory, and branding is drawn from marketing. While they are related, they are different. Framing refers to a theme that dominates how something is described and understood. Branding is associated with an artifact, an idea, or a commercial product.

Shih et al. [108] summarized the concept nicely. Entman's [109] widely cited definition stated, "to frame is to select some aspects of a perceived reality and make them more salient in a communicating text." [108].

> Frames, then, define problems - determine what a causal agent is doing with what costs and benefits, usually measured in terms of shared cultural values, diagnose causes - identify the forces creating the problem; making moral judgments - evaluating causal agents and their effects; and suggest remedies - offer and justify treatments for the problems and predict their likely impact. [109]

Gitlin [110] conceptualized frames as "the principles of selection, emphasis, and presentation composed of little tacit theories about what exists, what happens, and what matters." Framing is a way to draw attention to certain features of an issue while minimizing attention to others.

The different frames embedded in the media are important because they are performative, simultaneously specifying appropriate actions and ultimately reflecting policy agendas and the concerns and interests of other groups [111]. Therefore, the prominence of specific frames has implications for public health policy and a general understanding of (and responses to) epidemics [112]. Gislason made the case that frames serve as a launching point of resource mobilization in their study of the West Nile Virus. The "war" frames mask socioeconomic inequalities.

Framing public health issues in clever but deceptive ways is a successful tactic against implementing public health interventions. Genetically modified mosquitoes and soda bans illustrate the importance of studying how opposition groups frame their arguments [107]. Ribeiro et al. argue that eradicating mosquitos and taking preventative measures are only a few possible strategies to consider when coping with this and other disease outbreaks. One of the most important lessons taken from the case of ZIKV is that the absence of socioeconomic equality is also an underlying factor of disease emergence, and it may have the potential to influence its eradication, as has been the case in other mosquito-borne diseases [113].

Framing is a political tool interested parties use to establish a valence toward a subject. Frames are designed by those empowered to communicate and situate an issue within the public sphere. The public sphere is a quasi-political artifact traditionally associated with those people who contributed to political and policy decision-making, primarily white male landowners who read pamphlets and what newspapers happened to exist during the eighteenth and nineteenth before heading to a pub or town hall to discuss who and what to support as policy. More people were allowed into the meetings, politicking became more standardized, and the public sphere became actuated every two or four years, especially in the U.S. More newspapers surfaced, television became a commodity found in every home, the World Wide Web introduced the Internet, and the public sphere became more symbolic. Economic culture remains a powerful component of the public sphere. People participate as consumers of things, especially information. Framing things genetically engineered under the theme of the unknown and the unknown as inherently suspicious, if not frightening, has been an approach by actual opponents, the apprehensive, the self-interested, and the purveyors of public understanding, from media to government. While framers often use metaphors to describe their frames, such as Pandora's box and Frankenstein's monster, the theme is runaway science. The frame is "runaway science."

Traditionally, phenomena like pandemic influenza have been framed as a security issue, threatening the functioning of both state and society [114]. Intensive global attention to infectious diseases may require the pandemic to threaten global health security. Securitizing pandemic lifts it from the quirky public health area to the higher national and international security politics [115]. "It is by labeling something a security issue that becomes one." [116]. The term "threat" can be used in a securitizing move and is extensively used about pandemic influenza [117].

Successfully labeling an issue as "security" takes it beyond everyday political discourse and allows exceptional actions to be undertaken [118]. Virulent influenza may be more devastating to human life than war itself [119]. In 2016, the Commission on a Global Health Risk Frame for the Future released its report on infectious disease and global security.

> Framed as an issue of human security, the current level of investment in countering this threat to human lives looks even more inadequate. Very few threats can compare with infectious diseases in terms of their potential to result in catastrophic loss of life.
>
> The dynamics of infectious disease and the actions taken to counteract it can cause immense damage to societies and economies. And in a globalized, media-connected world, national borders are no barriers to real or perceived threats. Fears, whether rational or unwarranted, spread even more quickly than infections. Moreover, while economic or financial problems in fragile or failed states pose minimal direct risk to the rest of the world, infectious disease outbreaks in such states represent an immediate threat. The lack of health care and public health capacity in these countries is both a disaster for their populations and an acute vulnerability for the world. [120]

Ribeiro undertook a specific framing study on ZIKV. The analysis of 186 articles published in Brazil between December 2015 and May 2016 reveals a dominant "war" frame supported by two subframes: One focused on eradicating the vector (mosquito) and another on controlling microcephaly, placing the burden of prevention on women [113].

Abraham (2011) states that national and global security has a "high politics." This global mobilization through securitization can be deemed necessary because pandemics emerge from elsewhere, probably in Southeast Asia, and threaten the United States and the rest of the world [121].

The opponents of the Oxitec test in Key Haven have as their primary argument the potential and unknown environmental impact of the novel genetic technology being used in this project. Moreover, they believe that supporters of the genetically modified mosquito intervention manipulate public opinion through the ZIKV epidemic, town hall meetings, and public surveys on the issue [107].

The framing of ZIKV as a risk by government officials, the media, and other amplification stations [122] has been noteworthy. An amplification station takes a message, in this case, an outbreak message, and crafts around it an amplified, often exaggerated, set of letters within which a version of the original outbreak message may be found. Fear is a powerful driver though it is not universally powerful. Under certain circumstances, it is ineffective, if not counterproductive [123].

The primary amplification station is the media: traditional radio, TV news, newspapers, magazines, documentaries, commercial film, and digital media, with its news reporting more personalized and social. Government officials at all levels, personalities, and interest groups are an alternative amplification station for health-related communication. Together these amplification stations set public agendas and prime and frame outbreaks to a broad class of stakeholders. The media employs exaggeration and hyperbole often to increase readership and viewership, turning nearly every episode into a pandemic. Severe acute respiratory syndrome (SARS) in 2002, Avian flu in 2005, and multidrug resistant tuberculosis (MDR-TB) in 2012. Some feel it is happening again; the ZIKV implicates genetic engineering this time.

17.4.4 *Risk Fatigue*

The ZIKV posed its unique challenges to the American public. Research from Nowak et al. then from the University of Georgia suggested that only one in three people in an October 2016 nationally representative survey said they would be willing to get a ZIKV shot if available and recommended.

Nowak said that formulating a recommendation for a mosquito-transmitted disease like ZIKV would likely be difficult because almost all currently recommended vaccines are for diseases primarily transmitted from person to person [124].

An Associated Press-NORC Center for Public Affairs Research poll found critical gaps in general knowledge of the virus and its risks. One woman told the AP she had grown "numb" to outbreak warnings over the years. At the same time, she does not plan to travel during her pregnancy. She commented. "Once people realize it's a problem, there's usually a quick response, so I'm not worried." [125].

"The word 'new' in front of a vaccine doesn't work as well as when you put 'new' in front of laundry detergent," Glen Nowack, Director of the Center for Health and Risk Communication at UGA, said. "Many people interpret 'new' consumer products as better and improved things, thus worth trying. When you put 'new' in front of 'vaccine,' people think experimental or have not enough experience, and they take a 'wait and see approach.'" [124].

The public is experiencing risk fatigue. And this sense of ZIKV fatigue is pervasive [126].

17.5 Research on Controlling the Messaging

Effective management of new epidemic infectious disease risks in the phase that no treatment or vaccination is yet possible is mainly dependent on the precautionary behavior of the population. Implementation of precautionary behavior largely depends on effective risk communication, i.e., communication that induces realistic risk perceptions, correct knowledge, and skills to promote and enable precautionary practices, [127] like restraining from travel to some locations and taking special precautions for those who are pregnant or intending to become pregnant.

Risk perceptions are a necessary but often not sufficient condition for engaging in such behaviors. Therefore, higher risk perceptions may only predict protective behavior when people believe practical defensive actions are available (response efficacy) and are confident that they can engage in such protective acts (self-efficacy) [127].

Compliant message receivers need to be able to control their travel with the capabilities or capacity to do that. Asking someone to delay their pregnancy or engage in specific precautionary behavior, such as the use of condoms, requires the complaint messages can do that. In some cultures, that decision-making is not the women alone.

Asking women to be responsible for a situation that usually goes beyond their control is unacceptable. The only productive way to avoid getting pregnant is by not having sexual relations. Advising women to refrain from pregnancy for two years is burdensome and unrealistic (especially if they are in a relationship). Yet even this may not suffice given the conditions presented above, such as the high incidence of rapes and no emergency contraception [128].

The CDC has begun awarding grants to universities and state and local governments for ZIKV-related research. One of their priorities is training to address a shortage of workers in public health entomology and impending retirement among mosquito control managers.

For example, the Southeast Regional Center of Excellence in Vector-Borne Disease's Gateway Program at the University of Florida was funded by a $10 million grant from the U.S. Centers for Disease Control and Prevention. Led by Disease Expert Rhoel Dinglasin, researchers from the University of Miami, Florida, International University, and the University of South Florida also will participate [129].

17.6 Amplification Stations

Take the case of an NPR report that reported that federal funds to combat ZIKV had finally been authorized into a Senate bill. Given an opportunity to write the percentage of babies born to women infected during pregnancy, the news reported both use the upper limit of susceptibility (13% rather than the range in the literature 1–13%). It failed to distinguish which stage of pregnancy the fetus is most vulnerable.

17.6.1 Government Officials

The primary amplification stations came from government officials associated with two major health organizations: the Pan American Health Organization (PAHO) and the World Health Organization (WHO). In the USA, they were amplified additionally by Fauci from the Center for Disease Control (CDC), Califf, and others from the Food and Drug Administration (FDA) and the NIH.

17.6.2 Other Interest Groups

A classic illustration of risk profile shifting occurred in New Jersey in 2016 and involved misclaims about ZIKV and Sand Dunes. Margate, New Jersey, is a shoreline community. They have tried everything to prevent protective dunes from being built along the beach, invoking lost views, wrongly seizing property rights, and damaging tourism prospects. Now that those big-picture issues have failed to kill the project, Margate uses a mosquito to make a questionable claim: The dunes will help spread the ZIKV. In a lawsuit, six homeowners sued the US Army Corps of Engineers [130].

They claim puddles would collect behind the dunes and standing water breeds mosquitoes. A coastal expert said he found it unlikely since sand drains well. In addition, the northeast has not been the breeding ground for the *Ae. aegypti* mosquitoes. Experts claim any outbreak in New Jersey is likely to be limited.

17.7 Digital Amplification

Americans, when left alone, turn to resources for information on infectious diseases. Unfortunately, there are two problems with digital media on the Internet. First, they may not even understand what they are reading online. For example, online resources about ZIKV are hardly optimate. Basch, Fera, and Garcia (2020) determined the readability of one hundred articles on the Internet relating to ZIKV. They used an online readability calculator called Readable.io to conduct the following readability tests: Coleman-Liau Index, Flesch-Kincaid Grade Level, Flesch-Kincaid Reading Ease (FRE), Gunning Fog Index (GFI), and the Simple Measure of Gobbledygook-grade level. Their findings were disheartening. The data in this study make a compelling case that websites containing information on ZIKV are being written beyond recommended levels. In addition, the findings further complicate the issue of Internet-based details in that they may be difficult to read once information is gathered.

According to the American Medical Association, National Institutes of Health, and Centers for Disease Control and Prevention, medical information for the public should be written at no higher than an eighth-grade reading level [131].

If those who access information online have difficulty reading, decisions may be compromised. It has been reported that roughly one in three Americans read at a fourth- to fifth-grade reading level [131]. According to the OECD, half of U.S. adults can't read a book written at the eighth-grade level [132]. The average literacy score of 270 (global literacy rate: 273) out of 500 puts U.S. adults at Level 2, or below-basic, literacy [131].

Secondly, messaging found on the Internet is fraught with inaccuracies of all sorts. Misinformation on health topics, particularly in social media, is abundant. Messages that are wild, sexy, or conspiracy-laden can often garner more attention than messages that can inform and possibly save lives. It is just this dichotomy between the carefully crafted, well-researched health advice and the less rigorous, specious theories that happened with the ZIKV virus epidemic.

The published literature that ZIKV has been misrepresented on social media and shared information on social media sites has been reportedly rife with rumors [133]. In addition, a social media bot can also be created that panders to spreading rumors during times of pandemic and can cause significant obstacles in dealing with the spread of disease [134].

Instead of careful research and well-crafted advice on protecting yourself from the bites of the Aedes mosquitoes, Twitter, Reddit, and Facebook enthusiasts are screaming that Oxitec is solely responsible for spreading the ZIKV virus. While scientists and public and private health agencies are scrambling to develop vaccines and targeted pest control treatments, there is a growing population that prefers to confuse the public with "reports" of intentional human population control through ZIKV virus delivery with "proof" in the form of maps with mismatched dates and locations of release and infection [14].

While amplification is easier to explain when the actors are identifiable because motive can be attributed to the amplifiers, amplification without apparent motivation has become a problematic extension of Kasperson's SARF. Some participate in online mal-information, intentional misinformation because it can be profitable for employees of a troll farm like the Internet Research Agency in St. Petersburg, Russia, implicated in the 2016 U.S. Presidential Election of Donald Trump, the 2016 Brexit campaign which opened the EU's exit doors for the United Kingdom, the 2017 French election rumors over Macron's offshore bank accounts (this time Russia was joined with U.S.-based accounts) and the 2020 Central African Republic presidential and parliamentary elections (this time France was implicated as well as Russia).

In addition, mal-information trolls can profit from advertisement revenue associated with their online accounts. For pandemics, inaccurate information can help sell snake oil cures, such as ivermectin and antibacterial. No pandemic seems to be immune from conspiracy theories and miracle cures.

In this age of digital technology, the Internet, and social media, individuals are increasingly subjected to an information and disinformation overload. This includes political and economic information and medical news, often presented as a "discovery," "miracle cure," or some other press hyperbole [135].

Once a medical report is posted on a new media platform, little can be done to retract the information thoroughly. On more than one occasion, it has been heard that "once it is out on the Internet, it is there forever." However careful and measured information authored by experts that goes out to the patient community can cause much good, the same media conduits can disseminate sensational sounding but inaccurate medical content [135].

Social media poses both opportunities and challenges in response to ZIKV. Social media sites are popular sources of health information related to pregnancy and children's health. The interactive nature of social media and its high penetration in industrialized countries may allow for more effective communication of health information than is feasible through traditional media. Social media interacts with conventional media and amplifies its impact. For example, previous research found that Ebola-related videos released online by news channels drove up Ebola-related Twitter traffic [136].

Of course, social media also has the potential to amplify unnecessary anxiety during critical times periods of infectious disease outbreaks [136]. While reading about others' personal health experiences on social media may enhance feelings of identification, and social support and, in some cases, may improve health literacy about ZIKV, distinguishing accurate information from misinformation or communications that evoke excessive fear remains a challenge [136].

17.7.1 Digital Amplification of ZIKV and Microcephaly

As more new cases of microcephaly and other congenital disabilities and GBS are reported in the Americas, the more sensationalism there is likely to be in the media: The media and social networks may be the first to announce these events. Unofficially, social networks will exponentially increase the pressure and demand for information. They will also increase the potential for inaccurate information and rumors, which will spread like wildfire [137].

An interesting study compared coverage between online and in-print amplification. Among the one hundred manually coded ZIKV-related YouTube videos, forty-three consumer-generated videos, 38 Internet-based news videos, 15 TV-based news videos, and four professional videos. Collectively, these videos were viewed nearly nine million times. Internet-based news videos and consumer videos accounted for 67.7 and 22.4% of the total. Compared with consumer-generated videos, Internet-based news videos were 6.25 times more likely to mention the impact of ZIKV impact on babies (95% CI, 1.64–23.76), 5.63 times more likely to say the number of cases in Latin America (95% CI, 1.47–21.52); and 2.56 times more likely to mention ZIKV in Africa (95% CI, 1.04–6.31). In contrast, compared with consumer-generated videos, TV-based news videos were 6.67 times more likely to express anxiety or fear of catching ZIKV (95% CI, 1.36–32.70), 7.45 times more likely to highlight the fear of ZIKV among members of the public (95% CI, 1.20–46.16), and 3.88 times more likely to discuss not becoming pregnant (95% CI, 1.13–13.25) [136].

17.7.2 Digital Amplification of GM Mosquitoes

"The internet is not helping, now that anyone can write a distorted picture of what's going on," adds Bart Knols from the University of Amsterdam, "I fear that trials with genetically modified mosquitoes could be threatened by public opposition, which could lead to politicians getting cold feet and a failure to explore potential public health benefits of such trials." [138].

The high adoption rates for Facebook extend past the United States to the regions most affected by the ZIKV. More specifically, 82% of the entire population in North America and 78% of the entire population in Latin America and the Caribbean used Facebook in 2022 [139].

To better understand the assignment of blame and perceptions of risk in new media environments, Wirz et al. [140] looked at three different facets of conversations surrounding ZIKV on Facebook and Twitter: the prominence of blame in each language, how specific groups were discussed throughout the ZIKV outbreak, and the sentiment expressed about genetically engineered (GE) mosquitoes. The team concluded social media and online communication platforms are other sources of media system variables that need to be considered within traditional categorizations." [140].

17.7.3 Fake News

In many cases, grossly exaggerated, untrue, or even fake news has infiltrated the mainstream and social media by publicizing false medical information without any accountability or concern for the safety of patients who are often grasping for straws in their desperation to find a cure. Although disinformation is not always apparent, the purveyors of fake news are motivated by a political agenda or a desire for sensationalism [135].

Fake news was used during World War II. While hardly a new phenomenon, it has received heightened attention from President Trump in response to a flurry of charges of impropriety and demonstration of suspect judgment.

There will be incomplete and inaccurate information, and rumors and misconceptions will circulate. The 2016 ZIKV epidemic was another example of misinformation on social media, including Facebook and Twitter. Unfortunately, increasing exposure to misinformation online and on social media can reinforce the incorrect beliefs of users, making misinformation challenging to eradicate [141].

People may act based on this information. Health guidelines and recommendations may change as more is known about the health impact of ZIKV—for example, the complications of congenital malformation syndrome, GBS, and modes of transmission. As a result, the information must be updated as it changes [137].

Cummings and Kong and many others have claimed that "disinformation has the capability to breed uncertainty and undermine confidence." Moreover, they added, along with Broniatowski et al.'s 2018 Twitter study [142] that "repeated message exposure increases the likelihood that fabricated information will subsequently be regarded as true, as this false content blends into and alters individuals' understanding of events." [143].

17.8 Public Understanding of Messaging

Despite most women believing ZIKV is an essential issue in their community, only one-fourth thought they could get ZIKV in Miami. The survey of eighty-five women from antenatal clinics at the University of Miami Hospital and the Jackson Memorial Hospital involved eighty-five women surveyed between January 27, 2016, and March 3, 2017.

Overall, 92.6% of women believed ZIKV was an essential issue in their community, while 85.9% reported changed behavior during pregnancy due to ZIKV. However, only 26.9% of women thought they could get ZIKV where they live, 13.9% considered moving away from Florida, and 80.8% became more cautious about travel because of ZIKV. Only 2.6% traveled to other affected countries in Central America and South America during pregnancy [144].

How the public respond to messaging is especially important; in one of the few reports on the effectiveness of ZIKV-related messaging, Abramson et al. in the NYU Zika Report provide data that indicate nearly two-thirds of women reported wearing long sleeves or using bug spray, a little under a third sprayed their home for mosquitoes. Fewer still changed their travel plans or used birth control measures to avoid the ZIKV. For the most part, these behaviors did not vary by the women's level of knowledge about ZIKV or by their sense of being personally at risk for contracting the virus. There were two exceptions: Women who believed that ZIKV could be sexually transmitted were much more likely to engage in birth control practices to avoid ZIKV infection. Women who felt at risk for ZIKV infection were more likely to change travel plans [6].

17.9 Post Epidemic Warnings

The CDC noted no areas with local mosquito-borne ZIKV transmission in the continental U.S. in 2022. Still, ZIKV transmission continues in many places, with nearly one hundred countries and territories still having an active risk of ZIKV. "For this reason, the CDC continues to urge pregnant women not to travel to areas with risk of ZIKV and recommends that men and women who travel to an area with risk of ZIKV wait before trying to conceive," the agency said in a news release [145].

Somewhat less misogynistically, the CDC now recommends that men with possible ZIKV exposure who are planning to conceive wait for at least three months after symptom onset (if symptomatic) or their last potential ZIKV exposure (if asymptomatic) before unprotected sex. Additionally, the agency now recommends that for couples who are not trying to conceive, men can consider using condoms or abstaining from sex for at least three months after symptom onset (if symptomatic) or their last possible ZIKV exposure (if asymptomatic) to minimize their risk for sexual transmission of ZIKV [145].

17.10 Conclusion

Suitable communications materials, translated into the appropriate language(s), will be vital for explaining the technology and, therefore, will underpin engagement efforts at all levels. Researchers should include experienced science communicators on their team and sociologists and linguists to help develop the necessary vocabulary to convey the technical aspects of the research accurately and understandably to each group of stakeholders [146].

There is no easy solution to the current dissemination of misleading medical information through social media. Professional organizations should establish guidelines for researchers to follow when communicating their findings to media outlets. But this may need to be taken an additional step further. All communication must be informative but also trust-building. The next chapter acknowledges the world of engagement as the new frontier in communication studies.

References

1. Caplan A (2016) How the Zika virus outbreak Foretold Donald Trump's Win. Forbes. December 5. https://www.forbes.com/sites/arthurcaplan/2016/12/05/how-the-zika-virus-outbreak-foretold-donald-trumps-win/2/#2de5c9103538. Accessed 9 Mar 2017
2. Watts J (2016) Zika hysteria is way ahead of research into virus, says expert. The Guardian. February 17. https://www.theguardian.com/world/2016/feb/17/zika-hysteria-health-expert-research-leslie-lobel. Accessed 25 Sept 2016
3. News Service of Florida (2016) Lawmakers call for use of modified mosquitoes in Zika fight. Health News Florida. August 30. http://health.wusf.usf.edu/post/lawmakers-call-use-modified-mosquitoes-zika-fight#stream/0. Accessed 5 Sept 2016
4. Barry J (2009) Pandemics: avoiding the mistakes of 1918. Nature 459:21. May 21. https://www.nature.com/articles/459324a. Accessed 27 June 2018
5. Chan M (2017) Zika: We must be ready for the long haul. Commentary by WHO Director General Dr. Margaret Chan. WHO Media Centre. February 1. http://www.who.int/mediacentre/commentaries/2017/zika-long-haul/en/. Accessed 28 Mar 2017
6. Abramson D, Pitch-Loeb R (2016) U.S. Public's perception of Zika risk: awareness, knowledge, and receptivity to public health interventions. NYU Zika Briefing Report #1. https://www.nyu-pir2.org/research. Accessed 19 June 2019
7. Berube D (ed) (2021) Pandemic communication and resilience. Springer Nature, Berlin
8. ASTHO (2015) Before the Swarm: guidelines for the emergency management of vector-borne disease outbreaks. http://www.astho.org/Programs/Environmental-Health/Natural-Environment/Before-the-Swarm/. Accessed 28 Aug 2018
9. Reynolds B (2016) Zika CERC discussion: summary: CERC transcript 08 16 2016. August 16. https://emergency.cdc.gov/cerc/zika-teleconferences.asp. Accessed 28 Aug 2018
10. Musso D, Gubler DJ (2016) Zika virus. Clin Microbiol Rev 29. May 9. http://cmr.asm.org/content/29/3/487.abstract. Accessed 18 Jul 2017
11. Infectious Disease Advisor (2017) Update on Zika virus-associated birth defects in US for 2016. Infectious Diseases Advisor. April 10. http://www.infectiousdiseaseadvisor.com/zika-virus/zika-virus-birth-defects-us-2016/article/648717/. Accessed 4 May 2017
12. Vielot NA et al (2018) United States travelers' concern about Zika infection and willingness to receive a hypothetical Zika vaccine. Am J Trop Med Hyg 98:6. June. https://www.ncbi.nlm.nih.gov/pubmed/29692314. Accessed 26 Jul 2018

13. See Sunstein CR, Kahneman D, Sibony O (2021) Noise: a flaw in human judgment. Little, Brown Spark, Hatchette Book Group, NY
14. Kadri SM, Trapp-Petty M (2016) The Zika virus fight on social media. Arch Clin Microbiol 7(3):22. https://doi.org/10.4172/1989-8436.100051
15. Commission on a Global Health Risk Framework for the Future (2016) The neglected dimension of global security: a framework to counter infectious disease crises. The National Academies Press, Washington
16. Heintze C, Barrido MV, Kroeger A (2007) What do community-based dengue control programmes achieve? A systematic review of published evaluations. Trans Royal Soc Trop Med Hyg 101. http://www.sciencedirect.com/science/article/pii/S0035920306002616. Accessed 17 Jul 2017
17. Fiske S, Taylor S (2008) Social cognition: from brains to culture. McGraw Hill, NY
18. Petty R, Cacioppo J (1986) Communication and persuasion: central and peripheral routes to attitude change. Springer, NY
19. Eagly AH, Chaiken S (1993) The psychology of attitudes. Belmont, CA: Wadsworth; and Chaiken S, Maheswaran D (1994) Heuristic processing can bias systematic processing: effects of source credibility, argument ambiguity, and task importance on attitude judgment. J Personal Soc Psychol 66:460–473
20. Kahneman D (2011) Thinking, fast and slow. NY: Farrar, Straus and Giroux; and Tversky A, Kahneman D (1974) Judgment under uncertainty: heuristics and biases. Science 185(4157):1124–1131
21. Orr M (2016) How battling new epidemics helps fight age-old killers. Xconomy. April 3. http://www.idri.org/how-battling-new-epidemics/. Accessed 27 May 2017
22. Russell RE et al (2017) A framework for modeling emerging diseases to inform management. Emerg Infect Dis 23:1. January. https://wwwnc.cdc.gov/eid/article/23/1/16-1452_article. Accessed 31 May 2017
23. Panko B (2017) To fight deadly dengue fever in humans, create dengue resistant mosquitoes. Smithsonian. January 12. http://www.smithsonianmag.com/science-nature/save-humans-dengue-fever-engineer-mosquitoes-180961792/. Accessed 28 May 2017
24. Troncoso A (2016) Zika threatens to become a huge worldwide pandemic. Asian Pacific J Trop Biomed 6:6. http://www.sciencedirect.com/science/article/pii/S2221169116302921. Accessed 4 June 2017
25. Green MS et al (2002) When is an epidemic an epidemic? Israeli Med Assoc J 4. January. https://www.ima.org.il/FilesUpload/IMAJ/0/55/27606.pdf. Accessed 18 Sept 2018
26. Collier R (2009) Revised WHO pandemic scale requires higher incidence of disease for most alert levels. Canadian Med Assoc J 180:12. June 9. https://www.ncbi.nlm.nih.gov/pmc/articles/PMC2691422/. Accessed 18 Sept 2018
27. Bonneux L, Van Damme W (2010) Preventing iatrogenic pandemics of panic. Do it in a NICE way. British Med J. June 9. https://www.bmj.com/content/340/bmj.c3065. Accessed 27 June 2018
28. Bonneux L, Van Damme W (2006) An iatrogenic pandemic of panic. British Med J. March 30. https://www.bmj.com/content/332/7544/786. Accessed 27 June 2018
29. Kowitt SD, Schmidt AM, Hannan A, Goldstein AO (2017) Awareness and trust of the FDA and CDC: results from a national sample of US adults and adolescents. PLoS ONE 12(5):e0177546. https://doi.org/10.1371/journal.pone.0177546
30. Pew Research Center (2016) Beyond distrust: how Americans view their government 2015. June 29. http://www.people-press.org/2015/11/23/beyond-distrust-how-americans-view-theirgovernment/. Accessed 26 June 2022
31. Kalichman SC, Shkembi B, Kalichman MO, Eaton LA (2021) Trust in health information sources and its associations with COVID-19 disruptions to social relationships and health services among people living with HIV. BMC Public Health 21:817. https://doi.org/10.1186/s12889-021-10856-z
32. Pollard MS, Davis LM (2021) Decline in trust in the centers for disease control and prevention during the COVID-19 pandemic. https://www.rand.org/pubs/research_reports/RRA308-12.html#fn4. Accessed 26 June 2022

33. CDC Foundation (2016) Zika risk communication, community engagement focus of new prevention efforts by CDC, CDC Foundation, PAHO in U.S. Territories and the Americas. May 18. https://www.cdcfoundation.org/pr/2016/zika-risk-communication-community-engagement-focus-new-prevention-efforts. Accessed 26 June 2022
34. CDC (2017) Zika: CDC interim response plan. May. https://www.cdc.gov/zika/public-health-partners/cdc-zika-interim-response-plan.html. Accessed 28 Aug 2018
35. Chakradhar S (2015) Buzzkill: regulatory uncertainty plagues rollout of genetically modified mosquitoes. Nature Med 21(5):416–418. May. http://www.nature.com/nm/journal/v21/n5/full/nm0515-416.html. Accessed 3 Apr 2017
36. CVM Updates (2017) FDA requests comments on documents related to certain biotechnology and mosquito related products. April 12. http://vetdxnewsfeed.blogspot.com/2017/04/april-12-2017-at-0250pm-fda-requests.html. Accessed 9 May 2017
37. US AID Press Release (2016) USAID announces $30 million grand challenge to combat Zika and future disease threats. US. AID. April 13. https://www.usaid.gov/news-information/press-releases/apr-13-2016-usaid-announces-30-million-grand-challenge-combat-zika-and-future-threats. Accessed 9 June 2017
38. Soper F (1963) The elimination of urban yellow fever in the Americas through the eradication of Aedes Aegypti. Am J Public Health. https://www.ncbi.nlm.nih.gov/pmc/articles/PMC1253857/. Accessed 2 June 2017
39. Gubler DJ (2002) Epidemic dengue/dengue hemorrhagic fever as a public health, social and economic problem in the 21st century. Trends Microbiol 10:2. February. https://www.ncbi.nlm.nih.gov/pubmed/11827812. Accessed 7 Aug 2018
40. Sullivan P (2016) Officials sound alarm on Zika funding. The Hill. April 10. http://thehill.com/policy/healthcare/275701-officials-sound-alarm-on-zika-funding. Accessed 2 June 2017
41. Constantine A (2016) Federal government awards $40 million to combat Zika in U.S. Territories. NBC News. December 14. http://www.nbcnews.com/news/asian-america/federal-government-awards-40-million-combat-zika-u-s-territories-n696011. Accessed 4 Apr 2017
42. Purcell MY (2016) NACCHO Urges congress to spend $1.8 billion for Zika virus disease. https://www.naccho.org/uploads/downloadable-resources/Support-For-Zika-Funding-2016-Press-Release-Final.pdf. Accessed 10 Sept 2018
43. Hope M (2017) Texas medicaid to Thwart Zika pregnancies among low-income homes. Breitbart. April 26. http://www.breitbart.com/texas/2017/04/26/texas-medicaid-thwart-zika-pregnancies-among-low-income-homes/. Access 14 May 2017
44. Burwell S (2016) Zika supplemental funding spend plan. Department of Health and Human Services. October 26. https://www.naccho.org/uploads/downloadable-resources/HHS-Zika-Spend-Plan-to-Congress.pdf. Accessed 27 Aug 2018
45. Squiers L et al (2018). Zika virus prevention: U.S. travelers' knowledge, risk perceptions, and behavioral intentions—a national survey. Am J Trop Med Hyg 98(6). June. https://www.ncbi.nlm.nih.gov/pubmed/29737272. Accessed 26 Jul 2018
46. Carney T, Bennett B (2014) Framing pandemic management: new governance, science or culture? Health Sociol Rev 23(2):136–147
47. Aaltola M (2012) Understanding the politics of pandemic scares: an introduction to global politosomatics. Routledge, London, England
48. Relief Web (2017) Zika virus infection step-by-step guide to risk communication and community engagement report. Relief Web. January 13. http://reliefweb.int/report/world/zika-virus-infection-step-step-guide-risk-communication-and-community-engagement. Accessed 29 May 2017
49. Chen HL, Tang RB (2016) Why Zika virus infection has become a public health concern? J Chinese Med Assoc 79:174–178. http://www.sciencedirect.com/science/article/pii/S1726490116300065. Accessed 10 Apr 2017
50. Baird C (2017) Protect yourself! Zika virus: prevention, symptoms and treatment guide. Middletown, DE: Amazon
51. Soni N (2016) A new looming of Zika virus. Asian Pacific J Reprod 5:3. April 19. http://www.sciencedirect.com/science/article/pii/S2305050016300392. Accessed 19 Jul 2017

52. PAHO (2016) Zika virus infection step-by-step guide to risk communication and community engagement. November. http://iris.paho.org/xmlui/handle/123456789/18599. Accessed 29 May 2017
53. Whittemore K et al (2017) Zika Virus knowledge among pregnant women who were in areas with active transmission. Emerg Infect Dis 23:1. January. https://wwwnc.cdc.gov/eid/article/23/1/16-1614_article. Accessed 7 Feb 2017
54. Diniz D, Medeiros M, Madeiro A (2017) Brazilian women avoiding pregnancy during Zika epidemic. J Family Plan Reprod Health Care. 43:80. https://doi.org/10.1136/jfprhc-2016-101678. http://jfprhc.bmj.com/content/43/1/80.2.full. Accessed 11 Apr 2017
55. McNeil DG (2016) Zika: the emerging epidemic. W. W. Norton & Co, NY
56. Grenoble R, Schumaker E, Almendrala A (2016) WHO declares public health emergency around Zika Virus. The Huffington Post. February 3. http://www.huffingtonpost.com/entry/world-health-org-zika-virus-emergency_us_56af781ae4b077d4fe8ec2ac. Accessed 2 Oct 2016
57. Papenfuss M (2016) Activists in Florida block release of GM mosquitoes which could suppress Zika outbreak. Int Business Times. August 18. http://www.ibtimes.co.uk/activists-florida-block-release-gm-mosquitoes-which-could-suppress-zika-outbreak-1576655. Accessed 4 Sept 2016
58. CDC (2016) Zika virus case counts in the US. http://www.cdc.gov/zika/geo/united-states.html. Accessed 5 Sept 2016
59. AFP (2017) Angola records first virus cases. News 24. January 9. http://www.news24.com/Africa/News/angola-records-first-zika-cases-20170109. Accessed 1 Feb 2017
60. Liu A (2018) Takeda's Zika vaccine candidate wins FDA fast track status. FiercePharma. January 29. https://www.fiercepharma.com/vaccines/takeda-s-zika-vaccine-gets-fda-fast-track-though-virus-no-longer-emergency. Accessed 10 Jul 2018
61. Steenhuysen J (2017) Zika vaccine shows promise in early human trial. Eyewitness News. October 5. https://www.reuters.com/article/us-health-zika-vaccine/zika-vaccine-shows-promise-in-early-human-trial-idUSKBN1C92Y0. Accessed 25 Jul 2018
62. Colli G (2017) Zika virus threat not over yet. My Statesman. January 27. http://www.mystatesman.com/news/national/zika-virus-threat-not-over-yet/MVf. Accessed 3 Apr 2017
63. Louis D (2016). Zika still on researchers' hit list; Mohave County refining response plan. Havasu News. December 17. http://www.havasunews.com/news/zika-still-on-researchers-hit-list-mohave-county-refining-response/article_26b55210-c4ce-11e6-9093-3f38d7f4cf7e.html. Accessed 14 Apr 2017
64. McNeil D (2016) Zika is no longer a global emergency, W.H.O. Says. The New York Times. November 18. https://www.nytimes.com/2016/11/19/health/who-ends-zika-global-health-emergency.html. Accessed 22 May 2017
65. Krisch J (2016) WHO declares Zika global emergency over. The Scientist. November 21. http://www.the-scientist.com/?articles.view/articleNo/47564/title/WHO-Declares-Zika-Global-Emergency-Over/. Accessed 17 May 2017
66. Condliffe J (2016) In the battle against Zika, researchers prepare for a Marathon. Technol Rev. November 21. https://www.technologyreview.com/s/602945/in-the-battle-against-zika-researchers-prepare-for-a-marathon/. Accessed 4 Apr 2017
67. Isern S (2017) Dengue virus antibodies may worsen a Zika infection. The Conversation. January 4. http://theconversation.com/dengue-virus-antibodies-may-worsen-a-zika-infection-63980. Accessed 16 May 2017
68. CIDRAP (2017) Zika Scan for Jun 01, 2017: US Zika birth defects; WHO vector control program. Press Release. June 1. http://www.cidrap.umn.edu/news-perspective/2017/06/zika-scan-jun-01-2017. Accessed 17 Jul 2017
69. Corbel V et al (2017) International workshop on insecticide resistance in vectors of arboviruses, December 2016, Rio de Janeiro, Brazil. Parasites Vectors 10:278. June 2. https://parasitesandvectors.biomedcentral.com/articles/; https://doi.org/10.1186/s13071-017-2224-3. Accessed 7 Aug 2018

70. PAHO (2015). Neurological syndrome, congenital malformations, and Zika virus infection, malformations, and implications for public health in the Americas. Epidemiological Alert. December 1. https://www.paho.org/hq/dmdocuments/2015/2015-dec-1-cha-epi-alert-zika-neuro-syndrome.pdf. Accessed March 13, 2023.
71. Jamaica Observer (2017) PAHO says Zika virus outbreak continues a year after global emergency. Jamaica Observer. February 5. http://www.jamaicaobserver.com/news/PAHO-says-Zika-virus-outbreak-continues-a-year-after-global-emergency. Accessed 17 May 2017
72. CDC Foundation (2016) Zika risk communication, community engagement focus of new prevention efforts By CDC, CDC Foundation, PAHO. In U.S. Territories and The Americas. May 18. https://www.cdcfoundation.org/pr/2016/zika-risk-communication-community-engagement-focus-new-prevention-efforts. Accessed 19 June 2019
73. Salazar D (2016) FDA unveils tighter rules on e-cigarettes, other products. Drug Store News. May 6. https://www.drugstorenews.com/pharmacy/fda-unveils-tighter-rules-e-cigarettes-other-products/. Accessed 11 Aug 2018
74. CORDIS (2017) EU: community research and development information service. Projects Res Ser. http://cordis.europa.eu/project/rcn/207455_en.html. April 19. Accessed 23 May 2017
75. Medical Xpress (2017) Martinican blood donors help shed new light on the Zika virus. Medical Xpress. February 15. https://medicalxpress.com/news/2017-02-martinican-blood-donors-zika-virus.html. Accessed 23 May 2017
76. Wells MF et al (2016) Genetic ablation of AXL does not protect human neural progenitor cells and cerebral organoids from Zika virus infection. Cell Stem Cell 19:6. December 1. https://www.ncbi.nlm.nih.gov/pubmed/27912091. Accessed 29 June 2017. and Di Guardo G, Baleeiro Beltrão Braga P, Schatzmann Peron JP (2016) Zika virus-associated brain damage: animal models and open issues. Emerg Microbes Infect 21(5):e106. https://doi.org/10.1038/emi.2016.103
77. Gallain P et al (2017) Zika virus in asymptomatic blood donors in Martinique. Blood 129:2. January 12. http://www.bloodjournal.org/content/bloodjournal/129/2/263.full.pdf. Accessed 23 May 2017
78. Dart T (2017) 'It's going to hit the poorest people': Zika outbreak feared on the Texas border. The Guardian. April 23. https://www.theguardian.com/world/2017/apr/23/zika-outbreak-rio-grande-valley-texas-border-health. Accessed 9 May 2017
79. McGrory K (2017) Will Zika return to Florida this summer? Yes, and it could be worse. Tampa Bay Times. April 27. http://www.tampabay.com/news/health/will-zika-return-to-florida-this-summer/2321851. Accessed 5 June 2017
80. Geiling N (2016) House passes Zika bill that won't fight the virus but may put more pesticides in your water. Think Progress. May 25. https://thinkprogress.org/house-passes-zika-bill-that-wont-fight-the-virus-but-may-put-more-pesticides-in-your-water-b2d0fb78c119. Accessed 5 June 2017
81. Garcia-Navarro L (2016) Is the risk of catching Zika greater in poor neighborhoods? NPR WUNC. June 16. https://www.npr.org/sections/goatsandsoda/2016/06/16/482345540/is-the-risk-of-catching-zika-greater-in-poor-neighborhoods. Accessed 26 June 2022
82. Ribeiro B, Hartley S (2018) Why Brazil's Zika virus requires a political treatment. The Conversation. February 16. http://theconversation.com/why-brazils-zika-virus-requires-a-political-treatment-91955. Accessed 10 Jul 2018
83. Riggirozzi P (2017) The campaign to eradicate Zika has trampled over women's rights. The Independent. February 11. http://www.independent.co.uk/voices/the-campaign-to-eradicate-zika-has-trampled-over-womens-rights-a7573581.html. Accessed 30 May 2017
84. OHCHR (2015) Zika virus: "Improved water and sanitation services are the best answer"—UN experts note. OHCHR.org. March 15. http://www.ohchr.org/EN/NewsEvents/Pages/DisplayNews.aspx?NewsID=17212&LangID. Accessed 21 Jul 2017
85. McNeil D (2016) Growing support among experts for Zika advice to delay pregnancy. The New York Times. February 5. https://www.nytimes.com/2016/02/09/health/zika-virus-women-pregnancy.html?_r=0. Accessed 22 May 2017

86. Surugue L (2016) How Brazilian women are avoiding pregnancy over fear of Zika. Reuters. December 23. http://uk.reuters.com/article/us-health-zika-brazil-idUKKBN14B2JO. Accessed 2 June 2017; and Diniz D, Medeiros M, Madeiro A (2017) Brazilian women avoiding pregnancy during Zika epidemic. J Family Plan Reprod Health Care 43:80. https://doi.org/10.1136/jfprhc-2016-101678. http://jfprhc.bmj.com/content/43/1/80.2.full. Accessed 11 Apr 2017
87. Stone J (2017) How Brazil's Zika epidemic women's everyday plight: Watch. Forbes. July 15. https://www.forbes.com/forbes/welcome/?toURL=https://www.forbes.com/sites/judystone/2017/07/15/how-brazils-zika-epidemic-highlights-womens-every-day-plight-human-rights-watch/&refURL=https://www.google.com/&referrer=https://www.google.com/. Accessed 21 Jul 2017
88. Surugue, Léa. (2016). How Brazilian women are avoiding pregnancy over fear of Zika. Reuters. December 23. http://uk.reuters.com/article/us-health-zika-brazil-idUKKBN14B2JO. Accessed 2 June 2017
89. Davies S, Bennett B (2016) A gendered human rights analysis of Ebola and Zika: locating gender in global health emergencies. Int Aff. August 31. https://www.chathamhouse.org/publication/ia/gendered-human-rights-analysis-ebola-and-zika-locating-gender-global-health. Accessed 11 Apr 2017
90. Hotez PJ (2013) NTDs V.2.0: "Blue Marble Health"—neglected tropical disease control and elimination in a shifting health policy landscape. PLOS Neglected Trop Dis 7:11. http://journals.plos.org/plosntds/article?id=. https://doi.org/10.1371/journal.pntd.0002570. Accessed 17 Jul 2017
91. dos Santos T et al (2016) Zika virus and the Guillain–Barré syndrome—case series from seven countries. New England J Med. August 31. https://doi.org/10.1056/NEJMc1609015. Accessed 12 Apr 2017
92. Honein M (2018) Recognizing the global impact of Zika virus infection during pregnancy. New England J Med 378. March 15. https://doi.org/10.1056/NEJMe1801398. Accessed 10 Jul 2018
93. Air Safety Week (1997) Customer assurance: which is safer, flying or driving? Air Safety Week 11(15):1. April 14
94. Gigerenzer G (2004) Dread risk, September 11, and fatal traffic accidents. Psychol Sci 15:286–287
95. Gaissmaier W, Gigerenzer G (2012) 9/11, Act II: a fine-grained analysis of regional variations in traffic fatalities in the aftermath of the terrorist attacks. Psychol Sci 23(12):1449–1454
96. Busby JS, Onggo S (2013) Managing the social amplification of risk: a simulation of interacting actors. J Oper Res Soc 64(5): 638–653. May
97. Chang C et al (2016) The Zika outbreak of the 21st century. J Autoimmunity. February 28. 68. 1–13. https://www.ncbi.nlm.nih.gov/pubmed/26925496. Accessed 27 Oct 2016
98. Busby J, Duckett D (2012) Social risk amplification as an attribution: the case of zoonotic disease outbreaks. J Risk Res 15(9):1049–1074. October
99. Vasterman PLM (2005) Media-hype: self-reinforcing news waves, journalistic standards and the construction of social problems. Eur J Commun 20(4):508–530
100. Robledo I, Jankovic J (2017) Media hype: patient and scientific perspectives on misleading medical news. Movement Disorders 32:9. https://www.ncbi.nlm.nih.gov/pubmed/28370445. Accessed 18 Sept 2018; and Salzberg S (2017) Fake Medical journals are spreading, and they are filled with bad science. Forbes, January 3. https://www.forbes.com/sites/stevensalzberg/2017/01/03/fake-medical-journals-are-spreading-and-they-are-filled-with-bad-science/#400b129930c9. Accessed 18 Sept 2018
101. Conrow J (2018) Debate over "GMO-free" branding intensifies. Alliance for Science. https://allianceforscience.cornell.edu/blog/2018/10/debate-gmo-free-branding-intensifies/. Accessed 26 June 2022
102. Editor (2016) Genetically modified mosquitoes rub keys residents the wrong way. Health News Florida. May 22. http://health.wusf.usf.edu/post/genetically-modified-mosquitos-rub-keys-residents-wrong-way#stream/0. Accessed 5 Sept 2016

103. Staletovich J (2016) Unleash GMO mosquitoes against Zika under emergency rule, Gulf Coast leaders urge. Miami Herald. August 26. http://www.miamiherald.com/news/local/environment/article98185027.html. Accessed 5 Sept 2016
104. Carroll M (2017) Florida keys voters approve "Frankenskeeters". AMI Newswire. November 17. https://americanmediainstitute.com/investigations/511046021-florida-keys-voters-approve-frankenskeeters/. Accessed 28 Mar 2017
105. JW (2016) Comment. In d'Albissin, Alexis & Girard, Arnaud. Oxitec: Lord of the mosquitoes. The Blue Paper: Key West. June 19. http://thebluepaper.com/lord-of-the-mosquitoes/. Accessed 5 Sept 2016
106. d'Albissin A, Girard A (2016) Oxitec: Lord of the mosquitoes. The Blue Paper: Key West. June 19. http://thebluepaper.com/lord-of-the-mosquitoes/. Accessed 5 Sept 2016
107. Piotrowski H (2017) What genetically modified mosquitoes and soda bans can teach us about framing health issues. Baylor College of Medicine Blog Network. March 31. https://blogs.bcm.edu/2017/03/31/framing-health-issues/. Accessed 28 May 2017
108. Shih T-J, Wijaya R, Brossard D (2008) Media coverage of public health epidemics: linking framing and issue attention cycle toward an integrated theory of print news coverage of epidemics. Mass Commun Soc 11(2):141–160
109. Entman RM (1993) Framing: toward clarification of a fractured paradigm. J Commun 43(4):51–58
110. Gitlin T (1980) The whole world is watching: mass media in the making and unmaking of the new left. University of California Press, Berkeley
111. Greer SL, Singer PM (2017) The United States confronts Ebola: suasion, executive action, and fragmentation. Health Econ Policy Law 12(1):81–104
112. Gislason MK (2013) West Nile virus: the production of a public health pandemic. Sociol Health Illn 35(2):188–199
113. Ribeiro B, Hartley S, Nerlich B, Juspal R (2018) Media coverage of the Zika crisis in Brazil: the construction of a 'war' frame that masked social and gender inequalities. Soc Sci Med 200:137–144
114. Kamradt-Scott A, McInnes C (2012) The securitisation of pandemic influenza: framing, security and public policy. Glob Public Health 7:S2. December. https://www.ncbi.nlm.nih.gov/pubmed/23039054. Accessed 18 Sept 2018
115. Fidler DP (2003) Public health and national security in the global age: infectious diseases, bioterrorism and realpolitik. The George Washington Int Law Rev 35(4):787–856. https://www.repository.law.indiana.edu/facpub/416/. Accessed 12 Sept 2018
116. Waever O (2004) 'Aberystwyth, Paris, Copenhagen: New "Schools" in Security Theory and Their Origins between Core and Periphery'. Paper presented at the International Studies Association, Montreal, March 17. https://www.scribd.com/doc/40010349/Ole-Waever-Aberystwyth-Paris-en-New-Schools-in-Security-Theory-and-Their-Origins-Between-Core-and-Periphery. Accessed 12 Sept 2018
117. Kamradt-Scott A, McInnes C (2012) The securitisation of pandemic influenza: Framing, security and public policy. Global Public Health 7:S2. December. https://www.ncbi.nlm.nih.gov/pubmed/23039054. Accessed 18 Sept 2018
118. Kamradt-Scott A, McInnes C (2012) The securitisation of pandemic influenza: framing, security and public policy. Glob Public Health 7:S2. December. https://www.ncbi.nlm.nih.gov/pubmed/23039054. Accessed 18 Sept 2018, and Buzan B, Waever O, de Wilde J (1998) Security: a new framework for analysis. Boulder, CO: Lynne Rienner
119. Kamradt-Scott A, McInnes C (2012) The securitisation of pandemic influenza: framing, security and public policy. Glob Public Health 7:S2. December. https://www.ncbi.nlm.nih.gov/pubmed/23039054. Accessed 18 Sept 2018, and Francis T (1958) Influenza. Preventive Medicine in World War II, vol IV. Washington, DC: US Government Printing Office. Available from: http://history.amedd.army.mil/booksdocs/wwii/PM4/CH04.Influenza.htm. Accessed 18 Sept 2018
120. GHRF Commission (Commission on a Global Health Risk Framework for the Future) (2016) The neglected dimension of global security: a framework to counter infectious disease crises. http://nam.edu/GHRFreport. https://doi.org/10.17226/21891

121. Abraham T (2011) The chronicle of a disease foretold: pandemic H1N1 and the construction of a global health security threat. Political Studies. 59. https://onlinelibrary.wiley.com/. https://doi.org/10.1111/j.1467-9248.2011.00925.x. Accessed 18 Sept 2018
122. Kasperson R et al (1988) The social amplification of risk: a conceptual framework. Risk Anal 8(2):177–187
123. Witte K (1992) Putting the fear back into fear appeals: reconciling the literature. Commun Monogr 59:329–349
124. Beeson L (2016) Public skepticism would likely greet a new Zika vaccine, study says. UGA News Service. December 13. http://news.uga.edu/releases/article/public-skepticism-would-likely-greet-zika-vaccine/. Accessed 7 Feb 2017
125. Neergaard L, Swanson E (2016) Poll: some key gaps in Americans' knowledge about Zika virus. AP. April 8. https://apnews.com/45724cc8e53f4c68bd7a05fa53f1bebc/poll-some-key-gaps-americans-knowledge-about-zika-virus. Accessed 19 May 2017
126. McNeil DG (2017) Houston braces for another brush with the Peril of Zika. The New York Times. July 17. https://www.nytimes.com/2017/07/17/health/zika-virus-houston-texas.html. Accessed 21 Jul 2017
127. Brug J, Aro AR, Richardus JH (2009) Risk perceptions and behaviour: towards pandemic control of emerging infectious diseases. Int J Behav Med 16:3–6
128. Luna F (2017) Public health agencies' obligations and the case of Zika. Bioethics 31:575–581
129. AP (2016) Florida researchers will lead new Zika research program. 12CBS12.com. December 26. http://miami.cbslocal.com/2016/12/26/florida-researchers-to-lead-program-on-stopping-zika-spread/. Accessed 12 Jan 2017
130. Parry W (2016) Sand dune opponents invoke Zika fears in bid to stop project. ABC News. December 11. http://abcnews.go.com/Technology/wireStory/sand-dune-opponents-invoke-zika-fears-bid-stop-44123367. Accessed 22 Dec 2016
131. Wylie Communications (2019) U.S. literacy rate: can you read me now? https://www.wyliecomm.com/2019/03/us-literacy-rate/. Accessed 13 June 2022
132. Strauss V (2016) Hiding in plain sight: the adult literacy crisis. The Washington Post. November 1. https://www.washingtonpost.com/news/answer-sheet/wp/2016/11/01/hiding-in-plain-sight-the-adult-literacy-crisis/?noredirect=on. Accessed 13 June 2022
133. Wood MJ (2017) Conspiracy theories on twitter during the 2015–2016 Zika virus outbreak. Cyberpsychol Behav Soc Netw 21(8). August 1. https://doi.org/10.1089/cyber.2017.0669; and Safarnejad L, Xu Q, Ge Y, Krishnan S et al (2020) Contrasting misinformation and real-information dissemination network structures on social media during a health emergency. Am J Public Health. 110(S3):S340–S347. October. https://doi.org/10.2105/AJPH.2020.305854; and Sharma M, Yadav K, Yadav N, Ferdinand KC (2017) Zika virus pandemic—analysis of Facebook as a social media health information platform. Am J Infect Contr 45:301–302
134. Sharma M, Yadav K, Yadav N, Ferdinand KC (2017) Zika virus pandemic—analysis of Facebook as a social media health information platform. Am J Infect Control 45:301–302
135. Robledo I, Jankovic J (2017) Media hype: patient and scientific perspectives on misleading medical news. Move Dis 32:9. https://www.ncbi.nlm.nih.gov/pubmed/28370445. Accessed 18 Sept 2018
136. Basch CH et al (2017) Zika virus on YouTube: an analysis of English-language video content by source. J Prevent Med Public Health 50. https://www.jpmph.org/journal/view.php? https://doi.org/10.3961/jpmph.16.107. Accessed 13 June 2019
137. PAHO (2016) Zika virus infection step-by-step guide to risk communication and community engagement. November. http://iris.paho.org/xmlui/handle/123456789/18599. Accessed 28 May 2017
138. Christodoulou M (2011) Biological vector control of mosquito-borne diseases. The Lancet. February. 11. http://www.thelancet.com/journals/laninf/article/PIIS1473-3099(11)70017-2/abstract. Accessed 10 Apr 2017
139. Internet World Stats (2022) Facebook users in the world. https://www.internetworldstats.com/facebook.htm. Accessed 26 June 2022

140. Wirz CD, Xenos MA, Brossard D, Scheufele D et al (2018) Rethinking social amplification of risk: social media and Zika in three languages. Risk Anal 38(12):2599–2624. November 8
141. Safamejad L, Xu Q, Ge Y, Chen S (2021) A multiple feature category data mining and machine learning approach to characterize and detect health misinformation on social media. IEEE Internet Comp 25(5):43–51
142. Broniatowski DA et al (2018) Weaponized health communication: Twitter bots and Russian trolls amplify the vaccine debate. Am J Public Health 108(10):1378–1384
143. Cummings CL, Kong WY (2019) Breaking down 'fake news': differences between misinformation, disinformation, rumors and propaganda. In: Linkov I et al (eds) Resilience and hybrid threats. IOS Press
144. Chandrasekaran N, Marotta M, Taldone S, Curry C (2017) Perceptions of community risk and travel during pregnancy in an area of Zika transmission. Cureus 9(7):e1516. https://doi.org/10.7759/cureus.1516
145. AAFP (2018) Children exposed to Zika in Utero need long-term monitoring. August 17. https://www.aafp.org/news/health-of-the-public/20180817mmwr-zika.html. Accessed 14 May 2019
146. James S et al (2018) Pathway to deployment of gene drive mosquitoes as a potential biocontrol tool for elimination of Malaria in Sub-Saharan Africa: recommendations of a scientific working group. Am J Trop Med Hyg 98(Supp 6). https://www.ncbi.nlm.nih.gov/pubmed/29882508. Accessed 10 Sept 2018

Chapter 18
Pandemic Engagement

The ZIKV became a concern to many Americans, and its emergence fits into a broader public concern that the number of infectious disease threats to people's health has grown in the past generation. Some 51% of adults say there are more contagious disease threats today than 20 years ago [1].

According to the 2016 Harvard Poll, only about half (52%) said that the spread of the ZIKV is a public health threat in the U.S., with under a quarter (22%) saying it is a "major" public health threat. About the same fraction (24%) said they are concerned that they or their family members will get sick from the ZIKV in the next 12 months [2].

A (2016) survey from Pew Research Center shows the public is paying attention to these threats. Fully 82% of Americans say they pay "a lot" or "some" awareness to news of any kind about dangers from infectious diseases. There are notable demographic differences tied to people's views about the threat posed by the ZIKV.

Those who are most concerned include older Americans (especially older women) and those who report "fair" or "poor" health status. For instance, 44% of women ages 50 and older think ZIKV is a significant threat to the U.S. population, compared with 33% of older men. By contrast, 23% of adults under age 50 consider ZIKV a significant threat to the population. Additionally, 42% of those who report being in "fair" or "poor" health feel it is a considerable threat, vs. 27% of those who say they are in "excellent" or "very good" health. People of color and those who live in the South and Northeast also tend to be more concerned about ZIKV [3].

There is a lot known from public surveys and focus groups. Over 100 articles have been published since 2016 discussing knowledge, attitudes, and practices (KAP studies) associated with the ZIKV endemic. The few found here were selected because specific datasets enhanced the story being built and told.

Engagement should be proactive. Researchers may be confronted with well-organized dissent within or outside the community where the messaging is conducted. Execution of a robust, bold engagement plan may help mitigate negative messaging [4].

D. M. Berube, *Pandemics and Resilience: Lessons we should have learned from Zika*, Risk, Systems and Decisions, https://doi.org/10.1007/978-3-031-25370-6_18

Engagement during contained studies conducted within the laboratory provides an opportunity to explain project goals and operations, develop a relationship with the community, and build trust.

Community engagement is best managed by in-context social scientists who understand local value systems and can easily interface with the people. While still working in the laboratory, researchers should consider their obligations to the community near a facility, for example, and plan their interactions with the larger public. At each subsequent phase of testing, broader outreach and engagement will be required [4].

From the beginning, researchers should have a plan for interacting with those who disagree with the conduct of research on mosquitoes in their community. Some who disagree may hold deep-seated objections that limit compromise. In contrast, others may seek changes or have concerns that could be addressed and would, therefore, be amenable to dialogue if engaged [4].

Given the lack of good data on the effectiveness of these engagement activities, it must be asked whether the role of engagement is necessarily related to program success. The number of articles on engagement includes a set of sentences like the following: "Many educational efforts do not always have an immediate effect." However, "community participation can be essential for developing long-term, low-cost, sustainable programs [5]."

Public participation, engagement, enablement, and authorization are critical to the success of field trials and the future use and acceptability of new strategies. In cultural anthropology and public health, interventions often fail because of a lack of understanding of the local context, lay knowledge of the disease, and community, local or regional expectations around engagement [6].

> The benefits of public engagement include creating an informed citizenry, generating new ideas from the public, increasing the chances of research being adopted, increasing public trust, and answering ethical research questions. There can be significant gains in effectiveness and insight of decisions when the distributed intelligence of the public is combined with that of policymakers. Public engagement fosters global communication, enables shared experiences and methodology, standardizes strategy, and generates global viewpoints. This is especially pertinent to the developing world as it encourages previously marginalized populations to participate on a worldwide stage.
>
> From public engagement in the developing world, citizens of industrialized countries gain exposure to the challenges most humanity faces. Also, beyond benefiting society, public engagement in science helps the scientific enterprise itself.
>
> Given that the world economy is based on capital interests, public engagement can complement market signals in the research plan, resulting in a better match between research priorities and social needs. It can address distrust in science, potentially translating into a lack of research support. Finally, informed public questioning can probe excessive optimism and help prevent unintended consequences of such optimism. [7]

18.1 Experts

The public sometimes elects to understand differently from the scientifically rigorous process of bench scientists and engineers. At times, credentials may have mattered as a feature of credibility and trustworthiness. Still, those days have changed, especially with social media, where people believe in "friends" more than experts. As experts found the body of knowledge about the world blossoming at an incredible rate, no one expert could develop broad enough expertise to be a "Renaissance person." Experts seemed to have become fewer, less reliable, and less able to situate themselves within the social and intellectual milieu of the public [8].

There are three interrelated phenomena.

First, the expert community does not communicate well with the public. The first reaction tends to be to educate the public about the science. This has been called the "deficit theory" in science communication and has been demonstrated to be nearly worthless.

> Identifying the deficit model of the public with survey research per se and critical inquiry with qualitative protocols is fallacious and not helpful beyond temporary rhetoric. The critique of the public deficit model as a common-sense prejudice among experts is valid, but its identification with the protocol of survey research is dysfunctional.
>
> Converting a public that has turned away from science for many reasons and trying to win them back with a finely hewn scientific argument is ineffective. When the public hears labels like outbreak, epidemic, or pandemic, they react emotionally. Of course, jargon is a significant component of this problem. Other concerns orbit around the expert's attitude toward events and how they are framed for technical consumption without many concerns for the public. [9]

Second, there are mental shortcuts the public takes to make sense of complex and confusing things. These heuristics, ways of making sense, become biases, and one of the most powerful is the affirmation or confirmation bias, whereby the public attaches particular value to ideas that affirm or confirm what they always believe is true. This powerful bias tends to quadrant off the knowledge that may rebut thoughts that may have developed that are mostly incorrect.

Third, the public learns from the media, and the media amplifies risk events like infectious diseases because it produces dramatic headlines that increase readership and viewership. Decades ago, editors would make some effort to shape and fine-tune reportage; the World Wide Web has eliminated the editor per se.

Others amplify risks as well. Government officials and non-governmental organizations boost events to improve their standing and generate support for their causes, including attending rallies, voting occasionally, and sending money to a nonprofit.

The fear and anxiety generated by excessive and highly harmful reportage of infectious disease events may fuel apprehension sufficiently to convince some to take **necessary and reasonable** precautions. On the other hand, there is an evaluation of the risks associated with phenomena like genetically engineered mosquitoes. That needs to be more effectively communicated to all stakeholders if there is any chance to maximize wise and responsible engineering to improve the health and welfare of us and the ecosystems inhabited.

Effective programs had two common elements consistent with a policy of determined leadership that stressed community responsibility: (1) a vertical component, usually governmental, that initiated, planned, and oversaw the program, and (2) a horizontal component, usually householders, who helped execute control measures and permitted access to their property [10]. Experienced teams of engagers draw from the populations of these two groups converting the experience for the public into a more organic one. Community leaders meet with the public where they usually come together, such as in a church basement, engagement becomes more natural and ecological.

18.2 Design

Take, for example, efforts to reduce exposure to mosquitoes that might be used in the home. Ideal tools should be "user-friendly," requiring little additional work or behavior change by householders. They should be affordable, safe, and effective in reducing vector densities' indeed, as might be the case with insecticide-treated curtains in many societies, they could even be viewed as desirable by householders. There must be a goal for the ideal situation to have a suite of proven effective, safe, and environmentally friendly tools available for intervention. The most appropriate device or combination of tools can be selected to suit each community's specific biological and cultural needs [11].

Consider this outstanding example. North Texas has a five-D program with much potential. All North Texans are asked to adhere to the five D's when controlling mosquitoes' presence.

- DRAIN—all areas of standing water, including changing the water in wading pools and birdbaths and cleaning out gutters;
- DEET—Use bug spray and protective clothing with repellents containing permethrin or DEET;
- DRESS—Dress in light-colored clothing with long sleeves and wear long pants;
- DUSK/DAWN—Limit outdoor exposure at dusk and dawn;
- DOORS—Keep door and window screens in good repair [12].

This set of activities is simple, easy to recall, and doable.

Another set of measures comes from an American Mosquito Control Association (AMCA) article to eliminate and control mosquitoes. Individuals can act by eliminating standing water, introducing mosquito-eating fish, encouraging predators such as bats, birds, dragonflies, and frogs, and using least-toxic larvicides like bacillus thuringiensis israelensis (Bti) [13]. Community-based programs should encourage residents to employ these effective techniques, focus on eliminating breeding sites on public lands, and promote monitoring and action levels to determine what, where, and when control measures might be needed [14]. For example, Karen Cheng, a third-year medical student at Boston University School of Medicine, has developed an affordable automated device that safely delivers larvicide and insecticide that kills the larvae of mosquitoes in rooftop water collection tanks standard in developing societies [15].

Public engagement can mean participation. It can involve public stakeholders as partners rather than as audiences. Calling a public meeting where a set of speakers present an interpretation of a problem, and some proposed solution is hardly ideal public participation.

Let us begin by backing up and attempting to answer the question: Who are the stakeholders? The lesson of early control programs throughout the 1980s and early 1990s was that community-based programs need to incorporate a sense of ownership to be sustainable [16].

For example, one benefit of the Wolbachia project is how it involves residents participating in "citizen science" projects. Students study the mosquito life cycle and help maintain "Mozzie Boxes," raising Wolbachia-infected eggs until they mature into adult mosquitoes that go out and breed with others, furthering the spread of the Wolbachia-infected mozzies.

It is great to see the community engagement and support of projects like the "Wolbachia Warriors" program. This innovative technology will be increasingly important to the U.S. and Latin America because of the rapid spread of *Ae. aegypti*, a.k.a. the yellow fever mosquito, which transmits the disease [17].

18.3 Stakeholders

The stakeholders exist on many levels [18]. Delineating between those individuals and organizations with a bona fide interest, vested or otherwise, and those with merely a passive interest has been challenging in public engagement [19].

When it comes to public health issues such as epidemics or pandemics, there is a broad swath of stakeholders, a term of art that includes all individuals and institutions with a stake in some phenomenon like a pandemic.

In the case of vector control, there is a plethora of stakeholders, many of whom have no voices per se: humans of all sorts, especially those with diminished capacity caused by socioeconomic status, animals up and down food chains, agriculture, and precious natural environments.

According to the Association of State and Territorial Health Officials (ASTHO), vector control programs benefit from a broad view of essential stakeholders in public education efforts. Organizations such as schools, faith-based organizations and churches, community groups, and businesses can serve as valuable distribution pathways for relevant health information. Vector control program staff can work with elected and non-elected community leaders to coordinate releasing crucial public information [20].

Stakeholders often control channels for engagement. This would include public congregates and potential locations for outreach events such as schools, clubs, churches, and other organizations. Some individuals are at the front end of the pandemic, such as municipal works departments, green organizations, public health organizations, extension programs, and even citizen scientists. Social influencers are necessary (bloggers, newspapers, and local radio/TV stations that can do periodic stories or provide 30-s reminders and PSAs) [13].

The stake must be sufficiently large to constitute bona fide interest. A third party who may not be likely to affect the pandemic directly is highly problematic as a stakeholder. For a stakeholder, as a concept, to be meaningful, there must be non-stakeholders.

The stakeholders in the 2016 ZIKV pandemic included those infected and their offspring, their immediate families and related providers, the first-order health and care providers, and the people and institutions who were responsible for responding to the pandemic, including insurers and healthcare providers, and facilities, those impacted by the direct social responses to the pandemic, including those economically affected such as travel industry and probably many more. This distinction is not simply an academic exercise. Still, it will be addressed below when considering the concern over responses to the pandemic by a mostly unaffected population could have profound and life-threatening implications to an affected stakeholder, such as pregnant women (Fig. 18.1).

PAHO [21] produced a chart (above) that identified some key stakeholders. It is used here as a point for discussion and analysis (p. 26). As this project developed, it became clear this was a story about women and children of color, often impoverished and marginalized in places far from the suburbs of South Florida.

A broad range of community interests and stakeholders would be necessary for authentic community engagement, such as media, schools, municipal government committees, universities, and local non-government organizations (NGOs), making the mix of community stakeholders at any given research site unique [22].

It may not even be understood what constitutes a community. There is still considerable debate about what constitutes a "community" for research and what is meant by a community's decision to participate in a study.

General public	• People living in areas where there is a risk of infection • Pregnant women and women of reproductive age • Patients and persons with symptoms
Health workers	• Doctors and other health workers • Associations of medical professionals • Laboratory personnel
Travelers and the tourism industry	• Travelers to areas where there is a risk of infection • Airline companies • Professional airline organizations • Cruise lines
Community organizations	• Schools • Religious organizations • Businesses • Civic organizations

Fig. 18.1 PAHO chart of stakeholders. These groups have concerns and problems, some shared, others differing, which require specially targeted communications efforts to maintain trust and manage the expectations of the target audiences. PAHO/WHO (2016) Zika virus infection: step by step guide on risk communications and community engagement. Pan American Health Organization

> Suppose shared research risk is used as a defining criterion for the community during site selection. It becomes clear that even the specific field sites within the candidate countries or regions must be identified for community authorization to have an effective force. However, getting some prior consent before a final field site selection has been made may impose burdens unnecessarily on specific groups of people who may ultimately not be involved in the research while ignoring others who may become involved as the details of field site selection unfold. [22]

On a different level, stakeholders are often motivated to understand differently. To the question: Why do good people tolerate evil? Many responses extend from stupidity to perversion.

Michael Seifert, a community organizer in Brownsville, Texas, suggested that until people see local cases of babies with microcephaly–a congenital disability caused by ZIKV–the virus will be viewed as a sort of "urban legend," distant and unlikely [23].

18.4 Outreach

Public outreach occurs on at least two levels. First, there are the mosquito control districts, all of whom seem to have their approaches. For example, in Harris County (Texas), the mosquito control division is planning outreach—on social media and at community events like sports games and town halls—to instruct people about *Ae. aegypti.* The hype around ZIKV could, for once, help the cause [24]. Public meetings often serve as a minimal outreach exercise.

For example, in 2009, Oxitec's first trial in the Cayman Islands was largely unknown until Oxitec shared its results. It stirred criticism that the company had rushed a GM organism into the field without adequately consulting the public [25]. However, on September 8, some senior citizen residents in West Bay, Cayman seemed to have a Senior's Fellowship meeting at the John Gray Memorial Church with presentations from Mosquito Research and Control Unit Director Petrie and Renaud Lacroix from Oxitec; there were few other meetings reported in the media.

In Piracicaba, Oxitec engaged in 13 weeks of public engagement work. Oxitec's technicians, supported by the Secretary of Health, explained to Piracicaba's citizens what Friendly ™ *Aedes* is and how it works. An advertising campaign runs in newspapers, billboards, bus door posters, radio spots, and an information kiosk at Piracicaba's largest shopping mall [26].

Two lessons here. Some outreach activities are better than others and tend to be richer, deeper, and more extensive. Second lesson: There will always be members of the public who will be dissatisfied because of the confirmation bias mentioned earlier. They hear what they want and listen to what they already believe is true. The more extensive the outreach activities, the more likely some may be reached by them because they cannot look away.

The Oxitec issue was mishandled from the get-go, deciding unofficially to go with what the company was pitching without any public outreach—until the Florida mosquito board realized the public backlash. Only then did it combine with Oxitec for public meetings [27]. So, a set of meetings were held for Key Haven residents on April 11 and 12, 2016. These two three-hour meetings were less formal, and Oxitec representatives were there and allowed the public to speak with them, asking whatever questions they may have. The Florida Keys Project referring to the Key Haven trial had information on their Oxitec website.

Policy and management related to the release of organisms generated for pest management, whether bacterial, biotechnologically engineered, or the product of gene drive technology should be informed through public engagement and outreach. In the case of emerging genetically modified insect technologies, regulatory decisions can be conceptually distinguished into the development of frameworks, the assessment of the release of a specifically modified organism, and implementation decisions such as location and timing [28].

Some approaches to public engagement are more appropriate for different purposes and situations, and it is not always obvious how to match the path to the goal. And that is the rub.

18.5 Rationale

Engagement related to general guidance may enable a broader range of interests or knowledge to be considered and included. Still, it may make it challenging to articulate risks and benefits in terms that cover the full range of complexity and public concerns without ambiguity or overly technical language.

Shifting the needle of democratic reform discourse from "participation" and "involvement" in government decision-making to "engagement" and "empowerment"—a shift from information exchange models involving citizens to information processing models that help citizens make meaning of policy alternatives and share with them a real stake in the decision-making process [29].

Voting for regulators is not enough; even voicing approval or disapproval on an activity instant is also inadequate. The vote provides decision-makers with information about the tolerability of the proposal and about how this tolerability is distributed geographically among the county's voting districts. Still, it does not allow the public to express a choice among alternatives. Referenda cannot offer complex choices or provide balanced information on the ballot paper and often leave a divided public with different interpretations of the options perceived to be relevant [28].

[I]t seems curious that millions of dollars are invested in product development, clinical training, design, the building of facilities, etc., but often leave vital community engagement processes primarily to trial and error [30].

[T]aking science (and scientists) out of their laboratories and into the environment has consequences for residents, whose homes and places of work, leisure, education, and worship become outdoor laboratories for testing the efficacy of these methods or a forthcoming release site for a strategy. The ethical complexities of these field trials and environmental releases are considerable. Regulatory approval, collaborative partnerships, public engagement, and authorization are seen by many as essential ethical requirements that should be secured before undertaking open field releases [6].

Engagement related to insect control trials or experimental releases makes the range of relevant risks and potential benefits more specific. Still, many interests or concerns raised by the public may be beyond the scope of a narrow environmental assessment specified in regulations [28].

For a dynamic cohort, community engagement needs to be an ongoing process. The community engagement process has likely helped facilitate the current response rate of 85% in the research communities [31].

18.6 Justification

Why? Why go to all this trouble? McNaughton examined outreach activities in 2009–2010 associated with the release of Wolbachia-infected mosquitoes in Cairns, Australia. She reached this conclusion.

> [E]fforts to embrace the call for more ethical public participation and engagement in science need to develop engagement strategies and communication materials tailored for and understandable to the multiple publics at a given field site. Suppose projects plan to use open field trials. In that case, they need to be aware of and responsive to the needs, expectations, concerns, desires, and knowledge of the communities whose backyards they hope to use as open laboratories.
>
> Using long-term social research methods and attempting to understand lay knowledge rather than dismissing it as non-scientific or wrong can significantly aid the transmission of learning about new scientific endeavors, such that residents are enabled to participate, critique, assess, and determine whether they want these strategies to be trialed or implemented in their backyards and communities. [6]

Informed community mobilization adds effectiveness to government-run infectious disease control, according to Andersson et al. [32].

The Camino Verde project in Mexico and Nicaragua reduced *Ae. aegypti* larvae and pupae and protected against dengue virus infection. While there is not an expectation for community participation in dengue control to be accessible or easily sustainable, the intervention protocol that engages leadership and community members in discussing evidence and defining local strategies is a good starting point for various settings. Each site implementing the intervention has the advantage of local customization and strong community engagement [32].

The release of modified organisms is controversial, raising concerns about secondary ecological effects resulting from the sustained elimination or reduction of the harmful insect populations, the possibility of genome-level changes that may have unpredictable effects, and consequences for organic agriculture producers and other stakeholders within and outside the area where modified insects are released [28].

There is a growing consensus that the scarce resources available for mitigating tropical public health problems should be evidence-based and cost-effective. Some suggest public monies might be spent better.

Baly et al. [33] conducted an economic appraisal of two strategies for *Ae. aegypti* control: a vertical versus a community-based approach. Costs were calculated for 2000—2002 in three pilot areas of Santiago de Cuba, where a community intervention was implemented and compared with three control areas with routine vertical program activities. Reduction in *Ae. aegypti* foci were chosen as the measure of effectiveness. The pre-intervention number of foci (614 vs. 632) and economical costs for vector control (US$243,746 vs. US$263,486) were comparable in the intervention and control areas. During the intervention period (2001—2002), a 13% decrease in recurrent costs for the health system was observed. Within the control areas, these recurrent relative costs remained stable. The community-based approach was more cost-effective from a health system perspective (US$964 vs. US$1406 per focus) and a society perspective (US$1508 vs. US$1767 per focus) [33].

In the 1970s, the WHO conducted SIT releases for multiple species, including *Ae. aegypti* and *Culex spp*., in Delhi, India. However, the fate met by the project is a reminder of the importance of government and community consultation. Untruths reported by the media included the idea that the U.S. was using the WHO-associated research project to test dangerous chemosterilization methods in India and that the unstated goal of the program was to develop biological weapons. The undercurrent of these accusations was the subtext of scientific imperialism. Although the project might also have been a victim of the geopolitics of the time, it would undoubtedly have benefited from an active and effective community engagement campaign [34].

From a health system perspective, the community-based intervention can produce savings that might finance other control program activities (dengue-related or not) or address direct causes of vector proliferation such as water supply problems [33].

Primarily, it is disrespectful of communities to import a novel and potentially threatening set of technologies into a country without any legitimate state interest in these activities. Without a clear expression of such a claim, it is conceivable that insufficient resources could be devoted to appropriating review and oversight, including the necessary capacity for risk assessment. There might also be inadequate attention to how shared interests between the investigators and host communities should be negotiated and realized [22].

In some cases, they believe that citizens affected by decisions have the right to participate in those decisions, primarily when the research is funded by their tax contributions (this would be what social scientists term a normative justification) [35].

At the heart of Republican democracy is the idea that citizens exercise control over policies or actions that could affect their activities or lifestyles—this includes measures that would release genetically modified insects into their airspace [36].

Deliberative democracy theorists suggest that the deliberation enables participants to generate conclusions that the wider public would cause if they were informed and civic-minded [28]. Democratic deliberation augments participants' knowledge about issues, cultivates trust, builds civic capacity, and, over the long term, may increase general civic engagement and political participation [29]. In addition to contributing to greater citizen awareness of issues and the competing points of view surrounding those issues, citizen involvement through policy deliberation helps cultivate rational dialogue, active listening, and problem-solving skills [29].

In other cases, they reflect a desire to reduce conflict, help (re)build trust, and smooth the way for innovations (in other words, the reason is instrumental) [35]. A critical factor in reducing conflict is raising trust among the parties involved; getting work done without trust can be challenging and costly [29].

Deliberative public engagement can provide decision-makers with advice or recommendations from an informed, diverse group of citizens that is oriented to negotiating how to make decisions given controversy and diversity. This kind of public engagement is more likely to produce trusted choices that are seen as legitimate and may open opportunities for compensatory actions that balance gains and losses to groups [28].

And in still others they reflect the assumption that such participation from people who will use and be affected by technology will raise questions about the real-life functioning of developments when they leave the laboratory, perhaps leading to innovations that perform better in complex real-world conditions or that may be more socially, economically, and environmentally viable (these could be termed "substantive justifications") [35].

However, policymakers should realize that community involvement in vector control is not a "free ride" and carries a substantial opportunity cost for volunteer time spent on management. It is vitally important to acknowledge the significant investment in health made by the community in the form of unpaid labor contributed to vector control activities [33].

18.7 Typologies

Public engagement models developed from information exchange models, public hearings, or media broadcasts. Anytime a government initiative needs public participation, unidirectional information exchange models fall short. Information process models such as deliberative forums were introduced to scale the process.

This begins with participatory technology assessment: PTA incorporates similar approaches given different names by their respective authors, including constructive technology assessment [37], interactive technology assessment [38], real-time technology assessment [39], upstream public engagement [40], and technology appraisal [41]. PTA has been implemented through various methods, including consensus conferences, citizens juries, stakeholder workshops, deliberative polling, and public dialogues [35].

Dialogic approaches are a mixed bag. According to McNaughton's Cairns project review: "The most popular mechanism was face-to-face presentations, like those used in the focus groups, which also provided time for those present to ask scientists' questions, reflect on their answers, and hear other community members' views.

Residents expected multiple contact points with the program, including community presentations, website, newsletter, and general media coverage that would keep them informed, update them on new results, and allow time for people to digest and consider their responses. Identifying their needs and expectations and incorporating them into the engagement strategy was crucial in establishing trust and showing respect. Participants identified that collaborating directly with residents, leaders, and civic groups that were respected and trusted was the most appropriate way to engage and build awareness in this context" [6].

However, many involved with these initiatives—social scientists, representatives of NGOs, and public members recruited as participants—are skeptical about the value of these "public dialogues." They often merely describe the beliefs and attitudes of groups of people selected because they do not have specialized knowledge or strong opinions about the topic. And the knowledge gained from these initiatives often seems directed toward anticipating controversy to ward it off, rather than giving the public any actual role in decisions about research trajectories [35].

Longer-term research using mixed methods is essential to identifying concerns about a program. This, in turn, provides opportunities to develop precisely, considered, educationally and culturally appropriate responses and resources. This helped us create a consistent message, address concerns, and lay understandings sensitively, confidently, and effectively, build trust, and show respect because residents' concerns were taken seriously and not simply dismissed based on current knowledge [6].

South Africa's Public Understanding of Biotechnology (PUB) offers an introductory biotechnology course, profiles role models from the biotechnology field, conducts public opinion surveys, provides materials to educators, and provides space on its website to debate critical issues. It also offers free posters explaining biotechnology in several of South Africa's eleven national languages and develops games, crosswords, videos, and puzzles, which unconventionally promote understanding of biotechnology [7].

What constitutes public engagement in the developing world is often a collaboration of outreach and engagement? These have "included public events in bars and other venues outside the academic circuit, dramas, soap operas, comic books, poetry, games, storytelling, science fairs, and even science-based participation in Peru's parades and Brazil's annual Carnival." The Carnival initiatives aim to "put science on the street" [7] (Fig. 18.2).

Researchers have begun to take faith-based engagement activities more seriously, especially in the U.S. During demanding situations, people often turn to trusted leaders for advice. Trusted leaders can include **a** community or religious leaders, such as pastors, priests, rabbis, and imams [42]. Health communication and health promotion professionals can work with influencers or people who can persuade others to promote a concept through one-on-one communications or social media. Influences are trusted by the target population and can offer guidance on strategic health communication plans.

When influencers are religious leaders, while appropriate for many reasons, they confront a particular challenge in the case of the ZIKV. Preventing the sexual transmission of ZIKV is a specific challenge because, although condoms can help

Fig. 18.2 Araújo HRC, Carvalho DO, Ioshino RS, Costa-da-Silva AL, Capurro ML (2015) Aedes aegypti control strategies in Brazil: incorporation of new technologies to overcome the persistence of dengue epidemics. Insects 6:576–594. https://doi.org/10.3390/insects6020576

reduce the risk of ZIKV transmission, not all faith-based or community-based groups promote their use.

Efforts to engage faith communities in public health messaging must recognize that messages shared by a faith-based organization with its members must be consistent with the values of that organization. If the message is inconsistent, the organization might choose to deliver only partial, and possibly more limited, information [42].

18.8 Challenges

First, it is not easy to measure the benefits of engaging citizens. Second, citizens question whether their voices will be listened to at these events and whether they will, in turn, be affected by their input. Third, cost and time can limit the scope of engagement. Fourth, the engagement process can require time-consuming and continuous monitoring and evaluation. Fifth, an appropriate response to public needs may demand institutional reform. Broad participation and deep deliberation may also be at odds, mainly when substantial differences and feelings run high [7]. Protocols and planning for engagement activities become vitally important to help alleviate some of these misgivings.

Researchers must identify an appropriate method for obtaining community endorsement to conduct the studies. What constitutes acceptance will be culturally determined and may be identified through advanced discussions with the community.

Engagement is ongoing, and community authorization/endorsement must be continually confirmed as innovative studies are undertaken. Institutional ethics committees, regulators, and community advisory boards likely will play a role in defining the requirements for authorization. Appropriate ethical approvals must be obtained in studies involving human research within the field cage [4].

Online platforms: It has been seen that building an online platform is not enough; identifying opportunities for users to start projects, keeping barriers to participation low enough for anyone to participate, and launching an aggressive marketing campaign are keys to success for a public engagement program [7].

Engagement must be an iterative process that continues throughout the development pathway, understanding that opinions can change over time. However, consideration must be given to mechanisms to monitor for and avoid stakeholder fatigue throughout lengthy trials [4].

Today's most successful citizen participation efforts are those that understand engagement as a series of interrelated, developmental choices that have more to do with "what level of involvement" along the policy development-implementation continuum than any single technique for "one-off" events that fulfill statutory requirements.

The public comment period—the only legally mandated public engagement activity in these cases—is a one-size-fits-all approach to public engagement. It does not distinguish community engagement from outreach to the public.

No single "one-size-fits-all" approach to public participation or engagement can be applied universally; in other words, what works in one place may not work in another. Perhaps the most compelling reason for this is that different communities have divergent expectations, concerns, political biases, structures, and cultural sensibilities that need to be understood, respected, and considered to engage sensitively, ethically, and effectively [6].

Engagement activities involve information overload, asynchronous dialogue (especially online), institutional skepticism, and unrepresentativeness. While there are no easy solutions to these problems, an experienced engagement scholar should be able to reduce their impact on the participants.

For example, most participants in a Zika Contraception Access Network (Z-CAN) focus group in Puerto Rico preferred to receive information on contraception, potential side effects, and where to access contraceptive services via Internet-based channels and healthcare providers. Z-CAN provided same-day access to the full range of FDA-approved reversible contraception methods (i.e., condoms, contraceptive pills, contraceptive patches, vaginal rings, and long-acting reversible contraceptives (LARC) procedures at no cost to women in Puerto Rico during the ZIKV outbreak. Based on these findings, the Ante La Duda, Pregunta [When in Doubt, Ask] campaign was launched to promote awareness of Z-CAN services among those who chose to prevent pregnancy during the ZIKV outbreak. They discovered that ZIKV infection might not be a motivator to initiate contraceptive use; this type of campaign could have failed to resonate with the intended audience and could have failed to encourage women to seek Z-CAN services [43]. Data showed that there was community awareness regarding ZIKV in Puerto Rico. However, it was not a motivating factor in contraception decision-making; economic factors were the significant drivers.

Financial and time constraints have led to an over-reliance on "Rapid Assessment" techniques [44]. The actual cost of this "is the loss of contextual information, and the probable oversimplification of behavior due to the brevity of fieldwork and the lack of participant observation which enhances, and indeed is the one means to ensure, the validity of data." In other words, while quick and cheap (and sometimes the only options available because of budgetary and time constraints), these techniques produce a limited and less reliable understanding of the sociopolitical context and of lay understandings of disease, both of which are critical to improving the success of interventions and engagement [6].

There are multiple models for citizen participation: ChoiceWork Dialogue, Consensus Conferences, Citizens Juries, Deliberative Polling, Issue Forums, Study Circles, Citizen Assemblies, Neighborhood America, and more yet to be invented [29].

However, each seems to share four components: communication, consultation, engagement, and collaboration. Each model has a stage involving engagement.

Cohen's [45] definition of public engagement is clear and concise. Public engagement is a process that provides people with trustworthy information on key policy issues, elicits their input, and integrates it into decision-making and social action [45].

Public engagement strategies provide decision-makers with opportunities to improve the substance of public input, cultivate trust through the process, raise the legitimacy of decisions in the public eye, and lay the groundwork for lasting implementation. Engagement recognizes that the public has a right to influence policy and program development and creates opportunities for exchange between experts and the people in ways that yield balanced recommendations that affect policy [29].

How to engage a stakeholder depends on the context. The level of interest functions along a continuum against stakeholder efficacy. In other words, the more the stakeholder is empowered to function as an active stakeholder (their influence matters) and the level of interest and attention associated with the subject at hand, the more complicated the engagement becomes. There are many different matrices across the literature that visualized this relationship. One designed by the author is found below but highly influenced by the many that came before (Fig. 18.3).

Of relevance is one designed by Mike Depew of Modern Mosaic Consulting [46]. While his format is similar, this diagram emphasizes listening and efficacy. The outputs of an engagement might be current information to be considered by the organizers or clients, or it could be a collaboration on recommendations or advice. Community members may have different ideas about risk mitigation and might also aid in developing ecological risk assessments. Engaging this kind of information improves the experiment by simultaneously encouraging mutual engagement and building trust.

[redacted]

HIGH	Inform & Consult GOAL: Keep in tune with their attitudes. • Listen. • Challenging because vulnerable to fake news and conspiracy theories. • Engage and consult on interest area and attempt to raise interest or depress influence (if necessary).	Inform, Consult & Collaborate GOAL: Engage very closely & regularly • Listen. • Involve them in authentic governance and decision making as much as possible. • Engage and consult regularly organically (where they congregate).
STAKE HOLDER INFLUENCE	Inform GOAL: Increase level of interest. • Listen. • Inform using analog and digital media as well as town meetings. Remember you may be their first source. • Attempt to spark interest by emphasizing salience and move interest into attention.	Inform & Consult GOAL: Show consideration. • Listen. • Attempt to match interest with efficacy emphasize and discover paths to empowerment. • Maintain communication and add content continually. • Convert the interested into proponents and sometimes critics.
LOW	**STAKEHOLDER INTEREST**	**HIGH**

Fig. 18.3 Stakeholder engagement designed by the author (2022)

Outputs could also be aggregates of individual opinions, themes identified by analysts, or collective conclusions of the participants. A successful public engagement will have demonstrable outcomes and decisions, which should derive enhanced legitimacy from the public engagement [28].

18.9 The Stages

The issue of vector control demands has engagement on four levels: (1) companies interested in providing an alternative remedy to mosquito vectored infectious disease need to find some location to test their products (in this case, there are two companies using Wolbachia and Oxitec with its genetically engineered mosquito); (2) once demonstrated, the next step is to deploy the product and not only does someone have to pay for the product but also have community approval, sometimes not an option but a requirement; (3) companies offering a genetically engineered option, such as Oxitec, has a final hurdle to clear, releasing an engineered product into the wild; (4) maintaining engagement over the history of the interface.

Whenever possible, population-based researchers should include community members on the study team. Members of the ethnic population being studied should be involved in the design and implementation of the research and the analysis and interpretation of results, particularly in ethnically diverse communities that have experienced discrimination or stigmatization. Again, this instills trust and collaboration instead of fostering an image of researchers as biomedical "colonialists" who extract the information necessary to complete studies without considering the needs and priorities of the population [47].

There are some daunting questions. When researchers and potential participants speak a different language, that barrier profoundly impacts informed consent. Even when the same language is spoken, studies repeatedly show it is difficult for people to understand the complicated language of consent forms used in clinical or epidemiological research [47].

For small-scale releases, options for responding to concerns within the hosting community may include a project agreement to avoid releasing in the immediate location of the residence, or if that is unsatisfactory, at some mutually agreed on distance from the household; provision of mosquito repellent and provision of traps to remove mosquitoes from home [4].

Whereas some serious options can be presented, such as not allowing releases or monitoring at their home or place of work and not participating as human subjects through the vision of personal identifying information or specimens, it likely will be increasingly difficult to prevent some degree of exposure to gene drive mosquitoes as large-scale testing proceeds. The extent of exposure is likely to differ according to whether a suppression or replacement strategy is tested. Expectations for the long-term spread of the gene drive construct within the local mosquito population must be conveyed realistically in the engagement process. A survey of public understanding of these expectations is advised to be undertaken before releases [4].

This claim has never been more accurate. Public engagement involving genetically engineered mosquitoes requires the whole armor of engagement activities.

18.10 Stage One: The Field Test

The release area for small-scale trials should be sufficiently small to allow for intense community engagement through personal interactions [4].

It is advisable to survey to judge the level of community awareness before asking for endorsement of a release. A community liaison or reference group could help provide feedback on community satisfaction [4].

18.11 Stage Two: The Initial Deployment

Semifield testing provides additional opportunities for further engagement with the local community. Before and during semifield testing, engagement activities will include stakeholders living in the region of the field cage and potential release site(s). Although interaction with key opinion leaders should have begun earlier, it will become a critical engagement component at this point. It will be essential to ensure that the community understands the concept and goals of semifield testing, including information on safety studies that have already been performed. Observing research staff working inside the field cage with gene drive mosquitoes may help boost public confidence about safety. However, researchers will need to explain that cage trials are not aimed at evaluating safety but are for obtaining information that will optimize planning for field release [4].

18.12 Stage Three: The Ramp-Up Deployment

Ethically responsible population-based studies must seriously consider community needs and priorities, and researchers should work collaboratively with local populations to implement study goals. Investigators working on population-based studies confront unique ethical challenges in articulating these principles due to the community context of their research, their methods of inquiry, and the implications of their findings for social groups [47].

Researchers need to work collaboratively with the study population to ensure that participants and the communities they represent benefit from and are empowered by the research, not devalued or harmed [47].

18.13 Stage Four: The Long-Term Association

Different issues come into play when scientists move from the individual to the community, the process of complete collaboration. The potential exists for research results to be misapplied in health policy development or misinterpreted by the public media, thus promoting racist or discriminatory practices [47].

Community engagement practices remain as much art as science, and what makes them effective is still primarily determined by a combination of intuition, experience, and opinion. Remarkably, there is no explicit body of community engagement knowledge to which researchers can turn for guidance about approaches most likely effective in different contexts [48].

Lavery et al. [22] recommend an active, collaborative, and iterative approach to site selection—involving site visits, capacity assessment, regulatory analyses, and a range of expert consultations to avoid some of the consequences of weak engagement.

In 2010, he made unparalleled recommendations about the critical first phase: site location for research.

1. Early initiation helps avoid putting communities in situations in which they are pressured to make hurried and perhaps ill-considered decisions solely to meet the timelines of investigators.
2. Define your community probably as those who share identified risks.
3. Clearly indicate the purpose and goals stressing community involvement and ownership.
4. Disseminate information to permit a reasoned judgment about whether research warrants community support and participation.
5. Incorporate and involve the community; the involvement of authentic leaders and institutional collaborations and assets provides the community with a sense of familiarity, ownership, and security and establishes the basis for mutual trust; Participation by community members in meaningful ways helps limit the disruptive effect, to some degree, by naturalizing the research.
6. Review, assess, and modify [48].

Lavery et al. [48] tap into many significant concerns engagement scholars have about this process.

Consider the challenges involved with testing site selection. This is way before considering deployment of the technology, whether Wolbachia mosquitoes, an Oxitec engineered one, or a gene drive one.

Stakeholder engagement for testing site locations has been criticized for its general efficiency. For example, current approaches to community consent often involve large-scale activities that aim to canvas community knowledge, attitudes, and opinions about research with GM insects. This is time-consuming and costly, and although efficiency is not a compelling ethical goal in site selection when the choice is among several sites in various parts of the world, and accountable agencies provide funding within defined time horizons, time and cost need to be considered [22].

On a second level, there does not seem to be a handle on community consent. There is minimal empirical evidence about the effectiveness of community consent and the added complexity of interacting with communities as opposed to individuals in the consent process; it may not be realistic to rely entirely on surveys of communities' knowledge, attitudes, and opinions collected before the site selection decision as a reliable basis for inferring authentic community authorization [22].

Third, most of the work in this area centers on surveys, and undertaking large-scale surveys of knowledge, attitudes, and opinions in communities that will not be selected as research sites is not an ethically neutral matter. By its very nature, health research suggests a commitment to improving the lives of individuals and communities affected by various health problems. When researchers withdraw helpful interventions from individuals or gatherings after clinical trials, it has given rise to concerns about abandonment and a sense of loss for the individuals and communities that have helped make the research possible [22].

Southeast Asia Community Observatory (SEACO) took time and effort to design an exciting engagement process in Malaysia. SEACO used community engagement to recruit and retain subjects for a broad-based demographic health status study. This example was selected because it illustrates community engagement scholars often overlooked cultural variables. The community engagement strategy was based on three key activities: creating the representative structures to enable community consultation and participation; establishing mechanisms for information exchange, particularly for individuals in the community to engage directly with SEACO; and establishing processes for community-directed involvement in SEACO research and activities [31].

For example, after a group boycotted one of the initial community meetings, the team learned that this was because their representative had not been approached personally. The invitation had come via other social networks. The cultural protocols had been undervalued.

> Led by the core community engagement team, SEACO initiated the local community engagement process through critical community networks reflecting government and other political structures, ethnic leaders, leaders within the business communities, NGO and charity organizations, and social clubs such as the Lions' that support those who for various reasons, fell outside the more formal support structures. These alternative entry points were critical to ensuring representation from groups as diverse as the plantation workers and residents in federal land development (FELDA) communities, orphanages, and women's shelters. [31]

A critical part of any engagement is assessment. Set objectives and assess how they are being met and how they could better be met. It is astonishing how much government money is spent on activities to benefit the public that is not subjected to rigorous assessment beyond some external review done by "experts" who spend a few hours each year evaluating how successfully funds are being spent in the public interest.

18.14 Vector Control

Engagement is essential for the researchers to test their approach to vector control experimentally on one level. Participation in non-commercial scientific research is generally voluntary; therefore, unsurprisingly, recruitment is challenging. Many study protocols are abandoned due to the inability of researchers to recruit enough participants to power the study [49]. Participation in commercial, scientific research remains voluntary, and recruitment may be even more challenging. Engagement is essential to meeting ethical obligations of informed consent, building trust, and gaining acceptance of the research [4].

> A common principle is that communities should be provided with sufficient opportunity to interact with the project team to learn about the research and its implications to formulate reasoned positions about whether to host a trial. Information about the study and investigational product must be provided and discussed with communities and other stakeholders in a way that strives for all voices to be heard. The expectation is for community engagement with the open and honest exchange of ideas and information. Researchers should seek to learn from these groups about ways to improve the project or the product. Researchers and

funders must be open to the possibility that research plans may need to change in response to community input or that an ongoing project must be halted or moved. [4]

Often, the vector control experiments may require long-term participation. Still, research ethics generally requires that experimenters maintain the option for the research subject to elect to leave the investigation. However, once a genetically engineered mosquito is released, it cannot be called back; hence, agreeing to serve as a sample is a singular event. At best, a subject may elect to leave the experimental setting by physically leaving.

At all levels of engagement, co-ownership of the entire product development and testing process by in-country scientists and government authorities will be critical for acceptability [4].

Sustainability of disease control programs thus requires a sense of community ownership, including the use of community resources, ideas, and leadership in providing program design and direction; it must become unacceptable to allow mosquito larval habitats to exist in the community; homeowners must be convinced that it is in their best health interests to control *Ae. aegypti* on their premises and in their community. They will only develop effective, sustainable Ae in partnership with government agencies. *Aegypti* control programs. Community-based, integrated *Ae. aegypti* control is currently viewed by many public health officials as the only cost-effective approach that will provide sustainable and effective disease control over the long term [50].

As *Ae. aegypti* and *Ae. albopictus* thrive in stagnant water collections like those in peridomestic water supplies used in the absence of piped water provision; proliferation may be encouraged by human population growth or the migratory waves from areas of civil upheaval and the possible subsequent formation of uncontrolled slums. Therefore, consistent public awareness about the significance of eliminating any peridomestic stagnant water is of critical public health importance [51].

For example, Cuba initiated a program encouraging artificial flowers in cemeteries, mosquito-proof containers for water storage, and controlling unused containers that retain water and provide larval habitat for *Ae. aegypti* [50]. Singapore reduced exposure between 1973 and 1981 by serving 54,297 orders and 7047 summonses. These resulted in fines totaling approximately $800,000 U.S. dollars [50].

Interestingly, attitudes about reducing ZIKV-infected mosquitoes seem to depend on a handful of variables, the foremost of which appears to be whether there are reported cases of ZIKV infection nearby. In the eyes of some individuals, the mosquito project is an expression of horrifying corporate greed that treats the public as lab rats [52].

Watkins added social media would have helped in public outreach on that issue. She says the district never got ahead of it, instead of reacting rather than being proactive with residents. She says she would vote against the Key Haven release with her information [27].

We are in the dry season throughout the southeast, certainly in Florida. And so, we do not see mosquitoes as much now. But scientists say they believe ZIKV is here to stay in South Florida. And the Ae. aegypti mosquito, the primary carrier of this disease, is found throughout the southeast. The local transmission also happened last year in south Texas. So, it's not just Florida [53].

Evidence-based approaches, such as Communication for Behavioral Impact (COMBI) [54] and the Socializing Evidence for Participatory Action (SEPA) program based on CIET methods [55], are proving more successful in effecting behavioral change and reduction. Of entomological indices and hold promise as community-based vector control programs [16].

18.15 Review of Some Engagement Activities

A successful long-term sustainable control program will require top-down and bottom-up approaches. The top-down component involves integrating several new tools and is preferably managed over the long term by the community [56]. Public trust and complacency in such an ineffective approach have only increased the challenge of explaining the need for community involvement in controlling larval habitats [57].

Notably, a study (2018) by Ophir and Jamieson [58] associated the misbelief between MMR and ZIKV vaccines and also drew some conclusions predicting the importance of education in modifying well-being behavior. The study suggests that accurately communicating about the risks of ZIKV can help lessen the detrimental effects of the misbelief.

The study also found people who believe that ZIKV causes the congenital disability microcephaly (which is accurate) and those who think ZIKV is likely to cause death (which is inaccurate) were more likely to intend to vaccinate. People engaged in behaviors to protect against ZIKV were less likely to intend to get the vaccination, which "may be the result" [59].

"Even if we can't change what people think about the MMR vaccine, if we can give them an accurate picture of how vulnerable they are to a disease such as Zika, they can make a more informed decision about it," Ophir said [59].

"When a new disease arises, people who lack understanding of the new threat may extrapolate from their knowledge of other diseases," said Yotam Ophir, who co-authored the study with APPC Director Kathleen Hall Jamieson. While the authors found intentions were positively influenced by perceived severity of and vulnerability to ZIKV and belief in science's efficacy. As ZIKV vaccines are being developed and tested, the scientific understanding of the nature of the virus remains incomplete, and the public's knowledge of what science does know is lagging. As such, well-constructed and executed outreach activities will be necessary.

People who believe that ZIKV causes congenital disability microcephaly (which is accurate) and believe ZIKV is likely to cause death (which is inaccurate) are more likely to intend to vaccinate. People engaged in behaviors to protect against ZIKV were less likely to intend to get the vaccination, which "may be the result of their confidence that their actions preempt the need to be vaccinated," the researchers said [60].

Despite incontrovertible benefits, vaccinations have been the subject of misinformation about alleged risks. Resulting misbeliefs can attenuate vaccination rates, risking a resurgence in diseases considered under control [58].

Many programs focus only on changing people's knowledge and raising awareness, believing that behavior will change; when it does not (and usually does not), the standard response is to bombard people with even more entomological and epidemiological facts, often using sophisticated advertising techniques. But more information, fancy posters, colorful T-shirts, glossy pamphlets, and stylish TV features rarely lead to behavioral responses if they are not behaviorally focused [57].

Researchers suggest more outreach and community-level education efforts are needed to inform the public about the threat of ZIKV and promote preventative action. "During the height of the ZIKV outbreak in 2016, the department used a variety of digital, print, TV, and radio advertisements to inform Floridians on how to protect themselves and their loved ones from the threat of ZIKV, reaching an estimated 31.9 million people," said Devin Galetta, Interim Communications Director for the Florida Department of Health [61].

"People need to understand that by protecting themselves from the virus, they're protecting everyone from the virus," said lead author Kenneth M. Winneg, the managing director of survey research at the Annenberg Public Policy Center of the University of Pennsylvania. "It's not enough to have the most at risk protecting themselves. You need the entire community involved" [61].

Pérez et al. [62] calls for a "social learning process" implying a transfer of power and responsibilities to local people. In this sense, the actions undertaken must be oriented toward creating local capabilities, strengthening existing structures and organizations, and promoting group work for learning participation from participation itself.

- From a disease control point of view, this concept will better insert the *Ae. aegypti* vector controls social, cultural, political, and economic environments.
- From a scientific point of view, this concept opens a large field of research: participatory, qualitative, and emphasizing a process of learning from past errors [62].

Achieving community participation is a dynamic and complex process that entails continuous learning from present and past experiences. This learning process should concern stakeholders (local people, implementers, researchers, etc.). The ability to work in teams is essential. Therefore, these actors should have the necessary skills. Indeed, attitudes toward participation and communication skills of the researchers/implementers are critical [62]. In addition, Pérez et al. [62] warn that "attaining successful community participation implies adapting strategies to local situations and the communities' conditions and capabilities". Therefore, there can be no blueprints or simple recipes for achieving community participation. Each case is unique [62].

Even though each situation may be unique, some principles underlying successful cases can be learned from them.

No government and health system can solve the problem without every person's active and aware participation and organized community action. A remarkable fact is that community participation is generally high in epidemic times. This emphasizes the need for continuous surveillance by health personnel to ensure the effectiveness and sustainability of community participation initiatives. Sustainability over time will be enhanced by the availability of easy tools based on evidence [63].

On an international level, PAHO said a crucial tool in the long-term fight against ZIKV is community education and participation in eliminating mosquito vectors and breeding sites [64]. Prevention implies knowing about the mosquito's behavior and breeding sites and understanding every individual's environmental responsibility in the ongoing elimination of breeding sites. This should be a continuous practice in homes and communities [21]. Parks and Lloyd [57] detailed four examples of successful international awareness engagement campaigns. This article is highly recommended for anyone interested in engagement theory and practice. Sorry for the long quote, but it comes from one of the best works in the literature.

- In Bucaramanga, **Colombia**, high school students were trained in *Ae. aegypti* biology and control and assist as community-based health educators. Results over 11 years (1992–2001) showed a steady decrease (with occasional increases) in the number of houses with *Ae. aegypti* larvae present.
- In Purwokerto, Central Java, **Indonesia**, a partnership has been established between the local government, the Rotary Club, the Family Welfare Empowerment Organization (PKK), and municipal health services. This project operates at the neighborhood association level. Each neighborhood consists of between 25 and 50 households. Within each area, houses are grouped into sets of ten, known as "Dasa Wisma." Each dasawisma has a leader, usually a woman cadre from the PKK. The dasawisma arrange schedules, where one house inspects the other nine houses. Known as "Piket Bersama" (Picket Together), these house-to-house inspections are conducted every week, so each household takes its turn every ten weeks. The success of this project can be measured by reducing the house index from 20% before activities began to 2% once activities were running well.
- In Johore State, **Malaysia**, an integrated social mobilization and communication campaign motivated householders in Johor Bahru District to seek a prompt diagnosis for any fevers, to destroy any larval breeding sites found around their premises, Dengue volunteer inspection teams (DeVIT) were formed in forty-eight localities. Some 615 volunteers came forward to join DeVIT teams. During the three-month campaign, DeVIT teams advised 100,956 people, distributed 101,534 flyers, and inspected 1440 vacant lots. The campaign resulted in a dramatic drop in the occurrence of dengue in the district; three months after the campaign, tracking surveys revealed that 70% of householders were still checking their household premises.
- In El Progreso, **Honduras**, a method for cleaning large cement washbasins and metal drums (promoted as "the Untadita") was designed using existing household

behaviors. The Untadita is a homemade ovicide consisting of mixing chlorine bleach with detergent, applying this mixture to the walls of the containers, waiting a few minutes before scrubbing the walls with a brush, and then rinsing the walls. Householders' use of low-cost and easily obtainable materials weekly, the Untadita significantly impacted *Aedes* larval populations in the project area (it lowered the number and the age of larvae in washbasins and drums) and became one of the recommended control methods in the national program [57].

There are increasing numbers of U.S. awareness campaigns as well. For example, with fifty-seven reported travel-related cases, the Arizona Department of Health Services launched a new public awareness campaign encouraging residents and visitors to take precautions to prevent the spread of ZIKV in Arizona. According to Dr. Cara Christ, director of the Arizona Department of Health Services: "The Arizona Department of Health Services is proactively working with our partners to prevent the spread of disease in our state, and we need the community's help." "This new campaign ("Fight the Bite") will help build awareness about the steps people can take to prevent the spread of ZIKV, such as using insect repellent, wearing long-sleeve shirts and long pants when you're outdoors, and removing standing water from around your home." The outreach initiative includes billboards, radio public service announcements, and digital ads targeted at those most at risk, such as people traveling to ZIKV-affected areas and women who are pregnant or planning to get pregnant. The campaign also includes prevention materials that healthcare providers can share with patients on the best way to prevent ZIKV [65].

The WHO learned a brutal lesson in the 1970s in Delhi, India. "Untruths reported by the media included the idea that the United States was using the WHO-associated research project to test dangerous chemosterilization methods in India and that the unstated goal of the program was to develop biological weapons. The result was that the Mosquito Control Group had begun to have some successes and had its program prematurely terminated. The lesson, of course, is that even with great scientific success, such programs can fail if the correct relationships are not formed with the public and the government" [34].

Engagement with stakeholders is an essential component of pest management. Consider two reasonably successful cases.

Eliminating mosquito breeding sites remains the most crucial strategy for preventing and controlling ZIKV (dengue and chikungunya) infection. Therefore, communication plans for the response to ZIKV should include intersectoral action and community engagement to modify behaviors and encourage sustained practices to eliminate breeding sites and control the mosquito, as well as to inform and educate target audiences about the steps they can take to prevent ZIKV transmission [21].

18.16 Conclusion

Engagement is neither an option nor is it something other people do. Engagement is communication. It is complex and often chaotic. It must be tailored to the problem at hand. There is no silver bullet. Engagement activities are creative and personalized. This is a world where trust issues problematize communication. As such, communication effectively demands an engagement component as the trust-building component. Engagement activities are difficult and expensive, and they should be done for reasons other than wish fulfillment.

Entomologists and physicians must direct the control efforts and evaluate intervention strategies to determine program efficacy [50]. However, Gubler and Clark from the National Institute for Infectious Diseases recommended that a staff of community organizers include social scientists who can conduct the proper ethnographic studies needed to understand the community's values and diversity better, use this information to develop and target education messages, and to design intervention strategies. Marketing experts should be involved as community members to help develop educational materials/programs and to help market the program to the community.

In the last chapter, the title will be returned to and then how resilience is enhanced by lessons learned from ZIKV is examined.

References

1. Funk C, Tyson A (2021) Growing share of Americans say they plan to get a COVID-19 vaccine—or already have, March 5. https://www.pewresearch.org/science/2021/03/05/growing-share-of-americans-say-they-plan-to-get-a-covid-19-vaccine-or-already-have/. Accessed June 27, 2022
2. Harvard School of Public Health (2016) Zika virus and the election season, August. https://cdn1.sph.harvard.edu/wp-content/uploads/sites/94/2016/08/STAT-Harvard-Poll-August-2016-Zika.pdf. Accessed July 17, 2017
3. Rainie L (2016) Half of Americans say threats from infectious diseases are growing. PEW Research Center, July 8. https://www.pewresearch.org/science/2016/07/08/half-of-americans-say-threats-from-infectious-diseases-are-growing/. Accessed June 11, 2022
4. James S et al (2018) Pathway to deployment of gene drive mosquitoes as a potential biocontrol tool for elimination of malaria in Sub-Saharan Africa: recommendations of a scientific working group. Am J Trop Med Hygiene 98(Supp. 6). https://www.ncbi.nlm.nih.gov/pubmed/29882508. Accessed September 10, 2018
5. Bartlett-Healy K, et al (2011) Source reduction behavior as an independent measurement of the impact of a public health education campaign in an integrated vector management program for the Asian tiger mosquito. Inter J Environ Res Public Health 8:5, May. https://www.ncbi.nlm.nih.gov/pubmed/21655124. Accessed August 27, 2018
6. McNaughton D (2012) The importance of long-term social research in enabling participation and developing engagement strategies for new dengue control technologies. PLOS Neglected Trop Dis 6:8, August. E1786. https://doi.org/10.1371/journal.pntd.0001785. Accessed August 9, 2018
7. Cohen ERM et al (2008). Public engagement on global health challenges. BMC Pub Heal 8:168, May 20. https://doi.org/10.1186/1471-2458-8-168. Accessed September 4, 2018

8. Berube D (2018) How social science should complement scientific discovery: lessons from nanoscience. J Nanopart Res 20:120. https://doi.org/10.1007/s11051-018-4210-x
9. Bauer MW et al (2007) What can we learn from 25 years of PUS survey research? liberating and expanding the agenda. Pub Understand Sci 16. https://doi.org/10.1177/0963662506071287. Accessed June 13, 2019
10. Morrison A et al (2008) Defining challenges and proposing solutions for control of the virus vector Aedes aegypti. PLOS Med 5:3, March. https://doi.org/10.1371/journal.pmed.0050068. Accessed May 26, 2017
11. McCall, P. J. & Kittayapong, Pattamaporn. (2006). Control of dengue vectors: tools and strategies. Report of the Scientific Working Group on Dengue. TDR/World Health Organization, Geneva. http://www.who.int/tdr/publications/documents/swg_dengue_2.pdf. Accessed July 20, 2017
12. CBSDFW (2017) North Texas mosquito season means battling West Nile & Zika, May 5
13. American Mosquito Control Association (2017) Best practices for integrated mosquito management: a focused update, January. https://www.researchgate.net/publication/315924484_Best_Practices_for_Integrated_Mosquito_Management_A_Focused_Update. Accessed August 27, 2018
14. ABC News (2017) Beyond pesticides daily news blog, May 1. http://beyondpesticides.org/dailynewsblog/2017/05/wolbachiamosquitoesreleasednewfightzikamosquitobornediseases/. Accessed June 26, 2017
15. BU Press Release (2018) BU medical student develops new solutions to stop the spread of Zika virus. Eureka Alert! Press Release, January 11. https://www.eurekalert.org/pub_releases/2018-01/buso-bms_1011118.php. Accessed July 10, 2018
16. Kyle J, Harris EVA (2008) Global spread and persistence of dengue. Annual Rev Microbiol 62. https://www.ncbi.nlm.nih.gov/pubmed/18429680. Accessed May 18, 2017; Gubler DJ, Clark GG (1994) Community-based integrated control of Aedes aegypti: a brief overview of current programs. Am J Trop Med Hygiene 50:6, Supplement. http://www.ajtmh.org/docserver/fulltext/14761645/50/6_Suppl/TM05006S0050.pdf?expires=1495122720&id=id&accname=12094&checksum=9113072426049234C93BBCC9C118E0B6. Accessed May 18, 2017
17. Stone J (2016) Smart science confirms Wolbachia's value in fighting Zika as well as dengue. Forbes, May 4. https://www.forbes.com/forbes/welcome/?toURL=https://www.forbes.com/sites/judystone/2016/05/04/smart-science-confirms-wolbachias-value-in-fighting-zika-as-well-as-dengue/&refURL=https://www.google.com/&referrer=https://www.google.com/. Accessed June 2, 2017
18. Berube D (2008) Stakeholder participation in nanotechnology policy debates. In: Bennett D (ed) Nanotechnology: ethics and society. CRC Press, London, pp 225–229
19. Berube D (2006) The rhetoric of 'stake-holding. In: Alhoff F, Lin P, Moor J, Weckert J (eds) Nanoethics: Examining the societal impact of nanotechnology. Wiley and Sons, Hoboken, NJ, pp 225–240
20. ASTHO (2015) Before the swarm: guidelines for the emergency management of vector-borne disease outbreaks. http://www.astho.org/Programs/Environmental-Health/Natural-Environment/Before-the-Swarm/. Accessed August 28, 2018
21. PAHO (2016) Zika virus infection step-by-step guide to risk communication and community engagement, November. http://iris.paho.org/xmlui/handle/123456789/18599. Accessed May 29, 2017
22. Lavery JV, Harrington LC, Scott TW (2008) Ethical, social, and cultural considerations for site selection for research with genetically modified mosquitoes. Am J Trop Med Hygiene 79:3, September. https://www.ncbi.nlm.nih.gov/pubmed/18784220. Accessed August 8, 2018
23. Dart T (2017) 'It's going to hit the poorest people': Zika outbreak feared on the Texas border. The Guardian, April 23. https://www.theguardian.com/world/2017/apr/23/zika-outbreak-rio-grande-valley-texas-border-health. Accessed May 9, 2017
24. Zhang S (2015) Ride with the mosquito hunters protecting the US against Zika. Wired, February 8. https://www.wired.com/2016/02/mosquito-control-story/. Accessed October 24, 2016

25. Servick K (2016) Brazil will release billions of lab-grown mosquitoes to combat infectious disease. will it work? Science, October 13. http://www.sciencemag.org/news/2016/10/brazil-will-release-billions-lab-grown-mosquitoes-combat-infectious-disease-will-it. Accessed May 31, 2017
26. PR Newswire (2016) Oxitec and Piracicaba City Hall start release of Friendly™ *Aedes*. Market Watch, September 8. http://www.oxitec.com/oxitec-piracicaba-city-hall-start-release-friendly-Aedes-additional-10-downtown-neighborhoods/. Accessed September 18, 2016
27. Editorial (2016) Our recommendations in the Nov. 8 election. Florida Keys News, October 3. http://www.flkeysnews.com/opinion/editorials/article105619031.html. Accessed October 4, 2016
28. Burgess MM, Mumford JD, Lavery JV (2018) Public engagement pathways for emerging GM insect technologies. BMC Proceed 12(Suppl 8):12. https://doi.org/10.1186/s12919-018-0109-x
29. Lukensmeyer CJ, Torres LH (2006) Public deliberation: a manager's guide to citizen engagement. IBM Center for the Business of Government, Washington, DC
30. Newman PA (2006) Towards a science of community engagement. The Lancet 367:9507, January 28. https://www.thelancet.com/journals/lancet/article/PIIS0140-6736(06)68067-7/abstract?code=lancet-site. Accessed August 9, 2018
31. Allotey P et al (2014) Cohorts and community: a case study of community engagement in the establishment of a health and demographic surveillance site in Malaysia. Global Heal Action 7:23176. https://www.ncbi.nlm.nih.gov/pmc/articles/PMC4013487/. Accessed August 27, 2018
32. Andersson N et al (2015) Evidence based community mobilization for dengue prevention in Nicaragua and Mexico (Camino Verde, the Green Way): cluster randomized controlled trial. British Med J, July 8. https://www.bmj.com/content/351/bmj.h3267. Accessed August 27, 2018
33. Baly A et al (2007) Cost effectiveness of Aedes aegypti control programmes: participatory versus vertical. Trans Royal Soc Trop Med Hyg 101:6, March 21. https://www.ncbi.nlm.nih.gov/pubmed/17368696. Accessed August 27, 2018
34. McGraw EA, O'Neill SL (2013) Beyond insecticides: new thinking on an ancient problem. Nature: Microbiol 11:3, March. https://www.ncbi.nlm.nih.gov/pubmed/23411863. Accessed August 9, 2018
35. Marris C, Rose N (2010) Open engagement: exploring public participation in the biosciences. PLOS Biol 8(11):E1000549. https://doi.org/10.1371/journal.pbio.1000549. Accessed August 9, 2018
36. Neuhaus CP, Caplan AL (2017) Ethical lessons from a tale of two genetically modified insects. Nature Biotech 35(8):713–726, August
37. Rip A, Misa TJ, Schot J (eds) (1995) Managing technology in society. Pinter, London
38. Grin J, van de Graaf H, Hoppe R (eds) (1997) Technology assessment through interaction: a guide. Rathenau Institute, Den Haag
39. Guston DH, Sarewitz D (2002) Real-time technology assessment. Technol Soc 24:1–2. https://www.sciencedirect.com/science/article/pii/S0160791X01000471. Accessed August 9, 2018
40. Wilson J, Willis R (2004) See-through science. why public engagement needs to move upstream. DEMOS, London
41. Stirling A (2008) '"Opening up"' and '"closing down"': power, participation, and pluralism in the social appraisal of technology. Sci Technol Human Values 33:262–294
42. Santibañez S, Lynch J, Paye YP, McCalla H et al (2017) Engaging community and faith-based organizations in the Zika response, United States, 2016. Public Health Rep 132(4):436–442
43. August EM, Rosenthal J, Torrez R, Romero L et al (2020) Community understanding of contraception during the Zika virus outbreak in Puerto Rico. Current Topics Heal Educ 21(1):133–141
44. Manderson L, Abay A (1992) An epidemic in the field? Rapid assessment procedures and health research. Soc Sci Med 35:7, October. https://www.ncbi.nlm.nih.gov/pubmed/1411684. Accessed August 9, 2018

45. Cohen ERM et al (2008) Public engagement on global health challenges. BMC Pub Heal 8:168, May 20. /https://doi.org/10.1186/1471-2458-8-168. Accessed September 4, 2018
46. Depew M, Modern Mosaic Consulting (2022) 3 ways to leverage stakeholder engagement beyond strategic planning. https://modernmosaicconsulting.com/leveraging-stakeholder-engagement/. Accessed July 3, 2022
47. Marshall P, Rotimi C (2001) Ethical challenges in community-based research. Am J Med Sci 322:5, November. https://www.ncbi.nlm.nih.gov/pubmed/11721794. Accessed September 10, 2018
48. Lavery JV et al (2010) Towards a framework for community engagement in global health research. Trends Parasitol 26:6, June. https://www.ncbi.nlm.nih.gov/pubmed/20299285. Accessed August 8, 2018
49. Treweek S et al (2018) Strategies to improve recruitment to randomized trials. Cochrane Database Syst Rev, Issue 2. Art. No.: MR000013. https://www.ncbi.nlm.nih.gov/pubmed/17443634. Accessed August 27, 2018
50. Gubler D, Clark GC (1996) Community involvement in the control of Aedes aegypti. Acta Tropica 61. https://www.ncbi.nlm.nih.gov/pubmed/8740894. Accessed June 26, 2017
51. Verou R (2016) Zika virus, vectors, reservoirs, amplifying hosts, and their potential to spread worldwide: what we know and what we should investigate urgently. Inter J Infectious Diseases, 48, July. https://www.ncbi.nlm.nih.gov/pubmed/27208633. Accessed July 26, 2018
52. Brown K (2016) If genetically modified mosquitoes freaked you out, you will not like what's coming. GIZMODO, December 13. http://gizmodo.com/if-genetically-modified-mosquitoes-freaked-you-out-you-1790021060. Accessed March 15, 2017
53. Allen G (2017) Bacteria-infected mosquitoes tested as a way to control population. NPR Now, April 20. http://www.npr.org/2017/04/20/524833658/bacteria-infected-mosquitoes-tested-as-a-way-to-control-its-population. Accessed May 3, 2017
54. Parks W, Lloyd L (2004) Planning social mobilization and communication for dengue fever prevention and control: a step-by-step guide. Geneva: WHO. http://apps.who.int/iris/bitstream/10665/42832/1/9241591072.pdf. Accessed May 18, 2017
55. Andersson N (1996) Evidence-based planning: the philosophy and methods of sentinel community surveillance. World Bank Econ. Dev. Inst. Washington, DC
56. Gubler DJ (2011) Dengue, urbanization and globalization: the unholy trinity of the 21st century. Trop Med Heal 39:4. Supplement. https://www.ncbi.nlm.nih.gov/pmc/articles/PMC3317603/. Accessed July 17, 2017
57. Parks W, Linda L (2008) Planning social mobilization and communication for dengue fever prevention and control. WHO, Geneva. http://www.who.int/immunization/hpv/communicate/planning_social_mobilization_and_communication_for_dengue_fever_prevention_and_control_who_cds_wmc_2004.pdf. Accessed July 19, 2017
58. Ophir Y, Jamieson KH (2018) Intentions to use a novel Zika vaccine: the effects of misbeliefs about the MMR vaccine and perceptions about Zika. J Pub Heal 40(4):e531-e537, December. https://doi.org/10.1093/pubmed/fdy042
59. Annenberg Public Policy Center of the University of Pennsylvania (2018) False beliefs about MMR vaccine found to influence acceptance of Zika vaccine: Study finds spillover effect from misbeliefs about MMR vaccine and autism. ScienceDaily. 15, March. https://www.sciencedaily.com/releases/2018/03/180315130717.htm. Accessed September 10, 2018
60. Annenberg Public Policy Center (2018) Does a scientific breakthrough increase confidence in science? news of a Zika vaccine and trust in science. The Communication Initiative Network, May 15. http://www.comminit.com/global/content/does-scientific-breakthrough-increase-confidence-science-news-zika-vaccine-and-trust-sci. Accessed July 10, 2018
61. Garrett A (2018) Floridians took Zika virus seriously-but less than 50% tried to protect themselves. ClickOrlando.com, June 14. https://www.clickorlando.com/author/ashleygarrett. Accessed July 24, 2018
62. Pérez D et al (2007) Community participation in Aedes aegypti control: a sociological perspective on five years of research in the health area "26 de Julio", Havana, Cuba. Trop Med Inter Heal. 12:5, May. https://www.ncbi.nlm.nih.gov/pubmed/17445134. Accessed July 19, 2017

63. Troncoso A (2016) Zika threatens to become a huge worldwide pandemic. Asian Pacific J Trop Biomed 6:6. http://www.sciencedirect.com/science/article/pii/S2221169116302921. Accessed June 4, 2017
64. Jamaica Observer (2017) PAHO says Zika virus outbreak continues a year after global emergency. Jamaica Observer, February 5. http://www.jamaicaobserver.com/news/PAHO-says-Zika-virus-outbreak-continues-a-year-after-global-emergency. Accessed May 17, 2017
65. De Hoff F (2017) A new campaign launched to fight the Zika virus in Arizona. KVOA.cpm, April 24. http://www.kvoa.com/story/35228649/a-new-campaign-launched-to-fight-the-zika-virus-in-arizona. Accessed May 10, 2017

Chapter 19
Learning from ZIKV

> In 2015, the Association of State and Territorial Health Officials (ASTHO) presciently wrote in a report, "Before the Swarm," assessing vector (that is, not just mosquitoes, but ticks and other insects) control efforts: The unpredictable nature and severity of vector-borne disease outbreaks demonstrate the urgent need for careful preparation and the incorporation of vector-control emergency management activities into overall public health preparedness efforts. Since climate change is altering temperature and precipitation patterns across the country, it is critical that public health professionals also prepare for a potential increase in the geographic spread of existing vectors, such as *Ae. albopictus* or *Ae. aegypti*, and the potential for new vector-borne diseases. [1]

It seems like the crisis is over. Why then should there be concern about ZIKV? The answer: There is an excellent opportunity to use ZIKV as a learning moment without having it be clouded by gruesome death, much like what was seen during the brief Ebola outbreak in the U.S. in 2014.

Resiliency involved learned behavior as well as institutional investments into equipment and technology. When an outbreak fades, we thank our fortune and learn from the experience. However, we are not particularly good at learning lessons after outbreaks. Generally, a deep breath is taken, and people move on to the next virus.

At least in the U.S., there has been a constant challenge to try and situate to situate ourselves in the center of a pandemic involving zoonotic diseases. Those who suffer most are not Americans per se and do not look like most of us. How approaching the ZIKV outbreaks in Puerto Rico and other US possessions that U.S. citizens inhabit has been despicable.

Ziika/Zika is both a jungle in Uganda and a state of mind. There is a unique otherness toward zoonotic diseases moving into human populations. Not only does the developed world perceive the conditions as alien, but it has dissociated itself from its occurrence and implications. For them, the outbreaks are anomalies rather than a daily occurrence.

For example, one of the most concerning consequences of ZIKV was its effect on the development of children. Traditionally, diseases that could cross into the mother's unborn child were sexually transmitted, such as syphilis, gonorrhea, and chlamydia. However, new research by Miner et al. (2018) [2] indicated that both a West Nile

D. M. Berube, *Pandemics and Resilience: Lessons we should have learned from Zika*, Risk, Systems and Decisions, https://doi.org/10.1007/978-3-031-25370-6_19

virus, a mosquito-borne virus (As of January 9, 2018, 47 states and the D.C. have reported cases of West Nile virus to the United States CCD that has been detected across the continental U.S.), and the Powassan virus, a tick-transmitted disease a tick-transmitted disease, most commonly reported in the Great Lakes region, crossed the placenta and caused fetal death during infections in pregnant mice. The pathogens were also replicated in samples of both maternal and fetal tissues obtained from healthy pregnant human donors in their second trimesters [3].

The ability of the ZIKV to induce congenital disabilities did not come to light until the 2015–2016 Latin American epidemic, which sickened more than 1.5 million people. There may have been better preparation for the ZIKV had there been attention to the consequences of the West Nile and the Powassan viruses.

Some studies suggest that the ZIKV evolved and acquired the ability to infect the placenta and cause microcephaly. The history between the health services industry and the outbreak might have been less problematic had closer attention been paid to ZIKV infections prior to the outbreak in 2015–2016.

Yet, the events that led to the pathogen becoming a global public health concern were more challenging than they might have been. Miner argues, "It might be that ZIKV has always caused congenital disabilities, but it just hadn't been noticed until 2015. The outbreak in the Americas was so widespread that it generated many simultaneous cases, which probably made it easier for epidemiologists to identify the association between congenital disabilities and ZIKV" [3].

A mandate to learn from what experiences have come from the past in infectious disease control is imperative to foster resiliency to zoonotic diseases. Since originally thought to affect only tropical regions, mosquito-transmitted viruses are increasingly becoming a worldwide health challenge due to increased global movement and insecticide resistance [4]. According to Braack et al. [5], "…[i]t is, therefore, incumbent upon nations and governments to maintain and indeed increase vigilance against the insidious infiltration of vectors and viruses. All it takes is for a few infected mosquitoes, people, or animals to establish the vector or virus, which slowly simmers and grows until it emerges as a full-blown outbreak and public health emergency. This is the historical pattern of most of these diseases, be it yellow fever, dengue, chikungunya, West Nile virus, ZIKV, etc." [5].

Decisions regarding resource allocations to control the outbreak underscore the need for a tool to weigh policies according to their cost and the health burden they could avert. ZIKV was a case in point. According to Alfaro-Murillo et al. [6], estimates suggest that the health burden from microcephaly and Guillain–Barre syndrome (GBS) warrants substantial expenditures focused on ZIKV control. The research team identified the lifetime direct medical cost from ZIKV-related microcephaly, assuming a 79.7% probability of survival through the first year of life and an optimistic life expectancy of 35 years; they estimate $180,000 in lifetime expenses in the US as well as $57,000 as the direct media cost per case of GBS. However, indirect costs can be significant [6].

> The ZIKV outbreak impacted in Puerto Rico, and it has been estimated total direct medical costs for microcephaly and GBS of $104 million in our base case and from $39 million in our most conservative scenario up to $159 million in our less traditional design. However,

upon excluding direct non-medical costs and productivity losses, the values rise to $736 million in the base case, with $280 million in the most conservative scenario and as high as $1.13 billion in the less traditional design. These estimates do not include other indirect costs such as specialized child-care support, parental productivity losses, or the psychological toll of families with children with microcephaly, which are all substantial yet challenging to quantify. [6]

PAHO warned for hemispheric eradication to be completed, all the countries of the Americas, as well as European authorities responsible for territories in the Americas, must agree to eradicate. Countries that do not launch *Ae. aegypti* eradication programs would become possible sources of reinfestation, as happened during the early campaigns. Also, *Ae. aegypti* is present in significant numbers in Asia, posing the risk of reinfestation from that region [7].

Infectious disease health strategy cannot be unidimensional. Still, it will need an array of approaches, and it cannot be national but global, and those are the significant challenges waiting. It will not be easy, but it is possible. It will take a lot of understanding and no small a fraction of public investment of effort and resources.

19.1 Public Health Strategies

Another challenge to pandemic management is associated with recent findings by Platt et al. in 2018. The new study tests the effects of four other viruses: the two flaviviruses and two alphaviruses, chikungunya and Mayaro, which have led to outbreaks in ZIKV-affected areas. Pregnant mice infected with West Nile or Powassan virus—flaviviruses like ZIKV—also showed fetal harm. Over 40% of these infected fetuses died [8].

ZIKV, West Nile, and Powassan share similarities in their genetic information, Miner says. "So, there may be certain features of those virus genes and proteins in that family that confers this ability to infect certain cell types." But scientists don't understand what those features are yet, he says [8].

The goal is public mobilization in responding to infectious disease pandemics. Public health strategies include **environmental** tactics focused on controlling mosquito populations through aerial and indoor spraying, larvicide deployment, elimination of mosquito breeding grounds, and the potential introduction of genetically modified mosquitoes; **behavioral** strategies focused on promoting physical and chemical barriers to mosquitoes (insecticides and window screens) and on reproductive-related decision-making (delaying pregnancy, using contraceptives, or avoiding travel to areas with ZIKV infections); and **clinical** interventions, including screening and testing for disease and the availability of pregnancy termination services [9].

To protect ourselves against emerging diseases, the essential first step is practical global disease surveillance to warn early about emerging infections. This must be tied to incentives, such as national development, and eventually, be backed by a system for an appropriate rapid response. World surveillance capabilities are critically deficient [10].

There are as many as 1.6 million viruses with nothing known about lurking in mammals and birds, and as many as half might have the potential to jump to and infect humans. That is an estimate based on mathematical models, but the threat is clear. Six out of 10 infectious diseases that strike come from animals.

The list includes HIV/AIDS, Ebola, MERS, SARS, and in all probability, COVID-19. Scientists have identified about 260 viruses that infect people—a tiny fraction of what is out there. Viruses are ubiquitous. Add them all up, and they would weigh more than all the plants and animals, though they are tiny. About 100 million particles of the new coronavirus, SARS-CoV-2, can fit on a pinhead, virologist Peter Piot told a TEDMED interviewer earlier this year [11].

Last year, researchers projected in Nature Microbiology that by 2050, nearly half the world's population would be at risk of contracting diseases like these—caused by microbes called arboviruses—due to climate change and urbanization. An increasingly warm and densely populated planet will be more hospitable to the mosquito species that spread the disease, *Ae. aegypti* and *Ae. albopictus* [12]. Two brief concerns follow.

19.2 Rift Valley Fever

The mosquito-borne virus that causes Rift Valley fever may severely harm human fetuses if acquired by mothers during pregnancy, according to a 2018 Pittsburgh study. There are no vaccines or treatments for Rift Valley fever.

The hatching dynamic of *Aedes* mosquitoes, Africa's main reservoir of RVF, strongly depends on the rainfall pattern. Heavy rainfall results in a massive hatching episode and, consequently, the development of a large vector population.

ZIKV must take the "side roads" into the placenta to infect a fetus, while the Rift Valley fever virus can take the "expressway," Dr. Hartman, one of the study's lead authors, said [13].

> "If doctors had known about ZIKV's birth effects, they could have done more to protect pregnant women and babies. With Rift Valley fever, we're trying to get ahead of the curve." Rift Valley fever primarily occurs in livestock in sub-Saharan Africa, where outbreaks cause 90 to 100 percent of pregnant cows in a herd to miscarry or deliver stillborn calves, often a significant economic loss. But hundreds of cases also occur yearly in humans, causing flu-like symptoms and severe liver problems. According to Dr. Hartman, the outbreaks have moved beyond Africa: In late 2000, an attack in Saudi Arabia infected more than 100,000 people and led to at least 700 deaths. The mosquito that carries the disease is also found in Europe and the Americas. [13]

19.3 Disease X

Phylogenetic trees suggest that Africa was the ancestral origin of all mosquito and tick-transmitted flaviviruses. Many other arboviral infections—like the ZIKV in Africa—are circulating within immunologically adapted Indigenous African populations with a history of severe symptoms [5].

Given the preponderance of zoonoses of African origin (phylogenetic trees suggest that Africa was the ancestral origin of all mosquito and tick-transmitted flaviviruses) that have escaped previous endemic African settings to make geographical jumps to other regions through anthropogenic processes that are likely to escalate, Braack et al. [5] reported on mosquito-borne arboviral zoonoses that are already known to exist in Africa, either with potential for the continuing expansion of range or of which very little is known except that they infect humans or have genetic affinities which suggest they may infect humans. Public health consequences are difficult to predict when entering immunologically naïve populations in new geographical settings [5].

Competent vectors, commonly already present in the areas, provide opportunities for infection by exotic pathogens introduced by travel and trade. At the same time, the correct combination of environmental conditions (both abiotic and biotic) makes many far-flung parts of the world latently and predictably, but differentially, permissive to continuous transmission cycles. Socioeconomic factors and nutritional status determine human exposure to disease and resistance to infection, respectively, so disease incidence can vary independently of biological processes. The emergence of exotic vector-borne pathogens often depends on economic, commercial, or social events, commonly involving global movements of people and trade goods [14].

According to the 2018 R&D Blueprint for the World Health Organization, the infectious disease perhaps most unprepared for may be one that has never been heard of before. Hence, the WHO added "Disease X" to their list.

Disease X represents the knowledge that a serious international epidemic could be caused by a pathogen currently unknown to cause human disease. So, the R&D Blueprint explicitly seeks to enable cross-cutting R&D preparedness that is also relevant for an unspecified "Disease X" [15].

Disease X is not yet known to exist, but health officials recognize that the next significant outbreak could come from a virus they have not yet encountered, so they want to be ready. If the unknown virus hits public health, scientists say there needs to be a system that could rapidly develop a vaccine to fight it [16].

19.4 Review of a An Approach

Following are fundamental reviews of some more outstanding programs adopted to reduce exposure to the disease vector, provide treatment, and engage the public.

For example, Li et al. [17] reviewed Zika Contraception Access Network (Z-CAN). Z-CAN was adopted to overcome barriers to contraception access in Puerto

Rico. The Z-CAN intervention cost $26.1 million, including costs for the full range of reversible contraceptive methods, contraception-related services, and programmatic activities. The program is estimated to have prevented 85% of estimated ZIKV-associated microcephaly cases (26) and unintended pregnancies in the absence of Z-CAN.

Economically, it made high marks. The intervention cost was projected to have been more than offset by $79.9 million in ZIKV–associated costs avoided, 96% of which were lifetime ZAM-associated costs, and $137.0 million from avoided unintended pregnancies, with total net savings in one year of $216.9 million. The results were consistent with the previous CEA study [18].

19.5 What Should We Have Learned?

Following each outbreak, the public health community bemoans a lack of prescience, but after decades of reacting to each event with little focus on mitigation, there remains marginally better protection against the next epidemic. Ability to mitigate disease emergence is undermined by poor understanding of the diversity and ecology of viral threats and the drivers of their emergence [19].

Management of vector-borne disease outbreaks can quickly become an extremely complicated activity. A communication gap often develops between the people with technical expertise in government, industry, and academia and the control agencies with logistical training and experience on the ground. Only proactive preparation and continuous interagency communication can ensure that all these groups benefit from working together [20].

This last chapter makes the case that it is necessary to create comprehensive protocols so that we do not trip over each other as we reinvent the wheel when we confront an outbreak. There is a strong bias to relinquish pandemic events into historical tomes. No one enjoys pondering especially disconcerting events. Most avoid terrible news at nearly every opportunity. While there may be less anxiety, this behavior does not bode well for building sets of procedures to be followed when first noticing an outbreak to reduce the chance of the attack becoming an epidemic and, subsequently, a pandemic. While needing to retain flexibility, approaching a pandemic event with a blank slate is simply unacceptable, given the propensity that pandemics will become more a feature of most lives than ever before moving into the twenty-first century.

While most areas in the United States do not confront exotic vector-borne diseases frequently, lessons learned from recent outbreaks remain essential and should be revisited to factor in the many drivers of pandemics, especially human migration patterns and the potential impact of climate change on vector-borne disease incidence [20]. There may be a reluctance to focus on a zoonotic disease during a coronavirus pandemic. However, learning to do more than one thing at once is the key to optimal pandemic management.

19.5.1 *Lesson One: How May We Reduce the Development Driver?*

The current concept of pandemics must be reconsidered. It should be accepted that the spread and severity of infectious diseases are generally more dependent on the social conditions of populations than on the properties of the infectious agent. The most effective way to prevent any contagious disease pandemic is to invest in improving social conditions [21].

Establishing shared services agreements, equipment pools, regional districts, and standard contracts for services can benefit mosquito control programs with limited or no funding. Public health laboratories in other areas may assist with testing and surveillance during an emergency. Contractors can provide immediate help but may be already committed elsewhere or prohibitively expensive for many communities [20].

Approaches to such a potential global health security threat should be consistent and proactive. They should involve coordinated, multipronged, multilateral collaborative efforts that actively engage local, regional, national, and global agencies and resource pools [22].

Some health departments use electronic health alerts or notification systems to notify known medical providers of the discovery of new vector-borne pathogens or an increased risk of infection with endemic vector-borne pathogens. Health alerts can also inform medical professionals about what diseases are expected to appear in patients during a particular season. These notification systems can be particularly effective when targeting physicians through large insurance networks or hospitals. Alert systems that can quickly notify the public about vector-borne diseases and outbreaks can also be beneficial, as communication is particularly critical in emergency vector-borne disease conditions [20].

For example, the actual risk of ZIKV to the unborn remains challenging to quantify and remediate. Accurate, portable, and inexpensive point-of-care testing is required to identify cases better and manage the current and future outbreaks of ZIKV, including optimization of preventive approaches and identifying more effective risk reduction strategies [22].

Given the rapid geographic spread of ZIKV in recent years, coordinated local, regional, and global efforts are needed to generate sufficient resources and political traction to effectively halt and contain the further expansion of the current outbreak [22].

Zoonotic diseases bring with them a panoply of challenges. As humans move closer to animals who serve as reservoirs for viruses and as animals, such as mosquitoes, move closer to human populations, more occurrences can be expected where exotic diseases cross over into the human population and wreak their havoc. The United States is already home to competent hosts and vectors for many of the world's most vector-borne severe diseases, including Ae. albopictus and Ae. aegypti, which can carry West Nile, yellow fever, dengue, and chikungunya viruses, underscores creating and sustaining vector control programs [20].

19.5.2 Lesson Two: How May We Reduce Uncertainty?

Much may be able to be learned from new virus species, especially ZIKV. Mosquito-borne viruses are tricky to predict, and experts say it is premature to dismiss the threat entirely. The unpredictable nature and severity of vector-borne disease outbreaks demonstrate the urgent need for careful preparation and vector control emergency management activities in public health preparedness efforts [20].

Harvard's Viswanath explains that public health professionals need to operate from a place of total and utter transparency. When something isn't known, or it's unclear, say so. "Audience members won't be happy if you don't know everything, but they'll appreciate that you're being completely open with them," he says. This might be challenging, and might lead to backlash, but it's the only way to maintain long-standing trust. "It's not about us. It's about understanding what the challenges are in their lives. Our message should solve their problems."

There are two levels of effective communication for any type of critical public health information, particularly when the information includes uncertainty. Information needs to be assessed at both the systems level and the message level, and each requires its own strategy [23].

19.5.3 Lesson Three: How May We Adopt Viable Vector Control Approaches?

Local health officials believe that emerging infectious disease threats, including the ZIKV, require ongoing vigilance. Since local health departments are on the front lines of community preparation and response, funding is needed for public education, mosquito eradication, investigation and vector control, refinement of diagnostics and vaccines, and expanded capacity to test people suspected of having contracted the disease [24].

With states having cut back on mosquito surveillance, active surveillance for human disease, and laboratory testing for WNV and other arboviruses, the ability to rapidly detect emerging and outbreak-threshold threats and to rapidly initiate prevention measures to minimize human morbidity and mortality (e.g., public notification and killing adult mosquitoes) might be compromised [25].

The Council of State and Territorial Epidemiologists (CSTE) assessment demonstrates that critical state-level monitoring capacity built for West Nile Virus (WNV) has eroded since 2004, despite states eliminating less critical activities such as avian mortality surveillance. The study compares capacity between 2004 and 2012.

Since the first assessment, 22% of jurisdictions had stopped conducting active human surveillance, 13% had stopped mosquito surveillance, 70% had reduced mosquito trapping and testing, and 64% had eliminated avian mortality surveillance. Reduction in early detection capacity compromises local and national ability to rapidly detect changes in WNV and other arboviral activity and initiate prevention measures. Each authority is encouraged to review its current surveillance systems considering the local threat of WNV and emerging arboviruses (e.g., dengue and chikungunya) and ensure it can rapidly detect and respond to critical changes in arbovirus activity [25].

Jurisdictions were asked how they managed to reduce ELC funding for WNV surveillance during the past five years. Among respondents to specific questions, 30 of 47 (64%) eliminated dead bird surveillance, 32 of 48 (67%) decreased the number of mosquito trap sites, 35 of 50 (70%) reduced the number of mosquito pools tested, and 23 of 51 (45%) decreased the number of WNV tests done on human specimens [25].

Unsurprisingly, resource allocations are often skewed toward reactive vector control measures, and proactive approaches are often underfunded and poorly executed [26]. On top of this, there is a systemic complacency about infectious diseases in general, vector-borne diseases, and a lack of public health resources for research, surveillance, prevention, and control programs [27].

> The extent and methods of vector and arbovirus surveillance and control vary widely between jurisdictions in the United States. This likely leads to the patchy implementation of control regimens that lack the urgency and uniformity of the more effective scenarios simulated here. This lack of uniformity also permeates the research that has been done on the effectiveness of various vector control approaches. Thus, while it would be helpful to compare our results with more real-world studies, the current literature contains little overlap in study design, making it difficult to compare the results of these disparate approaches. Increased standardization in methods, investment in proactive strategies, and communication about vector population dynamics locally, nationally, and internationally could significantly reduce the public health risks of ZIKV and other current and future vector-borne infectious diseases. [28]

19.5.4 Lesson Four: How May We Be More Resilient and Responsive?

The term appeared in psychology in the 1980s. It was a metaphor for how an organism recovers from stress. The ultimate goal of resiliency is homeostasis. Resiliency has also become the new foci of risk that fosters success instead of failure and typifies the modern movement toward positive psychology [29]. "Resilience is described as a set of cognitive and behavioral strategies of individuals who enact resilience within an organizational context" [30].

"The emerging field of resilience assessment and management and its implementation could thus evaluate cross-domain alternatives to identify a policy design that enhances the system's ability to (i) plan for such adverse events; (ii) absorb stress; (iii) recover; and (iv) predict and prepare for future stressors through necessary adaptation" [31].

The most resilient health systems have been those that understand how to keep people well and take risks—instead of getting stuck in the fee-for-service environment. It is one where complete responsibility clinically and financially for a population can be accepted. And then some super interesting, elegant, innovative decisions can be made, like telehealth, hospital at home, and all those other great tools that volume-based systems are not prepared to use [32].

Humanity has tended to react to emerging diseases as they occur, using understanding of epidemiology to mitigate the damage done. If the Stockholm Paradigm reflects a fundamentally correct explanation of the evolution of interspecific associations, then reactive management policies for dealing with emerging diseases cannot be economically sustainable [33]. It is unaffordable not to plan.

Some organizations have sought to improve public health surveillance in partnership with governments, universities, and private organizations. The Global Virus Network is a good example of what can be done to bring resilience into epidemic planning.

For example, San Francisco, CA–Global Viral Forecasting Initiative, a nonprofit organization, and Global Viral Forecasting Inc., a for-profit company, have worked in tandem since 2009 to conduct in-depth field research and services better to understand the control of the spread of microbial threats [34]. On the for-profit side, GVF Inc. will be known as Metabiota. Taking inspiration from the scientific term, which describes the stable relationship between a group of hosts and their microbial inhabitants, Metabiota will continue to provide services to governments and the private sector as GVF Inc. has for years.

In 2012, they united to become the company Metabiota. This company specializes in microbial threat detection, evaluation, and response by integrating field and lab research. Metabiota products and services include global field-based biological threat research, pathogen discovery, outbreak response, and clinical trials [34].

In 2011, Global Virus Network (GVN) was launched. It is a group of virologists and medical researchers associated with the University Of Maryland School Of Medicine. Robert Gallo started it. Reinhardt Kurth and William Hall plan to serve as a clearinghouse for rapid data collection and disease containment. In response to an outbreak, the organization would send researchers from one of its many international academic centers into the field to collect samples and help local officials diagnose and treat infected people [35].

The primary mission of the Global Virus Network is to strengthen medical research and the public health response to emerging viruses and persisting viral threats. The GVN currently consists of 38 Centers and 6 Affiliates in 24 countries, focusing on all aspects of medical virology. Nathan Wolfe, a Stanford University virologist who directed the Global Viral Forecasting Initiative, a San Francisco–based infectious disease outbreak research and advocacy organization, welcomes the arrival of another group with a similar mandate. "What's being done now is just a drop in the bucket of what needs to be done to stop pandemics before they occur," he says [35].

Whether or not there are future ZIKV epidemics, certainly the emergence of new pathogens with unanticipated consequences will be seen. In applying the lessons of ZIKV to the next epidemic, research funding should be rapidly deployed and prioritize ethical reviews of emerging infection research. There should work to continue to strengthen surveillance systems and be ready to integrate efforts to prevent transmission with maternal, reproductive, and pediatric health services [36].

Weaver [37] concisely highlights five strategies or frames of reference where public health authorities can intervene.

1. Interrupt enzootic cycles. In this approach, the field's control of the vector and host/reservoir infections is needed.
2. Prevent enzootic spillover. Here, an attempt would be made to reduce the disease introduction to human population centers using bed nets and vaccines.
3. Limit urban disease introductions or prevent the disease through mosquito control by modulating the *Aedes* "vectorial capacity."
4. Actively intervene in the urban area. The interruption would involve vector control, elimination of standing water (e.g., natural, or artificial collections of water, pools, and containers), and enforcement of adequate garbage management and disposal.
5. Prevent spillback into the enzootic cycles. Under these conditions, human hosts become a source for reinfection of nonhuman primates [22].

19.5.5 Lesson Five: How May We Productively Engage in Animal Studies?

Mice and rats have long served as the preferred species for biomedical research animal models due to their anatomical, physiological, and genetic similarity to humans. Exposing them to zoonotic diseases has long been the preferred method for experimentation. Advantages of rodents include their small size, ease of maintenance, short life cycle, and abundant genetic resources. Alternatives to animal testing include sophisticated tests using human cells and tissues (also known as in vitro methods), advanced computer-modeling techniques (often referred to as in silico models) and very occasionally studies involving human volunteers (micro-dosing) though that is unlikely when the experiments will involve infants. The debate over alternatives to animal models should continue.

In the interim, a remarkable amount of pandemic research is undertaken using mice and rat models and will probably continue. As alternative to animal testing develop the experimental designs can be expected to adjust. However, a significant effort to make pandemic disease management resilient will probably demand an increase in animal studies rather than a reduction in the short term.

Most of what has been learned about infectious diseases and the effects it may have on fetuses and young infants has come from animal studies. For example, pregnant mice infected with West Nile (WNV) or Powassan virus (POWV)—both flaviviruses, like ZIKV—also showed fetal harm. Over 40% of these infected fetuses died. But among pregnant mice infected with one of two other mosquito-borne viruses. unrelated to ZIKV, all the fetuses survived, scientists report online [8].

Platt, Miner, and others explained: "Based on the susceptibility of human maternal and fetal tissue explants and pathogenesis experiments in immunocompetent mice, other emerging neurotropic flaviviruses may share with ZIKV the capacity for transplacental transmission, as well as subsequent infection and injury to the developing fetus" [2].

> Some have speculated that ZIKV is unique among flaviviruses in its capacity to cross the placental barrier to cause congenital malformations. Our data show that WNV and POWV replicate efficiently in human placental explants and cause similar placental infection, transplacental transmission, and fetal demise in immunocompetent mice, suggesting that ZIKV may not be the only flavivirus with the ability to cause congenital disease. [2]

The research underscores that "many viruses, including some similar to ZIKV, can infect the placenta and the baby's cells," says George Saade, an obstetrician–gynecologist and cell biologist at the University of Texas Medical Branch at Galveston. "This list keeps growing and highlights the risks from viruses we are unfamiliar with" [8].

> The new study tests the effects of four other viruses: the two flaviviruses and two alphaviruses, chikungunya, and Mayaro, which have led to outbreaks in ZIKV-affected areas. ZIKV, West Nile, and Powassan share similarities in their genetic information, Miner says. "So, certain features of those virus genes and proteins in that particular family may confer this ability to infect certain cell types." But scientists don't understand what those features are yet, he says… [8]

As more is learned about zoonotic disease, many generations of lab animals will forego themselves and their short lives for the human good and their sacrifices should not go unnoticed.

19.6 Thoughts on Surveillance

A major component of resilience lies with knowing what has not happened yet. The answer lies in improved surveillance. Given the drivers that suggest an increasing prevalence of infectious outbreaks, it may be time to reconsider how things are done, how records are made, how they are kept, and how they are used.

It was thought by Mlakar and others that "the rapid spread of ZIKV around the globe will be a strong impetus for collaborative research on the biological properties of the virus, particularly since the risk of neurotropic and teratogenic virus infections places a high emotional and economic burden on society [38]." That did not happen.

One would think that an adult and pediatric neurologist network with the US CDC to establish a surveillance system that could track ZIKV-induced GBS, and central nervous system (CNS) disease would be in order. This would facilitate the identification and characterization of disorders, the formation of a registry, and the mounting of comprehensive epidemiological studies [39].

Prompt alerts and rapid dissemination of relevant data are essential to effective outbreak response. Yet too often, signals are delayed, or data suppressed by officials fearful of the economic and political consequences [40]. For a compelling examination of this phenomenon, read Birk's book on COVID-19 during her experience in the Trump Administration.

Meanwhile, between epidemics and before a new pathogen emerges with potential for congenital infection, international agencies should devise mechanisms for accelerated development of non-invasive diagnostics and promote facilities for longer-term storage of routinely collected antenatal and neonatal samples, which can be linked to maternity and pediatric records. No matter how rapid the response to an emerging pathogen is, retrospective studies based on stored samples will always be needed [41].

Roos et al. wrote: "Now is the time to make plans and strategies regarding ZIKV studies." However, these plans never came to fruition. Maybe, it is time to reconsider how surveillance and preparedness are approached [39].

19.6.1 Anticipatory Surveillance

Sound, infectious disease epidemiology must be applied to the surveillance of influenza epidemics, e.g., define the target population, and draw appropriate random samples from the respective population to obtain unbiased estimates of the incidence of flu-like symptoms and the viral status of the model [21]. It is essential to soon have the toolsets to anticipate epidemics.

In 2018, a simulation study by Schwab et al. demonstrated a potential logical fallacy of reactive, surveillance-driven vector control: measures stop being implemented as soon as they are deemed adequate. This false sense of security leads to patchier control efforts that will do little to curb the size of future vector-borne disease outbreaks. More investment should be placed in collecting high-quality information to trigger early and uniform implementation. At the same time, researchers work to discover more informative metrics of human risk to start more effective control [28].

Critically, control efforts at these various spatial scales are frequently implemented reactively, only after a certain surveillance threshold is reached. [L]ittle attention has been paid to the reactive nature of many control efforts and the differences caused by focusing on different triggers for control [28].

Mosquito control efforts are often implemented only after human infections have been detected, or mosquito populations have peaked. Although proactive control of mosquito populations before introducing a pathogen into the landscape reduces outbreak size and public health costs more effectively than reactive control, the funds necessary to implement these measures often diminish in the absence of an outbreak [28].

> Scenarios in which control began before disease introduction achieved much more significant reductions in human infection than scenarios in which power was only implemented after arbovirus was already circulating. Surveillance information on vector ecology and population dynamics may thus provide more effective triggers for control than surveillance information on epidemiological dynamics that, by nature, only trigger control after disease introduction. However, the resources needed for vector surveillance are often only available when the risk of disease is known and acknowledged and may only be provided after confirmed active transmission. This creates an impossible situation for underfunded mosquito control

agencies, which cannot enact control without surveillance information to trigger it and cannot acquire surveillance information without the resources to collect it. [28]

Data science may also hold the key to making systems more resilient through the availability of massive amounts of data. While Roberts [42] may be discussing applications involving power grids, weather events, and supply chain problems, he also mentions emerging diseases. While some early applications, like Google Flu were less than wholly successful, novel approaches with stronger sets of tools have surfaced and the field of digital epidemiology has begun to blossom.

19.6.2 Internet/Digital Bio-Surveillance

Today's abundance of publicly available data from Internet-based sources has inspired ambitious disease surveillance efforts. Online news articles, Internet search terms, and user-generated content provide a wealth of information indicating disease occurrence, syndromes, and transmission patterns. Scientists can now collect and translate these data sources into useable surveillance platforms [43] (Fig. 19.1).

Online trends (i.e., the distribution of online behavior and interactions) are an essential health surveillance tool for detecting and tracking outbreaks [45]. In the past decade, the near-real-time availability of novel and disparate Internet-based data sources has motivated the development of complementary methodologies to track disease incidence and spread. These approaches exploit near-real-time information from Internet search engines, news reports, clinician's search engines, crowd-sourced participatory disease surveillance systems, Twitter microblogs, electronic health records, and satellite images to estimate the presence of a disease in each location [46].

These platforms have had mixed reviews. However, their utility is not as much in question as to their capabilities at the time. As better surveillance systems are developed, there are substantial benefits from early surveillance using the Internet.

Early warning is the sine qua non of a functional pandemic management system. It must be asked where the most premature infectious disease reporting is likely reported. Some of the first information received about an epidemic on the move is from the media. In a broad sense, this means much more than it did a few decades ago when media was defined by journalists and editors in print and television. Today, more is learned about ongoing events from social media than any other resource. When receiving messages from more traditional sources, these messages are framed.

Traditional media tell stories and construct related health events in associations, such as disease affecting an associated group of people, connected lifestyle, associated with God or nature, and government or medical institutions. Sometimes the health issue itself is framed metaphorically, such as in warfare. Many people have examined past expressions of mediated health outbreaks, which need not be repeated

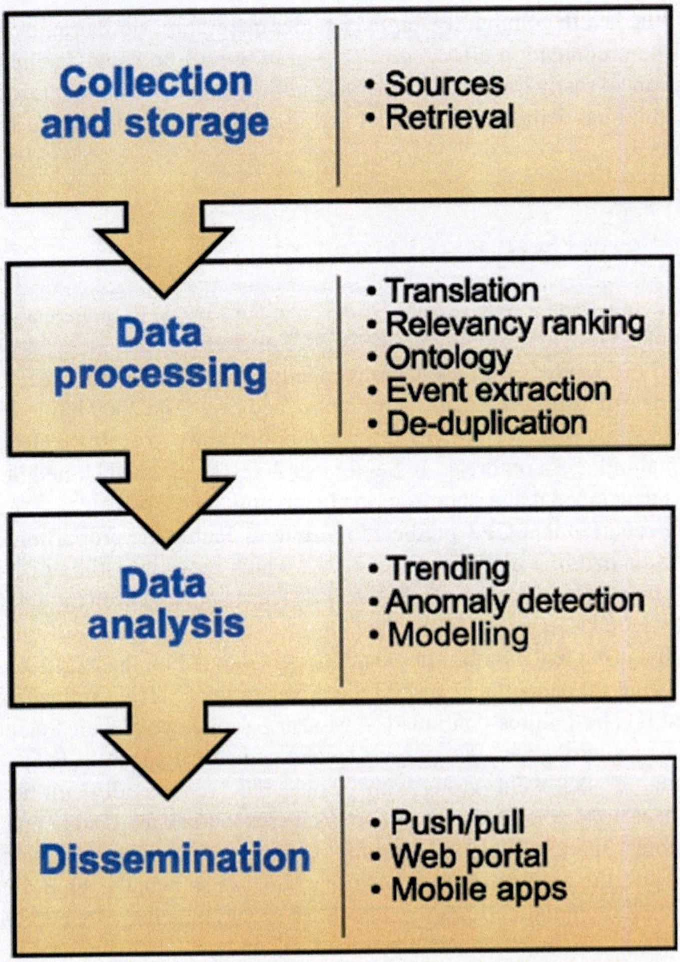

Fig. 19.1 General process of Internet-based bio-surveillance. Human input from information technology, public health, and other experts can occur at any step [44]

here. Instead, it is essential to determine if there are better approaches to characterizing an infectious attack beyond a pronouncement of an international crisis contradicted less than a year later. The de facto primary resource for planetary health information, the WHO, was surprised by the ZIKV outbreak. How can that be?

Alternative media functions as extra-institutional channels to challenge and contradict official media, communicate imminent risks to the public, and allow for more participatory communication and civic action [47].

New tools to enable responsibility digital bio-surveillance are arriving daily. For example, Ghenai and Mejova [48] developed a tool that combines crowdsourcing and machine learning techniques to identify ZIKV-related rumors and misinformation and clarify campaigns by public health agencies [48].

The public health community might consider integrating similar innovations into its digital communication efforts. Such integration will be better facilitated if the new tools can be easily linked with existing public health information infrastructures and place minimal demands in terms of additional resources, be they technological or human [49].

19.6.2.1 Google Flu Trends (GFT)

An experiment using Google searches didn't pan out very well. Indeed, after a round of predictions, Google discontinued the service altogether.

Google Flu Trends (GFT) used anonymized, aggregated Internet search activity to provide near-real-time estimates of influenza activity. "The 2009 influenza virus A (H1N1) pandemic [pH1N1] provided the first opportunity to evaluate GFT during a non-seasonal influenza outbreak. In September 2009, an updated United States GFT model was developed using data from the beginning of pH1N1."

"Nature reported that GFT predicted more than double the proportion of doctor visits for influenza-like illness than the CDC, which bases its estimates on surveillance reports from laboratories across the U.S. This happened even though GFT was built to predict CDC reports" [50].

Ironically, just a few months after announcing Google Flu, the world was hit with the 2009 swine flu pandemic, caused by a novel strain of H1N1 influenza. Google Flu missed it. The failures continued. As Lazer et al. show in their Science study, Google Flu was wrong for 100 out of 108 weeks since August 2011 [51].

Determining what went wrong with Google Flu Trends is difficult because the company does not disclose what search terms it uses to track flu. If someone searches "flu symptoms," they will likely be prompted to try a search for "flu vaccines," for example. Thus, the number of flu-related searches can snowball even if flu cases do not. They took more than 50 million search terms and matched them up with about 1100 data points on flu prevalence from the CDC [52].

Google Flu Trends is no longer available. Instead of running Google Flu Trends, the company will pass along the data gleaned from user searches to heath organizations that study infectious diseases so they can develop their prediction models. Those organizations include the Columbia University's Mailman School of Public Health, Boston Children's Hospital, and the Centers for Disease Control and Prevention Influenza Division, Google said on August 20, 2015 [53].

Initially, GFT used many Web search queries that correlate well with physician visits for influenza-like symptoms to estimate current weekly levels of influenza activity at regional and state levels and made claims of some significant breakthroughs. For example, Google Flu Trends (an online tool based on searches for influenza-related topics) detected influenza outbreaks in the USA 7–10 days before conventional surveillance systems. The CDC uses laboratory and clinical data to publish national and regional weekly statistics, typically with a 1–2-week(s) lag in reporting [54].

The decision to relinquish GFT is primarily speculative. However, Google has had some public pushback on its recent forays into health care. For example, in 2019, it was learned Goggle had accessed a collection of individuals' medical records after partnering with Ascension in a project code-named "Project Nightingale." At a government hearing in 2020, Ascension was asked for more details about its deal with Google. The controversy swirls around one key aspect of the partnership to pilot an electronic health records search tool that pulls patient electronic health records into an interface to help clinicians more easily find helpful information [55].

Just three years earlier, Google was burned when it was revealed that the Google-owned artificial intelligence company DeepMind had partnered with a group of London hospitals called the Royal Free London NHS Foundation. An investigation by UK regulators subsequently determined that the agreement breached data-protection laws [56].

19.6.2.2 ARGUS [57]

It seems Google has moved on from Google Flu Trends into other ventures. It would seem, however, that should be some system that combines their strengths. While this subject deserves a book of its own, here is ARGUS. Maybe what is needed in surveillance is some composite system using the best of Argus and the best from the GFT system.

According to an unpublished 2013 dissertation by Emily A. Iarocci entitled "Employing quantitative and qualitative analytic methods for Internet bio-surveillance of pandemic influenza." Using as a launching pad some work by Hartley et al [44].

One of the Internet bio-surveillance systems, **Argus**, hosted at Georgetown University Medical Center (GUMC), focused its efforts on monitoring social disruption caused by infectious disease events because human social reactions to disease often appear in news media before laboratory and official confirmation. Simply put by Adebayo et al. [45], online trends can aid in pandemic surveillance.

Project Argus is designed to report and monitor the evolution of biological events threatening human, plant and animal health globally, excluding the U.S. Argus collects, in an automated process, several thousand local, native-language Internet media articles daily.

This is how it works. Bayesian software tools and Boolean search strings, based on a taxonomy of infectious disease, identify candidate relevant articles. Regional experts, collectively fluent in roughly 40 languages, review these articles manually. Relevant media articles are identified based on direct indicators (reports of disease) and indirect indicators (socially disruptive events or precursors to disease, such as preventative measures or adverse enviroclimatic conditions). Regional experts write Argus reports based on these media articles; reports are posted to a password-protected Internet portal for users to view [59]. Nelson et al. [59] reviewed Argus against comparable WHO epidemiological data for the 2009 H1N1 epidemic. Argus reported first confirmed cases on the same day as WHO for 21 of the first 64 countries reporting cases, and 1–16 days (average 1.5 days) ahead of WHO for 42 of those countries [59].

19.7 Thoughts on Preparedness

Are we prepared for another ZIKV epidemic? Probably not. The ZIKV pandemic identified significant gaps in the public health system readiness and the need for sustained investments in the national public health emergency preparedness and response capacity [60]. What follows are a half dozen approaches toward preparedness with their own strengths and weaknesses. The best system will be a composite of all of the following.

19.7.1 The Global Virome Project (GVP)

From an analysis of recent viral discovery data [61], approximately 1.67 million yet-to-be-discovered viral species from critical zoonotic viral families and between 631,000 and 827,000 of these unknown viruses have zoonotic potential. The spillover rate is accelerating, leading to a nonlinear rise in pandemic risk and an exponential growth in their economic impacts.

The Global Virome Project was aimed to launch by Dennis Carroll, a former director of the U.S. AID's Pandemic Influenza and Other Emerging Threats Unit, in 2018 with an emphasis on large-scale sampling and viral discovery. Stakeholders from Asia, Africa, the Americas, and Europe, spanning industry, academia, intergovernmental agencies, non-governmental organizations, and the private sector, began meeting in 2016 to design a framework for the governance, management, technical operation, and scope of the GVP.

Critical efforts include developing finance streams; establishing a transparent, equitable implementation strategy; designing data- and sample-sharing protocols; developing laboratory and metadata platforms; targeting host taxa and sampling sites; analyzing return on investment; forming collaborative field and laboratory networks; developing risk characterization frameworks for viruses discovered; designing a strategy to assess and mitigate risk behaviors that facilitate viral emergence; and planning in-country capacity-building for sustainable threat mitigation [19].

Carroll et al. [19] predicted remarkable success.

> The GVP will likely accelerate the development of pathogen discovery technology, diagnostic tests, and science-based mitigation strategies, which may also provide unexpected benefits. Like the Human Genome Project, the GVP will provide a wealth of publicly accessible data, potentially leading to discoveries that are hard to anticipate, perhaps viruses that cause cancers and chronic physiological, mental health, or behavioral disorders. It will provide orders of magnitude more information about future threats to global health and

biosecurity, improve our ability to identify vulnerable populations, and more precisely target mitigation and control measures to foster an era of global pandemic prevention.

The project was raising funds from governments and foundations and was supposed to begin sampling wild animal populations in 2020 but was delayed due to the COVID-19 pandemic [62].

19.7.2 One Health

'One Health' represents a call for health researchers and practitioners at the human, animal, and environmental interfaces to work together to mitigate the risks of emerging and re-emerging infectious diseases (EIDs). A One Health approach emphasizing interdisciplinary cooperation is increasingly necessary for effective EID control and prevention.

One Health is grounded in recognizing that human, animal, and environmental health are interdependent, that animal species provide a shared reservoir for pathogen exchange and spread, and that varied and dynamic human–animal interactions drive many EIDs. One Health's response is to deconstruct the disciplinary silos that have separated biomedical and social sciences devoted to studying human disease from those committed to nonhuman disease and ecological concerns [63].

The threats posed by EIDs are comprised of complex and contingent sets of relations that involve socioeconomic and sociopolitical drivers and consequences, with the latter extending beyond the impact of the disease. The One Health approach must address a range of sociopolitical, ethical, and legal challenges that arise due to the spread of infection within and between species [63].

This approach is more thoughtful, if not philosophical, in its approach to infectious diseases. Its adherents argue that one of the reasons technological solutions are deferred to solve some of these most pressing problems, like climate change is because the technical approaches do not require behavioral change and especially do not demand a re-examination of the underlying value systems what One Health demands are a more integrated approach.

An example of this approach is MediLabSecure. MediLabSecure fosters intersectoral collaboration, expertise, and sharing of information. The resulting exchanges (methodological, communication, and operational) across disciplines and countries, dedicated research on intersectoral collaboration, and increasing diagnostic capacities are providing new paths and tools to public health professionals to face emerging viral threats such as a ZIKV epidemic in the Euro-Mediterranean region [64].

The MediLabSecure network, created in 2014, is a continuation of a prior network developed by the EpiSouth and EpiSouth Plus Projects [65]. It comprises 55 virology and medical entomology laboratories and 19 public health institutions in 19 countries in the Balkans, North Africa, the Middle East, and the Black Sea regions. It aims to set up awareness, risk assessment, monitoring, and control of emerging and re-emerging vector-borne viruses across disciplines and countries.

This objective is addressed by holding multidisciplinary "One Health" activities such as joint meetings, technical transversal workshops, expert visits where representatives of the four networks interact at national and international levels to foster cooperation and intersectoral risk assessment exercises and tailored research to explore how this intersectoral collaboration is implemented within partner countries. To date, the MediLabSecure human virology subnetwork, under the leadership of the Institut Pasteur in Paris, conducted three workshops on scientific knowledge and technical skills (real-time RT-PCR and serology by ELISA) for the detection of arbovirus infections [64].

19.8 Mosquito Alert and Citizen Science

What follows is a citizen science approach that seems to have worked well in various locations. Citizen science is scientific research conducted, in whole or in part, by amateur scientists. It is quintessentially collaborative and empowering.

19.8.1 Mosquito Alert

Spain's Mosquito Alert [66] is led by ICREA Movement Ecology Laboratory, Centre for Advanced Studies of Blanes-Consejo Superior de Investigaciones Científicas (CEAB-CSIC) and funded by the Spanish Foundation for Science and Technology and others. It encourages the public to systematically report tiger mosquito sightings, facilitates collaboration between citizen scientists, researchers, and public health administrations, and raises awareness about steps everyone can take to reduce the risk of mosquito-borne diseases.

Of 274 Spanish municipalities in which tiger mosquitoes were detected for the first time in 2014–2015, Mosquito Alert is alone credited with detections in 108 (39%). In total (including the municipalities in which both sources are credited), Mosquito Alert is credited with detections in 175 municipalities (64%).

19.8.2 Mosquito Stoppers

A 2-year citizen science program called Mosquito Stoppers was initiated in West Baltimore, Maryland, in 2014–2015. Mosquito Stoppers (hereafter MS) is part of Take Back the Block, a community beautification and science program established for West Baltimore. Nearly a dozen citizen scientists participated in yard surveys of potential mosquito habitats and evaluated mosquito nuisance. Jordan et al. [67] found that citizen scientists with minimal education and training could accurately collect data that reflect trends found in a comparable researcher-generated database [67].

Before engaging in this project, participants reported levels of social responsibility regarding social–ecological issues like that of individuals not engaged. After participation, however, differences resulted between the not engaged and the engaged group, with those who were engaged reporting higher levels of social responsibility [68].

Citizen science gives the same information and predictive capabilities as traditional surveillance tools. Early warning and invasion-front mapping are tasks for which citizen science is exceptionally well suited [69]. Palmer et al. [70] suggest that citizen science is positioned to revolutionize surveillance of mosquito-borne diseases. Palmer et al. reviewed similar citizen science projects in Germany, France, Portugal, The Netherlands, and the United Kingdom.

19.9 Pandemic Influenza Preparedness (PIP) Framework

The WHO reached an agreement on a pandemic influenza preparedness (PIP) framework for sharing influenza viruses and access to vaccines and other benefits in April 2011. While the framework encourages member states to share "PIP biological materials," it is not binding. It was built to calm concerns that large pharmaceutical firms would use biological materials, such as viruses, to research, patent, and market vaccines and antiviral medications back to countries that contributed the natural materials at prices beyond affordable [71].

This effort was predicated on the avian influenza A(H5N1) outbreaks in late 2006 when Indonesia refused to share virus specimens with WHO, claiming it was unfair to give pharmaceutical companies access.

The benefit-sharing system contains many components, but the most significant ones require the industry to pay half of GISRS's annual operating costs and provide benefits under the second standard material transfer agreement (e.g., vaccine donations); the framework does not direct developed member states to provide specific benefits for developing countries (e.g., vaccine donations) [71].

19.9.1 Integrated Pest Management (IPM)

How can the process of IPM be democratized so those who are directly affected, e.g., whence a vector experiment such as the release of a genetically engineered mosquito occurs in their neighborhoods, understand what will be happening and input into whether the experiment should be conducted at all? Two approaches dominate the psychology of IPM engagement: those that appeal to participants' sense of being a part of and contributing to knowledge and society; and those that provide monetary and other tangible incentives [72].

Deciding to participate in an experimental release vector control study involves complicated issues. Other strategies include reducing the effort required for participation through simplified consent processes and using opt-out rather than opt-in invitations. However, there remain concerns that simplification may underestimate societal consequences, and potential subjects may not fully understand opting-out.

Reviews to synthesize the evidence on the effectiveness of various approaches to recruitment and retention in longitudinal studies are inconclusive and have highlighted the difficulty in identifying the best practice because researchers often use multiple methods [72].

19.9.2 Integrated Pest and Vector Management (IPvM)

This may be the gold standard in preparedness. It combines many features of the programs and projects from above.

IPM is a synergistic, ecosystem-based strategy that focuses on the long-term suppression of pests or their damage through techniques, including biological control, trapping, habitat manipulation, and chemical control [73]. The AMCA identifies a subset of IPM as Integrated Mosquito Management (IMM).

The core of IMM includes four critical tactics: (1) surveillance, mapping, and rational setting of action thresholds; (2) physical control through manipulation of mosquito habitat; (3) larval source reduction and adult mosquito control; and (4) monitoring for insecticide efficacy and resistance [73].

Eisen et al. [26], in their study on dengue eradication, advocate a reallocation of resources from reactive and demonstrably ineffective vector control strategies, especially vehicle-mounted space-spraying, to more promising proactive vector control and prevention strategies; and the use of management models, such as the continuous improvement cycle, that emphasize monitoring and evaluation of outcomes for specific control strategies and adaptive modification to ensure steady improvement in control program performance and progress toward locally appropriate control strategies.

Emergency mosquito control is usually initiated at peak transmission time, but these efforts are generally misdirected and are too little and too late to impact the epidemic [27]. The AMCA calls for proactive needs assessment: (1) determining changes in the geographic distribution and abundance of mosquito species; (2) evaluating control efforts by comparing presurveillance and post-surveillance data; (3) obtaining relative measurements of the vector populations over time and accumulating a historical database; and (4) facilitating appropriate and timely decisions regarding interventions [73].

In addition, the effects of integrated vector management strategies are likely to be more sustainable as interventions are usually community-based rather than vertically implemented by specialized teams. Community-based interventions tend to be sustainable as they aim to change behavior and induce social mobilization [74].

IVM is an evidence-based, adaptive, multisectoral vector control approach. It involves a range of vector control tools used alone or combined based on knowledge of the local vector ecology and disease epidemiology [75]. Sometimes it is called IVC (control). Though integrated, it is not necessarily centralized. In some instances, centralization is preferable.

Decentralization of vector control services is considered positive because of faster operations and adjustments by the local epidemiology. However, centralization is often preferred in countries with successful malaria eradication campaigns, e.g., Guatemala and The Philippines [76].

Speaking on dengue, Erlanger, Kaiser, and Utzinger reported, "over the past three decades, concerns about environmental contamination and the emergence of vector resistance to insecticides have resulted in a shift in policy from outdoor space-spraying to a more targeted and longer-lasting use of insecticides, such as water-container treatment with temephos and indoor-residual spraying and the use of insecticide-treated curtains" [74].

When Erlanger, Kaiser, and Utizinger studied dengue in 2008, they reviewed data on the effectiveness of vector indices of all vector control methods. He concluded that integrated vector control was the most effective, while environmental management had minimal impact. "Dengue vector control effectively reduces vector populations, particularly when interventions use a community-based, integrated approach tailored to local eco-epidemiological and sociocultural settings and combined with educational programs to increase knowledge and understanding of the best practice" [74].

An extreme case can be made regarding that the most effective mosquito management programs seem to be a multilevel and multifaceted integrated approach. Eisen and his team [26] argued for a stronger focus on partnerships where researchers and public health professionals work together to (1) conceptualize promising, locally appropriate control program strategies; (2) implement these strategies in pilot programs; and (3) determine entomological and, most importantly, epidemiological outcomes [26].

> We advocate for the increased use of proactive vector control approaches. These approaches ideally should have the following characteristics, which will favor support from public health and political leadership as well as individual homeowners: 1) potential for implementation not only by vector control programs but also by individual homeowners; 2) low cost of implementation; and 3) minimal effort for long-term maintenance. [26]

This will require high public participation levels and sustained engagement activities. Cuba and Singapore are two success stories deserving mention. It is important to note both these programs are somewhat coercive involving fines.

> However, effective *Ae. aegypti* control can be achieved using an integrated approach targeting larval mosquitoes. This species was eliminated from most countries in tropical America during the 1950s and 1960s, effectively preventing both epidemic dengue and yellow fever. Unfortunately, these programs were disbanded in the early 1970s after success had been achieved. This change was followed by the rapid reinvasion by Ae. aegypti of most tropical American countries, putting them at high risk for epidemic dengue. In modern times, only

> Cuba and Singapore have successfully controlled *Ae. aegypti*. A combination "top-down–bottom-up" approach has been used in both countries. [77]

Recall that *Ae. aegypti* prefers to rest inside houses, so a truck or aerial insecticide spraying does not reach mosquitoes resting in hidden places such as cupboards. There have been several reports of homeowners in various countries refusing entry to household spraying teams or tightly shutting windows and doors to prevent outdoor insecticidal fogs from entering their house, thus reducing the potency of this intervention [77].

> Unfortunately, in today's world of uncontrolled urbanization–especially in tropical developing countries–the "top-down" methods used successfully in the past are no longer feasible because of a lack of resources. A city of 10 million people will have approximately 2 million households that need to be visited and checked for mosquitoes weekly; this is not feasible. Moreover, the vertically structured programs of the past had no sustainability. That sustainability will come only through community participation in mosquito control programs. The people living in those 2 million houses must assume responsibility for *Ae's* weekly inspection and control. *Aegypti* in and around their homes. Sustainability requires that this be an ongoing program that never ends as long as the threat of epidemic dengue transmission exists. *Ae. aegypti* control, therefore, must be an ongoing environmental management program. [77]

Abramides et al. [78] studied multiple intervention strategies to control *Ae. albopictus* in Spain. The process involved larvicide treatments, source reduction, adulticide treatments, and cleaning up uncontrolled landfills. It resulted in a significant decrease in eggs. They concluded both "The combination of the four integrated vector management (IVM) strategies was effective in reducing the number of eggs in the intervention areas compared with the control ones and offered it up as a model for controlling the populations of *Ae. albopictus* in the Mediterranean region" [78]. The source reduction involved a community-based approach.

Pérez et al. [79] studied 16 areas in La Lisa, a municipality of Havana, Cuba. They examined: organization and management, capacity-building, community work, and surveillance. A participatory assessment of process data was performed to determine whether the components and subcomponents were implemented, not implemented, or modified. They reported surveillance was the most implemented component (72.9%), followed by capacity-building (54.7%). Community work and organization and management were less implemented or modified with 50% and 45%, respectively. The implementation of strategy components was all correlated, apart from surveillance and capacity-building. If one part is implemented in a circumscription, the other members are also likely to be implemented. Notably, areas that did not undergo organizational changes commonly did not implement community work activities. Within the whole strategy, few activities were added. Scarcely implemented subcomponents were the most innovative ones. The difficulties encountered during implementation were related to appropriate training and skills, available time, lack of support and commitment to the strategy, lack of motivation of local leadership, coordination of actors, and mobilizing resources. The study showed a wide variability of fidelity in implementing the intervention [79].

Pérez et al. [80] made a powerful argument for the community participation process involving learning and access/control. Empirical evidence from 5 years of research in the context of Cuba showed that moves toward community-based *Ae. aegypti* control are feasible. However, to be successful, community-based dengue prevention should be a social learning process, implying a transfer of power and responsibilities to local people. Actions must be undertaken to create local capabilities, strengthen existing structures and organizations and promote group work for learning participation from participation itself [80].

Insect Resistance Management (IRM) strategies must be integrated within vector control programs early before resistance occurs. IRM should include regular resistance monitoring and careful implementation, monitoring, and evaluation of resistance-breaking interventions. Such interventions include rotations, mosaics, combinations of unrelated insecticides, and source reduction. Developing insecticide resistance management plans for *Aedes* mosquitoes will require improving resistance monitoring systems, engaging populations to IRM, developing novel diagnostic tools, evaluating operational insecticide resistance risk levels, enlarging the panel of available insecticides, and promoting the use of non-insecticidal control tools as part of integrated vector control management [75].

19.10 Conclusion

In 2010, Hostick, Runge-Ranzinger, Nathan, and Kroger examined dengue response protocols. It was a literature review and some case studies (interviews and questionnaires). Their case studies in Brazil, Guatemala, The Philippines, and Vietnam were notable.

The case studies confirmed most of this information: (1) a lack of personnel (entomologists, social scientists, operational vector control staff); (2) a lack of technical expertise at decentralized levels of services; (3) insufficient budgets; (4) inadequate geographical coverage; (5) interventions relying primarily on insecticides; (6) difficulties in engaging communities; (7) little capacity-building; (8) almost no monitoring and evaluation. Stakeholders noted a need for operational standards, evidence-based selection/delivery of combinations of interventions; development/application of monitoring and evaluation tools; needs-driven capacity-building [76].

These are not small challenges, but they are also possible to surmount. As the American Mosquito Control Association (AMCA) put it, budget often drives structure and implementation, resulting in inadequate mosquito control programs being funded at levels that provide comprehensive surveillance or control. Ultimately, such an approach may decrease the effectiveness of interventions and increase long-term costs [73].

To improve the efficiency and effectiveness of vector control programs through compliance with the principles of integrated vector management, the following tools and approaches are needed: the development of operational standards for vector control services, including minimum financial and personnel requirements by the geographical area(s) to be covered, their demography and the vector control methods to be implemented; evidence-based selection and delivery of different interventions or combinations of interventions, adapted to other settings; development and application of monitoring and evaluation tools for vector control service delivery; needs-driven capacity-building, especially in public health entomology and communication [76].

> Despite these threats' fears, the urge to respond must be tempered by reality and based on sound evidence. In the large urban zones where these vectors increase, to continue to use what has always been used, for that reason alone, or to pursue new approaches without sound supporting evidence would be wrong and potentially an extravagant waste of resources. Hence, there is an argument for instituting a global independent advisory body to guide decisions regarding selecting approaches and tools to control or prevent infections transmitted by urban *Aedes sp.* vector populations and the design of appropriate multi-center trials to evaluate their effectiveness. [81]

Finally, the 2016 Commission on a Global Health Risk Framework's analysis suggests that expected economic losses from potential pandemics could amount to around $60 billion per year. Implementing many of the aforementioned recommendations, by contrast, would cost about $4.5 billion per year. This figure has three elements: the cost of upgrading public health systems in low- and middle-income countries, which this report puts close to $3.4 billion per year; the cost of enhancing the WHO's pandemic prevention and response capabilities and of financing the WHO and World Bank contingency funds, which is assumed to be $130 million to $155 million per year; and proposed incremental investment in research and development of $1 billion per year [40].

19.11 Closing Remarks

> There are significant parallels between ZIKV disease and COVID-19 in terms of limited diagnostic techniques, therapeutics, and prognostic uncertainties. Both infections are associated with a substantial risk of adverse outcomes for the pregnant individual or the fetus. Existing social and economic inequalities amplify the risk burden of ZIKV disease and COVID-19 in vulnerable communities. Although each pathogen has unique features, there are underlying common principles regarding the recognition, communication, and mitigation of the risk of infection. Misinformation spread by social media platforms has undermined public health efforts and patient adoption of recommended mitigation strategies. Healthcare providers can provide partnership, social support, and evidence-based information to enhance health-seeking behaviors, thereby minimizing the risks for pregnant and reproductive-aged persons. [82]

While the world awaits the epidemiology and pathology of these unique viral infections to help explain the ZIKV epidemic in Brazil, now is the time, first and foremost, to update common thinking and approaches to studying teratology and the role of the placenta. Doing so might provide large scientific rewards for this outbreak and future similar epidemics [83].

More than half a decade later, the story about ZIKV can be read. Reading nearly all the publications on almost every conceivable media platform took a half a decade. Many of these reports were not in the English language and had to be translated from

Portuguese, Spanish, and Chinese. Only references that made corroborated claims were included unless it was an important outlier comment moved the text forward.

Experts in these fields vetted technical issues. My role was to serve as a distinct set of eyes in trying to make sense of the popular and technical reportage on ZIKV. Then COVID. While COVID is a zoonotic vector disease like ZIKV whereby the source is primarily contacted with an infected mosquito, they share human transmission potential, ZIKV through sexual contact and COVID through inhalation.

Learning to deal with infectious diseases will involve building emergency response measures alone and cannot begin to provide the same level of response as an organized, established vector control program. Both will need to be considered viable options in the future.

Association of State and Territorial Health Officials (ASTHO) reported that emergencies could strain an agency's staffing, equipment, and budget resources. State and local vector control programs cannot rely on federal agencies to supply timely financial aid or comprehensive emergency assistance when a disaster strikes. Only the sustainable dedication of resources for vector control, surveillance, and personnel can help advance understanding and capacity to respond to these diseases in a timely and effective fashion. The success of a mosquito control program depends on its ability to use multiple surveillance methods to provide data on disease threats, including novel threats that may spread into the United States and become endemic [20].

Vector control programs with limited or no funding can work with universities or colleges to access such institutions' experts and agricultural extension services. State agriculture and public health departments may facilitate access to surveillance data from veterinary diagnostic laboratories, veterinary clinic networks, zoos, equine, falconry, and raptor rehabilitation organizations. Additionally, vector control programs may contract with private companies or associations to conduct surveillance [20].

The target audience must also see the recommended actions as practical and feasible. The public must also believe that the risk is substantial enough to act. Several West Nile virus studies suggest that many people who ignore the advice to wear repellent or adopt other preventive measures do so because they do not perceive the risk of contracting the mosquito-borne disease to be high enough to warrant such actions [20].

Peter Sands frames investment into pandemic disease as a form of Pascal's wager on the existence of God.

> If we overinvest, we will spend more on shoring up our defenses against infectious-disease outbreaks than strictly necessary. Yet it is hard to see that as wasted money since these investments will help us achieve other critical health objectives. For example, strengthening national public health and primary care systems will help us tackle endemic diseases such as tuberculosis and malaria more effectively and detect the emergence of antimicrobial-resistant pathogens more swiftly. On the other hand, if we underinvest, we open the door to potential catastrophe. [40]

References

1. McKenna M (2016) Disorganized mosquito control will make US vulnerable to Zika. National Geographic, February 29. https://www.nationalgeographic.com/science/phenomena/2016/02/29/zika-mosquito-control/. Accessed August 8, 2018
2. Platt D et al (2018) Zika virus–related neurotropic flaviviruses infect human placental explants and cause fetal demise in mice. Sci Trans Med 10:eaao7090, January 31. https://www.ncbi.nlm.nih.gov/pubmed/29386359. Accessed July 6, 2018
3. Million-Weaver S (2018) Relatives of Zika virus could infect fetal tissue. AAAS, January 30. https://www.aaas.org/news/relatives-zika-virus-could-infect-fetal-tissue. Accessed July 10, 2018
4. Keating A (2018) New app identifies mosquito species and Wolbachia. Eureka Alert, August 30. https://www.eurekalert.org/news-releases/671992. Accessed May 23, 2022; Bhadra S, Riedel TE, Saldaña MA, Hegde S, Pederson N et al (2018) Direct nucleic acid analysis of mosquitoes for high fidelity species identification and detection of Wolbachia using a cellphone. PLOS Neglected Trop Dis 12(8):e0006671 https://doi.org/10.1371/journal.pntd.0006671
5. Braack L et al (2018) Mosquito-borne arboviruses of African origin: review of key viruses and vectors. Parasites Vect 11:29. https://www.ncbi.nlm.nih.gov/pubmed/29316963. Accessed July 7, 2018
6. Alfaro-Murillo J et al (2016) A cost-effectiveness tool for informing policies on Zika Virus control. PLOS Neglect Trop Dis 10:5. https://doi.org/10.1371/journal.pntd.0004743. Accessed July 10, 2018
7. PAHO (1997) The feasibility of eradicating Aedes aegypti in the Americas. Revista Panamericana de Salud Pública 1(1):68–72, January
8. Cunningham A (2018) Zika may not be the only virus of its kind that can damage a fetus. Science News, January 31. https://www.sciencenews.org/article/zika-may-not-be-only-virus-its-kind-can-damage-fetus. Accessed June 27, 2018
9. CDC (2018) Zika and pregnancy, August 22. https://www.cdc.gov/pregnancy/zika/testing-follow-up/reproductive-planning.html. Accessed June 19, 2019
10. Morse SS (1995) Factors in the emergence of infectious diseases. Emerg Infect Dis 1:1, January–March. https://wwwnc.cdc.gov/eid/article/1/1/95-0102_article. Accessed September 18, 2018
11. Smith F (2020) On the hunt for the next deadly virus. National Geograp, June 16. https://www.nationalgeographic.com/science/article/coronavirus-on-the-hunt-for-the-next-deadly-virus. Accessed May 25, 2022
12. Costly D (2020) Genetically engineered mosquitoes could soon be unleashed in the U.S. OneZero Medium, May 6. https://onezero.medium.com/genetically-engineered-mosquitoes-could-soon-be-unleashed-in-the-u-s-61896cc647b4. Accessed May 28, 2022
13. Baumgartner E (2019) A virus even more dangerous than Zika to pregnant women. The New York Times, January 7. https://www.nytimes.com/2019/01/07/health/rift-valley-pregnancy-zika.html. Accessed June 11, 2019; McMillen C et al (2018) Rift valley fever virus induces fetal demise in Sprague-Dawley rats through direct placental infection. Science Adv 4:12, December 5. https://advances.sciencemag.org/content/4/12/eaau9812. Accessed June 11, 2019
14. Randolph S, Rogers D (2010) The arrival, establishment and spread of exotic diseases: patterns and predictions. Nature Rev: Microbiol, 8, May. https://www.ncbi.nlm.nih.gov/pubmed/20372156. Accessed June 27, 2018
15. WHO (2018) List of blueprint priority diseases, February. http://www.who.int/blueprint/priority-diseases/en/. Accessed June 27, 2018
16. Dangerfield K (2018) 'Disease X' listed as potential global endemic by World Health Organization—but what is it? Globalnews.ca, March 13. https://globalnews.ca/news/4079589/disease-x-world-health-organization-global-pandemic/. Accessed June 27, 2018
17. Li R, Ellington S, Galang MR, Grosse S et al (2022) Economic evaluation of Zika contraception access network in Puerto Rico during the 2016–17 Zika virus outbreak. Contraception 107:68–73

18. Li R, Simmons KB, Bertolli J, Rivera-Garcia BL et al (2017) Cost-effectiveness of increasing access to contraception during the Zika virus outbreak, Puerto Rico. Emerging Infect Dis 23(1):74–82
19. Carroll D, Daszak P, Wolfe NF, Gao GF, et al (2019) The global virome project. Science 359(6378):872–874, January 12
20. ASTHO (2015) Before the swarm: guidelines for the emergency management of vector-borne disease outbreaks. http://www.astho.org/Programs/Environmental-Health/Natural-Environment/Before-the-Swarm/. Accessed August 28, 2018
21. Keil U, Schönhöfer P, Spelsberg A (2011) The invention of the swine-flu pandemic. European J Epidemiol 26:3. https://www.jstor.org/stable/41474227?seq=1#metadata_info_tab_contents. Accessed September 18, 2018
22. Sikka V et al (2016) The emergence of Zika virus as a global health security threat: A review and a consensus statement of the INDUSEM Joint working Group (JWG). J Global Infect Dis 8:1. http://www.jgid.org/article.asp?issn=0974-777X;year=2016;volume=8;issue=1;spage=3;epage=15;aulast=Sikka. Accessed September 18, 2018
23. Igoe K (2021) How do you communicate uncertainty and promote public health—during COVID-19 and beyond? July 20. https://www.hsph.harvard.edu/ecpe/how-to-communicate-uncertainty-and-promote-public-health-during-covid-19-and-beyond/. Accessed June 27, 2022
24. Purcell MY (2016) NACCHO urges congress to spend $1.8B for Zika virus disease. https://www.naccho.org/uploads/downloadable-resources/Support-For-Zika-Funding-2016-Press-Release-Final.pdf. Accessed September 10, 2018
25. Hadler JL et al (2014) National capacity for surveillance, prevention, and control of West Nile Virus and other Arbovirus infections—United States, 2004 and 2012. Morbidity Mortality Week Report 63:13, April 4. https://www.cdc.gov/mmwr/preview/mmwrhtml/mm6313a2.htm. Accessed September 4, 2018
26. Eisen L et al (2009) Proactive vector control strategies and improved monitoring and evaluation practices for dengue prevention. J Med Entomol 46:6, November. https://www.ncbi.nlm.nih.gov/pubmed/19960667. Accessed August 7, 2018
27. Gubler DJ (2002) Epidemic dengue/dengue hemorrhagic fever as a public health, social and economic problem in the 21st century. Trends Microbiol 10:2, February. https://www.ncbi.nlm.nih.gov/pubmed/11827812. Accessed August 7, 2018
28. Schwab SR et al (2018) The importance of being urgent: The impact of surveillance target and scale on mosquito-borne disease control. Epidemics, 23, June. https://www.ncbi.nlm.nih.gov/pubmed/29279187. Accessed August 11, 2018
29. Rutter M (2012) Resilience: causal pathways and social psychology. In: Unger M (ed) The social ecology of resilience: a handbook of theory and practice. Springer Science + Business Media
30. Berg SH, Aase K (2019) Resilient characteristics as described in empirical studies on health care In: Wiig S, Fahlbruch B (eds) Exploring resilience Springer briefs in safety management. https://doi.org/10.1007/978-3-030-03189-3_10
31. Massaro E, Ganin A, Perra N, Linkov I, Vespignani A (2018) Resilience management during large-scale epidemic outbreaks. Scient Reports 8(1859). https://doi.org/10.1038/s41598-018-19706-2
32. Health and Science Desk (2021) 14 lessons for the next pandemic. The New York Times, March 15. https://www.nytimes.com/interactive/2021/03/15/science/lessons-for-the-next-pandemic.html. Accessed June 11, 2022
33. Hoberg EP, Brooks DR (2015) Evolution in action: climate change, biodiversity dynamics and emerging infectious disease. Philos Trans R Soc B 370:20130553. https://doi.org/10.1098/rstb.2013.0553
34. Press Release (2012) GVFI now known as global viral and GVF Inc. Now Known as Metabiota. https://www.gvfi.org/assets/globalviral-metabiota-press-releases.pdf. Accessed May 15, 2019
35. Mann A (2011) New organization pledges scientific expertise for viral outbreaks. Nat Med 17:4, April. https://www.nature.com/articles/nm0411-394a. Accessed May 15, 2019

36. Becker-Dreps S, Stringer E, Bucardo F, Bowman N, Boivin M (2020) Is there a silver lining to the Zika virus epidemic in the Americas? The Lancet 20:14–15, January
37. Weaver SC (2013) Urbanization and geographic expansion of zoonotic arboviral diseases: mechanisms and potential strategies for prevention. Trends Microbiol 21:8, August. https://www.ncbi.nlm.nih.gov/pubmed/23910545. Accessed September 18, 2018
38. Mlakar J et al (2016) Zika virus associated with microcephaly. The New England J Med 34(10):951–958, March 10. http://www.nejm.org/doi/full/10.1056/NEJMoa1600651. Accessed May 23, 2017
39. Roos R (2016) Zika virus—a public health emergency of international concern. JAMA Neurol 73(12). http://jamanetwork.com/journals/jamaneurology/fullarticle/2557229?widget=personalizedcontent&previousarticle=2618389. Accessed May 30, 2017
40. Sands P, Munduca-Shah C, Dzau VJ (2019) The neglected dimension of global security—a framework for countering infectious-disease crises. The New England J Med 374:13. https://doi.org/10.1056/NEJMsr1600236
41. Ades AE et al (2020) Researching Zika in pregnancy: lessons for global preparedness. Lancet Infect Dis 20:e61-68
42. Roberts FS (2021) Data science and resilience. In: Roberts FS, Sheremet IA (eds) Resilience in the digital age. Springer Nature, Cham
43. Kluberg S, Mekaru S, McIver D, Madoff L et al (2016) Global capacity for emerging infectious disease detection, 1996–2014. Emerging Infect Dis 22(10), October. https://doi.org/10.3201/eid2210.151956
44. Hartley DM et al (2013) An overview of internet biosurveillance. Clinical Microbiol Infect, June.https://doi.org/10.1111/1469-0691.12273
45. Adebayo G, Neumark Y, Gesser-Edelsburg A, Ahmad WA, Levine H (2017) Zika pandemic online trends, incidence and health risk communication: a time trend study. Br Med J Glob Health 3:e000296. https://doi.org/10.1136/bmjgh-2017-000296
46. McGough SF, Brownstein JS, Hawkins JB, Santillana M (2017) Forecasting Zika incidence in the 2016 Latin America outbreak combining traditional disease surveillance with search, social media, and news report data. PLoS Negl Trop Dis 11(1):e0005295. https://doi.org/10.1371/journal.pntd.0005295
47. Deniz D (2016) Zika: From the Brazilian backlands to global threat. Zed books, London
48. Ghenai A, Mejova Y (2017) Catching Zika fever: application of crowdsourcing and machine learning for tracking health misinformation on Twitter. arXiv preprint:170703778
49. Vijaykumar S, Nowak G, Himelboim I, Jin Y (2018) Managing social media rumors and misinformation during outbreaks. Am J Infect Control 46:846–850
50. Lazer D, Kennedy R, King G, Vespignani A (2014) The parable of Google flu: traps in big data analysis. Science 343(6176):1203–1205. http://www.jstor.org/stable/24743402
51. Saltzberg S (2014) Why Google flu is a failure. Forbes, March 23. https://www.forbes.com/sites/stevensalzberg/2014/03/23/why-google-flu-is-a-failure/?sh=fa13cf555352. Accessed June 14, 2022
52. Pappas S (2014) Data fail! how Google flu trends fell way short. LiveScience, March 13. https://www.livescience.com/44089-google-flu-trends-problems.html. Accessed June 14, 2022
53. O'Connor F (2015) Google flu trends calls out sick, indefinitely. PCWorld, August 20. https://www.pcworld.com/article/423173/google-flu-trends-calls-out-sick-indefinitely.html. Accessed June 14, 2022
54. Carneiro HA, Mylonakis E (2009) Google trends: a web-based tool for real-time surveillance of disease outbreaks. Clin Infect Dis 49:1557–1564
55. Landl H (2020) Google defends use of patient data on Capitol Hill among scrutiny of Ascension deal. Fierce Healthcare Newsletter, March 4. https://www.fiercehealthcare.com/tech/senators-pressing-ascension-google-data-deal-as-tech-giant-defends-its-use-patient-records. Accessed June 15, 2022
56. Ledford H (2019) Google health-data scandal spooks researchers. Nature, November 19. https://www.nature.com/articles/d41586-019-03574-5. Accessed June 15, 2022

57. Collmann J, Robinson A (2011) Designing ethical practice in biosurveillance. The project Argus doctrine. In: Infectious disease informatics and biosurveillance, series: integrated series in information systems, 27; Zeng D, Chen H, Castillo-Chavez C, Lober WB, Thurmond M (eds). Springer, New York, pp 23–44. http://www.springer.com/public+health/book/978-1-4419-6891-3
58. Hartley DM et al (2010) Landscape of international event-based biosurveillance. Emerging Heal Threats J 3:e3. https://doi.org/10.3134/ehtj.10.003
59. Nelson NP, Yang L, Reilly AR, Hardin JE, Hartley DM (2012) Event-based internet biosurveillance: relation to epidemiological observation. Emerg Themes Epidemiol 9(4). http://www.ete-online.com/content/9/1/4. Accessed June 27, 2022
60. Goodman A (2020) The global impact of the Zika virus pandemic: the importance of emergency preparedness. Health 12:132–140. https://doi.org/10.4236/health.2020.122012
61. Anthony SJ, Epstein JH, Murray KA, Navarrete-Macias I et al (2013) mBio 4(5):e00598-13. https://doi.org/10.1128/mBio.00598-13
62. Robbins J (2020) Before the next pandemic, an ambitious push to catalog viruses in wildlife. Yale Environ 360, April 22. https://e360.yale.edu/features/before-the-next-pandemic-an-ambitious-push-to-catalog-viruses-in-wildlife. Accessed June 13, 2022
63. Degeling C, Johnson J, Kerridge I, Wilson A et al (2015) Implementing a One Health approach to emerging infectious disease: reflections on the socio-political, ethical and legal dimensions. BMC Public Health 15:1307. https://doi.org/10.1186/s12889-015-2617-1
64. Escadafal C, Gaayeb L, Riccardo F, Perez-Ramirez E et al (2016) BMC Publ Heal 16:1219. https://doi.org/10.1186/s12889-016-3831-1
65. Dente MG, Fabiani M, Gnesotto R, Putoto G et al (2009) EpiSouth: a network for communicable disease control in the Mediterranean region and the Balkans. EuroSurveillance 14(5), February 5. https://pubmed.ncbi.nlm.nih.gov/19215714/. Accessed June 12, 2022
66. Kampen H, Medlock JM, Vaux AGC, Koenraadt CJM, van Vliet AJH et al (2015) Approaches to passive mosquito surveillance in the EU. Parasit Vectors 8(9):13. https://doi.org/10.1186/s13071-014-0604-5
67. Jordan R, Sorensen A, Ladeau S (2017) Citizen science as a tool for mosquito control. J Am Mosq Control Assoc 33(3):241–245
68. Jordan R, Gray S, Sorensen A, Newman G et al (2016) Studying citizen science through adaptive management and learning feedbacks as mechanisms for improving conservation. Conservation Biology. 30:487–495
69. Dickinson JL, Zuckerberg B, Bonter DN (2010) Citizen science as an ecological research tool: challenges and benefits. Annu Rev Ecol Evol Syst 41:149–172
70. Palmer JRB, Oltra A, Collantes F, Delgado JA et al (2017) Citizen science provides a reliable and scalable tool to track disease-carrying mosquitoes. Nat Comm 8(916). https://doi.org/10.1038/s41467-017-00914-9
71. Fidler D, Gostin L (2011) The WHO pandemic influenza preparedness framework: a milestone in global governance for health. J Am Med Assoc 306:2, July 13. https://www.ncbi.nlm.nih.gov/pubmed/21750298. Accessed July 10, 2018
72. Allotey P et al (2014) Cohorts and community: a case study of community engagement in the establishment of a health and demographic surveillance site in Malaysia. Global Heal Action 7:23176. https://www.ncbi.nlm.nih.gov/pmc/articles/PMC4013487/. Accessed August 27, 2018
73. American Mosquito Control Association (2017) Best practices for integrated mosquito management: a focused update, January. https://www.researchgate.net/publication/315924484_Best_Practices_for_Integrated_Mosquito_Management_A_Focused_Update. Accessed August 27, 2018
74. Erlanger TE, Keiser J, Utzinger J (2008) Effect of dengue vector control interventions on entomological parameters in developing countries: a systematic review and meta-analysis. Medical Veterin Entomol 22. https://www.ncbi.nlm.nih.gov/pubmed/18816269. Accessed August 7, 2018

75. Corbel V et al (2017) International workshop on insecticide resistance in vectors of arboviruses, December 2016, Rio de Janeiro, Brazil. Parasites & Vectors, 10, 278, June 2. https://parasitesandvectors.biomedcentral.com/articles/; https://doi.org/10.1186/s13071-017-2224-3. Accessed August 7, 2018
76. Hostick O, Runge-Ranzinger S, Nathan MB, Kroeger A (2010) Dengue vector-control services: how do they work? A systematic literature review and country case studies. Trans Royal Soc Trop Med Hygiene 104:6, June 1. https://www.ncbi.nlm.nih.gov/pubmed/20400169. Accessed August 7, 2018
77. Parks W, Linda L (2008) Planning social mobilization and communication for dengue fever prevention and control. WHO, Geneva. http://www.who.int/immunization/hpv/communicate/planning_social_mobilization_and_communication_for_dengue_fever_prevention_and_control_who_cds_wmc_2004.pdf. Accessed July 19, 2017
78. Abramides GC, Roiz D, Guitart R, Quintana S et al (2011) Effectiveness of a multiple intervention strategy for the control of the tiger mosquito (Aedes albopictus) in Spain. Trans Royal Soc Trop Med Hygiene 105:281–288
79. Pérez D et al (2007) Fidelity research assists in evaluation, adjustment and scaling up of community-based interventions in Cuba. In: Abstracts of the 6th European Congress on Tropical Medicine and International Health and 1st Mediterranean Conference on Migration and Travel Health. Tropical Medicine and International Health, 14, 2, September. https://insights.ovid.com/tropical-medicine-international-health/tmih/2009/09/002/fidelity-research-assists-evaluation-adjustment/474/00060771. Accessed June 9, 2017
80. Pérez D (2007) Community participation in Aedes aegypti control: a sociological perspective on five years of research in the health area "26 de Julio", Havana, Cuba. Trop Med Inter Health 12:5, May. https://www.ncbi.nlm.nih.gov/pubmed/17445134. Accessed July 19, 2017
81. Bowman L, Donegan S, McCall PJ (2016) Is dengue vector control deficient in effectiveness or evidence? systematic review and meta-analysis. PLOS Neglect Trop Dis 10(3):e000455. https://www.ncbi.nlm.nih.gov/pubmed/26986468. Accessed August 7, 2018
82. McBroom K (2021) A comparison of Zika virus and COVID-19: clinical overview and public health messaging. J Midwifery Women's Health 66:334–342
83. Adibi J et al (2016) Teratogenic effects of the Zika virus and the role of the placenta. The Lancet, 387, April 9. http://thelancet.com/journals/lancet/article/PIIS0140-6736(16)00650-4/fulltext. Accessed October 14, 2016